What the Experts are Saying about the Real Goods *Solar Living Sourcebook:*

At the corner of Wasteful Living and Solar Way sits a store called Real Goods. Take a tour of the store and walk down this road apiece with *The Solar Living Sourcebook*. This is a fabulous compendium of hows, whys, and widgets for making a sensible home under the sun (and wind and micro-hydro and healthy living and sustainable transportation, too)."

—**Amory B. Lovins,** CEO, Rocky Mountain Institute

"Today, the most vital way to take control of our lives and feel 'powerful' is to take responsibility for our energy footprint. *The Solar Living Sourcebook* continues to be the bible for everyone wanting to get off carbon and on the solar grid. It is scrupulously honest, technically flawless, and steeped in experience and practical wisdom. There is nothing like it in the world."

—**Paul Hawken,** author of *Natural Capitalism* and *The Ecology of Commerce*

By 2025, most of what is being written today about energy will seem ridiculous. *The Solar Living Source Book* is one of the rare exceptions. John Schaeffer has had as much experience with the practicalities of sustainable living as anyone on the planet. With zest and intelligence, this book describes his vision, his tools, and his pragmatic problem-solving for those of us who expect to follow a parallel path.

—**Denis Hayes,** Chair of the Earth Day Network.

The Solar Living Source Book is a critical resource, to supply both information and inspiration, in this time when the US dependence on foreign oil and the US foreign policy that supports that thirst are wreaking havoc across the globe. From the Timor Gap to the Niger Delta, oil is the source of so much pain in the world. Thanks to John Schaeffer and the Solar Living Institute for making energy alternatives real, viable and accessible."

—**Amy Goodman,** Democracy Now! and author of *The Exception to the Rulers, Exposing Oily Politicians, War Profiteers, and the Media that Love Them.*

"Epic in scope, comprehensive in content, hopeful in spirit, the Real Goods *Solar Living Sourcebook* might be the most indispensable book in your library. "

—**Thom Hartmann,** author, *The Last Hours of Ancient Sunlight*

More of What the Experts are Saying about the Real Goods *Solar Living Sourcebook:*

"The priceless Solar Living Source Book is a gold mine of information on products, renewable energy, and sustainable living. I spend hours browsing through this book, reading about ideas that will help us forge an enduring human presence."

—**Dan Chiras,** Ph.D., author of *The New Ecological Home, The Solar House, The Natural House* and *The Natural Plaster Book.*

"As John Schaeffer points out in the Introduction to this extraordinarily useful book, solar living is not just an eco-groovy lifestyle choice for the environmentally hip; it is the only survivable option for humanity as fossil fuels dribble away during the remainder of this century. I bring every class of students from New College to the Solar Living Center in Hopland—a sustainability theme park—where they can see, hear, touch, and taste a way of life that uses dramatically less energy, leaves a reduced human footprint on the landscape, and yet opens space for enhanced human creativity and enjoyment. This book is the Solar Living Center on paper—an essential reference for anyone interested in practical sustainability. Study it carefully. And apply what you learn to your home, office, and daily routine."

—**Richard Heinberg,** author of *The Party's Over: Oil, War and the Fate of Industrial Societies* and *Powerdown: Options and Actions For a Post-Carbon World.*

"More than ever—*The Solar Living Sourcebook* is the world's best guide in the great transition to The Solar Age."

—**Hazel Henderson,** author, *The Politics of the Solar Age* and *Building A Win-Win World.*

"As the world crests the peak of petroleum supply, *The Solar Living Source Book* offers both a wealth of tools and skills for our communities to be more self-reliant and ways we can use much less of the world's resources."

—**Julian Darley,** Author *High Noon for Natural Gas* and Founder of the Post Carbon Institute.

"*The Solar Living Sourcebook* is *the* one-step guide to how to live wisely, sustainably and with optimism. A beacon to light our way in the twenty-first century— solar powered, of course."

—**David Pearson,** Author of *The New Natural House Book, The Natural House Catalog, Earth to Spirit: In Search of Natural Architecture* and *The House that Jack Built* series

TWELFTH EDITION

SOLAR LIVING SOURCE BOOK

 GAIAM   REAL  GOODS

THE REAL GOODS SOLAR LIVING BOOK SERIES

Real Goods Trading Company in Hopland, California—now Gaiam Real Goods—was founded in 1978 to make available new tools to help people live self-sufficiently and sustainably. Through seasonal catalogs, a biannually updated *Solar Living Sourcebook,* a website (www.realgoods.com), and a flagship retail store located at the 12-acre permaculture oasis called the Solar Living Center, in Hopland, California, Gaiam Real Goods provides a broad range of tools for independent living. Real Goods is the original solar pioneer, having sold the very first photovoltaic (solar electric) module in the world at retail in 1979. Twenty-six years later, Real Goods can boast of having solarized over 60,000 homes and businesses in the United States.

"Knowledge is our most important product" is the Real Goods motto. To further its mission, Gaiam Real Goods is a strong promoter of books for renewable energy and sustainable living, and has for the past ten years pioneered a co-publishing venture with Chelsea Green Publishing Company to bring important writings by relatively unknown authors to a larger market. Many of these books (listed above) are quoted extensively throughout this *Sourcebook*. The titles in this series are written by pioneering individuals who have firsthand experience in using innovative technologies to live lightly on the planet. Our co-published books are both practical and inspirational and they enlarge our view of what is possible. Beginning with this twelfth edition, we are partnering with New Society Publishers, who is distributing the *Sourcebook* for us and representing it at special events throughout North America to popularize the ideas, products, and practices of sustainable living to the nooks and crannies of America. The time has come.

John Schaeffer
President & Founder, Real Goods

GAIAM REAL GOODS

TWELFTH EDITION

SOLAR LIVING SOURCE BOOK

Your Complete Guide to
Renewable Energy Technologies
and Sustainable Living

Written and edited by **John Schaeffer**
Technical Editor **Doug Pratt**

A REAL GOODS SOLAR LIVING BOOK

Distributed by

NEW SOCIETY PUBLISHERS

ISBN 0-916571-05-X
The Real Goods Solar Living Sourcebook is the twelfth edition of the book originally published as the *Alternative Energy Sourcebook*, with over 500,000 in print, distributed in forty-four English-speaking countries.

Distributed by:
NEW SOCIETY PUBLISHERS LIMITED
PO Box 189, Gabriola Island, BC V0R 1X0 Canada
Telephone: 1.800.567.6772
www.newsociety.com

Gaiam Real Goods
Real Goods Solar Living Center
13771 Highway 101
Hopland, CA 95449
Business Office: 707.744.2100
To Order: 1.800.919.2400 or fax 800.482.7602
International Orders: 707.744.2100
International Fax: 707.744.2104
For Technical Information: 800.919.2400
Renewable Energy email: techs@realgoods.com
Website: http://www.realgoods.com

FOR THE EARTH

May we preserve and nurture it in our every action.

Contents

Acknowledgments

Thanks are due to everyone who has contibuted to the twelve editions of our *Sourcebook*. In the twenty two years since it was first published in 1982, more than 500,000 copies have been sold in 44 English-speaking countries. Thanks go first to Doug Pratt for his many technical writing contributions to this twelfth edition and several editions before it, and to his many overtime and weekend hours dedicated to finishing this on time while ensuring meticulous accuracy. Alan Berolzheimer has helped tremendously with polishing existing text and creating lots of new text, particularly for the Land and Shelter, Emergency Preparedness, Energy Conservation, and Sustainable Transportation chapters. Jill Shaffer, our book designer and production artist, has been a true pleasure to work with and her patience and diligence are greatly appreciated.

Many thanks go to Jeff Oldham, Real Goods Design and Consulting Manager, for providing insight into technologies, to Steve Rogers, Solar Residential Manager, and to all of our Real Goods Technicians who have shared their research and inspiration with us. And thanks also go to Lynn Powers, President of Gaiam, for making the book financially possible. Stephen Morris, former publisher of Chelsea Green Publishing, was of invaluable help in advising me on the direction for the book and how it fits into the book trade, and he contributed much to previous editions of the *Sourcebook*. Chris Plant, our new book distributor, from New Society Publishers, has been a huge help in bringing our new twelfth edition to market. I thank my loving wife, Nancy Hensley, my daughters, Sara and Ashley, and my stepson, Cy Hensley, who all have put up with many long hours of my concentration on this and previous *Sourcebooks*. And finally, for their financial support in making Real Goods possible and for assisting in the purchase and development of the Solar Living Center, I want to heartily thank all of the 10,000 former Real Goods shareowners, who have made this entire exhilarating adventure possible.

The Nonprofit Real Goods Solar Living Institute

The Real Goods Solar Living Institute is a 501(C)3 nonprofit organization that educates about and promotes the knowledge of renewable energy and other sustainable living practices through its hands-on workshops; through SolFest, an annual renewable energy educational event now in its tenth year, which happens in August each year; and through displays and demonstrations at the solar- and wind-powered 12-acre permaculture oasis called the Solar Living Center in Hopland, California. More information about the Institute and its classes and programs can be found in the Appendix of this book, on their website; www.solar living.org, or by calling 707.744.2017. The Institute has an innovative membership program and encourages tax-deductible donations to further its programs.

Introduction

I BEGAN WRITING INTRODUCTIONS to the *Solar Living Sourcebook* in the early 1980s. At that time I was convinced that the behavior of the human species was exceeding the capacity of the Earth to support us, and that we only had ten to twenty years to reverse a collision course with environmental collapse. In 1979 I wrote "According to the U.S. Congress's own Office of Technology, all known oil reserves will be exhausted by 2037." That was twenty-five years ago; 2037 seemed very far away. The defining work of the 1990s for me was the late Donella Meadows' *Beyond the Limits*, which confirmed that we *really did* only have a decade left to mend our unsustainable ways before oil production would decline and environmental decay would destroy the futures of our descendants. I expressed these views to *Sourcebook* readers, not that they needed to be persuaded. Even the newly elected Vice President of the United States seemed to get the message, reinforced by the Rio Earth Summit in 1992, though, unfortunately, he later squandered his mandate.

One would think that by now, halfway through the first decade of the twenty-first century, everyone in the world would understand that we're living on borrowed time—that we simply cannot afford another decade of denial because it will be too late to change course. Let's face it—the era of cheap energy is over. The human population of the Earth, which inched up over thousands of centuries, suddenly exploded from 1.7 billion to 6.4 billion in a single century, due to the abundance of new energy sources—mainly oil. That period of human history is soon coming to an end. It's now crystal clear that our species cannot survive the continued loss of biodiversity, the decimation of forests, shrinking ocean fisheries, falling water tables, foul water and air, and the relentless increase in carbon dioxide emissions that is changing the Earth's climate forever.

The exhaustive work done by Marion King Hubbert (the Hubbert Oil Peak), geologist Colin Campbell, and Richard Heinberg (author of *The Party's Over: Oil, War and the Fate of Industrial Societies* and *Power Down*), persuasively demonstrates that the "oil peak"—the time when 50% of all known oil reserves have been consumed—will occur sometime between 2006 and 2015. Even the CEOs of Exxon-Mobil and Royal Dutch Shell, the world's largest oil companies, concur. Thirteen megaprojects (oil fields which contain more than 500 million barrels of oil) were discovered in 2000, six in 2001, two in 2002, and none in 2003. Geologists believe that we'll probably never find another one. The world currently uses about 80 million barrels of oil per day and is expected to need 120 million barrels per day in 2020.

> It's now crystal clear that our species cannot survive the continued loss of biodiversity, the decimation of forests, shrinking ocean fisheries, falling water tables, foul water and air, and the relentless increase in carbon dioxide emissions that is changing the Earth's climate forever.

Such an increase in consumption would mean that only about one-quarter of our demand will be filled by new supply.

The "age of oil," in the geologic history of the world, will have proved to be only a small blip, from the time that Edwin L. Drake drilled the very first oil well in northwestern Pennsylvania in 1859 until that last drop is consumed sometime in the mid-twenty-first century.

The oil peak will profoundly affect all of us, particularly in the industrialized countries where we have become accustomed and addicted to exponential energy growth. The first and most obvious consequence, which we've already begun to experience, is skyrocketing oil and gasoline prices, which will likely never decline significantly from today's rates of $50/barrel and $2.30/gallon (in California). Indeed, soon those prices will sound very cheap. Decreasing supplies and rising prices will lead to economic contractions, because fossil fuels permeate all sectors of the

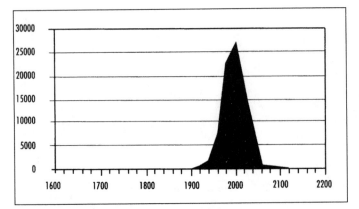
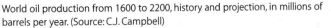

World oil production from 1600 to 2200, history and projection, in millions of barrels per year. (Source: C.J. Campbell)

world economy. Perhaps most significantly, as oil and natural gas become more scarce and costly, so will chemical fertilizers, thereby compromising the ability of industrial agriculture to feed the 8-plus billion people forecast to be alive in 2050. That system is totally dependent on fossil fuels—it takes 400 gallons of oil equivalents annually to feed each American—and therefore by definition it is unsustainable. Furthermore, most population experts agree that the planet's sustainable carrying capacity is around 2 billion people. For a thorough discourse on the subject of planetary carrying capacity, check out www.dieoff.org.

The most challenging legacy of the fossil fuel revolution is global warming. Carbon emissions exceed the capacity of the Earth's natural systems to "fix" carbon dioxide. Since scientists began recording average annual temperatures in 1866, the sixteen warmest years on record have all occurred since 1980, with 1998 and 2002 ranking number one and number two. Parts per million (ppm) carbon dioxide levels have increased from 280 ppm (where they were for centuries) to 360 ppm in 1998, and are forecast to hit 560 ppm by 2050. There are 775 million cars on the planet today, and with China making a concerted effort to get its population off of bicycles and into cars, that number could easily hit one billion by 2025. An increase of this magnitude will hasten the depletion of the remaining oil, and substantially increase the carbon dioxide emissions that are primarily responsible for climate change and global warming.

Believe it or not, some people still claim that global warming does not exist. In addition to the other evidence, consider the following:

• We have evidence that ice is melting at alarming rates in the world's major mountain ranges, including the Rockies, Andes, Alps, and Himalayas.

My father rode a camel. I drive a car. My son flies a jet airplane. His son will ride a camel.
—Saudi proverb

There are 775 million cars on the planet today, and with China making a concerted effort to get its population off of bicycles and into cars, that number could easily hit one billion by 2025.

- There is evidence that the Greenland ice sheet is melting exponentially faster than ever before. If it were to melt entirely, sea level would rise 23 feet.
- Swiss Re, the world's second largest insurance company, warned in March 2004 that the economic costs of disasters directly attributed to global warming likely will double to $150 billion per year in ten years.
- Pentagon defense advisor Andrew Marshall was commissioned to study global warming and concluded that climate change—not terrorism—is the number one immediate threat to world peace. He warned of impending violent storms, a much colder Europe, and the Greenland ice sheet melting and submerging London.

In spite of the evidence, our government continues to subsidize fossil fuel technologies, and the energy playing field is not level. The White House and Congress enacted incentives in 2003 giving the owner of a 10-mpg Hummer a whopping tax deduction of $34,000 but the owner of a 50-mpg hybrid vehicle just $4,000. For every 100,000 SUVs sold this year, American taxpayers will pay a subsidy of $1 billion, which ironically is the same amount that the federal government spent in the 1990s to encourage American car companies to build a hybrid car, and about the same amount we're spending each week to keep our troops in Iraq. After the first Gulf War in 1991 and with bitter memories of the oil shocks of the 1970s, both the Japanese and American governments declared major reductions in oil imports to be an urgent matter of national security. Both countries launched major initiatives to reduce their dependence upon Middle East oil. The Japanese government aggressively helped develop the Toyota Prius and the Honda Insight, assisting the automakers to sell below cost to get the cars out to the public. Ten years later, hybrid vehicles are a major success story. The Prius, which was voted *Motor Trend* Car of the Year in 2004, is in such heavy demand that it takes a year to get one; Toyota expects to sell 350,000 by 2006. Meanwhile, back in the USA, Ford, GM, and Chrysler abandoned hybrid concept cars in 2000, claiming that Americans would not buy them with SUVs on the market. Instead, vehicle fuel efficiency remains worse than it was twenty-five years ago, due to a complete lack of incentives or vision on the part of government and the car companies.

But enough doom and gloom. Even though I'm worried, I know what needs to be done, what *can* be done, and how we can have fun doing it. What has set Real Goods and the Solar *Living Sourcebook* apart since 1978 is our mission to educate by demonstration and positive example. All of the steps we've been advocating since our first-edition *Sourcebook* are still necessary to begin to repair the damage we've done to the planet through excessive consumption and reliance on fossil fuels. Those steps include:

- Transitioning to renewable energy sources in our homes and businesses on a massive scale.
- Cultivating an energy conservation consciousness and practicing it in our homes and businesses.
- Employing green building techniques that promote energy efficiency in all of our new buildings.

Pentagon defense advisor Andrew Marshall was commissioned to study global warming and concluded that climate change—not terrorism—is the number one immediate threat to world peace.

We must face the prospect of changing our basic ways of living. This change will either be made on our own initiative in a planned way, or be forced on us with chaos and suffering by the inexorable laws of nature.
—Jimmy Carter, 1976

We need an energy bill that encourages consumption.
—George W. Bush, 2002

- Using solar and appropriate technology for water heating, water pumping, and both on-grid and off-the-grid living.
- Using composting toilets and greywater systems in our homesteads.
- Developing transportation systems and vehicles that run on alternative fuels, or if they must use fossil fuels, that achieve maximum fuel economy.
- Expanding home and community-based gardens to provide as much food locally as possible.

Renewable energy is the foundation of Real Goods. We made the very first retail sale in the world of a photovoltaic (solar panel) module in 1979. In the 26 years since, we have solarized over sixty thousand homes and businesses. Worldwide, more than one million homes now get their electricity from solar cells. For the 1.7 billion people in the world still not connected to an electric grid, solar cells are by far the cheapest source of electricity. In the developing world, the monthly payment on a three- to five-year loan for a small PV system is less than what is typically spent on candles and kerosene for lamps.

Photovoltaic technology is totally reliable and our solar industry is now mature. In what other industry can you find a twenty-five-year warranty like the one that comes with almost every solar panel sold? Computers? Cars? With new home construction, you can easily wrap the cost of the solar system into your thirty-year mortgage, making PV cost-effective from day one. For a retrofit on an existing home, payback comes in six to twelve years which translates into a return on investment of 8 to 16%, far better than what you can earn in the stock, bond, or long-term CD markets. In the commercial sector, the numbers are even better. With federal tax incentives and accelerated depreciation, payback typically comes in two to five years, for a return on investment of 20 to 40%.

Solar power will increase property value, too. According to the (October 1999) *Appraisal Journal* of the National Appraiser's Association, home value increases $20 for every $1 achieved in annual energy savings. This means that a 2.5 kW system (the average-size residential system Real Goods sells) increases your property value by $18,000, while only costing you $16,000 (with California rebates). You're money ahead from day one. More importantly, a PV system allows you to lock in utility rates of $0.10 per kWh for 30 years—and we all expect that utility electric rates will skyrocket over that same period.

Is enough solar energy available to really make a dent in our fossil fuel dependence? Absolutely! There is 1,000 MW (megawatts) of power in every square kilometer of sunlight—the equivalent of one large nuclear power plant. Even if only 10% of the sun's energy is extracted as electricity, the area required to satisfy 100% of America's electrical energy needs would fit on the existing roofs of our buildings and would be smaller than the land area currently devoted to nuclear, gas, and coal infrastructure and extraction sites. The 4,000 billion kWh that America uses every year could be produced by photovoltaics in a square area 100 miles on a side, which is only half the area of the Nevada nuclear test site.

Solar is growing by leaps and bounds. In 2003, 675 MW of PV power was produced worldwide, up 33% from 2002. The photovoltaic industry experienced a compound annual growth rate of 38% between 1998 and 2003. Unfortunately, the United States' share is dwindling. In 1997, the United States accounted for 42% of all PV in the world, Europe for 18%, and Japan for 25%. But in 2003, the U.S. share declined by two-thirds to only 14%, while Europe increased to 26% and Japan doubled its share to 52% of the PV produced worldwide. The good news is that the price of PV continues to tumble. Solar is following the same pattern as the computer, electronics, and cell phone industries: For every doubling of supply the price declines about 20%.

Why has the United States been relegated to an "also ran" in the PV industry in the last five years? While Japan and Germany have been spending $300 million in research and development on PV, the United States spends less than $50 million. While the Germans guarantee a "feed-in tariff" (they buy your solar power) of $0.50/kWh, in the United States only thirty-eight states even allow "net metering" (where the utility has to buy power from your solar intertie system for the same price as the power you buy from them) As a result of German incentives, their PV market grew from 100 MW in 2001 to more than 400 MW in 2004. Japan also began an aggressive PV development strategy in the mid-1990s, providing an initial 75% government cash subsidy to spur demand (then ratcheted down to 10% over seven years). As a result of this program, Japan went from 2 MW of residential PV in 1994 to over 300 MW by 2002. The installed cost of PV was reduced by 62% in seven years—even during a period of very poor performance by the Japanese economy.

If only we had a government as visionary as the Germans and the Japanese. Similar programs in the United States would provide significant energy benefits, not the least of which would be alleviating some pressure on our overtaxed transmission grid. Equally important, encouraging the expansion of renewable energy in the United States would stimulate the American economy by generating hundreds of thousands of jobs in the manufacturing, sales, and installation of photovoltaics. Even our little company, Gaiam Real Goods, increased residential solar sales fifteen times from 2002 to 2003, almost solely as a result of the California Energy Commission incentives, and we almost doubled our workforce. Just imagine what a well-thought-out national program could do.

Wind energy is growing even faster than solar PV. The three windiest states in the United States (North Dakota, Kansas, and Texas) have enough usable wind energy to satisfy all of our national energy needs. Worldwide, enough wind-generating potential exists to produce twice the projected world electricity demand in 2020. Wind is the world's fastest-growing energy source, having increased sixfold from 4,800 MW in 1995 to 31,000 MW in 2002. Wind turbines now supply enough electricity to satisfy the residential needs of 40 million Europeans. Financially, wind is even better than PV: The cost of wind-generated electricity has dropped from $0.38/kWh in 1980 to less than $0.04/kWh today. This compares to photovoltaics at about $0.10/kWh.

The photovoltaic industry experienced a compound annual growth rate of 38% between 1998 and 2003.

What about the hydrogen panacea we've been hearing about from all levels of government lately? The "hydrogen economy" appears to be a thinly disguised attempt to keep the automobile industry in business-as-usual mode. Hydrogen is an energy carrier, not a primary source of energy. While it is abundant, it is also locked up in water and plants and does not have a favorable net energy balance. It takes more energy to extract the usable power from hydrogen than is generated by using that hydrogen as fuel. The Bush and Schwarzenegger plans of 2004 would make hydrogen by reforming fossil fuels or by nuclear-powered electrolysis, rather than from renewable energy sources. Bush has in fact asked for $16 billion to subsidize the construction of six new nuclear power plants. But under any foreseeable circumstances, from a carbon emissions point of view it's more sensible to use renewable energy sources to generate electricity rather than to generate hydrogen for fuel cells. Hybrid cars currently on the market are nearly as efficient as proposed fuel cell cars, and hybrids haven't even yet scratched the surface of their potential efficiency. And right now a hybrid costs $20,000 while a fuel cell car costs over $1 million! Read former DOE undersecretary Joseph Romm's new book, *The Hype about Hydrogen*, where he concludes that less than 5% of all cars will be powered by hydrogen in 2030.

Even if the "hydrogen economy" is more hype than substance right now, there is much to celebrate on the renewable energy and sustainability fronts. When we opened our first Real Goods store in Willits, California, in June 1978, our mission was to demonstrate and provide renewable energy alternatives—and it still is. In 2004, with the publication of this twelfth edition *Solar Living Sourcebook*, we are better positioned than ever to help you realize your dreams of transforming your lifestyle toward sustainability, whether your goal is to buy land and build a totally self-sufficient solar dream home, or simply to switch to rechargeable batteries and buy a water filter. Over the years, we've gathered an unbeatable team of renewable energy experts with over two hundred years of combined experience in solar, most of whom live with the products that we sell. Our commercial consulting division (Real Goods Design and Consulting Group) works with ecotourism resorts, developers, architects, builders, and green businesses to design, procure, install, and maintain renewable energy systems. Our Real Goods residential solar division specializes in residential renewable energy design and installation. Our Real Goods catalog division produces two 100-plus-page color catalogs every year that feature the latest products for energy conservation, healthy living, renewable energy, and environmental education, as well as the most thorough sustainable living library on the planet.

We are headquartered at the Real Goods Solar Living Center in Hopland, California, our 12-acre permaculture oasis where all of our products, ideas, and concepts come alive, not only in the interactive displays on site but in the persons of the 200,000 kindred spirits who visit the site annually. The Solar Living Center is operated by the nonprofit Solar Living Institute, which nurtures the site and offers over 2,000 class days per year in renewable energy, green building, permaculture, and sustainable living workshops. The SLI's mission is providing inspirational environmental education. They put on the annual SolFest energy fair and celebration for up to 10,000 like-minded folks. The Center is 100% solar powered, with over 140 kW of PV on site, and educational opportunities abound. Read more about the SLI's offerings in the Appendix, on pages 416-427.

The solar industry has been a lifelong adventure for me. I try to walk my talk, living with my wife in our solar home built of recycled and green materials, powered by solar (passive and active) and hydroelectric energy, with gorgeous gardens that provide most of our food. Our tractor and VW Jetta are 100% biodiesel powered. I'm so gratified to see the fruits of all our labors—my home even overlooks the Solar Living Center. As the solar industry continues to grow and as our cultural consciousness continues to evolve, it gives me hope that once and for all we will get things right in our homes, our communities, our country, and on our planet. We are indeed living on borrowed time. Let's turn it all around now, while we have this very last chance.

For the Earth,

John Schaeffer
Real Goods Founder and President

CHAPTER 1

Land and Shelter

LOOK OUT YOUR WINDOW. Chances are, the view you see presents a combination of the natural environment and the built environment. Wherever we live, we humans interact with the land to build our homes, with more or less impact on Mother Earth depending upon how we act. That, of course, is the crux of the matter: How we act, what we do, the choices we make. Shelter is a basic necessity of life, providing us with protection from the elements and serving as the nucleus of our family life. The choices we make about how we treat the land upon which we live, and how we create our sheltering places, are absolutely fundamental to any concept of sustainability.

It all begins and ends with land. So, not surprisingly, the craving to be intimately connected with land seems to be hard wired into human nature. Unfortunately, as we have become modern, industrialized creatures immersed in a culture that celebrates consumption, convenience, comfort, and personal gratification above nearly all other values, the challenge of honoring the land and being its stewards has become immense. Most of you reading these words probably consider yourselves enlightened to the goals of stewardship and sustainability; nevertheless, these challenges remain difficult and require vigilance. It always helps to refresh our understanding.

"Owning land" is a relatively modern notion in the scheme of human history, and by no means a universal practice. To many of the world's peoples who live intimately with the land, including the indigenous Indians of North America, the idea of land ownership is incomprehensible. By now the Native American notion of considering the consequences of our interaction with the land down to the seventh generation has achieved wide currency among open-minded, critically thinking people. But how often do we stop and contemplate the real meaning of that idea? How seriously do we integrate that principle into our daily lives? Consider, for example, the impact of human habitation on the Earth.

The choices we make about how we treat the land upon which we live, and how we create our sheltering places, are absolutely fundamental to any concept of sustainability.

■ The following is adapted from *The New Ecological Home* by Daniel D. Chiras, Chelsea Green Publishing.

Shelter, like many other elements of human existence, comes at an extraordinary cost to the planet and its inhabitants. Construction of the approximately 1.2 million new homes annually in the United States results in a massive drain on the Earth's natural resource base. In the U.S., 85% of all new homes are framed with wood, and wood is used elsewhere in construction, for example to manufacture exterior sheathing, doors, and floors. Today, nearly 60% of all timber cut in the U.S. is used to build houses.

According to the National Association of Home Builders, a typical 2,200-square-foot home requires 13,000 board feet of framing lumber; if laid end to end, this lumber would stretch two and a half miles. If all the dimensional lumber used to build these 1.2 million new homes each year was laid end to end, it would extend 3 million miles—to the moon and back six and a half times.

Humanity's demand for wood contributes significantly to problems of deforestation in many parts of the world. Deforestation, in turn, leads to other environmental problems, such as soil erosion, sediment pollution in streams, loss of wildlife habitat, and species extinction. While forests can be replanted and managed sustainably, deforestation continues to be a problem worldwide and will probably worsen as the human population's insatiable demand for wood and wood products continues to grow.

Home construction is also an enormous source of trash. Building an average 2,200-square-foot house generates three to seven tons of waste that includes scrap wood, cardboard, plastic, glass, and drywall. This small mountain of refuse typically is hauled off to local landfills: In the United States, home construction, remodeling, and demolition are responsible for approximately 25 to 30% of the nation's annual municipal solid waste stream.

Building homes clearly depletes natural resources and produces a huge amount of trash. However, the environmental problems created by shelter continue long after the construction of a home ends. Huge amounts of water, electricity, oil, natural gas, food, and countless household products stream into our homes throughout their life span. Although our individual needs may not seem like much, they add up quickly. According to the U.S. Department of Energy, each year America's homes consume approximately one-fifth of the nation's fossil-fuel energy, for heating, cooling, lighting, run-

ning appliances, and a host of other purposes. This fossil-fuel consumption is responsible for one-fifth of the nation's annual carbon dioxide emissions, which many scientists believe is contributing to dramatic changes in the Earth's climate with severe economic and ecological repercussions worldwide.

The day-to-day operation of conventional homes is also a source of other types of pollution. In industrial nations, billions of gallons of waste water are released each day from our homes. This waste, containing human excrement, toxic cleaning agents, and other household chemicals, is routed to septic tanks in rural areas or sewage systems in cities and towns. From septic tanks, these wastes seep into the ground, where they may trickle into and pollute underground water supplies. Although sewage treatment plants remove much of the waste, collectively they release millions of gallons of potentially harmful pollutants into surface waters each day—the lakes and streams from which we obtain our drinking water.

Furthermore, researchers have found that our homes contain many building materials and products, including furniture, furnishings, appliances, paints, stains, and finishes, that release toxic fumes into the indoor air (for example, formaldehyde).

Finally, and not coincidentally, what has come to be our conventional approach to shelter fosters a problem of extreme dependency. Most of our homes—and those of us living within them—are totally dependent on human-made life-support systems for food, energy, water, and waste disposal. As with a patient in an intensive care unit, cutting or disrupting the lines to the outside world, even for a day, causes suffering. The personal economic cost of this dependency, of course, is significant.

When we think of our homes, we don't tend to think about resource depletion, environmental pollution, habitat destruction, dangerous exposure to toxic chemicals, or dependency. We imagine our homes as comfortable refuges, not as sources of personal and environmental harm. Fortunately, many of our basic needs can be provided at a fraction of the typical economic and ecological costs. A new generation of architects, builders, and energy consultants is emerging, intent on creating homes that meet human needs in ways that minimize damage to people and the planet and protect the Earth's own life-support systems that sustain people and economies. ■■

According to the U.S. Department of Energy, each year America's homes consume approximately one-fifth of the nation's fossil-fuel energy, for heating, cooling, lighting, running appliances, and a host of other purposes. This fossil-fuel consumption is responsible for one-fifth of the nation's annual carbon dioxide emissions.

Choosing Land

■ The following is adapted from *The New Ecological Home* by Daniel D. Chiras, Chelsea Green Publishing.

No matter where you decide to build your home, you will need to assess each option by a variety of criteria. Proximity factors are perhaps the primary ones: how close a place is to work, schools, stores, recreation, and services such as fire protection, electricity, and mass transit. Although at first glance proximity factors may seem entirely anthropocentric, serving people only, they do have significant environmental implications. Living close to work, schools, and grocery stores, for example, can reduce dramat-ically time spent behind the steering wheel, thereby reducing gasoline consumption and air pollution.

Other practical factors also must be taken into consideration when shopping for land; for example, easements, zoning restrictions, covenants, and future land development plans.

Whether you're looking for land in the countryside or in already developed urban or suburban areas, remember that a good site provides many "free services," such as natural drainage or access to the sun for heating your home in the winter. ■ ■ ■

Shelter and Passive Solar Design

Sustainability and restoration are the keys to the future. To live sustainably, humans must find ways to live in harmony with other species, indefinitely, within the limits of the Earth's capacity to support us all. To accomplish this goal, we must savor every drop of our resources, so they can be used to their maximum potential. We must use energy and materials as efficiently as possible, and slow our consumption of nonrenewable resources. We must curb the rate at which we put pollution into the atmosphere, landfills, and surface water, before we reach the point at which our planet's natural systems can no longer handle the contamination. And before we get to sustainability, we must restore and regenerate what we have already destroyed.

As already discussed, building and living in houses is one of the most energy-intensive activities that humans do. Best practices toward sustainability mean striving to minimize construction waste, to avoid the use of nonrenewable materials, and, most importantly, to design structures that require minimal energy input to maintain indoor comfort. Today's common building and design practices are far from sustainable, as forests and fossil fuels dwindle and global warming becomes more evident. Most houses are built with little regard for their natural sites and microclimates, so that we have to outfit them with huge air conditioners and heaters. Then we pour thousands of dollars and tons of fossil fuels into these machines, at untold cost to the environment and our heirs. Many of the best sustainable building methods are thousands of years old, and by combining these traditional techniques with modern technology, we can help open the door to a truly sustainable future. This chapter on shelter provides information on some of the best and most innovative design and construction methods, ones that will carry us through the twenty first century and beyond.

When it comes to shelter, Real Goods will serve you better by providing ideas, concepts, and how-to information than by providing you with major hardware. For the most part, this is a simple case of bulky materials making the shortest possible journey. You will find Real Goods to be an incredibly rich source of information on all manner of energy-saving and energy-efficient technologies. After we have made you the local expert on passive solar and straw bale or rammed earth, buy your building materials locally as much as possible. A well-established industry is out there waiting to provide all your basic bulky supplies, and a good relationship with your local lumberyard will be mutually beneficial while you build your energy-efficient home.

With the accent on ideas more than hardware, we'll begin with a discussion of sustainable design principles and effective planning strategies, present a primer on passive solar design, and then discuss various building materials and styles. If you pay attention to energy and resources at each step, you will build a house that enables you to live lightly on the Earth.

Sustainable Design

The principles of ecological or sustainable design are holistic. They ask you to consider efficient energy and resource use at every stage of the home-building process: choosing a site, planning and design, choosing of materials,

Best practices toward sustainability mean striving to minimize construction waste, to avoid the use of nonrenewable materials, and, most importantly, to design structures that require minimal energy input to maintain indoor comfort.

When it comes to shelter, Real Goods will serve you better by providing ideas, concepts, and how-to information than by providing you with major hardware.

Building and living in houses is one of the most energy-intensive activities that humans do.

construction, setting up home systems, landscaping, and even the end of a structure's life cycle. Almost a century of cheap energy has led to habits of use and patterns of settlement completely divorced from the inherent qualities of the land.

Dan Chiras, author of *The Natural House*, *The Solar House*, and *The New Ecological Home*, among other books on sustainable living, offers these basic tenets of sustainable building:

1. Practice conservation. Use what you need and use it efficiently. This concept speaks to a number of issues, including building small, using passive solar design, emphasizing energy efficiency, reducing wood use, and using materials with low embodied energy.

2. Recycle, recycle, recycle. More and more building materials and products are available that are manufactured with recycled or "waste" materials.

3. Use renewable resources. The more the better. 'Nuff said.

4. Promote environmental restoration and sustainable resource management. Humans need to take an active role in restoring natural ecosystems that have been devastated by

human actions, thereby helping to reinstate the ecological services that ecosystems provide, such as flood control, air purification, water pollution control, oxygen production, and pest control. Reclaiming spoiled land, revegetating building sites with native plant species, facilitating conditions that produce biodiversity—these practices of sensitive stewardship can help to heal the planet, at least in the corner you occupy.

5. Create homes that are good for people, too. Sustainable design considers the health of the occupants of a house, as well as the health of the larger natural environment. These days it's easy, and affordable, to use materials, products, and components that are nontoxic. It's also sensible to design a house to be adaptable and accessible, because needs change with age and over the course of a family's life cycle.

CHOOSING A SITE

Natural builders, green architects, and environmentally aware designers of all stripes agree on the crucial importance of becoming intimately acquainted with the land and your building site as the first step in creating shelter. Learning a site takes time, and requires immersion in

The principles of ecological or sustainable design are holistic. They ask you to consider efficient energy and resource use at every stage of the home-building process.

Can You Afford a Green-Built House?

The future of green building depends, in large part, on green homes being affordable to build, furnish, and operate. Contrary to the myth that building an environmental home is a costly venture reserved only for the well-to-do, this option can be quite cost-competitive. Even with some unfortunate cost overruns, my home cost $5 to $15 per square foot less than most other new spec homes in my area. And its operating costs are much lower. But that is not to say that every aspect of building and furnishing the house was cheaper. A few environmentally friendly products, such as my super-efficient SunFrost refrigerator and compact fluorescent light bulbs, cost more, sometimes a lot more, than their less-efficient counterparts. However, may other green building products, such as recycled tile, carpeting, carpet pad, and insulation, cost the same or less that their counterparts.

Green builders I've talked to say that green building costs from 0 to 3 % more. But remember that investments in some green products and materials may end up provid-

ing additional comfort and saving money right away. Adding extra insulation, for example, keeps a house warmer in the winter and cooler in the summer. This not only makes living space more comfortable, it dramatically lowers fuel bills, potentially saving tens of thousands of dollars over the lifetime of a house. Furthermore, because a well-insulated house requires less heat, you'll be able to install a smaller and less expensive furnace, saving money right from the start. Smaller air-conditioning units are also needed, and in some cases measures to make a home more energy efficient make air conditioning unnecessary. Installing water-efficient toilets and showerheads reduces the demand for water, decreases electricity used for pumping water from wells, and could reduce the size of the leach field required for homes on septic systems. Because these options are no more expensive that standard fixtures, you end up saving money in the short and the long run.

Used with permission from *The Natural House*, by Dan Chiras.

a place through all seasons and under all conditions. The best way to get to know a site and understand its characteristics and potentialities is to spend time there. Site planning, building design, systems design, and the construction process itself all flow from deep knowledge of the site.

■ The following is adapted from *The Hand-Sculpted House: A Practical and Philosophical Guide to Building a Cob Cottage*, by Ianto Evans, Michael G. Smith, and Linda Smiley, Chelsea Green Publishing.

It's easy, but misguided, to think of a building as *something we put on a site*, rather than as *a creation growing out of a place* that already exists. Ecstatic architecture grows like a tree, responding to the nuances of the place where it develops and to the evolving needs of the dweller.

Selecting a building site is one of the most critical design decisions you will make, and should precede the rest of the design process. *You can't have a good building on a bad site.* For the magic to flow, the building and site need to grow together, each improving the other, like an excellent marriage. A good site will make a house easier to build and satisfying to inhabit.

The wrong site can have long-lasting negative effects that are difficult or impossible to mitigate, both during construction and throughout the building's lifetime.

Writers on natural building and sustainable design generally share a common set of guidelines about siting. These are adapted from *The New Ecological Home* by Dan Chiras.

1. Choose a site with good solar access. Solar energy is a vital component of sustainable shelter.

2. Consider wind currents and air drainage. These factors influence heating requirements, the potential for electricity generation, and gardening prospects.

3. Choose a sloping site for earth sheltering, which offers benefits ranging from energy efficiency to good air and water drainage to retention of flat land for other uses.

4. Seek favorable microclimates. Climate can vary dramatically over a site, with different microclimates advantageous or disadvantageous for different homesteading purposes.

5. Select a dry, well-drained site. Good drainage is important to prevent moisture problems in a building. Well-drained sites require little if any grading, which minimizes land disturbance, habitat loss, and the energy required to build a home.

6. Consider the soils on site. Select a site with stable subsoils. Like well-drained soils, stable subsoils minimize the risk of foundation and wall cracking. If you're considering an earthen building, such as cob, adobe, or rammed earth, you'll want to be sure that your site contains soils suitable for these purposes.

7. Avoid marshy areas. Wetlands are precious vanishing resource that need to be preserved. Even building near wetlands can damage them.

8. Select a site suitable for growing food. Providing at least some of one's own food is an important choice that has health, financial, and environmental benefits.

9. Select a site that offers building resources, including earth, sand, stones, trees, straw, or water. Any materials harvested from the site decrease the energy required to build, and embodied in, your home.

10. Choose a site with a good water supply, because water is of course a vital need that you will have to supply one way or another.

11. Minimize ecological disruption. Any kind of construction damages the land, at least temporarily, creates havoc for the plants and animals already there, and can cause erosion problems. Such damage is often obvious and dramatic, but the damage caused by the ongoing existence and use of the building after it is finished cumulatively may be even worse (or it may be healing, if done right). Think through the lifetime of the building, how it will affect, destroy, alter, or improve its site ecology over several hundred years.

12. Don't destroy beauty in your search for it. When siting a home, consider *not* placing it in the most beautiful spot on the property. As Christopher Alexander, coauthor of *A Pattern Language,* advises, "Leave those areas that are the most precious, beautiful, comfortable, and healthy as they are, and build new structures in those parts of the site which are least pleasant now." Create beauty; don't be an agent of its destruction. ■■

OFF-THE-GRID DESIGN

Living off the grid, disconnected from the conventional systems in our society that produce and deliver energy and other utilities, is the ultimate frontier of sustainable design.

There is nothing quite so satisfying as living completely off the electric grid. Particularly in these times when we expect that oil and fossil fuels won't be available for all that much longer, we seek solace from the utility company's hold on us. When there is a power outage in the area and your own house is lit up in unfettered splendor, there is a tendency to gloat, which often gives way to a magnanimous yearning to teach these wondrous new technologies to your entire community.

Taking responsibility for your own power isn't always easy, especially in the beginning. For me it is a happy ritual: every morning I walk to my barn where my power center is located, faithfully record the number of kilowatt hours my solar and hydro systems delivered the day before, check the voltage of my battery system, see if my batteries need equalizing, and occasionally check the water level. But far from drudgery, this daily ritual never ceases to be wondrous and satisfying. Living off-the-grid just doesn't seem to get old.

After the ritual becomes automatic, you never take energy for granted again. Each light bulb left on, each "phantom load" left plugged in and draining milliwatts, and each energy drain takes on new meaning. Do we really need that TV left on all night just in case we wake up and need to use the remote? Does the microwave really need to blink incessantly?

Monitoring becomes critical but is always fun. I have a voltmeter and an accumulating ammeter in my kitchen that is 300 feet from my energy system. I can glance in the middle of cooking a meal to see if the voltage is low, if the battery percentage of full goes below the critical 80% or on a sunny day see if my batteries are being overcharged due to a long equalization cycle.

Feeling connected to your energy source changes everything. You are suddenly empowered to create your own destiny and no longer beholden to the vicissitudes of the local power company. Once you've experienced it you'll never go back.

GREEN BUILDING

Striving to use environmentally friendly building materials as much as possible is another important dimension of sustainable building. "Green" building materials can be defined in terms of many criteria that take account of environmental and social responsibility. Examples include engineered lumber, biomass-based building panels, natural earthen materials, low- or no-VOC (volatile organic compound) paints and stains, and any number of products manufactured by companies with a strong commitment to protecting both the environment and its workers. The bottom line is, the greener your home, the less impact it has on people and the planet.

■ The following is adapted from *The Natural House*, by Dan Chiras, a Real Goods Solar Living Book, Chelsea Green Publishing.

In recent years, many manufacturers have climbed onto the green building bandwagon. Hardly a product in use today can't be replaced by a more environmentally sound alternative, and the variety of green building materials only promises to get better. If you are building a natural home, be sure to consider alternative materials for foundations, walls, floors, and windows. If you are going to build a more conventional wood-frame house or just an addition, give careful thought to ways that you can use green materials.

The products you buy should be produced by companies that care about the environment and take actions to manifest their commitment. Pollution prevention programs are an important sign of environmental responsibility. Production of materials from recycled scrap or sustainably harvested forests is another. Also purchase from companies that treat their employees well; it indicates they truly care about the health and well-being of the people who are making them wealthy—or at least keeping them in business.

Decisions about products are complex. Some materials may meet the criteria, but cost too much—either to purchase or to ship to your site. Others may not be as aesthetically appealing as standard building materials. Still others may not provide the structural strength you require. (In some cases they could actually be stronger than standard materials.) Sometimes a product may meet only one or two sustainable criteria; nonetheless, it is an improvement over conventional building materials.

My advice is to select the products and materials that meet the greatest number of green criteria possible. Don't expect perfection. You won't find it. Virtually all products have some shortcoming. Choose materials that offer the greatest gain for the environment and the health and welfare of the occupants of the house. One useful strategy is to concentrate on products that are used in great quantity, such as framing lumber, insulation, tile, concrete, and drywall. Purchase as many green building products as you can afford. This way, you will have a larger impact. ■ ■

THE HEALTHY HOME

One consequence of the growing emphasis on energy-efficient homes is that more and more houses are being built to be airtight. Adequate ventilation is a critical feature of any house, and this is especially true for airtight ones. Furthermore, without a holistic design, the functions of keeping heat in (or out, in the case of cooling) and ensuring sufficient ventilation can work at cross-purposes. The main reason ventilation is so important is that, unbeknownst to many people, most buildings suffer from some degree of indoor air pollution that can be harmful to our health. Exposure to indoor air pollution has been implicated in health problems ranging from annoying maladies like headaches, chronic bronchitis, asthma, and allergies, to debilitating illnesses such as multiple chemical sensitivity, and even cancer.

■ The following is adapted from *The Solar House,* by Daniel D. Chiras, Chelsea Green Publishing.

Indoor air pollution arises from a variety of sources. Conventional building materials are one common source, as are combustion appliances, cleaning products and disinfectants, household pesticides, pets, and even people themselves. The more airtight a home is, the more problems these and other chemicals create.

Many modern building materials contain toxic chemicals that are released into the room air, a process commonly referred to as outgassing. These chemicals can outgas for months, even years, after construction is complete. The list of offenders includes carpeting, drapery, vinyl floor tiles, some types of ceiling tiles, upholstery, vinyl wallpaper, particle board, plywood, oriented strand board, caulking compounds, paints, stains, and solvents. Fossil fuel-burning appliances such as stoves, ovens, gas fireplaces, furnaces, and water heaters all can contribute harmful gases such as carbon monoxide to indoor air. Wood stoves and fireplaces also release these gases and potentially harmful particulates into the air inside a home. ■ ■

As author Dan Chiras says, "What good is an energy-efficient, passive solar home if it poisons it occupants and wreaks havoc in the planet's life-support systems?"

PLANNING, DESIGN, AND COLLABORATION

If sustainability is the key to the future, planning and collaboration are the keys to successfully building a sustainable home. Whatever techniques, styles, materials, and systems you choose, careful planning throughout the process, and effective collaboration among all

Characteristics of Green Building Materials

- Produced by socially and environmentally responsible companies
- Produced sustainably-harvested, extracted, processed, and transported efficiently and cleanly
- Low embodied energy
- Locally produced
- Made from recycled waste
- Made from natural or renewable materials
- Durable
- Recyclable
- Nontoxic
- Efficient in their use of resources
- Reliant on renewable resources
- Nonpolluting

the various participants, will enhance your ability to obtain local and recycled products, minimize waste, reduce costs, and end up with the home you really want. It's well worth the time and effort spent before you start to build.

■ The following is adapted from *The Real Goods Independent Builder,* by Sam Clark, a Real Goods Solar Living Book, Chelsea Green Publishing.

To be in control of your building project, and have fun doing it, you need realistic expectations of what your house will cost and how much work it will take to make it livable. You especially will make your home-building experience more efficient, sane, and pleasurable if you follow a logical procedure for turning your ideas into designs and plans. Start with the information you have about your land, your needs, and your budget. Draw a site plan, a map that shows the characteristics of the land and locates important features such as the well, septic system, gardens, and roads. Make overlapping lists of the general features, qualities, functions, and activities you want in your house; then see how they can be combined in different patterns. Include everything. This is not the time to be realistic: Things can be eliminated later if necessary. But underline those activities and features that are central. To me, the layout or plan,is the heart of a house design. It largely should dictate the structural system, the heating design, and other features.

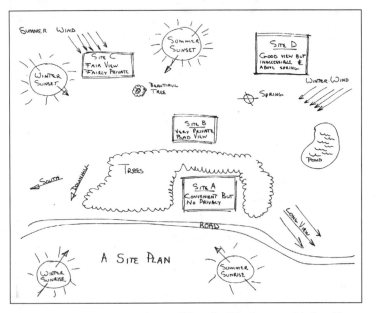

Inside the image:
SUMMER WIND — SUMMER SUNSET — Site D GOOD VIEW BUT INACCESSIBLE & ABOVE SPRING — WINTER SUNSET — Site C FAIR VIEW FAIRLY PRIVATE — BEAUTIFUL TREE — SPRING — WINTER WIND — Site B VERY PRIVATE BAD VIEW — POND — TREES — SOUTH — DRIVEWAY — Site A CONVENIENT BUT NO PRIVACY — LONG VIEW — ROAD — A SITE PLAN — WINTER SUNRISE — SUMMER SUNRISE

Beginning Site Plan.

It is possible to build a house with few if any drawings. But designing your house will be easier and more thorough if you learn to make and use scale floor plans, elevations, cross sections, and details. Making drawings is not just a way to put ideas on paper. It is a way to develop ideas and make them work for you. Basically, you make a scale drawing, such as a floor plan, and then systematically ask questions about it. Is the sun orientation good? Is the circulation efficient? Does the layout give enough privacy? Then you revise the drawing, until you have solved as many of the problems as possible. This labor may seem excessive when your real interest is not drawing but building, yet every hour you spend drawing will save you five hours of building. You can solve problems in advance on paper and avoid big and costly mistakes later.

The builder and designer should both be hired early. They should be asked to collaborate from start to finish. My suggested team model is this: During the design phase, the builder is supporting and consulting with the designer, checking costs, and suggesting those details that he or she prefers to execute. Later, the roles reverse: During construction, the designer supports and consults with the builder on detailing and other problems that arise. Other team members, such as an energy consultant, plumber, or electrician, should also be involved early in the process, so they can collaborate with the designer and the contractor. And, of course, as the owner, you are a key team member, whether or not you do any building. Most likely, you will be coordinating the process.■ ■■

A Passive Solar Design Primer

Warm in the Winter— Cool in the Summer

That's how we want our homes, right? And if it takes a minimum amount of heating or cooling energy to keep them that way, everybody wins. Passive solar design is the way to get there—and as we write this, in 2004, it's becoming clearer all the time that passive solar design has become an acceptable, almost mainstream, building concept. Using exactly the same pile of building materials and labor costs, you can have an energy-efficient, sunny, easy-to-maintain house, or an energy-sucking, expensive, cave-like house. Obviously the warm, sunny, low-maintenance house is going to be a lot nicer to live in, will generate less environmental impact, and will be worth far more if and when you decide to sell.

Don't Fear Solar

■ The following is adapted from *The Natural House*, by Dan Chiras, a Real Goods Solar Living Book, Chelsea Green Publishing.

The potential for solar design in most parts of the world is enormous. Even in those cold, cloudy winter regions like the northeastern United States, solar energy can support an efficient household nicely. Unfortunately, solar energy too often is thought to be the sole domain of homeowners in sun-drenched regions like the southwestern United States But nothing could be further from the truth. Solar energy is just as viable a solution in Ottumwa, Iowa, and Rochester, New York, as it is in West Palm Beach, Florida, or Tucson, Arizona. In areas with long, cold winters, capitalizing on the sun's free heat can result in major savings in fuel bills and a substantial decrease in human impact on the planet.

The first rule of sustainable building, then, is: Don't exclude solar from your plans simply because your home site is not as sunny as the residence of a friend who lives in Phoenix or the Australian outback. Think solar. Go solar. Plan to get as much solar energy as you can. You can take advantage of solar energy for heat with little, if any, extra cost to your building project, by properly orienting your house and rearranging items, notably windows, that you are going to buy anyway. It isn't going to cost you an arm and a leg to go solar. And over the lifetime of your home, it could save you thousands of dollars in fuel bills.

Solar has become an economically viable option in part because building specifications have changed. In the U.S., materials and techniques that once added to the cost of a solar home now are considered standard building practice. For example, double-pane, high-performace glass is found in virtually all new windows and patio doors. Insulation standards also have improved greatly. Vapor barriers, which prevent insulation from getting wet and thereby becoming less effective, are now commonplace in new construction. Solar homes aren't much different from nonsolar homes anymore. Because of these and other changes, you won't be spending more to build a passive solar home. ■■

Energy-Efficient Design

The concept of energy efficiency is integral to passive solar design. Indeed, the ability of a structure to use the least amount of energy, with minimal waste, to provide the desired degree of comfort, is the key to making passive solar design work. This is true for cooling as well as heating. Most energy-efficiency measures—such as wall and ceiling insulation—that contribute to the goal of passive solar heating also contribute to passive solar cooling.

By reducing heating and cooling loads, energy-efficient construction provides another significant advantage. Because you will need less fuel to heat and cool your home, you probably will be able to reduce the size of your back-up heating system. This, in turn, will result in savings that will offset the extra costs of some of your other sustainable design choices.

The Passive Solar Concept

■ The following is an excerpt from *The Passive Solar House,* by James Kachadorian, a Real Goods Solar Living Book, Chelsea Green Publishing.

A French engineer named Felix Trombe is credited with the simple idea of building a solar collector comprising a south-facing glass wall with an air space between it and a blackened concrete wall. The sun's energy passes through the glass, and is absorbed by the blackened wall. As the concrete warms, air rises in the space between the glass and the concrete. Rectangular openings at the bottom and top of the Trombe wall allow this warm air to flow to and from the living space. This movement of air is called thermosiphoning. At night the blackened concrete wall will radiate, or release, its heat.

Unfortunately the process can reverse at night—pulling warm air from the living space over to the cold glass. As this warmer air is cooled by the glass, it drops to the floor, which in turn pulls more warm air from the living space. The colder it is outside, the more energetically

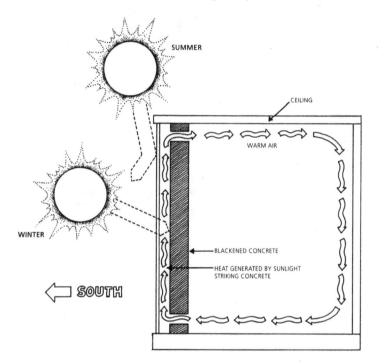

the Trombe wall will reverse thermosiphon. One way to control this heat loss is to mechanically close the rectangular openings at night, then reopen them when the sun comes out.

The Trombe wall is the "Model A" of passive solar design; that is, it is elegant in its simplicity and dependability, but largely has been supplanted by improved modern techniques. But even if Trombe walls rarely are used in contemporary buildings, they illustrate some essential characteristics of applied passive solar design: the system requires no moving parts, no switches to turn motors on or off, and no control systems, yet it will collect and store solar energy when it is functioning properly, and then radiate heat back into the living space, even after the sun has gone down. ■■

Principles of Passive Solar Design

Let's consider ten basic principles of solar design, none of which will seem controversial. What's striking about solar design is how logical and even "obvious" its tenets really are.

SOLAR PRINCIPLE #1

Orient the house properly with respect to the sun's relationship to the site.

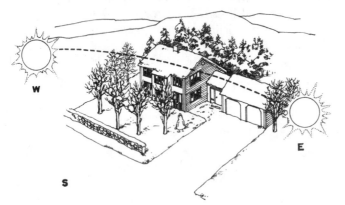

Use a compass to find true south, and then by careful observation site the house so that it can utilize the sun's rays from the east, south, and west during as much of the heating season as possible. Take into account features of the landscape, including trees and natural land forms, that can buffer the house against harsher weather or winds from the north in winter, and shade the house from too much sun in the summer.

SOLAR PRINCIPLE #2

Design on a twelve-month basis.

A home must be comfortable in summer as well as winter. When designing a solar home, carefully plan to accommodate and benefit from

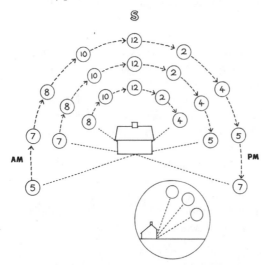

the sun's shifting patterns and other natural, seasonal cycles. Before finalizing a building plan, spend time at the site at different times of the day and year, and pay close attention to the sun, wind, and weather.

SOLAR PRINCIPLE #3

Provide effective thermal mass to store free solar heat in the daytime for nighttime use.

Design the home's thermal mass to effectively absorb the sun's free energy as it enters the building in winter, thereby avoiding overheating. Achieve thermal balance by sizing the storage capacity of the thermal mass to provide for the heating needs of the building through the night. In summer, properly sized thermal mass will serve to cool the building by providing "thermal lag"—that is, excess heat will be absorbed during the daylight hours; by the time

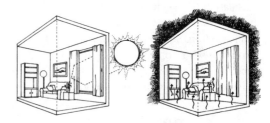

the mass has heated up, the day is over and that stored heat can be discharged by opening windows and increasing circulation during the night.

SOLAR PRINCIPLE #4

Insulate thoroughly and use well-sealed vapor barriers.

Build tightly constructed, properly insulated walls and roofs. Contemporary standards for wall and roof insulation are very compatible with solar design. Carefully install and seal discrete (or "positive") vapor barriers on the living-space side of walls, ceilings, and/or roofs to prevent moisture from migrating into the insulation along with heat, which tends to travel outward toward the cooler exterior. Incorporate an air-lock entrance for primary doors.

SOLAR PRINCIPLE #5

Utilize windows as solar collectors and cooling devices.

This idea sounds obvious, but many people overlook the obvious and spend large amounts of money purchasing, fueling, and maintaining furnaces and air conditioners to address needs that high-quality, operable windows also can address. Vertical, south-facing glass is especially effective for collecting solar heat in the winter, and these windows will let in much less heat in summer, because the sun's angle is more horizontal in winter and steeper in summer. This difference in seasonal angle can be exploited in very sunny locales by using awnings or overhangs that shade windows from the steep sun in summer, yet not from winter sun, which will penetrate further into rooms, supplying solar heat when it's most needed. Provide insulated window and patio door coverings to decrease nighttime heat loss in winter, and to control solar gain in spring, summer, and fall. Operable windows can be used to release excess heat and to direct cooling breezes into the house.

SOLAR PRINCIPLE #6

Do not overglaze.

Incorporate enough windows to provide plenty of daylight and to permit access to cooling breezes for cross-ventilation, but do not make the common mistake of assuming that solar design requires extraordinary allocations of wall space to glass. An overglazed building will overheat. As emphasized in Principle #5, locate your windows primarily on the home's south side, with fewer windows on the east- and west-facing sides, and only enough windows on the dark north side to let in daylight and fresh air.

SOLAR PRINCIPLE #7

Avoid oversizing the back-up heating system or air conditioner.

Size the conventional back-up systems to suit the small, day-to-day heating and cooling needs of the home. A well-insulated house, with appropriately sized and located thermal mass and windows, will need a back-up heating system

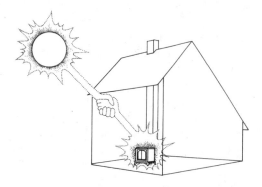

much smaller than conventional wisdom might dictate. Take into account the contribution of solar energy, breezes, and shade to the heating and cooling of the home. Do not oversize back-up oil, gas, or electric furnaces, as these units are inefficient, cycling on and off when not supplying heat at their full potential. Do not oversize air conditioners, as they are likewise expensive and wasteful when operated inefficiently. Remember that a heating and cooling contractor will tend to approach the problem of sizing from a "worst-case" perspective, and may not have the skills or experience to factor in the contributions of solar energy and other natural forces.

SOLAR PRINCIPLE #8

Provide fresh air to the home without compromising thermal integrity.

To maintain high-quality indoor air, a well-insulated and tightly constructed home needs a continual supply of fresh air, equivalent to replacing no less than two-thirds of the building's total volume of air every hour. This air exchange should occur through intended openings (such as exterior-wall fans) in both the

kitchen and bathroom, rather than through leakage around poorly sealed doors and windows. You should also investigate whole-house ventilation systems.

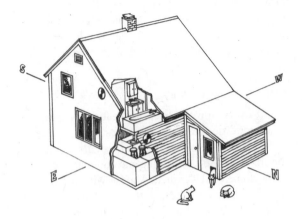

SOLAR PRINCIPLE #9

Use the same materials you would use for a conventional home, but in ways that maximize energy efficiency and solar gain.

With exactly the same construction materials, it is possible to build either an energy-efficient, sunny, and easy-to-maintain solar house, or an energy-gluttonous, dark, and costly-to-maintain house. When designing a solar home, rearrange and reallocate materials to serve dual functions—adding solar benefits as well as addressing architectural or aesthetic goals. Placing a majority of the home's windows on the south side is an example. The carefully designed and constructed solar home need not cost any more to build than a comparably sized nonsolar conventional home.

SOLAR PRINCIPLE #10

Remember that the principles of solar design are compatible with diverse styles of architecture and building techniques.

Solar homes need not look weird, nor do they require complicated, expensive, and hard-to-maintain gadgetry to function well and be comfortable year-round. In solar design, good planning and sensitivity to the surrounding environment are far more crucial than special technologies or equipment. Over thousands of years, many traditional cultures have used the sun and other natural forces to facilitate year-

round heating and cooling. Ancient examples come readily to mind, including the pueblos of the American Southwest and the rammed earth structures of North Africa and the Middle East. Build your home in the style you like, but use the solar principles for siting, glazing, thermal mass, and insulation to get the greatest possible benefit from free solar energy.

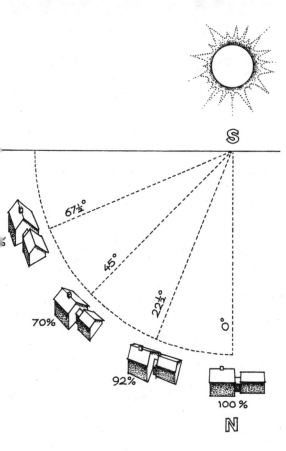

oriented to the sun at an angle other than true south is indicated at left.

As you can see, the solar benefit decreases exponentially as you rotate the home's orientation away from true south. Within roughly 20 degrees of true south, the reduced solar benefit is minimal, which allows some latitude in placing the house on a site that presents obstacles such as slopes and outcroppings.

Ideally, the north side of the site will provide a windbreak, with evergreen trees and a protective hillside. These natural features will protect the home from the harsher northerly winds and weather. Deciduous trees on the east, south, and west will shade the home in summer, yet drop their leaves in winter, allowing sunlight to reach the home.

KNOW YOUR SITE

Spend some time on your proposed homesite. Camp there to learn about its sun conditions in different seasons. Make a point of being on the site at sunrise and sunset at different times of the year. Develop a sense for which direction the prevailing wind comes from. Mark the footprint of your new home on the ground using stakes and string, and use your imagination to picture the view from each room. In addition to solar orientation, consider access, view, weather patterns, snow removal, power, septic, and of course, water.

The long axis of a solar home should run east to west, presenting as much surface area to the sun as possible. If your new home measures 24

LET NATURE HELP YOU

Consider a south-facing house site. For the sake of discussion, let's locate this house in Hopland, California, which is at north latitude 40 degrees. If the home faces true south, you will get the maximum solar benefit, but as you rotate your home off true south, the solar benefit will be reduced accordingly. For example, at solar noon in February in Hopland, the cost of being

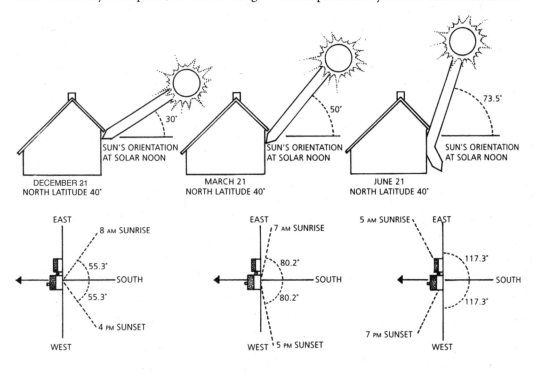

by 48 feet, maximize the amount of surface that the sun will strike by siting your home with the 48-foot dimension running east-west.

USE WINDOWS AS SOLAR COLLECTORS

If you locate the majority of the windows and patio doors on the east, south, and west walls of your home, they can function as solar collectors as well as ventilators, gathering warm solar energy when you need heat, and letting in breezes when you need fresh and cool air. One often sees pictures of solar homes with huge expanses of south-facing glass, tilted to be perpendicular to the sun's rays. Let's remember that you want your home to be comfortable all year round. Tilted glass, though technically more efficient at heat-gathering during winter months, is very detrimental in summer, and will result in overheated living spaces.

It's very important to design on a twelve-month basis, and to understand where the sun is in each season. The figure on page 21 shows the

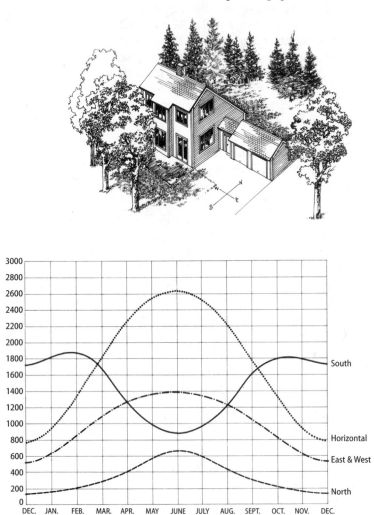

Insolation Year-Round for Latitude 40 Degrees North

sun's angles at three different times of the year at north latitude 40 degrees—December 21, March 21, and June 21. In December, the sun's low altitude almost directly strikes the south-facing vertical glass, and we can see again the importance of facing a home true south.

The March 21 and June 21 illustrations show that as the days grow longer, the breadth of solar aperture widens. Meanwhile, the altitude of solar noon rises to 50.0 degrees on March 21 and 73.5 degrees on June 21.

DESIGN ON A TWELVE-MONTH BASIS

In a northern state such as Vermont at the winter solstice (December 21), the sunlight shining through a south-facing patio door will penetrate 22 feet into the home. On the summer solstice (June 21), the sun will only enter the building a few inches.

Let's imagine south-facing glass at solar noon. If you plant a deciduous tree on the south side of your home, the sun will shine through the canopy in winter when the leaves are gone. Yet in summer, the tree's canopy will absorb almost all of the sun's heat. Plant deciduous trees at a distance from the home, based on the height to which the tree is expected to grow and the size of the anticipated canopy. If deciduous trees already exist on your site, cut down only those that directly obstruct the clearing needed to build the home. Thin adjacent trees' branches after you have gained experience with their shading patterns in both winter and summer.

Remember that because of the high arc of the summer sun, its heat will mostly bounce off vertical south-facing glass, unlike the almost direct horizontal hit your solar collectors will get in winter. This "device" called a solar home will "automatically" turn itself on during the coldest months and shut itself off during the summer months, so that solar collection is maximized for heat gain when you need the extra heat, and minimized when heat would be uncomfortable. If you can grasp these basic dynamics, you have started to let nature work for you.

The table at left shows the amount of energy received by south-, north-, and east/west-facing vertical glass, plus horizontal glass at solar noon at 40 degrees north latitude (measured in Btus, or British Thermal Units, the standard unit of measurement for the rate of transfer of heat energy). As you can see, the amount of energy received by vertical south-facing glass in December or January is over double the amount received in June. Note how the south-facing glass graph lines run counter to all other directions.

What about east- and west-facing glass? We've been singing the praises of south-facing glass, but at the beginning and end of the heating season, east- and west-facing glass can be effective solar collectors too. However, because the angle of the sun is perpendicular to east-facing glass as the sun rises, and perpendicular to west-facing glass as the sun sets, east- and west-facing glass do not "turn off" as solar collectors in summer. For this reason, location is a critical factor in deciding how to allocate east- and west-facing glass. For example, a solar home located in central California, which has a summer cooling load, should have less east- and west-facing glass than a home located in northern Washington state.

You may want to provide retractable nighttime insulation for at least some of the windows, since the same glass that admits heat during the daytime will steadily lose heat during the cooler nighttime hours. Window insulation can also keep heat out at times of overly intense sun. Another technique for reducing the amount of sunlight that reaches the windows during the hottest months of year is to design overhanging eaves (or movable awnings, which can be added later) to shade your glazing.

THE SOLAR SLAB

Passive Solar Principle #3 addresses thermal mass, the heat-storage system in passive solar design. As described below, many natural building techniques—such as cob, adobe, and

Passive Solar Checklist

- ✔ Small is beautiful
- ✔ East-west axis
- ✔ South-facing glazing
- ✔ Overhangs
- ✔ North-side earth berming
- ✔ Thermal mass inside building envelope
- ✔ High insulation levels
- ✔ Radiant barriers in roof
- ✔ Open airways to promote internal circulation
- ✔ Tight construction to reduce air infiltration
- ✔ Air-to-air heat exchanger
- ✔ Best high-tech windows available
- ✔ Reduced glazing on north and west sides
- ✔ Day lighting
- ✔ Invest in any energy-saving features possible
- ✔ Pay attention to the little details
- ✔ K.I.S.S. (keep it simple, stupid)

rammed earth—utilize the building material itself as thermal mass. Stones and concrete slabs also can be used to provide thermal mass. One passive solar design that makes innovative use of thermal mass is the Solar Slab, invented by Jim Kachadorian.

The Solar Slab consists of concrete blocks installed underneath a concrete slab. Oriented north-south, the cavities in the blocks create an airflow pathway. Vents on the north and south sides of the house allow room air to circulate underneath the floor. As heat is transferred into the home by the south glass or by the heat transfer through the wall, air inside the south wall

The Solar Slab provides thermal inertia to the home so that the home "wants" or tends to stay at a steady temperature, using very little purchased fuel in the process.

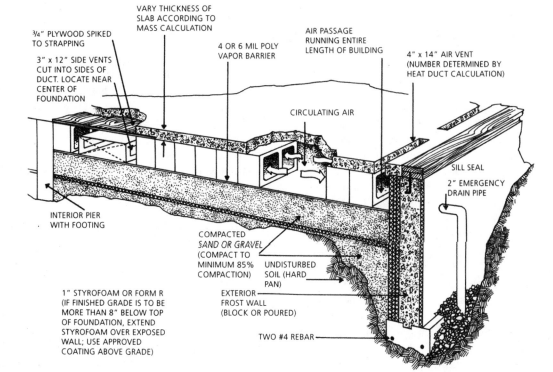

VARY THICKNESS OF SLAB ACCORDING TO MASS CALCULATION

¾" PLYWOOD SPIKED TO STRAPPING

3" x 12" SIDE VENTS CUT INTO SIDES OF DUCT. LOCATE NEAR CENTER OF FOUNDATION

4 OR 6 MIL POLY VAPOR BARRIER

AIR PASSAGE RUNNING ENTIRE LENGTH OF BUILDING

4" x 14" AIR VENT (NUMBER DETERMINED BY HEAT DUCT CALCULATION)

CIRCULATING AIR

SILL SEAL

2" EMERGENCY DRAIN PIPE

INTERIOR PIER WITH FOOTING

COMPACTED SAND OR GRAVEL (COMPACT TO MINIMUM 85% COMPACTION)

UNDISTURBED SOIL (HARD PAN)

1" STYROFOAM OR FORM R (IF FINISHED GRADE IS TO BE MORE THAN 8" BELOW TOP OF FOUNDATION, EXTEND STYROFOAM OVER EXPOSED WALL; USE APPROVED COATING ABOVE GRADE)

EXTERIOR FROST WALL (BLOCK OR POURED)

TWO #4 REBAR

The most obvious way to conserve land, energy, and building resources, especially wood, is to build smaller houses. A smaller house uses less materials, and causes less pollution, at every stage. Its economies are more than proportional to size, because a small house needs not just fewer timbers, but smaller ones, since the spans are shorter.

rises. Warmed air is then pulled out of the ventilated slab, and the cooler air along the north wall drops into the vents along the north wall. This thermosiphoning effect will continue to pull air naturally through the Solar Slab. As the air circulates, it also releases heat that is stored in the concrete blocks and the slab. At night, as the ambient temperature decreases, that stored heat radiates into the house. This passive solar process reduces the need to run a furnace or boiler to keep the house at a comfortable temperature. A Solar Slab, properly designed, can enable a home to achieve thermal balance every day.

Jim Kachadorian compares the Solar Slab to the mechanical inertia achieved by the flywheel on an old John Deere tractor: "The tractor's small engine slowly got the huge flywheel spinning. Once up to speed, very little energy was needed to keep the tractor moving.... Likewise, the Solar Slab provides thermal inertia to the home so that the home 'wants' or tends to stay at a steady temperature, using very little purchased fuel in the process. With this kind of thermal inertia built into the solar home, we can undersize the back-up heater, instead sizing equipment for those rare occasions when the tremendous thermal stability of the house isn't sufficient to cope with unusually hot or cold conditions outside."

Concerns have been raised about the potential for moisture to build up in the block cavities and promote the growth of mold and mildew.

Some builders have begun to adapt the Solar Slab design and use pipes instead of concrete blocks. But there is little doubt that the Solar Slab works as a simple, elegant method of passive solar heating.

EVERY LOCATION IS A POTENTIAL SOLAR HOME SITE

All too frequently we hear someone say, "Solar won't work here." How can solar energy not work? Although in some locales, as gardeners know, more sunlight is available for greater portions of the year, we all live in solar locations. Does our emphasis on ideal orientation and siting of a home mean that only if the solar conditions are ideal should you plan on building a passive solar house? No. Any contribution made by the sun and natural breezes to heating and cooling a home will be money not spent on fossil fuels, wood, and electricity.

The basic premises for a good solar home are simply the premises of good home design:

• Make the most of what's available to you, in terms of both your environment and the materials that you are planning to use in your home construction.

• Let the tendencies of nature work for you and not against you.

• Work toward the goal of keeping the conventional furnace and air conditioner switched off, and also try to minimize your reliance upon alternative back-up fuels such as wood. Only sunlight is free.

Architecture and Building Materials

Depending upon where we are in the world, we will find different climates, as well as different resources and methods of construction. Working with the materials and techniques most appropriate to the local area can save both money and energy. In most cases there may be a variety of appropriate responses to environmental and personal economic situations, so it is helpful to consider, research, and prioritize your options. Let's look at the advantages and disadvantages of several alternatives, to help you make an appropriate choice for your particular endeavor. Our emphasis will be on methods of what is known as "natural building."

As we consider alternatives, remember that the specific nature of a region and site will exert a great deal of influence in your choice. What materials exist on site, or are available locally? How much sunshine does the site receive, what

spots are shaded, where do the prevailing winds come from? Also important are issues of timing, financial constraints, and your own personal capabilities as a builder or those of builders in your area. Before you select any building material, familiarize yourself with the indigenous homes in your area. Such homes best exemplify the traditional and successful uses of materials and building strategies.

Wood

■ The following is adapted from *The Real Goods Independent Builder*, by Sam Clark, a Real Goods Solar Living Book, Chelsea Green Publishing.)

Most people build houses that consist largely of wood. But even a stone house, or a straw bale house, will be partly wood. Wood is an easy and forgiving material—but it is also a precious,

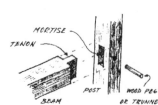

Mortise and tenon joint. From *A Shelter Sketchbook.* Used with permission.

often mismanaged resource. If you want to use wood responsibly, consider cutting wood from your own land and hiring a portable sawmill, see if native timber is available at local sawmills, and plan to incorporate recycled lumber in your structure.

The most obvious way to conserve land, energy, and building resources, especially wood, is to build smaller houses. A smaller house uses less materials, and causes less pollution, at every stage. Its economies are more than proportional to size, because a small house needs not just fewer timbers, but smaller ones, since the spans are shorter. This is the point most often ignored by people who talk sustainability. ■■■

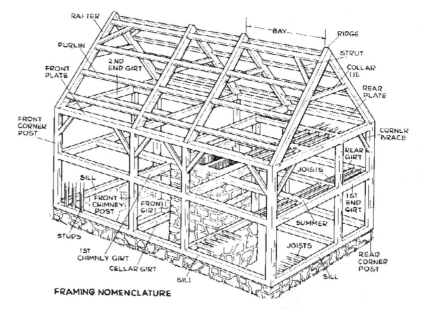

FRAMING NOMENCLATURE

Framing nomenclature.

Stick-Building

Change and innovation have become the constants in our building environment. Managing them is a critical skill. Over the years, I have learned that traditional approaches to building were often smarter than they appeared.

My starting point is a respect for the way most American houses have been built since 1860, which is often called stick-building, or light timber framing. In a stick-built house, the load-carrying frame consists of small, closely spaced framing members concealed inside floors, walls, and roof. Though it is often criticized, standard stick-building systems are particularly durable and adaptable. No system is as easy to build, fix, and change as the ordinary stick-built system: It is a people's technology. Unless it burns down, a house built this way will last forever if the roof is maintained and the basement is ventilated. It is every bit as strong as a timber building, and can be changed more easily.

Post and Beam

Post-and-beam framing has considerable aesthetic appeal. However, this time-honored method is wood-intensive, and utilizes mostly high-quality, large, and sometimes long timbers, which are increasingly hard to find and expensive. If you have a good source of indigenous wood that fits these criteria, post and beam can be an excellent and attractive way to build.

Post-and-beam structures are prevalent in the historic communities of the eastern United States. The frame of the building typically is prefabricated and can be worked on during the nonbuilding season, then brought to the site and erected efficiently.

As with any wood structure, the frame must be detailed properly with rigid geometry, lateral knee-bracing, and integration of frame and exterior sheathing to make a strong, resilient, and secure unit. Because wood is susceptible to rot, similar care must be taken in detailing the vapor barriers and insulation. Heavy timbers require heavy lifting and teamwork to erect, and skill in constructing joints that fit. Special tools and huge saws are required for production work, but these structures also can be built using only hand tools.

The infill walls can be conventional balloon framing, or more commonly now, stress-skin panels, which are a sheet of solid foam insulation with waferboard skins or drywall on the interior side. Interior finish can be paneling, drywall, or plaster. If wood timbers are planed, sanded, and finished with a pickling stain, the wood tends to stay lighter in color, and brighten the interior of the home. Chamfering the edges of the timbers softens the corners and is friendlier to the touch and eye. It is tempting to hurry, but taking care to finish exposed timbers before they go into place is well worth the time and effort.

Straw Bale

The idea of building anything permanent out of straw may seem laughable. After all, we all grew up with the tale of the Three Little Pigs. Yet traditional cultures throughout the world have used straw, grasses, and reeds as building materials, usually in combination with earth and timber, to create shelter for thousands of years. It must work!

Straw bales are made from the leftover stems of harvested grasses after the seed heads have been removed. Hay bales are made from grasses

Inexpensive, owner-builder friendly, fire resistant, highly insulative, and ecologically benign, straw bales are revolutionizing home construction methods.

Typical straw bale construction. Like bricks, only bigger.

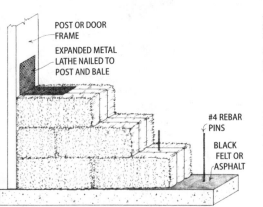

POST OR DOOR FRAME

EXPANDED METAL LATHE NAILED TO POST AND BALE

#4 REBAR PINS

BLACK FELT OR ASPHALT

A straw bale foundation and door frame ready to begin stacking. From *The Straw Bale House.* Used with permission.

Stacking walls is the easy part! Stucco application takes a bit of elbow grease. From *The Straw Bale House.* Used with permission.

that are harvested green with the seed heads included. Both can be used for building, but because hay is valued as an animal feed, it is much more expensive than straw. The embedded seed heads in hay may also attract vermin. Straw, with no nutritional value, is quite unattractive to mice and rats. Straw bales are super energy efficient, environmentally safe, inexpensive, and easy to work with. So long as they are protected from moisture, and their structural characteristics are taken into account, straw bales can be used to build structures that are attractive, safe, and durable.

Modern straw bale buildings first appeared shortly after the invention of mechanical baling equipment in the late 1800s. Pioneers in the timber-poor regions of Nebraska started using bales as expedient short-term shelters. When these temporary structures turned out to be both durable and comfortable in the extremes of Nebraska weather, they were plastered and adopted as permanent housing. A number of these 80- to 90-year-old buildings are still standing and in good repair.

Rediscovery of this century-old building technique is proving to be one good answer to our current search for low-cost, energy-efficient, sustainable housing. Inexpensive, owner-builder friendly, fire resistant, highly insulative, and ecologically benign, straw bales are revolutionizing home construction methods. As lumber prices rise and quality declines, environmentally conscious builders are looking to less expensive alternative building materials. At the same time, plagued by poor air quality, California and other states are banning agricultural burning of straw because it creates a major air pollution problem. If straw is baled rather than burned, a waste product can be turned into a very promising alternative building material: straw bales. Because straw matures in a matter of months, millions of tons of these sustainable building materials are produced annually, com-

pared to the decades it takes for trees to grow large enough for the sawmill. Straw bale construction is a win/win solution for farmers and builders.

A straw bale wall built with three-string bales (tight construction-grade bales are typically 16" x 23" x 46") will produce a superinsulated wall of R-50. That's two to three times more insulation than most kinds of new construction, without the environmental hazards of formaldehyde-laced fiberglass. Straw bale walls are highly fire-resistant because, with stucco on each side of the wall, there is too little oxygen inside to support combustion. And even without stucco protection, burn-through times greatly exceed building code requirements. Two-string bales can be used for construction if they are sufficiently compacted. But they are smaller in size, have a lower insulation rating, and do not have the structural strength of three-string bales. Good, tight, three-string bales are really preferable.

Straw bale construction is quick, inexpensive, and results in a highly energy-efficient structure. The technique is easy to learn, and the wall system for a whole building can be stacked in just a day or two. This method also allows for some interesting building shapes. To change the shape of a wall, or put a friendly looking radius on a sharp corner, use a weed whacker to trim the bales to the desired shape. Bales can also be bent in the center to form curved walls, and interesting insets and artwork can easily be carved into the straw before the stucco is applied. The 2-foot-thick walls leave room for attractive window seats like those found in old stone castles in Europe.

Modern load-bearing, plastered straw bale wall construction goes like this: A concrete slab is poured with short pieces of rebar (metal concrete-reinforcement rods) sticking out of the concrete about 12 inches high on 2-foot centers around the slab perimeter. After the top of the foundation is waterproofed, the bottom bales of the wall are wrapped with a polyethylene sheet (to prevent moisture migration from below) and impaled on the rebar. Subsequent bales are stacked on this bottom layer like bricks, with each new layer of bales offset by half over the bales below. Rebar pieces are driven into selected bale layers pinning them to the bales below. Plumbing generally is placed in interior frame walls. Electrical wiring is done on (or recessed into) the bales after the wall is up. Electrical outlet boxes are attached to wooden stakes driven into the bales. Preassembled window and door frames are set in place as the walls are con-

Straw bale interior finished in traditional Southwestern style. From *The Straw Bale House.* Used with permission.

structed. The frames are then pinned to the surrounding bales with dowels to hold them in place before the stucco is applied.

When the walls are finished and have completely compressed under the weight of the finished roof and ceiling for six weeks or so, the bales are wrapped with stucco mesh or chicken wire. Stucco is then troweled onto the wire, coating the walls inside and out. Plaster can be used on interior walls if desired. This forms a bug-proof, fire-resistant envelope around the straw, and securely attaches the electrical fixtures to the wall.

A large roof overhang and gutters help keep rain off the exterior walls and lessen the possibility of moisture migration. Straw bales can take some moisture on the exposed bale-ends, since the gaps between the straws are too big for capillary action to carry moisture into the bale. Of course, painting the stucco also goes a long way toward keeping the bales dry. If the bales are kept dry the building will last for many decades.

If the current trend continues, construction-grade straw bales will appear at your local lumberyard before too long. If you want to reduce your heating and cooling costs, and help prevent the current overcutting that is destroying our nation's forests and watersheds, straw bale construction is definitely an excellent sustainable building choice. See our product section for books, hardware, and specialized tools for straw bale construction.

Adobe

Adobe is one of the oldest and most widely used building methods. It is most economical in regions with prevalent adobe (clay) soils and a relatively dry climate, with bricks made and cured on site. Building with adobe is laborious. However, if the house is owner built, using materials found on site, it can be inexpensive.

In the United States adobe is primarily found in the Southwest and California. In regions that have hot, dry summers and cold winters, adobe can be utilized without insulation, since its mass provides a "thermal flywheel" that moderates temperature swings and ensures a comfortable home. In less sunny areas, it is advisable to provide insulation, such as foamboard applied to the exterior of the walls, then plastered over with stucco cement or adobe mud plaster.

The major concern with adobe is its seismic resistance. Classic adobe buildings have withstood centuries of temors because their buttressed walls are relatively thick for their height, and are topped with sturdy wood bond beams. Modern, thin-walled adobe structures have proven to be unsafe in an earthquake, unless carefully reinforced with steel and concrete, braced frequently, tied together with a bond beam, and securely linked to their foundations. The mix for the bricks (sand/clay content and waterproofing compounds used to keep the blocks from deteriorating in the rain), the detailing at the foundation and roof, and the integration of reinforcing frames are all quite important to the shelter's solidity and longevity.

In regions that have hot, dry summers and cold winters, adobe can be utilized without insulation, since its mass provides a "thermal flywheel" that moderates temperature swings and ensures a comfortable home.

Adobe lends itself to fantastic and wonderful building shapes. From *Build with Adobe.* Used with permission.

Rammed earth requires serious form work during construction. From *The Rammed Earth House.* Used with permission.

Rammed earth walls contain tremendous thermal mass, so rammed earth homes are highly energy efficient.

Think of rammed earth as a sort of "instant rock." It is magic, watching soil become stone beneath your feet.

Rammed earth produces walls with beautiful color variations. From *The Rammed Earth House.* Used with permission.

Rammed Earth

■ The following is adapted from *The Rammed Earth House* by David Easton, a Real Goods Solar Living Book.

Rammed earth construction combines many of adobe's best qualities with the advantages of slipformed walls. Soil from the site (some soil types are more suitable than others) is mixed with proper proportions of clay, sand, water, and cement. The earth is then tamped into reusable forms to infill walls between concrete frames, foundations, and bond beams, which tie the structure into an earthquake-resistant frame. If attention is paid to detailing at window and door openings, the rammed earth dwelling can have the same gracious and solid feeling as adobe, for less labor. Rammed earth walls contain tremendous thermal mass, so rammed earth homes are highly energy efficient. Further insulation may not be necessary, depending on your climate and site orientation.

It is possible to apply this material using a gunnite spray rig (Impacted Stabilized Earth or PISE), as Real Goods has done to cover the straw bale walls in our Solar Living Center. For a solid PISE wall the material is sprayed against a plywood form from the outside to a thickness of 18 to 24 inches. In mild climates, such as parts of California, Arizona, and New Mexico, this thickness may be sufficient insulation. With this technique, the walls need to be reinforced with steel to make a seismically safe home.

Think of rammed earth as a sort of "instant rock." The earth rammer plies his trade in an environment filled with the dust of soil and cement, the roar of diesel engines, and the staccato thump thump thump of backfill tampers. To me, this is magic . . . watching soil become stone beneath your feet, and knowing that, when the forms are removed, a well-built wall will be here that will survive the test of centuries.

To recombine into a strong and durable rammed earth wall, a soil should be a well-graded blend of different-sized particles. Large particles provide the bulk of the matrix, while the smaller particles fill in the spaces. With ever-decreasing particle sizes, virtually all of the air space within the wall matrix can be eliminated, resulting in the densest wall possible. Density is one of the contributors to ultimate wall strength. ■■■

Cob

Cob is a technique of building with earth that offers unique aesthetic possibilities in combination with the thermal mass benefits common to all earthen structures (such as adobe and rammed earth). Cob (the word comes from the Old English root meaning "a lump or rounded mass") is a composite made of earth, water, straw, clay, and sand that is hand worked into monolithic walls while still pliable. Because this technique does not use bricks or forms, hand-sculpted cob lends itself to organic shapes, curved walls, arches, and vaults. Cob homes are curvaceous rather than rectilinear. And, unlike adobe, cob can easily be built in cool, wet climates.

Cob has been revived in North America in the last decade or so primarily by Ianto Evans, Michael G. Smith, and the Cob Cottage Company, which has created a system called "Oregon Cob."

■ The following is adapted from *The Hand-Sculpted House: A Practical and Philosophical Guide to Building a Cob Cottage*, by Ianto Evans, Michael G. Smith, and Linda Smiley, Chelsea Green Publishing.

It was not until we were working on our fifth or sixth cottage that we realized how differently we

Cob lends itself to organic shapes, curved walls, arches, and vaults. Cob homes are curvaceous rather than rectilinear.

were building from either English or African cob and that our method offered significant advantages over what we know of traditional techniques. Our system, which we termed "Oregon cob," has evolved rapidly toward an even greater range of benefits and continues to improve as its popularity spreads and as experience builds on experience.

Historically, most cob contained very little or no straw, and any straw used was generally short and of poor quality. In contrast, the Oregon cob technique uses the longest, strongest straw available. And unlike traditional cob, Oregon cob uses a carefully adjusted proportion of sand, with just enough clay to bond the mix together. The cob therefore shrinks very little as it dries. This emphasis on precisely adjusting the mix according to the quality of the ingredients produces surprising strength. The builder can sculpt strong but thin earthen shelves, and interior partitions as thin as two inches.

Structural strength is also added by building walls selectively thicker as needed, and by curving walls whenever possible. The process feels like *growing* a building instead of violently forcing parts together. We pay particular attention to the elegant geometries of Nature—how strength is attained with a minimum of material—and we use those shapes. A common pattern that results is a series of walls curving in loose spirals connected by short, tight curves.

Oregon cob allows ad hoc design changes. Because building is incremental, by the handful, continuous adjustments can be made. Changes suggested by site, materials, the builders' skills, or the evolving building itself nearly always raise quality and are sometimes inspired and magical. Last-minute improvements can be tried, assessed, rejected, or adjusted. The medium melds the edges of sculpture and construction, using responsive, adaptable mixes of the basic ingredients to create the furniture, fixtures, and features with which we live.

As noted, cob is wonderfully suited to energy-efficient, passive solar design. It is also durable, fire resistant, wind resistant, and earthquake resistant. ■ ■

Earthships

Earthships—a sort of cross between rammed earth and adobe—employ opportunistic resources (used tires, aluminum cans) in a clever passive-solar strategy, often sunk into a hillside, or "earth-integrated." This innovative refuse-disposal and home-building concept was created by Michael Reynolds, a Taos architect.

A hole is excavated into the slope, then the tires are laid in a bricklike pattern, laboriously filled with soil, and compacted. The tires swell and interlock under the pressure of manually rammed earth, becoming very thick and resilient. Chinks between tires are stuffed with used and partially crushed aluminum cans. Like an adobe wall, integrity is further secured by a bond beam atop the wall. Roofing consists of the classic vigas (large wooden girders) and latillas, or modern laminated beams, along with plywood and foam sheathings.

A sloping glass wall along the front, oriented generally to the south, exposes the thermal mass of the tire-and-earth frame to direct solar gain. Exterior walls and rounded, sculpted interior surfaces are plastered and painted to look like adobe and rammed earth homes.

Earthships are often designed to be completely self-sufficient: Water from roof catchment, photovoltaic electricity, and innovative

Earthships give new meaning to "sweat equity." Where do you suppose all the cans came from?

The process feels like *growing* a building instead of violently forcing parts together.

indoor waste disposal are all common features. Effective passive solar design can keep a well-balanced earthship hovering around 65°F with no expenditure of energy, winter and summer.

We can judge earthship longevity only by their short (twenty-five-year) history, but they incorporate the benefits and share the risks of rammed earth and adobe construction: They are fireproof, earthquake resistant, thermally massive, made of appropriate materials, inexpensive, and indigenous. Earthships can be built for very little money by owner-builders, although building one will give you an intimate understanding of the term "sweat equity." Recently, the folks at Solar Survival Architecture have created smaller types of Earthships (the Nest and the Woodless Hut) that lend themselves to modular or kit construction and are less expensive. Along with straw bale and modern rammed earth techniques, Earthships represent a refreshing and optimistically innovative approach to shelter, with a strong emphasis on self-sufficient living.

Emerging Natural Building Technologies

The field of natural building continues to expand and diversify, as the movement gains in popularity. Innovative builders and designers are experimenting with new techniques and concepts for creating ecologically sustainable shelter. Here is a survey of some of the most promising emerging natural building technologies. For more information, see *The Natural House* and *The New Ecological Home*, by Daniel D. Chiras.

EARTHBAGS

This technique, pioneered by architect Nader Khalili, involves filling polypropylene bags with an appropriate earth mixture and using them as the building blocks for walls. The method produces good thermal mass and lends itself to round structures. It is also very labor intensive.

CAST EARTH

The evocative phase "cast earth" suggests just what it sounds like: Walls cast from liquid earth. In this process, a slurry of a composite earth mixture is poured into forms set on a foundation, and the forms are moved around as the mixture dries. Like other earth-building techniques, cast earth is ideal for passive solar design.

STRAW-CLAY

Straw-clay is more familiar in Europe than in the United States. This is an in-fill method of making walls. Chopped straw is mixed together with a water-clay mixture known as a slip, until the straw is evenly coated. Then the straw-clay mixture is poured into forms and tamped by hand. The process results in a natural wall with high insulative value and, especially when covered with earthen plaster, good thermal mass.

RASTRA BLOCK OR INSULATED CONCRETE FORMS

Insulated concrete forms (ICFs) are building systems that combine concrete or cement with some type of plastic foam to create foundation and/or wall components. Various manufacturers use different combinations that contain more or less cementitious and recycled-content material.

> Earthships are often designed to be completely self-sufficient: Water from roof catchment, photovoltaic electricity, and innovative indoor waste disposal are all common features.

Real Goods founder John Schaeffer's Rastra block house, "Sunhawk" in Hopland , California. See story on pages 37-40.

But ICFs are generally more resource and energy efficient than conventional building methods, provide good insulating value, and are easy to work with.

One example is Rastra, which I (John Schaeffer) have recently used to build my own house. Rastra blocks are made from cement (about 15%) and 100% recycled expanded polystyrene (EPS) foam (about 85%). In addition to providing an environmentally friendly alternative material for foundations and walls, Rastra blocks are also lightweight and easy to handle—you can even cut them with a saw! According to *Environmental Building News*, "the material is insect-proof, creates a very strong wall, and is extremely fire-resistant. Recent structural testing found a Rastra wall to be seven times better under earthquake-type stresses than a wood-framed shear wall, according to marketing director Richard Wilcox" (*EBN*, July/August 1996). A newer entry in the ICF market, called Tech Block, uses more recycled foam and less concrete than Rastra; however, it is nonstructural and must be installed with permanent sheathing (*EBN*, May 2001). Another huge advantage to Rastra is that it lends itself beautifully to being shaped and sculpted into just about any design.

PAPERCRETE

Can you imagine building a solid, durable home out of recycled newspaper? It is possible, with papercrete. Papercrete is a slurry made from chopped newspaper, sand, cement, and water. This slurry can be formed into blocks or poured into wall forms. Papercrete walls provide both good insulation and good thermal mass—not as good as primarily earthen materials like adobe or cob, but better than straw bale or straw-clay.

HYBRIDS

■ The following is adapted from *The Natural House*, by Daniel D. Chiras, A Real Goods Solar Living Book, Chelsea Green Publishing.

Today, many of the innovators who brought us straw bale homes, cob cottages, packed tire homes, and other alternative building technologies are wandering outside the conventional limits of their unconventional fields. They are mixing and matching building techniques to produce a new generation of hybrid natural homes. A home like mine, for example, combines packed tires and straw bales. The bedrooms, kitchen, and dining room are made from tires. The living room is straw bale. Other builders use packed tires and earthbags to build foundations for straw bale homes. The list goes on. The possibilities are only limited by the imagination.

Bill and Athena Steen, best-known for their work in straw bale construction, increasingly find themselves incorporating traditional clay building systems with straw bale. They use clay for floors, walls, and plaster. Once the straw bales have been placed in the wall, they often use traditional English cob to fill odd spaces and to sculpt window and door openings, as well as shelves, bancos for seating, and free-form fireplaces. Larger spaces in the wall that are too small for bales are packed with loose straw coated with clay slip. They also know that while straw bale offers excellent insulation, it is lacking in adequate thermal mass necessary for efficient passive solar construction. Therefore, they are now using adobe blocks to build interior mass walls that come in direct contact with winter sunlight pouring in through south-facing windows. Rammed earth and cob walls could also be used such cases.

"By viewing building as a process of combining different, but complementary materials rather than adhering to a particular building system," write the Steens, "we have given ourselves the freedom to create structures that respond to a wide variety of contexts and circumstances. They can be elegant or simple, quick or detailed, inexpensive or costly, and probably most importantly, they can be built from predominantly local materials in whatever combination best matches the local climate."

But also heed the words of Matts Myhrman and S. O. MacDonald: "By its nature, a hybrid structure often requires extra thought during the design process. Draw it, model it, get a 'second opinion,' and still expect to have to think on your feet once you get started." ■■

Yurts

Round houses have a particular kind of appeal. A yurt—the traditional structure invented by the nomads of central Asia—is a circular house that actually is composed of two cones, one inverted on top of the other, for walls and roof. This ingenious design, which uses a tension band to take the outward thrust of the roof, achieves a "spacious, economical structure—one of the most efficient surface-to-volume structures ever devised," in the words of Bill Coperthwaite, American yurt pioneer and advocate.

Today, many of the innovators who brought us straw bale homes, cob cottages, packed tire homes, and other alternative building technologies are wandering outside the conventional limits of their unconventional fields. They are mixing and matching building techniques to produce a new generation of hybrid natural homes.

In addition to their simplicity and efficient use of space, ecologically conscious people are drawn to yurts because they are portable. Some people build yurts as permanent or semipermanent structures; but many use them as temporary shelter or for nonliving spaces such as playhouses, studios, and sheds. Easily made from indigenous and local materials and requiring a minimum of them, yurts practically define "living lightly on the land."

Stone

Many people are attracted to the beauty and durability of stone, an ancient building material. Stone is fire resistant, and a well-engineered stone home can be made earthquake resistant too. Stonework buildings take a long time to construct, and they demand a lot of physical labor, but the work is rewarding and these homes can be very cost-effective if owner built. Because of its large thermal mass, a well-designed stone structure insulated with sprayed foam or batts in a frame can be very energy efficient.

There are two common ways to work with stone: formed wall construction, in which stones and mortar are built up in courses within reusable slipforms, and laid stone, where each stone is nestled and mortared atop the stones below. Laid stone requires great skill and much time, which usually means a costly finished product.

A stone building requires particular attention to wall footings and reinforcement, and careful choice and sizing of the stones, especially at the corners and around openings. These issues are more significant for laid stone than formed stone work. Pointing and fastidious cleanup can produce a beautiful, almost ageless structure. For guidance and inspiration, look to indigenous buildings in your area and also research the stone-building techniques of Helen and Scott Nearing, Frank Lloyd Wright, Ernest Flagg, and Ken Kern.

Log Cabins

Although log construction is neither resource efficient nor particularly energy efficient, it has its place in areas with a good source of logs, such as the Northwest and Southeast United States. Many people are drawn to the beauty of a log home, including the scale, proportion, and elegance of the corners and openings, and the use of small logs for railings, floor beams, and the like. This method of construction lends itself to prefabrication and off-site construction. Work can be done during the off-season and prepared so that, when the weather is right, the logs, already cut, notched, kerfed, and fitted, can be assembled, usually with a crane. As with timber frames, it is important to choose logs free of wood-eating insects and creatures. Log homes seem to work best in areas with cold, dry, snowy winters and mild, wet summers.

A log cabin requires a foundation adequate to support the enormous weight of the structure.

Laid stone construction. From *Step-by-Step Outdoor Stonework.* Used with permission.

Stonework buildings take a long time to construct, and they demand a lot of physical labor, but the work is rewarding and these homes can be very cost-effective if owner built.

Formed wall stone construction. From *The Independent Builder.* Used with permission.

While not particularly resource or energy efficient, log cabins offer a homespun charm many folks find irresistible. From *How to Build Your Own Log Home for Less than $15,000.* Used with permission.

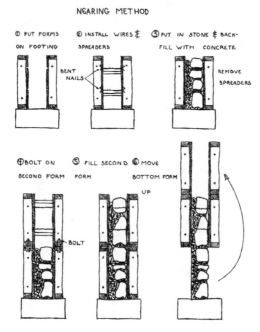

Room sizes and proportions need to be oversized due to the often massive thickness of the logs. The sealant used between the logs and the detailing at corners, windows, and doors are of great importance in providing a tightly sealed interior, to prevent cold air infiltration and heat loss. Detailing at door and window openings also needs to allow for settling, since the logs shrink as they dry. Increasing passive solar gain, by using lots of south-facing windows, requires special engineering to ensure the continuity of the structure's lateral resistance.

Prefabricated log structures—made with smaller-dimension softwood that has been totally milled and prepackaged—seldom make satisfactory homes. Their walls are too thin, the wood quality is often poor, and their thermal properties are marginal; these countrified, cookie-cutter structures are very expensive to heat and are resource-inefficient.

Cordwood construction, an unusual but promising log-building technique, uses smaller-dimensioned wood to good effect. It involves laying up short pieces of logs perpendicular to the wall like bricks. For more information, see Rob Roy's book *The Sauna* and other information provided through his Earthwood Building School: 518-493-7744; www.cordwoodmasonry .com.

Domes

Domes continue to exercise a strong appeal over our cultural imagination. However, while domes do have some unique advantages, these must be balanced against some equally unique disadvantages.

The great thing about domes is that, as Buckminster Fuller observed, the dome uses a minimum of materials to enclose a maximum of space. Also, domes are easy to erect, because all the pieces are light and the structure can go up quickly. Heavy lifting is not required, material use can be efficient (although waste is unavoidable because you'll be turning square materials into triangles), and materials are affordable. The structure of a dome is composed of a few standardized and repeated strut and hub sizes of small dimension. These components can be fabricated off site, indoors, and without a great deal of space. Because of the unusual shape and the angles, fabrication requires extreme accuracy, which can be achieved by using jigs. Sheathing the dome can be fast as well, since standardization is the name of the game. These structures are extremely strong and can withstand almost

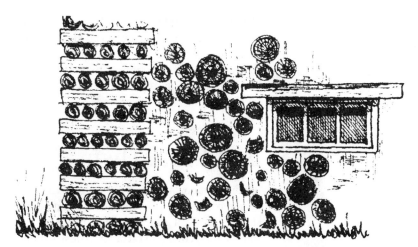

any force acting upon them if they are firmly anchored to a strong foundation.

However, once the shelter goes up and is sheathed, work proceeds very slowly. There is no free dome lunch: The equalizer comes in the detailing of windows and doors, roofing, interior finishing, and cabinetry. Interiors are slow to build and tedious to finish, as all shapes are unusual and either round or triangular. Placement of windows and doors can compromise the integrity of the dome shape, resulting in slow sagging, fatigue of the adjacent members, and, ultimately, collapse. The removal of any struts is discouraged, but if that's necessary, additional engineering is required. Aesthetically pleasing, weathertight roofing is difficult to achieve, particularly if it also needs to be fire resistant. High-quality asphalt shingles seem to have the best track record with existing domes.

Perhaps more important, it is difficult to create an energy-efficient dome. Domes have low thermal mass, and they are difficult to insulate well because they have such a thin frame (4 to 6 inches) and usually no attic space. Loose foam insulation could be used but infiltration losses

Cordwood construction. From *A Shelter Sketchbook*. Used with permission.

Many people are drawn to the beauty of a log home, including the scale, proportion, and elegance of the corners and openings, and the use of small logs for railings and floor beams. This method of construction lends itself to prefabrication and off-site construction.

Domes enclose maximum space with minimal material, but are very difficult to finish internally.

likely would be quite high. Nevertheless, the efficient shape can be easily vented and heated.

Domes are probably best employed as workshops, barns, and garages—in other words, buildings that require a minimum of interior finishing, minimal heating, and don't mind a drip or two when it rains. For more information, take a look at Buckminster Fuller's works and *The Dome Book*.

Build Your Own

Building your own house is a courageous undertaking, but it can also be extremely gratifying, and can save you a lot of money. Those of you who have experience in carpentry, construction, electrical work, or plumbing may be able to do some, most, or all of the work yourselves, with a little help from your friends. Even those with limited building experience can succeed admirably if you're committed and dedicated, as Carolyn Roberts documents in her book *A House of Straw: A Natural Building Odyssey* (Chelsea Green Publishing). Natural building methods, as discussed throughout this chapter, are especially well suited to owner building.

Whatever kind of structure you are going to build, you will have to deal with a variety of challenges. Many of us here at Real Goods have built our own houses, and earned some good down-to-earth advice from the school of hard knocks in the process. The following tips are a distillation of lessons we learned along the way. We hope that some of the mistakes we have made can help you not make the same ones. (New mistakes are always more interesting.)

QUOTES FROM *A HOUSE OF STRAW*, BY CAROLYN ROBERTS

■ This would be a home that would reflect my love of the desert. It would be made of the Sonoran sands and clays, blending with the desert's light hues and delicate textures, using nature's own elements to protect my family from the harsh extremes. It would be a home built of my enthusiasm and sweat, with the assistance of my two teenaged sons whom I loved dearly, and with friends who build straw bale houses because they love them and all they stand for. It would be a home to embrace and inspire all who pass within its doors.

■ The thought of a house as a living entity that worked with the elements and blended with the land was completely new to me. The idea of a house that could be built at least partially by regular, unskilled people like me and my friends appealed to my independent and thrifty nature.

■ As I drove home, Athena's words of warning haunted me. Somewhere in the middle of one of the casual discussions, Athena had looked firmly at us all. She said, "I can't stress enough how much work this is. Nobody understands until they've done this. It's going to take way longer and cost three times more than you think it will. Whatever you do, design a house with the minimum amount of square feet that you can live in. These people that think they're going to build 1,800-square-foot houses don't understand what they're getting into!" She had picked the exact square footage of my house.

■ As I imagined myself living in this smaller house, I noticed a change come over much of my activities at home. I started to simplify everything. I followed my urge to purge my closet of all the clothes I'd been hanging onto just in case I might want to wear them someday. If I hadn't worn it in a year, I put it in a bag for the homeless. I looked at the art on my walls with a new eye. My new house wouldn't have much spare wall space. Which ones did I really like? . . . Then I started going through the pots and pans, visualizing myself fitting into a much smaller kitchen . . . I even simplified our dinners. We did not need gourmet meals every night . . . Awakened energy surged through me. I didn't need caffeine to keep me awake in meetings at work. I greeted each day with enthusiasm for what I would learn and whom I would meet next. The house was becoming the joyful project that I had intended it to be. But of course, I hadn't begun the hard physical labor yet. . . . As one carpenter said to me, "Those two words, simplify and straw bale, should always go together." And so they should. The texture I was seeking in my life wouldn't come from merely having straw behind the stucco of my home. It would come from changing something inside me and the choices I made each day.

■ The house was built; I had made it. My mind wasn't prepared to have passed an inspection on the first attempt. My thoughts and emotions tumbled over each other. . . . Now, I could cry. Tears of fatigue streamed down my cheeks, and tears for all the fears I had faced without crying. Tears for all the days nobody had shown up to help, and tears for the days when I wanted to spend time with my boys but couldn't. There were tears for all the things that could have gone wrong and tears of self-pity for the days I had worked so hard, when other people were lounging by their pools. Then there were just tears. . . . As all the tears cleansed my Being, a peace radiated through me. The tears turned to those of joy that I had done it despite all the pressures, tears of amazement at all the people who had come to help, tears for all the miracles that made the impossible possible, tears of relief that the strain was behind me now and I had this wonderful house. . . . The house and I were bonded for life. It was a reflection of me, as I had made all these choices for its construction and colors. I knew every bump and crack in the bales, its weak points and strengths, its beauty and dark corners. My Spirit lined the walls and crevices; my love of the desert was reflected in its Soul. I was sure it had one.

■ What an amazing, challenging, soul-searching, exhausting, adventurous project it had been! I doubt if any period of my life will be richer or more fulfilling. I stood now with a strength and confidence that I'm not sure I really understood. My muscles had grown along with my inner strength.

■ One of the main gifts of this house to me, besides its warmth and protection, has been the change its construction brought about in my thought processes. The project was so enormous and unfamiliar, and I had so much at stake, that failure wasn't an option. The only way I found to survive was to look no further than the coming weekend. I was always planning for the upcoming phases of the house, but I didn't allow myself to question whether I'd be able to accomplish them; I simply moved forward with as much certainty as I could muster that when the time came, the solution would be there.

Words of Wisdom from Real Goods Staffers

ON WHERE TO BUILD

■ "Stay away from north-facing slopes. They're very cold."

—Terry Hamor

■ "I would have chosen a building site nearer to the main road, and with better gravity-flow for water instead of building a water tower."

—Debbie Robertson

■ "Don't buy too far out in the boonies. Elbow room is great, but there are practical limits.

"My property was five miles out a steep, rough dirt road. That's fine if you're independently wealthy, and don't have kids or friends. The difficult commute was my primary reason for eventually selling the property."

—Doug Pratt

ON BEING REALISTIC WITH TIME AND MONEY PLANS

■ "Be patient. You're going to be living in your house hopefully for many decades. If you can do it right by spending a few extra months, stop rushing, and pay attention to detail."

—John Schaeffer

■ "If I had to do it over again, I'd realize that it takes three times as long and costs three times as much as expected. I'd have my finances together so it could be built within a year's time instead of fifteen years!"

—Debbie Robertson

ON WHEN TO MOVE IN

■ "Get a cheap portable living space initially. I had a refurbished school bus that allowed me to move onto the property with a minimum of development. This saved rent and allowed me to check out solar access and weather patterns before choosing a building site and designing a house. A house trailer can do the same thing. Don't move in until it's finished. It's real tempting to move in once the walls and roof are up. Resist if at all possible."

—Doug Pratt

■ "I lived in a school bus until it was finished. It was good to be able to move into the shade in the summer and into the sun in the winter. Buses are cheaper than trailers, and come with a motor and charging system. One of my biggest mistakes was moving into the house before it was 100% completed. Most plugs and switches are still not done thirteen years later! Finish it before you move in, or you never will."

—Jeff Oldham

■ "I wouldn't have moved into an unfinished structure."

—Debbie Robertson

ON GENERAL BUILDING PLANS

■ "Use passive solar design. This was one of the things I did right. My house was cool in the summer and warm in the winter, and, with an intelligent design, yours can be too. Avoid the temptation to overglaze on the south side. You'll end up with a house that's too warm in the winter and cools off too quickly at night. Buy quality stuff. Cheap equipment will drive you crazy with frustration and eat up your valuable time."

—Doug Pratt

■ "Hire a carpenter and become his apprentice."

—Terry Hamor

■ "Take an honest assessment of your skills. Decide up front which tasks you can accomplish through your own efforts and which will require assistance. Assign a dollar amount and time required for each. Establish a working budget, which you should update throughout the project."

—Robert Klayman

■ "Money is no object when it comes to insulation! Don't use aluminum door or window frames. Aluminum is too good a conductor of heat. Use wood, vinyl, or fiberglass instead.

Pay attention to little details. It's not a matter of which roofing or siding system you choose; it's the small details like flashing, corner joints, drainage runoffs that really count.

—Jeff Oldham

■ "Design a compost bucket into your kitchen counters that's flush with the top. Design recycling chutes into the kitchen walls with 6" PVC so you can use for bottles and cans—it makes recycling fun for kids and a breeze to keep clean and organized."

—John Schaeffer

■ "Design your wood storage with the ability to load from the outside and retrieve from the inside. Build a laundry chute and a dumbwaiter. Have a small utility bathroom accessible from a back door. Plan for, or install, a dishwasher."

—Debbie Robertson

■ "If you're hiring a contractor, get him or her involved from the beginning of the design phase. Get his/her buy-in for the project before you start. Always find a contractor you can trust and pay him/her time and materials instead of doing a contract. With a contract one of you loses every time—with time and materials, it's a win-win."

—John Schaeffer

Spirit of the Sun

Green living pioneers John Schaeffer and Nancy Hensley have built a home that honors the land, wildlife, and pioneering spirit of their company, Real Goods.

■ Reprinted from *Natural Home Magazine,* November/December 2004.

Robyn Griggs Lawrence
Photography by Barbara Bourne

Between the two of them, John Schaeffer and Nancy Hensley have forty years of experience researching green building, off-the-grid living, and permaculture. John founded Real Goods, the nation's first solar retail business, in 1978 in Hopland, California. Nancy has lived off the grid since 1973 and joined Real Goods in 1989. The couple has realized their dream of employing their collective wisdom to create an energy-independent, nontoxic, environmentally gentle home that promotes sustainability—while also being tastefully beautiful and soul soothing.

They were right on track with that vision as construction on their 2,900-square-foot roundhouse—oriented to the cardinal directions and patterned after a red-tailed hawk ready to take flight—began in early 2001. The home is set on 320 acres overlooking the Hopland Valley that's richly landscaped with gardens, orchards, ponds, a lake, and a grotto with a waterfall. In 2002, John and Nancy moved into a barn on their property where they could watch the building progress. That summer, they noticed several large, black birds pecking mercilessly at the windows. How cute, they thought, until they realized that weeks later the birds—ravens—continued to attack the home's sixty-nine windows with a vengeance.

"We went into major raven research mode," Nancy says. "We talked to biologists, ornithologists, and shamans. People told us to dance around in circles with corn, build altars, give them offerings. We eventually learned that when the birds are nesting, they're very territorial. They saw their reflections and tried to scare off those 'other ravens.'"

"A shaman told us the ravens were upset because we were calling the house Sunhawk—they have a natural animosity toward hawks," John says. "So we made an altar and for thirty days we brought them tobacco, fish, and meat." In marked contrast to locals who thought a shotgun was the answer to their raven problem, John and Nancy's attitude was that the birds had inhabited this piece of land first and that they, as the human "intruders," should strive to live in harmony with the ravens. That open-mindedness translated into every step of building their ecologically friendly home.

BETTER LIVING THROUGH TECHNOLOGY

Built from Rastra blocks, which are made from 85% recycled polystyrene beads and 15% cement, Sunhawk is a masterful example of sustainability. Nancy spent months researching building materials and appliances. Her finds include recycled-tire roof shingles and repurposed granite countertops from a Berkeley café. Roof decking, fascia, barge rafters, and beams were made from reclaimed redwood, Douglas fir, and walnut from an area winery, vinegar plant, warehouse, and a converted orchard.

John and Nancy's house is, of course, entirely off the grid. (Who would expect less from a renewable

ered by rooftop solar hot-water panels and water heated by the excess voltage from the solar and hydroelectric systems.

Living off the grid entails absolutely no sacrifice, John quickly points out. In fact, there are unexpected bonuses. "We got a satellite system for Internet access and it's two to three times as fast as DSL or cable," he says. "The fun part is we get to use our house as a laboratory for the technology and products we order for Gaiam Real Goods."

NOW THIS IS PERMACULTURE

After John and Nancy bought their acreage in 1998, they spent numerous nights camping in various places around their property to find the ideal spot for their house—in the end, exactly the location where Nancy's intuition had initially told her it should be.

Building the Real Goods Solar Living Center in Hopland had taught the couple the importance of landscaping as an integral part of designing a homestead, so they made that a priority—even before finding a designer for the house. "We knew early on that this would be much more than just a house-building project," John says.

Their first major project after grading the road was to dig a ten-acre-foot lake flanked by a thirty-foot-wide

energy pioneer?) A 17-kilowatt solar system—recycled from a Gaiam Real Goods installation in Belize blown down in a hurricane—provides ample power in summer months, and a hydroelectric turbine produces 1.5 kilowatts per hour from a seasonal creek that runs through the property from December through May. "Our hydro system provides thirty-six kilowatt hours per day—almost twice the national average for electricity use," John points out. "And it cost only $1,500—it's way more cost effective than the electric company or any other source of electricity. It's a powerful feeling knowing we'll never have to pay an electric bill for the rest of our lives."

John is most proud of the home's innovative heating and cooling systems, which work so well largely because of the passive heating and cooling design. Cooling the house without air conditioning is no small feat in Hopland, where summer temperatures consistently rise to three digits. In Sunhawk's central core, rocks are buried 9 feet into the earth, where the temperature is a consistent 67 degrees. Two solar-powered fans pull this cool air up culverts and into the central core, where it travels by convection to the rest of the house. Even on the hottest day, the home's interior has never exceeded 76 degrees Fahrenheit. The home is heated primarily by radiant floor tubing pow-

grotto overflowing with waterfalls from the property's three natural springs. Three-and-a-half acres of lush permaculture landscaping include native grasses, a coastal redwood grove, Mediterranean foliage, lavenders, and a corridor of swamp cypress by the pond, which attracts a variety of wildlife including herons, egrets, ducks, coots, and giant bull frogs. To complement the landscaping, the couple added fruit and nut trees and Italian olive trees from which they hope to make their own olive oil. The orchards and the pond are key to the home's comfort; prevailing winds from the northwest flow across them and bring evaporative cooling inside.

The grounds continue to be a work in progress. Four Rastra window boxes on the home's south side provide enough growing area to keep the couple well fed. Vegetable beds are located six feet from the kitchen door so John and Nancy can step outside in any kind of weather to harvest arugula, cilantro, and exotic lettuces. Recently, as part of a Real Goods Solar Living Center permaculture workshop, students spent four days building an herb spiral on the home's north side and installing a composting system, worm bin, and drip watering systems.

LETTING THE HAWK SOAR

With the major landscaping in place, John and Nancy faced the task—and privilege—of creating a house that could live up to their ideals. "A home is far more than a shelter," John says. "It's an expression of our values and commitment, and it enables us to put our convictions into action. We wanted ours to promote not just the principles of sustainability, but to engender restoration and regeneration of the environment, while also nourishing the spirit."

They searched until they found architect Craig Henritzy of Berkeley, who understood that vision. However, they were taken aback when he showed them a set of plans for a Rastra house based loosely on the Cali-

fornia Native American roundhouse (in the style of the indigenous Pomo Indians) and symbolic of the hawk—which he declared John and Nancy's "house totem." "We thought it was visionary and unique, but a real challenge to pull off in a practical sense," John admits. Being open minded,

John and Nancy visited another roundhouse Henritzy had designed in Napa, and they knew they'd found their home. "That house was unlike any other we'd ever seen," John says. "It felt Native American yet twenty-second century—both ancient and futuristic."

A Conversation with the Homeowners

What do you love most about this house?
NANCY: I love the feeling of permanence and belonging. European houses are frequently homes to families for a thousand years. I see no reason ours won't be here to see our newly planted redwood trees turn 1,000 years old.

JOHN: There's nothing better than walking with bare feet on the warm floors on a cold winter day and curling up in the cushions in the south-facing window seats.

What's your favorite room?
NANCY: In the dining room we can sit at the table and enjoy 360-degree views of the pond and the wildlife. We keep a bird guide and binoculars on the dining room table.

JOHN: I love having a great "dishwashing" window in the kitchen where I can look out over the valley while I work. And of course there's the tower where we hold great solstice rituals.

What would you do differently?
NANCY: I would have created a hazardous materials collection station when we began building and introduced all the subcontractors to it. You don't want even nontoxic paints and sealers ending up in the landfill or your future garden.

JOHN: I would be more patient. When you're going to live in a home for decades, who cares if it takes an extra six months to complete? I'd spend more time and think out every detail. It's easier to change things before the house is done.

What advice can you offer new home builders?
NANCY: Include the outside areas in your planning. We just recently poured the sidewalks around the house, so we had a year of mud and dust. Don't wait until the house is done—do it at the same time.

JOHN: Erect your solar system before you start construction. That way you build with solar energy and don't have to listen to and smell generators—and you save money in the long run.

ABOVE: The master bedroom. BELOW: Detail of the kitchen. RIGHT: The floor plan.

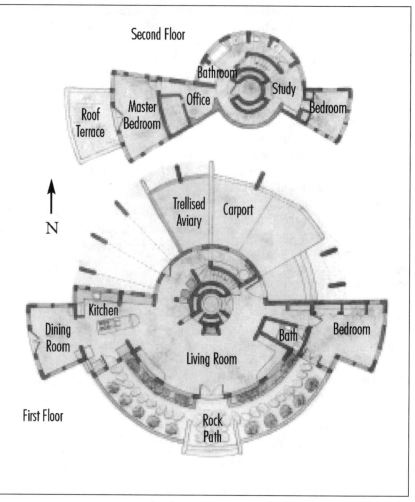

Second Floor

Roof Terrace

Master Bedroom

Office

Bathroom

Study

Bedroom

N

First Floor

Trellised Aviary

Carport

Dining Room

Kitchen

Living Room

Bath

Bedroom

Rock Path

"For Native Americans, the hawk symbolizes 'vision,' which has been important in John and Nancy's work," Henritzy explains. "Also, in the 'green architecture' field we often use just shed-type designs, which I feel has limited the progress of alternative designs being accepted in the housing market. This design explores passive solar with a geometry that celebrates the sun's cycles and playfully and beautifully assumes a hawk shape."

Because Rastra's Styrofoam beads give it a fluid quality, it's easily cut and sculpted, making the hawk shape and orientation with the cardinal directions possible. Sunlight falling on a solar calendar running from north to south on the living room floor marks the passing seasons. On the winter solstice, sunbeams stream through a stained-glass hawk above the south-facing French doors, causing the bird to "fly" across the floor from west to east. At exactly solar noon, the sun illuminates a slate hawk in the floor in front of the living room woodstove.

"I appreciate always knowing the position of the sun," John says. And that, Nancy adds, is really just a fringe benefit of a good passive solar design. "The very basic, most important thing is good southern exposure—taking advantage of sun and light," she points out. "What I love most is that the sun comes in at the right time and doesn't come in at the wrong time." ■■

Sunhawk: The Basics

HOUSE CONSTRUCTION:
Rastra block (85% recycled Styrofoam, 15% cement)
Foundation and Rastra block grout: 6-sack concrete and fly ash mix reinforced with rebar

ENERGY SYSTEM:
—Seventeen-kilowatt photovoltaics (4 kW AstroPower 110w modules and 13 kW Siemens 75w modules)
—Harris hydroelectric turbine

LAND AND SHELTER PRODUCTS

BOOKS

Gaviotas
A Village to Reinvent the World

By Alan Weisman. twenty-five years ago, the village of Gaviotas was established in one of the most brutal environments on Earth, the eastern savannas of war-torn, drug-ravaged Colombia. Read the gripping, heroic story of how Gaviotas came to sustain itself, agriculturally, economically, and artistically, eventually becoming the closest thing to a model human habitat since Eden. These incredibly resourceful villagers invented super-efficient pumps to tap previously inaccessible sources of water, solar kettles to sterilize drinking water, wind turbines to harness the mildest breezes—they even regrew a rain forest! To call this book inspiring is akin to calling the Amazon a winter creek. You won't be able to put it down. 240 pages, softcover. USA.

82306   Gaviotas   **$14.95**

The Solar House: Passive Heating and Cooling

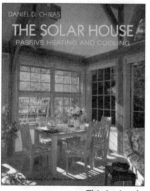

By Daniel D. Chiras. Passive solar design is basically simple, the devil is in the details. This book admits, even highlights, ideas from the 1970s that didn't work, but mostly highlights the good design ideas that *do* work. Chiras explains in detail how homebuilders can design excellent solar buildings attuned to regional climates, buildings that are warm in the winter, cool in the summer, and require a minimum of energy to stay that way. This is simply the best passive solar design book that's ever been available. Includes experienced reviews of current software programs for solar design, and extensive resource lists and appendixes. 284 pages, softcover. USA.

21-0335 The Solar House   **$30**

Finding & Buying Your Place in the Country

By Les and Carol Scher. If you've never bought rural land before, you need this book before you start your search. This is the "bible" on buying rural property, now in its fifth edition. Even for those who've bought land, this well-organized book has plenty of cautions, evaluations, and checklists to help you make informed development decisions. For practical dreamers, this book covers all the things you should consider before you buy rural land. 417 pages, softcover, USA.

82620   Finding & Buying Your Place in the Country, 5th ed.   **$28**

Superbia!

By Dan Chiras and Dave Wann. Practical ideas for creating more socially, economically and environmentally sustainable neighborhoods. Discussions include transformation of a fictitious neighborhood that adopts the authors' 31 precepts. Examples from all over North America and the world provide evidence that citizens can create Superbia! Includes a remarkably comprehensive resource list. 229 pages, softcover. Canada.

21-0356 Superbia   **$19.95**

The New Ecological Home

By Daniel D. Chiras. Shelter, while crucial to our survival, has come at a great cost to the land. But, as author Dan Chiras shows in his follow-up book to *The Natural Home,* this doesn't have to be the case. With chapters ranging from green building materials and green power to landscaping, he covers every facet of today's homes. A good starting point for decreasing the environmental impact of new construction. 352 pages, softcover. USA.

21-0346 The New Ecological Home   **$35**

More Small Houses

In *More Small Houses,* you'll find that smaller is beautiful—and more intimate and affordable. Twenty homes, each less that 2,000 square feet, each a study in craft and efficiency, are used to explore space-saving design ideas. 159 pages, hardcover. USA.

82-449   More Small Houses $24.95

The Not So Big House
A Blueprint for the Way We Really Live

By Sarah Susanka, with Kira Obolensky. This is one of the best, most inspiring home design books that's ever been written. Sarah's sense of what delights us, what

makes us feel secure, and what simply works well for all humans is flawless. Is it a surprise that people gather in the kitchen? It shouldn't be; human beings are drawn to intimacy, and this book proposes housing ideas that serve our spiritual needs while protecting our pocketbook. Commonsense building ideas are combined with gorgeous home design and dozens of small simple touches that delight the senses. This is an inspiring book on building houses that make us feel safe, protected, and comfortably at home. 199 pages, softcover. USA.

82-375   The Not So Big House $23

Ecological Design

By Sim Van der Ryn and Stuart Cowan. This groundbreaking book profiles how the living world and humanity can be reunited by making ecology the basis

for design. The synthesis of nature and technology, ecological design can be applied to create revolutionary forms of buildings, landscapes, and cities—as well as new technologies for a sustainable world. Some examples of innovations include sewage treatment plants that use constructed marshes to purify water and industrial "ecosystems" in which the waste from one process becomes fuel for the next. The first part of the book presents an overview of ecological design. The second part devotes chapters to five design principles fundamental to ecological design. A resource guide and annotated bibliography are included. 201 pages, softcover. USA.

80-512   Ecological Design $24

Green Remodeling
Changing the World One Room at a Time

By David Johnston and Kim Master. Millions of North Americans are renovating their homes every year. How do you remodel in a healthy, environmentally friendly way?

Green Remodeling is a comprehensive guide. Buildings are responsible for 40% of worldwide energy flow and material use; so how you remodel can make a difference. Green remodeling is more energy-efficient, more resource-conserving, healthier for occupants and more affordable to create, operate and maintain.

The book discusses simple green renovation solutions for homeowners, focusing on key aspects of the building including foundations, framing, plumbing, windows, heating and finishes. Room by room, it outlines the intricate connections that make the house work as a system. Then, in an easy-to-read format complete with checklists, personal stories, expert insights and an extensive resource list, it covers easy ways to save energy, conserve natural resources, and protect the health of loved ones. Addressing all climates, this is a perfect resource for conventional homeowners, as well as architects and remodeling contractors. 368 pages, softcover. Canada.

21-0365 Green Remodeling $29.95

Architectural Resource Guide

From the Northern California Architects/Designers/ Planners for Social Responsibility. Ever wondered where to find all those nontoxic, recycled, sustainable, or just plain better building materials? Well, here it is. This absolute gold mine of information has a huge wealth of products listed logically and intelligently following the standard architect's Uniform Specification System (e.g., concrete; thermal and moisture protection; windows and doors; finishes; etc.). Each section is preceded by a few pages explaining the pros and cons of particular materials, appliances. Each listing says what it is, gives the brand name, a brief explanation of the product or service, and access information, both national and

NorCal local, for the manufacturer. Although written for the professional architect or builder, there's plenty for everyone here. Updated continuously, we order small quantities to ensure shipping the latest edition. 152 pages (approx), softcover. USA.

80-395   Architectural Resource Guide $35

Circle Houses

By David Pearson. The natural world is filled with circles, yet circular houses are unusual in industrialized societies. Among nomadic peoples, round-shaped homes have since time immemorial served to shelter families, echoing natural forms. A round home represents the universe in microcosm: the floor (Earth), the roof (sky) and the

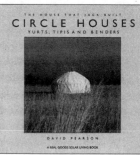

hole in the roof (sun). *Circle Houses* is a fascinating glimpse of tradition meeting timelessness, with stories of twenty-first century nomads, and basic instructions for designing and constructing your own tipi, yurt or bent-frame tent. 95 pages, hardcover. USA.

82616   Circle Houses $17

Treehouses

By David Pearson. This eclectic and delightful collection of tree-dwellers and their dwellings will inspire even the diehard couch potato to grab a hammer and head for the backyard. Featuring real projects from all around the globe, this book depicts twenty distinctive tree dwellings, each one a masterwork of inventiveness. Lavishly illustrated with color photography, this book portrays not only the finished project, but also the construction process, giving enough information to complete a simple project. The extensive resource section offers places to visit, both real and virtual. David Pearson, noted author of *The Natural House Book,* has captured the combined spirits of innocence and inspiration in this awesome little book. 120 pages, hardcover. USA.

82-614   Treehouses $16.95

Serious Straw Bale

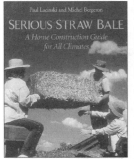

By Paul Lacinski and Michel Bergeron. Straw bale construction is a delightful form of natural, ecological building. But until the publication of *Serious Straw Bale: A Home Construction Guide for All Climates,* it was assumed appropriate only for dry, arid climates. This book does not advocate universal use of straw bale, but provides practical solutions to the challenges brought on by harsh climates. This is the most comprehensive, up-to-date guide to building with bales ever published. Builders, designers, regulators, architects, and owner-builders will benefit from learning how bale building technique now includes climates and locales once thought impossible. 371 pages, softcover. USA.

82-485   Serious Straw Bale $29.95

The Beauty of Straw Bale Homes

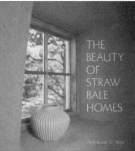

By Athena and Bill Steen. This book is a visual treat, a gorgeous pictorial celebration of the tactile, timeless charm of straw bale dwellings. A diverse selection of building types is showcased—from personable and inviting smaller homes to elegant large homes and contemporary institutional buildings. The lavish photographs are accompanied by a brief description of each building's unique features. Interspersed throughout the book are insightful essays on key lessons the authors have learned in their many years as straw bale pioneers. 128 pages, softcover.

82-484   The Beauty of Straw Bale Homes $22.95

Straw Bale Details

By Chris Magwood and Chris Walker. Nothing conveys a wealth of information like a good drawing by a professional architect, which is what this book is all about. It's the nitty-gritty of how to assemble a well-built, long-lasting strawbale structure. All practical foundation, wall, door and window, and roofing options are illustrated, along with notes and possible modifications. Has a clever ring binding that lays flat for easy worksite reference. 68 pages, softcover. Canada.

21-0341 Straw Bale Details $32.95

Buildings of Earth and Straw
Structural Design for Rammed Earth and Straw Bale

By Bruce King. Straw bale and rammed earth construction are enjoying a fantastic growth spurt in the United States and abroad. When interest turns to action, however, builders can encounter resistance from mainstream construction and lending communities unfamiliar with these materials. *Buildings of Earth and Straw* is written by structural engineer Bruce King, and provides technical data from an engineer's perspective. Information includes special construction requirements of earth and straw; design capabilities and limitations of these materials; and most importantly, the documentation of testing that building officials often require.

This book will be an invaluable design aid for structural engineers, and a source of insight and understanding for builders of straw bale or rammed earth. 190 pages, softcover with photos, illustrations, and appendices. USA.

80-041   Buildings Of Earth And Straw $25

Building with Earth
A Guide to "Dirt Cheap" and "Earth Beautiful"

By Paulina Wojciechowska. Building with Earth is the first comprehensive guide to the re-emergence of earthen architecture in North America. Even inexperienced builders can construct an essentially tree-free building, from foundation to curved roof, using recycled tubing or textile grain sacks. Featuring beautifully textured earth- and lime-based finish plasters for weather protection, "earthbag" buildings are being used for retreats, studios, and full-time homes in a wide variety of climates and conditions. This book tells (and shows in breathtaking detail) how to plan and build beautiful, energy-efficient earthen structures. Author Paulina Wojciechowska is founder of Earth, Hands and Houses, a nonprofit that supports building projects that empower indigenous people to build shelters from locally available materials. 200 pages, softcover. USA.

82-613   Building with Earth $24.95

Building with Structural Insulated Panels

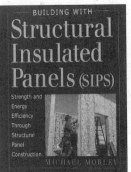

By Michael Morley. Structural Insulated Panels (SIPs) are in the process of replacing the post-war norm of stick-framed, fiberglass-insulated houses and light commercial buildings. SIPs produce a structurally superior, better insulated, faster to erect, and more environmentally friendly house than ever before possible. Experienced SIP builder Michael Morley covers choosing the right panels for the job, tools and equipment, wall and roof systems, and mechanical systems. This is an indispensable reference for anyone who wants to work with this innovative, field-tested construction system. 186 pages, hardcover, USA.

82621   Building with Structural Insulated Panels $35

Stone House

By Tomm Stanley. A great resource if you dream of building your own home. Topics range from traditional stonemasonry and slip forming to concrete making and restoring recycled wooden doors and windows. Wonderfully illustrated with photographs and supported by an online photo gallery. Includes a glossary and FAQ. 208 pages, softcover. USA.

21-0351 Stone House $33

The Hand-Sculpted House

By Ianto Evans, Michael G. Smith, and Linda Smiley. Learn how to build your own home from a simple earthen mixture. Derived from the Old English word meaning "lump," cob has been a traditional building method for millennia. Authors Ianto Evans, Michael G. Smith and Linda Smiley teach you how to sculpt a cozy, energy-efficient home by hand without forms, cement or machinery. Includes illustrations and photography. 384 pages, softcover. USA.

21-0331 The Hand-Sculpted House $35

Earthbag Building

The Tools, Tricks and Techniques

By Kaki Hunter and Donald Kiffmeyer. *Earthbag Building* is the first comprehensive guide for building with bags filled with earth—or earthbags. The authors developed this "Flexible Form Rammed Earth Technique" over the last decade. A reliable method for constructing homes, outbuildings, garden walls and much more, this enduring, tree-free architecture can also be used to create arched and domed structures of great beauty in any climate.

This profusely illustrated guide first discusses the many merits of earthbag construction, and then leads the reader through the key elements of an earthbag building, including: Special design considerations; Foundations, walls, and floors; Electrical, plumbing, and shelving; Lintels, windows and door Installations; Roofs, arches and domes; Exterior and interior plasters.

With dedicated sections on costs, making your own specialized tools, and building code considerations, as well as a complete resources guide, Earthbag Building is the long-awaited, definitive guide to this uniquely pleasing construction style. 288 pages, softcover. Canada.

21-0367 Earthbag Building $29

The Natural House

A Complete Guide to Healthy, Energy-Efficient Environmental Homes

By Daniel D. Chiras. Aimed at both builders and dreamers, this extraordinary book explores environmentally sound, natural, and sustainable building methods including rammed earth, straw bale, earthships, adobe, cob, and stone. Besides sanctuary to shelter our families, today we want something more— we want to live free of toxic materials. Chiras is an unabashed advocate of these new technologies, yet he openly addresses the pitfalls of each, while exposing the principles that make them really work for natural, humane living. This book is devoted to analysis of the sustainable systems that make

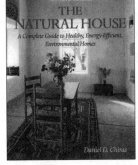

these new homes habitable. This is the short course in passive solar technology, green building material, and site selection, addressing more practical matters—such as scaling your dreams to fit your resources—as well. 469 pages, softcover. USA.

82-474   The Natural House $35

Pocket Reference to the Physical World

Compiled by Thomas J. Glover. This amazing book, measuring just 3.2 by 5.14 inches, is like a set of encyclopedias or Internet search engine in your shirt pocket! It's a comprehensive resource of everything from the physical properties of air to pipe-sizing information, and from calculating joist spacing to converting a broad range of weights and measures. A big chunk of knowledge in a very small package. No desktop is complete without one. 544 pages, softcover. USA.

80-506   Real Goods Pocket Ref $12.95

Sunshine to Electricity

BETTER THAN 99% of the world's energy comes from the sun. Some is harvested directly by plants, trees, and solar panels, much is used indirectly in the form of wood, coal, or oil, and a tiny bit is supplied by the non-solar sources, geothermal and nuclear power. Solar panels receive this energy directly. Both wind and hydro power sources use solar energy indirectly. The coal and petroleum resources that we're so busy burning up now represent stored solar energy from the distant past, yet every single day, enough free sunlight energy falls on the Earth to supply our energy needs for four to five years at our present rate of consumption. Best of all, with this energy source, there are no hidden costs, and no borrowing or dumping on our children's future. The amount of solar energy we take today in no way diminishes or reduces the amount we can take tomorrow and tomorrow and tomorrow.

Solar energy can be harnessed directly in a variety of ways. One of the oldest uses is heating domestic water for showering, dishwashing, or space heating. At the turn of the century, solar hot water panels were an integral part of 80% of homes in Southern California and Florida until gas companies, sensing a serious low-cost threat to their businesses, started offering free water heaters and installation. What we're going to cover in this chapter is one of the more recent, cleanest, and most direct ways to harvest the sun's energy: photovoltaics or PV.

Give me the splendid silent sun with all his beams full-dazzling.

—Walt Whitman

What are Photovoltaic Cells?

Photovoltaic cells were developed at Bell Laboratories in the early 1950s as a spin-off of transistor technology. Very thin layers of pure silicon are impregnated with tiny amounts of other elements. When exposed to sunlight, small amounts of electricity are produced. They were mainly a laboratory curiosity until the advent of spaceflight in the 1960s, when they were found to be an efficient, long-life, although staggeringly expensive, power source for satellites. Since the early 1960s PV cells have slowly but steadily come down from prices of over $40,000 per watt to current prices of around $5 per watt, or in some cases as low as $3 per watt in very large quantities. Using the technology available today, we could equal the entire electric production of the United States with photovoltaic power plants using about 10,000 square miles, or less than 12% of the state of Nevada. See map on page 57 for details. As the true environmental and societal costs of coal and petroleum become more apparent, PVs promise to be a major power resource in the future. And with worldwide petroleum sources close to peak, we are rapidly running out of "cheap" oil, making renewable energy all the more critical. Who says that space programs have no benefits for society at large?

In 1954, Bell Telephone Systems announced the invention of the Bell Solar Battery, a "forward step in putting the energy of the sun to practical use."

Fifty Years of Photovoltaics

Solar Energy Still Strikes a Powerful Chord

Hearing the hit songs of Rosemary Clooney or Perry Como was an ordinary occurrence in the mid-1950s. But it turned extraordinary on Sunday, April 25, 1954.

It was then that Bell Laboratories executives electrified members of the press with music broadcast from a transistor radio—powered by the first silicon solar cell. Bell called its invention "the first successful device to convert useful amounts of the sun's energy directly and efficiently into electricity."

The discovery was, indeed, music to peoples' ears. *The New York Times* heralded it as "the beginning of a new era, leading eventually to the realization of one of mankind's most cherished dreams—the harnessing of the almost limitless energy of the sun for the uses of civilization."

A press release dated April 25, 1954, announces the discovery of Bell's revolutionary solar cell.

A Single Cell Gives Birth to Millions

This fifty-year-old prophecy has, in many ways, come to fruition, with a billion watts of solar-electric modules now powering satellites, telescopes, homes, water pumps and even Antarctic research stations.

Today's solar cells are a vast improvement on their ancestors—achieving 16 to 19% efficiency compared to the 6% efficiency of Bell Labs' foray into the solar industry. Silicon is still used—though in slightly different forms to achieve higher efficiency rates as well as lower costs. Lab tests using single-crystal silicon modules, for example, have produced efficiency rates of up to 24%.

The two- to three-fold increase in efficiency is only part of the equation in the success of photovoltaics—an industry that has grown more than 20% compounded per year since 1980. A tenfold price drop since 1954, along with tax incentives and rebates, have made the move to solar power economically feasible for many homeowners and businesses, with typical payback periods of less than ten years.

The first advertisement for the silicon solar cell, appearing in the August 1954 issue of *National Geographic*.

All black and white images: property of AT&T Archives. Reprinted with permission of AT&T.

Chemist Calvin Fuller gets ready to diffuse boron into negative silicon. The addition of boron resulted in the first solar cell capable of producing useful amounts of electricity from the sun.

Where Will Solar Energy Be in 2054?

Of course, the satisfaction of using a renewable energy source may be the biggest motivator of all—and that's what fuels our excitement most. As Margaret Mead said so eloquently, "Never doubt that a small group of thoughtful, committed citizens can change the world. Indeed, it is the only thing that ever has."

Real Goods has now solarized more than 60,000 homes and businesses, and we can't wait to see what will happen as that number gets bigger over the next fifty years. You can count on Real Goods to continue leading the way.

A Brief Technical Explanation

A single PV cell is a thin semiconductor sandwich, with a layer of highly purified silicon. The silicon has been slightly doped with boron on one side, and with phosphorous on the other side. Doping produces either a surplus or a deficit of electrons depending on which side we're looking at. Electronics-savvy folks will recognize these as p- and n-layers, the same as transistors use. When our sandwich is bombarded by sunlight, photons knock off some of the excess electrons. This creates a voltage difference between the two sides of the wafer, as the excess electrons try to migrate to the deficit side. In silicon, this voltage difference is just under half a volt. Metallic contacts are made to both sides of the wafer. If an external circuit is attached to the contacts, the electrons find it easier to take the long way around through our metallic conductors, than to struggle through the thin silicon layer. We have a complete circuit and a current flows. The PV cell acts like an electron pump. There is no storage capacity in a PV cell, it's simply an electron pump. Each cell makes just under half a volt, regardless of size. The amount of current is determined by the number of electrons that the solar photons knock off. We can get more electrons by using bigger cells, or by using more efficient cells, or by exposing our cells to more intense sunlight. There are practical limits however to size, efficiency, or how much sunlight the cell can tolerate.

Since 0.5-volt solar panels won't often do us much good, we usually assemble a number of PV cells for higher voltage output. A PV "module" consists of many cells wired in series to produce a higher voltage. Modules consisting of about thirty-six cells in series have become the industry standard for large power production. This makes a module that delivers power at 17 to 18 volts, a very handy level for 12-volt battery

In 1992, the University of South Florida developed a 15.89% efficient thin-film cell, breaking the 15% barrier for the first time.

A Technical Step Back in Time

Before the invention of the silicon solar cell, scientists were skeptical about the success of solar as a renewable energy source. Bell Labs' Daryl Chapin, assigned to research wind, thermoelectric and solar energy, found that existing selenium solar cells could not generate enough power. They were only able to muster 5 watts per square meter—converting less than 0.5% of incoming sunlight into electricity.

THE SOLAR CELL THAT ALMOST NEVER WAS

Chapin's investigation may have ended there if not for colleagues Calvin Fuller and Gerald Pearson, who discovered a way to transform silicon into a superior conductor of electricity. Chapin was encouraged to find that the silicon solar cell was five times more efficient than selenium. He theorized that an ideal silicon solar cell could convert 23% of incoming solar energy into electricity. But he set his sights on an amount that would rank solar energy as a primary power source: 6% efficiency.

NUCLEAR CELL TRUMPS SOLAR SCIENTISTS

While Chapin, Pearson and Fuller faced challenges in increasing the efficiency of the silicon cell, archrival RCA Laboratories announced its invention of the atomic battery—a nuclear-powered silicon cell. Running on photons emitted from Strontium-90, the battery was touted as having the potential to run hearing aids and wristwatches for a lifetime.

That turned up the heat for Bell's solar scientists, who were making strides of their own by identifying the ideal location for the P-N junction—the foundation of any semiconductor. The Bell trio found that shifting the junction to the surface of the cell enhanced conductivity, so they experimented with substances that could permanently fix the P-N junction at the top of the cell. When Fuller added arsenic to the silicon and coated it with an ultra-thin layer of boron, it allowed the team to make the electrical contacts they had hoped for. Cells

The three inventors of the first silicon solar cell, Gerald Pearson, Daryl Chapin, and Calvin Fuller, examine their cells at their Bell lab.

were built using this mixture until one performed at the benchmark efficiency of 6%.

SOLAR BURSTS THE ATOMIC BUBBLE

After releasing the invention to the public on April 25, 1954, one journalist noted, "Linked together electrically, the Bell solar cells deliver power from the sun at the rate of 50 watts per square yard while the atomic cell recently announced by the RCA Corporation merely delivers a millionth of a watt" over the same area. The solar revolution was born.

Special thanks to: John Perlin, author of *From Space to Earth: The Story of Solar Electricity.*

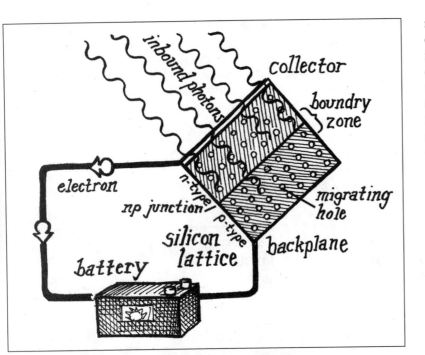

In 2002, 500 million photovoltaic cells were shipped around the world, totaling 500 megawatts of electricity.

Sharp 123 PV Module.

reliable power source, and it's usually easier to obtain at the gate site. More than 100,000 homes in the United States, largely in rural sites, depend on PVs as a primary power source, and this figure is growing rapidly as people begin to understand how clean and reliable this power source is, and how deeply our current energy practices are borrowing from our children. Worldwide, over one million homes currently derive their primary power from PV. Because they don't rely on miles of exposed wires, residential PV systems are more reliable than utilities, particularly when the weather gets nasty. PV modules have no moving parts, degrade very, very slowly, and boast a lifespan that isn't fully known yet, but will be measured in multiple decades. Standard factory PV warranties are ten to twenty-five years. Compare this to any other power generation technology or consumer goods.

charging. In recent years as PV modules and systems have grown larger, 24-volt modules consisting of seventy-two cells have also become standardized. The module is encapsulated with tempered glass (or some other transparent material) on the front surface, and with a protective and waterproof material on the back surface. The edges are sealed for weatherproofing, and there is often an aluminum frame holding everything together in a mountable unit. A junction box, or wire leads, providing electrical connections usually is found on the module's back. Truly weatherproof encapsulation was a problem with the early modules assembled twenty years ago. We have not seen any encapsulation problems with glass-faced modules in many years.

Many applications need more than a single PV module, so we build an "array." A PV array consists of a number of individual PV modules that have been wired together in series and/or parallel to deliver the voltage and amperage a particular system requires. An array can be as small as a single pair of modules, or large enough to cover acres.

PV costs are down to a level that makes them the clear choice for remote *and* grid intertie power applications. They are routinely used for roadside emergency phones and most temporary construction signs, where the cost and trouble of bringing in utility power outweighs the higher initial expense of PV, and where mobile generator sets present more fueling and maintenance trouble. It's hard to find new gate-opener hardware that isn't solar-powered. Solar with battery backup has proven to be a far more

Construction Types

There are currently four commercial production technologies for PV cells.

SINGLE CRYSTALLINE

This is the oldest and most expensive production technique, but it's also the most efficient sunlight conversion technology commercially available. Complete modules have sunlight to wire output efficiency averages of about 10 to 12%. Efficiencies up to 22% have been achieved in the lab, but these are single cells using highly exotic components that cannot economically be used in commercial production.

Boules (large cylindrical cylinders) of pure single-crystal silicon are grown in an oven, then sliced into wafers, doped, and assembled. This is the same process used in manufacturing transistors and integrated circuits, so it is very well-developed, efficient, and clean. Degradation is very slow with this technology, typically 0.25 to 0.5% per year. Silicon crystals are characteristically blue, and single crystalline cells look like deep blue glass. Examples are Shell and Sharp single crystalline products.

POLYCRYSTALLINE OR MULTICRYSTALLINE

In this technique, pure, molten silicon is cast into cylinders, then sliced into wafers off the large block of multicrystalline silicon. Polycrystal is very slightly lower in conversion efficiency compared to single crystal, but the manufacturing process is less exacting, so costs are a bit lower. Module efficiency averages about 10 to

11%, sunlight to wire. Degradation is very slow and gradual, similar to single crystal discussed above. Crystals measure approximately a centimeter (two-fifths of an inch) and the multicrystal patterns can be seen clearly in the cell's deep blue surface. Doping and assembly are the same as for single-crystal modules. Examples are Shell, Solarex, and Kyocera polycrystalline products.

STRING RIBBON

This clever technique is a refinement of polycrystalline production. A pair of strings are drawn through molten silicon pulling up a thin film of silicon like a soap bubble. It cools, crystallizes, and you've got ready-to-dope wafers. The ribbon width and thickness can be controlled, so there's far less slicing, dicing, or waste, and production costs are lower. Sunlight to wire conversion efficiency is about 8 to 9%. Degradation is the same as ordinary slice and dice polycrystal. Examples are Evergreen modules.

AMORPHOUS OR THIN-FILM

In this technique, silicon material is vaporized and deposited on glass or stainless steel. This production technology costs less than any other method, but the cells are less efficient, so more surface area is required. Early production methods in the 1980s produced a product that faded up to 50% in output over the first three to five years. Present-day thin-film technology has dramatically reduced power fading, although it's still a long-term uncertainty. Uni-Solar has a "within 20% of rated power at 20 years" warranty, which relieves much nervousness, but we honestly don't know how these cells will fare with time. Sunlight to wire efficiency averages about 5 to 7%. These cells are often almost black in color. Unlike other modules, if glass is used on amorphous modules it is not tempered, so breakage is more of a problem. Tempered glass can't be used with this high-temperature deposition process. If the deposition layer is stainless steel and a flexible glazing is used, the resulting modules will be somewhat flexible. These are often used as marine or RV modules. In the mid-1990s it appeared that amorphous modules could deliver the magic $2 per watt that would make solar sprout on every rooftop. There was a rush to build assembly plants. Oddly, it turned out that few homeowners wanted to cover every square foot of their roof with an unproven and still fairly expensive solar product. We're seeing fewer examples of thin-film technology. Uni-Solar makes examples of flexible, unbreakable modules.

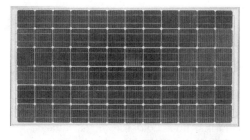

Shell Single Crystal Module

Solarex Multicrystalline Module

Evergreen String Ribbon Module

Uni-Solar Amorphous Module

Putting It All Together

The PV industry has standardized on 12-, 24-, or 48-volts for battery systems. These are moderately low voltages, which makes working with them somewhat safer. Under most circumstances, this isn't enough voltage to pass through your body. While not impossible, it's pretty difficult to hurt yourself on such low voltage. Batteries, however, where we store enormous quantities of accumulated energy, can be very dangerous if mishandled or miswired. Please see chapter 3, which includes batteries and safety equipment, for more information.

Multiple modules can be wired in parallel or series to achieve any desired output. As systems get bigger, we usually run collection and storage at higher DC voltages because transmission is

A Mercifully Brief Glossary of PV System Terminology

AC—Alternating Current. This refers to the standard utility-supplied power, which alternates its direction of flow 60 times per second, and for normal household use has a voltage of approximately 120 or 240 (in the United States). AC is easy to transmit over long distances, but it is impossible to store. Most household appliances require this kind of electricity.

DC—Direct Current. This is electricity that flows in one direction only. PV modules, small wind turbines, and small hydroelectric turbines produce this type, and batteries of all kinds store it. Appliances that operate on DC very rarely will operate directly on AC and vice versa. Conversion devices are necessary.

Inverter—An electronic device that converts the low voltage DC power we can store in batteries that to conventional 120-volt AC power as needed by lights and appliances. This makes it possible to utilize the lower cost (and often higher quality) mass-produced appliances made for the conventional grid-supplied market. Inverters are available in a wide range of wattage capabilities. We commonly deal with inverters from 50- through 10,000-watts capacity.

PV Module—A "solar panel" that makes electricity when exposed to direct sunlight. PV is shorthand for photovoltaic. We call them PV modules to differentiate from solar hot water panels, which are a completely different technology, and are often what folks think of when we say "solar panel." PV modules do not make hot water.

easier. Small systems processing up to about 2,000 watt-hours per day are fine at 12 volts. Systems processing 2,000 to 7,000 watt-hours per day will function better at 24 volts, and systems running more than 7,000 watt-hours per day should probably be running at 48 volts. These are guidelines, not hard and fast rules! The modular design of PV panels allows systems to grow and change as needs change. Modules from different manufacturers, different wattages, and various ages can be intermixed with no problems, so long as all modules have a rated voltage output within about 1.0 volt of each other. Buy what you can afford now, then add to it in a few years when you can afford to expand.

Efficiency

By scientific definition, the sun delivers 1,000 watts (1 kilowatt) per square meter at noon on a clear day at sea level. This is defined as a "full sun" and is the benchmark by which modules are rated and compared. That is certainly a nice round figure, but it is not what most of us actually see. Dust, water vapor, air pollution, seasonal variations, altitude, and temperature all affect how much power your modules actually receive. For instance, the 1991 eruption of Mt. Pinatubo in the Philippines reduced available sunlight worldwide by 10 to 20% for a couple of years. It is reasonable to assume that most sites actually will average about 85% of full sun, unless they are over 7,000 feet in elevation, in which case they probably will receive more than 100% of full sun.

PV modules do not convert 100% of the energy that strikes them into electricity (we wish!). Current commercial technology averages about 10% to 12% conversion efficiency for single- and multicrystalline modules, and 5% to 7% for amorphous modules. Conversion rates slightly over 20% have been achieved in the laboratory by using experimental cells made with esoteric and rare elements, but these elements are far too expensive to ever see commercial production. Conversion efficiency for commercial single and multicrystalline modules is not expected to improve; this is a mature technology. There's better hope for increased efficiency with amorphous technology, and much research is currently underway.

How Long Do PV Modules Last?

PV modules last a long, long time. How long we honestly don't yet know, as the oldest terrestrial modules are barely thirty-five years old, and still going strong. In decades-long tests, the fully developed technology of single and polycrystal modules has shown to degrade at fairly steady rates of 0.25 to 0.5% per year. First-generation amorphous modules degraded faster, but there are so many new wrinkles and improvements in amorphous production that we can't draw any blanket generalizations for this module type. The best amorphous products now seem to match closely the degradation of single-crystal products, but there is little long-term data. All full-size modules carry ten- to twenty-five-year warranties, reflecting their manufacturers' faith in the durability of these products. PV technology is closely related to transistor technology. Based on our experience with transistors, which just fade away after twenty years of constant use, most manufacturers have been confidently predicting twenty-year or longer lifespans. However,

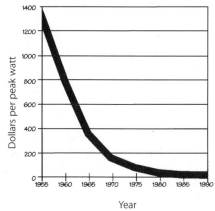

PHOTOVOLTAIC PRICES

Photovoltaic prices have decreased dramatically since 1955. Prices continue to drop 10 to15% per year as demand and production increase. We have seen prices as low as $3 per watt for very large systems in mid 2004.

keep in mind that PV modules are only seeing six to eight hours of active use per day, so we may find that lifespans of sixty to eighty years are normal. Cells that were put into the truly nasty environment of space in the late 1960s are still functioning well. The bottom line? We're going to measure the life expectancy of PV modules in decades—how many, we don't know yet.

Payback Time for Photovoltaic Manufacturing Energy Investment

In the early years of the PV industry, a nasty rumor was circulating that said PV modules would never produce as much power over their lifetimes as it took to manufacture them. During the very early years of development, when transistors were a novelty, and handmade PV modules costing as much as $40,000 per watt were being used exclusively for spacecraft, this was true. The truth now is that PV modules pay back their manufacturing energy investment in 1.4 to 10 years time (less than half of the typical warranty period), depending on module type, installation climate, and other conditions. Now, in all honesty, this information comes to us courtesy of the module manufacturers. The National Renewable Energy Labs have done some impartial studies on payback time (see the results at www.nrel.gov/ncpv/pdfs/24596.pdf). They conclude that modules installed under average U.S. conditions reach energy payback in three to four years, depending on construction type. The aluminum frame all by itself can account for six months to one year of that time.

Quicker energy paybacks, down to one to two years, are expected in the future as more "solar grade" silicon feedstock becomes available, and simpler standardized mounting frames are developed.

Maintenance

It's almost laughable how easy maintenance is for PV modules. Having no moving parts makes them practically maintenance-free. Basically, you keep them clean. If it rains irregularly or if the birds leave their calling cards, hose the modules down. Do not hose them off when they're hot, since uneven thermal shock theoretically could break the glass. Wash them in the morning or evening. For PV maintenance, that's it.

Control Systems

Controls for PV systems are usually simple. When the battery reaches a full-charge voltage, the charging current can be turned off, or directed elsewhere. Open-circuited, PV module voltage rises 5 to 10 volts and stabilizes harmlessly. It does no harm to the modules to sit at open-circuit voltages, but they aren't doing any work for you either. When the battery voltage drops to a certain set-point, the charging circuit is closed and the modules go back to charging. With the newer solid-state PWM controllers (Pulse Width Modulated), this opening and closing of the circuit happens so rapidly that you'll simply see a stable voltage. The most recent addition to PV control technology is Maximum Power Point Tracking, or MPPT controls. These sophisticated solid state controllers allow the modules to run at whatever voltage produces the maximum wattage. This is usually a higher voltage than batteries will tolerate. The extra voltage is downconverted to amperage that the batteries can digest comfortably. MPPT controls extract an average of about 15% more energy from your PV array, and do their best work in the wintertime when most residential systems need all the help they can get. Most controllers offer a few other whistles and bells, such as nighttime disconnect and LED indicator lights. See the Charge Controllers section in the chapter 3 for a complete discussion of controllers.

Powering Down

The downside to all this good news is that the initial cost of a PV system is still high. After

The truth now is that PV modules pay back their manufacturing energy investment in 1.4 to 10 years time (less than half of the typical warranty period), depending on module type, installation climate, and other conditions.

decades of cheap, plentiful utility power, we've turned into a nation of power hogs. The typical American home consumes about 20 to 30 kilowatt hours daily. Supplying this demand with PV-generated electricity can be costly; however, it makes perfect economic sense as a long-term investment. Fortunately, at the same time that PV-generated power started becoming affordable and useful, conservation technologies for electricity started becoming popular, and given the steadily rising cost of utility power, even necessary. The two emerging technologies dovetail beautifully. Every kilowatt-hour you can trim off your projected power use in a stand-alone PV-based system will reduce your initial setup cost by $3,000 to $4,000. Using a bit of intelligence and care in your lighting and appliance selection will allow you all the conveniences of the typical 20 kWh per day American home, while consuming less than 6 kWh per day. At $4,000 per installed kilowatt-hour that's $56,000 we just shaved off the initial system cost! With this kind of careful analysis applied to electrical use, most of the full-size home electrical systems we design come in between $10,000 to $20,000, depending on the number of people and intended lifestyle. Simple weekend cabins with a couple of lights and a boom box can be set up for $1,000, or less. With the renewable energy rebates and buydowns available in an increasing number of states, PV can be very cost-effective. Typical payback times in California run five to twelve years (8 to 20% return on investment!). Commercial paybacks with tax incentives typically pay back in half that time.

> Every kilowatt-hour you can trim off your projected power use in a stand-alone PV-based system will reduce your initial setup cost by $3,000 to $4,000.

Other chapters in this *Sourcebook* present an extensive discussion of electrical conservation, for both off and on grid (utility power), and offer many of the lights and appliances discussed. We strongly recommend reading these sections before beginning system sizing. We are not proposing any substantial lifestyle changes, just the application of appropriate technology and common sense. Stay away from 240-volt watt-hogs, cordless electric appliances, standard incandescent light bulbs, instant-on TVs, and monster side-by-side refrigerators, and our friendly technical staff can work out the rest with you.

PV Performance in the Real World

Okay, here's the dirt under the rug. Skeptics and pessimists knew it all along: PV modules could not possibly be all that perfect and simple. Even the most elegant technology is never quite perfect. There are a few things to watch out for, beginning with . . . Wattage ratings on PV modules are given under ideal laboratory conditions. As most of us figured out a long time ago, the world isn't a perfect place. Assuming you can avoid or eliminate shadows, the two most important factors that affect module performance out in the real world are percentage of full sun and operating temperature.

SHADOWS

Short of outright physical destruction, hard shadows are the worst possible thing you can do to a PV module. Even a tiny amount of shading dramatically affects module output. Electron flow is like water flow. It flows from high voltage to low voltage. Normally the module is high and the battery or load is lower. A shaded portion of the module drops to very low voltage. Electrons from other portions of the module and even from other modules in the array will find it easier to flow into the low-voltage shaded area than into the battery. These electrons just end up making heat, and are lost to us. This is why bird droppings are a bad thing on your PV module. A fist-sized shadow will effectively shut off a PV module. Don't intentionally install your modules where they will get shadows during the prime midday generating time from 10 A.M. to 3 P.M. Early or late in the day when the sun is at extreme angles there's little power, so don't sweat shadows then. Sailors may find shadows unavoidable at times, just keep them clear as much as practical.

Photovoltaic Summary

ADVANTAGES

1. *No moving parts*
2. *Ultra-low maintenance*
3. *Extremely long life*
4. *Noncorroding parts*
5. *Easy installation*
6. *Modular design*
7. *Universal application*
8. *Safe low-voltage output*
9. *Simple controls*
10. *Long-term economic payback.*

DISADVANTAGES

1. *High initial cost*
2. *Only works in direct sunlight*
3. *Sensitive to shading*
4. *Lowest output during shortest days*
5. *Low-voltage output difficult to transmit*

FULL SUN

As previously mentioned in the Efficiency paragraph, most of us seldom see 100% of full sun conditions. If you are not getting full, bright, shadow-free sun, then your PV output will be reduced. If you are not getting bright enough sun to cast fairly sharp-edged shadows, then you do not have enough sun to harvest much useful electricity. Eighty to 85% of a "full sun" (defined as 1000 watts per square meter) is what most of us actually receive on a clear sunny day. High altitude and desert locations will do better on sunlight availability. On the high desert plateaus, 105 to 110% of full sun is normal. They don't call it the "sunbelt" for nothing.

TEMPERATURE

The power output from all PV module types fades somewhat at higher temperatures. This is not a serious consideration until ambient temperatures climb above 80°F, but that's not uncommon in full sun. The backs of modules should be as well-ventilated as practical. Always allow some airspace behind the modules if you want decent output in hot weather. On the positive side of this same issue, all modules increase output at lower temperatures, as in the wintertime, when most residential applications can use a boost. We have seen cases when the modules were producing 30 to 40% over specs on a clear, cold winter morning with a fresh reflective snow cover and hungry batteries.

As a general rule of thumb, we usually derate official PV module output by about 15 to 25% for the real world. For panel-direct systems (where the modules are connected directly to a pump without any batteries), derate by 20%, or even by 30% for really hot summer climates if you want to make sure the pump will run strongly in hot weather.

Module Mounting

Modules will catch the maximum sunlight, and therefore have the maximum output, when they are perpendicular (at right angles) to the sun.

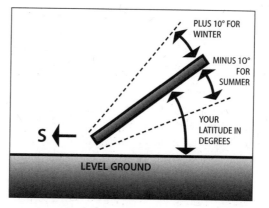

Proper PV mounting angle.

This means that tracking the sun across the sky from east to west will give you more power output. But tracking mounts are expensive and prone to mechanical and/or electrical problems, and PV prices have been coming down. Unless you've got a summertime high-power application such as water pumping, tracking mounts don't make a good investment any more.

PV systems are most productive if the modules are approximately perpendicular to the sun at solar noon, the most energy-rich time of the day for a PV module. The best year-round angle for your modules is approximately equal to your latitude. Because the angle of the sun changes seasonally, you may want to adjust the angle of your mounting rack seasonally. In the winter, modules should be at the angle of your latitude plus approximately 10 degrees; in the summer, your latitude minus a 10-degree angle is ideal. On a practical level, many residential systems will have power to burn in the summer, and seasonal adjustment may be unnecessary.

Generally speaking, most PV arrays end up on fixed mounts of some type. Tracking mounts are rarely used for residential systems anymore. Small water pumping arrays are the most common use of tracking mounts now. This rule of thumb is far from ironclad; there are many good reasons to use either kind of mounting. For a more thorough examination, see the PV mounting section beginning on page 70, which includes a large selection of mounting technologies.

System Examples

Following are several examples of photovoltaic-based electrical systems, starting from very simple and working up to very complex. All the systems that use batteries can also accept power input from wind or hydro sources as a supplement, or as the primary power source. PV-based systems constitute better than 90% of our renewable energy sales, so the focus here will be mostly on them.

A SIMPLE SOLAR PUMPING SYSTEM

In this simple system, all energy produced by the PV module goes directly to the water pump.

A utility intertie system without batteries is the simplest and most cost-effective way to connect PV modules to regular utility power.

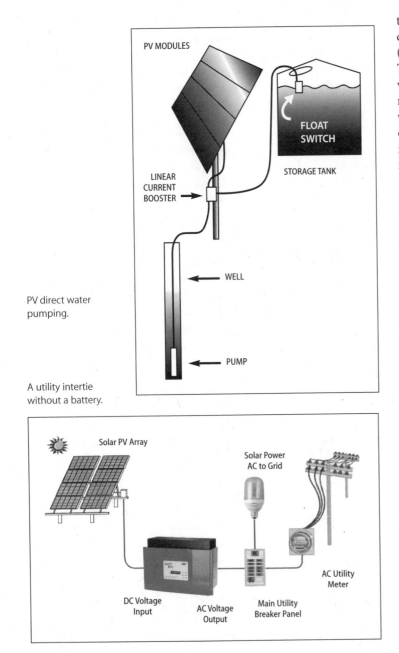

PV MODULES

FLOAT SWITCH

STORAGE TANK

LINEAR CURRENT BOOSTER

WELL

PUMP

PV direct water pumping.

A utility intertie without a battery.

Solar PV Array

Solar Power AC to Grid

DC Voltage Input

AC Voltage Output

Main Utility Breaker Panel

AC Utility Meter

There's no storage of the electrical energy; it's used immediately. Water delivered to the raised storage tank is our stored energy. The brighter the sun, the faster the pump will run. This kind of system (without battery storage) is called PV-direct, and is the most efficient way to utilize PV energy. Eliminating the electrochemical conversion of the battery saves about 20 to 25% of our energy, a very significant chunk! However, PV-direct systems only work with DC motors that can use the variable power output of the PV module, and of course this simple system only works when the sun shines.

There is one component of a PV-direct system we won't find in other systems. The PV-direct controller, or Linear Current Booster (LCB) is unique to systems without batteries. This solid-state device will downconvert excess voltage into amperage that will keep the pump running under low light conditions when it would otherwise stall. An LCB can boost pump output by as much as 40%, depending on climate and load conditions. We usually recommend them for PV-direct pumping systems.

For more information about solar pumping, see the extensive Water Development chapter in this *Sourcebook*.

A UTILITY INTERTIE SYSTEM WITHOUT BATTERIES

This is the simplest and most cost-effective way to connect PV modules to regular utility power. All incoming PV-generated electrons are converted to household AC power by the intertie inverter, and delivered to the main household circuit breaker panel, where they displace an equal number of utility-generated electrons. That's power you didn't have to buy from the utility company. If the incoming PV power exceeds what your house can use at the moment, the excess electrons will be forced backwards through your electric meter, turning it backwards. If the PV power is insufficient, that shortfall is automatically and seamlessly made up by utility power. It's like water seeking its own level (except it's really fast water!). When your intertie system is pushing excess power backwards through the meter, the utility is paying you regular electric rates for your excess power. You sell power to the utility during the daytime, it sells it back to you at night. This treats the utility grid like a big 100% efficient battery. But if utility power fails with one of these systems, even if it's sunny, the PV system will be off for the safety of utility workers.

There is a minimum of hardware for these intertie systems: a power source (the PV modules), an intertie inverter, a circuit breaker, and some wiring to connect everything. See the separate section on Utility Intertie Systems in chapter 3 for more information.

A SMALL CABIN SOLAR ELECTRIC SYSTEM

Most PV systems are designed to store some of the collected energy for later use. This allows you to run lights and entertainment equipment at night, or to temporarily run an appliance that takes more energy than the PV system is delivering. Batteries are the most cost-effective energy storage technology available so far, but batteries are a mixed blessing. The electrical/chemical conversion process isn't 100% efficient, so you have to put back about 20% more energy than you took out of the battery, and the storage capacity is finite. Batteries are like buckets, they can only get so full, and can only empty just so far. A charge controller becomes a necessary part of your system to prevent over- (and sometimes under-) charging. Batteries can also be dangerous. Although the lower battery voltage is generally safer to work around than conventional AC house current, it is capable of truly awe-inspiring power discharges if accidentally short-circuited. So fuses and safety equipment also become necessary whenever one adds batteries to a system. Fusing ensures that no youngster, probing with a screwdriver into some unfortunate place, can burn down the house. And finally, a monitoring system that displays the battery's approximate state of charge is essential for reliable performance and long battery life. Monitoring could be done without, just as you could drive a car without any gauges, warning lights, or speedometer, but this doesn't encourage system reliability or longevity.

The Real Goods Weekend Getaway Kit is an example of a small cabin or weekend retreat system. It has all the basic components of a residential power system: a power source (the PV module), a storage system (the deep-cycle battery), a controller to prevent overcharging, safety equipment (fuses), and monitoring equipment. The Weekend Getaway Kit on page 66 is supplied as a simple DC-only system. It will run 12-volt DC equipment, such as RV lights and appliances. An optional inverter can be added at any time to provide conventional AC power, and that takes us to our next example.

A FULL-SIZE HOUSEHOLD SYSTEM

Let's look at an example of a full-size residential system to support a family of three or more. The power source, storage, control, monitoring, and safety components have all been increased in size from the small cabin system, but most importantly, we've added an inverter for conventional AC power output. The majority of household electrical needs are run by the inverter, allowing conventional household wiring, and a greater selection in lighting and appliance choices. We've found that when the number of lights grows above five, AC power is much easier to wire, plus fixtures and lamps cost significantly less due to mass production, so the inverter pays for itself in appliance savings.

Often, with larger systems like this, we combine and preassemble all the safety, control, and inverter functions using an engineered, UL-approved power center. This isn't a necessity, but we've found that most folks appreciate the tidy appearance, fast installation, UL approval, and ease of future upgrades that preassembled power centers bring to the system.

Small cabin systems to run a few lights and an appliance or two can start at under $1000.

The typical Real Goods Fulltimer system takes only 2' x 5' of floor space inside your utility room.

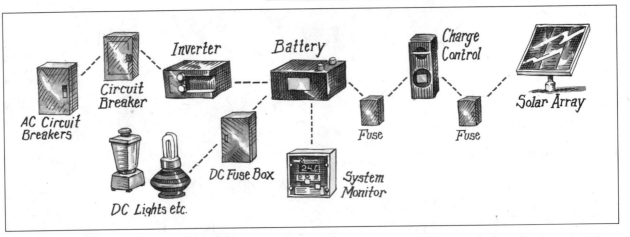

Labels in illustration: Inverter, Battery, Charge Control, Circuit Breaker, AC Circuit Breakers, Solar Array, Fuse, Fuse, DC Fuse Box, System Monitor, DC Lights etc.

A full-size household system has all these parts.

Everyone's needs, expectations, budget, site, and climate are individual, and your power system, in order to function reliably, must be designed with these individual factors accounted for.

Because family sizes, lifestyles, local climates, and available budgets vary widely, the size and components that make up a larger residential system like this are customized for each individual application. System sizing is based on the customer's estimate of needs, and an interview with one of our friendly technical staff. See the system sizing hints and worksheets starting on page 117.

A UTILITY INTERTIE SYSTEM WITH BATTERIES

In this type of intertie system, the customer has both a renewable energy system and conventional utility-supplied grid power. Any renewable energy beyond what is needed to run the household and maintain full charge on an emergency back-up battery bank is fed back into the utility grid earning money for the system owner. If household power requirements exceed the PV input, for example, at night or on a cloudy day, the shortfall is made up automatically and seamlessly by the grid. If the grid power fails, power will be drawn instantly from the backup batteries to support the household. Switching time in case of grid failure is so fast that only your home computer may notice. This is the primary difference between intertie systems with and without batteries. Batteries will allow continued operation if the utility fails. They'll provide emergency power backup, and will allow you to store and use any incoming PV energy.

In order to hasten this emerging technology, a number of federal and state sponsored programs exist, and an increasing number of them have real dollars to spend! These dollars usually appear as refunds, buydowns, or tax credits to the consumer. That's you. Programs and available funds vary with time and state. For the latest information, call your State Energy Office, listed in the Appendix, or check the Database of State Incentives for Renewable Energy on the Web at www.dsireusa.org.

System Sizing

We've found from experience that there's no such thing as "one size fits all" when it comes to energy systems. Everyone's needs, expectations, budget, site, and climate are individual, and your power system, in order to function reliably, must be designed with these individual factors accounted for. Our friendly and helpful technical staff, with over 69,000 solar systems under their collective belts, has become rather good at this. We don't charge for this personal service, so long as you purchase your system components from us. We do need to know what makes your house, site, and lifestyle unique. So filling out the household electrical demands portion of our sizing worksheets is the first, and very necessary step, usually followed by a phone call (whenever possible), and a customized system quote. Worksheets, wattage charts, and other helpful information for system sizing, are included in chapter 3—which also just happens to cover batteries, safety equipment, controls, monitors, and all the other bits and pieces you need to know a little about to assemble a safe, reliable energy system.

How Much PV Area to Equal U.S. Electric Production?

For those who just want to know the answer, without getting dragged through all the math, here it is:

A solar-electric array, using today's off-the-shelf technology, sited in nice sunny (largely empty) Nevada, that is big enough to deliver all the electricity the United States currently uses, would cover a square almost exactly 100 miles per side.

Want proof? Glad you asked … here we go …

According to the Energy Information Administration of the U.S. Department of Energy, www.eia.doe.gov/emeu/aer/txt/ptb0802a.html, the United States produced 3,838.6 billion kilowatt-hours of electricity in 2002. Let's round that up and aim for an even 4,000 billion kWh, which is a production figure we'll hit sometime between the years 2005 and 2010. Note that this is "production," not "use." Transmission inefficiencies and other losses are covered. This is the amount of power we need to stuff into the pipeline inlet.

We'll want our PV modules in a good sunny area to make the best of our investment. Nevada, thanks to climate and military/government activities, has a great deal of almost empty, and very sunny land. So looking at the National Solar Radiation Data Base for Tonopah, Nevada, http://rredc.nrel.gov/solar/pubs/redbook/, we see that a flat-plate collector on a fixed-mount facing south at a fixed-tilt equal to the latitude, 38.07° in this case, saw a yearly average of 6.1 hours of "full sun" per day in the years 1961 through 1990. A "full sun" is defined as 1,000 watts per square meter.

For PV modules, we'll use the large Evergreen Cedar-series 110-watt module, which the California Energy Commission, www.energy.ca.gov/greengrid/ certified_pv_modules.html, rates at 98.4 watts output, based on observed, real-world performance. 98.4 watts times 6.1 hours equals 600 watt-hours or 0.60 kilowatt-hours per day per module at our Tonopah site. At 26" x 62.5" this module presents 11.3 square feet of surface area. We'll allow some space between rows of modules for maintenance access, and for sloping wintertime sun, so let's say that each module will need 15 square feet.

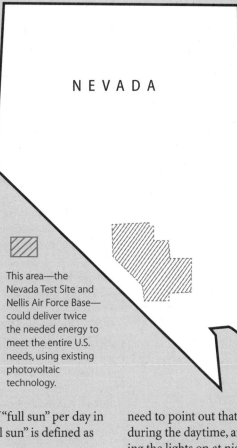

N E V A D A

This area—the Nevada Test Site and Nellis Air Force Base—could deliver twice the needed energy to meet the entire U.S. needs, using existing photovoltaic technology.

Conversion from PV module DC output to conventional AC power isn't perfectly efficient. Looking at the real-world performance figures from the California Energy Commission again, www.energy.ca.gov/greengrid/certified_inverters.html, we see that the Trace Engineering 20kW model PV-20208 is rated at 96% efficiency. We'll probably be using larger inverters, but this is a typical efficiency for large intertie inverters. So our 0.60 kWh per module per day becomes 0.576 kWh by the time it hits the AC grid.

A square mile, 5,280 feet times 5,280 feet equals 27,878,400 square feet. Divided by 15 square feet per module, we can fit 1,858,560 modules per square mile. At 0.576 kilowatt-hours per module per day, our square mile will deliver 1,070,530 kWh per day on average, or 390,743,654 kWh per year. Back to our goal of 4,000,000,000,000 kWh, divided by 390,743,654 kWh per year per square mile, it looks like we need about 10,237 square miles of surface to meet the electrical needs of the United States That's a square area about 102 miles on a side. This is a bit over half of the approximate 16,000 square miles currently occupied by the Nevada Test Site and the surrounding Nellis Air Force Range (www.nv.doe.gov/nts/ and www.nellis.af.mil/environmental/default.htm).

As a practical measure, we need to point out that PV power production happens during the daytime, and so long as we persist in turning the lights on at night, there will continue to be substantial power use at night. Also, so long as we're out in the desert, solar thermal collection may be a more efficient power generation technology. But, however you run the energy collection system, large solar-electric farms on what is otherwise fairly useless desert land could add substantially to the electrical independence and security of any country. The existing infrastructure of coal, nuclear, and hydro power plants could continue to provide reliable power at night, but nonrenewable resource use and carbon dioxide production would be greatly reduced.

—Doug Pratt

PHOTOVOLTAIC PRODUCTS

System Design

Power With Nature

Solar and Wind Energy Demystified

Author Rex Ewing is right up to date, explaining all the latest offerings like utility intertie systems and maximum power point tracking charge controls—all with extensive drawings, pictures, sidebars and witticisms. You'll be enlightened by the wealth of practical, hands-on information for solar, wind, and micro-hydro power; plus commonsense explanations of the necessary components such as batteries, inverters, charge controllers, and more. 255 pages, softcover. USA.

21-0339 Power with Nature   **$24.95**

There Are No Electrons

By Kenn Amdahl. Do you know how electricity works? Want to? Read this book and learn about volts, amps, watts and resistance. So much fun to read, you won't even notice you're learning. And if you already "know" about all this, you'll enjoy it even more—just be prepared to rethink everything. 222 pages, softcover. USA.

82646   There Are No Electrons   **$16**

The Solar Electric House

By Steven Strong. The author has designed more than 75 PV systems. This fine book covers all aspects of PV, from the history and economics of solar power to the nuts and bolts of panels, balance of systems equipment, system sizing, installation, utility intertie, stand-alone PV systems, and wiring instruction. A great book for the beginner. 276 pages, softcover. USA.

80800 The Solar Electric House   **$21.95**

Photovoltaics: Design and Installation Manual

By Solar Energy International staff. Producing electricity from the sun using PV systems has become a major industry worldwide. Designing, installing and maintaining such systems requires knowledge and training. This manual delivers the critical information to successfully design, install and maintain PV systems. The book contains an overview of photovoltaic electricity and a detailed description of PV system components, including PV modules, batteries, controllers and inverters. It also includes chapters on sizing photovoltaic systems, analyzing sites and installing PV systems, as well as detailed appendices on PV system maintenance, troubleshooting and solar insolation data for over 300 sites around the world. Topics covered also include: Stand-alone and PV/Generator hybrid system sizing; Utility-

Interactive PV systems; Component specification; system costs and economics; Case studies and safety issues.

Those wanting to learn the skills of tapping the sun's energy can do so with confidence with this book in their library. 317 pages, softcover. Canada.

21-0366 PV: Design & Installation Manual   **$59**

Renewable Energy with the Experts

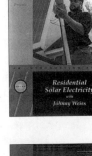

A professionally produced video series that covers basic technology introductions as well as hands-on applications. Each tape is presented by the leading expert in that field, and tackles a single subject in detail. Includes classroom explanations of how the technology works with performance and safety issues to be aware of. Then action moves into the field to review actual working systems, and finally assemble a real system on camera as much as possible. Other tapes in this series cover storage batteries with system sizing, and water pumping. You'll find these tapes in appropriate chapters. This is the closest to hands-on you can get without getting dirt under your fingernails. All tapes are standard VHS format. USA.

80360   Set of Three
**   Renewable Experts Tapes $100**

Johnny Weiss has been designing and installing PV systems for over 15 years. He teaches hands-on classes at Solar Energy International in Colorado. 48 minutes.

80361   Renewable Experts Tape—PV $39

Don Harris, the owner of Harris Turbines, practically invented the remote microturbine industry, and has manufactured over 1,000 in the past 15 years. 42 minutes.

80362   Renewable Experts Tape—Hydro $39

Mick Sagrillo, the former owner of Lake Michigan Wind and Sun, has been in the wind industry for over 16 years with over 1,000 turbine installation/repairs under his belt. 63 minutes.

80363   Renewable Experts Tape—Wind $39

The Solar Design Studio v5.0

This CD-ROM contains a suite of eleven solar energy programs for professionals, involved educators, and interested individuals. These detailed simulations are practical and easy enough to be used by non-technical individuals. Programs will design stand-alone, grid-tied, or water pumping PV systems, as well as solar hot water systems. Accessory programs deliver world climate info, sun path simulations, and output curves for all types of PV modules. Windows-based on CD-ROM. USA.

63-0018 The Solar Design Studio v5.0   **$149**

Photovoltaic Products

Evergreen Cedar Series PV Modules

Innovative, low-cost, string ribbon silicon production

Evergreen Solar is a rapidly growing U.S. company with an innovative String Ribbon™ polycrystalline production technology. String Ribbon produces a silicon wafer by drawing up a thin film of molten silicon between two strings, like a soap bubble. Production is a simple, nearly continuous process. With none of the usual slicing or dicing required to get thin blanks of silicon on which to build PV cells, there's very little waste, just fast, efficient cell production. That means PV modules that use less energy and waste less raw materials in the process. Like all polycrystalline modules, this is a very stable power production material with a life expectancy measured in decades.

As they've gained experience with String Ribbon production, Evergreen has been producing larger cells, resulting in higher wattage modules. Their latest offering, the Cedar Series, is available with either 36 cells, producing 50 to 60 watts, or with 72 cells, producing 100 to 120 watts. Features include standard 36-cells in series configuration, tempered-glass covers, clear anodized aluminum frames with standardized 24-inch mounting width, Tedlar backsheets, and roomy watertight junction boxes that are equipped with both terminal strips for lower voltage battery-based systems, and quick-connect MC cables for higher-voltage utility intertie systems.

The 100-watt module series features dual-voltage output. With 72-cell construction, it can be wired at the junction box to deliver either 12- or 24-volt output. This makes for less wiring and interconnection requirements on 24-volt systems, which are increasingly the standard voltage for household and pumping systems.

All Evergreen modules are UL listed and feature a great 25-year warranty. USA. *Please check availability before ordering. Production varies.*

EC-100-watt module series specs below are listed at 12-volt output. 24-volt output is the same wattage, but double the voltage, and half the amperage.

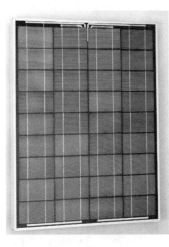

Evergreen EC-55 PV Module

Rated Power @ 25°C: 55 watts
Volts @ max power: 16.1 volts
Amps @ max power: 3.42 amps
Open Circuit Voltage: 20.0 volts
Short Circuit Current: 3.9 amps
Series Fuse: 15 amps
L x W x D: 32.1" x 25.7" x 2.2"
Construction: polycrystalline, tempered glass
Weight: 15.0 lb/6.8 kg
Warranty: within 20% of rated output for 25 years.

11582   Evergreen EC-55 Module $299

Tech note: 24v output from a single EC-100 series module requires a 24v jumper kit. Available at no charge.

41-0180 Evergreen 24v Jumper Kit $0

Evergreen EC-51 PV Module

Rated Power @ 25°C: 51 watts
Volts @ max power: 15.6 volts
Amps @ max power: 3.28 amps
Open Circuit Voltage: 20.0 volts
Short Circuit Current: 3.73 amps
Series Fuse: 15 amps
L x W x D: 32.1" x 25.7" x 2.2"
Construction: polycrystalline, tempered glass
Weight: 15.0 lb/6.8 kg
Warranty: within 20% of rated output for 25 years.

11002   Evergreen EC-51 Module $288

Evergreen EC-115 PV Module

Rated Power @ 25°C: 115 watts
Volts @ max power: 16.5 volts
Amps @ max power: 6.97 amps
Open Circuit Voltage: 20.0 volts
Short Circuit Current: 7.71 amps
Series Fuse: 15 amps
L x W x D: 62.4" x 25.7" x 2.2"
Construction: polycrystalline, tempered glass
Weight: 27.0 lb/12.2 kg
Warranty: within 20% of rated output for 25 years.

41-0124 Evergreen EC-115 Module $585

Evergreen EC-110 PV Module

Rated Power @ 25°C: 110 watts
Volts @ max power: 16.4 volts
Amps @ max power: 6.72 amps
Open Circuit Voltage: 20.0 volts
Short Circuit Current: 7.60 amps
Series Fuse: 15 amps
L x W x D: 62.4" x 25.7" x 2.2"
Construction: polycrystalline, tempered glass
Weight: 27.0 lb/12.2 kg
Warranty: within 20% of rated output for 25 years.

11581   Evergreen EC-110 Module $559

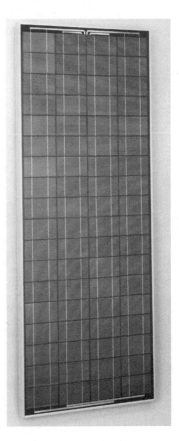

Sharp PV Modules

Low prices and now made in the United States

Sharp PV modules for all US sales are produced at their new Tennessee plant, which went into production in 2003. All Sharp modules currently feature watertight MC output cables for quick, foolproof series connections, making these modules a good choice for grid-tie systems. There are no junction boxes however, which make these modules a difficult choice for battery-based systems.

The 185-watt modules are single crystal and pack more power delivery into less space than any other module available. The 167-watt, 165 watt, 140-watt, and 70-watt are all polycrystalline modules with black anodized frames for lower visibility. The 70-watt modules are the PV industry's first triangular modules, and come in left or right configurations. The 70- and 140-watt modules match, and are designed to be used together. All Sharp modules feature extremely stout frames, tempered, low-reflection glass covers, built-in bypass diodes, and 25-year warranties.

167-watt model

Sharp 80- and 123-Watt PV Modules

With J-boxes for battery-based systems

Using weather-tight J-boxes, these high-efficiency polycrystalline modules are a great choice for battery-based systems. Featuring over 14% cell efficiency, which keeps the footprint smaller and the wattage higher, these 12-volt nominal modules also have an anti-reflective coating to increase light absorption, a textured cell surface to reduce reflection, terminal strips for easy electrical connections, and built in bypass diodes. Construction is conventional with tempered glass, EVA encapsulation, and full aluminum frames. Uses the standard 36 cells in series, so these will intermix with any other 36-cell modules. The 80-watt module is the same size as the 75-watt offerings from Shell (Siemens) and Astropower. The 123-watt module is the same width and within an inch lengthwise of the Astropower and Evergreen 100-watt series, making it easier to add to existing arrays and pick a mounting structure. UL listed. 25-year manufacturer's warranty. Assembled in USA.

Model	80 Watt	123 Watt
Rated Power @ 25°C:	80 watts	123 watts
Volts @ max power	17.1 volts	17.2 volts
Amps @ max power	4.67 amps	7.16 amps
Open Circuit Volts	21.3 volts	21.3 volts
Short Circuit Amps	5.3 amps	8.1 amps
Series Fuse	10 amps	15 amps
L x W x D	47.3" x 20.9" x 1.4"	59" x 26.1" x 1.8"
Weight	18.7 lb/8.5 kg	30.8 lb/14 kg
Price per watt	$4.94/watt	$4.84/watt
Item #	41-0177	41-0117
Price	$395	$595

Model	185 watt	167 watt	165watt	140 watt	70 watt
Rated Power @ 25°C:	185 watts	167 watts	165 watts	140 watts	70 watts
Volts @ max power	36.2 volts	23.5 volts	34.6 volts	19.95 volts	9.98 volts
Amps @ max power:	5.11 amps	7.1 amps	4.77 amps	7.02 amps	7.02 amps
Open circuit voltage	44.9 volts	29.0 volts	43.1 volts	24.85 volts	12.43 volts
Short circuit current	5.75 amps	7.91 amps	5.46 amps	7.81 amps	7.81 amps
Series fuse	10 amps	15 amps	10 amps	15 amps	15 amps
L x W x D	62" x 32.5" x 1.8"	52.3" x 39.5" x 1.8"	62" x 32.5" x 1.8"	45.9" x 39" x 1.8"	45.9" x 39" x 1.8"
Weight	37.5 lb.	35.3 lb.	37.5 lb.	32 lb.	26.9 lb.
Price per watt	$5.00/watt	$5.08/watt	$5.00/watt	$4.89/watt	$6.64/watt
Item #	41-0100	41-0118	11-612	41-0116	41-0115 L or R
Price	$925	$849	$825	$685	$465

70 and 140 watt modules

Shell Photovoltaic Modules

This company was formerly Siemens, and before that, Arco. Despite all the name changes, this experienced American crew has been making PV modules longer, and in most cases, better, than anyone. With production facilities in Vancouver, Washinngton, and Camarillo, California, the Arco/Siemens/Shell modules have long been the industry benchmark. Shell modules feature the most efficient single crystal silicon, low-iron tempered glass, EVA encapsulation, Tedlar backsheets, anodized aluminum frames, and weather-tight junction boxes with simple terminal strips and built-in bypass diodes. Shell modules also feature a TOPS (Textured Optimized Pyramidal Surface) coating, which reduces light reflection from the upper surface and gives the cells a fairly uniform charcoal-gray color.

SHELL POWERMAX™ PV MODULES

In mid-2004, Shell introduced their new PowerMax series of modules. These replace the former SQ and SP lines, but (thank you Shell!) retain the same footprint and will fit the same mounting structures. PowerMax Ultra series are single crystal products for the most watts within the available footprint. All Shell modules benefit from over 30 years of experience, a wide range of fine-tuning technical innovations to maximize power, and very tight +/- 5% power tolerances. Additional grounding holes are now provided, two of them facing outward on the end frames. The robust weather-tight ProCharger junction boxes continue. Series fuse ratings are raised to 20 amps for all modules, which will eliminate the need for combiner boxes in some applications.

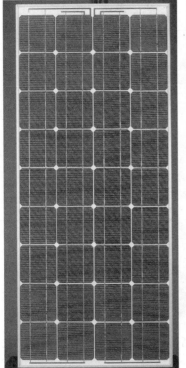

Shell PowerMax 80 and 85

The PM80 or 85 is a 12-volt, single crystal power production module with 36 cells in series. This module has the same footprint as the venerable SQ75, SP75, or the PC-4-JF. It's also the same size as the Astropower AP75, or GE 70-watt modules. So if you're looking to add to your older array, these will match nicely. Has the robust ProCharger junction box with terminal strip connections, anodized aluminum frame, and all the other features we've come to expect from the industry leader. UL listed. USA.

Model	PM80	PM85
Rated Power @ 25°C:	80 watts	85 watts
Volts @ max power:	16.9 volts	17.2 volts
Amps @ max power:	4.76 amps	4.95 amps
Open Circuit Volts	21.8 volts	22.2 volts
Short Circuit Amps	5.35 amps	5.45 amps
Series Fuse	20 amps	20 amps
L x W x D	47.3" x 20.8" x 2.2"	47.3" x 20.8" x 2.2"
Construction	Single crystal, tempered glass	Single crystal, tempered glass
Warranty	within 20% @ 25 years	within 20% @ 25 years
Weight	16.7 lb/7.6 kg	16.7 lb/7.6 kg
Item #	**41-0171**	**41-0172**
Price	**$449**	**$475**

Shell PowerMax Ultra Modules

The most powerful of the 24-volt PM-series, the 165- and 175-watt Ultra series uses single-crystal technology for the highest wattage within the available footprint. Has 72 cells in series and all the usual PowerMax features. These large modules reduce wiring and mounting effort on site. The robust ProCharger junction boxes provide both conventional terminal strips for battery-based systems, and pre-installed (removeable) MC cables for higher voltage utility intertie system series wiring. Cables are #12AWG, negative is 51", positive is 39" to support either side by side or top to bottom layout. This oversize module requires premium shipping. UL listed. USA.

Model	PM165	PM175
Rated Power @ 25°C:	165 watts	175 watts
Volts @ max power:	35.0 volts	35.4 volts
Amps @ max power:	4.72 amps	4.95 amps
Open Circuit Volts	44.5 volts	44.6 volts
Short Circuit Current	5.4 amps	5.43 amps
Series Fuse	20 amps	20 amps
L x W x D	63.9" x 32.1" x 2.2"	63.9" x 32.1" x 2.2"
Construction	Single-crystal, tempered glass	Single-crystal, tempered glass
Warranty	within 20% @ 25 yrs	within 20% @ 25 yrs
Weight	38 lb/17.3 kg	38 lb/17.3 kg
Item #	**41-0175**	**41-0176**
Price	**$940**	**$995**

Kyocera Photovoltaic Modules

Kyocera is a Japanese-based company with many, many years of experience in polycrystalline solar cells. Their cells have some of the highest efficiency found in commercial production. All Kyocera modules feature standard 36-cell construction, tempered, low-glare glass, EVA encapsulation, polyvinyl backsheets, anodized and clear-coated aluminum frames, with large, excellent, weatherproof junction boxes. Bypass diodes are installed halfway through the 36-cell series to reduce the effects of partial shading. The KC-series modules we're offering here are all UL, and cUL listed, and CE certified. A 25-year limited warranty is standard.

The Kyocera J-box is large, watertight, and simple, but requires wire terminals.

Kyocera KC120

Bigger wattage modules have some real advantages. You buy fewer modules, which means less bolting down and wiring up; thus, a faster, easier installation. Larger modules usually offer some economies of scale; it costs less to build one big module than two smaller ones. Kyocera's top-of-the-line module, the KC120, delivers a powerful 120 watts. No other module we offer delivers as much power in as small a footprint. The multicrystal KC120 features standard 36-cell construction, tempered, low-glare glass, EVA encapsulation, polyvinyl backsheet, anodized and clear-coated aluminum frame, with a large, excellent, weatherproof junction box. The J-box has four standard 1/2" knockouts. Bypass diodes are installed halfway through the 36-cell series to reduce the effects of partial shading. UL and cUL listed. CE certified. 25-year warranty. Japan.

Rated Power @ 25°C: 120 watts
Volts @ max power: 16.9 volts
Amps @ max power: 7.1 amps
Open Circuit Voltage: 21.5 volts
Short Circuit Current: 7.45 amps
Series Fuse: 11 amps
L x W x D: 56.1" x 25.7" x 2.0"
Construction: polycrystalline, tempered glass
Weight: 26.3 lb/11.9 kg
Warranty: within 20% of rated output for 25 years.

11568   Kyocera KC120 Module   **$685**

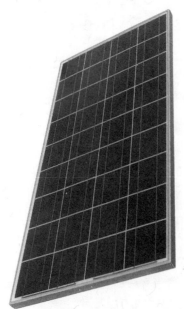

Kyocera KC80

Sporting the same features and construction as the KC120, the KC80 is assembled with 2/3-size cells. Features include standard 36-cell construction, tempered, low-glare glass, EVA encapsulation, polyvinyl backsheet, anodized and clear-coated aluminum frame, with a large, excellent, weather-proof junction box. The J-box has four standard 1/2" knockouts. Bypass diodes are installed halfway through the 36-cell series to reduce the effects of partial shading. UL and cUL listed. CE certified. 25-year warranty. Japan.

Rated Power @ 25°C:: 80 watts
Volts @ max power: 16.9 volts
Amps @ max power: 4.73 amps
Open Circuit Voltage: 21.5 volts
Short Circuit Current: 4.97 amps
Series Fuse: 7 amps
L x D x W: 38.4" x 25.7" x 2.0"
Construction: polycrystalline, tempered glass
Weight: 17.7 lb/8.0 kg
Warranty: within 20% of rated output for 25 years.

11567   Kyocera KC80 Module   **$465**

Kyocera 40W and 50W Modules

Kyocera's advanced cell-processing technology and automated production produces very highly efficient polycrystalline modules. These conventional 36-cell modules feature tempered glass covers, EVA encapsulant, polyvinyl backsheets, with a rugged full-perimeter anodized aluminum frame, and a weather-tight junction box with three 1/2" knockouts. They will intermix with other 36-cell modules for battery charging. These mid-size 40- and 50- watt modules fill a need for modest power requirements at reasonable prices. Twenty-five year manufacturer's warranty. Japan.

Model	KC40	KC50
Rated Power @ 25°C:	40 watts	50 watts
Volts @ max power:	16.9 volts	16.7 volts
Amps @ max power:	2.34 amps	3.0 amps
Open Circuit Coltage	21.5 volts	21.5 volts
Short Circuit Current	2.48 amps	3.1 amps
Series Fuse	5 amps	5 amps
L x W x D	20.7 x 25.7 x 2.0	25.2 x 25.7 x 2.0
Weight	9.9 lb/4.5 kg	11.0 lb/5.0 kg
Warranty	within 20% of rated output for 25 years	within 20% of rated output for 25 years
Item #	**41-0128**	**41-0129**
Price	**$240**	**$310**

Uni-Solar Triple Junction PV Technology

Uni-Solar delivers a nice breakthrough in lower cost, higher output amorphous film technology. For higher efficiency and lower production costs, each cell is composed of three spectrum-sensitive semiconductor junctions stacked on top of each other: blue light sensitive on top, green in the middle, red on the bottom. Like all Uni-Solar products, these flexible, unbreakable modules are deposited on a stainless steel backing with DuPont's Tefzel glazing and, due to their unique construction, have the ability to continue useful output during partial shading. The unbreakable construction, good resistance to heat-fading, and shade tolerance makes these modules an excellent choice for RVs, or any site where vandalism is a possibility. The new triple-junction technology brings the price down to a level equal to or better than single-crystal technology.

Uni-Solar has crafted their basic flexible amorphous PV material into several product lines, including conventional aluminum-framed modules; marine modules with flexible frames; and several roofing products.

Like all amorphous modules, power output will fade approximately 15% over the first month or two. Initial power will be as much as 18% higher than rated. This power bonus must be considered when designing the power, control, and safety systems. Rated power is post-fade, and is warranted for three to twenty years, depending on the model.

CONVENTIONAL FRAMED UNI-SOLAR MODULES

Frames are anodized aluminum. Large weatherproof junction boxes are supplied on 64-, 42- and 32-watt modules with a 20-year warranty. Wire pigtails are supplied on 21-, 11-, and 5-watt sizes with a 10-year warranty. These are all excellent choices for RVs, vandalism-prone sites, or systems that need to move easily without threat of breakage. UL and CUL listed. USA.

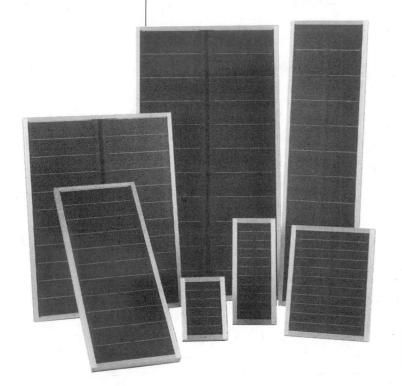

Uni-Solar 5-Watt Module

Rated Power @ 25°C: 5.0 watts
Volts @ max power: 16.5 volts
Amps @ max power: 0.3 amps
Open Circuit Voltage: 23.8 volts @ 25°C
Construction: triple junction, Tefzel glazing
Warranty: within 20% of rated output for 10 years.
Weight: 2.5 lb/1.13 kg
L x W x D: 19.3" x 8.5" x 0.9"

11230	Uni-Solar US5 Module	$89

Uni-Solar 11-Watt Module

Rated Power @ 25°C: 10.3 watts
Volts @ max power: 16.5 volts
Amps @ max power: 0.62 amps
Open Circuit Voltage: 23.8 volts @ 25°C
Construction: triple junction, Tefzel glazing
Warranty: within 20% of rated output for 10 years.
Weight: 3.6 lb/1.63 kg
L x W x D: 19.3" x 15.1" x 0.9"

11231	Uni-Solar US11 Module	$149

Uni-Solar 21-Watt Module

Rated Power @ 25°C: 21.0 watts
Volts @ max power: 16.5 volts
Amps @ max power: 1.27 amps
Open Circuit Voltage: 23.8volts @ 25°C
Construction: triple junction, Tefzel glazing
Warranty: within 20% of rated output for 10 years.
Weight: 6.6 lb/3.0 kg
L x W x D: 36.5" x 15.1" x 1.25"

11232	Uni-Solar US21 Module	$195

Uni-Solar 32-Watt Module

Rated Power @ 25°C: 32.0 watts
Volts @ max power: 16.5 volts
Amps @ max power: 1.94 amps
Open Circuit Voltage: 23.8 volts @ 25°C
Series Fuse: 4 amps
Construction: triple junction, Tefzel glazing
Warranty: within 20% of rated output for 20 years.
Weight: 10.6 lb/4.8 kg
L x W x D: 53.8" x 15.1" x 1.25"

11227	Uni-Solar US32	$239

Uni-Solar 42-Watt Module

Rated Power @ 25°C: 42.0 watts
Volts @ max power: 16.5 volts
Amps @ max power: 2.54 amps
Open Circuit Voltage: 23.8 volts @ 25°C
Series Fuse: 6 amps
Construction: triple junction, Tefzel glazing
Warranty: within 20% of rated output for 20 years.
Weight: 13.8 lb/6.3 kg
L x W x D: 36.5" x 29.1" x 1.25"

11233	Uni-Solar US42	$289

Uni-Solar 64-Watt Module

Rated Power @ 25°C: 64.0 watts
Volts @ max power: 16.5 volts
Amps @ max power: 3.88 amps
Open Circuit Voltage: 23.8 volts @ 25°C
Series Fuse: 8 amps
Construction: triple junction, Tefzel glazing
Warranty: within 20% of rated output for 20 years.
Weight: 20.2 lb/9.2 kg
L x W x D: 53.8" x 29.1" x 1.25"

11228	Uni-Solar US64	$419

FLEXIBLE MARINE MODULES

With a flexible, floating, marine module design, this series uses triple-junction film for greater efficiency in tight spaces. Can be rolled down to an 8-inch diameter without damage. Features blocking diodes within the potted junction block to prevent battery discharge and nickel-plated brass grommets at corners. Includes 8 feet of cable with built-in fuse holder and a 2-pin SAE connector. We've seen these flexible modules cleverly velcroed to the bottom of cockpit cushions. Just flip the cushions over for charging! UL listed. USA.

Uni-Solar Marine Flex USF-5

A 5-watt module that's ideal for battery maintenance charging on boats or RVs. Can be used without charge control on all but the smallest batteries. USA.

Rated Power @ 25°C: 5.0 watts
Volts @ max power: 16.5 volts
Amps @ max power: 0.30 amps
Open Circuit Voltage: 23.8 volts @ 25°C
Short Circuit Current: 0.37 amps @ 25°C
L x W x D: 21.8" x 9.7" x 0.3"
Construction: triple-junction, Tefzel glazing, flex frame
Warranty: within 20% of rated output for 3 years.
Weight: 1.18 lb/0.54 kg

11234   Uni-Solar Flex 5-Watt Module $109

Uni-Solar Marine Flex USF-11

An 11-watt module for modest battery charging needs. Good for small boats, laptop computers, or other mobile uses. USA.

Rated Power @ 25°C: 10.3 watts
Volts @ max power: 16.5 volts
Amps @ max power: 0.62 amps
Open Circuit Voltage: 23.8 volts @ 25°C
Short Circuit Current: 0.78 amps @ 25°C
L x W x D: 21.8" x 16.7" x 0.3"
Construction: triple-junction, Tefzel glazing, flex frame
Warranty: within 20% of rated output for 3 years.
Weight: 2.0 lb/0.91 kg

11221   Uni-Solar Flex 11-Watt Module $179

Uni-Solar Marine Flex USF-32

The largest flexible module available, this full-power 32-watt model is the ticket for extended travel, cruising, or independent living. USA.

Rated Power @ 25°C: 32.0 watts
Volts @ max power: 16.5 volts
Amps @ max power: 1.94 amps
Open Circuit Voltage: 23.8 volts @ 25°C
Short Circuit Current: 2.40 amps @ 25°C
L x W x D: 56.3" x 16.7" x 0.3"
Construction: triple-junction, Tefzel glazing, flex frame
Warranty: within 20% of rated output for 3 years.
Weight: 4.70 lb/2.14 kg

11235   Uni-Solar Flex 32-Watt Module $395

Options for Uni-Solar Marine Flex Modules

The StayPut mounting kits are noncorrosive, weatherproof, and almost indestructible. These mounts maintain a strong secure grip through the brass grommets, while still offering easy snap on and off operation. Choose canvas or hard base models, and the size appropriate for your module. Extension cord and conversion cords for the Marine modules also available.

StayPut Canvas Mount Kits
11597   6-piece for 32-Watt $17

StayPut Hard Mount Kits
11599   6-piece for 32-Watt $11

Extension Cords
11601   SAE to SAE, 10-ft. $17
11602   SAE to Male Lighter Plug, 1-ft. $14

Small PV Panels

13-Watt General Purpose Module

Delivering a peak 13.5 watts, this single-crystal, tempered glass glazed module is useful for any 12-volt charging applications. Features include weather-tight construction, a sturdy anodized aluminum frame with four adjustable mounting tabs, a standard 36 cells in series, built-in reverse-current diode, and 7-foot wire pigtail output with color-coded clips. Good for trickle battery recharging on RVs, farm equipment, boats, and a perfect match with the Aquasolar 200 pump for backyard fountains. China.

Rated Power @ 25°C: 13 watts
Volts @ max power: 17.5 volts
Amps @ max power: 0.71 amps
Open Circuit Voltage: 18.0 volts @ 25°C
Short Circuit Current: 750 mA @ 25°C
L x W x D: 15.1" x 14.4" x 1.0"
Construction: single crystal, tempered glass
Warranty: 90 days
Weight: 3.5 lb/1.6 kg

17-0170 13-Watt Module with Clips   **$139**

Lightweight Travel-Friendly Modules

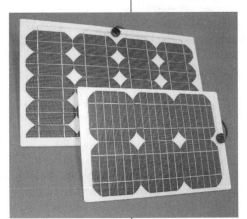

Featuring a lightweight "no-frame" design with fiberglass back, Tedlar cover, and grommet mounting holes in each corner, these unbreakable 36-cell modules are travel-friendly. A good choice for camping, expeditions, or just backyard battery charging. We use these versatile modules for a wide variety of small charging applications, such as laptop computers, digital cameras, and camp lighting. Comes standard with a blocking diode and a bare 6-foot output cord, or optionally with a preassembled female lighter socket on the 10-watt module. Mexico.

Model	10 watt	20 watt
Rated Power @ 25°C:	10 watts	20 watts
Volts @ max power:	16.8 volts	16.5 volts
Amps @ max power:	0.58 amps	1.21 amps
L x W x D:	16.5" x 10.6" x 0.2"	21" x12.75" x 0.2"
Construction:	single crystal, Tedlar coating	single crystal, Tedlar coating
Warranty:	3 years	3 years
Weight:	1.9 lb/0.86 kg	3.25 lb/1.47 kg
Item #	41-0164	41-0168
Price	$129	$249

06-0384 10W Lightweight Module with Lighter Socket   **$149**

NotePower 20-Watt Charger

A stylish, durable and protected way to provide 12-volt charging power in remote areas. This pair of modules zips open to deliver 20 watts of power for laptops, cell phones, camp lights or to recharge battery packs. Has a 10-foot power cord with lighter socket. Weighs only 3 lbs. 16" x 11" x 0.5" when closed. The case expands to 1.5" to carry paper or books. Two-year manufacturer's warranty.

41-0126 NotePower 20-Watt Charger   **$299**

Battery Saver Plus

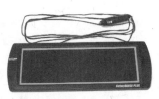

To keep your battery fully charged, even after long-term parking, place our solar-powered, 12-volt trickle charger on your dashboard in the sun, and plug it into your cigarette lighter. Output of this charger is a powerful 100 mA. (Most dash chargers are only 40-50 mA.) The 14.5" x 4.8" module is weather/rust/shock/UV resistant, and stores a generous 8-foot cord on the back. Not intended for outdoor use. Canada.

17-0147 Battery Saver Plus   **$35**

Deluxe Solar Charger

There's nothing quite like the frustration felt when your vehicle won't start. Don't let a weak battery ruin your fun or keep you from missing your next appointment. This Deluxe Clip-on Solar Charger is a 250-milliamp, 12-volt solar module that keeps your boat, RV, tractor, golf cart, or motorcycle battery topped off, even when it has been sitting unused for extended periods of time. This version is also water, weather, and rustproof, corrosion-resistant, and extremely durable. Its built-in, adjustable stand allows the panel to receive the maximum sunlight. Measures 10"W x 7.25"H x .5"D, and comes with a 6-foot connecting cable and alligator clips. Intended for outdoor use. China.

17-0144 Deluxe Solar Charger with Clips   **$89**

Solar Electric Packages for RVs, Cabins, and Sign Lighting

Solar Electric Cabin and Marine Kits

For less money than you think, you can have dependable solar power at your remote cabin or vacation camp. Our kits provide everything you need to produce, regulate and store DC power safely. We have over twenty-six years' experience designing solar systems, so we've already done the hard work for you.

Run lights, stereos, computers, water pressure pumps and other modest appliances without having to start—and listen to—that noisy, polluting generator. You provide wire and small parts as required by your particular installation, plus DC appliances such as lights, pumps or entertainment gear. An optional inverter, to provide conventional 120-volt AC, can be added to any of these kits. If you don't see exactly what you need, our experienced staff can help you create the perfect system at a reasonable price.

Weekend Getaway Kit for Cabins or RVs

Our smallest kit provides power for limited energy needs. The simple Battery Monitor shows system state of charge at a glance with red/yellow/green LEDs. The PV array can be doubled in the future without upgrading the controller. The no-maintenance sealed battery is optional (not needed for RVs). 12-volt DC output. 500+ watt hours.

INCLUDES:
1 Evergreen EC-115 PV Module
1 RV-type flush mount rack
1 Phocos CML 15 Charge Controller
1 Universal Battery Monitor w/fuseholder
1 30A Fuseblock w/fuse & spare
1 ATC-Type DC Fusebox w/20A fuses

BATTERY OPTION:
1 74Ah Sealed Gel Storage Battery

53-0106 Getaway Kit   **$740**
53-0107 Getaway Kit w/Battery   **$880**

Extended Weekend Kit for Cabins or RVs

Twice the size of the Getaway Kit, the Extended Kit can handle families for longer periods than weekends. The PV array can be doubled in the future without upgrading the charge controller, which also provides digital monitoring of battery voltage, PV amps, and load amps. The no-maintenance sealed battery pack is optional (not needed for RVs). 12-volt DC output. 1,000+ watt hours.

INCLUDES:
2 Evergreen EC-115 PV Modules
2 RV-type flush mount rack
1 Lyncom N35M Charge Controller
1 60A 2-Pole Safety Disconnect w/fuses & spare
1 ATC-Type DC Fusebox w/20A fuses

BATTERY OPTION:
2 98Ah Gel Storage Batteries w/interconnect cables

53-0108 Extended Weekend Kit   **$1,465**
53-0109 Extended Weekend Kit w/Battery   **$1,855**

Deluxe Full-Time Cabin Kit

For medium-sized weekend places or modest full-time cabins. It will power lights, stereos, TV, laptops and, with an optional inverter, other small appliances like a microwave or desktop computer. The controller provides system control and monitoring with battery voltage

and PV amps. You provide a 4-inch steel pipe for the PV array mount. The battery pack is optional. Can be set up initially as a 12- or 24-volt system. 2,000+ watt hours. See schematic below.

INCLUDES:
4 Evergreen EC-115 PV Modules
1 Pole-top fixed rack
1 TriStar 45A controller with monitor
1 6-Circuit DC Load Center
4 Circuit Breakers for above, 1-30A, 3-15A
1 110A Class T Fuse
BATTERY OPTION:
4 98Ah Gel Storage Batteries w/interconnect cables

53-0110 Full-Time Cabin Kit   **$2,830**
53-0111 Full-Time Cabin Kit w/Battery   **$3,610**

Marine Solar Kit for Boats or RVs

With solar power, your marine life can be truly independent. Solar power can keep your galley appliances, communications, navigation gear, running lights, and pumps all running happily. The marine kit employs an unbreakable, shade-tolerant Uni-Solar module with anodized aluminum frame. Flexible, floating versions of these modules can be substituted. The simple Battery Monitor shows system state of charge at a glance with red/yellow/green LEDs. The PV array can be doubled in the future without upgrading the controller. Mounting structure and DC fusebox are not included with this kit. 12-volt DC output.

INCLUDES:
1 Uni-Solar US64 PV Module
1 Phocos CML 10 Charge Controller
1 Universal Battery Monitor w/fuseholder
1 30A Fuseblock w/fuse & spare

53-0114 Marine Solar Kit   **$520**

SOLAR TRAVELER RV & MARINE KITS

Easy-Install Solar Traveler RV Kits

If you're feeling a little unsure of your hardware-store savvy, or you're a long way from the hardware store, these kits are for you. They include everything to add solar charging to an RV, right down to the wire, small parts, and instructions. The PV panels are RV-friendly Uni-

Solar panels, which are unbreakable and shade-tolerant —great features for RV use. Charge controllers have headroom for substantial future expansion.

These kits come in 32-watt or 64-watt starter packages with a Uni-Solar PV module, flush mounting hardware, a charge controller, wiring, small parts and instructions. Each starter kit can accept up to two expansion kits, which consists of another 32- or 64-watt module, mount, and interconnect wiring. Start small and see how it goes. USA

12202	**Basic Kit 64**	**$755**
12204	**Basic Kit 32**	**$445**
12203	**Expansion Kit 64**	**$629**
12205	**Expansion Kit 32**	**$360**

Full-Time Cabin Kit Schematic (at 24-volt)

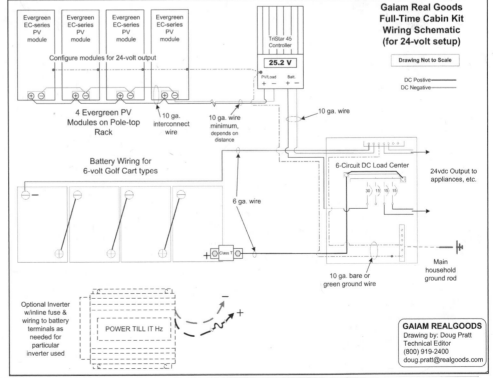

Gaiam Real Goods Full-Time Cabin Kit Wiring Schematic (for 24-volt setup)

Drawing Not to Scale

DC Positive ———
DC Negative ———

Evergreen EC-series PV module (×4)
Configure modules for 24-volt output
4 Evergreen PV Modules on Pole-top Rack
10 ga. interconnect wire

TriStar 45 Controller
25.2 V
PV/Load   Batt.

10 ga. wire minimum, depends on distance
10 ga. wire

Battery Wiring for 6-volt Golf Cart types

6 ga. wire
Class T

6-Circuit DC Load Center
30   15   15   15

24vdc Output to appliances, etc.

10 ga. bare or green ground wire
Main household ground rod

Optional Inverter w/inline fuse & wiring to battery terminals as needed for particular inverter used
POWER TILL IT Hz

GAIAM REALGOODS
Drawing by: Doug Pratt
Technical Editor
(800) 919-2400
doug.pratt@realgoods.com

Solar Floodlights for Signs and More

Our solar-powered sign-lighting kits share a core group of proven off-the-shelf components that are combined to deliver the run time you need. These are maintenance-free systems—set them up, turn them on and they'll take care of themselves.

We provide: outdoor floodlight(s), solar module(s), pole-top mount, charge/light controller, sealed battery, and safety fuse. You provide: mounting for the floodlight, box or shelter for the battery and control, 2.5-inch steel pipe for the PV module mount, and wiring/small parts as needed for your particular installation.

Kits state run time for a single 13-watt floodlight based on 3.5 hours sunlight, about the national wintertime average. Adding a second floodlight will halve your run time.

The Morningstar Sunlight 10 controller provides daytime charge control, turns on the light at dusk for a selected time period and shuts it off if the battery gets too low. The sealed, maintenance-free battery provides four to five days of backup power with no sun, and will last five to ten years. The solar module(s) and pole-top mount provide sufficient power to run the specified number of hours. All components USA.

24-Hour Solar Sign Lighting Kit

We're obviously not running 24 hours a day. This kit runs a pair of 13-watt floodlights for 12 hours a day. Provides a pair of Sharp 80-watt modules, a pole-top mount, a pair of 98AH sealed batteries, Morningstar Sunlight 10 controller, fusing and a pair of 13-watt floodlights.

12254   24-Hour Solar Sign Lighting Kit $1,550

12-Hour Solar Sign Lighting Kit

Our basic all-nighter lighting kit. Provides a Sharp 80-watt module with pole-top mount, 98AH sealed battery, Morningstar Sunlight 10 controller, fusing, and 13-watt floodlight.

12238   12-Hour Solar Sign Lighting Kit $885

8-Hour Solar Sign Lighting Kit

Provides a Kyocera 50-watt module with pole-top mount, 74AH sealed battery, Morningstar Sunlight 10 controller, fusing and 13-watt floodlight.

12253   8-Hour Solar Sign Lighting Kit $760

4-Hour Solar Sign Lighting Kit

Provides a Uni-Solar US32-watt module with side-of-pole mount, 51AH sealed battery, Morningstar Sunlight 10 controller, fusing, and 13-watt floodlight.

12252   4-Hour Solar Sign Lighting Kit $569

Portable Power Packages

Small PV/Battery Packages for Portable Electronics

These small PV/battery packages are intended to provide power and/or recharging for cell phones, GPS, laptops, or small battery-powered appliances with modest power uses. For larger systems see the various portable battery packs covered under Large Storage Batteries in chapter 3.

PowerDock Power Station for Mobile Gadgets

The PowerDock has everything you need to run laptops, cell phones, GPSs, camp lights, and other modest-power gear in remote sites. This slick and sturdy power station has a zip-out, unbreakable 15-watt PV panel and 9.2 amp/hrs of sealed 12-volt batteries. Also included are a pair of fused lighter socket outlets, a power meter to show state of charge, and a water-resistant, canvas carrying case with five roomy compartments for converters and accessories. Will accommodate laptop computers up to 9" x 15" x 2". Want to double your charging power? Add the 15-watt Solar Boost. Plugs into one of the lighter sockets and folds

into the PowerDock for travel. The quality construction is delightfully robust, with easy no-tools access to the batteries (lots of Velcro!). Has four batteries, which currently cost $8 each and typically need to be replaced in two to four years. Weighs 14.5 lbs, yet is comfortable to carry with the wide, well-padded shoulder strap. This product is quality all the way! Comes with an AC wall-watt charger for when the sun don't shine. One-year manufacturer's warranty, 10 years on PV array. USA & Mexico.

53-0112 PowerDock $349
41-0127 PowerDock 15W Boost $179

iSun and Battpak

This 2.2-watt folding unit can be switched for 6- or 12-volt output, comes with two power cables and seven different DC plugs that cover more than 90% of cell phones, discmans, MP3 players, or GPS units. For more power, up to five iSun units can be plugged into each other with the included spine connector plug. Measures 7.25" x 4.25" (8.25" unfolded) x 1.25". Weighs 11 oz. Delivers 145mA @ 15.2v, or 290mA @ 7.6v. China.

The Battpak docks mechanically and electrically to the back of the iSun, or it can be used stand-alone with the included AC and DC charging cords. It will charge up to ten rechargeable AA or AAA size batteries. ON and FULL lites indicate charging activity. Two DC output plugs will deliver 1.5v to 12v depending on how many batteries are installed. Measures 7.5" x 2.9" x 1.25". Weighs 6.5 oz. China.

11613	iSun Solar Charger	$80
50267	Battpak Charger	$30
50268	iSun with Battpak	$99

Mobile DC Power Adapter

Use a 12-volt source to run and recharge any battery-powered device using less than 12 volts DC. Our adapter delivers stable DC power at 3, 4.5, 6, 7.5, 9, or 12 volts at up to 2.0 amps. Voltage varies less than 0.5 volt in our tests. Perfect for your camera, DVD, or MP3 player, camcorder, Game Boy®, cell phone, or whatever. The 6-foot fused input cord with LED power indicator plugs into any lighter socket; the 1-foot output cord has six adapter plugs. A storage compartment in the power adapter holds your spare plugs to prevent loss. Slide switch selects voltage. China.

| 53-0105 | Mobile DC Power Adapter | $24 |

12v DC PC Power Supplies

Run and recharge your PC from any 12-volt DC source. Available in 65- or 100-watt output with stabilized voltage adjustable from 15 to 21 volts. Output voltage varies less than 0.5 volt from zero to full load in our tests. External voltage adjustment is recessed for safety. Two-foot fused input cord with LED power indicator plugs into any lighter socket. Four-foot output cord has six different adapters to fit almost any PC laptop with a round power socket. Will not recharge newer Macintosh products requiring 24-volt input. 65-watt unit has maximum 4.0 amp output. 100-watt unit has maximum 7.0 amp output. 3.8" x 2.3" x 1.5", weighs 8 oz. Taiwan.

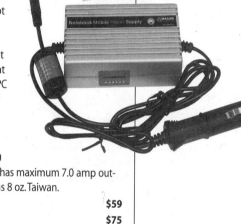

| 53-0103 | 65-watt | $59 |
| 53-0104 | 100-watt | $75 |

Solar e-Power

The pocket-size, folding Solar e-Power charges most models of Nokia, Ericsson, Motorola, or Samsung cell phones. Power cord adapter included. It can also charge 4 AA or AAA batteries in the attached battery holder. Phone charging can be solar direct or energy stored in the batteries. The folding design keeps it small and protects the cells during transport. Delivers 250 mA @ 6.0 volt. 4.5" x 2.6" x 1.3". China.

| 50266 | Solar e-Power | $69 |

Photovoltaic Mounting Hardware

A PV mounting structure will secure your modules from wind damage, and lift them slightly to allow some cooling air behind them. Mounts can be as small as one module for an RV, or big enough to carry hundreds of modules for a utility intertie system. PV systems are most productive if the modules are approximately perpendicular to the sun at solar noon, the most energy-rich time of the day for a PV module. If you live in the Northern Hemisphere, you need to point your modules roughly south. The best year-round angle for your modules is approximately equal to your latitude. For better winter performance, raise that angle about 10°, for better summer performance, lower that angle about 10°.

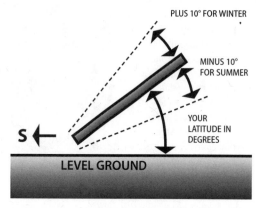

Proper PV mounting angle.

Off-the-grid systems probably want to orient their modules for best wintertime performance, as this is typically when off-the-grid systems are most challenged for power delivery. On-grid, utility-intertied systems are usually set up for best summer performance. Most utilities allow credits to be rolled over from one month to the next. We'll make the most of those long summer days to deliver the maximum kilowatt-hours for the year.

You can change the tilt angle of your array seasonally as the sun angle changes, but on a practical level, many residential systems will have power to burn in the summer. Most folks have found seasonal adjustments to be unnecessary.

The perfect mounting structure would aim the PV array directly at the sun, and follow it across the sky every day. Tracking mounts do this, and in years past Real Goods has enthusiastically promoted trackers. But times change. The electrical and/or mechanical complexity of tracking mounts assures you of ongoing maintenance chores (ask us how we know . . .), and the

falling cost of PV modules makes trackers less attractive. For most systems now, an extra module or two on a simple fixed mount is a way better investment in the long run.

In the following pages you'll find numerous mounting structures, each with its own particular niche in an independent energy system. We'll try to explain the advantages and disadvantages of each style to help you decide if a particular mount belongs in your system.

In ascending order of complexity, your choices are:
- RV Mounts
- Home-Built Mounts
- Top-of-Pole Fixed Mounts
- Fixed Roof or Ground Mounts
- Passive Trackers
- Active Trackers

RV MOUNTS

Because of wind resistance and never knowing which direction the RV will be facing next, most RV owners simply attach the module(s) flat on the roof. RV mounts raise the module an inch or two off the roof for cooling. They can be used for small home systems as well. Simple and inexpensive, most of them are made of aluminum for corrosion resistance. Obviously, they're built to survive high wind speeds. These are a good choice for systems with a module or two. For larger systems, the fixed or pole-top racks are usually more cost-effective.

HOME-BUILT MOUNTS

Want to do it yourself? No problem. Small fixed racks are pretty easy to put together. Anodized aluminum or galvanized steel are the preferred materials due to corrosion resistance, but mild steel can be used just as well, so long as you're willing to touch up the paint occasionally. Slotted steel angle stock is available in galvanized form at most hardware and home-supply stores, and is exceptionally easy to work with. Wood is not recommended because your PV modules will last longer than any exposed wood. Even treated wood won't hold up well when exposed to the weather for over forty years. Make sure that no mounting parts will cast shadows on the modules. Adjustable tilt is nice for seasonal angle adjustments, but most residential systems have power to spare in the summer, and seasonal adjustments usually are abandoned after a few years.

FIXED ROOF OR GROUND MOUNTS

This is easily the most popular mounting structure style we sell. These all use the highly adaptable SolarMount extruded aluminum rails as their base. This a la carte mounting system offers the basic rails in various lengths. Buy enough rails to fit your particular array, then add optional roof standoffs and/or telescoping back legs for seasonal tilt and variable roof pitches. This mounting style can be used for flush roof arrays, low-profile roof arrays with a modest amount of tilt, high-profile arrays that stand *way* up, ground mounts in either low or high profile, or even flipped over and used on south-facing walls. Use concrete footings or a concrete pad for ground mounts. The racks are designed to withstand wind speeds up to 100 mph or more. They don't track the sun, so there's nothing to wear out or otherwise need attention. Getting snow off of them is sometimes troublesome; it can pile up at the base. Ground mounting can leave the modules vulnerable to grass growing up in front, or to rocks kicked up by mowers. A foot or so of elevation, or a concrete pad is a good idea for ground mounts.

TOP-OF-POLE FIXED MOUNTS

A very popular and cost-effective choice, pole-top mounts are designed to withstand winds up to 80 mph; in some cases up to 120 mph. This mounting style is a very good choice for snowy climates. With nothing underneath it, snow tends to slide right off.

For small or remote systems, pole-top mounts are the least expensive and simplest choice. We almost always use these for one- or two-module pumping systems. Tilt and direction can be adjusted easily. Site preparation is easy—just get your steel pipe cemented in straight. The pole is common schedule 40 steel pipe, which is not included (pick it up locally to save on freight). Make sure that your pole is tall enough to allow about 33% burial depth, and still clear livestock, snow, and weeds. Ten feet total pole length is usually sufficient. Taller poles sometimes are used for theft deterrence. Pole diameter depends on the specific mount and array size. Pole sizes listed are for "nominal pipe size." For instance, what the plumbing industry calls "4-inch," is actually 4.5-inch outside diameter. When a mount says it fits "4-inch," it's actually expecting a 4.5-inch diameter pipe.

PASSIVE TRACKERS

Tracking mounts will follow the Sun from east to west across the sky, increasing the daily power output of the modules, particularly in summer and in southern latitudes. Trackers are most often used on water pumping systems with peak demands in summer. See the "To Track or Not to Track" sidebar on the next page for a discussion of when tracking mounts are appropriate.

Passive trackers follow the sun from east to west using just the heat of the sun and gravity. No source of electricity is needed; a simple, effective, and brilliant design solution. The north-south tilt axis is seasonally adjustable manually. Maintenance consists of two squirts with a grease gun once every year.

Do Tilt Angle or Orientation Matter?

Much has been made in years past of PV tracking and tilt angle. You would think that it's nearly a life or death matter to point your PV modules *directly* at the sun during all times of the day. We're here to shake up this belief. Tilt angle and orientation make a lot less difference than most folks believe.

From the rigorously scientific and completely trustworthy Sandia National Labs comes this very interesting chart that details tilt angle versus compass orientation, and the resulting effect on yearly power production. South-facing, at a 30° angle (7:12 roof pitch) delivered the most energy. They labeled that point 100%. All other orientations and tilt angles are expressed as a percentage of that number. Note that we can face southeast or southwest, a full 45° off due south, and only lose 4%. Our tilt angle can be 15° off, and we only lose 3%. In reality, we can face due east or west, and only lose about 15%! A couple of bird droppings could cost you more energy.

Ask Dr. Doug

Welcome to Dr. Doug's (aka Doug Pratt, Real Goods Sales Technician and Technical Editor) down home technical folk wisdom.

ROOF SLOPE AND ORIENTATION (Northern California data)

	Flat (0°)	4:12 (18.4°)	7:12 (30°)	12:12 (45°)	21:12 (60°)	Vertical (90°)
South	.89	.97	1.00	.97	.89	.58
SSE, SSW	.89	.97	.99	.96	.88	.59
SE, SW	.89	.95	.96	.93	.85	.60
ESE, WSW	.89	.92	.91	.87	.79	.57
East, West	.89	.88	.84	.78	.70	.52

The moral of this story? Shadows, bird droppings, leaves, and dirt will have far more effect on your PV output than orientation. Keep your modules clean and shade-free, but don't worry if they aren't perfectly perpendicular to the sun at noon every day. You'll do fine.

To Track or Not To Track

Photovoltaic modules produce the most energy when situated perpendicular to the sun. A tracker is a mounting device that follows the sun from east to west and keeps the modules in the optimum position for maximum power output. At the right time of year, and in the right location, tracking can increase daily output by more than 30%. But beware of the qualifiers, trackers are often *not* a good investment.

Trackers work best during the height of summer when the sun is making a high overhead arc. They add very little in winter unless you live in the extreme southern United States, or even further south. Trackers need clear access to the sun from early in the morning until late in the afternoon. A solar window from 9 A.M. to 4 P.M. is workable; if you have greater access, more power to you (literally).

Tracking mounts are expensive, and PV power is getting cheaper. If you have a project that peaks in power use during the summer, such as water pumping or residential cooling, then tracking may be a very good choice. For many water pumping projects the most cost-effective way to increase daily production is to simply add a tracking mount.

Ask Dr. Doug

If your project peaks in power use during winter, such as powering a typical house, then tracking doesn't offer you much. In most of North America, winter tracking will add less than 15%. One of the new generation of MPPT charge controls is a much better investment in this situation. They add 15 to 30%, and do their best work in the winter. See the Controllers section for more information.

Tracking will boost daily output by about 30% in the summer and 10 to 15% in the winter. The two major problems with passive technology are wind disturbances, and slow "wake-up" when cold. The tracker will go to "sleep" facing west. On a cold morning, it may take more than an hour for the tracker to warm up and roll over toward the east. In winds over 15 mph, the passive tracker may be blown off course. These trackers can withstand winds of up to 85 mph (provided you follow the manufacturer's recommendations for burying the pipe mount), but will not track at high wind speeds. If you have routine high winds, you should have a wind turbine to take advantage of those times, but that's a different subject.

ACTIVE TRACKERS

Active trackers use photocells, electronics, and linear actuators like those on giant old-fashion satellite TV dishes to track the Sun very accurately. A small controller bolted to the array is programmed to keep equal illumination on the photocells at the base of an obelisk. Power use is minuscule. Active trackers average slightly more energy collection per day than a passive

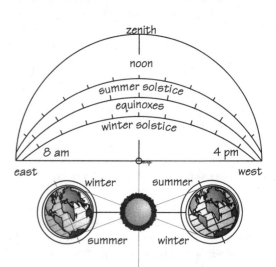

tracker in the same location, but historically they also have averaged more mechanical and electrical problems. Based on our experience, the experience of hundreds of customers, and the dropping price of PV power, we no longer recommend active trackers. Their high initial cost and continuing maintenance problems just aren't worth the investment any longer. Want more power? Add some more PV. It's hard to beat the reliability of *no moving parts*.

PV MOUNTING PRODUCTS

Solar Pathfinder (Rent or Buy)

It only takes a fist-sized chunk of shade to effectively turn off a solar module. Wondering if that tree or building will shade your array during certain times of the year? In less than two minutes on site, the Pathfinder

will show exactly what times of day and months of the year shade will fall on any site. This wonderfully simple tool only requires you to stand on the site, face south (there's a built-in compass with magnetic declination adjustment), level the Pathfinder (built-in bubble level), and read the shade reflections on the dome. The chart under the dome shows the hours and months that bit of shade will be a problem. Every professional needs to own one of these. Homeowners probably only need it once, so we've got a nice $25/week rental program. Credit card required for deposit. The handheld model is just the Pathfinder head. The Professional model adds an adjustable tripod mount and protective custom metal carrying case. USA.

11605	Weekly Pathfinder Rental	$25
11603	Handheld Pathfinder	$250
11604	Professional Pathfinder w/ tripod & metal case	$360

UniRac Tracker 200

One and two module trackers for water pumping

Water needs usually peak in the summer, the same time a tracking mount delivers more hours of peak sunlight, delivering longer pump run times. For most of North America, a tracking mount will boost your mid-summer gallons per day by about 30%. This is an application where tracking mounts still make good sense.

These are simple, dependable, heat-driven passive trackers that follow the sun from east to west every day. They mount on standard 2.5" schedule 40 steel pipe (purchase pipe locally). About one-third of the total pipe length should be buried and cemented. Non-corrosive aluminum components with stainless steel hardware. These racks will accept one or two of the modules listed. 10-year manufacturer's warranty. USA.

T200/60	T200/68	T200/80
Astropower/GE 65 thru 75 Siemens/Shell 65 thru 85	Astropower/GE 100 thru 120 Evergreen 47 thru 115 Kyocera 35 thru 120 Siemens/Shell 90 thru 110	Astropower/GE 150, 165 Siemens/Shell 130 thru 175 Sharp 165 thru 185 UniSolar 64

14328	T200/60	$495
13996	T200/68	$520
14314	T200/80	$535

Solar mount rail
cross section

UNIRAC FIXED RACKS

UniRac makes a wide variety of high-quality, easy-to-install mounting options. Most are based on their incredibly adaptable extruded SolarMount™ aluminum rail. The clever bolt slots allow module and footing bolts to be placed at any points needed along the rail. All UniRacs feature the highest quality, non-corrosive components. Constructed of aluminum and galvanized steel with stainless steel fasteners, there are no painted or zinc-plated pieces. Durability is excellent, assembly is fast, and instructions are clear and complete. UniRacs ship directly from the manufacturer via UPS, usually within one week, thanks to standardized construction. There's no waiting for racks and no exorbitant shipping charges since everything packs small and tight. Engineered to sustain wind loads of 50 lbs. per sq.ft. (about 120 mph). USA.

Flush roof mounting.

RAIL SIZING CHART

Pick your module on the left. Read to right for number of modules carried per rail set, plus clamp size (A thru H).
Order rail and clamp sets as needed for your array.

	48" 14354 $83	60" 14355 $99	72" 14356 $115	84" 14357 $131	96" 14358 $145	106" 14359 $157	120"* 14360 $178	132"* 14361 $190	144"* 14362 $202	156"* 14363 $214	168"* 14364 $226	180"* 14365 $238	192"* 14366 $259	204"* 14367 $271	216"† 14368 $283
L-feet supplied	4	4	4	4	4	4	6	6	6	6	6	6	8	8	8
Number of PV Modules supported, followed by clamp kit depth code letter															
Evergreen all EC-100 series		2C	3C				4C		5C		6C		7C		8C
Kyocera KC80, 120, 125G		2C	3C				4C		5C		6C		7C		8C
Sharp 80w	2C		3C		4C		5C		6C	7C	8C				
123w		2F	3F				4F		5F		6F		7F		
140w, 167w			2H					3H			4H			5H	
165, 175, 185w			2F			3F			4F			5F		6F	
Shell 70w thru 85w	2C		3C		4C		5C		6C	7C	8C				
130w thru 175w		2D				3D			4D			5D		6D	
Uni-Solar US 42, 64			2B		3B			4B		5B			6B		7B

* 120" thru 204" rails and components normally are shipped via truck freight, as they're too long for UPS. For UPS shipping, add one item UPS-20 per rail set. Rails will be cut in half and a splice kit added. USA.

† 216" rails ship via truck freight only.

UniRac SolarMount™ Roof or Ground Mounts

These universal mounting systems are sold a la carte. SolarMount rails are sold in pairs. Buy enough length to carry all your modules (plus 1″ between modules and 1.5″ at each end), buy mounting clamp sets as required, then add optional tilt legs or optional standoff mounting points as needed.

The standard rails are supplied with L-bracket feet, which are acceptable for most asphalt shingle roofs. Standoff mounts are a must for tile, ceramic, shake, or flat roofs. Standoffs are a more leak-secure, roofer-friendly option for all roofs. SolarMounts are installed flush with your roof, unless you use the optional tilt legs, which are also used for ground mounting. Both high-profile and low-profile style legs are offered, depending on your module orientation. See example drawings. Stainless steel fastener hardware is supplied for everything. All components made in the USA with 10-year manufacturer's warranty.

14269 UniRac UPS-20   **$20**

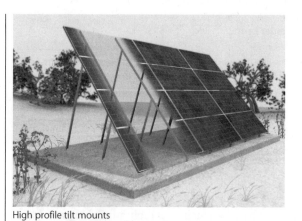
High profile tilt mounts

Low profile tilt mounts

UniRac Top-Mount Rail Clamp Sets

These easy-to-install clamps secure the PV modules from the top using aluminum clamps and stainless nuts and bolts. They require 1" between modules and 1.5" at row ends. Sold in sets with all the clamps and hardware to secure from two to eight modules. Module frame depths vary. Clamp depth is selected by the letter following the item number. Find the correct letter for your modules from the Rail Sizing Chart at left, and specify on your order. For instance, all Evergreen modules are "C" depth. USA.

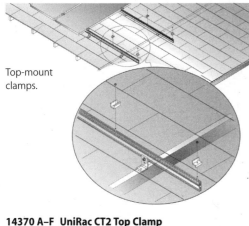

Top-mount clamps.

14370 A–F	UniRac CT2 Top Clamp Set for 2 Modules	$16
14370 G	UniRac CT2 Top Clamp Set for 2 Modules	$11
14370 H	UniRac CT2 Top Clamp Set for 2 Modules	$14
14371 A–F	UniRac CT3 Top Clamp Set for 3 Modules	$20
14371 G	UniRac CT3 Top Clamp Set for 3 Modules	$14.50
14371 H	UniRac CT3 Top Clamp Set for 3 Modules	$18
14372 A–F	UniRac CT4 Top Clamp Set for 4 Modules	$24
14372 G	UniRac CT4 Top Clamp Set for 4 Modules	$18
14372 H	UniRac CT4 Top Clamp Set for 4 Modules	$22
14373 A–F	UniRac CT5 Top Clamp Set for 5 Modules	$28
14373 G	UniRac CT5 Top Clamp Set for 5 Modules	$21.50
14373 H	UniRac CT5 Top Clamp Set for 5 Modules	$26
14374 A–F	UniRac CT6 Top Clamp Set for 6 Modules	$32
14375 A–F	UniRac CT7 Top Clamp Set for 7 Modules	$36
14376 A–F	UniRac CT8 Top Clamp Set for 8 Modules	$40

UniRac Standoffs

Standoffs are the connection between your roof framing and the PV racking. Standoffs penetrate your roof, and use a leak-proof flashing just like a plumbing vent. They're even the same size as a typical 1.25" vent pipe. This is familiar territory for your roofer. The flat base is lag-bolted to a framing member. Two $5/16$" x 3.5" stainless lag bolts are included, then a standard plumbing vent flashing slips over the top. Standoffs are chrome-plated steel for strength with corrosion resistance. Standoff mounts are required for tile, ceramic, shake, and flat roofs. They're a more leak-secure, roofer-friendly option for all roofs.

Raised flange-type standoffs replace the L-feet and fasten directly to the SolarMount rails. These are the most common type used. Flat-top standoffs support common uni-strut stringers that fasten to the standard L-feet, a common technique on flat roofs. All standoffs are available in 3", 4", 6", or 7" heights. Order as needed to raise rails above roofing material. Flashings are <u>NOT</u> included. USA.

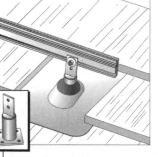

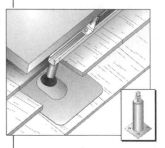

Optional standoff feet

14400	UniRac STR3 Raised Flange Standoff	$18.00
14401	UniRac STR4 Raised Flange Standoff	$18.50
14402	UniRac STR6 Raised Flange Standoff	$19.00
14400	UniRac STR7 Raised Flange Standoff	$19.50
14396	UniRac STR3 Flat Top Standoff	$17.00
14397	UniRac STR4 Flat Top Standoff	$17.50
14398	UniRac STR6 Flat Top Standoff	$18.00
14399	UniRac STR7 Flat Top Standoff	$18.50

UniRac Rail Splice Kits

For very long flush or low-profile installations, splice kits can be used to join rails end to end. Due to thermal expansion and contraction, no more than two rail sections should be spliced, and splice kits are never used with high-profile tilt legs. Each splice kit contains two splices plus stainless steel hardware. USA.

13666	UniRac SP-2 Splice Kit	$16

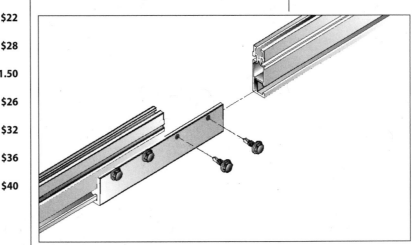

Rail splice kit

UniRac Low Profile Tilt Legs

Low-profile orientation minimizes the vertical height of your array. Useful to hide an array behind a parapet on a flat roof, or just to keep the visual impact to a minimum. These can be used for rooftop or ground mounting. The aluminum rear legs are telescoping and adjustable on site. Longer rail sets, 192" thru 216", require two sets of legs. USA.

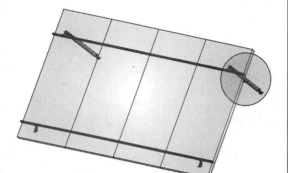

LOW PROFILE TILT LEGS SIZING CHART

Find your PV module on the left, then your required angle between the modules and the mounting surface (roof or ground), then read across until you find the number of modules you need to mount. Order leg kit(s) as needed from that column.

	TLL2-12 14390 $49	TLL2-30 14391 $59	TLL2-44 14392 $67	TLL3-12 14393 $68	TLL3-30 14394 $83	TLL3-44 14395 $94
Legs per kit:	2	2	2	3	3	3
Number of PV Modules per Rail Set						
Evergreen all EC-100 series and Sharp 123w						
07–16°	2–3, 7–8*			4–6		
16–40°		2–3, 7–8*			4–6	
24–60°			2–3, 7–8*			4–6
Kyocera KC80						
13–28°	2–3, 7–8*			4–6		
30–60°		2–3, 7–8*			4–6	
Kyocera KC120, 125G						
8–17°	2–3, 7–8*			4–6		
18–45°		2–3, 7–8*			4–6	
26–60°			2–3, 7–8*			4–6
Sharp 140w						
9–21°	2, 5*			3–4		
22–25°		2, 5*			3–4	
32–60+°			2, 5*			3–4
Sharp 167w						
8–18°	2, 5*			3–4		
19–48°		2, 5*			3–4	
28–60+°			2, 5*			3–4
Sharp 165, 175, 185w						
7–16°	2–3, 6*			4–5		
16–41°		2–3, 6*			4–5	
24–60+°			2–3, 6*			4–5
Shell 70 thru 85w and Sharp 80w						
9–20°	2–4			5–8		
21–54°		2–4			5–8	
31–60+°			2–4			5–8
Shell 130 thru 175w						
7–15°	2–3, 8			5–7		
16–39°		2–3, 8			5–7	
23–60°			2–3, 8			5–7
Uni-Solar US 42, 64						
8–18°	2–3, 6–7*			4–5		
19–47°		2–3, 6–7*			4–5	
27–60+°			2–3, 6–7*			4–5

* Long rails, 2 leg kits required per rail set.

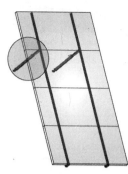

UniRac High-Profile Tilt Legs

High-profile orientation maximizes module density on a given footprint. High profile legs are telescoping and adjust on site. Sizing is based on the length and desired angle of your rails. Two legs are provided for rail sets up to 106". For 120" thru 180" rail sets, a short and a long leg are provided for each rail (four legs total). Rails longer than 180" are not recommended for high-profile mounts. Do NOT use rail splices with any high profile mounting. USA.

Rail Set & Tilt Angle	Item #	Price	Rail Set & Tilt Angle	Item #	Price
48", 10–23°	14384	$37	120", 5–10°	14387	$65
48", 33–60+°	14385	$54	120", 17–38°	14388	$99
60", 8–18°	14384	$37	120", 27–60°	14389	$129
60", 26–60+°	14385	$54	132", 6–10°	14387	$65
72", 7–16°	14384	$37	132", 17–37°	14388	$99
72", 22–60°	14385	$54	132", 26–60°	14389	$129
84", 5–12°	14384	$37	144", 6–9°	14387	$65
84", 17–47°	14385	$54	144", 16–33°	14388	$99
84", 28–60°	14386	$68	144", 24–56°	14389	$129
96", 5–11°	14384	$37	156", 5–8°	14387	$65
96", 16–43°	14385	$54	156", 14–30°	14388	$99
96", 26–60°	14386	$68	156", 22–49°	14389	$129
106", 4–10°	14384	$37	168", 5–7°	14387	$65
106", 15–39°	14385	$54	168", 13–28°	14388	$99
106", 24–60°	14386	$68	168", 20–46°	14389	$129
			180", 3–7°	14387	$65
			180", 12–26°	14388	$99
			180", 19–43°	14389	$129

UniRac Pole-Top Mounts

Pole-top mounts are the easiest for most home owners to install, and will shed snow better than any other mounting style. Tilt and direction adjust easily. You supply the Schedule 40 or 80 steel pipe locally. Approximately one third of the total pipe length should be buried and backfilled with concrete. Pole heights more than about 10 feet above ground are okay, but should have guy wires or be firmly attached to the side of a building.

Uni-Rac mounts use only non-corrosive aluminum with stainless steel hardware. All necessary nuts and bolts for modules are included. Modules mount with long dimension perpendicular to the rails, as shown in photo. 10-year manufacturer's warranty. USA.

UniRac 5001-series pole-top mount

UNIRAC POLE-TOP MOUNTS

Item #	Price	UniRac #	Pipe size	Evergreen EC100 series	Kyocera 80	Kyocera 120	Sharp 80	Sharp 123	Shell/Siemens 165–185	Shell/Siemens 70–85	Shell/Siemens 130–175	UniSolar 64
42-0177	$112	500023	2.5"							1		
42-0178	$118	500025	2.5"		1							
42-0179	$124	500027	2.5"						1		1	
42-0180	$131	500030	2.5"							2		
42-0181	$139	500031	2.5"		2							
42-0208	$115	500034	2.5"				1					
42-0182	$121	500036	2.5"	1								
42-0183	$125	500037	2.5"									1
42-0185	$134	500040	2.5"				2					
42-0186	$123	500043	2.5"			1		1				
42-0188	$225	500104	3"							3		
42-0189	$228	500105	3"								2	
42-0190	$231	500106	3"						2			
42-0191	$249	500107	3"		3							
42-0192	$258	500109	3"							4		
42-0193	$285	500113	3"		4							

UNIRAC POLE-TOP MOUNTS

Item #	Price	UniRac #	Pipe size	Evergreen EC100 series	Kyocera 80	Kyocera 120	Sharp 80	Sharp 123	Shell / Siemens 165–185	Shell / Siemens 70–85	Shell / Siemens 130–175	UniSolar 64
42-0194	$224	500114	3″	2								
42-0195	$233	500115	3″									2
42-0196	$239	500117	3″				3					
42-0197	$263	500121	3″	3								
42-0198	$268	500122	3″				4					
42-0199	$278	500124	3″									3
42-0200	$231	500128	3″			2		2				
42-0201	$265	500131	3″			3		3				
42-0202	$336	500219	4″								3	
42-0203	$339	500220	4″							3		
42-0205	$362	500224	4″	4								
42-0206	$373	500227	4″			4						
42-0207	$376	500228	4″					4				
42-0209	$390	500322	4″							6		
42-0210	$427	500325	4″		6							
42-0211	$489	500330	4″		8							
42-0212	$408	500331	4″									4
42-0250	$420	500333	4″				6					
42-0213	$630	500420	6″							8		
42-0214	$710	500424	6″							10		
42-0215	$774	500428*	6″							12		
42-0216	$779	500429*	6″		10							
42-0217	$867	500430*	6″		12							
42-0218	$571	500431	6″							4		
42-0219	$643	500433	6″				8					
42-0220	$659	500435	6″									6
42-0221	$698	500436	6″							6		
42-0222	$714	500437	6″				10					
42-0224	$778	500440*	6″									8
42-0225	$821	500441*	6″				12					
42-0226	$837	500443*	6″							8		
42-0227	$896	500445*	6″									10
42-0230	$965	500448*	6″							10		
42-0231	$1021	500451*	6″									12
42-0232	$576	500453	6″								4	
42-0233	$632	500456	6″	6		6		6				
42-0234	$698	500460	6″								6	
42-0235	$704	500461	6″			8						
42-0236	$706	500462	6″	8				8				
42-0237	$826	500466*	6″			10					8	
42-0238	$828	500467*	6″	10				10				
42-0239	$948	500470*	6″	12		12						
42-0251	$952	500471	6″					12				
42-0240	$956	500472*	6″								10	

* Requires truck shipping. All others go UPS.

GAIAM REAL GOODS

UniRac 5000 Series Pole-Top.

UniRac 5002 Series Pole-Top.

UniRac 5004 Series Pole-Top.

UniRac 5003 Series Pole-Top.

RV Mounting Structures

For RVs, we've got simple mounts, and even simpler mounts. The really, really simple solution is RV Mounting Feet. This set of four rustproof aluminum feet fits any single PV module and holds it just above the roof for junction box clearance and cooling. Stainless hardware for attachment to the module is included. You provide hardware to attach to the RV roof as needed. One set of feet mounts one module.

The slightly more elegant solution is the RV flush mount, which provides a pair of aluminum channels bolting down to four L-brackets. This provides more cooling clearance, and depending on assembly, mostly will hide under the module. Depending on your module size, these flush mounts can take one or two modules. Choose your size. You provide roof attachment hardware as needed. Both mounts have a 10-year manufacturer's warranty. USA.

42-0169	**RV Mounting Feet**	**$18**
42-0170	**RV Flush Mount 16"**	**$49**
42-0171	**RV Flush Mount 20"**	**$54**
42-0172	**RV Flush Mount 24"**	**$59**
42-0173	**RV Flush Mount 28"**	**$64**
42-0174	**RV Flush Mount 32"**	**$69**
42-0175	**RV Flush Mount 40"**	**$74**
42-0176	**RV Flush Mount 44"**	**$79**

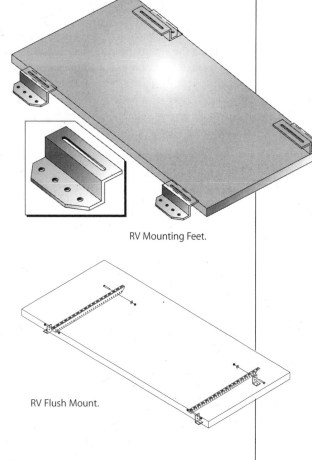
RV Mounting Feet.

RV Flush Mount.

Wind and Hydropower Sources

Wind and hydro energy sources are most often developed as a booster or bad weather helper for a solar-based system.

These hybrid systems have the advantage of being better able to cover power needs throughout the year, and are less expensive than a similar capacity system using only one power source.

Sunlight, wind, and falling water are the renewable energy big three. These are energy sources that are commonly available at a reasonable cost. Solar or sunlight, the single most common and most accessible renewable energy source, is well covered at the beginning of this chapter. We've found in our years of experience that wind and hydro energy sources most often are developed as boosters or bad-weather helpers for a solar-based system. These hybrid systems have the advantage of being better able to cover power needs throughout the year, and are less expensive than a similar capacity system using only one power source. When a storm blows through, the solar input is lost but a wind generator more than makes up for it. The short, rainy days of winter may limit solar gain, but the hydro system picks up from the rain and delivers steady power twenty-four hours a day. This is not to say that you shouldn't develop an excellent single-source power system if you've got it; like a year-round stream dropping 200 feet across your property, for instance. But for most of us, we'll be further ahead if we don't put all our eggs in one basket. Diversify!

Our experienced technical staff is well versed in supplying the energy needs of anything from a small weekend getaway cabin all the way up to an upscale state-of-the-art resort. We'll be glad to help put together a system for you. There is usually no charge for our friendly and personalized services.

Wind Systems

We generally advise that a good, year-round wind turbine site isn't a place that you'd want to live. It takes average wind speeds of 8 to 9 mph and up, to make a really good site. That's honestly more wind than most folks are comfortable living with. But this is where the beauty of hybrid systems comes in. Many, very happily livable sites do produce 8-plus mph during certain times of the year, or when storms are passing through. Tower height and location also make a big difference. Wind speeds average 50 to 60% higher at 100 feet compared to ground level (see chart on page 95). Wind systems these days are almost always designed as wind/solar hybrids for year-round reliability. The only common exceptions are systems designed for utility intertie; they feed excess power back into the utility, and turn the meter backwards.

Hydro Electric Systems

For those who are lucky enough to have a good site, hydro is really the renewable energy of choice. System component costs are much lower, and watts per dollar return is much greater for hydro than for any other renewable source. The key element for a good site is the vertical distance the water drops. A small amount of water dropping a large distance will produce as much energy as a large amount of water dropping a small distance. The turbine for the small amount of water is going to be smaller, lighter, easier to install, and vastly cheaper. We offer several turbine styles for differing resources. The small Pelton wheel Harris systems are well suited for mountainous territory that can deliver some drop and high pressure to the turbine. The propeller-driven Low Head Stream Engine, and smaller Jack Rabbit turbines are for flatter sites with less drop, but more volume, and the Stream Engine, with a turgo-type runner, falls in between. It can handle larger water volumes and make useful power from shorter vertical water drop distances.

Read on for detailed explanations of wind generators and hydro turbines. If you need a little help and guidance putting together a system or simply upgrading, our technical staff, with decades of hands-on experience in renewable energy, will be glad to help. Call us toll-free at: 800-919-2400.

Hydroelectricity

If you could choose any renewable energy source you wanted, hydro is the one. If you don't want to worry about a conservation-based lifestyle—always nagging your kids to turn off the lights, watching the voltmeter, basing every appliance decision on energy efficiency—then you had better settle next to a nice year-round mountain stream! Hydropower, given the right

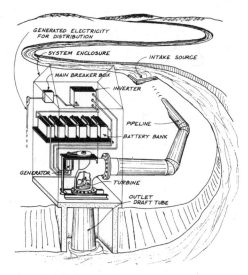

Low-head installation.

site, can cost as little as a tenth of a PV system of comparable output. hydropower users often are able to run energy-consumptive appliances that would bankrupt a PV system owner, like large side-by-side refrigerators and electric space heaters. hydropower probably will require more effort on site to install, but even a modest hydro output over twenty-four hours a day, rain or shine, will add up to a large cumulative total. Hydro systems get by with smaller battery banks because they only need to cover the occasional heavy power surge rather than four days of cloudy weather.

Hydro turbines can be used in conjunction with any other renewable energy source, such as PV or wind, to charge a common battery bank. This is especially true in the West, where seasonal creeks with substantial drops only flow in the winter. This is when power needs are at their highest and PV input is at its lowest. Small hydro systems are well worth developing, even if used only a few months out of the year, if those months coincide with your highest power needs. So, what makes a good hydro site and what else do you need to know?

What is a "Good" Hydro Site?

The Columbia River in the Pacific Northwest has some really great hydro sites, but they're not exactly homestead-scale (or low-cost). Within the hydro industry, the kind of home-scale sites and systems we deal with are called micro-hydro. The most cost-effective hydro sites are located in the mountains. Hydropower output is determined by water's volume times its fall or drop (jargon for the fall is "head"). We can get approximately the same power output by running 1,000 gallons per minute through a 2-foot drop as by running 2 gallons per minute through a 1,000-foot drop. In the former scenario, where lots of water flows over a little drop, we are dealing with a low-head/high-flow situation, which is not truly a micro-hydro site. Turbines that can handle thousands of gallons efficiently are usually large, bulky, expensive, and site-specific. But if you don't need to squeeze every last available watt out of your low-head source, the Low Head Stream Engine generator will produce very useful amounts of power from low-head/high-flow sites. In addition, the turgo runner used on the Stream Engine is good at high volume sites with 15 feet or more head.

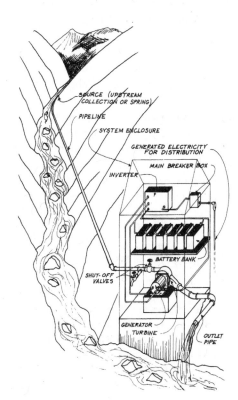

High-head Installation.

Hydropower, given the right site, can cost as little as a tenth of a PV system of comparable output.

Small hydro systems are well worth developing, even if used only a few months out of the year, if those months coincide with your highest power needs.

In general, any site with more than 100 feet of fall will make an excellent micro-hydro site, but many sites with less fall can be very productive also. The more head, the less volume will be necessary to produce a given amount of power.

For hilly sites that can deliver a minimum of 50 feet of head, Pelton wheel turbines offer the lowest-cost generating solution. The Pelton-equipped Harris turbine is perfect for low-flow/high-head systems. It can handle a maximum of 200 gallons per minute, and requires a minimum 50-foot fall in order to make useful amounts of power. In general, any site with more than 100 feet of fall will make an excellent micro-hydro site, but many sites with less fall can be very productive also. The more head, the less volume will be necessary to produce a given amount of power. Check the output charts in the products section for a rough estimate of what your site can deliver.

A hydro system's fall doesn't need to happen all in one place. You can build a small collection dam at one end of your property and pipe the water to a lower point, collecting fall as you go. It's not unusual to use several thousand feet of pipe to collect a hundred feet of head (vertical fall).

Our Hydro Site Evaluation service will estimate output for any site, plus it will size the piping and wiring, and factor in any losses from pipe friction and wire resistance. See the example at the end of this editorial section.

What If I Have a High-Volume, Low-Head Site or Want AC-Output?

Typically, high-volume, low-head, or AC-output hydro sites will involve engineering, custom metalwork, formed concrete, permits, and a fair amount of initial investment cash. None of this is meant to imply that there won't be a good payback, but it isn't an undertaking for the faint-of-heart or thin-of-wallet. AC generators are typically used on larger commercial systems, or on utility intertie systems. DC generators are typically used on smaller residential systems.

DC generation systems offer several advantages for small hydro. Control is easy and cheap. The batteries we'll use to store energy allow power output surges way over what the turbine is delivering. The DC to AC inverters we have available to us now will deliver far cleaner and more tightly regulated AC power than a small AC hydro turbine can manage, and the inverter will cost less than a small AC control system.

For the homestead with a good creek, but little significant fall, the Low Head Stream Engine turbine is a far less costly alternative. With just three or four feet of fall, and some site develop-

The Jack Rabbit is a drop-in-the-creek generator.

ment work, this simple turbine can provide for a modest homestead.

If you'd rather look into the typical low-head scenario, contact the DOE's Renewable Energy Clearinghouse at 800-363-3732, or use Internet access at www.eere.energy.gov/RE/hydropower .html for more free information on low-head hydro than you ever thought was possible.

How Do Micro-Hydro Systems Work?

The basic parts of micro-hydro systems are: the pipeline (called the penstock in the trade), which delivers the water; the turbine, which transforms the energy of the flowing water into rotational energy; the alternator or generator, which transforms the rotational energy into electricity; the regulator, which controls the generator or dumps excess energy, depending on regulator style; and the wiring, which delivers the electricity. Our micro-hydro systems also use batteries, which store the low-voltage DC electricity, and usually an inverter, which con-

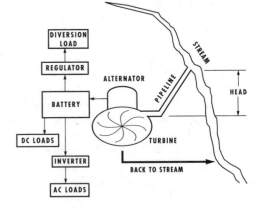

Typical micro-hydro system.

verts the low-voltage DC electricity into 120 volts or 240 volts AC electricity.

Most micro-turbine systems use a small DC alternator or generator to deliver a small but steady energy flow that accumulates in a battery bank. This provides a few important advantages. The battery system allows the user to store energy and expend it, if needed, in short powerful bursts (like a washing machine starting the spin cycle). The batteries will allow us to deliver substantially more energy for short periods than a turbine is producing, as long as the battery and inverter are designed to handle the load. DC charging means that precise control of alternator speed is not needed, as is required for 60 Hz AC output. This saves thousands of dollars on control equipment.

And finally, with the quality of the DC to AC inverters that are available to us now, you'll enjoy cleaner, more tightly controlled AC power through an inverter, than through a small AC turbine. The bottom line is that a DC-based system will cost far less than an AC system for most residential users, and will perform better.

DC Turbines

Several micro-hydro turbines are available with simple DC output. We currently offer the Harris, the Stream Engine, and the Jack Rabbit.

HARRIS TURBINE

The Harris turbine uses a hardened cast silicone bronze Pelton turbine wheel mated with a low-voltage DC alternator. Pelton wheel turbines work best at higher pressures and lower volumes. Life begins at about 50 feet of head for these turbines, and has no practical upper limit. Harris offers a couple of choices for the alternator. The standard Harris is based on common 1970s Motorcraft alternators with windings that are customized for each individual application. Bearings and brushes will require replacement at intervals from one to five years, depending on how hard the unit is working. These parts are commonly available at any auto parts store. Harris also offers an optional permanent-magnet alternator that is custom-made. The PM alternators deliver more power under almost all conditions, have no brushes to wear out, and are mounted on larger, more robust bearings with two to three times the life expectancy. For most customers, this is a very worthwhile $700 upgrade.

Depending on the volume and fall supplied, Harris turbines can produce from 1 kWh (1,000 watt-hours) to 35 kWh per day. Maximum alternator instantaneous output is about 2500 watts in a 48-volt system with cooling options required. The typical American home consumes 15 to 25 kWh per day with no particular energy conservation, so with a good hydro site, it is fairly easy to live a conventional lifestyle.

The Harris turbine can be supplied with one, two, or four nozzles. The maximum flow rate for any single nozzle can be from 20 to 60 gallons per minute (gpm), depending on the head pressure. The turbine can handle flow rates to about 120 gpm before the sheer volume of water starts getting in its own way. Many users buy two- or four-nozzle turbines with different sized nozzles, so that individual nozzles can be turned on and off to meet variable power needs or water availability. The brass nozzles are easily replaceable because they eventually wear out, especially if there is grit in the water. They are available in 1/16-inch increments, from 1/16 through 1/2 inch. The first nozzle doesn't have a shut-off valve, while all nozzles beyond the first one are supplied with ball valves for easy, visible operation.

STREAM ENGINE

The Stream Engine turbines use a unique brushless permanent-magnet alternator with three large, sealed shaft bearings. Permanent magnets mean there are no field brushes to wear out. Magnetic field strength is adjusted by varying the air gap between the magnet disk and the stationary alternator windings. Once the unit is set up, almost no routine maintenance is required. Setup does require some time with this universal design, however, and involves trial and adjustment: selecting one of four alternator wiring setups, and then adjusting the permanent-magnet rotor air gap until peak output is achieved. A precision shunt and digital multimeter are supplied to expedite this set-up process. Output voltage can be user-selected at 12, 24, or 48 volts. Maximum instantaneous output is 800 watts for this alternator.

This low-maintenance alternator is employed on two very different turbines. The original Stream Engine uses a cast bronze turgo-type runner wheel. This turgo wheel can handle a bit more water volume than the Harris Pelton wheel—up to about 200 gpm before it starts choking—and starts to deliver useful output at lower 15- to 20-foot heads. Nozzles are cone-shaped plastic casings; you cut them at the size desired, from 1/8 up to 1 inch. Two- or four-

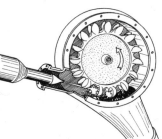

An 1880s vintage single-nozzle Pelton wheel. Only the generator technology has changed.

The bottom line is that a DC-based system will cost far less than an AC system for most residential users, and will perform better.

Two- and four-nozzle Harris turbines. The four-nozzle is upside down to show the Pelton wheel and nozzles.

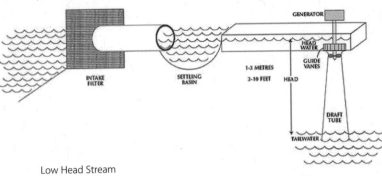

Low Head Stream Engine installation.

A typical installation places the batteries at the house on top of the hill, where the good view is, and the turbine at the bottom of the hill, where the water ends its maximum drop. Low-voltage power is difficult to transmit if large quantities or long distances are involved.

nozzle turbines are available, with replacement nozzles being a readily available bolt-in.

The Low Head Stream Engine uses the same alternator, but is packaged quite differently. It works on 2 to 10 feet of fall, which happens on the downstream end of the turbine for a change. Flows of 200 to 1,000 gpm can be accommodated through the 5-inch propeller turbine. The large draft tube on the output must be immersed in the tail water. See the illustration. This turbine requires more site preparation, but almost no maintenance or attention once it's installed and tuned.

JACK RABBIT

The Jack Rabbit is a 100-watt, low-speed alternator that originally was designed for towing behind seismic sleds for ocean-floor mapping. This drop-in-the-stream propeller-driven charger reaches maximum output at 9 to 10 mph stream speed. This is a relatively low-cost, low-impact hydro solution for folks in the flatlands. But it's also relatively low-output. It isn't worth installing unless you've got a choke point in your stream that delivers 9 mph or higher flow rates. The 24-volt version in particular seems to require higher flow rates to produce useful power output. It is available in 12- or 24-volt versions, and is not field adjustable or switchable.

Power Transmission

One disadvantage of lower-voltage DC hydro systems is the difficulty of transmitting power from the turbine to the batteries, particularly with high-output sites. A typical installation places the batteries at the house on top of the hill, where the good view is, and the turbine at the bottom of the hill, where the water ends its maximum drop. Low-voltage power is difficult to transmit if large quantities or long distances are involved. The batteries should be as close to the turbine as is practical, but if more than 100 feet of distance is involved, things will work better if the system voltage is 24 or even 48 volts.

Transmission distances of more than 500 feet often require expensive large-gauge wire or technical tricks. Don Harris has been working with Outback Power Systems to develop a hydroelectric version of their maximum power point tracking MX60 controller, which allows inputs up to 140 volts, while feeding the batteries whatever it is that makes them happy. At the time of this writing, the Hydro MX60 isn't ready for release, but beta units are in trial. Please consult with the Real Goods Technical Staff about this, or other transmission options.

Controllers

Hydro generators require special controllers or regulators. Controllers designed for photovoltaics may damage the hydro generator, and will very likely become crispy-critters themselves if used with a hydro generator. You can't simply open the circuit when the batteries get full like you can with PV. So long as the generator is spinning, the energy needs a place to go. Controllers for hydro systems take any power beyond what is needed to keep the batteries charged, and divert it to a secondary load, usually a water- or space-heating element. So extra energy heats either domestic hot water or the house itself. These diversion controllers are also used with some wind generators, and can be used for PV control as well if this is a hybrid system.

The Jack Rabbit turbine is a special case. This low-speed turbine breaks the usual control rules, and can be used with an inexpensive PV control, or often, no control at all. Because of the Jack Rabbit's slow speed, it won't generate high voltage if disconnected from the battery.

Site Evaluation

Okay, you have a fair amount of drop across your property, and/or enough water flow for one of the low-head turbines, so you think micro-hydro is a definite possibility. What happens next? Time to go outside and take some measurements, then fill in the necessary information on the Hydro Site Evaluation form. With the information on your completed form, the Real Goods technicians can calculate which turbine and options will best fill your needs, as well as what size pipe and wire and which balance-of-system components you require. Then we can quote specific power output and system costs so that you can decide if hydro is worth the installation effort. The Jack Rabbit doesn't need the fall or flow measurements; it just requires a minimum of 13 inches of water depth and a flow that's moving at 8 miles per hour (a jogging pace), or up.

DISTANCE MEASUREMENTS

Keep the turbine and the batteries as close together as practical. As discussed earlier, longer transmission distances will get expensive. The more power you are trying to move, the more important distance becomes.

You'll need to know the distance from the proposed turbine site to the batteries (how many feet of wire), and the distance from the turbine site to the water collection point (how many feet of pipe). These distances are fairly easy to determine; just pace or tape them off.

FALL (DROP OR HEAD) MEASUREMENT

Next, you'll need to know the fall from the collection point to the turbine site. If a pipeline is in place already, or if you can run one temporarily and fill it with water, this part is easy. Simply install a pressure gauge at the turbine site, make sure the pipe is full of water, and turn off the water at the bottom. Read the static pressure, and multiply your reading in pounds per square inch (psi) by 2.31 to obtain the drop in feet. If the water pipe method isn't practical, you'll have to survey the drop or use a fairly accurate altimeter or GPS device. A number of relatively inexpensive sports watches come with a built-in altimeter now. If the altimeter can read ±10 feet, that's close enough. Strap it on, take a hike, and record the difference.

The following instructions represent the classic method of surveying. You've seen survey parties doing this and if you've always wanted to attend a survey party, this is your big chance to get in on the action. You'll need a carpenter's level, a straight sturdy stick about eye-level tall, a brightly colored target that you will be able to see a few hundred feet away, and a friend to carry the target and to make the procedure go faster and more accurately. (It's really hard to party alone.)

Stand the stick upright and mark it at eye level. (Five feet even is a handy mark that simplifies the mathematics, if that's close to eye level for you.) Measure and note the length of your stick from ground level to your mark. Starting at the turbine site, stand the stick upright, hold the carpenter's level at your mark, make sure it is level, then sight across it uphill toward the water source. With hand motions and body English, guide your friend until the target is placed on the ground at the same level as your sightline, then have your friend wait for you to catch up.

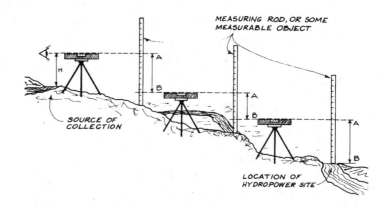

Measuring Fall

Repeat the process, carefully keeping track of how many times you repeat. It is a good idea to draw a map to remind you of landmarks and important details along the way. If you have a target and your friend has a stick (marked at the same height, please) and level, you can leapfrog each other, which makes for a shorter party. Multiply the number of repeats between the turbine site and the water source by the length of your stick(s) and you have the vertical fall. People actually get paid to have this much fun!

FLOW MEASUREMENT

Finally, you'll need to know the flow rate. If you can, block the stream and use a length of pipe to collect all the flow. Time how long it takes to fill a 5-gallon bucket. Divide 5 gallons by your fill time in seconds. Multiply by 60 to get gallons per minute. Example: The 5-gallon bucket takes 20 seconds to fill. 5 divided by 20 = 0.25 times 60 = 15 gpm. If the flow is more than you can dam up or get into a 4-inch pipe, or if the force of the water sweeps the bucket out of your hands, forget measuring: You've got plenty!

Conclusion

Now you have all the information needed to guesstimate how much electricity your proposed hydro system will generate based on the manufacturer's output charts on the next pages. This will give you an indication whether your hydro site is worth developing, and if so, which turbine option is best. If you think you have a real site, fill out the Hydroelctric Site Evaluation Form, and send it to the Technical Department at Real Goods, or just give us a call. We will run your figures through our computer sizing program, which allows us to size plumbing and wiring for the least power loss at the lowest cost, and a myriad of other calculations necessary to design a working system. You'll find an example of our Hydro Survey Report on the next page, followed by the form for the information we need from you.

CALCULATION OF HYDROELECTRIC POWER POTENTIAL

Copyright © 1988 By Ross Burkhardt. All rights reserved.

ENTER HYDRO SYSTEM DATA HERE:	Customer: Meg A. Power	
PIPELINE LENGTH:	1,300.00	FEET
PIPE DIAMETER:	4.00	INCHES
AVAILABLE WATER FLOW:	100.00	G.P.M.
VERTICAL FALL:	200.00	FEET
HYDRO TO BATTERY DISTANCE:	50.00	FEET (one way)
TRANSMISSION WIRE SIZE:	2.00	AWG #
HOUSE BATTERY VOLTAGE:	24.00	VOLTS
HYDRO GENERATION VOLTAGE:	29.00	VOLTS

Power produced at hydro:	*Power delivered to house:*	
49.78 AMPS	49.78	AMPS
29.00 VOLTS	28.20	VOLTS
1,443.53 WATTS	1,403.59	WATTS

Four-nozzle, 24V, high output w/cooling turbine required

PIPE CALCULATIONS

HEAD LOST TO PIPE FRICTION:	7.61	FEET
PRESSURE LOST TO PIPE FRICTION:	3.29	PSI
STATIC WATER PRESSURE:	86.62	PSI
DYNAMIC WATER PRESSURE:	83.33	PSI
STATIC HEAD:	200.01	FEET
DYNAMIC HEAD:	192.40	FEET

hydropower CALCULATIONS

OPERATING PRESSURE:	83.33	PSI
AVAILABLE FLOW:	100.00	GPM
WATTS PRODUCED:	1,443.53	WATTS
AMPERAGE PRODUCED:	49.78	AMPS
AMP-HOURS PER DAY:	1,194.65	AMP-HOURS
WATT-HOURS PER DAY:	34,644.83	WATT-HOURS
WATTS PER YEAR:	12,645,362.71	WATT-HOURS

LINE LOSS (USING COPPER)

TRANS. LINE ONE-WAY LENGTH:	50.00	FEET
VOLTAGE:	29.00	VOLTS
AMPERAGE:	49.78	AMPS
WIRE SIZE #:	2.00	AMERICAN WIRE GAUGE
VOLTAGE DROP:	0.80	VOLTS
POWER LOST:	39.95	WATTS
TRANSMISSION EFFICIENCY:	97.23	PERCENT
PELTON WHEEL RPM WILL BE:	2,969.85	AT OPTIMUM WHEEL EFFICIENCY

This is an estimate only! Due to factors beyond our control (construction, installation, incorrect data, etc.), we cannot guarantee that your output will match this estimate. We have been conservative with the formulas used here and most customers call to report more output than estimated. However, be forewarned! We've done our best to estimate conservatively and accurately, but there is no guarantee that your unit will actually produce as estimated.

Real Goods
Hydroelectric Site Evaluation Form

Name:_____

Address:_____

Phone #:_____ Date:_____

Pipe Length:_____ (from water intake to turbine site)

Pipe Diameter:_____(only if using existing pipe)

Available Water Flow:_____(in gallons per minute)

Fall:_____(from water intake to turbine site)

Turbine to Battery Distance:_____(one way, in feet)

Transmission Wire Size:_____(only if existing wire)

House Battery Voltage:_____(12, 24, ??)

Alternate estimate (if you want to try different variables)

Pipe Length:_____ (from water intake to turbine site)

Pipe Diameter:_____(only if using existing pipe)

Available Water Flow:_____(in gallons per minute)

Fall:_____(from water intake to turbine site)

Turbine to Battery Distance:_____(one way, in feet)

Transmission Wire Size:_____(only if existing wire)

House Battery Voltage:_____(12, 24, ??)

For a complete computer printout of your hydroelectric potential, including sizing for wiring and piping, please fill in the above information and send to Real Goods.

HYDROELECTRIC PRODUCTS

SUNSHINE TO ELECTRICITY

BOOKS

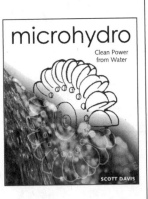

Microhydro

Clean Power from Water

By Scott Davis. Hydroelectricity is the world's largest source of renewable energy. But until now there were no books that covered residential-scale hydropower, or micro-hydropower. This excellent guide covers the principles of micro-hydro, design and site considerations, equipment options, plus legal, environmental and economic factors. Author Scott Davis covers all you need to know to assess your site, develop a workable system, and get the power transmitted to where you can use it. 157 pages, softcover. Canada.

21-0340 Microhydro $19.95

Microhydro Power Video

A program we highly recommend to anyone considering residential hydro. Covers basic technology and hands-on applications, with easy-to-follow explanations of micro-hydro power. Also covers performance and safety issues. With micro-hydro pioneer Don Harris of Harris Hydroelectric.

VHS, 42 min. USA.

80362 Microhydro Power Video $39

HYDRO GENERATION PRODUCTS

Harris Hydroelectric: *Bringing hydropower home*

When Don Harris moved to California's Santa Cruz Mountains in the mid-1970s, he was seven miles from the nearest power line—and direct sunshine wasn't plentiful enough to make power from the sun.

hydropower was his best bet, but the only hydro plants in operation were AC, and way too big for residential use. Then he visited a hydro museum and saw a way to make micro-hydro work.

In a machine shop from his old drag-racing days, Don created 50 Pelton wheels and launched Harris Hydroelectric, whose very first customer was Real Goods in 1981. Harris has manufactured over 3,000 micro-hydro plants since then—including the one that powers Real Goods President John Schaeffer's home all winter, for an original outlay of just $1,500 providing him with 36 kWh per day.

Harris Hydroelectric Turbines

The generating component of the standard Harris turbines is an automotive alternator equipped with custom-wound coils appropriate for each installation. Field strength, and therefore output, is adjustable with a large rheostat. The built-in ammeter shows the immediate effects of tweaking the field.

The new optional Permanent Magnet alternator eliminates brushes and annual brush replacement. It has bigger bearings that last two to three times longer than the standard automotive bearings, so overall maintenance is greatly reduced. Under most operating conditions the PM alternator delivers more wattage (see output chart). This is a highly recommended upgrade. (Also available for older machines in service. Please call.)

The rugged turbine wheel is a one-piece Harris casting made of tough silicon bronze. Hundreds of these turbine wheels have been in service for over a decade.

Four-nozzle turbine.

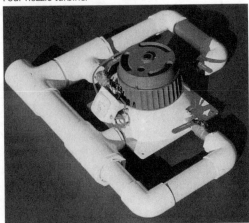

Even with dirty, gritty water, it takes at least ten years to wear out a Harris wheel. The cast aluminum wheel housing serves as a mounting for the alternator and up to four nozzle holders. It also acts as a spray shield, redirecting the expelled water down into the collection/drain box. Expelled water depends on gravity to drain.

Single-nozzle turbine with permanent magnet option.

Harris Hydroelectric turbines are available with one, two, or four nozzles to maximize the output of the unit. Having an extra nozzle or two beyond what is absolutely required makes it easy to change flow rates or power output for seasonal or household variations. Just turn different size nozzles on and off. The individual brass nozzles are screwed in from the underside of the turbine. They do wear with time, and will need eventual replacement. Any turbine delivering 20 amps or more (at whatever system voltage), needs the cooling fan option. USA.

17101	**1-Nozzle Turbine**	**$995**
17102	**2-Nozzle Turbine**	**$1,095**
17103	**4-Nozzle Turbine**	**$1,249**
17329	**Permanent Magnet Option**	**$700**
17132	**24-Volt Option**	**$0 no charge**
17136	**48-Volt Option**	**$149**
17134	**Extra Nozzles**	**$5**
17135	**Optional Cooling Fan Kit**	**$79**

All Harris turbines are custom-assembled to match your site. We need to know: head, flow rate, pipeline size & length, system voltage, and distance from turbine to batteries. Allow 2 to 4 weeks delivery time.

Harris Pelton Wheels

For do-it-yourselfers interested in a small hydroelectric system, we offer the same tough and reliable Pelton wheel used on the complete turbines above. Harris silicon bronze Pelton wheels resist abrasion and corrosion far longer than polyurethane or cast aluminum wheels. These are 5" diameter quality castings that can accommodate nozzle sizes of $^1/_{16}$" through $^1/_2$". Designed with threads for Delco or Motorcraft alternators. USA.

17202	**Silicon Bronze Pelton Wheel**	**$275**

The Stream Engine

The highly adjustable Stream Engine uses a turgo-type runner wheel that can produce useful power from lower heads. The permanent-magnet generator has no brushes to wear out, and uses a clever combination of alternator wiring changes and adjustable air gap to compensate for differing flow rates. Setup does require some time with this universal design, and involves trial and adjustment, selecting one of four alternator wiring setups, and then adjusting the permanent-magnet rotor air gap until peak output is achieved. A precision shunt and digital multi-meter are supplied to expedite this setup process. Output voltage is user-settable from 12 to 48 volts. Available in two- or four-nozzle turbines, the Stream Engine is supplied with universal cut-to-size nozzles of $^1/_8$" to 1". See the Nozzle Chart to determine flow rates at various heads. Head can be from 5 feet to 400 feet, and flows of 5 gpm to 300 gpm can be used (note that flow rates above 150 gpm have diminishing returns, as the sheer volume of water starts to get in its own way). Maximum output is approximately 800 watts. See performance chart. Canada.

17109	**Stream Engine, 2-nozzle**	**$1,895**
17110	**Stream Engine, 4-nozzle**	**$2,045**
17111	**Turgo Wheel only (bronze)**	**$620**
17112	**Extra Universal Nozzle**	**$40**

Harris Turbine Max Flow Rates in GPM

	Number of Nozzles			
Feet of Head	1	2	3	4
25	17	35	52	70
50	25	50	75	100
75	30	60	90	120
100	35	70	105	140
200	50	100	150	200
300	60	120	–	–

HARRIS TURBINE OUTPUT IN WATTS

Feet of Head	Gallons per Minute Flow (Permanent Magnet specs in bold.)														
	3		6		10		15		20		30		50		100		200
25	–		–		–		20	**25**	30	**40**	50	**65**	115	**130**	200	**230**	520	**580**
50	–		–		35	**40**	60	**75**	80	**100**	125	**150**	230	**265**	425	**500**	850	**900**
75	–		25	**30**	60	**75**	95	**110**	130	**160**	210	**250**	350	**420**	625	**750**	1,300	**1,300**
100	–		35	**45**	80	**95**	130	**150**	200	**240**	290	**350**	500	**600**	850	**1,100**	–
200	30	**45**	100	**130**	180	**210**	260	**320**	400	**480**	580	**650**	950	**1,100**	1,500	**1,500**	–
300	70	**80**	150	**180**	275	**300**	400	**450**	550	**600**	850	**940**	1,400	**1,500**	–	–

Maximum wattage @ voltage: 12V=750, 24V=1,500, 48V=2,500

STREAM ENGINE NOZZLE FLOW CHART (in gpm)

Head Pressure Feet	PSI	1/8	3/16	1/4	5/16	3/8	7/16	1/2	5/8	3/4	7/8	1.0	Turbine RPM
						Nozzle Diameter In Inches							
5	2.2					6.18	8.40	11.0	17.1	24.7	33.6	43.9	460
10	4.3			3.88	6.05	8.75	11.6	15.6	24.2	35.0	47.6	62.1	650
15	6.5		2.68	4.76	7.40	10.7	14.6	19.0	29.7	42.8	58.2	76.0*	800
20	8.7	1.37	3.09	5.49	8.56	12.4	16.8	22.0	34.3	49.4	67.3	87.8*	925
30	13.0	1.68	3.78	6.72	10.5	15.1	20.6	26.9	42.0	60.5	82.4*	107*	1140
40	17.3	1.94	4.37	7.76	12.1	17.5	23.8	31.1	48.5	69.9	95.1*	124*	1310
50	21.7	2.17	4.88	8.68	13.6	19.5	26.6	34.7	54.3	78.1*	106*	139*	1470
60	26.0	2.38	5.35	9.51	14.8	21.4	29.1	38.0	59.4	85.6*	117*	152*	1600
80	34.6	2.75	6.18	11.0	17.1	24.7	33.6	43-9.	68.6	98.8*	135*	176*	1850
100	43.3	3.07	6.91	12.3	19.2	27.6	37.6	49.1	76.7	111*	150*	196 *	2070
120	52.0	3.36	7.56	13.4	21.0	30.3	41.2	53.8	84.1*	121*	165*	215*	2270
150	65.0	3.76	8.95	15.0	23.5	33.8	46.0	60.1	93.9*	135*	184*	241*	2540
200	86.6	4.34	9.77	17.4	27.1	39.1	53.2	69.4	109*	156*	213*	278*	2930
250	106	4.86	10.9	19.9	30.3	43.6	59.4	77.6*	121*	175*	238*	311*	3270
300	130	5.32	12.0	21.3	33.2	47.8	65.1	85.1*	133*	191*	261*	340*	3590
400	173	6.14	13.8	24.5	38.3	55.2	75.2	98.2*	154*	221*	301*	393*	4140

STREAM ENGINE POWER OUTPUT (IN WATTS)

Net Head Feet	5	10	15	20	30	40	50	75	100	150	200*	300*
				Flow Rate In US Gallons Per Minute								
5				5	8	10	15	20	30	40		
10			7	12	18	23	30	45	60	80	100	
15	5	10	15	20	30	40	50	75	100	125	150	200
20	8	16	25	32	50	65	85	125	170	210	275	350
30	12	30	45	60	90	120	150	225	300	400	500	700
40	16	40	60	80	120	160	200	300	400	500	600	
50	20	50	75	100	150	200	250	375	500	600		
75	30	75	110	150	225	300	375	560	700			
100	40	100	150	200	300	400	500	650				
150	60	150	225	300	400	550	650					
200	80	200	300	400	550	700						
300	120	240	360	480	720							
400	160	320	480	640								

*Flow rates above 150 gpm are possible, but will suffer from diminishing returns.

Low Head Stream Engine

Using the same brushless, permanent-magnet generator as the regular Stream Engine, the Low Head model works with drops of 2 to 10 feet, and flows of 200 to 1,000 gallons per minute. Peak output is 1,000 watts. Wattage output at your site will be determined by available flow rate and drop. Generator is driven by a 5" low-head propeller turbine and includes the large 10-foot long tapered draft tube, which must be immersed in the tailwater. 12- to 48-volt output. Canada.

17328   Low Head Stream Engine $1,895

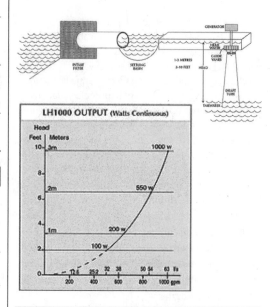

LH1000 OUTPUT (Watts Continuous)

Jack Rabbit Submersible Hydro Generator

No pipes or dams! Power from any fast-running stream!

The Jack Rabbit is a special low-speed alternator mounted in a heavy-duty, oil-filled cast aluminum housing with triple shaft seals. Originally designed for towing behind seismic sleds for oil exploration, this marine-duty unit is ideal for home power generation near a reasonably fast-moving stream. In a 9-mph stream (slow jog) the Jack Rabbit produces about 2,400 watt-hours daily. In a 6-mph stream (brisk walk) it produces over 1,500 watt-hours. The 12.5" propeller requires 13" of water depth. A rock or timber venturi often can be constructed to increase stream speed and power output.

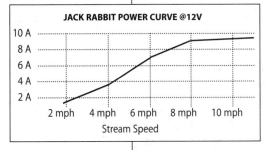

The Jack Rabbit turbine can be mounted in a variety of ways—on bridging logs, on a counter-weighted cantilever, or on an adjustable frame. For streams that are subject to flooding violent enough to dislodge logs or boulders, you should have a method of raising the generator.

For best output, the prop should face upstream, but in tidal flow applications output happens in either direction due to the rectified alternator output. Comes with three meters of output wire and a separate rectifier pack.

The Jack Rabbit is a very rugged machine that requires no maintenance. Small debris that would clog or slow a Pelton turbine passes through harmlessly. The prop is top-grade ductile aluminum, which can be hammered back into shape in case it really gets walloped by a log. Props are available as replacement parts.

The Jack Rabbit rarely requires a diversion controller and dummy load to prevent overcharging. In all but the highest output sites, it can be simply shut off with a PV controller. The slow speed prevents any high-voltage buildup.

The Jack Rabbit weighs 22 pounds, is 14.4" long, 8.4" diameter, has a 12.5" prop, and carries a 3-year manufacturer's warranty. Voltage not field switchable. Great Britain.

17104	Jack Rabbit Hydro Generator 12V	$1,195
17105	Jack Rabbit Hydro Generator 24V	$1,195
17108	Jack Rabbit Spare Prop	$169

Experience has shown that the Jack Rabbit needs high flow rates to function well. A choke point in your stream that delivers 9-mph or higher is good. The 24-volt version in particular seems to require higher flow rates to produce useful power output.

JACK RABBIT POWER CURVE @12V

Stream Speed (2 mph, 4 mph, 6 mph, 8 mph, 10 mph) vs (2 A, 4 A, 6 A, 8 A, 10 A)

HYDRO SYSTEM CONTROLS AND ACCESSORIES

Hydroscreen Water Intakes

Maintenance-free clean water

Crud in your hydro intake is nothing but trouble, reducing output, shortening life expectancy, and requiring regular attention. Self-cleaning Hydroscreens are robust, noncorrosive stainless steel wedge-wire water intake screens with a stainless frame. They're installed at about a 45° angle, so the water washes over them.

Complete assembly.

Flat screen.

Edge-on close-up of wedge wire.

The wires are flat on top and wedge-shaped. Anything that can't drop through the precise 1.25mm (0.05") gap between wires is flushed off the surface. No maintenance! No regular screen cleaning. High water floods pass right over. Available as either a flat screen with stainless ramp and tail (you build your own collection box under it), or as a complete assembly with screen, collection box, and 4" output pipe. Screen area is 12" x 12". These models are sized to work with any Harris or Stream Engine turbine installation up to 200 gpm. One-year manufacturer's warranty. USA.

47-0100	Hydroscreen Flat Screen	$269
47-0101	Hydroscreen Complete Assembly	$570

TriStar PV or Load Controllers

Best control for power diversion

The reliable, UL-listed MorningStar TriStar controller can provide either solar charging, load control, or diversion regulation for 12-, 24-, or 48-volt systems. This is a full-featured, modern, solid-state PWM controller. Features include reverse polarity protection, short circuit protection, lightning protection, high temperature current reduction, 4-stage battery charging, simple DIP switch controlled voltage setpoint setup, large 1"/1.25" knock-

outs on bottom, sides, and back, with extra wire bending room and terminals that accept up to #2 AWG wire. Circuit boards are fully conformal coated, the aluminum heat sink is anodized, and the enclosure is powder-coated with stainless fasteners to laugh at high humidity tropical environments.

This controller is particularly good for hydro diversion because the controller will allow up to 300 amp inrush currents to start motors or warm up cold resistive loads without the electronic short circuit protection interfering.

The optional 2-line, 16-character digital meter may be mounted to the controller in place of the cover plate, or remotely in a standard double-gang box using standard RJ-11 connectors (ethernet cables). Displays all system information, self-test results, and setpoints with intuitive up/down and left/right scrolling buttons. The remote temperature sensor option with 10-meter cable will automatically adjust voltage setpoints, and is recommended if your batteries will routinely be exposed to temperatures under 40°F or over 80°F. 10.1" H x 4.9" W x 2.3" D, weighs 4 lb. CE and UL listed. 5-year manufacturer's warranty. USA.

48-0010 TriStar 45-Amp Controller		**$169**
48-0011 TriStar 60-Amp Controller		**$218**
48-0012 TriStar Digital Meter		**$99**
48-0013 TriStar Remote Temperature Sensor		**$45**

Air Heater Diversion Loads

These resistive loads are enclosed in vented aluminum boxes for safety. They can be used on any DC system between 12 and 48 volts. Box needs at least 12" clearance to combustibles. Both units are shipped in the highest resistance mode, and can be easily reconfigured for lower resistance by changing connections in the terminal block. The HL-100 unit can be configured for 30 or 60 amps in nominal 12-volt mode. Two-year mfr.'s warranty. USA.

Item #	Price	Model	Diversion Amps @ Voltage Below			Resistance Setting
			15v	30v	60v	
48-0081	$235	HL-100	30/60	—	—	0.5/0.25 ohm
			15	30	—	1 ohm
			3.8	7.5	15	4 ohm
48-0082	$235	HL-75	20	40	—	0.75 ohm
			5	10	20	3 ohm

Diversion Water Heater Elements

Put your spare energy to work!

These industrial-grade DC heater elements give you someplace to dump extra wind or hydropower. Use with a diversion controller such as the MorningStar Tristar.

The water heater elements fit a standard 1" NPT fitting. The 12/24V model has a pair of 25A/12V elements. Can be wired for 25A or 50A @ 12V, or in series for 25A @ 24V. The 24/48V model has a pair of 30A/24V elements. Can be wired for 30A or 60A @ 24V, or in series for 30A @ 48V. USA.

25078	**12/24V Water Heating Element**	**$90**
25155	**48V/30A Water Heating Element**	**$116**

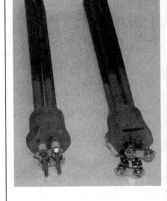

Wind Energy

■ This section was adapted from *Wind Energy Basics,* a Real Goods Solar Living Book by Paul Gipe. He is also the author of *Wind Power: Renewable Energy for Home, Farm and Business* (2004), *Wind Energy Comes of Age* (1995), and *Energía Eólica Práctica* (2000). Gipe has written and lectured extensively about wind energy for more than two decades.

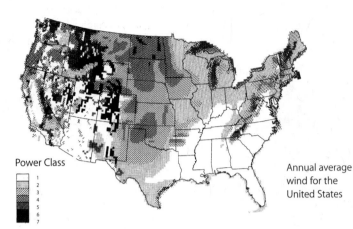

Power Class
1
2
3
4
5
6
7

Annual average wind for the United States

Small Wind Turbines Come of Age

The debut of micro wind turbines has revolutionized living off-the-grid. These inexpensive machines have brought wind technology within reach of almost everyone. And their increasing popularity has opened up new applications for wind energy previously considered off-limits, such as electric fence charging and powering remote telephone call boxes, once the sole domain of photovoltaics.

Micro wind turbines have been around for decades for use on sailboats, but they gained prominence in the 1990s as their broader potential for off-the-grid applications on land became more widely known. While micro wind turbines have yet to reach the status of widely available consumer commodities such as personal computers, the day may not be far off.

The use of wind power is "exploding," say Karen and Richard Perez, the editors of *Home Power* magazine. "There are currently over 150,000 small-scale RE (renewable energy) systems in America and they are growing by 30% yearly. [But] [t]he small-scale use of wind power is growing at twice that amount—over 60% per year." And a large part of that growth is due to Southwest Windpower of Flagstaff, Arizona.

Southwest Windpower awakened latent consumer interest in micro wind turbines with the introduction of its sleek Air series. Since launching the line in 1995, they've shipped thousands of the popular machines.

"What Americans, and folks all over the world, are finding out," the Perezes say, "is that wind power is an excellent and cost-effective alternative" to extending electric utility lines, and fossil-fueled backup generators.

Hybrid Wind and Solar Systems

You might say that joining wind and solar together is a marriage made in heaven. The two resources are complementary: in many areas wind is abundant in the winter when photovoltaics are least productive, and sunshine is abundant in the summer when winds are often weakest. The sun and wind together not only improve the reliability of an off-the-grid system, but are also more cost effective than using either source alone.

Hybrid systems include a DC source center (for DC circuit breakers), batteries, inverters, and often an AC load center. These components are necessary whether you're using just wind or solar. So, it's best to spread the fixed cost of these components over more kilowatt-hours by using photovoltaic panels in addition to a wind turbine.

Engineers have found that these hybrids perform even better when coupled with small

Both wind and PV can happily feed a common battery.

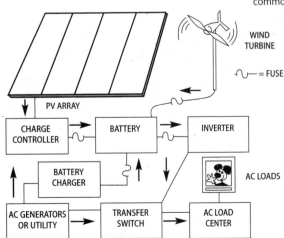

Installing a 1,500-watt wind turbine on a tilt-up tower using an electric winch.

©Paul Gipe

Bob and Ginger Morgan's Bergey Excel being installed by a crane near Tehachapi, California (see page 103).

©Paul Gipe

backup generators to reduce the battery storage needed. Many of those living off-the-grid reach the same conclusion by trial and error.

Typically a micro turbine, such as Southwest Windpower's Air series, or a small wind turbine, such as the Bergey XL1, a modest array of PV panels, batteries, and a small backup generator will suffice for most domestic uses. Though Pacific Gas & Electric Co. found that most Californians living off-the-grid had backup generators, they seldom used them. In a well-designed hybrid system, the backup generator provides peace of mind, but little electricity.

Size Matters

In wind energy, size, especially rotor diameter, matters. Nothing tells you more about a wind turbine's potential than its diameter—the shorthand for the area swept by the rotor. The wind turbine with the bigger rotor will intercept more of the wind stream and almost invariably will generate more electricity than a turbine with a smaller rotor, regardless of their generator ratings.

The area swept by a wind turbine rotor is equivalent to the surface area of a photovoltaic array. When you need more power in a PV array, you increase the surface area of the panels exposed to the sun by adding more panels. In the case of wind, you find a wind turbine that sweeps more area of the wind—a turbine with a greater rotor diameter.

Micro turbines range in size from 3 to 5 feet in diameter. The lower end of the range is represented by the Southwest Windpower's Air series. These machines are suitable for recreational vehicles, sailboats, fence-charging, and other low-power uses. Micro turbines will generate about 300 kilowatt-hours (kWh) per year at sites with average wind speeds typical of the Great Plains (about 12 mph at the height of the turbine).

Mini wind turbines are slightly larger than micro turbines and are well suited for vacation cabins. They span the range from 5 to 9 feet in diameter and include Southwest Windpower's Whisper H40 as well as Bergey Windpower XL1. Wind turbines in this class can produce 1,000 to 2,000 kWh per year at sites with an average annual wind speed of 12 mph.

Household-size turbines, as the name implies, are suitable for homes, farms, ranches, small businesses, and telecommunications. They span an even broader range than the other size classes, and encompass turbines from 10 to 23

feet in diameter. This class includes Bergey's Excel model and African Windpower's 3,6 meter (12-foot) diameter turbine. Household-size wind turbines can generate from 2,000 kWh to 20,000 kWh per year at 12 mph sites.

Generators

Most small wind turbines use permanent-magnet alternators. This is the simplest generator configuration and is nearly ideal for micro and mini wind turbines. There is more diversity in household-size turbines, but again nearly all use permanent-magnet alternators. Some manufacturers use Ferrite magnets, others use rare-earth magnets. The latter have a higher flux density than ferrite magnets. Both types do the job.

POWER CURVES

The power curve indicates how much power (in watts or kilowatts) a wind turbine's generator will produce at any given wind speed. Power is presented on the vertical axis; wind speed on the horizontal axis. In the advertising wars among wind turbine manufacturers, often the focus is the point at which the wind turbine reaches its "rated" or nominal power. Though rated power is just one point on a wind turbine's power curve, many consumers mistakenly rely on it when comparison shopping. But not all power curves are created equal. Some power curves are, to be diplomatic, more "aggressive" than others.

Manufacturers may pick any speed they choose to "rate" their turbine. In the 1970s, it was easy for unscrupulous manufacturers to manipulate this system to make it appear that their turbines were a better buy than competing products. By pushing "rated power" higher, they were able to show lower relative costs (turbine cost/rated power) or they were able to increase their price—and profits—proportionally.

Unlike wind farm turbines, the performance of many small wind turbines have not been tested to international standards. As a rule, don't place much faith in power curves, unless the manufacturer has stated clearly the conditions under which the curve was measured (usually in a detailed footnote), or in "rated power." Some power curves can be off by 40% in low winds, and as much as 20% at "rated" power. And it's the performance in lower winds that matters most.

Most homeowners seldom see their turbines operating in winds at "rated" speeds of, say, 28 mph. Small wind turbines operate most of the time in winds at much lower speeds, often from

10 to 20 mph. For size or price comparisons, stick with rotor diameter or swept area. Both are more reliable indicators of performance than power curves.

Robustness

Wind turbines work in a far more rugged environment than photovoltaic panels that sit quietly on your roof. You quickly appreciate this when you watch a small wind turbine struggling through a gale. There's no foolproof way to evaluate the robustness of small wind turbine designs.

In general, heavier small wind turbines have proven more rugged and dependable than lightweight machines. Wisconsin's wind guru Mick Sagrillo is a proponent of what he calls the "heavy metal school" of small wind turbine design. Heavier, more massive turbines, he says, typically last longer. Heavier in this sense is the weight or mass of the turbine relative to the area swept by the rotor. By this criteria, a turbine that has a relative mass of 10 kg/m2 may be more robust—and rugged—than a turbine with a specific mass of 5 kg/m2.

Siting

To get the most out of your investment, site your wind turbine to best advantage: well away from buildings, trees, and other wind obstructions. Install the turbine on as tall a tower as you're comfortable working with. Jason Edworthy's experience at Nor'wester Energy Systems in Canada convinces him that the old 30-foot rule still applies. This classic rule from the 1930s dictates that for best performance, your wind turbine should be at least 30 feet above any obstruction within 200 feet of the tower. Under the best of conditions, a tower height of 30 feet is the absolute minimum, says Edworthy.

Putting a turbine on the roof is no alternative. Seldom can you get the turbine high enough to clear the turbulence caused by the building itself. Imagine trying to mount a 30-foot tall tower on a steeply pitched roof. It's a recipe for disaster. Even if you could, turbine-induced vibrations will quickly convince you otherwise. (It's like putting a noisy lawn mower on your roof.) While many small wind turbines are relatively quiet, some are not—another good reason to put them out in the open, well away from buildings.

Towers

Most small wind turbines are installed on guyed, tubular masts that are hinged at the base. With an accompanying gin pole, these towers can be raised and lowered, simplifying installation and service. Some tilt-up towers use thin-walled steel tubing, others use thick-walled steel pipe. Household-size turbines also use guyed masts of steel lattice, as well as free-standing truss towers. With the advent of pre-engineered, tubular mast kits, there's now less excuse than ever for installing micro and mini wind turbines on inadequate towers.

INSTALLATION

Those with good tool skills—who work safely—can install a micro turbine themselves using a pre-engineered, tilt-up tower kit. Installing mini wind turbines, because of the greater forces involved, requires considerably more skill. Household-size turbines should be left only to professionals—and even the pros have made tragic mistakes. If you like doing the work yourself, start with a micro turbine and a tilt-up tower. Once you're satisfied you know what you're doing and you're comfortable with the technology, you can try your hand with a larger turbine.

The sleek Air 403 Industrial Turbine.

The Bergey Excel Turbine.

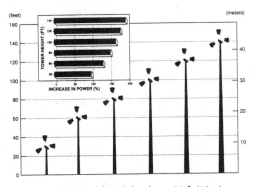

Increase in Power with Height above 30 ft (10m)

Chart adapted from *Wind Power for Home & Business.*

	ESTIMATED ANNUAL ENERGY OUTPUT					
	at Hub height in thousand kWH / Y					
Avg Wind Speed (mph)	**Rotor Diameter, m (ft)**					
	1 (3.3)	1.5 (4.9)	3 (9.8)	7 (23)	18 (60)	40 (130)
	thousands of kWh per year					
9	0.15	0.33	1.3	7	40	210
10	0.20	0.45	1.8	10	60	290
11	0.24	0.54	2.2	13	90	450

Bergey Excel and photovoltaic panels at home of Dave Blittersdorf, founder of NRG Systems.
©Paul Gipe

Typical Costs

The cost of a wind system includes the cost of the turbine, tower, ancillary equipment (disconnect switches, cabling, etc.), and installation. The total cost of a micro turbine can be as little as $1,000, depending upon the tower used and its height, while that of household-size machines can exceed $50,000. When comparing prices, remember that bigger turbines cost more but often are more cost-effective. For an off-the-grid power system, the addition of a wind turbine will almost always make economic sense by reducing the number of photovoltaic panels or batteries needed.

For grid-intertie systems, the economics depend upon the winds at your site and a host of other factors, including the average wind speed, the cost of the wind system, the cost of utility power, and whether your utility provides net billing.

Some Do's and Don'ts

Do plenty of research. It can save a lot of trouble—and expense—later.

Do visit the library. Books remain amazing repositories of information. (I do have a bias about books, since I write them. But I've always been a firm believer in "you get what your pay for" when it comes to free information, whether it's from the Web, manufacturers, or their trade associations.)

Do talk to others who use wind energy. They've been there. You can learn from them what they did right, and what they'd never do again.

Do read and, equally important, follow directions.

Do ask for help when you're not sure about something. The folks at Real Goods Renewables are there to help.

Do build to code. In the end it makes for a tidier, safer, and easier-to-service system.

Small Wind Turbines
Cuisinart for birds or red herring?

Avian mortality is an issue that always seems to come up with wind turbines. Do birds die from running into wind turbine blades, towers, and guy wires? Yes they do. In large numbers? Hardly. Let's look at the facts. Birds die all the time, in great numbers, due to human endeavors. The number one bird killer is glass-faced office buildings. These account for 100 million to 900 million bird deaths per year.[1,2] Some skyscrapers have been monitored and shown to kill as many as 200 birds per day. Cars and trucks kill another 50 to 100 million birds per year.[3,4] Cell phone, TV, radio, and other communications towers kill 4 to 10 million per year[5]. And finally, Muffy the housecat takes at least 100 million birds a year, possibly many more.[6,7]

In comparison, the vast majority of large wind plants report zero bird kills, and even the famous Altamont Pass area, in which wind turbine bird kills were first reported, has about 1000 bird deaths per year.[8]

This sounds like a lot, until we add there are 5,400 turbines on this site. That gives an average of 0.19 bird kills per turbine per year.[9] The problem seems to be the abundance of prey in the tall grasses around the towers, that this is a heavily used migratory route, and that early tower designs provided a wealth of good perches.

That's the true story for large power production turbines. Bird kills are a non-issue. In comparison, residential-size turbines and towers simply do not pose any significant risk to birds. The towers are too short, the blade swept area is too small. The chance of your residential turbine ever hitting a bird are vanishing small. If you or your neighbors are truly concerned about bird deaths, getting rid of Muffy would have much more real effect.

—Doug Pratt

1. Curry & Kerlinger, Dr. Daniel Klem of Muhlenberg College has done studies over a period of twenty years, looking at bird collisions with windows. www.currykerlinger.com/birds.htm
2. Defenders of Wildlife, "Bird mortality from wind turbines should be put into perspective." www.defenders.org/habitat/renew/wind.html
3. Curry & Kerlinger, "Those statistics were cited in reports published by the National Institute for Urban Wildlife and U.S. Fish and Wildlife Service." www.currykerlinger.com/birds.htm
4. Defenders of Wildlife, "Bird mortality from wind turbines should be put into perspective." www.defenders.org/habitat/renew/wind.html
5. Curry & Kerlinger, "U.S. Fish and Wildlife Service estimates that bird collisions with tall, lighted communications towers, and their guy wires result in 4 to 10 million bird deaths a year." www.currykerlinger.com/birds.htm
6. Curry & Kerlinger, "The National Audubuon Society says 100 million birds a year fall prey to cats." www.currykerlinger.com/birds.htm
7. Defenders of Wildlife, the American Bird Conservancy estimates that feral and domestic outdoor cats probably kill on the order of hundreds of millions of birds per year (Case 2000). One study estimated that in Wisconsin alone, annual bird kill by rural cats might range from 7.8 to 217 million birds per year (Colemen & Temple 1995). www.defenders.org/habitat/renew/wind.html
8. http://www.currykerlinger.com/studies.htm
9. http://www.nrel.gov/docs/fy04osti/33829.pdf

Do take your time. Remember, there's no rush. The wind will always be there.

Do be careful. Small wind turbines may look harmless, but they're not.

Don't skimp and don't cut corners. Taking short cuts is always a surefire way to ruin an otherwise good installation.

Don't design your own tower—unless you're a licensed mechanical engineer.

Don't install your turbine on the roof despite what some manufacturers may say!

And, of course, **don't** believe everything you read in sales brochures.

In general, doing it right the first time may take longer and cost slightly more, but you'll be a lot happier in the long run. ■■

SOURCES OF INFORMATION

For more on micro turbines see *Wind Energy Basics*, Chelsea Green Publishing, 1999. 222 pp., ISBN 1-890132-07-1, available from Real Goods. Describes a new class of small wind turbines, dubbed *micro turbines*. These inexpensive machines, when coupled with readily available photovoltaic panels, have revolutionized living off-the-grid.

For more on small wind turbine technology see *Wind Power: Renewable Energy for Home, Farm and Business*, Chelsea Green Publishing, 2004, 504 pp., ISBN 1-931498-14-8, also available from Real Goods. Explains how modern, integrated wind turbines work, and how to use them most effectively.

For information about the commercial wind power industry see *Wind Energy Comes of Age*, New York: John Wiley & Sons, 1995. 613 pages. ISBN 0-471-10924-X. A chronicle of wind energy's progress from its rebirth during the oil crises of the 1970s to its maturation on the plains of northern Europe in the 1990s. Selected

Real Goods Solar Living Center's Whisper 3000 wind turbine atop a hinged, tilt-up tower in Hopland, California.
©Paul Gipe

as one of the outstanding academic books published in 1995.

To determine your local wind resources visit http://rredc.nrel.gov/wind/pubs/atlas/ or search for the National Renewable Energy Laboratory's *Wind Energy Resource Atlas of the United States*.

For tips on installing micro and mini wind turbines using a Griphoist brand winch, see "Get a Grip!" in *Homepower Magazine #68*, December 1998/January 1999 or visit www.wind-works.org.

The Gipe Family Do-It-Yourself Wind Generator Slide Show

1. Taking delivery of a new Air turbine and 45-foot tower.

2. Checking the packing list against parts delivered.

3. Securing the tower's base plate.

4. Aligning the guy anchors.

5. Driving the screw anchors.

6. Unspooling the guy cable.

7. Assembling the mast.

8. Clevis and gated fitting hook.

9. Gin pole and lifting cables.

10. Strain relief for supporting power cables.

11. Final assembly of the turbine.

12. Disconnect switch and junction box.

13. Slowly raising the turbine with a Griphoist-brand hand winch.

14. Air turbine safely installed with a Griphoist-brand hand winch.

All photos ©Paul Gipe

WIND PRODUCTS

BOOKS AND SITE-EVALUATION TOOLS

Wind Power

Renewable Energy for Home, Farm and Business
New in 2004

By Paul Gipe. Wind energy technology has truly come of age—with better, more reliable machinery and a greater understanding of how and where wind power makes sense. In addition to expanded sections on gauging resources and turbine siting, this thoroughly updated edition contains a wealth of examples, case studies, and human stories. Covers everything from small off-grid to large grid-tie, plus water pumping. This is the wind energy book for the industry. Everything you ever wanted to know about wind turbines. Has an abundance of pictures, charts, drawings, graphs, and other eye candy, with extensive appendixes. 504 pages, softcover. USA.

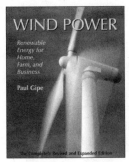

21-0343 Wind Power: Renewable Energy for Home, Farm, and Business   **$50**

Wind Energy Basics

A Guide to Small and Micro Wind Systems

By Paul Gipe. A step-by-step manual for reaping the advantages of a windy ridgetop as painlessly as possible. Chapters include wind energy fundamentals, estimating performance, understanding turbine technology, off-the-grid applications, utility intertie, siting, buying, and installing, plus a thorough resource list with manufacturer contacts, and international wind energy associations. A Real Goods Solar Living Book. 222 pages, softcover. USA.

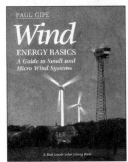

80928   Wind Energy Basics   **$19.95**

Residential Wind Power with Mick Sagrillo

Mick Sagrillo, the former owner of Lake Michigan Wind and Sun, has been in the wind industry for over sixteen years, with over 1,000 turbine installation/repairs under his belt.

This professionally produced video includes basic technology introductions, hands-on applications, and performance and safety issues. The action moves into the field to review actual working systems, including assembling a real system. Tape runs 63 minutes. Standard VHS format. USA.

80363   Renewable Experts Tape—Wind   **$39**

NRG WindWatcher

The next generation of wind loggers. Collects and records wind speed and power; calculates average speed and average power density for accurate power predictions; delivers monthly averages from constant 2-second data samples; and saves data at the end of each month. Runs one year on a single D-cell battery. Includes anemometer with weather boot, stub mast, 100' of sensor wire, ground wire, battery and Wind-Watcher with surface mounting box. 4.7' square. Designed for indoor or weather-protected mounting. Will track wind direction and deliver monthly average with optional direction vane (Includes small separate mounting mast and 100' sensor wire). Two-year manufacturer's warranty. USA.

43-0154 WindWatcher   **$395**
43-0159 Direction Vane   **$229**

Wind Speed Indicator

Test how much power you will get from a wind generator before buying one. Our accurate indicator includes a pick-up vane and 50' of connecting tube.

16285   Wind Speed Indicator   **$75**

Handheld Windspeed Meter

This is an inexpensive and accurate windspeed indicator. It features two ranges, 2 to 10 mph and 4 to 66 mph. A chart makes easy conversions to knots. Speed is indicated by a floating ball viewed through a clear tube. Many of our customers are curious about the wind potential of various locations and elevations on their property, but are reluctant to spend big money for an anemometer just to find out. Here is an economical solution. Try taping this meter to the side of a long pipe, (use a piece of tape over the finger hole for high-range reading) and have a friend hold it up while you stand back and read the meter with binoculars. This is very helpful in determining wind speeds at elevations above ground level. Includes protective carrying case and cleaning kit.

63205   Windspeed Meter   **$20**

Lightning Protection

Do you need lightning protection on a wind system? Is the Pope Catholic? You need serious industrial-grade protection here, and it's surprisingly simple and inexpensive.

The first level of protection depends on you to ground the tower, the guy wires, and make sure ALL the

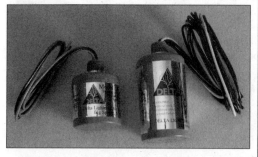

grounds in your wind and household system are connected together. Lightning wants to go to ground. Make it easy, and give it a path outside your system wiring. The second level of protection is needed when light-

ning decides to take a trip through your system wiring. Delta arrestors will clamp any high voltage spike, and direct most of the energy to ground. They will withstand multiple strikes until capacity is reached and then rupture, indicating the need for replacement. Delta arrestors can be installed at any convenient point, or points, between the turbine and the battery. Multiple arrestors are a good idea in lightning-prone areas. Lightning does strike twice!

The 3-wire DC model is for turbines such as the Air X with 2-wire DC output. Connect one wire to positive, one to negative, and one to ground. It can take up to 60,000 amps current, or 2,000 joules per pole. Response time to clamp is 5 to 25 nanoseconds depending on amp load.

The 4-wire model is for turbines like the AWP 3.6, Whisper, or Bergey with 3-wire wild AC output. Connect one wire to each of the output wires and one to ground. It can take up to 100,000 amps current, or 3,000 joules per pole. Same clamp time as above. USA.

| 25194 | DC 3-wire Lightning Protector | $45 |
| 25749 | AC 4-wire Lightning Protector | $60 |

WIND GENERATORS

Southwest Windpower Air X

The third generation of North America's most popular turbine

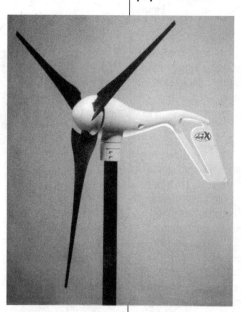

Southwest Windpower has introduced their third generation of the Air turbine series, and things just keep getting better for this simple, attractive wind generator. Proven features from previous models include a sleek noncorrosive cast aluminum alloy body, three flexible carbon-reinforced blades, a neodymium permanent-magnet alternator, and built-in regulation. What's improved is noise control and charge control. The sophisticated new microprocessor-based speed and charge control delivers better battery charging by optimizing the alternator output at all points of the power curve. It won't overcharge smaller battery packs, and it eliminates the blade "flutter" that was a noise problem at high wind speeds by actually controlling and limiting the blade speed.

Both standard and marine models are rated at 400 watts output. Standard models are unpainted, marine models are powder-coat white. Both feature a large LED in the underbelly to indicate output. Available in 12- or

24-volt versions, the voltage cannot be changed in the field. See chart below for full specifications. Three year manufacturer's warranty. USA.

AIR X TURBINE SPECIFICATIONS

	Standard	Marine
Swept Area Ft^2	11.5	11.5
Output	400 watts	400 watts
Available DC voltages	12–24V	12–24V
Rated wattage output	28 mph	28 mph
Max. wind speed	110 mph	110 mph
Cut-in wind speed	7 mph	7 mph
Number of blades	3	3
Rotor diameter	46"	46"
Tower top weight	13 lbs.	13 lbs.
Tower type	1.5" pipe*	1.5" pipe*
*Pipe size shown is for nominal steel pipe		

Air X Standard Turbine (specify voltage)		$649
16-277	Std. 12V	
16-278	Std. 24V	
Air X Marine Turbine (specify voltage)		$875
16-281	Marine 12V	
16-282	Marine 24V	

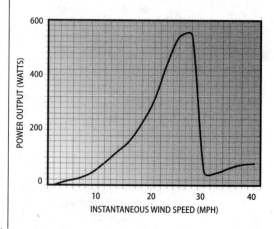

Installation Accessories for Air X

Between your battery and turbine, you need overcurrent protection. If your DC center isn't providing it, Southwest recommends the simple surface-mounted breakers below. Use the larger 70 amp breaker for 12-volt.

The stop switch disconnects the turbine from the battery, and then connects the positive and negative to each other to give maximum braking. It's highly recommended for marine applications, especially if you plan to sleep on the boat. Turbines DO make noise, and this will help a lot.

16153	Air 403/X 50-Amp Stop Switch	$20
25167	50-Amp Surface-Mount Circuit Breaker	$50
25823	70-Amp Surface-Mount Circuit Breaker	$50

Air 403 Industrial

The 400-watt Industrial model is specially designed for survival in the most extreme environments; on remote mountaintops and off-shore platforms. It will survive winds in excess of 100 mph and temperatures down to –60°F for extended periods without attention. Features of this fully powder-coated model include anodized

cooling fins for continuous high-power output, larger internal wiring, and special carbon fiber blades that are twice the strength of standard Air blades. Supplied as a package with industrial turbine, an external C40 diversion control, and an air-heater dump load. It's designed for hybrid solar/wind systems in telecommunication stations, offshore platforms, monitoring stations, lighthouses, and cathodic protection systems. Three-year manufacturer's warranty. Industrial models are made to order. Allow 8 weeks minimum delivery time. USA.
Specify voltage (12, 24, or 48) at time of order.

| 16276 | Air 403 Industrial w/External Control & Dump Load | $1095 |

I'd like to tell you that your company was instrumental in my getting started in this industry. When I was in school, I used to get the old RG "Hard Corps" catalog and would memorize the thing front-to-back. After college, I knew exactly what I wanted to do. So—my thanks to Real Goods.

—Steve Wilke, National Sales Manager
Bergey Windpower Company

Whisper Wind Turbines

Whisper currently offers three turbine models that are designated by the square footage of blade sweep. Nothing measures output potential better than blade size. Big generators can't deliver when connected to a small blade. Whisper is trying to start an honest industry trend here.

- The H40 model sweeps 40 square feet, and delivers a peak of 900 watts.
- The H80 model sweeps 80 square feet, and delivers a peak of 1,000 watts.
- The 175 model sweeps 175 square feet, and delivers a peak of 3,200 watts.

All Whisper models use modern, lightweight, brushless, direct-drive, low-speed alternators with start-up speeds of 7 to 7.5 mph. The 175 model provides overspeed protection by tipping the generator and blades back on their mount; the H models twist to the side. Both schemes simply move the blades at right angles to the incoming wind, return automatically when wind speed drops, and rely on nothing more complex than wind pressure and gravity. The fiber composite

Whisper H40 Turbine.

blades are long-lasting without repainting. The three-blade design of the H-models offers better balance and less machine strain in shifting winds.

H40 and H80 turbines are delivered with the capable and new-for-2004 Whisper Controller, which includes the power rectifier, a built-in diversion controller, and makes it possible to change the turbine output voltage on site. The optional Digital Display shows kWh of production, volts, amps, plus max volts and amps. The 175 is delivered with the E-Z wiring center that also supplies a complete monitoring package, but doesn't allow voltage changes.

H40 Turbine

The H40 model is specifically designed for sites with high wind speeds. It has a smaller 7-foot (2.1 meter) blade direct-connected to a fairly large 900-watt alternator. If you have 12 mph (5.4 m/s) or higher average winds, this turbine is for you. Got lower wind speeds? Check the H80 below.

The H40 features a neodymium permanent-magnet alternator combined with the high-efficiency composite airfoil 3-blade design. The H40 is available in 12-, 24-, or 48-volt output, and with the new Whisper Controller, can be easily switched in the field.

H80 Turbine

The H80 model is specifically designed for sites with low to medium wind speeds. It has a larger 10-foot (3.0 meter) blade connected to a 1,000-watt low-speed alternator. Got wind speeds averaging 12 mph (5.4 m/s) or less? This 1,000-watt turbine is for you. The large swept area does better at harvesting useful energy from low winds. The advanced 3-blade airfoil design is optimized for the lower wind speeds that most sites see daily. The H80 is available in 12-, 24- or 48-volt output, and with the new Whisper Controller, can be easily switched in the field.

Whisper H80 Turbine.

175 Turbine

The 175 is Whisper's most powerful turbine. The 14.8-foot (4.26 meter) 2-blade rotor features a hand-laid fiberglass construction with foam core for light weight and easy startup. This 3,000-watt turbine has been upgraded with a stronger, larger-diameter yaw shaft, new blade stabilizer straps, and a third spindle bearing for longer life. This powerful turbine does best in steady, moderate winds. The 2-bladed design isn't recommended for sites with exceptionally strong, or rapidly shifting winds. The 175 is shipped as 24- or 48-volt. Please specify.

Tower kits and anchors are offered below in Wind Accessories. All Whisper turbines feature a two-year manufacturer's warranty. USA.

	Model H40	Model H80	Model 175
Swept Area Ft^2:	38.5	78.5	172
Rotor Diameter:	7 feet	10 feet	14.8 feet
Tower Top Weight:	47 lb	65 lb	155 lb
Tower Type:	2.5" steel pipe	2.5" steel pipe	5" steel pipe
Max. Design Wind Speed:	120 mph	120 mph	120 mph
Cut-in Wind Speed:	7.5 mph	7.0 mph	7.1 mph
Number of Blades:	3	3	2
Rated Wattage (@mph):	900@28	1,000@24	3,200@27
Generator Type:	Brushless Permanent-Magnet Alternator		
Available Voltage:	12–48 VDC	12–48 VDC	24–48 VDC

16239 Whisper H40 w/Controller, 12 to 48V $1,895

16241 Whisper H80 w/Controller, 12 to 48V $2,365

16242 Whisper 175 w/wiring center, 24 or 48V (specify voltage) $6,450

48-0084 Digital Display for H40/H80 Controller $99

Whisper H-80 and H-40 power curve

Whisper H-175 Power Curve

African Wind Power

Strong and quiet for batteries or intertie

The AWP 3.6 is a robust, low-speed 1,000-watt turbine that has nothing but happy reviews from everyone who's flown it. It features an exceptionally large 11.8' (3.6m) blade diameter, giving it tremendous energy harvesting potential, especially in light winds of 10 mph or less—the kind of winds that most of North America enjoys most of the time. Independent private-party side-by-side testing has shown the AWP 3.6 delivering at least twice the power of competing 1,000 watt turbines in light winds. It has twice the swept area of the Bergey XL.1 ... draw your own conclusions.

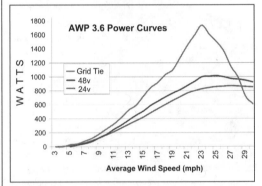

Not only is the AWP 3.6 the best-performing 1,000-watt turbine in the real world, it's also the quietest. Most owners report that they can only tell if the turbine is working by looking at it, or their monitoring equipment. It's rare to be able to hear it above ambient noise. It has a very low rotation and blade tip speed. If you're concerned about noise complaints from the neighbors, or even your own family, this turbine is your best choice.

This is a new slightly upgraded version of the AWP 3.6 turbine. It now has slip rings on the yaw shaft, a stainless steel tail, and lighter weight foam core blades. It hits peak output at 25 mph. Overspeed is governed by automatically twisting to the side to shed wind and reduce output. Maximum wind speed is 100 mph.

The AWP 3.6 is available in 24 or 48vdc, with adjustable diversion control/rectifier and DC air heater load, or in a direct-intertie configuration with the Windy Boy 1800 inverter. In intertie configuration, it's rated at 1,500 watts and comes as a complete package with turbine, rectifier/controller, diversion load, and inverter. The controller will send excess energy to the diversion load

The Whisper 175 Turbine that's been cranking out power at the Solar Living Center for the past eight years.

(a heater element) if the turbine exceeds the inverter capacity, or the grid is down. A complete selection of tilt-up guyed pipe tower kits from 43' to 127' is available in the Wind Accessories section following. Two-year manufacturer's warranty by North American importer. Made in South Africa.

43-0163	AWP 24-volt control/rectifier	$210
43-0164	AWP 24-volt diversion load	$105
43-0165	AWP 48-volt control/rectifier	$195
43-0166	AWP 48-volt diversion load	$95

AFRICAN WIND POWER 3.6 SPECIFICATIONS

Model	24 volt DC	48 volt DC	Utility-Intertie
Swept Area	109 ft²	109 ft²	109 ft²
Rated wattage @ mph	850 @ 25	1000 @ 25	1500 @ 25
Cut-in wind speed mph	6 mph	6 mph	6 mph
Max design wind speed	100mph+	100mph+	100mph+
Generator type		Brushless Permanent Magnet, 3-phase	
Rotor diameter in feet		11.8 ft.	
Tower top weight		250 lb.	
Tower		4" Sch. 10 steel pipe	
Item #	**43-0167**	**43-0168**	**43-0169**
Price	**$2,800**	**$2,780**	**$6,190**

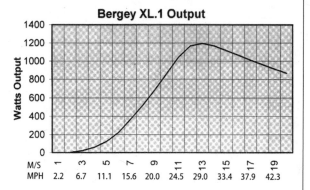

Bergey Wind Generators

Bergey wind generators are wonderfully simple: no brakes, pitch-changing mechanism, gearbox, or brushes. Bergey pioneered the automatic side-furling design that practically all wind turbines use now. This furling design forces the generator and blades partially out of the wind at high wind speeds, but still maintains maximum rotor speed. AUTOFURL uses only aerodynamics and gravity, there are no brakes, springs, or electromechanical devices to reduce reliability. These turbines are designed to operate unattended at wind speeds up to 120 mph.

Bergey XL.1 Wind Generator

The 1000-watt XL.1 is Bergey's newest turbine. With 24-volt output, this wind charger is intended for off-the-grid or remote-location battery charging. This 2.5-meter (8.2-foot) turbine is designed for high reliability, low maintenance, and automatic operation under adverse conditions. The XL.1 features Bergey's proven AUTOFURL storm protection, a direct-drive permanent-magnet alternator, excellent low-wind performance, and nearly silent operation thanks to the state-of-the-art, patent-pending, blade airfoils combined with the oversized, low-speed alternator. Low-wind-speed performance is enhanced by low-end boost circuitry that increases output at wind speeds down to 5.6 mph (2.5 m/s).

The BWC PowerCenter provides battery-friendly constant voltage charging, and will control up to 30 amps of PV input also. The simple controller shows battery voltage and which charging sources are active with at-a-glance LED indicators.

See the Wind Accessories section for complete XL.1 tower kits that include tubing, and anchors, and will ship by UPS. Excellent 5-year manufacturer's warranty. China & USA.

BERGEY XL.1 WIND TURBINE SPECIFICATIONS
Model: 24 volt DC
Swept Area Ft²: 52.8 ft²
Rated wattage @ mph: 1,000 @ 25 (11 m/s)
Cut-in wind speed mph: 6 (2.5 m/s)
Max design wind speed: 120 mph (54 m/s)
Generator type: Brushless Permanent Magnet, 3-phase
Rotor diameter in feet: 8.2 ft. (2.5 m)
Tower top weight: 75 lb. (34 kg)
Tower stub: Proprietary

16243	Bergey XL.1 w/Controller	$2,150

Bergey XL.1 Output

Graph — Watts Output (y-axis: 0 to 1400) versus wind speed.

M/S	1	3	5	7	9	11	13	15	17	19
MPH	2.2	6.7	11.1	15.6	20.0	24.5	29.0	33.4	37.9	42.3

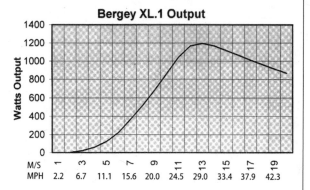

Proven Wind Turbines

The toughest and quietest turbine for residential use

Proven turbines are built in Scotland, where they have an intimate familiarity with some of the world's harshest weather and most brutal wind turbine conditions. Developed for the ferocious wind climates of northern Scotland, and built with marine quality using all non-corrosive components, Proven turbines have been put to the test in some of the world's worst weather locations, and will last a lifetime. When wind turbulence is unavoidable, this is the turbine you want to fly! These turbines belong to the "heavy metal" school of wind machines. They are heavy, tough, slow survivors. Proven turbines are all very slow speed, very quiet turbines. Most owners report they can't hear, or can just barely hear the turbine, even in the strongest winds. Proven turbines use an extremely low rpm brushless permanent magnet alternator. Low rpm means a low blade tip speed and almost silent operation. In an extremely

Assembling a WT 2500.

strong 40 mph wind, they make 55 to 65 decibels, about the same as a normal conversation. Lower wind speeds are even quieter. The smallest 600-watt turbine spins faster than the larger machines, and will be noisier.

These unique turbines are all "downwind machines." There's no tail, the blades run on the downwind end. As wind speeds increase, the very clever and robust polypropylene blades hinge inward, forming a progressively smaller cone area for the wind to push against. There is no wind speed at which these turbines stop producing, up to 145 mph!

Three models are available at 600-, 2500-, and 6000-watts. We have put together packages for each model that include the turbine, an appropriate charging/rectifier regulator with diversion control, and the required shipping box. Metering is standard on the 2,500- and 6,000-watt turbines; on the 600-watt turbine metering is only available as an option at 48 volt. The direct utility intertie packages for the 2,500 and 6,000 turbines

WT-6000 on mono-pole mast

include turbine, Windy Boy inverter(s) with display, an appropriate charging/rectifier regulator with diversion control (for those times the utility goes down), and the required shipping box. Voltage must be specified at time of purchase. Attractive self-supporting tilt-up mono-pole masts are available as well as conventional guyed towers. Please see Wind Accessories. Two-year manufacturer's warranty, with five-year option for utility-intertie units. Scotland.

Pricing subject to fluctuation w/International currency. Delivery is 5–6 weeks air freight, 10–12 weeks surface.

PROVEN TURBINE SPECIFICATIONS

Model	WT600	WT2500	WT6000
Swept Area ft²	55.4	96.7	254.0
Available DC voltages	12, 24, 48	24, 48, 120, 300	48, 120, 300
Rated wattage @ mph	600 @ 22.5	2,500 @ 26	6,000 @ 26
Cut-in wind speed mph	6	6	6
Max design wind speed	145	145	145
Generator type	Brushless Permanent Magnet		
Rotor diameter	8.4 ft	11.1 ft	18 ft
Tower top weight	154 lb	418 lb	1100 lb
Tower stub	Use Proven tower or adapter		

43-0176 Proven WT600/12-volt Turbine Package		**$5,250**
43-0177 Proven WT600/24 or 48-volt no-meter Turbine Package (specify 24 or 48v)		**$5,185**
43-0178 Proven WT600/48-volt w/meter Turbine Package		**$5,540**
43-0179 Proven WT2500/24 or 48-volt w/meter Turbine Package specify 24 or 48v)		**$10,590**
43-0180 Proven WT2500/Windy Boy Intertie Turbine Package		**$12,585**
43-0181 Proven WT6000/48-volt w/meter Turbine Package		**$20,360**
43-0182 Proven WT6000/Windy Boy Intertie Turbine Package		**$25,250**

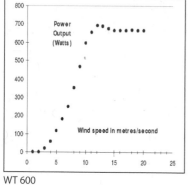

WT 600

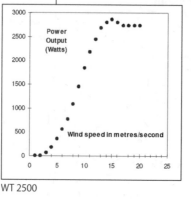

WT 2500

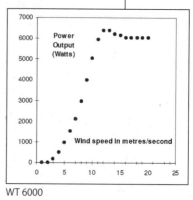

WT 6000

WIND ACCESSORIES

Griphoist Hand Winches

The Griphoist is an extraordinarily safe portable hand winch that can be used in any position, with any length of cable, going in or out. This is THE tool for safe, controlled raising and lowering of small residential wind-plants with tilt-up towers, such as we offer below. The Griphoist holds in position the instant the hand lever is released. It cannot get away from you like traditional drum winches. The cable is special, and must be ordered separately. You will need a lifting cable approximately twice as long as your tower height. The Pull-All model is acceptable for the lightest turbines, the Air X for instance. The Super Pull-All model is acceptable for turbines up to about 250 lb. with shorter towers. The Tirfor 508D is a safe choice for turbines 200 lb. and up, or turbines on towers 100' or taller.

	Pull-All	Super Pull-All	Tirfor 508D
Lifting Capacity:	700 lb	1,500 lb	2,000 lb
Weight:	3.85 lb	8.25 lb	18 lb
Supplied With:	32' of $^3/_{16}$" + 2 slings	32' of $^1/_4$" + 2 slings	30' of $^5/_{16}$"
Item #:	**43-0196**	**43-0202**	**43-0208**
Price:	**$165**	**$550**	**$595**
Add'l Cable, 75 Ft.	43-0197 $110	43-0203 $80	43-0209 $95
Add'l Cable, 100 Ft.	43-0198 $145	43-0204 $105	43-0210 $125
Add'l Cable, 150 Ft.	43-0199 $220	43-0205 $160	43-0211 $185
Add'l Cable, 200 Ft.	43-0200 $290	43-0206 $210	43-0212 $250
Add'l Cable, 250 Ft.	43-0201 $365	43-0207 $260	43-0213 $310

Tower Kits for Wind Turbines

The basic rule of thumb when installing any wind turbine: Be 30-feet higher than any trees, buildings, or other obstructions within a 200- to 300-foot radius. Anything that ruffles up the wind before it hits your turbine is going to reduce the power available, and it's going to beat up your turbine. You really want to be up in the clean, smooth airflow. Plus, there's more power available as you go up. At 100 feet, there's about 40% more energy than at ground level. "The taller the tower, the greater the power." Don't skimp on your tower!

Towers come in a variety of styles. The most basic division is between self-supporting towers, like the old water pumpers used, and guy-wire supported towers. Guyed towers are significantly less expensive, and are usually the tower of choice for smaller homestead turbines. Tubular fold-over guyed towers that use steel pipe or lighter weight steel tubing for the actual tower, with guy wire supports, have become the most common do-it-yourself tower type. This style completely assembles the tower, guy-wire system, and turbine while safely on the ground, then stands it all up. Rigging and setup takes awhile during initial assembly, but standing up or folding down the tower only takes a few minutes, and everything happens with your feet firmly planted on the ground.

Most of the tower kits offered below are tilt-up tubular guyed towers.

Roof Mounting Kit For Air 403 & Air X

We don't recommend this mounting option for any building that people plan to sleep (or lose sleep) in. Wind turbines make noise, and it transmits down the tower very nicely. This kit includes vibration isolators, which helps, but won't entirely cure the noise problem. This kit also includes clamps, straps, safety leashes, bolts, and roof seal. Mounting pipe and lag bolts are not included. USA.

16235 Air 403 Roof Mount w/Seal $99

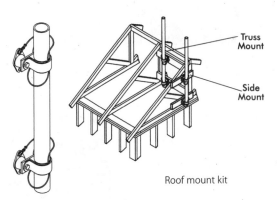

Truss Mount

Side Mount

Roof mount kit

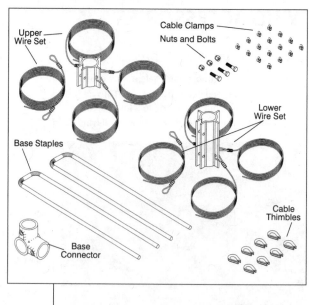

Tower Kits For Air 403 and Air X

These low-cost, guyed, tubular tower kits provide all the hardware, cabling, and instructions to assemble a stand-up tower. Pipe and anchors are not included. 1.5" nominal steel pipe can be purchased locally; guy anchors are offered below. 36" anchors recommended with 25´ tower, 48" anchors with 47´ tower. USA.

16233	Air 403 25´ Guyed Tower Kit	$189
16234	Air 403 47´ Guyed Tower Kit	$260

Mounting Kits For Air 403 and Air X Marine

A quality 9-foot mast mount that's built to withstand hurricane-force winds, and looks good too. Includes white powder-coated aluminum poles, all stainless steel hardware, self-locking nuts, and vibration dampening mounts. Main mast is 1.9" od (nominal 1.5" pipe), stays are 1" od. Masts and hardware kits are sold separately for those who want to supply their own masts. USA.

16148	Air Marine Tower Hardware Kit	$169
16149	Air Marine Tower Mast Kit	$179

Our powder-coated tower kits make it easy to install the **Air marine**

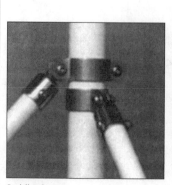

Saddle clamps

Tilting mast base

Stay base

Tower Kits For Whisper H40 or H80 Turbines

These guyed, tubular tower kits provide all the hardware, couplings, hinged base plate, cabling, and instructions to assemble a tilt-up tower. Pipe and anchors are not included. Steel pipe can be purchased locally; guy anchors are offered below. See chart below for anchor recommendations. See Tower Sizing graphic for layout and amount of ground space required. All these towers will fit either the H40 or the H80 turbine, and require nominal 2.5" pipe (which is actually 2.875" od). USA. All tower kits drop shipped from manufacturer.

Item #	Tower Height	Anchor Hole Dimensions	Guy Wire Radius	Auger Size	Price
16256	24'	1.5' dia. 3' deep, 12" concrete fill	15'	36"	**$290**
16257	30'	2' dia. 3' deep, 9" concrete fill	20'	48"	**$495**
16258	50'	2' dia. 3' deep, 10" concrete fill	25'	48"	**$619**
16259	65'	2' dia. 3' deep, 12" concrete fill	33'	60"	**$795**
16260	80'	2' dia. 4' deep, 12" concrete fill	40'	60"	**$949**

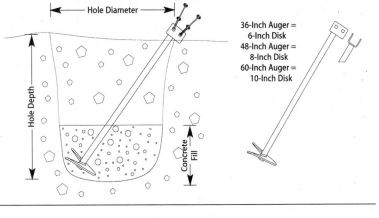

Anchors for Wind Towers

These anchor augers are sold in sets of four. With smaller turbines like the Air 403 or the Windseeker, you can simply screw them into place. No digging! With larger turbines, you need to dig anchor holes, per the recommendations above, and fill the bottom with concrete. USA.

16-253	36" Galvanized Anchor Augers (4)		**$199**
16-254	48" Galvanized Anchor Augers (4)		**$225**
16-255	60" Galvanized Anchor Augers (4)		**$270**

36-Inch Auger = 6-Inch Disk
48-Inch Auger = 8-Inch Disk
60-Inch Auger = 10-Inch Disk

Tower Kits for Whisper 175 Turbines

These guyed, tubular tower kits provide all the hardware, couplings, hinged base plate, cabling, and instructions to assemble a stand-up tower. Pipe and anchors are not included. Steel pipe can be purchased locally; guy anchors are offered below. See chart for anchor recommendations. See Tower Sizing graphic for layout and amount of ground space required. These towers will fit only the 175 turbine, and require nominal 5" pipe (which is actually 5.563" od). USA.

All tower kits drop-shipped from manufacturer.

Item #	Tower Height	Anchor Hole Dimensions	Guy Wire Radius	Auger Size	Price
16249	30'	2' dia. 3' deep, 12" concrete fill	20'	48"	**$980**
16250	42'	2.5' dia. 4' deep, 12" concrete fill	25'	48"	**$1,060**
16251	70'	2.5' dia. 5' deep, 16" concrete fill	35'	60"	**$1,455**

Tower Kits for African Wind Power Turbines

These are guyed tilt-up pipe towers. Kits are based on 4-inch schedule 10 pipe in standard 21-foot lengths. (Nominal 4-inch pipe is actually 4.5" O.D.) Kits include all couplers, end fittings, rigging, turnbuckles, and ground anchors. All hardware is galvanized. Note that these kits *include* anchors and even your hoisting cable. These are really high-quality kits that give you everything but the pipe. 43- and 64-foot towers have a guy radius of 21 feet. 85- and 106-foot towers have a guy radius of 42 feet. The 127-foot tower has a guy radius of 54 feet.

43-0170	43-foot AWP Tower Kit	$1,700
43-0171	64-foot AWP Tower Kit	$2,000
43-0172	85-foot AWP Tower Kit	$2,550
43-0173	106-foot AWP Tower Kit	$2,900
43-0174	127-foot AWP Tower Kit	$3,700

Towers for Bergey XL.1 Turbines

These are complete tilt-up, guyed tower kits that include galvanized tubular tower sections with associated hardware, guy anchors, guy wires, and gin pole. Only the tower electrical wiring is needed to complete your installation.

The 9- and 13-meter towers have 2-meter long, 3.5-inch diameter tower sections. These tower kits are shipped via UPS. 19-, 25-, and 32-meter towers have 3-meter long, 4.5-inch diameter tower sections, and are shipped via common carrier (semi truck).

16261	Bergey XL.1 Tower Kit, 9m (30′)	$790
16262	Bergey XL.1 Tower Kit, 13m (42′)	$1,060
16263	Bergey XL.1 Tower Kit, 19m (64′)	$1,250
16264	Bergey XL.1 Tower Kit, 25m (84′)	$1,490
16265	Bergey XL.1 Tower Kit, 32m (104′)	$1,995

Towers for Proven Wind Turbines

Proven provides aesthetically pleasing self-supporting mono-pole towers. These attractive, uncluttered pole mounts are equipped with a hinged base and come with a gin pole for easy raising and lowering. Equip yourself with the appropriate griphoist, and you'll be prepared to raise or lower your turbine anytime.

Conventional guyed tilt-up pipe tower kits are also offered. These include guy cables, galvanized turnbuckles, anchors, couplings, pivot assembly for base, hardware, and installation manual. You supply the 5" schedule 40 steel pipe. Also included with these kits is a Pulley Kit that contains the pulleys, cable, and hardware required to raise the tower. No other wind turbine manufacturer is offering such a complete package. Equip yourself with the appropriate griphoist, and you'll be prepared to raise or lower your turbine anytime.

If you're providing your own tower, rather than one of Proven's kits you'll need the appropriate tower adapter to connect your turbine.

Tilt-up Mono-Pole Towers for Proven Turbines

43-0183	5.5m (18′) Mono-Pole for WT600	$2,315
43-0184	6.5m (21′) Mono-Pole for WT2500	$3,600
43-0185	11m (36′) Mono-Pole for WT2500	$6,725
43-0186	9m (30′) Mono-Pole for WT6000	$6,500
43-0187	15m (49′) Mono-Pole for WT6000	$8,200

Tilt-Up Guyed Pipe Tower Kits for Proven Turbines

43-0188	13m (42′) Guyed Tower Kit for WT2500	$3,100
43-0189	20m (63′) Guyed Tower Kit for WT2500	$3,960
43-0190	27.5m (84′) Guyed Tower Kit for WT2500	$4,985
43-0191	34m (105′) Guyed Tower Kit for WT2500	$5,790
43-0192	41m (126′) Guyed Tower Kit for WT2500	$6,865

Please call for towers for other turbines.

Proven Tower Adapters

43-0193	Proven WT600 Tower Adapter	$245
43-0194	Proven WT2500 Tower Adapter	$345
43-0195	Proven WT6000 Tower Adapter	$465

Hydrogen Fuel Cells

Power Source of Our Bright and Shining Future?

Fuel cells seem to be the darlings of the energy world lately. If we believe all the hype, they'll bring clean, quiet, reliable, cheap energy to the masses, allowing us to continue an energy-intensive lifestyle with no penalties or roadblocks. Are fuel cells going to save our energy-hog butts? Maybe. They sure will help. Can you buy one yet? Certainly . . . for a price. It helps if you're the Department of Defense, or have a fat government grant or contract in your back pocket. This is still a very young, very rapidly developing technology. The first small commercial units started appearing in 2003 for relatively outrageous prices. Residential units that will run off natural gas or propane to provide backup power probably won't show up until at least 2006, and they're liable to be fairly expensive. Want a bit more background information before you order one? Good plan, read on.

Traditional energy production, for either electricity or heat, depends on burning a fuel source such as gasoline, fuel oil, natural gas, or coal to either spin an internal combustion engine, or to heat water, then piping the resulting steam or hot water to warm our buildings, or run the turbines to make electricity. Burning fuels produce some byproducts we'd be better off not letting loose into the environment, and every energy conversion costs some efficiency. The increasing industrialization of the world is requiring more and more fuel, creating our current unsustainable run on the world's energy reserves. The ultimate effect of all this fossil fuel burning is to create global climate change and global warming. Most scientists predict that if we fail to curb our fossil fuel consumption, the average world temperature will rise 4° to 10°C in the next fifty to one hundred years, and the oceans will rise 3 to 8 feet in the same time. Other, even less savory results of global warming are still speculative and debatable. What is known is that we're messing with things we don't understand, can't fix quickly, and almost surely won't enjoy.

Fuel cells are chemical devices that go from a source, like hydrogen or natural gas, straight to heat and electrical output, without the combustion step in the middle. Fuel cells increase efficiency by two- or three-fold, and dramatically reduce the unintended byproducts. A fuel cell is an electrochemical device, similar to a battery, except fuel cells operate like a continuous battery that never needs recharging. So long as fuel is supplied to the negative electrode (anode), and oxygen or free air is supplied to the positive electrode (cathode), they will provide continuous electrical power and heat. Fuel cells can reach 80% efficiency when both the heat and electric power outputs are utilized.

HOW FUEL CELLS OPERATE

A fuel cell is composed of two electrodes, sandwiched around an electrolyte material. Hydrogen fuel is fed into the anode. Oxygen (or free air) is fed into the cathode. Encouraged by a catalyst, the hydrogen atom splits into a proton

> Fuel cells are chemical devices that go from a source, like hydrogen or natural gas, straight to heat and electrical output, without the combustion step in the middle. Fuel cells increase efficiency by two- or three-fold, and dramatically reduce the unintended byproducts.

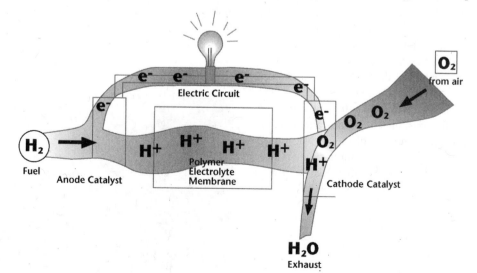

If you run down to your friendly local gas supply to buy a tank of hydrogen, what you get will be a byproduct of petroleum refining.

The dream is to build a fuel cell that accepts water input. But it takes more energy to split the water into hydrogen and oxygen than we'll get back from the fuel cell. One potential future scenario has large banks of PV cells splitting water.

and an electron. The proton passes through the electrolyte to reach the cathode. The electron takes a separate, outside path to reach the cathode. Since electrons flowing through a wire are commonly known as "electricity," we'll make those free electrons do some work for us on the way. At the cathode, the electrons, protons, and oxygen all meet, react, and form water. Fuel cells are actually built up in "stacks" with multiple layers of electrodes and electrolyte. Depending on the cell type, electrolyte material may be either liquid or solid.

Like other electrochemical devices such as batteries, fuel cells eventually wear out and the stacks have to be replaced. Stack replacement cost is typically 20 to 25% of the initial fuel cell cost. Current designs have run times of 40,000 to 60,000 hours. Under continuous operation that's 4.5 to 6.8 years.

WHAT'S "FUEL" TO A FUEL CELL?

In purest form, a fuel cell takes hydrogen, the most abundant element in the universe, and combines it with oxygen. The output is electricity, pure water (H_2O), and a bit of heat. Period. Very clean technology! No nasty byproducts and no waste products left over. There are some real obvious advantages to using the most abundant element in the universe. We've got plenty of it on hand, and, as most anyone can clearly see, a hydrogen-based economy is far more clean and sustainable than a petroleum-based economy.

On the downside, hydrogen doesn't usually exist in pure form on Earth. It's bound up with oxygen to make water, or with other fuels like natural gas or petroleum. If you run down to your friendly local gas supply to buy a tank of hydrogen, what you get will be a byproduct of petroleum refining.

Since hydrogen probably isn't going to be supplied in pure form, most commercial fuel cells have a fuel-processing component as part of the package. Fuel processors, or "reformers" do a bit of chemical reformulation to boost the hydrogen content of the fuel. This makes a fuel cell that can run on any hydrocarbon fuel. Hydrocarbon fuels include natural gas, propane, gasoline, fuel oils, diesel oil, methane, ethanol, methanol, and a number of others. Natural gas or propane are favored for stationary generators, and methanol or even gasoline have been used in experimental automotive fuel cells. But if you feed the fuel cell something other than pure hydrogen, you're going to get something more than pure water in the output, carbon dioxide usually being the biggest component. Fuel cells emit 40 to 60% less carbon dioxide than conventional power generation systems using the same hydrocarbon fuel. Other air pollutants such as sulfur oxides, nitrogen oxides, carbon monoxide, and unburned hydrocarbons are nearly absent, although you'll still get some trace byproducts we'd be better off without.

The dream is to build a fuel cell that accepts water input. But it takes more energy to split the water into hydrogen and oxygen than we'll get back from the fuel cell. One potential future scenario has large banks of PV cells splitting water. The hydrogen is collected, and used to run cars and light trucks. Efficiency isn't great, but nasty byproducts are zero.

Competing Technologies

Just like there's more than one "right" chemistry with which to build a battery, there are quite a number of ways to put together a functional fuel cell. As an emerging technology, several fuel cell chemical combinations are receiving experimental interest, substantial funding, and absolutely furious development. At the risk of boring you, or scaring you all away, we're going to list the major chemical contenders for fuel cell power, and their strong or weak points.

PHOSPHORIC ACID

This is the most mature fuel cell technology. Quite a number of 200-kilowatt demonstration/experimental units are in everyday operation at hospitals, nursing homes, hotels, schools, office buildings, and utility power plants. The Department of Defense currently runs about thirty of these with 150- to 250-kilowatt output. If you simply must have a fuel cell in your life right now, this is your baby. Currently that price is about $3,000 per kilowatt. These are large, stationary power plants, usually running on natural gas. Phosphoric acid cells do not lend themselves well to small-scale generators. Locomo-

The typical phosphoric acid fuel cell is about the size of a freight car. This one's in New York's Central Park.

tives and buses are probably about as small as they will go. Operating temperatures are 375° to 400°F, so they need thermal shielding. Efficiency for electric production alone is about 37%; if you can utilize both heat and power, efficiency hits about 73%. Stack life expectancy is about 60,000 hours.

PROTON EXCHANGE MEMBRANE

PEM is the fuel cell technology that is probably seeing the most intense development, and is the type of fuel cell you are most likely to see in your lifetime. Because it offers more power output from a smaller package, low operating temperatures, and fast output response, this is the favored cell technology for automotive and residential power use. Thanks to their low noise level, low weight, quick start-up, and simple sup-

PEM cells are smaller, run at lower temperatures, and are more consumer-friendly.

port systems, experimental PEM fuel cells have even been produced for cell phones and video cameras. Recent advances in performance and design raise the tantalizing promise of the lowest cost of any fuel cell system. Current prices are about $3,000 per kilowatt, but they will drop as mass-production economies begin to kick in. Operating temperature is a very low 175° to 200°F, so minimal thermal shielding is needed. Efficiency for electric production alone is about 36%, if you can utilize both heat and power, efficiency hits about 70%. PEM cells can start delivering up to 50% of their rated power at room temperature, so start-up time is minimal. Stack life expectancy is about 40,000 hours.

There are some small demonstration PEM fuel cells available for sale currently, but for something that's going to run your house you'll need some patience. A great many manufacturers have residential units under development,

many are in beta testing, and if you want it bad enough, you'll find a few for sale, although this technology isn't expected to fully mature till sometime after 2010. For current information visit the nonprofit Fuel Cells 2000 website at www.fuelcells.org.

MOLTEN CARBONATE AND SOLID OXIDE

These two fuel cell technologies aren't related, except that they are both large baseload-type cells that utility companies can use to supply grid power. These technologies have barely delivered their first prototypes, so operational plants are still far in the future. Very high operating temperatures, 1,200° to 1,800°F, are the norm for these cells. Electrical generation efficiencies are over 50%, and pollution output is way, way down, so these technologies are going to be very attractive to utility companies. The high operating temperatures mean these cells don't turn on or off casually. Start-up can take over ten hours. These aren't fuel cells for off-the-grid homes.

Cruisin' the City on Fuel Cells?

One of civilization's biggest fossil fuel uses, and the one we seem to have the hardest time weaning ourselves from, is automobiles. Do fuel cells offer any hope for hopeless driving addicts? Yes, they do. Every major auto manufacturer has a well-funded, active fuel cell development program. Several manufacturers, Daimler-Chrysler, Toyota, and Honda in particular, have several generations of prototypes behind them already. The stakes are high. The company that develops a workable fuel cell–powered auto holds the keys to the future. Several manufacturers are aiming for the first commercial models to be available by 2006 or 2007, with full-scale production a couple years later. These will probably be methanol-fueled PEM cells with a modest

Do fuel cells offer any hope for hopeless driving addicts? Yes, they do. Every major auto manufacturer has a well-funded, active fuel cell development program.

The stakes are high. The company that develops a workable fuel cell-powered auto holds the keys to the future.

As an electrical supplier, a fuel cell is closer kin to an internal combustion engine than to any renewable energy source. Think of a fuel cell as a vastly improved generator. It burns less fuel, it makes less pollution, there are no moving parts, and it hardly makes any noise. But unless you have a free supply of hydrogen, it will still use nonrenewable fossil fuels for power, which will cost something.

battery or capacitor pack to help meet sudden acceleration surges.

One of the primary problems has been the size of the fuel cell stack required. Mercedes' first prototype was a small bus; its second prototype was a van; and its latest, fifth, prototype is in the new compact A-series sedan. Each prototype just has room for a driver and passenger. So cell stacks are getting smaller and more efficient, but we could still stand some improvement (or an expectation adjustment).

Crash-safe hydrogen storage in a vehicle also continues to be a trying problem. We demand much safer vehicles than we used to. The first generations of fuel cell vehicles will be using on-board reformers with methanol or other liquid fuels, so this isn't a problem that requires an immediate solution.

How Do Fuel Cells Compare to Solar or Wind Energy?

As an electrical supplier, a fuel cell is closer kin to an internal combustion engine than to any renewable energy source. Think of a fuel cell as a vastly improved generator. It burns less fuel, it makes less pollution, there are no moving parts, and it hardly makes any noise. But unless you

have a free supply of hydrogen, it will still use nonrenewable fossil fuels for power, which will cost something.

Photovoltaic modules, wind generators, hydro generators, and solar hot water panels all use free, renewable energy sources. No matter how much you extract today, it doesn't impact how much you can extract tomorrow, the next day, and forever. This is the major difference between technologies that harvest renewable energy, and fuel cells, which are primarily going to continue using nonrenewable energy sources.

In the long run, we think that fast-starting PEM fuel cells will take the place of backup generators in residential energy systems. They'll be in place to pick up the occasional shortfall in renewable-powered off-the-grid systems, or to provide backup AC power for grid-supplied homes when the utility fails. If your home is off-the-grid, then solar, wind, or hydro is still going to be the most cost-effective, and environmentally responsible primary power source. In an ideal world, every backup power fuel cell would be supplied by a small PV-powered hydrogen extractor that cracks water all day and stores the resulting hydrogen for later use. Guess we'd better get busy designing hydrogen extractors. . . .

—Doug Pratt

CHAPTER 3

Balance of Systems

From Panel to Plug: How to Get Your Renewable Energy Source Powering Your Appliances

A Renewable Energy System Primer

HOW ALL THE PIECES FIT TOGETHER

IN OTHER CHAPTERS and sections of the *Sourcebook* we provide details about how individual components such as PV modules or inverters work, and why they're needed. This is the place where we explain how and why all the bits and pieces fit together. Read this Renewable Energy System Primer first to get an overall idea of how everything works, and then move on to details as needed.

Stand-Alone Systems (Battery Back-up)

Explanations in this section will revolve around systems that use batteries and are capable of stand-alone operation. If your interest lies in simple utility intertie systems, with no battery back-up, please see the separate section dealing with utility intertie, which immediately follows on pages 122 to 136.

IT ALL REVOLVES AROUND THE BATTERY

The battery, or battery pack, is the center of every stand-alone renewable energy system. Everything comes and goes from the battery, and much of the safety and control equipment is designed either to protect the battery, or to protect other equipment from the battery. The battery is your energy reservoir; it stores up electrons until they're needed to run something.

Battery

The size of your reservoir determines how many total electrons you can store, but has minimal effect on how fast electrons can be poured in or out, at least in the short term. A smaller battery will limit how long your system will run

without recharging. A larger battery will allow more days of run time without recharging. However, there are practical limits in each direction. Batteries are like the muscles of your body: They need some exercise to stay healthy, but not too much exercise or they get worn out at an early age. We usually aim for three to five days worth of battery storage. Less than three and your system will be cycling the battery deeply on a daily basis, which is bad for the battery. More than five and your battery starts getting more expensive than a back-up generator or other recharging source.

Once you start removing electrons from your battery, you'd better have a plan to replace them or you'll soon have an empty reservoir—what we call a dead battery—which brings us to . . .

THE CHARGING SOURCE

Batteries aren't the least bit fussy about where their recharging electrons come from. So long as

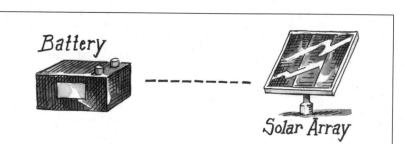
Battery — *Solar Array*

The battery is your energy reservoir; it stores up electrons until they're needed to run something.

the voltage from the charging source is a bit higher than the battery voltage, the electrons will flow "downhill." You can haul your battery into town to get it charged, or you can connect it to your vehicle charging system and charge it on the way to town and back. But batteries are heavy and the sulfuric acid in them just loves to eat your clothes, so hauling batteries around is no fun. You'll want an on-site charging source. This can be a fossil-fueled (gasoline or propane) generator, a wind turbine, a hydroelectric generator, solar modules (PVs), a stationary bicycle generator, or combinations of these.

As long as incoming energy keeps up with outgoing energy, your battery will remain happy. Up to about 80% of full charge, a battery will accept very large current inputs with no problem. As it gets closer to full, we have to start applying some charge control to prevent overcharging. Batteries will be damaged if they are severely or regularly overcharged.

BATTERIES NEED CONTROL

Someplace in your system there needs to be a charge controller. Charge controllers use various strategies, but they all have a common goal: to keep the battery from overcharging and toasting. A battery that is regularly overcharged will use excessive amounts of water, may run dry, and sheds flakes of active lead from the battery

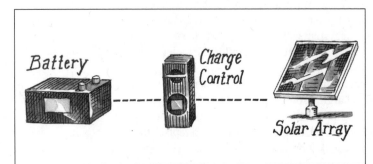

plates, reducing future capacity. A charge controller watches the battery voltage, and if it gets too high, will either open the charging circuit, or dump excess electrons into a heater element.

We use different strategies depending on the charging source. Solar modules can use simple controls between the solar array and the battery that just open the circuit if voltage gets too high. Rotating generators such as wind and hydroelectric turbines use a controller that diverts excess energy into a heater element. (Really bad things happen if you open the circuit between a generator and the battery while the generator is spinning!)

Now you have a battery, a way to recharge it, a controller to prevent overcharging, but no clue how full or empty it is. You need a system monitor.

MONITORING YOUR SYSTEM

System monitors all have a common goal: to keep you informed about your system's state of charge, and help you prevent the exceptionally

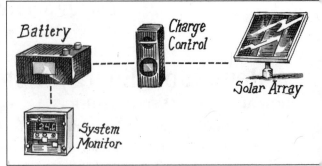

deep discharges that sap the life out of your batteries. Monitors can be as simple as a red or green light, or as complicated as a PC-linked system that logs dozens of system readings every few minutes. At the simplest level, displaying the battery voltage gives an approximate indication of how full or empty your battery is, and what's going on at the moment. More complete monitors may show how many amps are flowing in or out at the moment, and the best monitors keep tabs on all this information, then give it to you as a simple "% of full charge" reading. Many charge controllers offer some simple monitoring abilities, but the best monitors stand alone, and get mounted in the kitchen or living room where they're more likely to get some regular attention.

You can run a renewable energy system without a monitor, but it's not a smart idea. You could run your car without any gauges or meters too, but you'd be much more likely to exceed speed limits, run out of gas, or do accidental damage. System monitors are a real bargain, compared to batteries.

Now you have a battery, a way to recharge it, a controller to prevent overcharging, and a system monitor to tell how full or empty the battery is. Next, you need to be able to use the stored energy.

USING YOUR POWER

Batteries deliver direct current, or DC, when you ask them for power. You can tap this energy to run any DC lights or appliances directly. This is fine if you're running an RV or a boat; vehicles pretty much run on DC, thanks to their automotive heritage. With a DC breaker or fuse panel, and a wiring circuit or two, you're all set to run your DC light or pump.

A charge controller watches the battery voltage, and if it gets too high, it will either open the charging circuit, or dump excess electrons into a heater element.

The best monitors stand alone, and get mounted in the kitchen or living room where they're more likely to get some regular attention.

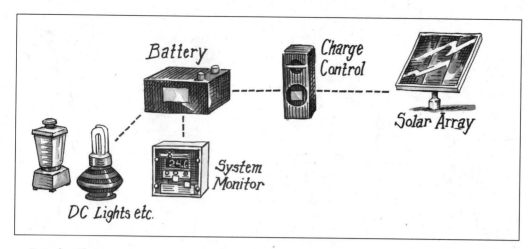

Battery Charge Control Solar Array System Monitor DC Lights etc.

But what if you want to run an entire household, or a microwave in your RV? Household appliances run on 120 volt AC, or alternating current. AC power is easier to transmit over long distances, so it has become the world standard. AC appliances won't run on DC from a battery. You could convert all your household lights and appliances to DC, but joining the mainstream offers some definite advantages. AC appliances are mass produced, making them less expensive, universally available, and generally of better quality, when compared to "RV appliances."

But AC power can't be stored—it has to be generated as demanded. To meet this highly variable demand, you need a device called an inverter. The inverter converts low-voltage DC from the battery into high-voltage AC as fast as your household lights and appliances demand it.

Inverters come in a variety of outputs from 50 watts up through hundreds of thousands of watts, and range in cost from under $30 to more than you want to know. You want one that will deliver enough wattage to start and run all the AC appliances you might turn on at the same time, but no bigger than necessary. Bigger inverters cost more, and use a bit more power on standby, waiting for something to turn on. However, lower-cost inverters usually leave out some

protection equipment, which means a shorter life expectancy. (We've seen really cheap inverters blow up the instant a compact fluorescent lamp was turned on.)

Many of the larger, household-size inverters come with built-in battery chargers that come on automatically anytime outside AC power becomes available by starting a generator or plugging into an RV park outlet. This provides a simple back-up charging source for times when the primary solar or wind charger isn't keeping up.

Now you have a battery, a way to recharge it, a controller to prevent overcharging, a system monitor to tell how full or empty it is, and a way to use your stored energy. Only one component is missing, and it's the most important one: Safety!

HOW TO AVOID BURNING DOWN YOUR HOUSE

For the most part, low-voltage DC systems and batteries are a lot safer than jittery high-voltage AC systems. Just keep your tongue out of the DC sockets and you'll be fine. However, if something does short-circuit a battery, look out! Because they're close by, batteries can deliver more amps into a short circuit than an AC system can, because the AC is strung out over many

The inverter converts low-voltage DC from the battery into high-voltage AC as fast as your household lights and appliances demand it.

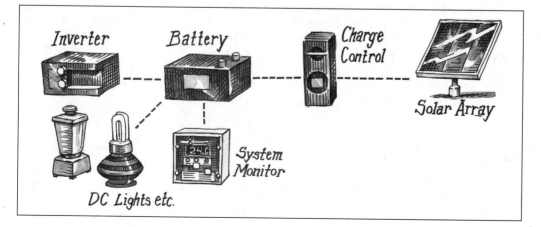

Inverter Battery Charge Control Solar Array System Monitor DC Lights etc.

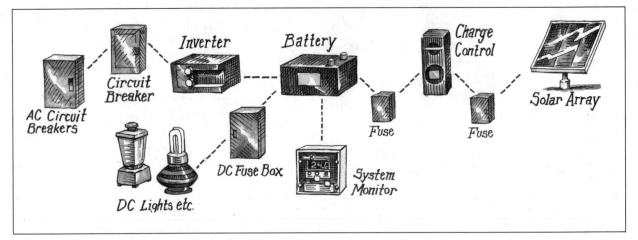

*The more energy
your system processes
per day, the higher your
system voltage is
likely to be.*

miles. So common sense, and the Electric Code, says you *must* put some overcurrent protection on any wire attached to a power source. "Overcurrent protection" is code-speak for a fuse or circuit breaker. "Power source" means batteries, or any charging source such as PV arrays or a wind turbine. These fuses or circuit breakers must be rated for DC use, and appropriately sized for the wires they're protecting. A DC short circuit is harder to interrupt than an AC short, so DC-rated equipment usually contains extra stuff and is therefore more expensive. DC-rated fuses and circuit breakers are built tougher, and you really, really want them that way. You only need to see wires glowing red with burning insulation dripping off them once, to appreciate the serious importance of fuses.

Now you have a battery, a way to recharge it, a controller to prevent overcharging, a system monitor to tell how full or empty it is, a way to use your stored energy, and the safety equipment to prevent burning down your house. That's a complete system!

WHAT SYSTEM VOLTAGE
SHOULD YOU USE?

Renewable energy systems can be put together with the battery bank running at 12, 24, or 48 volts. The more energy your system processes per day, the higher your system voltage is likely to be. As systems get bigger, it gets easier to push the DC power around if the voltage is higher. If you double the voltage, you only need to push half the amperage through the wires to perform a given job. There are no hard

and fast rules here, but in general, systems processing up to 2,000 watt-hours per day are fine at 12 volts. From 2,000 to about 7,000 watt-hours per day, a system will work better at 24 volts, and from 7,000 watt-hours on up we prefer to run at 48 volts. These are guidelines, not rules. We can make a system run at any of these three common voltages, but you'll buy less hardware and big wires if system voltage goes up as the watt-hours increase.

Ready . . . Set . . . Build!

Are you ready to put together a renewable energy system? Of course not. You probably still have a lot of questions about how big, how many, what brand, and more. These questions can be answered by consulting other sections in this chapter: Safety and Fusing, Controllers, Monitors, Batteries, Inverters, Wiring, and a System Sizing Worksheet. Each of these sections explains a particular renewable energy component in more detail, and helps you make the best choice for your individual needs.

And finally, if you aren't the technical type, and don't take discovery joy from figuring this stuff out yourself—or even if you do, but would appreciate a little guidance—there's our highly experienced technical staff, which we think is the best in the business. We're available for consultation, system design, and troubleshooting Monday through Saturday, 7:30 A.M. to 5:00 P.M. Pacific time.

—Doug Pratt

System Sizing Worksheets

Off-the-Grid Systems

The following System Sizing Worksheet provides a simple and convenient way to determine approximate total household electrical needs for off-the-grid systems. Once the worksheet is completed, our Gaiam Real Goods technicians can size the photovoltaic and battery system to meet your needs. Do you want to go with utility intertie? That's simpler—see the next paragraph. Ninety percent of the renewable energy systems we design are PV-based, so these worksheets deal primarily with PV. If you are fortunate enough to have a viable wind or hydro power source, you'll find output information for these sources in their respective sections of chapter 2 of this *Sourcebook*. Our technical staff has considerable experience with these alternate sources and will be glad to help you size a system. Give us a call.

Utility Intertie Systems

These are easy. Whatever your renewable energy system doesn't cover, your existing utility company will. So we don't need to account for every watt-hour beforehand. Either tell us how much you want to spend, or how many kilowatt-hours of utility power you'd like to displace on an average day. For direct-intertie systems without batteries, you'll invest about $2,000 for every kilowatt-hour per day your solar system delivers. For battery-based systems that can provide limited emergency back-up power, you'll invest about $3,500 for every kilowatt-hour per day. These are very general ballpark figures for initial system costs. Remember, state and other rebates can reduce these costs by up to 50%.

General Information

The primary principle for off-the-grid systems is: Conserve, conserve, conserve! As a rule of thumb, it will cost about $3 to $4 worth of equipment for every watt-hour per day you must supply. Trim your wattage to the bone! Don't use incandescent light bulbs or older standard refrigerators.

If you're intimidated by this whole process, don't worry. Our tech staff is here to hold your hand and help you through the tough parts. We do need you to come up with an estimate of Total Household Watt-Hours per Day, which this worksheet will help you do. We can pick up the design process from there. You are in the best position to make lifestyle decisions: How late do you stay up at night, are you religious about always turning the light off when leaving a room, are you running a home business or is the house empty five days a week, do you hammer on your computer for twelve hours a day, does your pet iguana absolutely require his rock heater twenty-four hours a day? These are questions we can't answer for you. So figure out your watt-hours to let us know what we're shooting for.

Determine the Total Electrical Load in Watt-Hours per Day

The form following allows you to list every appliance, how much wattage it draws, how many hours per day it runs, and how many days per week. This gives us a daily average for the week, as some appliances, like a washing machine, may only be used occasionally.

Some appliances may only give the amperage and voltage on the nameplate. We need wattage. Multiply the amperage by the voltage to get wattage. Example: A blender nameplate says, "2.5A 120V 60Hz." This tells us the appliance is rated for a maximum of 2.5 amps at 120 volts/60 cycles per second. 2.5 amps times 120 volts equals 300 watts. Be cautious of using nameplate amperage, however. For safety reasons, this must be listed as the highest amperage the appliance is capable of drawing. Actual running amperage is often much less. This is particularly true for refrigerators and entertainment equipment. The Watt Chart for Typical Appliances may help give you a more "real" wattage use.

AC device	Device watts	×	Hours of daily use	×	Days of use per week	÷	7	=	Average watt-hours per day
		×		×		÷	7	=	
		×		×		÷	7	=	
		×		×		÷	7	=	
		×		×		÷	7	=	
		×		×		÷	7	=	
		×		×		÷	7	=	
		×		×		÷	7	=	
		×		×		÷	7	=	
		×		×		÷	7	=	
		×		×		÷	7	=	
		×		×		÷	7	=	
		×		×		÷	7	=	
		×		×		÷	7	=	
		×		×		÷	7	=	
		×		×		÷	7	=	
		×		×		÷	7	=	
		×		×		÷	7	=	
		×		×		÷	7	=	
		×		×		÷	7	=	
		×		×		÷	7	=	
		×		×		÷	7	=	
		×		×		÷	7	=	

1 Total AC watt-hours/day

2 × 1.1 = Total corrected DC watt-hours/day

DC device	Device watts	×	Hours of daily use	×	Days of use per week	÷	7	=	Average watt-hours per day
		×		×		÷	7	=	
		×		×		÷	7	=	
		×		×		÷	7	=	
		×		×		÷	7	=	
		×		×		÷	7	=	
		×		×		÷	7	=	
		×		×		÷	7	=	
		×		×		÷	7	=	

3 Total DC watt-hours/day

3 (from previous page)	Total DC watt-hours/day	
4	Total corrected DC watt-hours/day from Line 2 +	
5	Total household DC watt-hours/day =	
6	System nominal voltage (usually 12 or 24) ÷	
7	Total DC amp-hours/day =	
8	Battery losses, wiring losses, safety factor × 1.2	
9	Total daily amp-hour requirement =	
10	Estimated design insolation (hours per day of sun, see map on p. 434) ÷	
11	Total PV array current in amps =	
12	Select a photovoltaic module for your system	
13	Module rated power amps ÷	
14	Number of modules required in parallel =	
15	System nominal voltage (from line 6 above)	
16	Module nominal voltage (usually 12) ÷	
17	Number of modules required in series =	
18	Number of modules required in parallel (from Line 14 above) ×	
19	Total modules required =	

BATTERY SIZING

20	Total daily amp-hour requirement (from line 9)	
21	Reserve time in days ×	
22	Percent of useable battery capacity ÷	
23	Minimum battery capacity in amp-hours =	
24	Select a battery for your system, enter amp-hour capacity ÷	
25	Number of batteries in parallel =	
26	System nominal voltage (from line 6)	
27	Voltage of your chosen battery (usually 6 or 12) ÷	
28	Number of batteries in series =	
29	Number of batteries in parallel (from line 25 above) ×	
30	Total number of batteries required	

LINE BY LINE INSTRUCTIONS

Line 1   Total all average watt-hours/day in the column above.

Line 2   For AC appliances, multiply the watt-hours total by 1.1 to account for inverter inefficiency (typical by 90%). This gives the actual DC watt-hours that will be drawn from the battery.

Line 3   DC appliances are totalled directly, no correction necessary.

Line 4   Insert the total from line 2 above.

Line 5   Add the AC and DC watt-hour totals to

get the total DC watt-hours/day. At this point you can fax or mail the design forms to us and after a phone consultation we'll put a system together for you. If you prefer total self-reliance, forge on.

Line 6 Insert the voltage of the battery system; 12-volt or 24-volt are the most common. Talk it over with one of our tech staff before deciding on a higher voltage as control and monitoring equipment is sometimes hard to find.

Line 7 Divide the total on line 5 by the voltage on line 6.

Line 8 This is our fudge factor that accounts for losses in wiring and batteries, and allows a small safety margin. Multiply line 7 by 1.2.

Line 9 This is the total amount of energy that needs to be supplied to the battery every day on average.

Line 10 This is where guesswork rears its ugly head. How many hours of sun per day will you see? Our Solar Insolation Map in the appendix (page 434) gives the average daily sun hours for the worst month of the year. You probably don't want to design your system for worst-possible conditions. Energy conservation during stormy weather, or a back-up power source can allow use of a higher hours-per-day figure on this line and reduce the initial system cost.

Line 11 Divide line 9 by line 10; this gives the total PV current needed.

Line 12 Decide what PV module you want to use for your system. You may want to try the calculations with several different modules. It all depends on whether you need to round up or down to meet your needs.

Line 13 Insert the amps of output at rated power for your chosen module.

Line 14 Divide line 11 by line 13 to get the number of modules required in parallel. You will almost certainly get a fraction left over. Since we don't sell fractional PV modules, you'll need to round up or down to a whole number. We conservatively recommend any fraction from 0.3 and up be rounded upward.

If yours is a 12-volt nominal system, you can stop here and transfer your line 14 answer to line 19. If your nominal system voltage is something higher than 12 volts, then forge on.

Line 15 Enter the system battery voltage. Usually this will be either 12 or 24.

Line 16 Enter the module nominal voltage. This will be 12 except for unusual special-order modules.

Line 17 Divide line 15 by line 16. This will be how many modules you must wire in series to charge your batteries.

Line 18 Insert the figure from line 14 and multiply by line 17.

Line 19 This is the total number of PV modules needed to satisfy your electrical needs. Too high? Reduce your electrical consumption, or, add a secondary charging source such as wind or hydro if possible, or, a stinking, noisy, troublesome, fossil-fuel–gobbling generator.

BATTERY SIZING WORKSHEET

Line 20 Enter your total daily amp-hours from line 9.

Line 21 Reserve battery capacity in days. We usually recommend about three to seven days of back-up capacity. Less reserve will have you cycling the battery excessively on a daily basis, which results in lower life expectancy. More than seven days capacity starts getting so expensive that a back-up power source should be considered.

Line 22 You can't use 100% of the battery capacity (unless you like buying new batteries). The maximum is 80%, and we usually recommend to size at 50 or 60%. This makes your batteries last longer, and leaves a little emergency reserve. Enter a figure from .5 to .8 on this line.

Line 23 Multiply line 20 by line 21, and divide by line 22. This is the minimum battery capacity you need.

Line 24 Select a battery type. The most common for household systems are golf carts at 220 amp-hours, or L-16s at 350 amp-hours. See the Battery Section for more details. Enter the amp-hour capacity of your chosen battery on this line.

Line 25 Divide line 23 by line 24; this is how many batteries you need in parallel.

Line 26 Enter your system nominal voltage from line 6.

Line 27 Enter the voltage of your chosen battery type.

Line 28 Divide line 26 by line 27; this gives you how many batteries you must wire in series for the desired system voltage.

Line 29 Enter the number of batteries in parallel from line 25.

Line 30 Multiply line 28 by line 29. This is the total number of batteries required for your system.

Typical Wattage Requirements for Common Appliances

Use the manufacturer's specifications if possible, but be careful of nameplate ratings that are the highest possible electrical draw for that appliance. Beware of appliances that have a "standby" mode and are really "on" twenty-four hours a day. If you can't find a rating, call us for advice (800.919.2400).

DESCRIPTION	WATTS
Refrigeration (*Refrigerators only are listed in watt-hours per day*):	
5–10-yr.-old 22-cu. ft. auto defrost	3,000
New 22-cu. ft. auto defrost (with Energy Star rating of 400 kwh/yr)	1,100
12-cu. ft. Sun Frost refrigerator	600
5–10-yr.-old standard freezer	3,500
Sundanzer 8-cu. ft. freezer	530
Dishwasher:	
cool dry	700
hot dry	1,450
Trash compactor	1,500
Can opener (electric)	100
Microwave (.5 cu. ft.)	900
Microwave (.8 to 1.5 cu. ft.)	1,500
Exhaust hood	144
Coffeemaker	1,200
Food processor	400
Toaster (2-slice)	1,200
Coffee grinder	100
Blender	350
Food dehydrator	600
Mixer	120
Range, small burner	1,250
Range, large burner	2,100
Water Pumping:	
AC Jet Pump (⅓ hp), 300 gal per hour, 20' well depth, 30 psi	750

DESCRIPTION	WATTS
AC submersible pump (½ hp), 40' well depth, 30 psi	1,000
DC pump for house pressure system (typical use is 1–2 hours per day)	60
DC submersible pump (typical use is 6 hours per day)	50
Shop:	
Worm drive 7¼" saw	1,800
AC table saw, 10"	1,800
AC grinder, ½ hp	1,080
Hand drill, ⅜"	400
Hand drill, ½"	600
Entertainment/Telephones:	
TV (27-inch color)	170
TV (19-inch color)	80
TV (12-inch black & white)	16
Video games (not incl. TV)	20
Satellite system, VCR	30
DVD/CD player	30
AC-powered stereo (avg. volume)	55
AC stereo, home theater	500
DC-powered stereo (avg. volume)	15
CB (receiving)	10
Cellular telephone (on standby)	5
Cordless telephone (on standby)	5
Electric piano	30
Guitar amplifier (avg. volume)	40
(Jimi Hendrix volume)	8,500
General Household:	
Typical fluorescent light (60W equivalent)	15
Incandescent lights	(as indicated on bulb)
Electric clock	4
Clock radio	5
Electric blanket	400

DESCRIPTION	WATTS
Iron (electric)	1,200
Clothes washer (vertical axis)	900
Clothes washer (horizontal axis)	250
Dryer (gas)	500
Dryer (electric)	5,750
Vacuum cleaner, average	900
Central vacuum	1,500
Furnace fan:	
¼ hp	600
⅓ hp	700
½ hp	875
Garage door opener: ¼ hp	550
Alarm/security system	6
Air conditioner: per 1 ton or 10,000 BTU/hr capacity	1,500
Office/Den:	
Computer	55
17" color monitor	100
17" LCD "flat screen" monitor	45
Laptop computer	25
Ink jet printer	35
Dot matrix printer	200
Laser printer	900
Fax machine (plain paper):	
standby	5
printing	50
Electric typewriter	200
Adding machine	8
Electric pencil sharpener	60
Hygiene:	
Hair dryer	1,500
Waterpik	90
Whirlpool bath	750
Hair curler	750
Electric toothbrush: (charging stand)	6

Utility Intertie Systems

How to Sell Power to Your Electric Utility Company (*and get yours for free!*)

Utility electric costs are rising, blackouts are increasing, and the power grid we once took for granted is becoming increasingly unreliable. Many of our customers are turning to the reliability and consistency of solar power and utility intertie systems. Solar is clean, reliable, nonpolluting energy that is quickly becoming cost-effective, with payback periods approaching six years where utility rates are high and rebate programs are in place. State and federal government rebate programs may reduce initial costs by as much as 50%. The time for solar is now.

A utility intertie system makes it possible to generate your own solar power and sell it back to your utility company. Your utility electric meter will run backwards anytime your renewable energy system is making more power than your house needs at the moment. If the meter runs backwards, your utility company is buying power from you, and, if you live in a state with net metering, they will buy it at the same retail cost that you pay them. It feels great! Even if your RE system isn't making more energy than your house needs at the moment, any watt-hours your RE system delivers will displace an equal number of utility watt-hours, directly reducing your utility bill, and reducing your exposure to the fluctuating and often increasing utility rates that are becoming the norm as utility deregulation proceeds.

THE LEGALITIES OF ELECTRICITY AND NET METERING

The 1978 federal Public Utilities Regulatory Policies Act (PURPA) states that any private renewable energy producer in the United States has a right to sell power to their local utility company. The law doesn't state that utilities have to make this easy, or profitable, however. Under PURPA, the utility usually will pay its "avoided cost," otherwise known as wholesale rates. This law was an outgrowth of the 1973 Arab oil embargo and was intended to encourage renewable energy producers. But two to four cents per kilowatt-hour was not sufficiently encouraging to have much of an effect on home utility intertie installations.

The next wave of encouragement started arriving in the mid-1980s with net metering laws. There is no blanket federal net metering law (yet). As of early 2004, thirty-three states have statewide net metering laws, and an additional six have individual utilities that support net metering. These policies allow small-scale renewable energy producers to sell excess power to their utility company, through the existing electric meter, for standard retail prices. Details of these state laws vary widely. For an up-to-date summary see the Database of State Incentives for Renewable Energy website: http://dsireusa.org. Or call your State Energy Office; there's a complete directory in the Appendix.

HOW UTILITY INTERTIE SYSTEMS WORK

An intertie system uses an inverter that takes any available renewable power generated on-site, turns it into conventional AC, and feeds it into your electrical circuit breaker panel. Electricity flow is much like water flow; it runs from higher to lower voltage levels. When you turn on an electrical appliance the voltage level in your house slightly decreases, letting current flow in from the utility. The intertie inverter carefully monitors the utility voltage level, and feeds its power into the grid at just slightly higher voltage. As long as the voltage level in your house is being maintained by the intertie inverter, no current flows in from the utility. You are displacing utility power with renewable power (in most cases solar power). If your house demands more current than the intertie system is delivering at the moment, utility power compensates seamlessly. If your house is using less current than the intertie system is delivering, the excess pushes out through your meter, turning it backwards, back into the grid. This treats the utility like a big 100% efficient battery. You sell it power during the day, it sells power back to you at night for the same price (assuming you have a PV system that is operating during the daytime). If you live in one of the majority of states that allow net metering, then everything goes in and out through a single residential meter. Very simple, and extremely encouraging to renewable energy producers.

A utility intertie system makes it possible to generate your own solar power and sell it back to your utility company. Your utility electric meter will run backwards anytime your renewable energy system is making more power than your house needs at the moment.

For an up-to-date summary of state net metering laws, see the Database of State Incentives for Renewable Energy website: http://dsireusa.org.

KEEP IT SAFE!

Utility companies are naturally concerned about any source that might be feeding power into their network. Is this power clean enough to sell to your neighbors? Are the frequency, voltage, and waveform within acceptable specifications? And most importantly, what happens if the utility power goes off? Utility companies take a very dim view of any power source that might send power back into the grid in the event of a utility power failure, as this could be a serious threat to power line repair workers.

All intertie inverters have precise output specifications, voltage limits, and multi-layered automatic shut-off protection features as required by UL specification 1741. This specification details all the necessary output and safety requirements as agreed upon by utility and industry representatives, and is a specific UL listing for this new family of intertie inverters. If the utility fails, the inverter will disconnect in no more than 30 milliseconds! Most utility companies agree this is a faster response than their repair crews usually will manage. Even an "island" situation will be detected and shut off within two minutes. An island happens (theoretically) when your little neighborhood is cut off, isolated, and just happens to require precisely the amount of energy that your intertie inverter is delivering at the moment. Safety has been the top concern during development of intertie technology. These inverters just can't cause any problems for the utility. The worst problem they could cause the homeowner is to shut off the intertie system temporarily because utility power drifts outside specifications.

Because small-scale intertie is a new and evolving development, some smaller utility companies may not yet have worked out their official policies and procedures. Many utilities simply follow the lead of Pacific Gas & Electric, the largest utility company in the United States. If PG&E approves a particular product or procedure, it's probably safe enough for the rest of us.

Be aware that not all electric meters will support net metering. Some meters are equipped with a ratchet that will only allow them to turn forward. Some of the newer digital remote-reading data collection meters simply count how many times the hash mark on the wheel goes by; they just assume it's moving forward. Also remember that not all states have net metering laws for small-scale renewable energy producers. Your utility company will make sure you're equipped with the right kind of meter. There's usually no charge if a change is required.

DO YOU NEED BATTERIES OR NOT?

Utility intertie inverters come in two basic configurations: those that support back-up batteries, and those that don't. This is a major fork in the road of system design, as the two system types operate very differently, require completely different equipment, and have big price differences. To help you choose wisely, read on.

Battery-based systems are basically large stand-alone renewable energy systems, using a sophisticated inverter that also can intertie with the utility. Several manufacturers offer battery-based intertie inverters. The Xantrex SW is the venerable choice, with almost ten years of experience. The SW-series inverters are large 2,500- to 5,500-watt units with 24- or 48-volt battery packs. OutBack's FX and VFX series is a newer offering with units from 2,000 to 3,600 watts and 24- or 48-volt battery packs. Beacon Power has a 5,000-watt all-in-one-box solution that's easy to install for residential customers. And in the near future we expect a battery-based unit from SMA, which will work along with their existing Sunny Boy series of direct-intertie inverters. All these inverters will allow selected circuits in your house to continue running on battery and solar power if utility power fails. They probably won't run your entire house, but they can keep the fridge, furnace, and a few lights going through an extended outage, and they'll do it automatically without noise or fuss. With inverter, safety, control, battery, and a minimal PV array, these systems start out at approximately $7,500. Average-size systems that will comfortably (partially) support a typical suburban home run about $20 to $25,000.

Direct-intertie systems are much simpler. They consist of PV modules, a mounting structure that holds the modules, the intertie inverter(s), a couple of safety switches, and a run of wiring that connects everything. Without batteries, less safety and control equipment is necessary and less physical space is required on-site. About 5% more of your PV power ends up as useful AC power, and more of your investment tends to end up buying PV power. But if utility power goes off, so does your intertie system, regardless of whether the sun is shining. Direct-intertie inverters come in modest to large sizes from 700 watts up to 6,000 watts, and are produced by an increasing number of companies, including Xantrex, SMA, and Sharp. Direct-intertie systems start at just under $5,000 for a system that delivers approximately 2.0 kilowatt hours per day. The average-size system we've sold over the past few years sells for about

If the utility fails, the inverter will disconnect in no more than 30 milliseconds! Most utility companies agree this is a faster response than their repair crews usually will manage.

Utility intertie inverters come in two basic configurations: those that support back-up batteries, and those that don't.

Looking out over the darkened neighborhood from inside your comfortable, well-lit home does wonders for your sense of well-being.

Real Goods has now solarized over 60,000 homes and businesses, more than any other company in the world.

$17,500 (before any rebates) and delivers approximately 9.5 kilowatt hours per day.

Which type of intertie system is right for you? That primarily depends on the reliability of your utility provider. If you live in an area where storms, poor maintenance, or mangled deregulation schemes tend to knock out power regularly, then a battery-based system is going to greatly increase your security and comfort level. Looking out over the darkened neighborhood from inside your comfortable, well-lit home does wonders for your sense of well-being. On the downside, adding batteries will raise your system cost by $3 to $5,000, and adds system components that will need some ongoing maintenance and eventual replacement. If utility power is reliable and well maintained in your area, then you have little incentive for a battery-based system. Those dollars could be better spent on PV wattage. You'll see more benefits and get to enjoy almost zero maintenance. Just be aware that if utility power does fail, your solar system will automatically shut off too for safety reasons.

REBATES CAN REDUCE YOUR TOTAL SYSTEM COST

Money is available out there for doing the right thing. A number of states have programs that will pay you an incentive in the form of a rebate, buydown, or grant to defray your renewable energy system costs. It's difficult to keep tabs on all of the rapidly changing programs in all of the states. The Database of State Incentives for Renewable Energy does a great job on its website: http://dsireusa.org. We also provide a list of State Energy Offices in the Appendix. Give yours a call to ask about net metering and money available for renewable energy projects.

California has a buydown program for intertie systems that our technicians are fully versed in, and we'll be happy to help you figure out how to navigate it. In the San Francisco/ North Bay and Los Angeles areas, we provide complete installation services. Our fully trained technicians have years of experience sizing these systems (1-800-919-2400). Real Goods has now solarized over 60,000 homes and businesses, more than any other company in the world.

DO IT FOR YOUR CHILDREN AND THE PLANET

Don't go into utility intertie expecting to cash in immediately. You won't. Intertie systems will pay for themselves in saved electric bills, but the payback point can be six to ten years or more depending on your electric rates, and if rebates are available. Still, that often translates into a return on investment of 10 to 20% annually, which is way better than the best bonds and CDs or other safe financial investments today. Most state intertie regulations won't let residential customers deliver more energy than their average electric use. The details vary from state to state, but if you make more power than you use, you'll either give it away, or sell it at wholesale rates. Don't convert your backyard into a PV farm just yet. Intertie is something you do because it feels good to be independent of the utility and to cover your own electricity needs directly with a clean, nonpolluting, renewable power source. It also feels good, of course, to increase the value of your home and make a healthy return on your investment. Solar modules last for decades, require almost no maintenance, and don't borrow from our kids' future, which is probably the best possible reason to invest in renewable energy.

Converting California Sunshine (or Anywhere Else's) into Electricity

All the Details on How You Can Solarize Your Home, Slash Your Electric Bill, and Spin Your Meter Backwards

We've talked a lot about how to solarize your home using clean, safe, reliable, and cost-effective power from the sun. It's easier than you think, and Real Goods is here to take care of all the details. Rebates and tax incentives currently available in California and some other states have made solar power affordable and cost effective, providing a 10% return on investment. There has never been a better time than now to go solar. In most cases, we can provide you with a system that will cost the same or less than what you currently pay for utility power—and in many cases solar will pay for itself almost instantly.

Real Goods has provided solar energy to over 60,000 homes and businesses in America since 1978. We are the world's oldest and largest supplier of renewable energy. Our customers are our first priority and we will take care of every detail, from the planning stages to the installation and deployment of your new solar energy system.

Depending upon the circumstances of individual installation, system payback times are as brief as five to ten years, with annual rates of return as high as 10 to 20%. The simple fact is that installing a solar electric system on your home is now cost effective for the first time. With your personal Real Goods Solar Expert walking you through every step of the process, it couldn't be easier. And you'll have peace of mind knowing that you've become part of the energy solution for future generations as we see our world's supply of oil continue to dwindle, and its price continue to rise.

Your Real Goods Solar Expert will come to your home (if you're in our California service territory) to review your electrical needs and your last twelve months of electric bills. He or she will then evaluate your home for solar potential, provide you with a financial analysis and payback period, and give you both a price quotation and installation time, should you decide to proceed. We'll even take care of the paperwork needed for rebates and tax credits.

TODAY'S SOLAR ECONOMICS— YOUR SYSTEM PAYS FOR ITSELF THE DAY IT'S INSTALLED!

At Real Goods, we've been selling solar since 1978, when the price was well over $300 per watt or more than $6 per kWh (kilowatt-hour). Today, with State of California incentives, that $6 per kWh cost has come down to around $0.12/kWh, a 98% drop! This means that you'll be guaranteeing yourself an electric rate of $0.10/kWh for the next thirty years— and we all know that over that time utility rates will be sharply increasing.

What do these lower prices and governmental incentives mean for the average homeowner? We frequently see 15% annual returns on investment (ROI)—far better than the stock market, bond market, money markets, and long-term CDs.

According to the National Appraisal Institute (*Appraisal Journal,* October 1999), your home's value increases $20 for every $1 reduction in annual utility bills. This means our 2.5 kW system increases your home's value by $17,520 while costing you only $16,000. You're $1,500 ahead from day one!

STATE AND FEDERAL INCENTIVES

In 1996, the State of California, through its California Energy Commission (CEC), decided to level the playing field for solar energy by providing cash rebates directly to customers who installed solar systems on their homes, provided the homes were connected to the utility company grid (because the money came from the utilities . . .).

Fetzer Vineyards, Hopland, California. "The installation of photovoltaic panels on the Fetzer Vineyards Administration Building marks another milestone in our continuing quest toward sustainability of our winery. We are especially pleased to have our neighbors and friends at Real Goods oversee the project."

—PAUL DOLAN,
PRESIDENT, FETZER VINEYARDS

The rebate provided to customers of Southern California Edison (SCE), Pacific Gas & Electric (PG&E), and San Diego Gas & Electric Company (SDG&E) is $3.00 per delivered watt, or about 35% of the total system cost. On January 1, 2005, the rebate goes down to $2.80 or less per watt.

In addition to this generous rebate, there is a further state tax credit of 7.5% of the cost of your renewable energy system (solar or wind), after rebates. This comes directly off of the amount of state income tax that you owe.

Other states, for example New York, have similar rebate and tax credit programs for renewable energy systems. The information we provide here is for California, but it serves as

California's Solar Incentives Made Easy: *Real Goods does it all for you!*

Based upon your projected installed system costs, Real Goods will maximize the rebate and tax credits you're eligible to receive on your new solar system. Our solar experts will fill out all the paperwork and collect the rebate on your behalf so you won't be responsible for any income tax. You only pay us the after-rebate price, without waiting for the rebate. In addition, we'll supply you with the paperwork for your 7.5% state tax credit.

Let Real Goods' Solar Experts
Solarize Your Home and Drastically
CUT Your Electric Bill
With Sharp
Solar Products
SHARP
be sharp

www.realgoods.com/calsolar
888.507.2561
California license no. 691556

Call 1-800-919-2400 for a free brochure.

an example of how similar incentives in your state may make the choice to go solar financially attractive. Check with your state energy office or a local renewable energy contractor to learn the details of your state's program. Then call Real Goods for further consultation.

Businesses can also get a 10% Federal Investment Tax Credit that can be carried forward fifteen years or back three tax years. The "MACRS" five-year accelerated depreciation schedule is also applicable. With these incentives, we have often seen paybacks as low as three years and internal rates of return in excess of 25%.

TIME-OF-USE METERING (TOU)

California's net metering laws (thirty-eight states have them now) mandate that your utility company cannot charge you more for your electricity than they pay you for the solar power you generate.

The physics of solar power make it the perfect energy source—it puts out the most power in mid-afternoon, which is exactly when California utilities have the greatest demand due to summer air conditioning. In order to decrease the summer mid-afternoon power load, the utilities have instituted "Time-of-Use" metering to encourage homeowners to conserve in the afternoon. This is great news for solar system owners!

Time of Use is an excellent choice for PV systems that meet a minimum of 50% of your electrical needs. This means you can sell your solar power to the utility between noon and 6 P.M. for as much as $0.325/kWh, and then you can buy that same power back at a rate as low as $0.095/kWh. This gives you huge financial benefit in the long run and also reduces the initial size of your solar system, and consequently your ultimate payback time.

YOUR PROPERTY TAX WON'T GO UP

A new law in California prohibits county property tax assessors from increasing your tax assessment

because of value increases from your installation of a solar system. This is great news for your cash flow, as you'll be saving hundreds or even thousands of dollars every year from your new solar system, increasing your property's resale value by $17,520 due to your utility bill savings, and not adding even one penny in property tax!

FINANCING YOUR SOLAR SYSTEM

Real Goods has researched numerous quality home equity lenders and other loan sources in California. If you choose not to self-fund your solar system, there are many cost-effective alternatives. Many lenders are open to financing solar systems for less than prime rate. Our favorites are solar homefinancing.com (800.216.0086) and solarfinancing.com, which fund home equity loans at prime less .05%. For many homeowners, this means instant positive cash flow.

HOW MUCH POWER CAN YOUR SOLAR SYSTEM PRODUCE?

If you've checked with any of our competitors, you'll see widely divergent opinions on how much real electricity a solar system can produce. Our industry is relatively new, so consistent standards of measurement haven't yet been adopted to level the playing field for everyone. In the meantime, all we can do is be completely honest with our customers about the true output they can expect from their solar systems.

STC or "Name Plate" Ratings: STC stands for Standard Test Conditions. It is the rated output in watts that the manufacturer puts on its photovoltaic (PV) modules under laboratory-perfect conditions.

PTC Ratings: PTC stands for Practical Test Conditions, or the ratings under the PVUSA Test Conditions. This is the standard used by the California Energy Commission, and in general runs about 17%-20% less than STC. PTC ratings generally are used by the solar industry, installers, our competitors, and the general public.

Real Life Expectations: Many industry professionals have studied the issue of photovoltaic ratings and are uncomfortable with both STC and PTC ratings because they seem over-rated to real-world conditions. Real Goods recently attended an excellent seminar at the national solar conference in Portland, Oregon, and the consensus was that to be conservative in your expectations, you should expect your solar system to yield in AC output to your electrical panel about 70% of STC (manufacturer's name plate) ratings (that is, multiply STC by 0.7).

In Summary: Throughout our literature, like all in the solar industry, we will use PTC ratings, but we recommend that you be conservative in your expectations and calculate your numbers for "real life" expectations. What this means is well summarized by the example at the right featuring a solar panel we commonly use for 2.5kW systems.

This is how a 3kW system at the outset becomes a 2kW system in actual electricity yield.	
STC rating	2,970 Watts
PTC rating	2,450 Watts
Real Life rating	2,080 Watts

HOW REAL GOODS' SOLAR PROFESSIONALS WORK TO SOLARIZE YOUR HOME

1. Feasibility and Cost/Benefit Analysis:

After you contact one of our solar experts, we go to work immediately to determine if solar is right for you. We ask you to provide us with:

- copies of your last twelve months of utility bills;
- a description or rough drawing of your home's directional orientation;
- a description of your roofing material, condition, and age;
- a description of your current electrical service and installation;
- your desire to be 100% energy independent, or what fraction thereof.

PROJECTED RATES OF RETURN ON YOUR 2.5KW RESIDENTIAL SOLAR SYSTEM

The following figures are calculated based on an electric bill "baseline" of 383 kWh per 30 days AND time-of-use (TOU) metering as previously described. Note: TOU metering may not be the best solar solution for all homes. Savings vary slightly from these numbers depending upon your utility company. These are averages for California.

If your monthly electric bill is usually:	A 2.5kW solar electric system will reduce it to:
$50	$0
$75	$20
$100	$40
$150	$78

If you choose to pay cash for your system, the payback period will range from six to thirteen years, depending upon how you value the discounted pre-tax cash flow and the anticipated inflation rate. Your return on investment will typically be 8 to 16%, far better than just about any investment available today … or you can choose to finance your solar system with a home equity loan at 4.2% (current rate from www.solarfinancing.com).

For electric bills in excess of $150 per month, contact one of our solar technicians at 888.212.5640 for a thoroug h financial payback analysis. Solar systems for homes with large utility bills are always cost-effective and have shorter payback times and higher returns on investment due to the higher base electricity rates.

THE BOTTOM LINE

Total installed cost of 2.5kW system (incl. sales tax) before rebate and tax credit	$25,031.34
CEC rebate ($3.00/W)	($7,350.00)
Amount paid to Real Goods after installation completed	$17,681.34
State tax credit (7.5% of post-rebate installed cost)	($1,326.10)
Total installed cost (incl. sales tax) after rebate and tax credit	$16,355.24
Estimated monthly cost based on PTC (CEC) ratings:	
Electricity cost savings per month (@ 2.45kW @ $0.185 @ 5.5 hrs/day for 30.4 days/month)	$75.78
Monthly payment on financing $16,355.00 for 30 years @ 4.2%	$79.00
Total Monthly Cost for Electricity	$3.22

Note: Again, the above pricing model does not include your potential benefits for time-of-use (TOU) metering, which could provide significant savings and quicker payback.

2.5KW RESIDENTIAL SOLAR SYSTEM COMPONENTS

Component	Qty
2,900 watts of solar electric panels	2.9 kW
SMA Sunny Boy 2500 DC-to-AC inverter with metering	1
Solar panel mounting kits and roof standoffs	4
Utility-required safety disconnect switch w/fuses	1
High-voltage solar array disconnect switch	1
Delta lightning arrestor	1
Weatherproof output cable (50 ft., male-female)	2–3
Splice Kits	2

YOUR "GRID CONNECTED" SOLAR ELECTRIC SYSTEM

1 Solar panels convert sunlight to electricity.

2 Inverter conditions electricity for use.

3 Solar electricity powers your home!

4 Surplus electricity is sold to your utility.

Next, our technician will discuss with you the possibility of making a personal visit to your home to assess your location, access to solar gain, and your electrical service. This service is provided for a $50 assessment fee to cover our expenses, refundable upon purchase. We'll get back to you with:

- a comprehensive recommendation of which solar system is right for your situation;
- a price quotation that includes all equipment, delivery to your home, building permit fees, installation, building inspections, and amount of rebate available from the state or your utility;
- a recommendation on whether time-of-use (TOU) metering will work for you;
- a financial analysis of return on investment for your system, years to payback, and internal rate of return, including any state or federal tax credits or incentives available to you;
- if you choose not to purchase the solar system outright, we'll provide a variety of financing options.

2. Contract Signing, Procurement, and Installation Scheduling: When you have made a decision to proceed with a solar installation, after personal consultation in your home, the next steps are:

- sign our installation contract and provide us with a $1,000 deposit;
- our solar experts will submit your application to the California Energy Commission or other relevant agency, qualifying you for the maximum rebate;
- procure a building permit. You will have approximately seven months to complete your installation before the CEC rebate runs out.

- if you are building a new home, additional information will be needed by our solar technician.

3. Installation of your Solar System: Our professionals will typically install your complete system in pne to three days.

- 50% of the payment is due the day we begin installation. The balance is due upon completion of the installation. We will collect the rebate on your behalf so you won't be responsible for any income tax. You only pay us the after-rebate price, without waiting for the rebate.
- Upon completion and payment, we'll schedule an inspection with the local building inspector and utility for final approval.
- You'll be making your own clean solar power and contributing to the energy solution!

UTILITY INTERTIE INVERTERS
Straight from the Sun to Your Utility Meter

PV modules that produce conventional AC power have long been a dream of the renewable energy industry. Small individual inverters that delivered AC power output from each PV module were introduced in the mid-1990s, turned out to be stunningly expensive, and have since disappeared in the United States. It costs a lot more to build many small inverters rather than a single large one. Using some reasonable economies of scale, we can offer several cost-effective ways to connect PV modules directly to your household AC system.

Trace/Xantrex provided the first generation of utility intertie inverters with their large multi-tasking sine wave series in the mid-1990s. The original full-featured SW-series can intertie and support emergency back-up batteries. The ability to intertie and provide battery backup is now shared by the new generation of highly adaptable OutBack FX- and VFX-series inverters, by the new Beacon M5 inverter, and other offerings are due to debut in the near future. This type of fairly large inverter is the proper choice if you absolutely must have backup batteries due to unreliable utility power, but they're expensive for smaller arrays, and the batteries with their associated control and safety equipment add considerable cost.

Direct-intertie inverters don't use batteries at all. They use the utility grid as if it was a bottomless, perfectly efficient battery. These are the best choice for systems that don't need back-up power. Incoming PV power is converted to conventional AC power, and fed to your household breaker panel, displacing utility use. When there's excess power, your meter runs backwards, giving you retail credit; when there's a deficit, power is purchased, running the meter forward again. This back and forth energy trading runs on a one-to-one basis, perfect efficiency, which sure beats batteries. Xantrex, SMA, Sharp, Fronius, and an increasing number of others make direct-intertie inverters.

An additional advantage of direct-intertie inverters is the high energy efficiency of the system. PV modules are controlled with power point tracking, which allows the modules to run at their maximum power output, rather than a lower battery-charging voltage. Batteries are eliminated, which typically costs 5 to 20% of charging power; and all transmission is done at high voltage, allowing smaller wire sizes, and less loss.

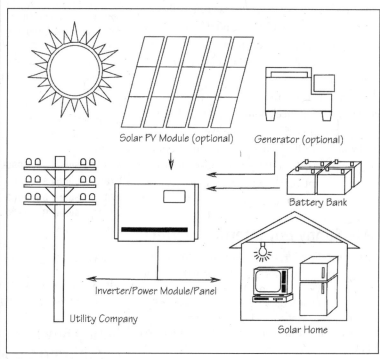

Typical back-up power grid interactive system.

A multiple number of intertie inverters can be paralleled for system expansion. Direct-intertie inverters will not operate without an AC power source to feed back into, and will not charge batteries. All inverters designed for utility intertie, whether they're battery-base or direct-intertie are UL-1741 listed.

Some states and utility companies are offering rebate or buydown programs for intertie systems. Please call your local utility or our technical staff for details.

UTILITY DIRECT-INTERTIE INVERTERS (Without Batteries)

SMA (SUNNY BOY) INVERTERS

The world's most experienced utility-intertie inverter manufacturer

SMA is a high-quality German manufacturer with more than 80,000 utility intertie inverters in service since 1996. The direct-intertie Sunny Boy series is their third-generation design. These are high-voltage input inverters that use one to three series-strings of PV modules. This eliminates the need for combiner boxes or large wire, and greatly eases long-distance transmission. Maximum power point tracking delivers the highest possible wattage under all conditions. Maximum inverter efficiency and safety have been the primary design goals during development of all SMA products.

All Sunny Boy inverters feature powder-coated, stainless steel, outdoor-rated, NEMA 4 cases that can be mounted inside or outside, and will deliver full-power operation at temperatures up to 60°C (140°F). Ground-fault protection is built into each inverter, and five-year warranties are standard.

SMA Sunny Boy Intertie Inverters

Direct-intertie with German engineering

Sunny Boy now offers several closely related models at 700, 1,100, 1,800, 2,500, and 6,000 watts. All models feature the best of high-quality German engineering. High efficiency, redundant processors, easy data communications, maximum power point tracking, an excellent reliability record, and powder-coated, stainless steel, outdoor-rated boxes are only a few of the quality design features. Outdoor installation in a shady location is preferred.

Sunny Boy inverters require high-voltage series-strings of PV modules. The number of modules varies with inverter model, PV modules used, climate in which they'll be installed, and the owner's budget and/or output requirements. Smaller inverters require shorter, lower-voltage, PV strings. Bigger inverters require longer strings. Call for details.

Output is delivered to your household main breaker panel by feeding through a dedicated circuit breaker. Conversion efficiency is an excellent 93.5 to 94%. Internal power use is less than 7 watts during operation, less than 0.1 watts during standby, and the inverter will disconnect from the utility at night to eliminate any phantom load.

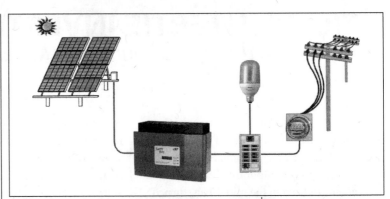

The display option is one of the slickest in the inverter industry. The 2-line, 40-character backlit display is acoustically activated by knocking on the lid and scrolls through: Instantaneous AC Watts; DC Volts; Daily kWh; Total kWh (since the inverter was turned on); Operating Hours; and Inverter Status. Several optional remote monitoring and professional data acquisition options are also available; please call for details.

All Sunny Boy installations will need a high-voltage, DC-rated raintight disconnect. Up to three Sunny Boy inverters can be connected to the high-voltage DC disconnect listed below. The lightning protection is optional, but highly recommended. All Sunny Boy inverters are warranted by the manufacturer for 5 years. Germany.

Warning! We strongly recommend that high-voltage PV systems, as used by the Sunny Boy and Sharp inverter series, only be installed by experienced and/or licensed electricians.

49-0168	**Sunny Boy 6000U w/Display**	**$6,155**
27857	**Sunny Boy 2500U w/Display**	**$2,845**
27856	**Sunny Boy 2500U Standard**	**$2,710**
27904	**Sunny Boy 1800U w/Display**	**$2,437**
27903	**Sunny Boy 1800U Standard**	**$2,330**
49-0116	**Sunny Boy 1100U w/Display**	**$2,098**
49-0117	**Sunny Boy 700U w/Display**	**$1,828**
24648	**Hi-Volt DC Disconnect**	**$165**
25775	**Hi-Volt Lightning Protector**	**$40**

Sunny Boy Installation and Monitoring Options

The Sunny Breeze fan clips to the top of the heat-sink fins, provides thermostatic-controlled cooling as needed, and includes an AC power supply. It's needed with indoor installations, and some heavily loaded outdoor installations.

Please call for help with remote monitoring. There are many options.

49-0107	**Sunny Breeze Fan Option**	**$150**
24687	**Sunny Boy GFI Fuse, 1A/600V**	**$19**
27902	**Sunny Boy RS-232 Comm. Module**	**$158**
49-0118	**Sunny Boy RS-485 Comm. Module**	**$158**
27858	**Sunny Boy Control**	**$549**
24710	**On-Board Digital Display (after purch)** (specify inverter model & color)	**$317**

Sharp Sunvista Intertie Inverter
Direct-intertie with Japanese engineering

The Sharp intertie inverter has proven itself over several years and tens of thousands of Japanese installations.

We've been selling it in North America for a year now and have had excellent results. Offering up to 3,500 watts of output, with maximum power point tracking, the Sharp is unique with three separate input channels. Each channel operates independently, making the most of the array attached to that particular input. PV arrays can face different directions, with different numbers or even brands or sizes of PV modules to make the best use of available space, and it will still harvest maximum output from each module. All modules within a single input channel must match and face the same direction. This inverter also lends itself to starting with a smaller system with lots of room for future expansion. With most 24 volt-nominal modules, strings of 5 to 7 modules are acceptable. With 12 volt-nominal modules, strings of 9 to 14 are usually acceptable.

The light gray Nema 3R case can be mounted indoors or outdoors. Inverter output lands on a 20A/240V circuit breaker in your main breaker panel. It has automatic active cooling fans to allow full power output under very hot conditions up to 105°F. During even hotter conditions, it will limit power output to the most efficient setting possible without overheating. An interior wall-mounted LCD display is included. It shows instantaneous output, cumulative output for one year, and total CO_2 abatement A great feature for bragging rights.

UL and IEEE approved. Requires a high-voltage DC disconnect between PV arrays and inverter for electric code compliance. Lightning protection is optional, but recommended. Size is 23.6"W x 17.7"H x 7.6"D, weight 62 lb. Five-year manufacturer's warranty. Japan.

49-0108	**Sharp 3500 Sunvista Inverter**	**$3,500**
24648	**Hi-Volt DC Disconnect**	**$165**
25775	**Hi-Volt Lightning Protector**	**$40**

Fronius Intertie Inverters

Fronius is one of the largest suppliers of direct-intertie inverters in Europe for good reason. They feature advanced high-frequency technology that delivers high efficiency, precision maximum power point tracking, and a lighter-weight design. Active cooling allows full power from -4° to 122°F, and a wide voltage input range of 150 to 450 volts per-

mits highly flexible array sizing. The backlit LCD display is standard, and shows instant AC output, daily output, output since installation, CO_2 emissions avoided, and much more, including fault displays for quick repairs. The integrated DC and AC disconnects mean a clean installation with less hardware on the wall. The NEMA 3R enclosure can be outside or inside mounted. Expansion slots allow future upgrades or remote displays. There are data collection options for those who want to connect to their PC. The IG2000 model delivers a nominal 1,800 watts of output, the IG3000 delivers 2500 watts. Both are 240 volt and land on a 2-pole 15A circuit breaker. UL 1741 approved. 18.5" x 16.5" x 8.8"; weight 26 lb. Five-year manufacturer's warranty. Germany.

49-0149	**Fronius IG3000 Inverter**	**$3,575**
49-0150	**Fronius IG2000 Inverter**	**$2,975**

DIRECT-INTERTIE INSTALLATION ACCESSORIES

MC Cable Extensions

Most PV modules used for intertie now are being supplied with the quick-connect MC connectors for fast series connections. Sometimes you need some extra length, without switching to standard wiring and back. Our MC extension cables

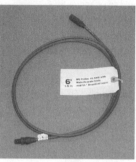

make it easy, with a male connector on one end and a female on the opposite end. We almost always use one to get from the far end of the array back to the wiring J-box. USA.

26662	PV MC-Type Output Cable 6' M/F	$18.00
26666	PV MC-Type Output Cable 15' M/F	$20.00
26663	PV MC-Type Output Cable 25' M/F	$23.00
26667	PV MC-Type Output Cable 30' M/F	$25.00
26664	PV MC-Type Output Cable 50' M/F	$28.00
26665	PV MC-Type Output Cable 100' M/F	$52.00

Distribution Block

Our 3-pole distribution block is handy for joining multiple PV strings before making the long run to the DC disconnect and inverter.

3-pole for positive, negative, and ground. Accepts one #14 to #2/0 wire on the primary side, and up to four #14 to #4 wires on the secondary side. Up to two of these distribution blocks will fit nicely in the raintight 10" x 8" x 4" J-box listed below.

24705	3-Pole Distribution Block	$39
24213	J-Box 10x8x4 R/T	$46

AC Combiner Panel

The AC Combiner Panel is necessary when you have more than one direct-intertie inverter on your house.

We need to combine all the power outputs before passing through the AC disconnect. The utility requires there only be one disconnect. Combiner is raintight for exposed outdoor use. Use a 15A/240V breaker for each Sunny Boy 2500, a 20A/120V for each Sunny Boy 1800. A pair of Sunny Boy 1100 or 700 models can land on each 15A/240V breaker. Combiner will accept up to four 240V, or up to eight 120V breakers (or combinations).

48-0018	AC 4/8 Raintite Combiner Panel	$87
24707	15A/240V Circuit Breaker	$16.50
24709	20A/120V Circuit Breaker	$8.50

AC Disconnect

Some unenlightened utilities require a lockable disconnect between your intertie inverter(s) and the main panel. This is not an electric code requirement. It allows your friendly local lineman to indulge his paranoia by positively locking your inverter off during nearby repairs. These disconnects are both raintight with visible blades and a handle that's lockable in the off position. The smaller 30A disconnect requires two fuses. USA.

24644	30A/2-Pole Fused Disconnect, Raintight (Square D brand # D221NRB)	$66
24501	30A FRNR30 Fuse	$5
24229	60A/3-pole Unfused Disconnect, Raintight (GE brand # TGN3322R)	$159

UTILITY INTERTIE INVERTERS WITH BACK-UP ABILITY

If you want back-up ability from your solar electric system, you'll need batteries. How big those batteries are depends on how much you want to run, and for how long. For some folks, a few hours would be just fine. Three days is probably about the practical limit. For longer outages, a back-up generator is the answer. Run a couple days. Start the generator. Recharge. Run a couple more days . . . repeat as needed. Talk with one of our tech staff to size your battery pack.

In all cases, we strongly recommend sealed batteries for intertie systems, partly because they're likely to be forgotten and won't get regular watering, and partly because sealed batteries are going to be a lot happier about floating along for weeks, months, and years between discharge cycles. Sealed batteries have their chemistry tweaked for "set it and forget it" service, and you'll get longer, more dependable life from them. And you'll never need to kick yourself for forgetting to water them.

There are at least three manufacturers with an inverter that's smart enough to manage utility intertie and a back-up battery pack. (And maybe more by the time you read this.) Let's give a quick introduction to the known prospects, with an honest review of their good and bad points. In chronological order . . .

TRACE/XANTREX

Trace started the whole business of small-scale everyday folks selling excess renewable power back to the utility in the mid-1990s with the SW inverter series. In the late 1990s, Trace was purchased by Xantrex and before long much of the engineering staff left to form OutBack Engineering. This wasn't a happy time for Xantrex products. Now, in 2004, Xantrex is fully staffed, and has been rolling out a nice upgrade to the well-proven SW series, called the SW-Plus. The shocker here is that Xantrex has announced that the SWP series will not have a grid-tie option! Apparently the original SW-series will still be produced, and will continue to be Xantrex's battery-based grid-tie option. On the plus side, this is equipment that is well known, proven, and understood by most installers and inspectors. On the negative side, the SW inverter is more expensive than the new SWP. The SW has a less-than-perfect multi-stepped approxi-mation of a sine wave that works fine for 98% of the AC hardware out there, but occasionally we run into a gizmo that just isn't happy with this waveform. Grid-tie requires a $500 extra-cost GTI option, and the single most annoying feature is that the SW holds all field programming in volatile memory. If the inverter loses battery power, during a battery swap or even just routine maintenance, it forgets all charging voltages, amperages, generator starting procedures. It even forgets to sell excess power to the utility! Very annoying.

Xantrex systems are basically a la carte packages. Although they usually are purchased as pre-assembled Power Panels, there are separate boxes for inverters, DC disconnect, charge control, AC disconnect, grid-tie interface, and possibly others that need to be wired into a whole. This is a techie product, and techies hate to leave any possible function behind.

Xantrex inverters and options are fully covered in the Inverters section of this chapter.

OUTBACK ENGINEERING

OutBack formed when Xantrex bought Trace and much of the engineering staff decided they didn't enjoy working for a big company. These folks probably have more experience with residential inverters than any engineering team on Earth. The OutBack inverter is a clean-sheet-of-paper next-generation product. As such, it has many good and a few bad points. On the plus side, it delivers an absolutely perfect sine wave, and does it quietly, from a package that's about half the size and weight of the Xantrex SWP. Access for repairs in the field is easy on this inverter, and components are bundled into just a few major pieces.

Much of the software is outside the inverter on the small, light, easily returned and upgraded Mate. If you have a problem, a call to the manufacturer likely will put you in touch with the engineer who built it. The level of service from these folks is really terrific. On the negative side OutBack is an engineer-owned and operated company, and engineers will *never* stop developing their product until someone takes it away. So there will always be new capabilities and functions on next year's model. These will probably always be developing products, and although OutBack has the most extensive beta-testing programs our industry has ever seen, we still find occasional bugs.

OutBack systems are also a la carte packages, with several separate boxes for inverters, DC, and AC centers, and the like, and are available as pre-assembled Power Centers. Hey, it's a techie product, after all.

OutBack inverters and options are fully covered in the Inverters section of this chapter.

Beacon Power

The Beacon M5 inverter is a fully integrated one-box package for utility intertie with battery backup. This is the non-techie solution when you don't want to play with it, or modify it, or have a raft of options to sort through. You just want it to work well, and take a minimum of space. The inverter, charge controller, DC and AC overcurrent protection, and all switchgear is housed in a single outdoor-rated enclosure. On the plus, side the M5 is tidy, compact, relatively quick to install, and delivers a big healthy 5,000 watts. On the negative side, it's 120-volt output only; it won't run 240-volt appliances. But then you shouldn't be running any 240v appliances during a utility outage anyway. On first glance, the M5 seems expensive. Actually, because all the bits and pieces are included, the M5 tends to cost less for a complete system in the 5,000-watt range.

Beacon M5 Inverter

BATTERY-BASED INTERTIE PRODUCTS

Doug Pratt with the Beacon M5 Inverter powering the Real Goods Solar Living Center.

Beacon Power M5 Inverter

The single box intertie solution

In 2004, Beacon Power introduced a new concept to the large household inverter market. The M5 is a single, compact outdoor-rated unit that handles all the necessary functions for solar-electric intertie with battery backup. Rated at a big healthy 5,000 watts, the M5 converts any incoming PV energy to conventional AC power and uses it to displace utility power. If more PV energy is available than your house can use, the extra is pushed back through your meter, giving you credit toward later use. The M5 provides three ground-fault protected 50-amp PV input breakers with state-of-the-art Maximum Power Point Tracking control of the PV array to extract the maximum amount of energy at any moment. If the utility fails, the M5 instantly switches to battery power. It will easily keep the fridge, furnace, computer, small kitchen appliances, household lights, and entertainment equipment running during utility outages.

The M5 has an easy hang-on-the-wall mount with intelligent wiring interfaces that speed and simplify installation. The 50A, 120V output will run a separate electrical subpanel for all the circuits you need to back up.

For true set-it-and-forget-it operation the M5 shows it's working with a simple set of LED indicators. For those who want to play with it a bit, the optional PC-based Smart Power Monitor lets you track a range of performance parameters plus total PV power generated, using common Ethernet plug and cable. Field reprogrammable setpoints provide complete system adjustment and control.

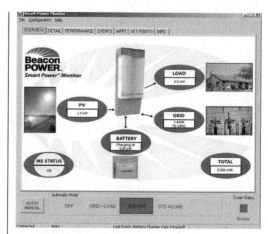

Home screen for the optional monitor software.

A complete system will require a 48-volt PV array with string combiners, a 48-volt sealed battery pack, 2/0 battery cables, and a 175A battery breaker. The M5 inverter is UL 1741 certified, measures 42"H x 16"W x 10"D, weighs 120 lb., and has a 5-yr. mfr.'s warranty. USA.

BEACON POWER M5 SPECIFICATIONS
Continuous output: 5,000 VA
Surge amps: 62.5 amps
DC voltage: 48 vdc
Charging rate: No AC charger
Preferred circuit breaker: 175 amp
DC cable size: 2/0 cu
Weight: 120 lb.

49-0151	Beacon Power M5 Inverter	$6,995
27098	2/0, 10-ft Inverter Cables, 2-lug pair	$140
48-0186	Beacon 175A Breaker Assembly	$250
49-0152	Beacon Smart Power Monitor software	$249

Sealed Batteries for Utility Intertie Systems

Solar Gel Batteries

These true deep-cycle gel batteries come highly recommended by the National Renewable Energy Labs. They are completely maintenance-free, with no spills or

fumes, and can be operated in any position. Because the electrolyte is gelled, no stratification occurs, and no equalization charging is required.

Self-discharge is under 2% per month. Because of thinner plate construction, gel batteries can be charged or discharged more rapidly than other deep-cycle batteries. All Solar Gel batteries now come with secure, bolt-up, flag-type terminals.

To avoid gassing, gel batteries must be charged at lower peak voltages; typically, 13.8 to 14.2 volts maximum is recommended. Higher-charging voltages will drastically shorten life expectancy. Life expectancy depends on battery size and use in service, but no sealed battery will perform better or longer than these. Typical life expectancy is five years or more. Amp/hour rating at standard 20-hour rate. USA.

15-210	**8GU1 12V 32Ah** (24.2 lbs., 7.75" x 5.2" x 7.25")	**$69**
15-211	**8G22NF 12V 51Ah** (37.6 lbs., 9.5" x 5.5" x 9.25")	**$119**
15-212	**8G24 12V 74Ah** (53.6 lbs., 10.8" x 6.75" x 9.8")	**$149**
15-213	**8G27 12V 86Ah** (63.2 lbs., 12.75" x 6.75" x 9.8")	**$159**
15-214	**8G31 12V 98Ah** (71.7 lbs., 12.9" x 6.75" x 9.4")	**$189**
15-215	**8G4D 12V 183Ah** (129.8 lbs., 20.75" x 8.5" x 10")	**$339***
15-216	**8G8D 12V 225Ah** (160.8 lbs., 20.75" x 11" x 10")	**$399***
15-217	**8GGC2 6V 180Ah** (68.4 lbs., 10.25" x 7.2" x 10.8")	**$189***

*Shipped via truck.

Hawker Envirolink Sealed Industrial Battery Packs

The best choice for emergency power backup systems

The Hawker Envirolink series offers true gel construction, not the cheaper, lower life-expectancy AGM construction. These batteries are the top-quality choice for long-life, low-maintenance emergency power back-up systems. No routine maintenance is required, and this battery type thrives on float service. Life expectancy is 10 to 15 years with proper care and a charger/controller that keeps voltage below 2.35v/cell. Delivered in pre-assembled 12- or 24-volt packs with smooth steel cases and no exposed electrical contacts. Cycle life expectancy is 1250 cycles to 80% depth of discharge. Interconnect cables are only needed between 12- or 24-volt trays; cell interconnects are soldered in place at the factory. Five-year manufacturer's warranty. USA.

Hawker Envirolink Sealed Industrial 12-Volt Battery Packs

Item #	Ah @ 20hrs	Weight lbs.	Size (") L x W x H	Price
17-0296	369	390	26.3 x 6.5 x 23	**$1,530**
15819	553	564	31 x 7.75 x 23	**$2,050**
15820	925	885	31.13 x 13 x 23	**$3,270**
15821	1,110	1,050	29.3 x 13 x 23	**$3,865**
15822	1,480	1,374	25.7 x 19.6 x 23	**$5,000**

Hawker Envirolink Sealed Industrial 24-Volt Battery Packs

Item #	Ah @ 20hrs	Weight lbs.	Size (") L x W x H	Price
17-0297	369	780	25.7 x 11 x 23	**$2,895**
15823	553	1,128	31 x 13 x 23	**$3,865**
15825	925	1,770	38.5 x 16.6 x 23	**$6,175**
15826	1,110	2,100	38.7 x 19.7 x 23	**$7,280**
15827	1,480	2,748	38.7 x 25.6 x 23	**$9,420**

COMPLETE INTERTIE SYSTEMS

Typical 2,500-Watt Residential Intertie System

Almost every intertie system is individual. Available mounting space and style vary, power needs and budgets vary, and we're happy to put together just what you need. But for information purposes, here's what an average-sized system consists of. This one is using the 2,500-watt Sunny Boy inverter to full potential, is supplied with flush roof-top mounts, the optional display for the Sunny Boy, an optional lightning arrestor, and the lockable disconnect required by some utilities. Installation and permit costs are estimates based on Northern California averages. Wiring is supplied locally as needed to get from roof to inverter to main breaker panel.

Based on the North American yearly average of 5.5 hours noon-equivalent sunlight per day, this system will deliver approximately 10.7 kWh/day. Rebates and tax credits vary. In California (where our residential techs design, provide, and install several hundred turn-key solar systems every year), the current rebate is $3/watt and a 7.5% state tax credit is available that together reduce the bottom line below by around 35% off the top.

Item #	Quantity	Description	Retail	Extended price
11612	18	SHARP NE-Q5E2U 165 WATT MODULE 24 VOLT	$ 825.00	$ 14,850.00
27857	1	SUNNY BOY 2500 W/DISPLAY	2,845.00	2,845.00
14362	2	UNIRAC SOLARMOUNT SM/144	202.00	404.00
14365	2	UNIRAC SOLARMOUNT SM/180	238.00	476.00
14372 F	2	UNIRAC CT4 SOLARMOUNT TOP MOUNT CLAMP SET F	24.00	48.00
14373 F	2	UNIRAC CT5 SOLARMOUNT TOP MOUNT CLAMP SET F	28.00	56.00
24644	1	2 POLE SAFETY DISCONNECT 30 AMP RAINTIGHT	66.00	66.00
24501	2	30 AMP RK5 FUSE	5.00	10.00
24648	1	SAFETY DISCONNECT, 30 AMP, 600 V., UNFUSED, R/T	165.00	165.00
25775	1	DELTA LIGHTNING ARRESTOR, 440-650 VOLT	40.00	40.00
26664	2	SHARP PV OUTPUT CABLE 50 FT. MALE/FEMALE	28.00	56.00
570101	1	LABEL WARNING ELECTRIC SHOCK HAZARD	2.20	2.20
13666	2	UNIRAC SPLICE KIT	16.00	32.00

System component subtotal	$ 19,050.20
Shipping	$ 300.00
Installation including permit fees and cost to pull permit (tax exempt)	$ 4,300.00
Total installed system price	$ 23,650.20

Fully Integrated Controls or Pre-Assembled Power Centers

Fully integrated controls, otherwise known as power centers, are how most renewable energy systems go out the door now. A power center puts the inverter(s), DC protection/distribution, charge controller(s), monitoring, and AC protection/distribution hardware all in one tidy, pre-wired, UL-approved package. You get *ONE* component to hang on the wall. The alternative is to buy all the bits and pieces a la carte, hang them on the wall, and run all the wiring between them. If you're good (and/or lucky), it will only take half a dozen trips to the hardware store for small parts, but you will save yourself a couple hundred dollars. Most of the remote home power systems we've sold in the past few years have been built around one of these integrated control packages.

Intelligent manufacturers have taken the safety, charge control, and monitoring functions and combined them onto one neat, fully assembled and wired, UL-approved board, or enclosed box. The technical staff at Real Goods endorses this concept wholeheartedly, as it generally results in safer, faster, better-looking installations that rarely, if ever, cause problems with building inspectors. Most integrated controls are easily adaptable for diverse system demands, and expandable in the future. So now, instead of dealing with lots of bits and pieces that you (or your paid electrician) have to wire together on site, you can have it all delivered pre-wired.

Power centers are partially custom assembled to match the particular needs of a renewable energy system. Everybody needs theirs set up a little differently, so these are usually built to order. You must communicate with one of our technical staff to order an integrated control. If there's anything better, cheaper, or safer for your application, we'll let you know during this interview period.

At this time, Xantrex and OutBack both are offering various Power Center assemblies with a wealth of options.

A typical residential power center using OutBack products

> Fully integrated controls, otherwise known as Power Centers, are how most renewable energy systems go out the door now.

OutBack Power Systems: Light Years Ahead

Fighting aliens isn't part of the job for Christopher Freitas and Robin Gudgel pictured with fellow Out-Back engineers. But their inverters are out of this world.

Since 2001, the team at OutBack Power Systems in Arlington, Washington, has been designing and building the finest power electronics in the world—made in the United States and competitive in the world market. With 28 employees and 400% growth last year, this young and eclectic engineering-driven, employee-owned company has a good thing going.

LIGHTENING UP

The OutBack team's sense of humor (their Men in Black marketing campaign speaks volumes) is topped only by its revolutionary designs, cultivated partly before they saw the light: Robin and Greg came to know the importance of tight specs and reliability while designing power supplies for nuclear-tipped cruise missiles.

NEXT STOP: LATE NIGHT

Christopher (who disdains titles, but reluctantly calls himself Marketing VP) has been asked so many times why the company is called OutBack that he wrote a David Letterman–style Top 10 List. Our favorite: "It doesn't start with an 'X' or sound like a drug."

OutBack Power Systems—fighting for your energy independence

PRE-ASSEMBLED POWER CENTERS

OUTBACK-BASED POWER CENTERS

*Pre-assembled, pre-wired, UL-approved, and
ready to hang on the wall*

This is how most residential and commercial systems are
installed now, even by the pros. Let a specialist with immediate access to all the small bits and pieces do the major component assembly. It costs a little more initially, but saves big-time
in on-site labor and trips to the hardware store.

Using quality OutBack components, each power center is
assembled in an ETL-approved shop. Each custom-built panel
is pre-wired, tested, and ready to go. You simply add batteries,
charging source(s), and output wiring on site. All power centers come with DC and AC enclosures, the inverter(s), controllers, and other options of your choice, on a sturdy Out-
Back mounting plate.

Power centers come in two basic sizes: the original large mounting
plate with full-size AC and DC enclosures and space for up to four Out-
Back inverters; and the new half-size rack with smaller AC and DC
enclosures and space for up to two OutBack inverters. The half-size racks
are usually more than adequate for most residential uses, and naturally
cost less. In addition to half- and full-size racks, OutBack inverters come
in sealed or vented models. Vented models are for indoor installations,
and are the right choice for most installations. Sealed models are for
marine, jungle, bug-infested, or other challenging sites. So four power
center styles are listed below.

Options? We have lots of options! All the controls, breakers, transformers, GFIs, and more, that OutBack makes, plus a few things they
don't make are available assembled on power centers. We've listed the
more common basic racks and options below; this is not a complete listing. Please work with one of our experienced technicians to specify a
panel that fits your current and future needs. Prices run from $3,200 to
$12,000 depending on size and options. Proudly made in USA.

See the Inverter section for a complete introduction to OutBack inverters.

A Full Rack with four inverters mounted.

A Half Rack with two inverters mounted.

Standard OutBack
Vented Half-Rack Power Centers

Base price for standard centers

All 20"H x 44"W x 12.75"D. Weight 145 lb. w/one
inverter, 215 lb. w/two inverters (2).

49-0119 OutBack Half-Rack Vented	
Power Center VFX2812	**$3,740**
49-0120 OutBack Half-Rack Vented	
Power Center VFX2812 (2)	**$6,400**
49-0121 OutBack Half-Rack Vented	
Power Center VFX3524	**$3,740**
49-0122 OutBack Half-Rack Vented	
Power Center VFX3524 (2)	**$6,740**
49-0123 OutBack Half-Rack Vented	
Power Center VFX3648	**$3,725**
49-0124 OutBack Half-Rack Vented	
Power Center VFX3648 (2)	**$6,555**

Standard OutBack
Sealed Half-Rack Power Centers

Base price for standard centers

All 20"H x 44"W x 12.75"D. Weight 145 lb. w/one
inverter, 215 lb. w/two inverters (2).

49-0138 OutBack Half-Rack Sealed	
Power Center FX2024	**$3,185**
49-0139 OutBack Half-Rack Sealed	
Power Center FX2024(2)	**$5,295**
49-0140 OutBack Half-Rack Sealed	
Power Center FX2548	**$3,560**
49-0141 OutBack Half-Rack Sealed	
Power Center FX2548(2)	**$6,030**

Standard OutBack
<u>Vented Full</u>-Rack Power Centers

Base price for standard centers

All 36"H x 50"W x 12.75"D. Weight 260 lb. w/two inverters (2); add add'l 70 lb. per inverter.

49-0153	**OutBack Full-Rack Vented Power Center VFX2812 (2)**	**$6,715**
49-0154	**OutBack Full-Rack Vented Power Center VFX2812 (3)**	**$9,525**
49-0155	**OutBack Full-Rack Vented Power Center VFX2812 (4)**	**$12,275**
49-0156	**OutBack Full-Rack Vented Power Center VFX3524 (2)**	**$6,715**
49-0157	**OutBack Full-Rack Vented Power Center VFX3524 (3)**	**$9,525**
49-0158	**OutBack Full-Rack Vented Power Center VFX3524 (4)**	**$12,275**
49-0159	**OutBack Full-Rack Vented Power Center VFX3648 (2)**	**$6,655**
49-0160	**OutBack Full-Rack Vented Power Center VFX3648 (3)**	**$9,425**
49-0161	**OutBack Full-Rack Vented Power Center VFX3648 (4)**	**$12,160**

Standard OutBack
<u>Sealed Full</u>-Rack Power Centers

Base price for standard centers

All 36"H x 50"W x 12.75"D. Weighst 260 lb. w/two inverters(2); add add'l 70 lb. per inverter.

49-0162	**OutBack Full-Rack Sealed Power Center FX2024 (2)**	**$5,595**
49-0163	**OutBack Full-Rack Sealed Power Center FX2024 (3)**	**$7,840**
49-0164	**OutBack Full-Rack Sealed Power Center FX2024 (4)**	**$10,050**
49-0165	**OutBack Full-Rack Sealed Power Center FX2548 (2)**	**$6,325**
49-0166	**OutBack Full-Rack Sealed Power Center FX2548 (3)**	**$8,565**
49-0167	**OutBack Full-Rack Sealed Power Center FX2548 (4)**	**$11,470**

Please call us for quotes on larger Full-Rack systems with up to eight inverters.

OUTBACK PANEL OPTIONS

A wide variety of options are available installed on your OutBack Panel.

Ground Fault Protection

If your PV array is installed on a residence, this protection is code-required. Uses three of the _" breaker spaces, provides two 60A breakers.

27931	**GFP Installed**	**$195**

OutBack MX60 Controller

The best charge control on the market, with MPPT accepts up to 140v input, delivers up to 60 amps at 12 to 60 volts. Installed price includes two 60A breakers.

27930	**MX60 Controller Installed**	**$735**

DC Circuit Breakers

Extra input or output breakers for wind, hydro, or any DC appliance.

27915 1A	**1 Amp DC Breaker Installed**	**$39**
27915 10A	**10 Amp DC Breaker Installed**	**$39**
27915 15A	**15 Amp DC Breaker Installed**	**$39**
27915 30A	**30 Amp DC Breaker Installed**	**$39**
27915 40A	**40 Amp DC Breaker Installed**	**$45**
27915 60A	**60 Amp DC Breaker Installed**	**$45**
27915 100A	**100 Amp DC Breaker Installed**	**$75**

AC Circuit Breakers

Extra input or output breakers for generators, AC appliances, or subpanels.

27922 single 15A	**AC Breaker Installed 15 Amp**	**$29**
27922 single 20A	**AC Breaker Installed 20 Amp**	**$29**
27922 single 50A	**AC Breaker Installed 50 Amp**	**$39**
27922 dual 15A	**AC Breaker Installed Dual 15 Amp**	**$45**
27922 dual 20A	**AC Breaker Installed Dual 20 Amp**	**$55**
27922 dual 50A	**AC Breaker Installed Dual 50 Amp**	**$69**

120/240 Transformer

Used for a variety of step-up, step-down, or balancing chores. Can output 240vac from a single inverter, or can balance two 120vac inputs from a generator for battery charging, or will allow two OutBack inverters to share a heavy 120vac load, while still wired for 240vac output. (Note, only OutBack inverters can do this trick.) Wiring varies with intended use, please let us know your intention at time of order.

27929	**X240 Transformer Installed**	**$335**

Xantrex SW Plus Power Panels

Xantrex is offering a completely new line of components for the new SW Plus Inverter/Chargers. All components will ship by UPS, or other courier to your installation site. No expensive truck service, and with smaller, lighter components, now systems can be installed without a crew or expensive lifting equipment.

The basic power panel requires 36" H x 48" W of wall space for an SW Plus inverter with DC and AC boxes. All pieces are powder-coated for continued good looks. The backing plate is optional; you could screw all the components into plywood backing, but it has all mounting holes pre-punched nice and square, and now has very handy hangers that will support the heavy components hands-free while you put in the bolts. The backing plate is shipped in two pieces.

The new taller AC and DC conduit boxes now have space for up to two SW-series inverters plus ground fault protectors, terminal and distribution strips, and all the incoming, outgoing, or bypass circuit breakers you'll want, including the big DC ones for the inverter(s).

Basic Power Panel Packages

Xantrex offers Basic Power Panel Packages with either one or two of the SW Plus inverters with AC and DC boxes. DC inverter input breakers, AC bypass breakers, and stacking controls are included as needed, but no other options. Add the backing plate, DC input breakers, charge controls, transformers, generator start modules, and other components. as needed for your particular installation. Options are described briefly below. These packages ship unassembled via UPS or other courier.

Item #	Model	AC Output	DC Input	Price
49-0169	SWP2524	2.5kW	24vdc	$3,450
49-0170	SWP2524 (2)	5.0kW	24vdc	$6,115
49-0171	SWP2548	2.5kW	48vdc	$3,450
49-0172	SWP2548 (2)	5.0kW	48vdc	$6,115
49-0173	SWP4024	4.0kW	24vdc	$3,950
49-0174	SWP4024 (2)	8.0kW	24vdc	$7,140
49-0175	SWP4048	4.0kW	48vdc	$3,450
49-0176	SWP4048 (2)	8.0kW	48vdc	$7,115
49-0177	SWP5548	5.5kW	48vdc	$4,650
49-0178	SWP5548 (2)	11.0kW	48vdc	$8,540

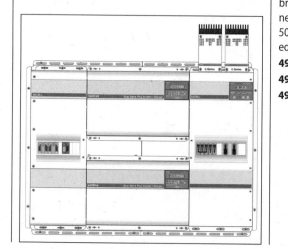

A Dual Inverter power Panel with two charge controllers and a battery status meter option.

XBP Backing Plate

Two-piece, powder-coated pre-punched steel backing plate with equipment hangers. 36" H x 48" W when assembled. Has conduit pass-thru holes so most wiring

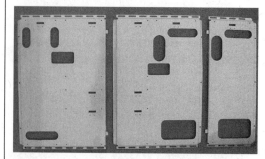

can disappear directly into the wall. Will accept up to two SWP inverters with AC and DC boxes, or one inverter with the 4 kW transformer. All screws and hardware for mounting included. USA.

49-0179 XBP Backing Plate   **$425**

XBP-DC Backing Plate Extension, DC Side

Adds 14" width to the backing plate for installations with heavy DC input/output and a second DC box. Powder-coated, pre-punched, with all hardware. Shown with XBP Plate above. USA.

49-0180 XBP-DC Backing Plate   **$150**

DCCB-L DC Conduit Box

One DC conduit box with appropriate inverter input breaker(s) comes with each power panel package, but sometimes we need more DC breaker or charge controller space. These 13.5"W x 33"H x 8.8"D boxes accept 3 large and 6 small breakers, and up to two C-series charge controls on top. The basic box has only a 12-hole grounding bar. The 175 and 250 models add a large inverter input breaker with inverter cabling, a DC negative bus bar, and a 500A/50mv shunt as standard equipment. USA.

49-0181 DCCB-L Basic   **$425**
49-0182 DCCB-L-175   **$675**
49-0183 DCCB-L-250   **$675**

CFMP Breaker Mounting Plate

In the DC conduit box, this replaces the standard 3 large/6 small breaker mounting plate with a mounting plate that will accept up to 10 small breakers. USA.

48-0190 CFMP Mounting Plate $35

ACCB-L AC Conduit Box

The long AC Conduit Box with space for up to two SW-series inverters is included with the Basic Power Panel Package. This 33"h x 11"w x 8.75"d box comes with a 60-amp bypass breaker assembly, ground, neutral, and hot terminal bus bars, and an input/output terminal block for each inverter. Additional Square D™ QOU DIN-rail mount circuit breakers can be installed as needed. These are standard breakers that can be purchased locally. 12 breaker slots are provided, each bypass assembly takes 3 slots. The second inverter space can also be used to mount the Autotransformer option for 240 volt output. The L2-PCK kit provides a second bypass breaker, L2 terminal strip, and wiring as needed to add a second inverter. USA.

49-0189 ACCB-L-L1 Conduit Box w/ L1
    Bypass Breaker $475
49-0190 ACCB-L2-PCK L2 Bypass Breaker Kit $125

PVGFP PV Ground Fault Breakers

If you mount your PV array on a residence, you need ground fault protection. These bolt into the DC conduit box and take 2, 3, 4, or 5 small breaker slots (one more slot than the number of poles). Each pole can handle one charge controller, up to 100 amps. USA.

3-Pole GFP; 1-Pole GFP

48-0191 PVGFP-CF-1
    1-Pole Grd. Fault Protector $250
48-0192 PVGFP-CF-2
    2-Pole Grd. Fault Protector $275
48-0193 PVGFP-CF-3
    3-Pole Grd. Fault Protector $300
48-0194 PVGFP-CF-4
    4-Pole Grd. Fault Protector $325

CF60 DC Circuit Breaker

A 60-amp DC circuit breaker for input or output power as needed. Takes one small breaker slot. This CF-type breaker is rated for up to 160vdc open circuit. USA.

48-0195 CF60 Circuit Breaker $40

60A DC Circuit Breaker

CC-PCK Charge Controller Wiring Kit

Provides all the bushings, fittings, and wire to install one charge controller up to 60 amps on the top of the DCCB box. Does not include controller or input/output breakers. USA.

Charge Controller Wiring Kit

48-0196 CC-PCK Controller Wiring Kit $35

GJ175F-PCK Large Breaker Add Package

Adding an inverter or a whole lot of DC input? You'll need one of these kits that provide either a 175- or 250-amp breaker with flag terminal adapters and a wiring kit. Each kit contains: 1 large circuit breaker, a #2/0 AWG, 20" long negative cable, a #2/0 AWG, 26" long positive cable, and a #6 (THHN) AWG, 7" long ground wire. USA.

48-0197 GJ175F-PCK Large Breaker Kit $120
48-0198 GJ250F-PCK Large Breaker Kit $120

BSM Battery Status Meter

A very good system monitor. Shows volts and amps at the moment, plus keeps a running total of the system

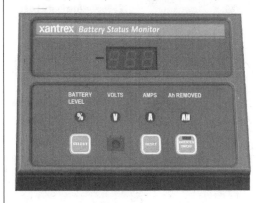

percent of full charge. Designed by the same engineer as the TriMetric monitor, but with a nicer user interface. Mounts on the front of the DC conduit box. USA.

48-0199 TM500A System Monitor, no shunt $195

TX4K and TX6K Autotransformer for 120/240VAC Conversion

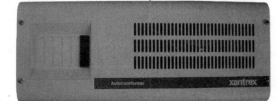

Need a bit of 240vac now and then, but only need one inverter? The auto-transformer takes 120vac from your inverter and delivers 240vac to your well pump or other modest appliance that requires higher voltage. Fits in the lower inverter space between the DC and AC boxes. Two models, choose 4,000 or 6,000 watts maximum as needed. USA.

49-0184	TX4K Autotransformer	$750
49-0185	TX6K Autotransformer	$800

GSM Generator Start Module

If your generator supports two- or three-wire remote start ability, this module will provide automated start and stop functions based on battery voltage. Program-mable thru the SWP inverter programming. USA.

27855	Generator Start Module	$159

ALM Auxiliary Load Module

Provides three relays that can turn stuff off or on based on battery voltage. Want to automatically turn some appliance(s) off if the battery gets too low, or start some other appliance once the battery gets mostly charged? Have it either or both ways, here's your answer. A nerd's dream! Programmed thru the inverter. USA.

25865	Auxiliary Load Module	$159

Code-Approved Inverter Cables

Our inverter and battery cables are all flexible, code-approved cable with color coding for polarity and shrink-

wrapped lugs for $3/8$" bolt studs. These two-lug versions (lugs on both ends) are great for Xantrex power centers and disconnects. Just cut them at the perfect point to be both the inverter cables as well as the battery cables. USA.

15851	2/0, 3' Inverter Cables, 2 lug, single	$30
15852	4/0, 3' Inverter Cables, 2 lug, single	$40
27311	4/0, 5' Inverter Cables, 2 lug, pair	$120
27312	4/0, 10' Inverter Cables, 2 lug, pair	$205
27099	2/0, 5' Inverter Cables, 2 lug, pair	$85
27098	2/0, 10' Inverter Cables, 2 lug, pair	$140

Safety and Fusing

Any time we have a power source and wiring, we need to ensure that the current flow will be kept within the capability of the wire. Too much current through too small a wire leads to melted insulation or even fires. Renewable energy systems have at least two power sources. The solar, wind, or hydro generator is one (or more), and the battery pack is the other. The battery requires particular concern. Even small battery packs store up what can be an awesome amount of energy for future use. So we'll cover battery safety first.

In many ways, batteries are safer than traditional AC power. It's not impossible, but it's pretty difficult to shock yourself with the low-voltage direct current (DC) in the 12- or 24-volt battery systems we commonly use. On the other hand, because batteries are right there in your house, they can deliver many more amperes into an accidental short circuit than a long-distance AC transmission system can. These high-amperage flows from battery systems can turn wires red hot almost instantly, burning off the insulation and easily starting a fire. So Real Goods' #1 safety rule is: **Any wire that attaches to a battery must be fused!** (The National Electrical Code says the same thing, although it spends a lot more words saying it.)

And if red-hot wires and melting insulation aren't enough to teach caution, it's difficult to stop a DC short circuit. In a really severe short circuit, popping open a circuit breaker or fuse may not stop the electrical flow. The current will simply arc across the gap and keep on cooking! Once a DC arc is struck, it has much less tendency to self-snuff than an AC arc. Arc-welders greatly prefer using DC for this very reason, but impromptu welding is not the sort of thing you want to introduce into your house. Fuses that are rated for DC use a special snuffing powder inside to prevent an arc after the fuse blows. AC-rated fuses rarely use this extra level of protection.

Many DC circuit breakers have limited current interruption ability, and require an upstream main fuse that's designed to stop very large (20,000-amp) current flows. All the Class T fuses we offer have this ability and rating. When large household-sized inverters started becoming widely available, large DC-rated circuit breakers, DC disconnects, and DC power centers started showing up in the industry. For household systems, these are the safest, and most aesthetically pleasing, solution.

The National Electrical Code requires "overcurrent protection and a disconnect means" between any power source and any appliance. A DC-rated circuit breaker covers both those requirements, and has become the first choice. For small systems, a breaker box and circuit breakers may get expensive and folks are tempted to cut corners. Do so at your own peril.

Large-inverter safety requires some specialized equipment. Full-size 3,000- to 5,000-watt inverters can safely handle up to 500 amps input at times. Systems of this size will use either an OutBack or Xantrex DC power center that provides all the fusing, safety disconnect, ground fault protection, and interconnection space for everything DC in one compact, pre-wired, engineered, UL-approved package.

For mid-size, 400- to 1,000-watt inverters a 110-amp Class T fuse assembly with an appropriate Blue Sea circuit breaker makes a cost-effective and legal safety disconnect. The small inverters of 100 to 300 watts simply plug or hardwire into a fused outlet.

Fuses are sized to protect the wires in a circuit. See the chart near the fuse listings for standard sizing/ampacity ratings. These ratings are for the most common wire types. Some types of wire can handle slightly more current, and all types of wires can safely handle much more current for a few seconds. If a particular appliance, rather than the wire, wants protection at a lower amperage, then usually a separate or in-line fuse is used just for that appliance.

Real Goods' #1
safety rule is:
**Any wire that
attaches to a battery
must be fused!**

BALANCE OF SYSTEMS

At the very least,
even tiny systems need
a catastrophic-overload fuse
at the DC positive battery
terminal.

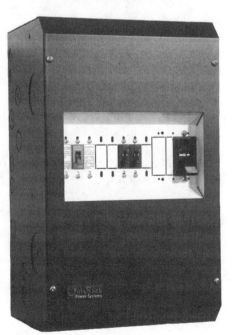

An OutBack DC Enclosure with large breaker for the inverter, ground fault protection for the PV array, smaller breakers for the charge controller, and lots of space left over still.

Lightning Protection and Grounding

Let's preface this subject with a warning: *Lightning is capricious.* That means it does whatever it wants to do, whenever it wants to. This does not mean you're defenseless, however. You can do plenty to avoid lightning damage 99% of the time.

The most important thing to keep in mind about lightning is that it *wants* to go to ground. Your job is to give it a clear, easy path, and then stay out of the way. Good, thorough grounding is your first, and most effective, line of defense. Solar arrays have a nice conductive aluminum frame around each module. But frames are usually anodized, which makes them act as an insulator. There will not be a good electrical path between the frames and the mounting structure. You need to run bare copper wire, no smaller gauge than your PV interconnect wiring, across the backs of all the modules, with a good mechanical connection to each module frame. Stainless steel self-tapping screws with stainless star washers are the best choice. Run this ground wire to your household ground rod, if it's within about 50 feet. If your array is more remote, install a ground rod at the array and ground to it. It's fine to have multiple ground rods in a single system . . . however . . . *all ground points must be connected to each other with #6-gauge copper wire*. This is probably the most important point about lightning

Ask Dr. Doug

safety. Connect all the grounds together!

Why? Let's say your array just took a hit, but the lightning took the easy path to ground you provided. Now the earth around your array is at several thousand volts potential, and that energy is looking for a way to spread out, the quicker and easier the better. It sees a nice fat ground rod over at your house, and if it just jumps into this highly conductive copper PV-negative wire that someone nicely provided, it can get over to the house ground rod real easy…so it does. Goodbye charge controller, and whatever else gets in the way. If your ground rods are all connected together with bare copper wire, however, the lightning is going to take the easy route, and pretty much stay out of your system wiring. If your bare ground wire is buried in direct contact with the ground, not inside a conduit, the

lightning will like it even better, because it has more ground contact to disperse into.

For wind turbine towers, which obviously are just begging for lightning strikes, you need a ground rod at the base of the tower, and a ground rod at each guy wire anchor. Connect them all together, and then connect back to the house ground.

Good grounding like this will save your system from damage through better than 90% of strikes. For that occasional capricious strike that decides to take an excursion through your wiring, you need to have lightning protectors installed. These $45 gizmos short out internally if voltage goes above 300, and route as much energy as possible to ground. You should have at least one on each set of input wires. If you live in a high-lightning zone, a couple of protectors on each input is cheap insurance.

Good grounding and lightning protectors are sufficient protection for residential systems. If you're installing in a really severe lightning climate, like a mountaintop (or Florida, according to some folks), seek professional lightning protection assistance! Much more can be done for extreme sites.

SAFETY EQUIPMENT

OutBack PV Combiner Box

OutBack's highly adaptable PV Combiner box provides code-compliant fusing for individual PV strings. The box is outdoor-rated, rainproof, powder-coated aluminum, and can be mounted vertically, or laid on a roof with as little as 14° slope. It will accept up to 12 DC breakers for circuits up to 125 VDC, or it can be set up with up to 7 touch-safe midget fuses for high-voltage systems up to 600 VDC. Breakers or fuse holders are sold separately as needed. Can be configured for one, or two separate PV output circuits. You heard right. Run two PV controller circuits from a single combiner box!

Set-screw output terminals will accept up to 1/0 gauge bare cable. Plenty of space for current sensors or lightning protection. The TBB (insulated terminal bus bar) option is required when running two RV Power Products Solar Boost controls, to provide a second isolated negative circuit. PV combiner box measures 9.5"W x 13.1"H x 3.5"D and weighs 6 lb. (plus options). ETL-listed, made in USA.

24-653	PV Combiner Box	$139
24-660	PV Array DC Breaker, 6-amp	$12
24-661	PV Array DC Breaker, 9-amp	$12
24-662	PV Array DC Breaker, 10-amp	$12
24-663	PV Array DC Breaker, 15-amp	$12
24-664	PV Array DC Breaker, 30-amp	$12
24-665	Hi-Volt Fuseholder	$18
24-666	Hi-Volt Fuse, 4-amp	$18
24-667	Hi-Volt Fuse, 6-amp	$18
24-668	Hi-Volt Fuse, 10-amp	$18
24-669	Hi-Volt Fuse, 15-amp	$18
24-670	OutBackTBB Option	$19

OUTBACK POWER GEAR

All the safety and connection pieces for a professional installation

OutBack supplies a full range of mounting, safety, control, and enclosure equipment to make your system safe and good looking (at least to techie types). If you're going the pre-assembled power center route, these necessary parts are already covered for you. If you're assembling on site, forge on (and don't hesitate to call us for help).

Power gear is available for full-size racks that will accept up to four OutBack inverters, or half-size racks that accept up to two OutBack inverters. All OutBack metal work is highest quality, fits perfectly, has a wealth of pre-punched knockouts and mounting points, and is powder-coated for great corrosion resistance and continued good looks.

OUTBACK AC POWER GEAR

OutBack AC Enclosures

Provides space for AC bypass switches, the incredibly useful transformer option, and for smaller systems it can function as the complete household AC breaker panel. Full-size enclosure comes with a pair of 60A bypass switch assemblies for two inverters, plus ground, neutral, and hot bus bars. Has space for thirteen additional OutBack, or eight additional Square D breakers. The half-size enclosure comes with a pair of 50A bypass breakers for two inverters, one general purpose 20A breaker, a 15A breaker with AC plug outlets, plus ground, neutral, and hot bus bars. Has space for eight additional OutBack, or six additional Square D breakers. USA.

48-0074	AC Enclosure, Half-Size w/50A Bypass	$385
24689	AC Enclosure, Full-Size w/60A Bypass	$369

OutBack AC Auto-Transformer

Can be used for step-up, step-down, or load balancing. Unique to OutBack, two FX inverters can share a heavy 120-vac load, while wired for 240-vac output with this transformer. Rated to 4kVA without the fan, or 6kVA with the cooling fan. Includes a 25A two-pole breaker. Mounts inside AC enclosure. USA.

24690	AC Auto-Transformer	$290
24691	Auto-Transformer Cooling Fan	$29

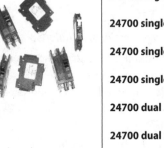

OutBack AC Circuit Breakers

Extra input or output breakers for generators, AC appliances, or subpanels.

24700 single 15A	**OutBack AC Circuit Breaker single 15A**	**$15**
24700 single 20A	**OutBack AC Circuit Breaker single 20A**	**$15**
24700 single 30A	**OutBack AC Circuit Breaker single 30A**	**$25**
24700 single 60A	**OutBack AC Circuit Breaker single 60A**	**$25**
24700 dual 25A	**OutBack AC Circuit Breaker dual 25A**	**$35**
24700 dual 30A	**OutBack AC Circuit Breaker dual 30A**	**$50**
24700 dual 60A	**OutBack AC Circuit Breaker dual 60A**	**$50**

Terminal Bus Bars

The AC enclosures come with all the bus bars most folks will need, but for the creative types, here you go. Isolated bus bars, with 12 holes. Choose your color. USA.

24670	**Terminal Bus Bar-Black**	**$19**
24702	**Terminal Bus Bar-Red**	**$19**
24654	**Terminal Bus Bar-White**	**$19**

OUTBACK DC POWER GEAR

OutBack Mounting Plates

A punched, powder-coated plate that lines up all your hardware perfectly. Plenty of holes for attaching securely to your wall. Available in half- or full-rack sizes for two or four inverters with AC and DC enclosures.

48-0072	**Half-Rack Mounting Plate**	**$129**
24685	**Full-Rack Mounting Plate**	**$179**

OutBack DC Enclosures

Both DC enclosures come supplied with a large DC breaker for one inverter, a 500A/50mV DC shunt, plus positive and negative bus bars and knockouts on five surfaces for inverter, conduit, and control connections. The Full-Size enclosure supports up to four inverters, and accepts ten 3/4" breakers, three 1" breakers, and two 1.5" breakers. The optional charge control bracket

Full-size DC enclosure.

takes two MX60s and mounts on either side. The Half-Size enclosure supports up to two inverters, and accepts eight 3/4" breakers and two 1.5" or three 1" breakers. The Half-Size *includes* a mounting bracket for one MX60 control, and a Mate.

Half-size DC enclosure.

48-0073	**DC Enclosure, Full-Size w/100A Breaker**	**$459**
24672	**DC Enclosure, Full-Size w/175A Breaker**	**$519**
24671	**DC Enclosure, Full-Size w/250A Breaker**	**$519**
24682	**Charge Control Bracket**	**$39**
48-0037	**DC Enclosure, Half-Size w/100A Breaker**	**$325**
48-0038	**DC Enclosure, Half-Size w/175A Breaker**	**$385**
48-0035	**DC Enclosure, Half-Size w/250A Breaker**	**$385**

OutBack DC Circuit Breakers

OutBack DC breakers can be used for additional inverters, input from DC charging sources, or output to DC appliances. They come in three widths.

24674	**250A DC Breaker, 1.5"**	**$119**
24675	**175A DC Breaker, 1.5"**	**$119**
24676	**100A DC Breaker, 1"**	**$60**
24677	**60A DC Breaker, 3/4"**	**$29**
24678	**40A DC Breaker, 3/4"**	**$29**
24679	**30A DC Breaker, 3/4"**	**$25**
24680	**15A DC Breaker, 3/4"**	**$25**
24681	**10A DC Breaker, 3/4"**	**$25**

OutBack Ground Fault Protection

Required for PV arrays mounted on a residence. This simple kit includes two 60A input breakers with an isolated negative bus bar and a ground bus bar. Takes three 3/4" breaker slots in the DC enclosure.

24673	**Grd Fault Protect/2**	**$129**

Trace DC Disconnects

To provide DC system overload/disconnect equipment for everything passing in or out of the battery bank, here's an alternative that meets all the code requirements. Best used with systems that only use DC power for one or two loads, the Trace DC Disconnects are available in two basic versions, 175 amps and 250 amps, with a variety of options.

No fumbling for expensive replacement fuses in the dark, these units feature easy to reset circuit breakers. Breakers are rated for 25,000-amps interrupting capacity at 65 volts, and are UL-listed for DC systems up to 125 volts. Main breaker lugs accept up to #4/0 AWG fine strand cable, no ring terminals required. .75", 2", and 2.5" knockouts are provided. Mates up with the conduit box option for DR and SW series Trace inverters. (Matching color scheme, too.) Space is available inside the disconnect for the DC current shunt(s) required by most monitors.

Options, all of which can be installed in the field, include a second main breaker for dual inverter systems, a DC bonding block for negative and ground wires, and up to four smaller DC breakers. 15-, 20-, and 60-amp circuit breakers are offered for PV or other charging input, or DC power output to loads. The negative/ground bonding block provides a necessary connection point for any negative wires. It accepts up to four 4/0, two #1/0, two #2, and four #4 cables. USA.

25-010	Trace DC175-amp Disconnect	$329
25-011	Trace DC250-amp Disconnect	$329
25-012	Trace 2nd 175-amp Breaker	$195
25-013	Trace 2nd 250-amp Breaker	$195
25-014	Trace 15-amp Auxiliary Breaker	$29
25-008	Trace 20-amp Auxiliary Breaker	$29
25-015	Trace 60-amp Auxiliary Breaker	$39
25-016	Trace Neg/Grd Bonding Block	$50

Code-Approved DC Load Centers

We find that a simple six-circuit center will handle the DC fusing needs for 99% of systems. This six-circuit center meets all safety standards and utilizes only UL-listed components. All current-handling devices have been UL approved for 12-volt through 48-volt DC applications.

These DC circuit breakers require a main fuse inline before the breaker box to provide catastrophic overload protection. If you already have a large Class T fuse or circuit breaker for your inverter, it can be used for this protection. If not, you must use the 110-amp class T fuse block on the input. USA.

Breakers must be ordered separately.

23119	6-Circuit Load Center	$43
24507	Fuse Block, 110A Fuse and Cover	$53
23131	10A Circuit Breaker	$19
23132	15A Circuit Breaker	$15
23133	20A Circuit Breaker	$15
23134	30A Circuit Breaker	$15
23140	60A Circuit Breaker	$19

Class T DC Fuse Blocks with Covers

For those who want to protect their mid-sized inverter or other DC systems, we offer the excellent DC-rated, slow-blowing Class T fuses with block, cover, and #2/0 lugs. Dimensions are 2" x 2.5" x 5.5", including the cover. Rated for DC voltage up to 125 volts.

For quick and easy basic catastrophic protection, just bolt the bare replacement fuse between your battery + terminal and your output cable. Has $^5/_{16}$" bolt holes. USA. *Use with surface-mount DC breakers below for a disconnect means and NEC compliance.*

24518	Fuse Block, 300A Fuse & Cover	$75
24212	Fuse Block, 200A Fuse & Cover	$55
24507	Fuse Block, 110A Fuse & Cover	$53
24217	300-Amp Class T Replacement Fuse	$38
24211	200-Amp Class T Replacement Fuse	$18
24508	110 Amp Class T Replacement Fuse	$18

Surface-Mount DC Circuit Breakers

High amperage protection, small, inexpensive package

Here is a compact DC circuit breaker at an affordable price. These sealed, surface-mounting breakers are water- and vaporproof. They are thermally activated, and manually resettable, with the reset handle providing a positive trip identification. The red trip button can also be activated manually, allowing the breaker to function as a switch. They are "trip-free," and cannot be held closed against an overload current. Thermal tripping allows high starting loads, without tripping. They have 1/4" studs for easy electrical connection.

These simple-to-install, surface-mount breakers are an excellent low-cost choice for small to mid-sized inverters, wind generators, control and power diversion equipment, or any other application with current flows of 50 to 150 amps. For use with DC power systems up to 30 volts. USA.

Rated at minimal 3,000-amps interrupting capacity. Requires a class T fuse inline before the circuit breaker for NEC compliance. See above.

25167	50A Circuit Breaker	$50
25823	70A Circuit Breaker	$50
25169	100A Circuit Breaker	$50
25184	150A Circuit Breaker	$50

Safety Disconnects

Safety disconnects provide overcurrent protection and a quick disconnect means between any power source and appliance. Both the batteries and the renewable generator (PV, wind, hydro turbine) are considered a power source. The 2-pole units have two separate paths for power, 3-pole units have three paths, so these can provide protection for different circuits. We carry several different sizes. Fuses should be sized according to the wire size you are using. Raintight boxes should be used if the box will be exposed to the weather. USA.

24201	30A 2-Pole Safety Disconnect	$45
24644	30A 2-Pole Safety Disconnect, Raintight	$66
24404	30A 3-Pole Safety Disconnect	$99
24202	60A 2-Pole Safety Disconnect	$65
24645	60A 2-Pole Safety Disconnect, Raintight	$109
24203	60A 3-Pole Safety Disconnect	$114
24204	100A 2-Pole Safety Disconnect	$190

Fuses sold separately below.

DC-Rated Fuses

These fuses fit the Safety Disconnects and the Minimum Protection Fuseblocks. Class RK5 fuses are rated for DC service up to 125 volts. They are time delay types that will allow motor starting or other surges. The 45A fits a 60A holder. Buy extras! You won't find these DC-rated fuses at your local hardware store (when you need one bad). USA.

24501	30A DC Fuse	$5
24502	45A DC Fuse	$11
24503	60A DC Fuse	$10
24505	100A DC Fuse	$20

Minimum Protection Fuseblocks

For simple, inexpensive protection, we offer larger diameter, lower resistance fuseblocks that offer more contact surface area than automotive-type fuses. Single fuseblocks with spring clips. 30A block takes up to 10-gauge wire, 60A block takes up to 2-gauge wire. Order fuses separately. USA.

24401	30A Fuseblock	$13
24402	60A Fuseblock	$18

ATC-Type DC Fusebox

Using the newer, safer, ATC fuse style, this covered plastic box has wire entry ports on the sides. Six individually fused circuits with a pair of main lugs for battery positive, and a negative bus bar. Will accept up to 10-gauge wire. Requires a Torx T-15 screwdriver. 8.75" x 5.5" x 2". USA.

24-418	ATC-Type DC Fusebox	$39

Inline ATC Fuse Holder

A simple inline holder for any ATC-type fuse. With 14-gauge pigtails. Not for exposed outdoor use.

25540	ATC Inline Fuse Holder	$2.95

ATC-Type DC Fuses

These are the newer plastic body DC fuses that most automotive manufacturers have been using lately. Compared to old-style round glass fuses, they're safer, as energized metal parts aren't exposed, they have more metal to metal surface contact, the amp rating is easier to read and color-coded, and with the different colors they look cool! These come in little boxes of five fuses. USA.

24219	ATC 2A Fuse	$2.95
24220	ATC 3A Fuse	$2.95
24221	ATC 5A Fuse	$2.95
24222	ATC 10A Fuse	$2.95
24223	ATC 15A Fuse	$2.95
24224	ATC 20A Fuse	$2.95
24225	ATC 25A Fuse	$2.95
24226	ATC 30A Fuse	$2.95

DC Lightning Protector

Here is simple, effective protection that can be used on any PV, wind, or hydro system up to 300 volts DC. These

high-capacity silicon oxide arrestors will absorb multiple strikes until their capacity is reached and then rupture, indicating need for replacement. Three-wire connection, positive, negative, and ground, can be done at any convenient point in the system. The AC unit is identical in operation and capacity; only the wire colors vary. The 4-wire unit is for wind turbines with 3-phase output (African, Proven, Whisper, Bergey turbines). USA.

25194	DC Lightning Protector LA302 DC	$45
25277	AC Lightning Protector LA302 R	$45
25749	4-Wire Lightning Protector LA603	$60

Raintight Junction Boxes and Power Distribution Block

Most battery-based PV systems will need a fused combiner box where the small array wiring joins the larger lead-in wiring, but small systems and utility intertie systems, which rarely require a fused combiner box, can use a distribution block and raintight box. This is the easy, secure way to join your lead-in cables to the smaller interconnect wires from the PV array. Insert cables and tighten set screws.

The 2-pole block accepts positive and negative, the 3-pole also accept ground. For every pole of both blocks, the primary side accepts one large cable, #14 to 2/0, the secondary side accepts four smaller cables, #14 to #4. For use with copper or aluminum conductors.

Raintight boxes are 16-gauge zinc-coated steel with gray finish. Removable cover is fastened at bottom by a screw. For noncorrosive environments. USA.

48-0014	2-Pole Power Distribution Block	$27
24705	3-Pole Power Distribution Block	$39
24213	Junction Box 10" x 8" x 4"	$49
24214	Junction Box 12" x 10" x 4"	$65

Automatic Transfer Switch

Transfer switches are designed as safety devices to prevent two different sources of voltage from traveling down the same line to the same appliances. This transfer switch will safely connect an inverter and an AC generator to the same AC house wiring. If the generator is not running, the inverter is connected to the house wiring.

When the generator is started, the house wiring is automatically disconnected from the inverter and connected to the generator. A time-delay feature allows the generator to start under a no-load condition and warm up for approximately one minute. Available in both 110-volt and 240-volt models. Each will handle up to a maximum of 30 amps. These switches are great for applications where utility power may be available only a few hours per day, or if frequent power outages are experienced. Housed in a metal junction box with hinged cover. Wires are clearly marked and installation schematic is included inside cover. Two-year warranty. USA.

23121	Transfer Switch, 30A, 110V	$99
23122	Transfer Switch, 30A, 240V	$149

Charge Controllers

If batteries are routinely allowed to overcharge, their life expectancy will be reduced dramatically.

Almost any time batteries are used in a renewable energy system, a charge controller is needed to prolong battery life. The most basic function of a controller is to prevent battery overcharging. If batteries routinely are allowed to overcharge, their life expectancy will be reduced dramatically. A controller will sense the battery voltage, and reduce or stop the charging current when the voltage gets high enough. This is especially critical with sealed batteries that do not allow replacement of the water that is lost during overcharging.

The only exception to the need for a controller is when the charging source is very small and the battery is very large in comparison. If a PV module produces 1.5% of the battery's ampacity or less, then no charge control will be needed. For instance, a PV module that produces 1.5 amps charging into a battery of 100 amp-hours capacity won't require a controller, as the module will never have enough power to push the battery into overcharge.

PV systems generally use a different type of controller than a wind or hydro system requires. PV controllers can simply open the circuit when the batteries are full without any harm to the modules. Do this with a rapidly spinning wind or hydro generator and you will quickly have a toasted controller, and possibly a damaged generator. Rotating generators make electricity whenever they are turning. With no place to go, the voltage will escalate rapidly until it can jump the gap to some lower-voltage point. These mini lightning bolts can do damage. With rotating generators, we generally use diversion controllers that take some power and divert it to other uses. Both controller types are explained below.

PV Controllers

Most PV controllers simply open or restrict the circuit between the battery and PV array when the voltage rises to a set point. Then, as the battery absorbs the excess electrons and voltage begins to drop, the controller will turn back on. With some controllers, these voltage points are factory preset and nonadjustable, while others can be adjusted in the field. Early PV controllers used a relay, a mechanically controlled set of contacts, to accomplish this objective. Newer solid-state controllers use power transistors and

PWM (Pulse Width Modulation) technology to turn the circuit rapidly on and off, effectively floating the battery at a set voltage. PWM controllers have the advantage of no mechanical contacts to burn or corrode, but the disadvantage of greater electronic complexity. Both types are in common use, with most new designs favoring PWM, as electronics have been gaining greater reliability with experience.

The latest generation of PV controls employs an electronic trick called Maximum Power Point Tracking, or MPPT. This allows the controller to run the PV array at whatever voltage delivers the highest wattage. This is often at a higher voltage level than the batteries would tolerate. The excess voltage is converted to amperage that the batteries can digest happily before it leaves the controller. So MPPT controllers usually push more amps out than they're taking in. On average, an MPPT control will deliver about 15% more energy per year than a standard control. And they do their best work in winter, when most off-grid homes need all the charging help they can get. Cold temperatures tend to elevate PV voltages, and long hours of darkness tend to lower battery voltages. We expect MPPT controls to gradually take over most of the PV controller market.

Controllers are rated by how much amperage they can handle. National Electric Code regulations require controllers to be capable of withstanding 25% overamperage for a limited time. This allows your controller to survive the occasional edge-of-cloud effect, when sunlight availability can increase dramatically. Regularly or intentionally exceeding the amperage ratings of your controller is the surest way to turn your controller into a crispy critter. It's perfectly okay to use a controller with more amperage capacity than you are generating. In fact, with this piece of hardware, buying larger to allow future expansion is often smart planning, and usually doesn't cost much.

A PV controller usually has the additional job of preventing reverse-current flow at night. Reverse-current flow is the tiny amount of electricity that can flow backwards through PV modules at night, discharging the battery. With smaller one- or two-module systems, the amount of power lost to reverse current is negligible. A dirty battery top will cost you far more power loss. Only with larger systems does

On average, a maximum power point controller will deliver about 15% more energy per year than a standard controller. And they do their best work in winter, when most off-grid homes need all the charging help they can get.

reverse-current flow become anything to be concerned about. Much has been made of this problem in the past, and almost all charge controllers now deal with it automatically. Most of them do this by sensing that voltage is no longer available from the modules, when the sun has set, and then opening the relay or power transistor. A few older or simpler designs still use diodes—a one-way valve for electricity—but relays or power transistors have become the preferred methods (see sidebar on Dinosaur Diodes).

Hydroelectric and Wind Controllers

Hydroelectric and wind controllers have to use a different strategy to control battery voltage. A PV controller can simply open the circuit to stop the charging and no harm will come to the modules. If a hydro or wind turbine is disconnected from the battery while still spinning, it will continue to generate power, but with no place to go, the voltage rises dramatically until something gives. With these types of rotary generators, we usually use a diverting charge controller. Examples of this technology are the Trace C-series, or the MorningStar TriStar series of controllers. A diverter-type control will monitor battery voltage, and when it reaches the adjustable set point, will turn on some kind of electrical load. A heater element is the most common diversion load, but DC incandescent lights can be used also. (Incandescent lights are nothing but heater elements that give off a little incidental light, anyway.) Both air and water heater elements are commonly used, with water heating being the most popular. It's nice to know that any power beyond what is needed to keep your batteries fully charged is going to heat your household water or hot tub. Diversion controllers can be used for PV regulation as well, in addition to hydro or wind duties in a hybrid system with multiple charging sources. This controller type is rarely used for PV regulation alone, due to its higher cost.

Controllers for PV-Direct Systems

A PV-direct system connects the PV module directly to the appliance we wish to run. They are used most commonly for water pumping and ventilation. By taking the battery out of the system, initial costs are lower, control is simpler, and maintenance is virtually eliminated.

Although they don't use batteries, and therefore don't need a controller to prevent overcharging, PV-direct systems often do use a device to boost pump output in low light conditions. While not actually a controller, these booster devices are closer kin to controllers than anything else, so we'll cover them here. The common name for these booster devices is LCB, short for Linear Current Booster.

LCBs are a solid-state device that helps motors start and keep running under low light conditions. The LCB accomplishes this by taking advantage of some PV module and DC motor operating characteristics. When a PV module is exposed to light, even very low levels, the voltage jumps way up immediately, though the amperage produced at low light is very low. A DC motor, on the other hand, wants just the opposite conditions to start. It wants lots of amps, but doesn't care much if the voltage is low. The LCB provides what the motor wants by down-converting some of the high voltage into amperage. Once the motor starts, the LCB will automatically raise the voltage back up as much as power production conditions allow without stalling the motor. Meanwhile, on the PV mod-

Regularly or intentionally exceeding the amperage ratings of your controller is the surest way to turn your controller into a crispy critter.

If a hydro or wind turbine is disconnected from the battery while still spinning it will continue to generate power, but with no place to go, the voltage rises dramatically until something gives.

Dinosaur Diodes

When PV modules first came on the market a number of years ago, it was common practice to use a diode, preferably a special ultra-low forward-resistance Schottky diode, to prevent the dreaded reverse-current flow at night. Early primitive charge controllers didn't deal with the problem, so installers did, and gradually diodes achieved a mystical must-have status. Over the years, the equipment has improved and so has our understanding of PV operation. Even the best Schottky diodes have 0.5 volt to 0.75 volt forward voltage drop. This means the module is operating at a slightly higher voltage than the battery. The higher the module voltage, the more electrons that can leak through the boundary layer between the positive and negative silicon layers in the module. These are electrons that are lost to us; they'll never come down the wire to charge the battery.

The module's 0.5 volt higher operating voltage usually results in more power being lost during the day to leakage than the minuscule reverse-current flow at night we're trying to cure. Modern charge controllers use a relay or power transistor, which has virtually zero voltage drop, to connect the module and battery. The relay opens at night to prevent reverse flow. Diodes have thus become dinosaurs in the PV industry. Larger, multi-module systems may still need blocking diodes if partial shading is possible. Call our tech staff for help.

An LCB (linear current booster) can boost pump output as much as 40% by allowing the pump to start earlier in the day and run later. Under partly cloudy conditions an LCB can mean the difference between running or not.

Bottom line: For most residential power systems, an MPPT control will do more for your power well-being than a tracking mount.

Tracking Mounts or Maximum Power Point Tracking?

Which delivers more power?

Until recently, tracking mounts have been the only way to increase the daily output from your PV array. By following the sun from east to west, trackers can increase output as much as 30%. Tracking works best in the summer, when the sun makes a high arc across the sky. Tracking is the best choice for power needs that peak in the summer, such as water pumping and cooling equipment. In the winter, tracking gains are much less, usually about 10 to15%.

What if you're looking for more winter output? The new MPPT charge controls can increase your cold-weather output up to 30%.

Ask Dr. Doug

MPPT controls are a refinement of linear current boosting, which allows the PV module to run at its maximum power point, while converting the excess voltage into amps. MPPT controls perform best with cold modules and hungry batteries, a wintertime natural! Some MPPT controllers also allow wiring and transmitting at higher voltage. A 72-volt array can charge a 24-volt battery, for instance. This saves additional costs in wire and reduces transmission losses.

Bottom line: For most residential power systems, an MPPT control will do more for your power well-being than a tracking mount.

ule side of the LCB, the module is being allowed to operate at its maximum power point, which is usually a higher voltage point than the motor is operating at. It's a constant balancing act until the PV module gets up closer to full output, when, if everything has been sized correctly, the LCB will check out of the circuit, and the module will be connected directly to the motor.

LCBs cost us something in efficiency, which is why we like to get them out of the circuit as the modules approach full power. All that fancy conversion comes at a price in power loss—usually about 10% to 15%—but the gains in system performance more than make up for this. An LCB can boost pump output as much as 40% by allowing the pump to start earlier in the day and run later. Under partly cloudy conditions, an LCB can mean the difference between running or not. We usually recommend them strongly with most PV-direct pumping systems. With PV-direct fan systems, where start-up isn't such a bear, LCBs are less important

PV, HYDRO, AND WIND CHARGE CONTROLLERS

Morningstar SunGuard

A small controller at the right price!

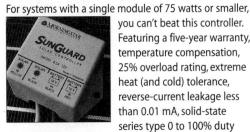

For systems with a single module of 75 watts or smaller, you can't beat this controller. Featuring a five-year warranty, temperature compensation, 25% overload rating, extreme heat (and cold) tolerance, reverse-current leakage less than 0.01 mA, solid-state series type 0 to 100% duty cycle PWM, epoxy encapsulated circuitry, 1500W transorb lightning protection, and more! For 12-volt systems only, it is rated for up to 4.5 amp PV and regulates at 14.1 volts. Okay for sealed batteries. USA.

**25739   Morningstar SunGuard
             4.5A/12V Controller**   **$30**

Phocos CML-Series PV Charge Controllers

Sophisticated German-engineered controllers at affordable prices, in 5-, 10-, 15-, and 20-amp sizes. These are 12- or 24-volt controls with automatic selection when you connect the battery. Packed with features, including

PWM 3-stage charging, a 3-LED graphic battery charge indicator, LED charging and overload indicators, electronic circuit and load protection, and automatic low voltage load disconnect with acoustic warning (25 beeps). Charging voltage temperature compensation is builtin.

Choose between sealed and wet-cell battery charging programs. Will accept up to #10 AWG wire. 3.9" x 3.2" x 1.4" Two-year manufacturer's warranty. German-engineered, made in China.

48-0069 Phocos CML5   **$25**
48-0070 Phocos CML10   **$35**
48-0071 Phocos CML15   **$45**
48-0086 Phocos CML20   **$55**

SunSaver 10-Amp Controller

The SunSaver 10-amp controller provides high-reliability, low-cost PWM charge control for small 12-volt systems. Featuring an easy-to-access, graphic terminal strip for connections, a weather-resistant anodized aluminum case, fully encapsulated potting, reverse-current protection, and field selection of battery type for flooded or sealed batteries. Has an LED to show charging activity, and features automatic temperature compensation, based on the temperature of the controller. Accepts up to #10 gauge wire. 6" x 2.2" x 1.3", 8 oz. Five-year manufacturer's warranty. USA.

25763   SunSaver 10A/12V Controller   **$55**

Sunlight 10- and 20-Amp PV and Light Controllers

A simple PV and light control

The Sunlight 12-volt solid-state PWM charge controller has automatic light control, and low voltage disconnect for battery protection. Connect a battery, a PV panel, and a 12V gizmo, and you've got a system that automatically turns on your sign lighting, stairway light, or other PV-powered task at dusk. If voltage falls below 11.7V, a low voltage disconnect occurs, and will not let the light run until battery voltage recovers to 12.8V. A "test" button overrides LVD for five minutes. Load control has ten settings from off through dusk to dawn, including three unique settings that run a few hours after dusk, turn off, then run an hour or two before dawn. Charge control can be set for either sealed batteries at 14.1V or wet-cell batteries at 14.4V with a simple jumper, and provides reverse-current protection at night. Automatic charge voltage temperature compensation is built in. Reverse polarity protected. Available in 10 or 20 amp models for 12-volt applications. Five-year manufacturer's warranty. USA.

25742   Sunlight 10 PV/Load control   **$109**
25743   Sunlight 20 PV/Load control   **$139**

Trace C-12

Charge/load/lighting controller

This 12-volt controller can do multiple functions. For starters, it's an electronic, 3-stage 12-amp PV charge controller with easily adjustable bulk and float voltages, nighttime disconnect, and automatic monthly equalization cycles.

It's also a 12-amp automatic load controller with 15-amp surge and electronic overcurrent protection, auto-

matically reconnecting at 12-second and then 1-minute intervals. Voltage set points for connect and disconnect are also easily adjustable. Loads will blink five minutes before disconnect, allowing the user to reduce power use. Even after disconnect, it allows one ten-minute grace period by pushing an override button.

Plus, it's an automatic lighting controller that will turn on at dusk for sign or road lighting. Run time is user adjustable from two to eight hours, or dusk to dawn. If voltage gets too low, low voltage disconnect will override run time to protect the battery.

The LED mode indicator will show the approximate state of charge and indicate if low voltage disconnect, overload, or equalization has occurred.

The C-12 is compatible with any 12V battery type including sealed or nicads. Interior terminal strip will accept up to #10 AWG wire. The box is rain-tight, and the electronics are conformal coated. Certified by ETL to UL specs. Manufacturer's warranty is two years. USA.

| 27315 | Trace C-12 Multi Controller | $110 |
| 27319 | Battery Temperature Probe | $29 |

BZ MPPT 200 Maximum Power Point Tracking Controller

New tracking technology for smaller systems

MPPT technology is finally available for smaller 12-volt systems at a reasonable price. Run your PV modules at their maximum power point, usually 16 to 19 volts, while downconverting that extra voltage into amps your battery can digest. Charge current can increase by up to 30%, with a 15% boost being about the yearly

average. This is far cheaper than an equivalent amount of PV wattage. MPPT controllers work best in the winter with cold PV modules and hungry batteries, when your system probably

appreciates any charging boost the most. A simple red/yellow/green battery voltage graphic appears on the face of the controller.

The MPPT200 accepts up to 200 watts of PV input. The PWM-type charge control float voltage is set at 14.1

volts, safe for all lead-acid battery types, and is adjustable from 13 to 15 volts. ATO-type automotive fuses are supplied on the input and output. Will accept up to 12-gauge wire. Circuit board is 100% conformal coated for moisture protection. For surface or flush mounting. Fits a 4^{11}/$_{16}$" standard electrical box. Size is 5.5" x 5.5" x 3"D. Five-year manufacturer's warranty. USA.

| 25820 | BZ Products MPPT200 Controller | $119 |

BZ Products Model 20

The Model 20 is a modern 12-volt, 20-amp, solid-state PWM controller with a large LCD digital meter displaying either battery voltage or PV charging current. For small systems, this is all the control and monitoring your system needs, and at a bargain price. Has an accurate three-digit display, including tenths of a volt and tenths of an amp. Features include easily user-adjustable float voltage, built-in temperature compensation, internal PV and battery fusing, a manual equalization switch, lightning protection, and LED indicators for float voltage or equalizing. Has two ^{1}/$_2$" knockout fittings in base, wire terminals will accept up to #6 gauge, and has reverse polarity protection. Measures 6.4"H x 4.7"W x 1.5"D, weighs 1.0 lb. Five-year manufacturer's warranty. USA.

| 25764 | BZ Model 20 Controller/Monitor | $99 |

Lyncom Charge Controllers

Lyncom makes a selection of high-quality, modest-cost controllers in 7- to 35-amp sizes. They all feature Pulse Width Modulation electronic control, reverse polarity protection, temperature compensation, lightning protection, automatic low voltage disconnect, an easy-access terminal strip, and can be set for sealed or wet-cell batteries.

The basic SR series for small 12-volt systems is available in 7- or 12-amp sizes. The 12- or 20-amp PV series adds over-voltage and over-temperature protection for 12-volt systems. The 35-amp N series adds short-circuit protection, automatic battery equalization, and can do 24 volt with 12/24 auto select software.

"T" units have a timer circuit for automatic lighting systems.

"M" units have a digital meter. Reads battery volts, PV amps, or load amps.

Sizes vary. All Lyncom controllers have a five-year manufacturer's warranty. USA.

25789	Lyncom SR7 Control	$47
25792	Lyncom SR12M Control	$98
25794	Lyncom PV12T Control	$88
25798	Lyncom PV20M Control	$152
25803	Lyncom N35M Control	$189

ProStar PV Charge Controllers

Control and monitoring in a single package

MorningStar is one of the largest and most reliable PV controller manufacturers in the world. The ProStar line, with models at 15 or 30 amps, features solid-state, constant-voltage PWM control, automatic selection of 12- or 24-volt operation, simple manual selection of sealed or flooded battery charging profile, nighttime disconnect, built-in temperature compensation, automatic equalize charging, reverse-polarity protection, electronic short circuit protection, moderate lightning protection, and roomy terminal strips that accept up to #6 AWG wire sizes. A built-in LCD display scrolls through battery voltage, charging current, and load current (if any loads are connected directly to the controller). An automatic low-voltage disconnect will shut off connected loads if batteries get dangerously low for more than 55 seconds. Maximum internal power consumption is a minimal 22 milliamps. Dimensions are 6.01" x 4.14" x 2.17". Five-year manufacturer's warranty. USA.

25771	ProStar 15-amp PV Controller w/ meter	$179
25729	ProStar 30-amp PV Controller w/ meter	$219

TriStar PV or Load Controllers

Full-featured 45- or 60-amp controls

The reliable, UL-listed MorningStar TriStar controller can provide either solar charging, load control, or diversion regulation for 12-, 24-, or 48-volt systems. This is a full-featured, modern, solid-state PWM controller. Features include reverse-polarity protection, short circuit protection, lightning protection, high temperature current reduction, 4-stage battery charging, simple DIP switch controlled voltage setpoint setup, large 1"/1.25" knockouts on bottom, sides, and back, with extra wire bending room and terminals that accept up to #2 AWG wire. Circuit boards are fully conformal coated, the aluminum heat sink is anodized, and the enclosure is powder-coated with stainless fasteners to laugh at high humidity tropical environments.

This controller is particularly good for hydro diversion because the controller will allow up to 300-amp inrush currents to start motors or warm up cold resistive loads without the electronic short circuit protection interfering.

The optional 2-line, 16-character digital meter may be mounted to the controller in place of the cover plate, or remotely in a standard double-gang box using standard RJ-11 connectors (ethernet cables). Displays all system information, self-test results, and setpoints with intuitive up/down and left/right scrolling buttons. The remote temperature sensor option with 10-meter cable will adjust voltage setpoints automatically, and is recommended if your batteries routinely will be exposed to temperatures under 40°F or over 80°F. 10.1" H x 4.9" W x 2.3" D, weighs 4 lb. CE and UL listed. Five-year manufacturer's warranty. USA.

48-0010	TriStar 45-Amp Controller	$169
48-0011	TriStar 60-Amp Controller	$218
48-0012	TriStar Digital Meter	$99
48-0013	TriStar Remote Temperature Sensor	$45

Xantrex C-Series Multi-Function DC Controllers

The versatile C-series of solid-state controllers can be used for PV charge control, DC load control or DC diversion. They only operate in one mode, so PV charge control and DC load control requires two controllers. The 40-amp controller can be selected manually for 12-, 24- or 48- volts. The 35- and 60-amp controllers can do 12 or 24 volts. All setpoints are field adjustable, with removable knobs to prevent tampering. They are protected electronically against short circuits, overload, overtemp, and reverse polarity with auto-reset. No fuses to blow! Two-stage lightning and surge protection is included.

The three-stage PV charge control uses solid-state pulse width modulation control for the most effective battery charging. Has an automatic "equalize" mode every 30

days, which can be turned off for sealed batteries. A battery temperature sensor is optional but recommended if your batteries will routinely see temps below 40°F or above 90°F. The diversion control mode will divert excess power to a dummy heater load and offers the same adjustments and features as the PV control. Your dummy load must have a smaller amperage capability than the controller. The DC load controller has adjustable disconnect and auto-reconnect voltages with a time delay for heavy surge loads.

The optional LCD display with backlighting replaces the standard front panel, or can be mounted remotely. It continuously displays battery voltage, DC amperage, cumulative amp-hours, and a separate resettable "trip" amp-hour meter. Note that this meter can only sense and display current flows passing through the charge controller.

9" x 5" x 2", weighs 4 lbs. Certified by ETL to UL standards. Two-year manufacturer's warranty. USA.

25-027	C-35	$119
25-017	C-40	$159
25-028	C-60	$199
27-319	Battery Temperature Sensor	$29
25018	Xantrex C-Series LCD Display	$99
25774	Remote C-Series LCD Display w/50" cord	$129

OutBack MX60 MPPT Charge Controller

Get the most from your PV array

This is the absolute queen of PV charge controls. The MX60 charge controller will deliver up to 70 amps continuous output to any battery system between 12 and 60 volts. The Maximum Power Point Tracking will run your PV array at its greatest power point, and down-convert the excess voltage to something your batteries can digest. Power conversion efficiency is an incredible 99.1% at 40 amps, or 97.3% at the full 70 amps output. MPPT technology will deliver about 15% more power to your batteries on a yearly average, and delivers up to 30% boost in the winter, when your batteries probably need all the help they can get. Any PV array with less than 140 volts open circuit output (usually 72-volt nominal) can be used to charge a lower battery voltage, reducing wire size and power loss. Set points are fully adjustable for any battery type, chemistry, and charging profile. Periodic equalization cycles may be programmed (for wet cells), or not (for sealed cells).

The MX60 comes with a four-line, 80-character backlit LCD display for monitoring and programming. It's easy to read, easy to use, and easy to understand. Our photo here is showing the home screen. From the top, it's displaying:

PV volts in, Bat volts out

Amps in, Amps out

Watts at the moment

Aux state

kWh delivered today, functional state

Performance data for the past 60 days is saved in the easily accessible logging file. The Auxiliary output is a 12-volt signal that can be programmed to turn on or off at specific voltages. It's always 12v regardless of system voltage so it can run an external relay or warning light. The optional OutBack Mate will plug into the MX60, and will allow control and monitoring of up to eight MX60 controls from up to 300 feet away. The Mate can also output to a PC computer for data logging or remote monitoring.

Standby power use is less than 1 watt. For installation in indoor or protected locations. Accepts 14- to 4-guage wire. Size: 14.5"H x 5.75"W x 5.75" D. Weighs 10 lb. Two-year manufacturer's warranty with extended option. Made in USA.

25812 OutBack MX60 Charge Controller $649

Working MX60 processing over 2,900 watts with room for more.

CONTROLLER ACCESSORIES AND SPECIALTY CONTROLLERS

Air Heater Diversion Loads

These resistive loads are enclosed in vented aluminum boxes for safety. They can be used on any DC system between 12 and 48 volts. Box needs at least 12" clearance to combustibles. Both units are shipped in the highest resistance mode, and can be reconfigured easily for lower resistance by changing connections in the terminal block. The HL-100 unit can be configured for 30 or 60 amps in nominal 12-volt mode. Two-year manufacturer's warranty. USA.

Item #	Price	Model	Diversion Amps @ Voltage Below			Resistance Setting
			15v	30v	60v	
48-0081	$235	HL-100	30/60	—	—	0.5/0.25 ohm
			15	30	—	1 ohm
			3.8	7.5	15	4 ohm
48-0082	$235	HL-75	20	40	—	0.75 ohm
			5	10	20	3 ohm

Diversion Water Heater Elements

Put your spare energy to work!

These industrial-grade DC heater elements give you someplace to dump extra wind or hydro power. Use with a diversion controller such as the MorningStar Tristar.

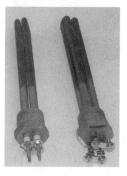

The water heater elements fit a standard 1" NPT fitting. The 12/24V model has a pair of 25A/12V elements. Can be wired for 25A or 50A @ 12V, or in series for 25A @ 24V. The 24/48V model has a pair of 30A/24V elements. Can be wired for 30A or 60A @ 24V, or in series for 30A @ 48V. USA.

25078 12/24V Water Heating Element $90

25155 48V/30A Water Heating Element $116

Voltage Controlled Switches

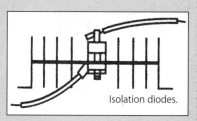

Got something you want to turn on, or off, based on your battery voltage? This is your ticket. Maybe you'd like to divert power to pumping or fans once the battery is charged? Or you'd like to shut off some appliances if the battery gets dangerously low? This 30-amp relay switch is adjustable for any DC voltage 10 to 63 volts. You set the high and low trip points. There's a normally open and a normally closed connection, so you can turn things on or off at your trip point. The "active high" unit turns ON the relay at the high trip point, and holds it until the voltage drops below the low trip point. The "active low" unit is the opposite. Quiescent power use is 17ma. The relay coil draws 75/47/32ma. at 12/24/48V. One-year manufacturer's warranty. Canada.

26657	Voltage Controlled Sw. Active High	$85
26658	Voltage Controlled Sw. Active Low	$85

About Diodes

Diodes are one-way valves for electricity. In PV systems, they help the electrons get where they're supposed to go, and stay out of places they aren't supposed to go. But even the best valves have some restriction to flow in the forward direction, so diodes are used for strictly limited jobs.

The most common diode is a blocking diode. These are usually already installed inside the J-box of most larger modules. Installed in the middle of the series string of PV cells that make up the individual module, a blocking diode prevents a shaded module from stealing power from its neighbors. Smaller modules can have blocking diodes added inline. But unless you have a multiple-module array that is so poorly positioned that it will get hard shade during prime midday sun hours, don't worry about blocking diodes. Large RV or sailboat arrays need blocking diodes (if they aren't already installed by the

Blocking or bypass diode.

module manufacturer).

Another common use is the bypass diode, used in series strings. If one module in the string has limited output due to shading or other problems, a bypass diode installed across the output terminals will allow the other modules in the string to continue output. They are usually only necessary on long, high-voltage series strings. You must have an amperage rating higher than the module string.

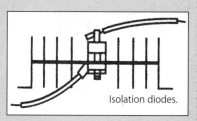

Isolation diodes.

Isolation diodes are used to prevent backflow from batteries, or when combining parallel strings. Reverse-current flow from batteries is usually better handled by the charge controller, but a few controllers don't do nighttime disconnect.

30-Amp Power Relays

These general purpose DPDT (double-pole double-throw) power relays have either a 12-volt/169-milliamp, a 24-volt/ 85-milliamp, or a 120vac/85-milliamp pull-in coil, depending on model chosen. Other controls like timers, load disconnects, or float switch, can be limited in their amperage capability. These relays can be used remotely to switch larger loads, up to 30 amps per pole. They can also be used to switch an appliance that's a very different voltage. Turn on an AC gizmo with a DC control, or vice-versa. Because these are highly adaptable DPDT relays, they can control multiple switching tasks simultaneously. Screw terminal connections. Approx. 2.5"H x 2.5"W x 3.5"L. Choose your system voltage. USA.

25112	12V/30A Relay	$40
25151	24V/30A Relay	$40
25160	120vac/30A Relay	$40

100-Amp Isolation Diode

There used to be controls that needed this big diode, but they're rare now. If you need it, you'll know it. USA.

48-0201 100-amp Diode with Heat Sink $59

Blocking or Bypass Diodes

These are flat-pack diodes with operating current de-rated 50% from manufacturer's specs for reliable operation and low voltage drop. 45-volt maximum. Use for 12- or 24-volt systems. May require the addition of shrink tubing or fork terminals for installation in module J-boxes. USA.

25807   10-amp Diode (set of 2) $9

Spring Wound Timer

Our timers are the ultimate in energy conservation; they use absolutely no electricity to operate. Turning the knob to the desired timing interval winds the timer. Timing duration can be from 1 to 12 hours (or 1 to 15 minutes), with a hold feature that allows for continuous operation. This is the perfect solution for

automatic shutoff of fans, lights, pumps, stereos, VCRs, and Saturday morning cartoons! It's a single-pole timer, good for 10 amps at any voltage, and it mounts in a standard single gang switch box. A brushed aluminum faceplate is included. One-year warranty.

25400	Spring-Wound Timer, 12 hour w/hold	$39
25399	Spring-Wound Timer, 15 minute, no hold	$35

12-Volt Digital Timer

This accurate quartz clock will control lights, pumps, fans, or other appliances from any 12VDC power source. The switched connections are isolated, so the switched voltage can be up to 36 volts DC or 240 volts AC, and up to 8 amps. Normally open and normally closed contacts are provided, as well as a manual on/off/auto button. The 7-day timer can accept up to eight on/off events per day. The internal battery will retain programming for one year after loss of 12V power (but won't switch on/off events). 4" x 4" x 2.5". One-year manufacturer's warranty. USA.

25064   12V Digital Timer $85

Battery Selector Switch

Few renewable energy homes use dual battery banks, but most boats and RVs do. This high-current switch permits selection between Battery 1 or Battery 2, or both in parallel. The off position acts as a battery disconnect. Wires connect to 3/8" lugs. Capacity is 250 amps continuous, 360 amps intermittent. Okay for 12- or 24-volt systems. 5.25" diameter x 2.6" deep. Marine UL-listed. USA.

25715   Battery Switch $34.95

Dual Battery Isolator

For RVs or boats, this heavy-duty relay automatically disconnects the household battery from the starting battery when the engine is shut off. This prevents inadvertently running down the starting battery. When the engine is started, the isolator connects the batteries, so both will be recharged by the alternator. The relay is large enough to handle heavy, sustained charging amperage. Easy hookup: Connect a positive wire from each battery to the large terminals. Connect ignition power to one small terminal, ground the other small terminal. USA.

25716   Dual Battery Isolator $23

DC-to-DC Converters

Got a DC appliance you want to run directly, but your battery is the wrong voltage? Here's your answer from Solar Converters. These bi-directional DC-to-DC converters work at 96% efficiency in either direction. For instance, the 12/24 unit will provide 12V @ 20A from a 24V input, or can be connected backwards to provide 24V @ 10A from a 12V input. Current is electronically limited and has fuse back-up. Four units are offered: The 12/24 unit described above; a larger 12/24 unit that delivers 25A or 50A depending on direction, a 24/48 unit that delivers 5A or 10A depending on direction, and a 12/48 unit that delivers 2.5A or 10A depending on direction. The housing is a weatherproof NEMA 4 enclosure measuring 4.5" x 2.5" x 2", good for interior or exterior installation (keep out of direct sunlight to keep the electronics happy). Rated for –40°F to 140°F (–40°C to 60°C). One-year manufacturer's warranty. Canada.

25709	**12/24 DC-to-DC Converter, 20A**	**$185**
25782	**12/24 DC-to-DC Converter, 50A**	**$349**
25710	**24/48 DC-to-DC Converter, 10A**	**$210**
25711	**12/48 DC-to-DC Converter, 10A**	**$210**

PV Direct Pump Controllers

These current boosters from Solar Converters will start your pump earlier in the morning, keep it going longer in the afternoon, and give you pumping under lower light conditions when the pump would otherwise stall.

Features include Maximum Power Point Tracking, to pull the maximum wattage from your modules; a switchable 12- or 24-volt design in a single package; a float or remote switch input; a user-replaceable ATC-type fuse for protection; and a weatherproof box. Like all pump controllers, a closed float switch will turn the pump off, an open float switch turns the pump on.

Four amperage sizes are available in switchable 12/24 volt. Input and output voltages are selected by simply connecting or not connecting a pair of wires. Amperage ratings are surge power. Don't exceed 70% of amp rating under normal operation. The 7-amp model is the right choice for most pumping systems, and is the controller included in our submersible and surface pumping kits. 7- and 10-amp models are in a 4.5" L x 2.25" W x 2" D plastic box. 15-amp model is a 5.5" L x 2.9" W x 2.9" D metal box. 30-amp model is supplied in a NEMA 3R rain-tight 10" x 8" x 4" metal box. One-year manufacturer's warranty. Canada.

25002	**7A, 12/24V Pump Controller**	**$106**
25003	**10A, 12/24V Pump Controller**	**$135**
25004	**15A, 12/24V Pump Controller**	**$220**
25005	**30A, 12/24V Pump Controller**	**$395**

Universal Generator Starter

Automatic start and stop based on voltage

The Solar Converters Genset starter will automatically control the starting and stopping of a backup generator based on the voltage of any 12- to 48-volt battery. Start and stop voltage points are user adjustable. Supports 2,

3, or 4 wire, and advanced start systems. There's a time delay to prevent false starts, and a 20-second precrank relay closure for diesel preheaters. A 30-second delay for engine warm up is provided before closing the load relay. (Optional external power relays will be required for some functions.) LEDs indicate operation. A manual start/stop switch is on the board, and an external one can be added. All connections land on a clearly labeled screw terminal strip that will accept up to #12-gauge wire. On-board relays rated at 5 amps. Available as either a bare, open frame, as pictured, measuring 5" x 5" x 1", or mounted in a NEMA 3R outdoor box, measuring 10" x 8" x 4". Quiescent power use less than 20ma. One-year manufacturer's warranty. Canada.

48-0017 Univ. Gen. Starter, bare   **$265**
48-0083 Univ. Gen. Starter w/Enclosure   **$340**

Monitors

The system monitor is the gizmo that allows you to peer into the electrical workings of your system and keep track of what's going on.

If you're going to own and operate a renewable energy system, you might as well get good at it. Your reward will be increased system reliability, longer component life, and lower operating costs. The system monitor is the gizmo that allows you to peer into the electrical workings of your system and keep track of what's going on.

The most basic, indispensable, minimal piece of monitoring gear is the voltmeter. A voltmeter measures the battery voltage, which in a lead-acid battery system can be used as a rough indicator of system activity and battery state of charge. By monitoring battery voltage, you can avoid the battery-killing over- and undervoltages that come naturally with ignorance of system voltage. No battery, not even the special deep-discharge types we use in RE systems, likes to be discharged beyond 80% of its capacity. This drastically reduces the life expectancy of the battery. Since a decent voltmeter is a tiny percentage of battery cost, there is no excuse for not equipping your system with this minimal monitoring capability.

Beyond the voltmeter, the next most common monitoring device we use is the ammeter. Ammeters measure current flow (amps) in a circuit. They can tell us how much energy is flowing in or out of a system. Small ammeters are installed in the wire. Larger ammeters use a remote sensing device called a shunt. A shunt is a carefully calibrated resistance that will show a certain number of millivolts drop at a specific number of amps current flow. For instance, our 100-amp shunt is rated at 100 millivolts. For every millivolt difference the meter sees between one side of the shunt and the other, it knows that one amp is flowing. By noticing which side is higher, it knows if the flow is into or out of the battery, and will display a + or – sign. A single shunt gives us the net amperage in or out for the entire system, but won't tell us, for example, that 20 amps are coming down from the PV array, but 15 amps are going straight into the inverter, and only 5 amps are passing through the shunt.

The most popular monitoring device is the accumulating amp-hour meter. This is an ammeter with a built-in clock and simple computing ability that will give a running cumulative total of amp-hours in or out of the system. This allows the system owner to monitor easily and accurately how much energy has been taken from the battery bank. Our most popular meter, the TriMetric, can display remaining energy as an easily understood percentage. When the meter says your battery system is 80% full, that's a number anyone can understand. Not every system or every person needs this kind of high-powered monitor/controller, but for those with an in-depth interest or an aversion to technology, it's available at modest cost.

INSTALL YOUR MONITOR WHERE IT WILL GET NOTICED

The best, most feature-laden monitor won't do you a bit of good if nobody ever looks at it. We usually recommend installing your monitor in the kitchen or living room so it's easy for all family members to notice and learn from. It's surprising how fast kids can learn given the proper incentive. A "No computer games or Saturday morning cartoons unless the batteries are charged" rule works wonders to teach the basics of battery management. Sometimes this even works on adults, too.

Elastic Voltmeters!

A voltmeter is a basic battery monitoring tool. It can give you a rough indication of how full or empty a lead-acid battery is, because the at-rest voltage will rise or fall slightly with state of charge. HOWEVER ... the most important thing to know about a voltmeter is ... this is a very *e l a s t i c* sort of indicator. Voltage stretches up or down when charging or discharging. When charging, a 12-volt battery can read over 14 volts. Almost the instant you stop charging, it will drop under 13 volts. Did you lose power when the voltage dropped? Not at all. It was just the voltage snapping back down to normal. Now put that battery under a heavy load like a toaster or microwave. The voltage will drop well under 12 volts, maybe even under 11 volts depending on how big the battery is, how well charged it is, and how heavy the load. But it'll bounce right back up as soon as the microwave goes off. The lesson here is that you need to know what else is happening when you read your voltmeter. A reading of 11.8 volts isn't a problem while the microwave is running, but if nothing is on, and the batteries have been sitting awhile to stabilize, then 11.8 volts is a very nearly dead battery. You'll learn with practice.

SYSTEM METERS AND MONITORS

Affordable Meter for 12- and 24-Volt Systems

This nifty little digital meter plugs into a lighter socket and auto-selects for 12- or 24-volt systems. The main 3-digit digital display shows 1/10-volt increments, plus there's a 3-LED red/yellow/green display that gives system state of charge at a glance. China.

57-0103 Affordable Meter for 12- & 24V Systems   **$16**

Analog Voltmeters and Ammeters

Use these low-cost ammeters to monitor either charging or discharging current flow. Meter reads in one

direction only, so full monitoring may require two meters. Voltmeter size is 2.5" square, designed for panel mounting in a 2" round hole. USA.

25313	0–30A Ammeter	$24
25778	11–16 Volt Analog Voltmeter	$27
25304	22–32 Volt Analog Voltmeter	$30

Universal Battery Monitor

Here's simple at-a-glance system monitoring that anyone can understand. Ten LEDs are labeled in 10% increments and range from green to yellow to red. Draws only 1/4 watt. Can be set for 12-, 24- or 48-volt operation. Easy mounting tabs, 2-foot cable that can be spliced and extended with small 18-gauge wire as needed. 5.2" x 2.75" x 1.1". 2-amp fuse recommended, listed below. Five-year manufacturer's warranty. USA.

57-0100	Universal Battery Monitor	$65
25540	ATC In-line Fuseholder	$2.95
24219	ATC 2A Fuse (5)	$2.95

Digital Meters

Our easy-to-read, surface-mount digital voltmeter monitors battery voltage up to 40 volts (for 12- and 24-volt systems), and reads three digits (tenths of a volt). Draws only 8 milliamps to operate.

Our Digital Ammeter works on 12- or 24-volt systems, reads four digits up to 199.9 amps, shows reverse flow with a negative sign, and includes a 100A shunt. 4" x 2" x 1.75". One-year manufacturer's warranty. USA.

25297	Voltmeter	$58
25298	Ammeter	$89

TriMetric 2020 Monitor

The TriMetric is a user-friendly 12- to 48-volt digital monitor that displays volts, amps or percentage of full charge, and tracks a number of battery management items. The bright 3-digit LED display is now 33% larger, and the face and label are larger, the deeper battery management data is easier to access and is listed right on the label. A "charging" light shows when positive amp flow is happening or flashes to show that "charged" criteria were met during the last day.

A new feature is the "battery reminder" light, which can be programmed to flash a warning for any of three conditions: 1) low battery voltage (you set the voltage); 2) equalize batteries (you set 1 to 250 days); 3) full charge batteries (you set 1 to 60 days). Every five seconds, the digital display will spell out what the reminder light is warning you about.

Fits a standard double-gang electric box turned sideways. Standard TriMetric works from 8 to 35 volts; adding the Lightning Protector/48V adapter, the range is 12 to 65 volts. 100A shunt reads to 250 amps in 0.01 increments; 500A shunt reads to 1,000 amps in 0.1 increments. Shunt must be purchased separately. Accuracy is ±1%. Draws 16 to 32 mA depending on display mode. One-year manufacturer's warranty. USA.

25371	TriMetric 2020 Monitor	$165
25303	Lightning Protector/48V Adapter	$26
23145	TriMetric Surface-Mount Box	$9.95
25351	100-Amp Shunt	$26
25364	500-Amp Shunt	$35
48-0188	Meter Wire, 18 ga. 3 twisted pairs	$1.20/ft

Please specify length.

Protect Your Monitor Investment

Monitors are expensive, protection is cheap. What more do we need to say? Inline fuse holder works with all monitors.

25540	Inline ATC Fuse Holder	$2.95
24219	2A ATC Fuse 5/box	$2.95

TM500 System Monitor

Batteries are a significant and expensive part of any independent energy system. They are also the part of your system that is most vulnerable to mistreatment. The Xantrex TM500 meter keeps track of the energy your system has available as well as energy consumed, ensuring adequate reserve power and proper treatment of your batteries. It measures DC system voltage (needs

the TM48 adapter for 48-volt systems), net amperage, cumulative amp-hours, days since full, peak voltage, minimum voltage, and more! Measuring 4.55 inches square by 1.725

inches deep, it may be surface or flush mounted. A 500-amp shunt is required (Xantrex and OutBack power centers, modules, and panels already have a shunt wired in). We strongly recommend purchasing the version with the shunt included. Their shunt is pre-wired for simply plugging in. Otherwise some tedious wiring is required. USA.

25740	TM500 Meter with Prewired Shunt	$245
25724	TM500 Meter without Shunt	$195
25725	TM48 48-Volt Adapter for TM500	$40

AC/DC Monitor

With quad display panel

Because it displays both AC and DC values, this is a great monitor for renewable energy systems, boats, RVs, and battery-based intertie systems. The large 4"x 2.5" LCD display with switchable backlighting shows four meas-

urements simultaneously, and can quickly cycle through everything it's monitoring with the clearly labeled front panel buttons. It measures two DC battery voltages, DC net current, DC net wattage, AC voltage, AC frequency, AC current (with optional sensor), plus time and temperature. Min/Max alarms can be set for all inputs. There is an audible alarm and an external alarm terminal to turn on a remote alarm. Pushing any button stops the audible alarm, and there's a master button to turn all alarms on or off. We are offering this meter with the optional AC current sensor and larger 250-amp shunt as a package. For 12- or 24-volt systems. Can be used on 48-volt systems so long as a 12V power source is connected to BAT 1 terminals. Will display up to 60 volts on BAT 2. Draws only 5.5ma with backlight off. For panel or wall mounting, wiring terminal strip protrudes out back. 6" x 4" x 0.6". Has shunt on positive side, requires 3 inline fuses, below. One-year manufacturer's warranty. USA.

57-0108	AC/DC Monitor	$365
25540	Inline ATC Fuse Holder	$2.95 (needs 3)
24219	2A ATC Fuse 5/box	$2.95

Brand Electronics ONE Meter

Multi-channel digital monitoring for utility intertie systems

Ever wonder how many watt-hours your battery-based intertie system is really contributing? Here's the ultimate meter to answer all your questions, and a few you hadn't even thought of yet. In the three-channel basic package we're offering below, the ONE Meter monitors DC power input from your PV, wind, or hydro system, plus AC power output through the inverter output terminals, plus AC power input/output through the inverter AC1 terminals. Results are displayed on the 4 x 20 backlit LCD display. Two displays are available, the flush-mounted oak-trimmed one we've shown, which mounts in a standard 3-gang electric box, or a fully enclosed stand-alone display. We'll supply the oak-trimmed flush display unless ordered otherwise. Additional displays (up to 3) can be run from a single base.

The ONE Meter is computer-ready with a common RS-232 serial-port connection for downloading to a PC, and comes with a Windows-compatible BASIC terminal program for logging, compiling, and display. The ONE Meter also has stand-alone logging capability. AC current sensors are simple snap-on types, DC current sensors are toroid hall effect types, cabling is standard 8-conductor CAT-5 LAN type with RJ-45 connectors, widely used for computer networking. Cabling is supplied with the meter.

Additional data channels (up to 9) can be monitored with a single base display. Optional extra channels must be specified for AC or DC. Our basic package is set up for a single Trace SW-series inverter. Each additional SW inverter in your system will require two more AC channels for complete monitoring. Made in USA. One-year manufacturer's warranty.

25766	Brand ONE Meter Basic Package	$799
25767	Brand Optional DC Channel	$150
25768	Brand Optional AC Channel	$150
25769	Brand Optional Oak Display	$150
25770	Brand Optional Stand-Alone Display	$150

Homeowner's Digital Multimeter

If you're going to be the owner/operator of a renewable energy system, you need one of these. A multimeter is your first line of help when things go wrong. This is a nice one with all the features you'll ever need, and a great price. LCD 0.5" display reads 31/2 digits. Reads DC volts, AC volts, amps up to 10, plus continuity and ohms. Has good quality, flexible test leads. Powered by an included 9-volt battery. China.

25000   Digital Multimeter $20

Appliance Cost Control Monitor

Ever wondered what that old fridge, freezer, or other appliance costs to run? Just plug the Cost Control between your appliance and outlet, and it will show current wattage plus track peak wattage and cumulative power use for as long as it's plugged in. Enter your power cost per kilowatt-hour, and it shows what your appliance costs per day, month, and year based on actual use. The internal battery stores all data until reset, so power failures won't erase your history. Simple, intuitive operation. Maximum load 1,800 watts. Accuracy is +/- 2%. Indoor, 120vac/60hz use only. China.

01-0447 Appliance Cost Control Monitor $30

Watt's Up Pro AC Power Measurement Tool

Determine power consumption of any AC appliance (up to 15 amps) in your home. Simply plug into this accumulating watt-hour meter for hours or months to see how much power it's using. Displays: watts, within 0.1; kWh; time; average monthly cost; voltage now; current now; power factor; max/min voltage; current; duty cycle (% of time the appliance runs). Plus memory storage and the ability to download the data to a PC (serial cable and software included). One thousand data points are stored, so years of usage can be accurately recorded. One-year manufacturer's warranty. USA.

25819   Watt's Up Pro $149

AC Kilowatt-Hour Meter

Keep track of your intertie system's output with this simple, familiar, rebuilt utility meter. Equipped with the easy-read dial, this analog digital meter displays kilowatt-hours, and will monitor both 120- or 240-volt circuits. Our round meter socket has 1" threaded outlets top and bottom. Both meter and socket are required for operation, and can be mounted inside or outside. USA.

25780   Meter $49
25781   Socket $25

Solar Pathfinder (Rent or Buy)

It only takes a fist-sized chunk of shade to effectively turn off a solar module. Wondering if that tree or building will shade your array during certain times of the year? In less than two minutes on site, the Pathfinder will show exactly what times of day and months of the year shade will fall on any site. This wonderfully simple tool only requires you to stand on the site, face south (there's a built-in compass with magnetic declination adjustment), level the Pathfinder (built-in bubble level), and read the shade reflections on the dome. The chart under the dome shows hours and months that bit of shade will be a problem. Every professional needs to own one of these. Homeowners probably only need it once, so we've got a nice $25/week rental program. Credit card required for deposit. The handheld model is just the Pathfinder head. The Professional model adds an adjustable tripod mount and protective custom metal carrying case. USA.

11605	Weekly Pathfinder Rental	$25
11603	Handheld Pathfinder	$250
11604	Professional Pathfinder w/ tripod & metal case	$360

Large Storage Batteries

Batteries provide energy storage, and are required for any remote, stand-alone, or back-up renewable energy system. Batteries accumulate energy as it is generated by various renewable energy devices such as PV modules, wind, or hydro plants. This stored energy runs the household at night or during periods when energy output exceeds energy input. Batteries can be discharged rapidly to yield more power than the charging source can produce by itself, so pumps or motors can be run intermittently. For personal safety and good battery life expectancy, batteries need to be treated with some care, and get properly recycled at the end of their life. If common-sense caution is not used, batteries can easily provide enough power for impromptu welding and even explosions.

Batteries 101

A wide variety of differing chemicals can be combined to make a functioning battery. Some combinations are very low cost, but they have very low power potential; others, like the lithium-ion batteries used in better laptops, can store astounding amounts of power, but have astounding costs to go with them. Lead-acid batteries offer the best balance of capacity per dollar, and are far and away the most common type of battery storage used in stand-alone power systems.

This battery type is also well known as the common automotive battery. Although storage batteries and starting batteries have very different internal construction, they're both lead-acid types, and that's what we're going to cover in this section of the *Sourcebook*.

Simplified Lead-Acid Battery Operation

The lead-acid battery cell consists of positive and negative lead plates made of slightly different alloys, suspended in a diluted sulfuric acid solution called an electrolyte. This is all contained in a chemically and electrically inert case. If we lower the voltage on the battery terminals by turning on a load such as a light bulb, the cell will discharge. Sulfur molecules from the electrolyte bond with the lead plates, releasing electrons, which flow out the negative battery terminal, through the light, and back into the positive terminal. This flow of electrons is what we call electricity. If we raise the voltage on the battery terminals by applying a charging source, the cell recharges. Electrons push back into the battery, bond with the sulfur compounds, and force the sulfur molecules back into the electrolyte solution. Electrons will always flow from higher voltage to lower voltage if a path is available.

This back-and-forth energy flow isn't perfectly efficient. Lead-acid batteries average about 20% loss. For every 100 watt-hours you put into the battery, you can pull about 80 watt-hours back out. Efficiency is better with new batteries, and drops gradually as batteries age. Still, this is the best energy storage medium for the price.

A single lead-acid cell produces approximately two volts, regardless of size. Each individual cell has its own cap. A battery is simply a collection of individual cells. A 12-volt automotive starting battery consists of six cells, each producing 2 volts, connected in series. Larger cells provide more storage capacity; we can run more electrons in and out, but the voltage output

> Lead-acid batteries offer the best balance of capacity per dollar, and are far and away the most common type of battery storage used in stand-alone power systems.

Why Batteries?

It's a reasonable question. Why don't renewable energy systems simply produce standard 120-volt AC power like the utility companies? It's simple. No technology exists to store AC power; it has to be produced as needed. For a utility company, with a power grid spread over half a state or more, considerable averaging of power consumption takes place. So it can (usually) deliver AC power as demanded. A single household or remote homestead doesn't have the advantage of power averaging. Batteries give us the ability to store energy when an excess is coming in, and dole it out when there's a deficit. Batteries are essential for remote or back-up systems. Utility intertie systems make do without batteries, by using the utility grid like a battery to make up shortfalls or overages. But this requires the presence of grid power, and doesn't provide any back-up ability in case of a power outage caused by grid failure.

stays in the 2.0- to 2.5-volt potential of the chemical reaction that drives the cell. Cells are connected in series and parallel to achieve the needed voltage and storage capacity.

Battery Capacity

Think of a battery as a bucket. It will hold a specific amount of energy, and no amount of shoving, compressing, or wishing is going to make it hold any more. For more capacity, you need bigger buckets or more buckets.

A storage battery's capacity is rated in amp-hours: the number of amps it will deliver, times how many hours. How fast or slow we pull the amps out will affect how much energy we get. A slower discharge will yield more total amps. So battery capacity figures need to specify how many hours the test was run. The 20-hour rate is the usual standard for storage batteries, and is the standard we use for all the storage batteries in our publications. A 220 amp-hour golf cart battery, for example, will deliver 11 amps for 20 hours. This rating is designed only as a means to compare different batteries to the same standard, and is not to be taken as a performance guarantee. Batteries are electrochemical devices, and are sensitive to temperature, charge/discharge cycle history, and age. The performance you will get from your batteries will vary with location, climate, and usage patterns. But in the end, a battery rated at 200 amp-hours will provide you with twice the storage capability of one rated at 100 amp-hours.

What if the battery you're looking at only has a rating for Cold Cranking Amps? That's a battery designed for engine-starting service, not storage. Stay away from it. Starting batteries suffer short, ugly lives when put into storage service.

Batteries are less-than-perfect storage containers. For every 1.0 amp-hour you remove from a battery, it is necessary to pump about 1.2 amp-hours back in, to bring the battery back to the same state of charge. This figure varies with temperature, battery type, and age, but is a good rule of thumb for approximate battery efficiency.

Advantages and Disadvantages of Lead-Acid Batteries

Lead-acid batteries are the most common battery type. Thanks to the automotive industry, they are well understood, and suppliers for purchasing, servicing, and recycling them are practically everywhere. Lead-acid batteries are the most recycled item in U.S. society, averaging better than a 95% return rate in most states. Of all the energy-storage mediums available to us, lead-acid batteries offer the most bang for the buck by a very wide margin. All the renewable energy equipment on the market is designed to work within the typical voltage range of lead-acid batteries. That's the good part.

Now the bad part: The active ingredients, lead and sulfuric acid, are toxins in the environment, and need to be handled with great respect. (See "Dr. Doug's Battery Care Class," pp. 170–72, for more info.) Acid can cause burns, and just loves to eat holes in your blue jeans. Lead is a danger if it enters the water cycle, is a strategic metal, and has salvage value, so it should be recycled. Lead is also incredibly heavy. More lead equals more storage capacity, but also more weight. Work carefully with batteries to avoid strains and accidents.

Lead-acid batteries produce hydrogen gas during charging, which poses a fire or explosion risk if allowed to accumulate. The hydrogen must be vented to the outside. See our drawings of ideal battery enclosures for both indoor and outdoor installations at the end of this section. Lead-acid batteries will sustain considerable damage if they are allowed to freeze. This is harder to do than you may imagine. A fully charged battery can survive temperatures as low as −40°F without freezing, but as the battery is discharged, the liquid electrolyte becomes closer to plain water. Electrolyte also tends to stratify, with a lower concentration of sulfur molecules near the top. If a battery gets cold enough and/or discharged enough, it will freeze. At 50% charge level, a battery will freeze at approximately 15°F. This is the lowest level you should intentionally let your batteries reach. If freezing is a possibility, and the house is heated more or less full time, the batteries should be kept indoors, with a proper enclosure and vent. For an occasional-use cabin, the batteries may be buried in the ground within an insulated box.

Batteries will slowly self-discharge. Usually the rate is less than 5% a month, but with dirty battery tops to help leak a bit of current, it can be 5% a week. Batteries don't fare well sitting around in a discharged state. Given some time, the sulfur molecules on the surface of the discharged battery plates tend to crystallize into a form that resists recharging, and will even block off large areas of the plates from doing active service. This is called sulfation, and it kills batteries far before

Batteries are less-than-perfect storage containers. For every 1.0 amp-hour you remove from a battery, it is necessary to pump about 1.2 amp-hours back in, to bring the battery back to the same state of charge.

Lead-acid batteries produce hydrogen gas during charging, which poses a fire or explosion risk if allowed to accumulate.

Unit of Electrical Measurement

Most electrical appliances are rated with wattage, a measure of energy consumption per unit of time. One watt delivered for one hour equals one watt-hour of energy. Wattage is the product of current (amps) times voltage. This means that one amp delivered at 120 volts is the same amount of wattage as 10 amps delivered at 12 volts. Wattage is independent of voltage. A watt at 120 volts is the same amount of energy as a watt at 12 volts. To convert a battery's amp-hour capacity to watt-hours, simply multiply the amp-hours times the voltage. The product is watt-hours. To figure how much battery capacity it will require to run an appliance for a given time, multiply the appliance wattage times the number of hours it will run to yield the total watt-hours. Then divide by the battery voltage to get the amp-hours. For example, running a 100-watt lightbulb for one hour uses 100 watt-hours. If a 12-volt battery is running the light, it will consume 8.33 amp-hours (100 watt-hours divided by 12 volts equals 8.33 amp-hours).

Once a bank of batteries has been in service for six months to a year, it generally is not a good idea to add more batteries to the bank.

their time. The cure is regular charge and discharge exercise, or lacking that, trickle charging that will keep the battery fully charged. Special pulse chargers that deliver high-voltage spikes of charging power have proven very effective at reducing and preventing sulfation in batteries that don't get much exercise.

Lead-acid batteries age in service. Once a bank of batteries has been in service for six months to a year, it generally is not a good idea to add more batteries to the bank. A battery bank performs like a chain, pulling only as well as the weakest link. New batteries will perform no better than the oldest cell in the bank. All lead-acid batteries in a bank should be of the same capacity, age, and manufacturer as much as possible.

Why Can't I Use Car Batteries?

Batteries are built and rated for the type of "cycle" service they are likely to encounter. Cycles can be "shallow," reaching 10 to 15% of the battery's total capacity, or "deep," reaching 50 to 80% of total capacity. No battery can withstand 100% cycling without damage, often severe.

Automotive starting batteries and deep-cycle storage batteries are both lead-acid types, but there are important construction and even materials differences. Starting batteries are designed for many, many shallow cycles of 15%

Much like the muscles of your body, wet-cell batteries need regular exercise to maintain good performance. This makes them less than ideal for emergency and back-up power systems.

discharge or less. They deliver several hundred amperes for a few seconds, and then the alternator takes over and the battery is quickly recharged. Deep discharges cause the lead plates to shed flakes and sag, reducing life expectancy. Deep-cycle batteries are designed to deliver a few amperes for hundreds of hours between recharges. Their lead plates are thicker, and use different alloys that don't soften with deep discharges. Neither battery type is well suited to doing the other's job, and will suffer a short ugly life if forced to do the wrong service.

Automotive engine-starting batteries are rated for how many amps they can deliver at a low temperature, aka cold cranking amps (CCA). This rating is not relevant for storage batteries. Beware of any battery that claims to be a deep-cycle storage battery and has a CCA rating.

Sealed or Wet?

Lead-acid batteries come in two basic flavors, traditional "wet cells" or the newer "sealed cells." Wet cells have caps that can be removed. In fact, you have to remove them every couple of months to add distilled water. Wet-cell batteries cost less, their problems are more easily diagnosed, and they're usually the right choice for remote homesteads. On the downside, they do require routine maintenance, they produce hydrogen gas during charging that needs to be vented outside, and they have to ship as hazardous freight, which makes them expensive to move. Much like the muscles of your body, wet-cell batteries need regular exercise to maintain good performance. Without regular cycling, they become "stiff" chemically, and won't perform well until they're limbered up. This makes them less than ideal for emergency and back-up power systems.

This is where sealed batteries come in. Sealed cells do better at sitting around for long periods waiting for activity. Plus, nobody has to remember to water them periodically, and they can be installed in places where hydrogen gassing couldn't be tolerated. Sealed batteries come in two basic styles. Gel batteries have the liquid electrolyte in a jellied form. AGM (absorbed glass mat) batteries use a fiberglass spongelike mat to hold liquid electrolyte between the plates. There is intense debate and rivalry as to which technology is better. AGM is less expensive and fussy to build, and the gels tend to be a bit more robust and longer lived. We offer small sealed batteries as AGM types, and larger sealed batteries as gel types.

Sealed batteries require special charge controls. To prevent gassing and the irreplaceable loss of water, charging voltage on sealed cells needs to be held to a maximum of 14.1 volts (or 2.35 volts per cell). Higher charging voltages, as commonly used for wet cells, will seriously reduce the life expectancy of sealed batteries. Almost all the charge controls we offer are either voltage adjustable, or have a selection switch for sealed or wet-cell batteries.

How Big a Battery Pack Do I Need?

We usually size remote household battery banks to provide stored power for three to five days of autonomy during cloudy weather. For most folks, this is a comfortable compromise between cost and function. If your battery bank is sized to provide a typical three to five days of back-up power, then it will also be large enough to handle any surge loads that the inverter is called upon to start. A battery bank smaller than three days' capacity will get cycled fairly deeply on a regular basis. This isn't good for battery life. A larger battery bank cycled less deeply will cost less in the long run. Banks larger than five days' worth start getting more expensive than a back-up power source (such as a modest-sized generator).

As a general rule of thumb, you'll be better off building your battery pack with the fewest number of cells. Larger cells with more ampacity means fewer interconnects, fewer cells to water, less maintenance time, and a smaller chance that one cell will fail early, causing the entire pack to be replaced before its time. Bigger batteries tend to be higher quality, and also have longer lives.

Occasionally we run into situations with 3/4-horsepower or larger submersible well pumps or stationary power tools requiring a larger battery bank simply to meet the surge load when starting. Call the Real Goods technical staff for help if you are anticipating large loads of this type. Our System Sizing Worksheet at the beginning of Chapter 3 (pp. 117-19) has a quick battery sizing section to help you out.

New Technologies?

Compared to the electronic marvels in the typical renewable energy package, the battery is a very simple, proven technology. Tremendous amounts of research have been directed lately into energy-storage technology. Auto manufacturers are searching desperately for a lightweight battery with high energy density and low cost—The Magic Battery. Several dozen battery technologies are currently under intense development in the laboratory. Several of these are bearing fruit now for cell phones, laptops, and hybrid vehicles, but they are still far too expensive for the amount of energy storage required in a renewable energy system. The nickel-metal hydride battery pack in the Toyota RAV4 EV is reportedly worth $30,000, and it would only make a very modest-sized remote home battery pack.

Lead-acid batteries are also in the laboratory. The possibilities of lead-acid technology are far from tapped out. This old dog is still capable of learning some new tricks. For now we must coexist with traditional battery technology—a technology that is nearly one hundred years old, but is tried and true and requires surprisingly little maintenance. The care, feeding, cautions, and dangers of lead-acid batteries are well understood. Safe manufacturing, distribution, and recycling systems for this technology are in place and work well. Could we say the same for a sulfur-bromine battery?

Higher charging voltages, as commonly used for wet cells, will seriously reduce the life expectancy of sealed batteries.

If your battery bank is sized to provide a typical three to five days of back-up power, then it will also be large enough to handle any surge loads that the inverter is called upon to start.

Dr. Doug's Battery Troubleshooting Guide

What Could Go Wrong?

Famous last words . . . Judging by the phone calls we get, plenty can go wrong. Folks have more trouble with batteries than with any other single component of their systems. Here's a troubleshooting guide.

GOT WATER?

Check the fluid level in all the cells. It should be above the plates, but not above the split ring about ½" below the top of the cell. It's bad for the plates to be exposed to air. Add distilled water as needed. In an emergency, clean tap water is better than no water, but not by much. Distilled water is really the right stuff.

Auto parts stores sell special battery-filler bottles with a nifty spring-loaded automatic shut-off nozzle. Just stick it in the cell, push down, and wait until it stops gurgling. The fluid level will be perfect and there will be no drips or mess. It's a tool worth having. Or consider automatic battery-watering systems, which make refills on large industrial batteries a snap.

CONNECTIONS TIGHT AND CLEAN?

Check battery connections by wiggling all the cables, snug up the hardware, and make sure no corrosion is growing.

Got corrosion? This is like cancer—you've got to clean it all out, or it will return. Disassemble the connection, wire brush and scrape off all the crud you can, then attack what's left with a baking soda and water solution until all traces of corrosion are gone. Clean the bare metal with sandpaper or a wire brush, assemble, and then coat all exposed metal with grease or petroleum jelly to prevent future corrosion.

IS IT CHARGED?

Every system needs some way to monitor the battery state of charge. A voltmeter is the basic minimum. Voltage can be tricky, however. This is a very elastic sort of indicator. Voltage will stretch upwards when a battery is being charged, and it will stretch downwards when a battery is being discharged. Only when the battery has been sitting for a couple hours with no activity can you get a really accurate sense of the battery's state of charge with a voltage reading. First thing in the morning is usually a good time.

A fully charged 12-volt battery will read 12.6 volts, or a bit higher. (Those with 24- or 48-volt packs multiply accordingly.) Lower voltage readings mean a lower state of charge, or that something is turned on. At 12.0 volts, you're in the caution zone, and at 11.6 volts any further electrical use is doing damage that will reduce the battery life expectancy. When charging, the voltage needs to get up to 14.0 volts or higher periodically. If your battery doesn't climb above 13.5 volts, it isn't getting close to being fully charged. You don't need to achieve a full charge every day, but at least once a week is good.

IS IT HEALTHY?

The very best tool for checking your battery's state of health is a hydrometer. This looks like an oversized turkey baster with a graduated float inside it. You must wait at least 48 hours after adding water before you run a hydrometer test. A sample of electrolyte is drawn up from a cell until the float rises. Note at what level it's floating. It measures the specific gravity of the electrolyte to three decimal places. A typical reading would be 1.220. The decimal point is universally ignored, so our example reads as "twelve

---

Voltage can be tricky. This is a very elastic sort of indicator. Voltage will stretch upwards when a battery is being charged, and it will stretch downwards when a battery is being discharged.

When charging a 12-volt battery, the voltage needs to get up to 14.0 volts, or higher periodically. If your battery doesn't climb above 13.5 volts, it isn't getting close to being fully charged.

Volt Readings for your Deep-Cycle Battery

DC Volts 12 · 13

Get the longest life from deep-cycle batteries by keeping the state of charge at 60% or more for lead-acid and 70% for gel cells. For accurate volt readings, don't charge or discharge batteries for 2 hours.

Voltage Reading	State-of-Charge
12.75	100
12.70	95
12.65	90
12.60	85
12.55	80
12.50	75
12.45	70
12.40	65
12.30	55
12.25	50
12.25	45

twenty." Under normal conditions, all the cells in a battery pack should read within 10 points of each other. That would be 1.215 to 1.225 in our example. As a general rule, it doesn't matter how high or low the readings are, so long as they're all about the same. We're only looking for differences between cells.

When a cell starts to go bad, it will read much higher or lower—usually lower—than its neighbor cells. At 20 to 30 points difference, you may only need a good equalizing charge to bring all the cells into line. At 40 to 50 points difference, you probably have a failing cell. Plan for replacement within two or three months. At 60 points or more difference, you have a failed cell that's sucking the life out of all the surrounding cells, and needs to be replaced as soon as possible.

Common Battery Questions and Answers

What voltage should my system run? I notice that components for 12-, 24-, and 48-volt battery banks are available.

Voltage is selected for ease of transmission on the collection and storage side of the system. Low voltage doesn't transmit well, so as home power systems get bigger and are pushing around greater amounts of energy, it's easier to do at higher voltages. Generally, systems under 2,000 watt-hours per day are fine at 12 volts, systems up to 7,000 watt-hours per day are good at 24 volts, and larger systems should be running at 48 volts. These aren't ironclad guidelines by any means. If you need to run hundreds of feet from a wind or hydro turbine, then higher voltage will help keep wire costs within reason. Or if you're heavily invested in 12-volt with a large inverter and 12-volt lights, you can still grow your system without upgrading to a higher voltage.

I'm just getting started on my power system. Should I go big on the battery bank assuming I'll grow into it?

The answer is yes and no. Yes, you should start with a somewhat larger battery bank than you absolutely need, perhaps sizing for four to five days of autonomy instead of the bare minimum three days. Over time, most folks find more and more things to use power for once it's available. But if this is your first venture into remote power systems and battery banks, then we usually recommend that you start with some "trainer batteries." You're bound to make some mistakes and do some regrettable things with your first set of batteries. You might as well make mistakes with inexpensive batteries. So no, don't invest too heavily in batteries your first couple of years. The golf cart–type deep-cycle batteries make excellent trainers. They are modestly priced, will accept moderate abuse without harm, and are commonly available. In three to five years, when the golf cart–type trainers wear out, you'll be much more knowledgeable about what you need and what quality you're willing to pay for.

The battery bank I started with two years ago just doesn't have enough capacity for us anymore. Is it okay to add some more batteries to the bank?

Lead-acid batteries age in service. The new batteries will be dragged down to the performance level and life expectancy of the old batteries. Your new batteries will be giving up two years of life expectancy right off the bat. Different battery types have different life expectancies, so we really need to consider how long the bank should last. For instance, it would be acceptable to add more cells to a large set of forklift batteries at two years of age because this set is only at 10% of life expectancy. But an RV/marine battery at two years of age is at 100% of life expectancy.

I keep hearing rumors about some great new battery technology like "flywheels" that will make lead-acid batteries obsolete in the near future. Is there any truth to this, and should I wait to invest in batteries?

The truth is that several dozen battery technologies are currently under intense development. Some of them, like nickel-metal hydride, look very promising, but none of them are going to give lead-acid a run for your money within the foreseeable future. Lead-acid is also in the laboratory. Lead-acid technology is going to be around, and is going to continue to give the best performance per dollar for a long time to come.

> Generally, systems under 2,000 watt-hours per day are fine at 12 volts, systems up to 7,000 watt-hours per day are good at 24 volts, and larger systems should be running at 48 volts.

Dr. Doug's Battery Care Class

Basic Battery Safety

1. Protect your eyes with goggles and hands with rubber gloves. Battery acid is a slightly dilute sulfuric acid. It will burn your skin after a few minutes of exposure, and your eyes almost immediately. Keep a box or two of baking soda and at least a quart of clean water in the battery area. Flush any battery acid contact with plenty of water. If you get acid in your eyes, flush with clear water for fifteen minutes and then seek medical attention.

2. Tape the handles of your battery tools or treat them with Plastic-Dip so they can't possibly short out between battery terminals. Even small batteries are capable of awesome energy discharges when short circuited. The larger batteries we commonly use in RE systems can easily turn a 10-inch crescent wrench red hot in seconds while melting the battery terminal into a useless puddle, and for the grand finale possibly explode and start a fire at the same time. This is more excitement than most of us need in our lives.

3. Wear old clothes! No matter how careful you are around batteries, you'll probably still end up with holes in your jeans. Wear something you can afford to lose, or at least have holes in.

4. Now stop thinking, "Oh, none of that will happen to me!" I used to think that too. I'm knowledgeable, professional, and work carefully. Yet I've managed to melt terminals, start fires, blow up batteries, and come near to blinding a roommate over the years. Don't take chances: Safety measures are easy, and the potential harm is permanent.

The Yearly Maintenance for Wet-Cell Batteries

1. Check the water level after charging and fill with distilled water. Batteries use more water when they're being fully charged every day, and they'll also use more water as they get older. Check the water level every couple of months. Fluid level should be above the plates, but not above the split ring about ½" below the

Simple Lead-Acid Battery Care

1. Take frequent voltage readings. The voltmeter is the simplest way to monitor battery activity and approximate state of charge.

2. Baking soda neutralizes battery acid. Keep at least a couple boxes on hand, inside the battery enclosure, in case of spills or accidents.

3. Batteries should be enclosed and covered to prevent casual access to terminals. Wet-cell batteries must be vented to prevent accumulation of hydrogen and other harmful gasses. A 2" pvc vent at the highest point in the battery box is sufficient.

4. Do not locate any electrical equipment inside a battery compartment. It will corrode and fail.

5. Check the water level of your batteries once a month until you know your typical usage pattern. Use distilled water only. The trace minerals in tap water kill battery capacity. Get a battery-filler bottle from the auto parts store to make this job easy.

6. Be extra careful with any metal tools around batteries. Tape handles or treat with Plastic-Dip so tools can't possibly short between terminals if dropped.

7. Protect battery terminals and any exposed metal from corrosion. After assembly, coat them with grease or Vaseline, or use one of the professional sprays to cover all exposed metal.

8. Never smoke or carry an exposed flame around batteries, particularly when charging.

9. Wear old clothes when working with batteries. Electrolyte loves to eat holes.

10. Chemical processes slow down at colder temperatures. Your battery will act as if it's smaller. At 0°F, batteries lose about 50% of their capacity.

In cold climates, install your batteries in the warmest practical location.

11. All batteries self-discharge slowly, and will sulfate if left unattended for long periods with no trickle charging. Clean, dry battery tops reduce self-discharging.

12. Batteries tend to gain a memory for typical use, and will resist wider discharge cycles initially. Like the muscles of your body, periodic stretching exercises increase strength and flexibility. It's good to do an occasional deep discharge and equalizing charge to maintain full battery capacity.

13. A few large cells are better than many small cells when building a battery pack. Larger cells tend to have thicker plates and longer life expectancies. Fewer interconnections and cells mean less maintenance time, and less chance of one cell failing early and requiring the entire battery pack to be replaced early.

top of the cell. It's bad for the plates to be exposed to air. Add distilled water as needed. In an emergency, clean tap water is better than no water, but not by much. Distilled water is really the right stuff.

Don't fill before charging; the little gas bubbles will cause the level to rise and spill electrolyte. Only pure water goes back into the battery. The acid doesn't leave the cell, so you never need to replenish it.

Auto parts stores sell special battery-filler bottles with a nifty spring-loaded automatic shut-off nozzle. Just stick it in the cell, push down, and wait until it stops gurgling. The fluid level will be perfect and there will be no drips or mess. It's a tool worth having. Consider an automatic battery watering device if you have a large industrial battery bank.

2. Clean the battery tops. The condensed fumes and dust on the tops of the batteries start to make a pretty fair conductor after a few months. Batteries will discharge significantly across the dirt between the terminals. That power is lost to you forever! Sponge off the tops with a baking soda and water solution, or use the battery cleaner sprays you can get at the auto parts store. Follow the baking soda cleaner with a clear water rinse. Make sure that cell caps are tight and that none of the cleaning solution gets into the battery cell! This stuff is deadly poison to the battery chemistry. Clean your battery tops once or twice a year.

3. Clean and/or tighten the battery terminals. Lead is a soft metal and will gradually "creep" away from the bolts. If you were smart and coated all the exposed metal parts around your battery terminals with grease or Vaseline when you installed them, they'll still be corrosion-free. If not, then take them apart, scrub or brush as much of the corrosion off as possible, then dip or brush with a baking soda and water solution until all fizzing stops and then scrape some more. Keep this up until you get all the blue/green crud off. (It's like cancer, if you don't get it all, it will return.) Then carefully cover all the exposed metal parts with grease when you put it back together. If your terminals are already clean, then just gently snug up the bolts.

4. Run a hydrometer test on all the cells. Wait at least 48 hours after adding water before you run this test. Your voltmeter is great for gauging battery state of charge; the hydrometer is for gauging state of health. Use the good kind of hydrometer with a graduated float (not the cheapo floating-ball type). We sell one for $8 (#15-702). A sample of electrolyte is drawn up

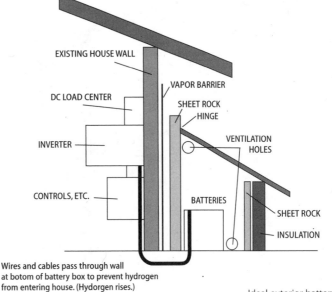

Wires and cables pass through wall at bottom of battery box to prevent hydrogen from entering house. (Hydorgen rises.)

Ideal exterior battery enclosure

from a cell until the float rises. Note at what level it's floating. It measures the specific gravity of the electrolyte to three decimal places. A typical reading would be 1.220. The decimal point is universally ignored, so our example reads as "twelve twenty." Under normal conditions, all the cells in a battery pack will read within 10 points of each other. That would be 1.215 to 1.225 in our example. What we're looking for in this test is not the state of charge, but the difference in points between cells. In a healthy battery bank, all cells will read within 10 points of each other. Any cell that reads 20 to 25 points lower is probably starting to fail, although you should run an equalizing charge before casting judgment. You may have three to six months to round up a replacement set. At 60 points difference or more, the bad cell is sucking the life out of your battery bank and needs to get out now! Don't pay any attention to the color-coded good-fair-recharge markings on the float. These only pertain to automotive batteries, which use a slightly hotter acid.

5. Run an equalizing charge (wet-cell batteries only). An equalizing charge is a controlled slight overcharge. This will even out any small differences among cells that have

BATTERY STATE-OF-CHARGE

Voltage Reading	Percent of Full Charge
12.6	100%
12.4–12.6	75–100%
12.2–12.4	50–75%
12.0–12.2	25–50%
11.7–12.0	0–25%

Ideal interior battery enclosure

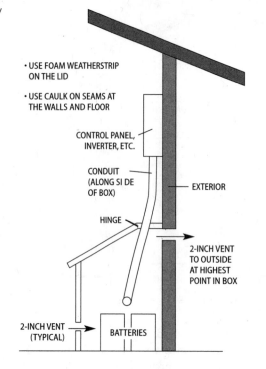

- USE FOAM WEATHERSTRIP ON THE LID
- USE CAULK ON SEAMS AT THE WALLS AND FLOOR

CONTROL PANEL, INVERTER, ETC.

CONDUIT (ALONG SI DE OF BOX)

HINGE

EXTERIOR

2-INCH VENT TO OUTSIDE AT HIGHEST POINT IN BOX

2-INCH VENT (TYPICAL)

BATTERIES

developed over time, and help push any sulfation back into the electrolyte. It's like a minor tune-up for your batteries. It's a good idea to do an equalization every one to six months. Equalizing is more important in the winter when batteries tend to run at lower charge levels than in the summer. *Do NOT equalize sealed batteries! Permanent damage will occur!*

Run your batteries up to about 15.0 volts (or 2.50 volts per cell) and hold them at between 15.0 and 15.5 volts (2.5 to 2.6 volts per cell) for two to four hours. You'll get fairly vigorous gassing and bubbling on all cells. Do not take the caps off or loosen them during the equalizing charge, you'll just lose more water and spatter acid over the top of the battery. Do NOT top up the water just before equalizing. All that gas production can raise the electrolyte level above the cell top and get messy. Top up the water a day or two later.

The Yearly Maintenance for Sealed Batteries

1. Clean the battery tops. The crud and dust on the tops of the batteries start to make a pretty fair conductor eventually, even on sealed batteries. Batteries can discharge significantly across the dirt between the terminals. That power is lost to you forever! Sponge off the tops with a baking soda/water solution, or use the battery cleaner sprays you can get at the auto parts store. Follow the baking soda cleaner with a clear water rinse. Cleaning the tops of your sealed batteries once a year is sufficient.

2. Clean and/or tighten the battery terminals. Lead is a soft metal and will gradually "creep" away from the bolts. If your terminals are already clean, then just gently snug up the bolts. Terminal corrosion is rare, but not impossible on sealed batteries. If needed, take them apart, scrub or brush as much of the corrosion off as possible, then dip or brush with a baking soda and water solution until all fizzing stops and then scrape some more. Keep this up until you get all the blue/green crud off. (It's like cancer, if you don't get it all, it will return.) Then carefully cover all the exposed metal parts with grease when you put it back together.

LARGE BATTERIES, CHARGERS, AND ACCESSORIES

Battery Book for Your PV Home

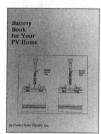

This booklet by Fowler Electric Inc. gives concise information on lead-acid batteries. Topics covered include battery theory, maintenance, specific gravity, voltage, wiring, and equalizing. Very easy to read and provides the essential information to understand and get the most from your batteries. Highly recommended by Dr. Doug for every battery-powered household. 22 pages, softcover. USA.

80104   Battery Book for Your PV Home $8

Battery Care Kit

Our kit includes the informative *Battery Book for Your PV Home,* and a full-size graduated battery hydrometer. The *Battery Book* explains how batteries work and shows how to care for your batteries and get the best performance. Every battery-powered household needs a copy. The hydrometer is your most

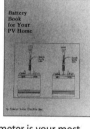

important tool for determining battery state of health. Detect and fix small problems before they require complete battery bank replacement. *Use Tip: The color-coded red, yellow, and green zones on the hydrometer are calibrated for automotive starting batteries: ignore them. Just look for differences among cells.*

15754   Hydrometer/Book Kit $14

POWER SUPPLY PACKAGES

Here is a handy way to carry a 12-volt power supply for laptops, cellphones, digital cameras, CD players, etc. Our power supply packages will accept solar, automotive, or 120vac charging, so no matter where you are, you should be able to rustle up enough power to keep all the equipment happy. Batteries are sealed, spill-proof, airline approved, and replaceable, with a three- to four-year life expectancy. LED indicators show battery state of charge. All have a one-year manufacturer's warranty. China.

Power Depot

Our smallest power supply is 12-volt DC only with a 7Ah battery. It has a nice detachable 13-watt fluorescent work light that will run over 10 hours on one charge. Weighs 7.0 lbs. 8.25"H x 5.8"W x 3.25"D.

15856   Power Depot $59

Power Zone

Our medium-size power supply delivers AC or DC and comes with a 12Ah battery. The built-in inverter delivers 150 watts AC. It has a single DC lighter socket and a DC charge input port. The pair of 6-watt fluorescent lights will run over 15 hours on one charge. The built-in AC charger stores its cord in a nice rear compartment, which also stores the DC adapter cord and the DC jumper cables. Weighs 17.6 lbs. 12"L x 9.5"H x 4.1"D.

15857   Power Zone $149

Power Depot 2000

Our larger power supply delivers AC or DC, features an 18Ah battery, a built-in 300-watt AC inverter with dual outlets, dual DC sockets, a built-in AC charger, and a small work light. Jumper cables with side storage pockets are included. Weighs 18.0 lb. 16"L x 14.5"H x 4.5"D.

15858   Power Depot 2000 $169

Sealed Replacement Batteries

Do not buy until your battery is ready for replacement! Batteries will die sitting on the shelf. We buy in small batches, or ship directly from the manufacturer to ensure fresh batteries.

15200	**12V/7Ah**	
	(6"L x 2.5"W x 3.9"H, 5.5 lbs.)	**$28**
15202	**12V/12Ah**	
	(6"L x 3.9"W x 3.9"H, 8.7 lbs.)	**$50**
15209	**12V/17.5Ah**	
	(7.2"L x 3"W x 6.6"H, 12.8 lbs.)	**$59**

Portable Power System

Our Portable Power Generator can supply lighting or run entertainment equipment at your occasional-use cabin or on your boat. Also ideal for emergency use and disaster relief, the system is housed in a rugged weathertight Rubbermaid box (21" x 15" x 12"). Already pre-wired are a 15-amp PV controller, a 150-watt Statpower Prowatt inverter for AC output, a pair of lighter plugs for DC output, an externally mounted DC voltmeter "fuel gauge," an input plug for PV module(s), and all the fusing to keep everything safe. All you need to add is the PV module(s) or other charging source of choice, and a 98Ah (or smaller) battery. All components except the battery and PV module(s) are pre-wired and ready to rock. The Portable Power Generator weighs 9 lbs. as delivered, or 81 lbs. with the maximum-capacity sealed battery listed below. One-year manufacturer's warranty. USA.

12107	Portable Power Generator	$699
15214	Solar Gel Battery 12V, 98 Ah	$189

PV module and battery sold separately; call for sizing.

XPower 1500 Powerpack

Our largest portable powerpack, the XPower 1500 delivers 120vac and 12vdc. It packs a large 60Ah sealed battery and a 1,500-watt inverter with dual outlets in a nice, rugged wheeled cart with a removable 38"-high waist handle for ease of movement. Includes a 5-amp battery charger for when AC power is available from a wall plug or generator, and a heavy-duty DC charging cord for lighter socket or a solar panel of your choice. Has a battery charge indicator. Weighs 60 lbs. Measures 14.8"H x 15.6"W x 12.3"D without handle. One-year manufacturer's warranty. Canada.

17-0283 XPower 1500 Powerpack $380

WET-CELL BATTERIES

"Golf Cart"-Type Deep-Cycle Batteries

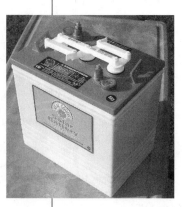

These batteries are excellent "trainer" batteries for folks new to battery storage systems. First-time users are bound to make a few mistakes as they learn the capabilities and limitations of their systems. These are true deep cycle batteries and will tolerate many 80% cycles without suffering unduly. Even with all the millions of dollars spent on battery research, conventional wisdom still indicates that these lead-acid "golf cart" batteries are the most cost-effective battery storage solution for smaller to medium-sized systems. Typical life expectancy is approximately three to five years. Cycle life expectancy is typically about 225 cycles to 80% depth of discharge. USA.

Battery Voltage: 6 volts
Rated Capacity: 220 amp-hours
L x W x D: 10.25" x 7.25" x 10.25"
Weight: 63 lb/28.6 kg

15101 Deep-Cycle Battery, 6V/220Ah $79
Free shipping in the continental U.S. for 10 or more.

L-16 Series Deep Cycle Batteries

For larger systems or folks who are upgrading from the "golf cart" batteries, these L-16s have long been the workhorse of the alternative energy industry. The larger cell sizes offer increased ampacity and lower maintenance due to larger water reserves. L-16s typically last from six to eight years. Cycle life expectancy is typically about 300 cycles to 80% depth of discharge. USA.

Battery Voltage: 6 volts
Rated Capacity: 370 amp-hours
L x W x D: 11.25" x 7" x 16"
Weight: 128 lb/58 kg

15102 Deep Cycle Battery, 6V/370Ah $199
Free shipping in the continental U.S. for 10 or more.

Industrial-Quality IBE Batteries

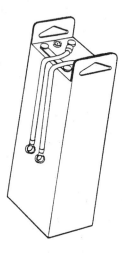

IBE industrial batteries are some of the best batteries available. With proper maintenance, they will easily last fifteen to twenty years or more. Each cell is individually packaged in a steel or plastic case with two lifting handles and coated with acid-resistant paint. These individual cells can be carried by two people, making them the best choice for sites without a forklift, or with difficult access. The industrial 2-volt cells provide increased performance, greater reliability, and less frequent maintenance. Every performance and life-enhancing trick known to the battery trade has been incorporated into these cells. Cycle life expectancy is 1,500 cycles to 80% depth of discharge, or 5,000 cycles to 20% depth of discharge.

Priced as 6-cell, 12-volt batteries. For 24-volt systems you must purchase two batteries. These batteries are totally recyclable. Interconnects and input/output leads are sold separately. USA.

15600	IBE Battery, 12V/1015Ah	$1,870
15601	IBE Battery, 12V/1107Ah	$1,960
15602	IBE Battery, 12V/1199Ah	$2,030
15603	IBE Battery, 12V/1292Ah	$2,240
15605	IBE Battery, 12V/1464Ah	$2,340
15606	IBE Battery, 12V/1568Ah	$2,520
15607	IBE Battery, 12V/1673Ah	$2,740

Specify system voltage and plastic or steel case when ordering. Allow 4–6 weeks for delivery. Plastic cases may require an additional 3–4 weeks.
Shipped freight collect.

Pre-assembled Industrial-Quality Hawker Solar Batteries

The lowest-cost industrial battery choice for those with forklift access

From Hawker Power, the world's largest battery manufacturer (Hawker recenty bought Yuasa General), these very high-quality industrial batteries consist of six propylene cells in what the industry calls a steel tray.

Most folks would call it a steel box with a flip top. Features include heat-sealed cell covers, molded lead cell interconnectors, and free freight! Yep, free freight in the continental United States to a business with a forklift, or to a trucking terminal. You must have a forklift to move these heavy, pre-assembled cells. Sold with six cells, or 12-volts, per tray. Purchase multiple trays for higher system voltages. With proper maintenance, these cells will last fifteen to twenty years. Cycle life expectancy is 1500 cycles to 80% depth of discharge, or 5000 cycles to 20% depth of discharge. Interconnect cables are only needed between 12-volt trays; no cell interconnects required. Five-year manufacturer's warranty. USA.

Hawker Solar 12-Volt Battery Packs (formerly Yuasa)

Item #	Ah @ 20hrs	Weight (lb.)	Size (") L x W x H	Price
15799	632	498	30.75 x 7.8 x 25	$1,230
15801	845	630	20.2 x 13 x 25	$1,520
15803	1,055	762	39 x 9 x 25	$1,775
15805	1,270	918	29.8 x 13.5 x 25	$2,050
15807	1,482	1,068	33.7 x 13 x 25	$2,475
15809	1,690	1,260	39 x 13.5 x 25	$2,895

SEALED BATTERIES

Sealed AGM-Type Batteries

These small, sealed batteries may be just what your next expedition needs to keep those laptops running and lights on after dark. The 6V/4Ah size is the replacement

for several brands of solar sensor lights. The 12V/7Ah size is a replacement battery for our Multi Power Supply, and is a useful 12-volt storage battery for small power systems. These smaller batteries use an absorbed glass mat construction. AGM costs a bit less than gel batteries, but is slightly less durable. Usual life expectancy is three to four years. USA.

15198	6V/4Ah Sealed Battery (2.8" L x 1.9" W x 4.25" H 1.95 lb.)	$14.50
15200	12V/7Ah Sealed Battery (6" L x 2.5" W x 3.9" H 5.5 lb.)	$28
15202	12V/12Ah Sealed Battery (6" L x 3.9" W x 3.9" H 8.7 lb.)	$50

Real Goods Solar Gel Batteries

These true deep cycle gel batteries come highly recommended by the National Renewable Energy Labs. They are completely maintenance-free, with no spills or fumes. Because the electrolyte is gelled, no stratification occurs, and no equalization charging is required. Self-discharge is under 2% per month. May be airline transported as approved by DOT (Department of Transportation), ICAO (International Commercial Airline Organization), and IATA (International Airline Transport Association).

To avoid gassing, gel batteries must be charged at lower peak voltages; typically, 13.8 to 14.2 volts maximum is recommended. Higher charging voltages will drastically shorten life expectancy. Life expectancy depends on battery size and use in service, but no sealed battery will perform better or longer than these. Typical life expectancy is five to ten years. Smaller batteries will ship by UPS; larger sizes are shipped freight collect. Free shipping on any order of ten or more batteries! Batteries will be shipped from the nearest warehouse in AZ, CA, CO, GA, IL, MD, MO, NJ, NV, TX, UT, or WA. Amp-hour rating at standard 20-hour rate. These batteries are all supplied with flag-type terminals for secure bolt up connections. Dimensions are L x W x H including terminals. USA.

15210	32AH/12V SU1 Solar Gel Battery, 23.4 lb. 7.75" x 5.2" x 7.25"	$69
15211	51AH/12V S22NF Solar Gel Battery, 37.0 lb. 9.4" x 5.5" x 9.25"	$119
15212	74AH/12V S24 Solar Gel Battery, 52.0 lb. 10.25" x 6.75" x 9.8"	$149
15213	86AH/12V S27 Solar Gel Battery, 62.7 lb. 12.75" x 6.75" x 9.25"	$159
15214	98AH/12V S31 Solar Gel Battery, 69.5 lb. 12.9" x 6.75" x 9.75"	$189
15215	183AH/12V S4D Solar Gel Battery, 127.0 lb. 20.75" x 8.5" x 10.6"	$339
15216	225AH/12V S8D Solar Gel Battery, 157.0 lb. 20.75" x 11" x 10.6"	$399
15217	180AH/6V S6VGC Solar Gel Battery, 68.4 lb. 10.25" x 7.2" x 10.8"	$189

Hawker Envirolink Sealed Industrial Battery Packs

The best choice for emergency power backup systems

The Hawker Envirolink series offers true gel construction, not the cheaper, lower life-expectancy AGM construction. These batteries are the top-quality choice for long-life, low-maintenance emergency power back-up systems. No routine maintenance is required, and this battery type thrives on float service. Life expectancy is ten to fifteen years with proper care and a charger/controller that keeps voltage below 2.35v/cell. Delivered in preassembled 12- or 24-volt packs with smooth steel cases and no exposed electrical contacts. Cycle life expectancy is 1,250 cycles to 80% depth of discharge. Interconnect cables are only needed between 12- or 24-volt trays; cell interconnects are soldered in place at the factory. Five-year manufacturer's warranty. USA.

Hawker Envirolink Sealed Industrial 12-Volt Battery Packs

Item #	Ah @ 20hrs	Weight (lb.)	Size (") L x W x H	Price
17-0296	369	390	26.3 x 6.5 x 23	$1,530
15819	553	564	31 x 7.75 x 23	$2,050
15820	925	885	31.13 x 13 x 23	$3,270
15821	1,110	1,050	29.3 x 13 x 23	$3,865
15822	1,480	1,374	25.7 x 19.6 x 23	$5,000

Hawker Envirolink Sealed Industrial 24-Volt Battery Packs

Item #	Ah @ 20hrs	Weight (lb.)	Size (") L x W x H	Price
17-0297	369	780	25.7 x 11 x 23	$2,895
15823	553	1,128	31 x 13 x 23	$3,865
15825	925	1,770	38.5 x 16.6 x 23	$6,175
15826	1,110	2,100	38.7 x 19.7 x 23	$7,280
15827	1,480	2,748	38.7 x 25.6 x 23	$9,420

Iota DLS-Series Battery Chargers

Efficient, reliable, and affordable

Iota chargers are solid-state, high-efficiency chargers, very similar to the now-out-of-business Todd charger we've offered for years. Iota features include light, compact construction; automatic fan cooling; reverse-polarity protection; easy terminal access; external, replaceable fuses; rugged reliability; and UL listing. The Iota is a safe constant-voltage charger, so as the battery voltage gradually rises with charging, the charger will gradually reduce amperage flow.

Chargers are available in 12-volt models at 30, 55, 75, and 90 amps; in 24-volt models at 25 and 40 amps; and in a 48-volt model at 10 amps. Standard charger models are safe to use with any battery type, or with any battery pack that is constantly float-charged. Each charger has switchable dual voltage output, with about a 0.6-volt difference. Use low voltage for float charging, or high voltage for the most rapid battery charging. Voltage range is selected by inserting a phone-type plug; a customer-supplied switch can be inserted easily for manual control, or Iota offers their optional IQ switch that delivers automatic 3-stage smart charging. We offer three custom high-volt 12-volt models for wet-cell battery charging designed to keep generator run time to a minimum. The hi-volt models only are set at 14.8/15.4-volt. Two-year manufacturer's warranty. USA.

15791	12V/30A Std. Charger	$135
15793	12V/55A Std. Charger	$170
15795	12V/75A Std. Charger	$335
17-0277	12V/90A Std. Charger	$380
15792	12V/30A Hi-Volt Charger	$150
15794	12V/55A Hi-Volt Charger	$185
15796	12V/75A Hi-Volt Charger	$350
17-0278	24V/25A Std. Charger	$260
17-0279	24V/40A Std. Charger	$390
17-0280	48V/10A Std. Charger	$315
15853	Iota IQ3 Switch	$35

Std. Charger Model	12V/30A	12V/55A	12V/75A	12V/90A	24V/25A	24V/40A	48V/10A
Max. Amp Draw (AC):	8A	15A	18A	21A	12A	20A	15A
Size: 6.5" x 3.5" x	7"	7"	10"	10"	7"	10"	10"
Weight:	5.5 lb.	7 lb.	10 lb.	8 lb.	5.5 lb.	8 lb.	8 lb.
DC Output Voltages:	13.6–14.2	13.6–14.2	13.6–14.2	13.6–14.2	27.2–27.7	27.2–27.7	54.0–55.0

BATTERY ACCESSORIES

DC-to-DC Converters

Got a DC appliance you want to run directly, but your battery is the wrong voltage? Here's your answer from Solar Converters. These bi-directional DC-to-DC converters work at 96% efficiency in either direction. For instance, the 12/24 unit will provide 12V @ 20A from a 24V input, or can be connected backwards to provide 24V @ 10A from a 12V input. Current is electronically limited and has fuse backup. Four units are offered: The 12/24 unit described above; a larger 12/24 unit that delivers 25A or 50A depending on direction, a 24/48 unit that delivers 5A or 10A depending on direction, and a 12/48 unit that delivers 2.5A or 10A depending on direction. The housing is a weatherproof NEMA 4 enclosure measuring 4.5" x 2.5" x 2", good for interior or exterior installation (keep out of direct sunlight to keep the electronics happy). Rated for –40°F to 140°F (–40°C to 60°C). One-year manufacturer's warranty. Canada.

25709	12/24 DC-to-DC Converter, 20A	$185
25782	12/24 DC-to-DC Converter, 50A	$399
25710	24/48 DC-to-DC Converter, 10A	$210
25711	12/48 DC-to-DC Converter, 10A	$210

Full-Size Battery Hydrometer

This full-size graduated-float-style specific gravity tester is accurate and easy to use in all temperatures. Specific gravity levels are printed on the tough, see-through plastic body. It has a one-piece rubber bulb with a neoprene tip. USA. *Note: Ignore the color-coded green/yellow/red zones on the float. These are calibrated for automotive starting batteries, which use a denser acid solution. Storage batteries will read usually in the red or yellow zones. Just watch for differences among individual cells.*

15702	Full-Sized Hydrometer	$7.95

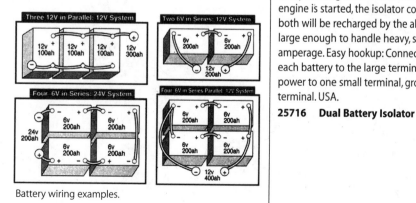

Battery wiring examples.

Code-Approved Inverter Cables

Our inverter and battery cables are all flexible code-approved cable with color coding for polarity and shrink-wrapped lugs for 3/8" bolt studs. These are two-lug versions (lugs on both ends). USA.

15851	2/0, 3' Inverter Cables, 2 lug, single	$30
15852	4/0, 3' Inverter Cables, 2 lug, single	$40
27311	4/0, 5' Inverter Cables, 2 lug, pair	$120
27312	4/0, 10' Inverter Cables, 2 lug, pair	$205
27099	2/0, 5' Inverter Cables, 2 lug, pair	$85
27098	2/0, 10' Inverter Cables, 2 lug, pair	$140

Heavy-Duty Insulated Battery Cables

Serious, beefy, 2/0 or 4/0 gauge battery interconnect cables for systems using 1,000-watt and larger inverters. These single code-approved cables have 3/8" lug terminals on both ends, and are shrink-wrapped for corrosion resistance. Pick your gauge, length, and color. USA.

15841	4/0, 20" Black	$32
15840	4/0, 20" Red	$32
15839	4/0, 13" Black	$27
15850	2/0, 20" Black	$24
15849	2/0, 20" Red	$24
15848	2/0, 12" Black	$16
15847	2/0, 9" Black	$14

Battery Selector Switch

Few renewable energy homes use dual battery banks, but most boats and RVs do. This high-current switch permits selection between Battery 1 or Battery 2, or both in parallel. The off position acts as a battery disconnect. Wires connect to 3/8" lugs. Capacity is 250 amps continuous, 360 amps intermittent. Okay for 12- or 24-volt systems. 5.25" diameter x 2.6" deep. Marine UL-listed. USA.

25715	Battery Switch	$34.95

Dual Battery Isolator

For RVs or boats, this heavy-duty relay automatically disconnects the household battery from the starting battery when the engine is shut off. This prevents inadvertently running down the starting battery. When the engine is started, the isolator connects the batteries, so both will be recharged by the alternator. The relay is large enough to handle heavy, sustained charging amperage. Easy hookup: Connect a positive wire from each battery to the large terminals. Connect ignition power to one small terminal, ground the other small terminal. USA.

25716	Dual Battery Isolator	$23

Rechargeable Batteries and Chargers

Americans toss over three billion small consumer batteries into the landfill every year.

If you've waded this deeply into the *Sourcebook,* we probably don't need to bore you with why you ought to be using rechargeable batteries. They're far less expensive, they don't add toxins to the landfill, and they don't perpetuate a throw-away society. You know all that already, right? Well, just in case you want a review, we'll provide a brief one. Already feeling secure in your knowledge and just want to choose the best rechargeable battery or charger for your purpose? Then skip ahead to "Choosing the Right Battery."

Why We Need to Use Rechargeable Batteries

Americans toss over three billion small consumer batteries into the landfill every year. The rest of the world adds another few billion to the total. The vast majority of these are alkaline batteries. Alkaline batteries are the common Duracell, Energizer, and Eveready brands of batteries you find in the grocery store checkout lane. While domestic battery manufacturers have

The latest generation of NiMH (nickel-metal hydride) AA-size batteries actually has *more* storage capacity than the premium name-brand alkalines!

refined their formulas in the past few years to eliminate small amounts of mercury, alkaline batteries are still low-level toxic waste. Casual disposal in the landfill may be acceptable to your local health officials, but what a waste of good resources! Our grandchildren, maybe even our children, will be part of the next gold rush, when we start mining our twentieth-century landfills for all the refined metals and other depleted resources that are waiting there.

So throw-away batteries are wasteful and a health hazard. They're also surprisingly expensive. Alkaline cells cost around $0.90 to $2.50 per battery, depending on size and brand. Use it once, and then throw it away. In comparison, rechargeable nickel-metal hydride cells—including the initial cost of the battery, a charger, and one or two cents worth of electricity—cost $0.04 to $0.10 per cycle, assuming a very conservative lifetime of 400 cycles. Chargers outlive batteries, so these rechargeable cost figures err on the high side. Typically, throw-away batteries cost $0.10/hour to operate, while rechargeable batteries cost only $0.001/(1/10 of one cent) per hour.

ALKALINE BATTERIES VERSUS RECHARGEABLES
What's the smart choice?

2,000 Alkaline Batteries @ $2.⁸⁵ / 4 pack

4,000 Ah for $1,425.⁰⁰

=

4 Nimh Batteries $12.⁰⁰
1 Charger $29.⁰⁰
2,000 Charges @ 1¢ ea. $20.⁰⁰

4,000 Ah for $61.⁰⁰

When *Not* to Use Rechargeable Batteries

Rechargeable batteries cost far less, they reduce the waste stream, and they usually present fewer disposal health hazards. What's not to love about them? It isn't a perfect world; rechargeable technology does have downsides. Specifically, lower capacity, lower operating voltage, self-discharge, and, in the case of nicad cells only, toxic waste. Let's look at each issue individually.

Lower capacity. This disadvantage is rapidly disappearing. The latest generation of NiMH (nickel-metal hydride) AA-size batteries actually has *more* storage capacity than the premium name-brand alkalines! As this trend continues and spreads to the larger C- and D-size cells, capacity simply won't be an issue. NiMH batteries will support high-drain appliances such as digital cameras or remote-control toys longer than any other battery chemistry, including throw-away batteries. And they will do it hundreds of times before hitting the trash.

Lower operating voltage. Nicad and NiMH batteries operate at 1.25 volts per cell. The alkaline cells that many battery-powered appliances are expecting to use operate at 1.50 volts per cell. This quarter of a volt is sometimes a problem for voltage-sensitive appliances, or devices such as large boom boxes that use many cells in series. The cumulative voltage difference over eight cells in series can add up to a problem if the device isn't designed for rechargeable battery use.

Self-discharge. All battery chemistries discharge slowly when left sitting. Some chemistries, like the alkaline and lithium cells used for throw-away batteries, have shelf lives of five to ten years. Rechargeable chemistry will self-discharge much faster, with shelf lives of two to four months. These batteries aren't the ones you want in your glove box emergency flashlight, and probably aren't the best possible choice for battery-powered clocks.

Toxic nicad cells. Nicad cells contain the element cadmium. Human beings do not fare well when ingesting even tiny amounts of cadmium. Real Goods stopped selling nicad batteries several years ago as soon as a better alternative was available. These cells absolutely must be properly recycled. We accept all nicad batteries returned to us for recycling. Radio Shack stores also accept nicads for recycling. Do not dispose of nicad cells casually! They are toxic waste! Now, that said, let us point out that the newer, higher-capacity nickel-metal hydride batteries are not toxic at all, and will recharge just fine in any nicad battery charger you might have already.

Rechargeable batteries aren't the ones you want in your glove box emergency flashlight, and probably aren't the best possible choice for battery-powered clocks.

RECHARGEABLE BATTERY COMPARISONS

	StandardAlkaline	Rechargeable Alkaline	NiMH
Volts	1.5	1.5	1.25
Capacity (compared to alkaline)	100%	90% initially Diminishes each cycle	70 to 110% No loss with cycles
Capacity in mAh AAA AA C D	750 mAh 2,000 mAh 5,000 mAh 10,000 mAh	750 mAh 1,800 mAh 3,000 mAh 7,200 mAh	750 mAh 2,200 mAh 3,500 mAh 7,000 mAh
Avg. Recharge Cycles	1	12–25	400–600
Self-Discharge Rate	Negligible; 5-year shelf life	Negligible; 5-year shelf life	Modest; 15% loss @ 60 days
Strengths	1.5-volt standard Long shelf life High capacity	1.5 volt standard Long shelf life Good capacity	Sustains high draw No voltage fade Many recharges Nontoxic
Weaknesses	Highest cost Throw-away	Loses capacity each cycle Special charger required	1.2 volt Self-discharges
Best Uses	Clocks Emergency stuff	Clocks Emergency stuff Toys, games, radios, flashlights	Digital cameras! Remote-control toys Flashlights; Any high-drain appliance

Choosing the Right Battery

There are three basic rechargeable battery chemistries to choose from: rechargeable alkaline, nickel-metal hydride (NiMH), and nickel cadmium (nicad). Each has strengths and weaknesses, although nicads compare so poorly now that we aren't even going to dignify them by charting. We've summarized everything and made the comparisons with standard throwaway alkaline batteries in the chart on page 179.

Regardless of your chemistry choice, buy several sets of batteries so you can always have one set in the charger while another set is working for you.

Choosing the Right Battery Charger(s)

Now that you've seen the choices in rechargeable batteries, you'll also need to decide which charger(s) to use. The three types of chargers are AC charger (plugs into a wall socket), solar charger (charges directly from the sun), and 12-volt charger (plugs into your car's cigarette lighter or your home's renewable energy system).

SOLAR CHARGERS

The great benefit of solar battery chargers is that you don't need an outlet to plug them into,

you just need the sun! They can be used anywhere the sun shines. However, most solar chargers work more slowly than plug-in types, and it may take several days to charge up a set of batteries. This needn't be a problem. We strongly recommend you buy an extra set of batteries with your solar charger, so you can charge one set while using the other. Given time and patience (prerequisites for a prudent, sustainable life in any case), solar chargers are the best way to go.

DC CHARGERS

The 12-volt charger will recharge your batteries from a car, boat, or from your home's 12-volt renewable energy system (if you're so blessed). It provides a variable charge rate depending on battery size. Most AC chargers put the same amount of power into every battery, regardless of size. Even though the 12-volt charger is a fast charger, it will not drain your car battery unless you forget and leave it charging for several days. Some 12-volt chargers can overcharge batteries, however, so pay close attention to how long it takes to charge, and don't leave them in there much longer. Leaving them in the 12-volt charger for one-and-a-half times the recommended duration will not harm the batteries.

AC CHARGERS

The AC (plug in the wall) chargers are by far the most convenient for the conventional utility-powered house. Most of us are surrounded by AC outlets all day long and all we have to do is plug in the charger and let it go to work. All these chargers turn off when charging is complete, and go into a trickle charge maintenance mode. Your batteries will always be fully topped up and waiting for you. Most AC chargers automatically figure out what chemistry and size each battery is, then charges it accordingly.

Frequently Asked Questions on Rechargeable Batteries

Is it true that rechargeable batteries develop a "memory?"

This was a nicad problem many years ago. If you only use a little of the battery's energy every time, and then charge it up, the battery will develop a "memory" after several dozen of these short cycles. It won't be able to store as much energy as when new. This was only true with nicads, not with NiMHs or alkalines. Improved chemistry has practically eliminated any memory problems. Short cycles have never been a problem for NiMH or alkaline batteries.

How do I tell when a NiMH (or nicad) battery is charged?

Because these batteries always show the same voltage from 10 to 90% of capacity, it's tricky to know when they're charged. A voltmeter tells little or nothing about the battery's state of charge. Either use one of the smart chargers like the AccuManager, or time the charge cycle. (Calculate the recommended charging time by dividing the battery capacity by the charger capacity and add 25% for charging inefficiency. For instance: A battery of 1,100 mAh capacity in a charger that puts out 100 milliamps will need eleven hours, plus another three hours for charging inefficiency, for a total of fourteen hours to recharge completely.) A cruder method to use is the touch method. When the battery gets warm, it's finished charging. (This method is only applicable with the plug-in chargers.)

My new NiMH (or nicad) batteries don't seem to take a charge in the solar-powered charger.

All batteries are chemically "stiff" when new. NiMH batteries are shipped uncharged from the factory, and are difficult to charge initially. A small solar charger may not have enough power to overcome the battery's internal resistance; hence, no charge. The solution is to use a plug-in charger for the first few cycles to break in the battery. Another solution is to install only one battery at a time in the solar charger. Once the batteries have been broken in, there is no problem. These batteries will give you many, many years of use for no additional cost with solar charging.

My tape player (or other device) doesn't work when I put NiMHs in. What's wrong?

Some battery-powered equipment will not function on the lower-voltage NiMH batteries, which produce 1.25 volts (versus 1.5 volts produced by fresh alkalines). Some manufacturers use a low-voltage cutoff that won't allow using rechargeables. If the cutoff won't let you use NiMHs, it's also forcing you to buy new alkalines when they're only 50% depleted! Sometimes with devices that require a large number of batteries (six or more), the ¼-volt difference between rechargeables and alkalines gets magnified to a serious problem. Even a fully charged set of NiMHs will be seen as a depleted battery pack by the appliance. You need a new appliance that's designed for rechargeable batteries.

Do I have to run the rechargeable battery completely dead every time?

No. For greatest life expectancy, it's best to recharge NiMH batteries when you first sense the voltage is dropping. You'll be down to 10 or 15% capacity at this point. Further draining is not needed or recommended.

Alkaline rechargeable batteries really hate deep cycles. They'll do best when they're only run down about 50%.

RECHARGEABLE BATTERIES AND CHARGERS

Batteries That Really Just Keep Going and Going!

Americans use nearly three billion disposable batteries every year; the rest of the world adds another few billion to that total. Beyond being expensive and a colossal waste of resources, this presents a huge low-level toxic disposal problem. Rechargeable batteries are the obvious, easy, money-saving solution. Buy an extra set. That way you have one set in your charger at all times and one set in the appliance. When the appliance gets weak, just swap batteries with the fresh ones. If you change batteries once a week, they will last for ten to fifteen years.

Rechargeable Batteries with MORE Energy Than Throwaway Batteries!

The capacity of NiMH batteries keeps increasing as the technology develops. Our high-capacity AA-size NiMH batteries are now equal to or better than throw-away alkaline cells. Our tests show name-brand alkaline AA batteries deliver approximately 1,800 to 2,100mAh, then you throw them away. Our AA NiMH cells deliver 2,200mAh, and will do that about 500 times before you throw them away. In addition, NiMH chemistry delivers at a steady voltage all the way down to 10% of charge. No voltage fade like standard alkaline cells! This means you get to actually *use* more of the available capacity with environmentally friendly rechargeable batteries.

ALKALINE BATTERIES VERSUS RECHARGEABLES
What's the smart choice?

2,000 Alkaline Batteries @ $2.⁸⁵ / 4 pack
4,000 Ah for $1,425.⁰⁰

=

4 Nimh Batteries $12.⁰⁰
1 Charger $29.⁰⁰
2,000 Charges @ 1¢ ea. $20.⁰⁰
4,000 Ah for $61.⁰⁰

Hi-Capacity NiMH Rechargeable Batteries

Equal to the best throw-away batteries and non-toxic

Nickel-metal hydride batteries just keep getting better and better. These batteries excel at hard use. NiMH bat-teries deliver steady voltage with no sag all the way down to 10% of charge. This makes them a better choice for heavy-dis-charge uses such as digital cameras or remote-control toys,  where you actually get to use more of the stored energy than any other battery type will deliver. There is no memory effect from short cycling, no toxic content, and average life expectancy is about 500 cycles. Because they self-discharge over 3-6 months, these cells aren't a good choice for emergency lights. China.

17-9275 4	AAA 4-pack (750mAh)	$12
17-9275 8	AAA 8-pack (750mAh)	$22
17-9276 4	AA 4-pack (2,200mAh)	$12
17-9276 8	AA 8-pack (2,200mAh)	$22
17-9175	C 2-pack (3,300mAh)	$11
17-9176	D 2-pack (7,000mAh)	$20
17-0284	9V single (160mAh)	$9
17-0285	9V 2-pack (160mAh)	$16

Hi-Speed Digital Charger

Our speedy, affordable battery charger comes with both AC and DC power cords, charges AA- or AAA-size NiMH or nicad batteries in pairs, and stops charging automatically when batteries are full, then switches to gently trickle-charge your batteries until you're ready for them. Charges most batteries in 5 hours or less. LEDs show red when charging, green when finished, and flashes red/green if a battery is defective. Taiwan.

17-0281 Hi-Speed Digital Charger $29

AccuManager 20

Using a pulse-charging technology that rechargeable batteries just love, the AccuManager 20 charger will recharge your NiMH, nicad, or Accucell batteries faster, and better, than any other charger we've ever seen. You can mix different battery types, sizes, and charging times. After charging, each cell is stored at the peak of readiness with a gentle trickle charge, until you're ready for it. There are four bays that accept AAA, AA, C or D cells, plus two bays that accept 9-volt cells. Charge times will run from 20 minutes to 12 hours depending on type and size. Supplied with both AC and DC charging cords and will run off standard 120vac, automotive 12 volts, or off a 10- to 15-watt solar panel with appropriate plug, such as our solar option sold below. Three-year manufacturer's warranty. Germany/China. For rechargeable batteries only.

50269	**AccuManager 20 Battery Charger**	**$69**
06-0384	**Lightweight 10W Solar Option**	**$149**

iSun and Battpak

This 2.2-watt folding unit can be switched for 6- or 12-volt output, comes with two power cables and seven different DC plugs that cover more than 90% of cell phones, discmans, MP3 players, or GPS units. For more power, up to five iSun units can be plugged into each other with the included spine connector plug. Measures 7.25" x 4.25" (8.25" unfolded) x 1.25". Weighs 11 oz. Delivers 145mA @ 15.2v, or 290mA @ 7.6v. China.

The Battpak docks mechanically and electrically to the back of the iSun, or it can be used stand-alone with the included AC and DC charging cords. It will charge up to 10 rechargeable AA- or AAA-size batteries. ON and FULL lights indicate charging activity. Two DC output plugs will deliver 1.5v to 12v depending on how many batteries are installed. Measures 7.5" x 2.9" x 1.25". Weighs 6.5 oz. China.

11613	**iSun Solar Charger**	**$80**
50267	**Battpak Charger**	**$30**
50268	**iSun with Battpak**	**$99**

Quickcharger Battery Charger

For recharging nickel-metal hydride (NiMH) and nickel cadmium (nicad) batteries, this sophisticated microprocessor-controlled charger will provide the fastest complete charge of any battery charger on the market without creating damaging heat. An automatic discharge circuit conditions each battery for full capacity and the longest life by running a controlled discharge at the beginning of the charge cycle. Works perfectly with the newer ultra-high capacity NiMH cells. A graphic LCD display shows what's happening, and displays battery voltage. After fully recharging, it will hold the cells on a gentle trickle charge, so they always come out at the peak of performance. Accepts one to four AAA- through D-size cells. Batteries on each charging cycle should be the same size and chemistry. Runs on 12vdc, comes with an AC adapter. One-year manufacturer's warranty. China.

17-0184 Quickcharger Battery Charger	**$66**

12-Volt Nicad Charger

This 12-volt charger will recharge AAA-, AA-, C-, or D-size NiMH or nicad batteries in 10 to 20 hours from your 12-volt power source. It will charge up to four batteries simultaneously. Batteries must be charged in pairs. Has a clever contact system on the positive end that will deliver different current flow depending on battery size. AAAs will receive 50 milliamps, AAs will receive 130 milliamps, Cs and Ds will receive 240 milliamps. It comes with a 12-volt cigarette lighter plug to go into any 12-volt socket. Make sure the cigarette lighter works with the key off when using it in your car. One or two charge cycles will not discharge the car's battery. China.

50214	**12-Volt Nicad Charger**	**$20**

Inverters

If you want to run conventional household appliances with your renewable energy system, you need a device to produce AC house current on demand. That device is an inverter.

What is an inverter, and why would you want one in your renewable energy system? This is a reasonable question. Some small renewable energy systems don't need an inverter.

The inverter is an electronic device that converts direct current (DC) into alternating current (AC). Renewable energy sources such as PV modules and wind turbines make DC power, and batteries will store DC power. Batteries are the least expensive and most universally applicable energy storage method available. Batteries store energy as low-voltage DC, which is acceptable, in fact preferable, for some applications. A remote sign-lighting system, or a small cabin that only needs three or four lights, will get by just fine running everything on 12-volt DC. But most of the world operates on higher-voltage AC. AC transmits more efficiently than DC and so has become the world standard. No technology exists to store AC, however. It must be produced as needed. If you want to run conventional household appliances with your renewable energy system, you need a device to produce AC house current on demand. That device is an inverter.

Inverters are a relatively new technology. Until the early 1990s, the highly efficient, long-lived, relatively inexpensive inverters that we have now were still a pipe dream. The world of solid-state equipment has advanced by extraordinary leaps and bounds. More than 95% of the household power systems we put together now include an inverter.

Benefits of Inverter Use

The world runs on AC power. In North America, it's 120-volt, 60-cycle AC. Other countries may have slightly different standards, but it's all AC. Joining the mainstream allows the use of mass-produced components, wiring hardware, and appliances. Appliances may be chosen from a wider, cheaper, and more readily available selection. Electricity transmits more efficiently at higher voltages, so power distribution through the house can be done with conventional 12- and 14-gauge romex wiring using standard hardware, which electricians appreciate and inspectors understand. Anyone who's wrestled with the heavy 10-gauge wire that a low-voltage DC system requires will see the immediate benefit.

Modern brand-name inverters are extremely reliable. The household-size models we sell have failure rates well under 1%. Efficiency averages about 90% for most models, with peaks at 95% to 98%. In short, inverters make life simpler, do a better job of running your household, and ultimately save you money on appliances and lights. But watch out for the flood of cheapo imported inverters! High failure rates and poor performance are common with these.

An inverter/battery system is the ultimate clean, uninterruptible power supply for your computer, too. In fact, an expensive UPS (uninterruptible power supply) system is simply a battery and a small inverter with an expensive enclosure and a few bells and whistles. Just don't run the power saw or washing machine off the same inverter that runs your computer. We often recommend a small secondary inverter just for the computer system. This ensures that the water pump or some other large, unexpected load can't possibly hurt you.

Electrical Terminology and Mechanics

Electricity can be supplied in a variety of voltages and waveforms. As delivered to and stored by our renewable energy battery system, we've got low-voltage DC. As supplied by the utility network, we've got house current: nominally 120 volts AC. What's the difference? DC electricity flows in one direction only, hence the name direct current. It flows directly from one battery terminal to the other battery terminal. AC current alternates, switching the direction of flow periodically. The U.S. standard is 60 cycles per second. Other countries have settled on other standards, but it's usually either 50 or 60 cycles per second. The electrical term for the number of cycles per second is Hertz, named after an early electricity pioneer. So AC power is defined as 50 Hz or 60 Hz.

Our countrywide standard also defines the voltage. Voltage is similar to pressure in a water line. The greater the voltage, the higher the pressure. When voltage is high, it's easier to transmit a given amount of energy, but it's more difficult to contain and potentially more dangerous. House current in this country is delivered at about 120 volts for most of our household appliances, with the occasional high-consumption appliance using 240 volts. So our power is defined as 120/240 volts/60 Hz. We usually use the short form, 120 volts AC, to denote this particular voltage/cycle combination. This voltage is by no means the world standard, but is the most common one we deal with. Inverters are available in international voltages by special order, but not all models from all manufacturers.

Description of Inverter Operation

If we took a pair of switching transistors and set them to reversing the DC polarity (direction of current flow) 60 times per second, we would have low-voltage alternating power of 60 cycles or Hertz (Hz). If this power was then passed through a transformer, which can transform AC power to higher or lower voltages depending on design, we could end up with crude 120-volt/60 Hz power. In fact, this was about all that early inverter designs of the 1950s and 1960s did. As you might expect, the waveform was square and very crude (more about that in a moment). If the battery voltage went up or down, so did the output AC voltage, only ten times as much. Inverter design has come a long, long way from the noisy, 50% efficient, crude inverters of the 1950s. Modern inverters hold a steady voltage output regardless of battery voltage fluctuations, and efficiency is typically in the 90 to 95% range. The waveform of the power delivered has also been improved dramatically.

Waveforms

The AC electricity supplied by your local utility is created by spinning a bundle of wires through a series of magnetic fields. As the wire moves into, through, and out of the magnetic field, the voltage gradually builds to a peak and then gradually diminishes. The next magnetic field the wire encounters has the opposite polarity, so current flow is induced in the opposite direction. If this alternating electrical action is plotted against time, as an oscilloscope does, we get a picture of a sine wave, as shown in our wave-form gallery above. Notice how smooth the curves are. Transistors, as used in all inverters, turn on or off abruptly. They have a hard time reproducing curvy sine waves.

Early inverters produced a square-wave alternating current (see the wave-form gallery again). While square-wave does alternate, it is considerably different in shape and peak voltage. This causes problems for many appliances. Heaters or incandescent lights are fine, motors usually get by with just a bit of heating and noise, but solid-state equipment has a really hard time with it, resulting in loud humming, overheating, or failure. It's hard to find new square-wave inverters anymore, and nobody misses them.

Most modern inverters produce a hybrid waveform called a quasi-, synthesized, or modi-

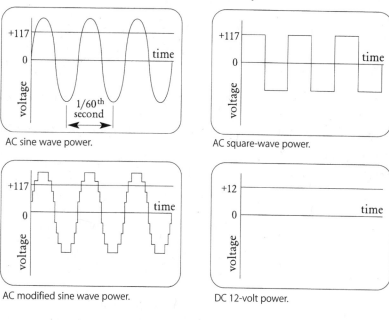

AC sine wave power.

AC square-wave power.

AC modified sine wave power.

DC 12-volt power.

fied sine wave. In truth, this could just as well be called a modified square-wave, but manufacturers are optimists. (Is the glass half full or half empty?) Modified sine wave output cures many of the problems associated with square-wave. Most appliances will accept it and hardly know the difference. There are some notable exceptions to this rosy picture, however, which we'll cover in detail below in the Inverter Problems section.

Full sine wave inverters have been available since the mid-1990s, but because of their higher initial cost and lower efficiency, they were only used for running very specific loads. This is changing rapidly. A variety of high-efficiency, moderate-cost pure sine wave inverters is available now. Xantrex, OutBack, and other manufacturers have unveiled a whole series of sine wave units, and it's obviously the wave of the future. True sine wave inverters have very nearly become the standard for larger household power systems already. With very rare exceptions, sine wave inverters will happily run any appliance that can plug into utility power. Motors run cooler and more quietly on sine wave, and solid-state equipment has no trouble. True sine wave inverters deliver top-quality AC power, and are almost always a better choice for household use.

Inverter Output Ratings

Inverters are sized according to how many watts they can deliver. More wattage capability will cost more money initially. Asking a brand-name

> True sine wave inverters deliver top-quality AC power, and are almost always a better choice for household use.

Most households end up with one of the full-size inverters because household loads tend to grow, and larger inverters are equipped often with very powerful battery chargers.

Reading Modified Sine Wave Output with a Conventional Voltmeter

Most voltmeters that sell for under $100 will give you weird voltage readings if you use them to check the output of a modified sine wave inverter. Readings of 80 to 105 volts are the norm. This is because when you switch to "AC Volts" the meter is expecting to see conventional utility sine wave power. What we commonly call 110- to 120-volt power actually varies from 0 to about 175 volts through the sine wave curve. 120 volts is an average called the Root Mean Square, or RMS for short, that's arrived at mathematically. More expensive meters are RMS corrected; that is, they can measure the average voltage for a complex waveform that isn't a sine wave. Less expensive meters simply assume if you switch to "AC Volts" it's going to be a sine wave. So don't panic when your new expensive inverter checks out at 85 volts: the inverter is fine, but your meter is being fooled. Modern inverters will hold their specified voltage output, usually about 117 volts, to plus or minus about 2%. Most utilities figure they're doing well by holding variation to 5%.

Battery chargers are the most useful inverter option, and for most folks, the cheapest and easiest way to add a powerful, automatic battery charger to their system.

inverter for more power than it can deliver will result in the inverter shutting down. Asking the same of a cheapo inverter may result in a crispy critter. All modern inverters are capable of briefly sustaining much higher loads than they can run continuously, because some electric loads, like motors, require a surge to get started. This momentary overcapacity has led some manufacturers to fudge on their output ratings. A manufacturer might, for instance, call its unit a 200-watt inverter based on the instantaneous rating, when the continuous output is only 140 watts. Happily, this practice is fading into the past. Most manufacturers are taking a more honest approach as they introduce new models and are labeling inverters with their continuous wattage output rather than some fanciful number. In any case, we've been careful to list the continuous power output of all the inverters we carry. Just be aware that the manufacturer calling the inverter a 200-watt unit does not necessarily mean it will do 200 watts continuously. Read the fine print.

How Big an Inverter Do You Need?

The bigger the inverter, the more they cost initially. So you don't want to buy a bigger one than you need. On the other hand, an inverter that's too small is going to frustrate you because you'll need to limit usage.

A small cabin with an appliance or two to run doesn't take much. For example, if you want to run a 19-inch TV, a DVD player, and a light all at once, total up all the wattages (about 80 for the TV, 25 for the DVD, and 20 for a compact fluorescent light, a total of 125 watts), pick an inverter that can supply at least 125 watts continuously, and you're all set.

To power a whole house full of appliances and lights might take more planning. Obviously not every appliance and light will be on at the same time. Mid-size inverters of 600 to 1,000 watts do a good job of running lights, entertainment equipment, and small kitchen appliances, in other words, most common household loads. What a 1,000-watt inverter will not do is run a mid- to full-size microwave, a washing machine, or larger handheld power tools. For those loads, you need a full-size 2,000-plus watt inverter. In truth, most households end up with one of the full-size inverters because household loads tend to grow, and larger inverters are equipped often with very powerful battery chargers. This is the most convenient way to add battery charging capability to a system (and the cheapest, too, if youre already buying the inverter).

Chargers, Lights, Bells, and Whistles

As you go from small plug-in-the-lighter-socket inverters to larger household-size units, you'll find increasing numbers of bells and whistles, most of which are actually pretty handy. At the very least, all inverters have an LED light showing that it's turned on. Mid-size units may feature graphic volt and amp meters; better units may have an LCD display and be able to signal generators and other remote devices to turn on or off.

Battery chargers are the most useful inverter option, and for most folks, the cheapest and easiest way to add a powerful, automatic battery charger to their system. Chargers are pretty much standard equipment on all larger household-size inverters. With a built-in charger, the inverter will have a pair of "AC input" terminals. If the inverter detects AC voltage at these terminals, because you just started the back-up generator, for instance, it will automatically connect the AC input terminals to the AC output terminals, and then go to work charging the batteries. Inverter-based chargers tend to be extremely robust and adjustable, and will treat your batteries nicely.

Many financially challenged off-the-grid systems start out with just a generator, an inverter/charger, and some batteries. You run the generator every few days to recharge the batteries. Add PV charging as money allows, and the generator gradually has to run less and less.

How to Cripple Your Inverter

Do you want to reduce your inverter's ability to provide maximum output? Here's how: Keep your battery in a low state of charge, use long, undersized cables to connect the battery and inverter, and keep your inverter in a small airtight enclosure at high temperatures. You'll succeed in crippling, if not outright destroying, your inverter. Low battery voltage will severely limit any inverter's ability to meet a surge load. Low voltage occurs when the battery is undercharged, or undersized. If the battery bank isn't large enough to supply the energy demand, then voltage will drop, and the inverter probably won't be able to start the load. For instance, a

Xantrex DR2412 will easily start most any washing machine, but not if you've only got a couple of golf cart batteries. The starting surge will demand more electrons than the batteries can supply, voltage will plummet, and the inverter will shut off to protect itself (and your washing-machine motor). Now suppose you have eight golf cart batteries on the DR2412, which is plenty to start the washer with ease, but decided to save some money on the hookup cables, and used a set of $16.95 automotive jumper cables. The battery bank is capable, the inverter is capable, but not enough electrons can get through the undersized jumper cables. The result is the same: low voltage at the inverter, which shuts off without starting the washer. Do not skimp on inverter cables or on battery interconnect cables. These items are inexpensive compared to the cost of a high-quality inverter and good batteries. Don't cripple your system by scrimping on the electron delivery parts.

All inverters produce a small amount of waste heat. The harder they work, the more heat they produce. If they get too hot, they will shut off or limit their output to protect themselves. Give the inverter plenty of ventilation. Treat it like a piece of stereo gear: dry, protected, and well ventilated.

How Big a Battery Do You Need?

We usually size household battery banks to provide stored power for three to five days of autonomy during cloudy weather. For most folks, this is a comfortable compromise between cost and function. With three- to five-day sizing, the battery will be large enough to handle any surge loads that the inverter is called upon to start. A battery bank smaller than three-days' capacity will cycle fairly deeply on a regular basis. This isn't good for battery life. A larger battery bank cycled less deeply will cost less in the long run, because it lasts longer. Banks larger than five-days' capacity are more expensive than a back-up power source (such as a modest-sized generator). However, we occasionally run into situations with ¾-horsepower or larger submersible well pumps or stationary power tools requiring a larger battery bank simply to meet the surge load when starting. Call the Gaiam Real Goods technical staff for help if you are anticipating large loads of this type.

Treat your inverter like a piece of stereo gear: dry, protected, and well ventilated.

Happy Inverters

1. Keep the inverter as close to the battery as possible, but not in the battery compartment. Five to ten feet and separated by a fireproof wall is optimal. The high-voltage output of the inverter is easy to transmit. The low-voltage input transmits poorly.
2. Keep the inverter dry and as cool as possible. They don't mind living outdoors if protected.
3. Don't strangle the inverter with undersize supply cables. Most manufacturers have recommendations in the owner's manual. If in doubt, give us a call.
4. Fuse your inverter cables and any other circuits that connect to a battery!

Inverter Safety Equipment and Power Supply

In some ways batteries are safer than conventional AC power. It's fairly difficult to shock yourself at low voltage. But in other ways batteries are more dangerous: They can supply many more amps into a short circuit, and once a DC arc is struck, it has little tendency to self-snuff. One of the very sensible things the National Electrical Code requires is fusing and a safety disconnect for any appliance connected to a battery bank. Fusing is extremely important for any circuit connected to a battery! Without fusing, you risk burning down your house.

Several products exist to cover DC fusing and disconnect needs for inverter-based systems. For full-size inverter fusing, you can use a properly sized OutBack or Xantrex DC power center. Our technical staff can answer your questions about

Ten feet is the longest practical run between the battery and inverter. This is true even for small 100- to 200-watt inverters.

which one is better for you. Mid-size inverters can use a Class T fuse with a small surface-mounted circuit breaker to provide safe and compliant connection. All this safety and connection equipment is covered in the Safety section of this chapter, beginning on page 143. We've recommended fusing or breaker sizing, and required cable sizing with all larger inverters.

The size of the cables providing power to the inverter is as important as the fusing, as we noted earlier. Do not restrict the inverter's ability to meet surge loads by choking it down with undersized or lengthy cables. Ten feet is the longest practical run between the battery and inverter. This is true even for small 100- to 200-watt inverters. Put the extension cord on the AC side of the inverter! Larger models will include cable and circuit breaker requirement charts. Batteries should either be the sealed type, or live in their own enclosure. Don't put your inverter in with the batteries!

Potential Problems with Modified Sine Wave Inverter Use—
and How to Correct Them

We have been painting a rosy picture up to this point, but now it is time for a little brutal honesty to balance things out. If you know about these problems beforehand it is usually possible to work around them when selecting appliances.

Wave-form Problems

Wave-form problems are probably the biggest category of inverter problems we encounter. We talked about sine wave versus modified sine wave in the technical description above. The problems discussed here are caused only by modified sine wave inverters. If you are planning to use a pure sine wave unit, you can skip this section.

Computers run happily on modified sine wave.

AUDIO EQUIPMENT

Some audio gear will pick up a 60-cycle buzz through the speakers. It doesn't hurt the equipment, but it's annoying to the listener. There are too many models and brands to say specifically which have a problem and which do not. We've had better luck with new equipment recently. Manufacturers are starting to put better power supplies into their gear. We can only recommend that you try it and see.

Some top-of-the-line audio gear is protected by SCRs or Triacs. These devices are installed to

guard against powerline spikes, surges, and trash (nasties that don't happen on inverter systems). However, they see the sharp corners on modified sine wave as trash and will sometimes commit electrical hara-kiri to prevent that nasty power from reaching the delicate innards. Some are even smart enough to refuse to eat any of that ill-shaped power, and will not power up. The only sure cure for this (other than more tolerant equipment) is a sine wave inverter. If you can afford top-of-the-line audio gear, chances are you've already decided that sine wave power is the way to go.

COMPUTERS

Computers run happily on modified sine wave. In fact, most of the uninterruptible power supplies on the market have modified sine wave or even square-wave output. The first thing the computer does with the incoming AC power is to run it through an internal power supply. We've had a few reports of the power supply being just a bit noisier on modified sine wave, but no real problems. Running your prized family computer off an inverter will not be a problem. What can be a problem is large start-up power surges. If your computer is running off the same household inverter as the water pump, power tools, and microwave, you're going to have trouble. When a large motor, such as a skilsaw, is starting, it will pull the AC system voltage way down momentarily. This can cause computer crashes. The fix is a small, separate inverter that only runs your computer system. It can be connected to the same household battery pack, and have a dedicated outlet or two.

LASER PRINTERS

Many laser printers are equipped with SCRs, which cause the problems detailed above. Laser printers are a very poor choice for renewable energy systems anyway, due to their high standby power use keeping the heater warm. Lower-cost inkjet printers can do almost anything a laser printer can do while only using 25 to 30 watts instead of 400 to 900 watts.

CEILING FANS

Many variable-speed ceiling fans will buzz on modified sine wave current. We've also had some reports of fan remote controls not working on modified sine wave. The fans spin up fine, but the noise is annoying, and you can't change speeds.

Potential Problems with ALL Inverters—
and How to Correct Them

You sine wave buyers can stop smirking here, the following problems apply to all inverters.

RADIO AND TV INTERFERENCE

All inverters broadcast radio noise when operating. The FCC put out new guidelines concerning interference, and as manufacturers come out with new models like the OutBack FX series or the Xantrex SW Plus series, this is much less of a problem. Most of this interference is on the AM radio band. Do not plug your radio into the inverter and expect to listen to the ball game; you'll have to use a battery-powered radio and be some distance away from the inverter. This is occasionally a problem with TV interference when inexpensive TVs and smaller inexpensive inverters are used together. Distance helps. Put the TV (and the antenna) at least 15 feet from the inverter. Twisting the inverter input cables may also limit their broadcast power (strange as it sounds, it works).

PHANTOM LOADS AND VAMPIRES

A phantom load isn't something that lurks in your basement with a half-mask, but it's close kin. Many modern appliances remain partially on when they appear to be turned off. That's a phantom load. Any appliance that can be powered up with a button on a remote control must remain partially on and listening to receive the "on" signal. Most TVs and audio gear these days are phantom loads. Anything with a clock—VCRs, coffee makers, microwave ovens, or bedside radio-clocks—uses a small amount of power all the time.

Vampires suck the juice out of your system. These are the black power cubes that plug into an AC socket to deliver lower-voltage power for your answering machine, electric toothbrush, power-tool charging stand, or any of the other huge variety of appliances that use a power cube on the AC socket. These villainous wastrels usually run a horrible 60 to 80% inefficiency (which means that for every dime's worth of electricity consumed, they throw away six or eight cents' worth.) Most of these nasties always draw power, even if no battery/toothbrush/razor/cordless phone is present and charging. It would cost their manufacturers less than 25 cents per unit to build a power-saving standby mode into the power cube, but since you, the consumer, are paying for the inefficiency, what do they care? The appliance might be turned off, but the vampire keeps sucking a few watts. Have you ever noticed that power cubes are usually warm? That's wasted power being converted to heat. By the way, cute and appropriate as it is, we can't take credit for the vampire name. That's official electric industry terminology (bless their honest souls).

So what's the big deal? It's only a few watts, isn't it? The problem isn't the power consumed by vampires and phantom loads, it's what is required to deliver those few watts twenty-four hours per day. When there's no demand for AC power, a full-size inverter will drop into a sleep mode. Sleep mode keeps checking to see if anything is asking for power, but it takes only a tiny amount of energy, usually 2 to 3 watts. If the cumulative phantom and vampire loads are enough to awaken the inverter, it consumes its own load plus the inverter's overhead, which is typically 15 to 20 watts. The inverter overhead is the real problem. An extra 15 to 20 watts might

Do not plug your radio into the inverter and expect to listen to the ball game; you'll have to use a battery-powered radio and be some distance away from the inverter.

Many modern appliances remain partially on when they appear to be turned off. That's a phantom load.

Have you ever noticed that power cubes are usually warm? That's wasted power being converted to heat.

The solution for phantoms and vampires is outlets or plug strips that are switched off when not in use.

sound like grasping at straws, but over twenty-four hours every day, this much power lost to inefficiency can easily add a couple $450 modules to the size of the PV array and maybe another battery or two to the system requirements. You want to make sure your inverter can drop into sleep mode whenever there's no real demand for AC power.

KEEPING THE VAMPIRES AT BAY

With some minimal attention to details, you can keep the vampires and phantoms under tight control and save yourself hundreds or thousands of dollars on system costs.

The solution for clocks is battery power. A wall-mounted clock runs for nearly a year on a single AA rechargeable battery. We have found a wide selection of good-quality battery-powered alarm clocks, and most of the other timekeepers anyone could possibly require, just by looking around. Clocks on house current are ridiculously wasteful.

The solution for phantoms and vampires is outlets or plug strips that are switched off when not in use. The outlets only get switched on when the appliance is actually being used or recharged. As a side benefit, this cures PMS—perpetual midnight syndrome—on your VCR, too!

If you're aware of the major energy-wasting gizmos we all take for granted, it's easy to avoid

or work around them. When you consider the minor inconvenience of having to flip a wall switch before turning on the TV against the $400 to $500 cost of another PV module to make up the energy waste, it all comes into perspective. A few extra switched outlets during construction looks like a very good investment. If your house is already built, then use switched plug strips.

Conclusions

Inverters allow the use of conventional, mass-produced AC appliances in renewable energy-powered homes. They have greatly simplified and improved life "off-the-grid." Modern solid-state electronics have made inverters efficient, inexpensive, and long lived. Batteries and inverters need to live within 10 feet of each other, but not in the same enclosure. Modified sine wave inverters will not be perfectly problem-free with every appliance. Expect a few rough edges. The best inverters produce pure sine wave AC power, which is acceptable to all appliances. Sine wave units tend to cost a bit more, but are strongly recommended for all full-size household systems. Keep the power-wasting phantom loads and power cube vampires under tight control, or be prepared to spend enough to overcome them.

INVERTERS AND RELATED PRODUCTS

MODIFIED SINE WAVE INVERTERS

Modified sine inverters represent the lower-cost end of residential inverters. It's easier to build and control this type of inverter, and they tend to be slightly more efficient than pure sine wave inverters. There are two basic design types. High-speed electronic switchers are the most common type of modified sine wave inverter. These are all the small to medium-sized, largely overseas-produced units. The Wagan inverters are an example. Electronic switchers are relatively lightweight. Cost is usually a pretty fair indicator of how long you can expect them to last. To make them cheaper, manufacturers will cut corners on protection circuitry. Buyer beware! The other basic design type is transformer-based. These units tend to come in larger wattages, be much more robust, and will withstand a lot more abuse. The Xantrex DR-series is an example.

WAGAN POWER INVERTERS

Wagan offers an extensive line of 12-volt inverters at attractive prices. You won't find built-in battery chargers or automatic transfer switches on these no-frills inverters. You will find dependable modified sine wave output, honest output ratings, modest no-load power use, and reasonable prices. Features on all models include automatic overload protection, input short circuit protection, and a one-year manufacturer's warranty. Taiwan.

Small Mobile Inverters for Battery-Powered Gizmos

This series of incredibly handy Wagan inverters are great for recharging laptops, PDAs, cameras, cell phones, or any small AC appliance up to the wattage rating of the inverter. Plugs into any standard 12-volt lighter socket. Overload, short-circuit, and low-battery protected. Shuts off at 10 volts. Surges to about twice the rated power. Single 3-prong outlet. Weighs 6.4, 9.6, and 14.5 oz. respectively. One-year manufacturer's warranty. China.

49-0130	75 Watt	$29
49-0131	110 Watt	$35
49-0132	150 Watt	$45

400-Watt Inverter

The 400-watt model can run small kitchen appliances, lights, tunes, or TV for a small cabin, boat, or RV. Includes battery clamps for heavier-duty use. Delivers 400 watts continuously, or up to 1,000 watts surge. On/off switch, automatic cooling fan, and dual AC outlets. Full specs above.

27878	400-watt Inverter	$89

700-Watt Inverter

The 700-watt model can comfortably run almost anything you need in your small cabin, RV, or boat. Must be hard-wired to the battery; cables are included. Delivers 700 watts continuously, or up to 1,800 watts surge. On/off switch, automatic cooling fan, and dual AC outlets. Full specs above.

27879	700-watt Inverter	$159

1,000-Watt Inverter

The first of Wagan's new Elite line, the 1,000-watt model has a smaller size, higher surge ability, lower idle current, and lower cost per watt. Cables are included for hard-wiring to the battery. Delivers 1,000 watts continuously, or up to 2,500 watts surge. On/off switch, automatic cooling fan, and dual AC outlets. China.

49-0133	1,000-watt Inverter	$195

WAGAN POWER INVERTERS SPECIFICATIONS			
	150-Watt	**400-Watt**	**700-Watt**	**1,000-Watt**
Continuous output:	150 watts	400 watts	700 watts	1,000 watts
Surge capacity:	500 watts	1,000 watts	1,800 watts	2,500 watts
Idle current:	0.2A (2.4 watts)	0.3A (3.6 watts)	0.95A (11.4 watts)	0.5A (6 watts)
Dimensions:	5"x 4.5"x 2"	7.5"x 5"x 2"	14"x 9.7"x 4.5"	13.5"x 4.8"x 3.1"
Weight:	0.9 lbs.	2.0 lbs.	4.7 lbs.	6.0 lbs.

THE XANTREX DR-SERIES INVERTERS

All DR inverters come equipped with high-performance battery chargers as standard equipment. They are available for 12- or 24-volt systems in a variety of wattages.

DR-series inverters are conventional, high-efficiency, modified sine wave inverters. All models are Electrical Testing Lab approved to UL specifications. These inverters will protect themselves automatically from overload, high temperature, high- or low-battery voltages; all the things we've come to expect from modern inverters. You just can't hurt them externally. Fan cooling is temperature regulated; low battery cut-out is standard and is adjustable for battery bank size. All models come with a powerful three-stage adjustable battery charger and all models have a 30-amp transfer switch built in. Charge amperage is adjustable to suit your batteries or generator. Charging voltages are tailored by a front panel knob. Pick your battery type, eight common choices are given plus two equalization selections, and you've automatically picked the correct bulk and float voltage for your batteries.

The optional battery temperature probe will allow the inverter to factor all voltage set points by actual battery temperature. This is a good feature to have when your batteries live outside and go through temperature extremes. Comes with a 15-foot cord. The conduit box option allows the battery cables to be run in 2-inch conduit to meet NEC.

The DR-series is stackable in series (must be identical models) for three-wire splitable 240-volt output. Stacking is now easier and less expensive, too; just plug in the inexpensive stacking cord, bolt up the battery cable jumpers, and it's done.

All models are designed for wall mounting with conventional 16-inch stud spacing. All Xantrex inverters carry a two-year manufacturer's warranty. USA.

Conduit Box Option

The optional conduit box allows the very heavy input wiring from the batteries to be run inside conduit to comply with the NEC. Has three (one on each side) ½", ¾", and 2" knockouts and plenty of elbow room inside for bending large wire. USA.

27-324   Conduit Box option-DR Series $69

DR Stacking Kit

The DR stacking kit includes the plug-in comm-port cable, like the SW, plus a pair of 2/0 battery cables to jumper from the primary to the secondary inverter. USA.

27-313   DR Stacking Intertie Kit $85

Battery Temperature Probe

Comes standard with the SW-series; recommended for the DR-series if your batteries are subject to temperatures below 40°F or above 90°F. Will adjust the battery charger automatically to compensate for temperature. 15-foot length, phone plug on inverter end, sticky pad on battery end. This same temp probe also works with all C-Series Trace controllers. USA.

27-319   Battery Temperature Probe, 15´ $29

Remote Control for DR, UX, & TS Series Inverters

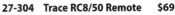

A 50-foot ON/OFF remote control with led indicator. Plugs into all Trace DR, UX, and TS series inverters. USA.

27-304   Trace RC8/50 Remote $69

Model	DR1512	DR2412	DR1524	DR2424	DR3624
Input Voltage:	10.8 to 15.5 volts DC	10.8 to 15.5 volts DC	21.6 to 31.0 volts DC	21.6 to 31.0 volts DC	21.6 to 31.0 volts DC
Output Voltage:	120 volts AC true RMS ± 5%	120 volts AC true RMS ± 5%	120 volts AC true RMS ± 5%	120 volts AC true RMS ± 5%	120 volts AC true RMS ± 5%
Wave-form:	Modified sine wave	Modified sine wave	Modified sine wave	Modified sine wave	Modified sine wave
Continuous Output:	1,500 watts	2,400 watts	1,500 watts	2,400 watts	3600 watts
Surge Capacity:	3,200 watts	6,000 watts	4,200 watts	7,000 watts	12,000 watts
Idle Current:	0.045A @ 12V (0.54 watt)	0.045A @ 12V (0.54 watt)	0.030A @ 24V (0.72 watt)	0.030A @ 24V (0.72 watt)	0.030A @ 24V (0.72 watt)
Charging Rate:	0 to 70 amps	0 to 120 amps	0 to 35 amps	0 to 70 amps	0 to 70 amps
World Voltages Available:	yes, call for details	no	yes, call for details	yes, call for details	no
Average Efficiency:	94% at half rated power	94% at half rated power	94% at half rated power	95% at half rated power	95% at half rated power
Recommended Fusing	200A Class T or Trace DC-175	300A Class T or Trace DC-175	110A Class T or Trace DC-175	200A Class T or Trace DC-175	300A Class T or Trace DC-250
Dimensions:	8.5" x 7.25" x 21"	8.5" x 7.25" x 21"	8.5" x 7.25" x 21"	8.5" x 7.25" x 21"	8.5" x 7.25" x 21"
Weight:	35 lb.	45 lb.	35 lb.	40 lb.	45 lb.
Item #	**27237**	**27245**	**27238**	**27242**	**27246**
Price	**$1,029**	**$1,395**	**$979**	**$1,395**	**$1,595**

PURE SINE WAVE INVERTERS

Pure sine inverters increasingly represent the future of residential inverters. They tend to be more expensive, but higher quality and longer lasting than modified sine units. Efficiency is slightly lower on pure sine wave inverters, but there are hardly ever compatibility problems with appliances. Just like modified sine units, we have high-speed switchers in smaller sizes for lower prices, like the Wagan or Samlux units, and transformer-based units in larger wattages at higher prices, like the OutBack or Xantrex units.

Wagan Pure Sine Wave Inverters

Every once in a while we run into an appliance that just isn't happy with anything less than pure, clean, sine wave power. Here's how to deliver a modest amount of pure sine wave power for a fussy appliance or two, without breaking the bank. Choose 150 or 300 watts continuous. Both inverters can double their rated output for a short surge. The 150-watt model is equipped with lighter plug for DC input, and AC outlets. The 300-watt model is equipped with both a lighter plug and/or battery clamps for input, and two AC outlets. Taiwan.

27880	Wagan 150W Sine Wave Inverter	$149
27881	Wagan 300W Sine Wave Inverter	$195

Pure Sine Wave Inverters

Affordable clean sine wave inverters

Just the thing for smaller off-the-grid homes, these mid-sized 600- and 1,000-watt inverters deliver clean sine wave power at an affordable price. Available in either 12v or 24v input, these compact lightweight units feature overload protection, thermostatic fan, low battery alarm, warning lights, on/off switch, and automatic shutdown. Output is a pair of 3-prong plugs on front, input is #2 gauge terminals on back.

The 600w unit will surge to 1,000 watts, is 13.2" x 9.3" x 3.3", weighs 6.6 lb.

The 1,000w unit will surge to 2,000 watts, is 15.5" x 9.3" x 3.3", weighs 8.8 lb.

One-year manufacturer's warranty. Taiwan.

49-0109	Samlex 12v 600w Pure Sine Inverter	$295
49-0110	Samlex 24v 600w Pure Sine Inverter	$319
49-0111	Samlex 12v 1000w Pure Sine Inverter	$589
49-0112	Samlex 24v 1000w Pure Sine Inverter	$599

Exeltech Pure Sine Wave Inverters

The purest sine wave on the market

The Exeltech 1,100-watt unit is our most reliable mid-sized pure sine wave inverter. When your video, audio, or other sensitive electronic components demand clean power, Exeltech has proven to be durable and reliable. Delivers 1,100 watts continuously, or up to 2,200 watts surge. Will automatically protect itself against overload, overheating, or battery voltage too high or low. Has a thermostatically controlled cooling fan, and runs almost silently. Features high operating efficiency (for sine wave inverters) of 87% to 90%. Dual AC outlets. Measures 7.7" x 3.6" x 13.8" and weighs 10 lb. Available in 12- or 24-volt versions, or 32, 36, 48, 66, or 108 volts by special order. USA.

27147	Exeltech XP 1100/12V Sine Wave Inverter	$799
27148	Exeltech XP 1100/24V Sine Wave Inverter	$899
27882	Exeltech XP 1100 (specify volts) Sine Wave Inverter	$1099

Fighting for Your Independence

OutBack Power Systems

Available in black only. We need to keep the chrome de-animizers for battling the aliens.

OUTBACK FX-SERIES INVERTERS

The most advanced and adaptable inverters available

OutBack is an engineer-owned, customer-focused power equipment manufacturer. That means they make stuff that *works* and they actively support their customers. OutBack has introduced several new models and increased the power output of their incredibly clean, efficient, sine wave FX series inverters in the past year. The FX is available either in vented models for protected, indoor installations, or in sealed models for installation in marine, jungle, bug-infested, or other challenging sites. The vented models cost a bit less per watt, and are going to be the choice for most folks. But if corrosion, moisture, or critters are a problem where you live, make your life problem-free with a sealed unit.

Grid-tie units are available in both sealed and vented cases for 24- and 48-volt units. The grid-tie software package has some substantial differences, so these are being offered as separate models, at the same pricing. No additional accessories or extra-cost options are required for code-approved grid-tie with OutBack FX-series inverters.

OutBack inverters deliver a very clean sine wave power, with strong surge abilities up to 50 amps on all models. A built-in 60-amp transfer switch with automatic battery charging takes over whenever an outside power source such as a generator or utility power is available. The optional Mate allows adjustment of battery charging and other operations, but is not required for daily operation. Any custom programming is retained even without battery power. (Yea! No more lost programming!) Multiple inverters can be connected for 120-, 240-, or even 3-phase 208-volt AC output. Multiple inverter stacks *do* require the Mate and Hub. All OutBack inverter installations will require additional parts for mounting and overcurrent protection. Call for assistance. Inverters listed below are North American standard 120v/60hz output. Export models with 230v/50hz output are available by special order, same pricing. All inverters are 16.25"L x 8.25"W x 11.5H and weigh 60 to 62 lb. USA of course. Two-year manufacturer's. warranty with optional extended warranty.

In the table below, FX models are sealed, VFX are vented, a G or GT indicates grid-tie. First two digits are rated wattage, second two digits are battery voltage. A GTFX2524 is a grid-tie, sealed unit, 2,500 watts output, 24-volt input, for instance.

Model	FX2012T*	VFX2812	FX2024	GTFX2524*	VFX3524 and GVFX3524	FX2548	GTFX3048*	VFX3648 and GVFX3648
Battery Voltage:	12 volts	12 volts	24 volts	24 volts	24 volts	48 volts	48 volts	48 volts
Cooling:	Sealed	Vented	Sealed	Sealed	Vented	Sealed	Sealed	Vented
Continuous Output:	2,000 W	2,800 W	2,000 W	2,500 W	3,500 W	2,500 W	3,000 W	3,600 W
Wave form:	Pure Sine	Pure Sine	Pure Sine	Pure Sine	Pure Sine	Pure Sine	Pure Sine	Pure Sine
Surge Capacity:	4,800 VA	4,800 VA	4,800 VA	4,800 VA	5,000 VA	4,800 VA	4,800 VA	5,000 VA
Battery Charging Rate Max.:	80 amps	125 amps	55 amps	55 amps	85 amps	35 amps	35 amps	45 amps
Recommended DC Breaker:	OBDC-250	OBDC-250	OBDC-175	OBDC-175	OBDC-250	OBDC-100	OBDC-100	OBDC-175
Recommended DC Cable:	4/0 Ga.	4/0 Ga.	2/0 Ga.	2/0 Ga.	4/0 Ga.	2/0 Ga.	2/0 Ga.	2/0 Ga.
Item #:	49-0134	49-0125	27905	27905	49-0126	27906	27906	49-0127
Price:	$1,995	$2,345	$1,795	$1,995	$2,345	$2,245	$2,345	$2,345

*Equipped with Turbo cooling fan

OUTBACK INVERTER OPTIONS

OutBack FX-Series Adapter Kit

The easy way to order what every inverter installation needs. The FX Adapter Kit contains one each of the DC Cover, DC Conduit Adapter, and AC Adapter described below. Order one Adapter Kit for each FX-series inverter, and you're covered.

27910   FXA Adapter Kit $129

OutBack FX2000 DC Cover

A cast aluminum cover for the DC cable connections and other accessories in the DC wiring area on top of the inverter. Required for a NEC compliant installation. USA.

27907   DC Cover $65

OutBack FX2000 DC Conduit Adapter

An aluminum conduit adapter required when mounting the FX2000 to an OutBack DC Center, or to a 2" conduit. Required for a NEC compliant installation. USA.

27908   DC Conduit Adapter $45

OutBack FX2000 AC Adapter

A cast polycarbonate AC wiring compartment extension. Provides a 2" conduit adapter, two 1" conduit knockouts, and an AC outlet knockout. Required when mounting an FX inverter to an OutBack AC Center. USA.

27909   AC Adapter $35

OutBack Mate

The brains of your OutBack system

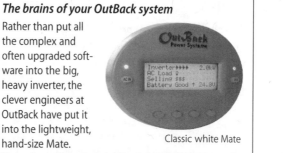

Rather than put all the complex and often upgraded software into the big, heavy inverter, the clever engineers at OutBack have put it into the lightweight, hand-size Mate.

Classic white Mate

Upgrades are simple now! Single inverter systems will run just fine without the Mate, but for larger multi-inverter systems, or folks who like to fine-tune, the Mate is a necessity. The classic oval Mate is available in white or black; the new square Mate2 is black. The 4-line, 80-character backlit display is also a highly detailed system monitor, and can

Classic black Mate

be 1,000' from the system. Runs on common Ethernet cable (Cat 5E cable with RJ45 connectors). An opto-isolated RS232 port will talk to PC computers. The Mate is supplied with a 50' cable and will talk to one appliance. Hubs let the Mate communicate with multiple inverters and charge controls. Four-

Mate 2

and ten-appliance Hubs available. 5.75"W x 4.25"H x 2" D, 1 lb. Two-year manufacturer's warranty. USA.

25813   OutBack Mate, White w/50' cable $295
48-0184 OutBack Mate, Black w/50' cable $295
48-0185 OutBack Mate2, Black w/50' cable $295

25814   OutBack Hub-4 w/2-3' & 2-6' cables $195
48-0039 OutBack Hub-10 w/2-3' & 4-6' cables $375

XANTREX SW PLUS INVERTER SERIES

New choices and lower prices for off-grid sine wave power

Xantrex introduces a new, improved version of their well-proven sine wave inverter technology. Available in 24- or 48-volt versions, the SWP-series offers continuous AC power output at 2.5, 4.0, and 5.5 kVA. This highly capable inverter can be used for independent households, or to provide automatic back-up power when utility power fails. Peak efficiency is 95%, and search mode power is under 2 watts! The SWP series is still using the basic SW operating system that was first introduced in the mid-1990s, and is now well known and respected. One of the most welcome improvements with the SW Plus is that all inverter programming is held in non-volatile memory so customized charge or control settings won't be lost if DC power is interrupted. The Plus series is also FCC Part B compliant, which means less potential interference with radio and telephone equipment.

Many of the standard equipment bells and whistles from the former SW series are now options. But then, the base inverter price is $700 lower, so it's hard not to save money on this new model. Options include matching AC and DC conduit boxes for code-compliant installations. The AC conduit box now contains bypass/disconnect circuit breakers. With the optional Generator Start Module, it will automatically start and stop any remote-start capable generator based on battery voltage and customer needs. The optional Auxiliary Load Module will automatically turn up to three external items on or off based on battery voltage. An Inverter Control Module w/25' cable can be installed for remote operation, monitoring, or connection to a PC. The Grid Tie Interface and utility intertie are not available for the SWP series. Powder-coated steel enclosure rated for indoor use. 120vac/60hz output. 22.5"W x 15"H x 9"D. CSA listed to UL specs. Two-year manufacturer's warranty. USA.

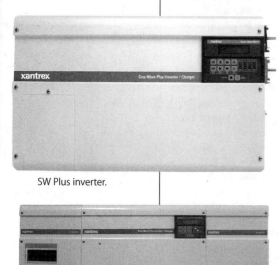

SW Plus inverter.

SW Plus inverter with short AC & DC conduit box options.

XANTREX INVERTER OPTIONS

SW Plus Generator Start Module

An add-on module that plugs into the SW Plus inverter for automatic generator control.

27855   **Xantrex Generator Start Module**   **$159**

SW Plus Auxiliary Load Module

An add-on module that allows battery voltage-based control of up to three circuits.

25865   **Xantrex Auxiliary Load Module**   **$159**

SWP Remote Control Panel Option

Offers a complete LCD control panel with all controls, indicators, and display. Includes 25 feet of cable. USA. Please specify voltage.

49-0139 **SWP Remote Inverter Control Module**   **$275**

TRACE SWP SERIES COMPARISON CHART

Model	SWP2524	SWP4024	SWP2548	SWP4048	SWP5548
Continuous Output:	2,500 VA	4,000 VA	2,500 VA	4,000 VA	5,500 VA
Wave Form:	Pure Sine	Pure Sine	Pure Sine	Pure Sine	Pure Sine
Surge Amps (5 sec):	80A	85A	80A	95A	105A
DC Input Voltage Range:	22–32 vdc	22–32 vdc	44–64 vdc	44–64 vdc	44–64 vdc
Charging Rate:	0–70A	0–110A	0–40A	0–60A	0–75A
Preferred Circuit Breaker:	175A	250A	175A	175A	250A
DC Cable Size:	2/0	4/0	2/0	2/0	4/0
Weight:	105 lb.	113 lb.	105 lb.	113 lb.	136 lb.
Item #:	49-0102	49-0135	49-0103	49-0136	49-0137
Price:	$2,299	$2,800	$2,299	$2,800	$3,500

SW/SWP Communications Adapter

This allows PC connection and monitoring of one SW- or SWP-series inverter with version 4.01 or later software. Includes adapter, DOS-based software, 50-foot cable, and DB9 connector. No picture.

27305   SW/SWP Communications Adapter $175

SW/SWP Stacking Interface

To stack a pair of SW or SWP inverters, this 42" comm-port cable is all you need. Simple phone-type plug-in connectors. 240V output from the stacked pair.

27314   SW/PS Stacking Interface $40

Conduit Box Options

The optional short conduit box allows the very heavy input wiring from the batteries to be run inside conduit to comply with the NEC. Has three (one on each side) ½", ¾", and 2" knockouts and plenty of elbow room inside for bending large wire. The SWP short AC conduit box has a 60A bypass breaker with additional breaker slots. Long AC and DC conduit boxes are also available with more room and many options. See the Power Panels section on page 140. USA.

27209   SW / SWP DC Conduit Box, short $94
49-0138 SWP AC Conduit Box, short $369

Trace T240 Autotransformer

A very handy 2:1 bi-directional transformer with a 4,500-watt continuous rating. It can be used to step up 120-volt inverter output to run the occasional 240-volt appliance, like a water pump. It can be used to balance the load on a 240-volt generator, by delivering 120 volts to an inverter/charger while evenly loading the generator. Or it can be used as a step-down transformer, allowing a stacked pair of inverters to share the start-up and running of a large 120-volt appliance. Supplied in a powder-coated enclosure with multiple ¾" and 1" knock-outs and a 2-pole 25-amp circuit breaker. Weighs 39.4 lb. Measures 6.3" x 7" x 21". Accepts 14 to 4 AWG wire sizes. USA.

27326   T240 Autotransformer $350

Code-Approved Inverter Cables

Our inverter and battery cables are all flexible code-approved cable with color coding for polarity and shrink-wrapped lugs for ⅜" bolt studs. These are two-lug versions (lugs on both ends) . USA.

15851	2/0, 3' Inverter Cables, 2 lug, single	$30
15852	4/0, 3' Inverter Cables, 2 lug, single	$40
27311	4/0, 5' Inverter Cables, 2 lug, pair	$120
27312	4/0, 10' Inverter Cables, 2 lug, pair	$205
27099	2/0, 5' Inverter Cables, 2 lug, pair	$85
27098	2/0, 10' Inverter Cables, 2 lug, pair	$140

Heavy-Duty Insulated Battery Cables

Serious, beefy, 2/0 or 4/0 gauge battery interconnect cables for systems using 1,000-watt and larger inverters. These single code-approved cables have ⅜" lug terminals on both ends, and are shrink-wrapped for corrosion resistance. Pick your gauge, length, and color. USA.

15841	4/0, 20" Black	$32
15840	4/0, 20" Red	$32
15839	4/0, 13" Black	$27
15850	2/0, 20" Black	$24
15849	2/0, 20" Red	$24
15848	2/0, 12" Black	$16
15847	2/0, 9" Black	$14

Wire, Adapters, and Outlets

It is important to use common sense when dealing with electricity, and this might be best done by acknowledging ignorance on a particular subject and requesting advice and help from experts when you are unsure of something.

Wires are freeways for electrical power. If we do a poor job designing and installing our wires, we get the same results as with poorly designed roads: traffic jams, accidents, and frustration. Big, wide roads can handle more traffic with ease, but costs go up with increased size, so we're looking for a reasonable compromise. Properly choosing wire and wiring methods can be confusing, but with sufficient planning and thought we can create safe and durable paths for energy flow.

The National Electrical Code (NEC) provides broad guidelines for safe electrical practices. Local codes may expand upon or supersede this code. It is important to use common sense when dealing with electricity, and this might be best done by acknowledging ignorance on a particular subject and requesting advice and help from experts when you are unsure of something. The Gaiam Real Goods technical staff is eager to help with particular wiring issues, although you should keep in mind that local inspectors will have the final say on what they consider the most appropriate means to the end of a properly installed electrical system. Advice from your local inspector or local electrician, who will be familiar with local code requirements and renewable energy systems, is probably your best advice.

Wire

Wire comes in a tremendous variety of styles that differ in size, number, material, and type of conductors, as well as the type and temperature rating of insulation protecting the conductor. One of the basic ideas behind all wiring codes is that current-carrying conductors must have at least two layers of protection. Permanent wiring should always be run within electrical enclosures, conduit, or inside walls. Wires are prone to all sorts of threats, including but not limited to abrasion, falling objects, and children using them as a jungle gym. The plastic insulation around the metal conductor offers minimal protection, based on the assumption that the conductors will be otherwise protected. Poor installation practices that forsake the use of conduit and strain-relief fittings can lead to a breach of the insulation, which could cause a short circuit and fire or electrocution. Electricity, even in its most apparently benign manifestations, is not a force to be managed carelessly.

A particular type and gauge (thickness) of wire is rated to carry a maximum electrical current. The NEC requires that we not exceed 80% of this current rating for continuous duty applications. With the low-voltage conditions that we run across in independent energy systems, we also need to pay careful attention to voltage drop. As voltage decreases, amperage must increase in order to perform any particular job. So lower-voltage DC systems are likely to be pushing higher amperages. This problem becomes more acute with smaller conductors over longer runs. Some voltage drop is unavoidable when moving electrical energy from one point to another, but we can limit this to reasonable levels if we choose the right size wire. Acceptable voltage drop for 120-volt circuits can be up to 5%, but for 12- or 24-volt circuits 2 to 3% is the most we want to see. We provide a good all-purpose chart and formula on page 200 for figuring wire size at any voltage drop, any distance, any voltage, and any current flow. Use it, or give us a call and we'll help you figure it out.

DC Adapters and Outlets

With the rising popularity and reliability of inverters that let folks run conventional AC appliances, demand for DC fixtures and outlets continues to diminish, which is mostly a good thing. The roots of the independent-energy movement grew from the automobile and recreational-vehicle industries, so plugs and outlets for low-voltage applications are based usually on that somewhat flimsy creation, the cigarette lighter socket. This is an unfortunate choice. The limited metal to metal contact in this plug configuration seriously limits the current-carrying capacity. Cigarette lighter plugs and outlets are usually rated for 15 amps surge, but 7 to 8 amps continuously is absolutely the maximum you can expect without meltdown. Lighter sockets are not approved by the National Electric Code. There is no plug-and-socket combo that is approved officially for 12-volt use. A large, open receptacle accessible to small children presents an obvious safety hazard. If you must use lighter socket outlets, be sure they're fused!

The conventional rules for DC outlets and plugs says the center contact of the receptacle (female outlet) and the tip of the plug (male adapter) should be positive, and the outer shell of the receptacle and side contact of the plug should be negative. However, the polarity can be reversed in a variety of ways during installation. For some appliances, such as incandescent or halogen lights, it doesn't make a bit of difference. But for those expensive 12-volt compact fluorescent lamps, or that costly DC television you just bought, it matters greatly. It may even be a matter of life or death. Don't assume. Check the polarity with a handheld voltmeter before plugging in.

Another convention is that the power source is presented by the female outlet. This way, it is more difficult to accidentally short out against something or electrocute yourself. Some small solar modules have DC male adapters on them, even though they are a power source, so they can plug into the dashboard lighter socket and trickle charge the battery. These are such a small power source that they present little to no hazard. It will be difficult to get a visible spark from these trickle-charger panels.

Some people would rather not use lighter sockets and plugs in their houses because they look dangerous . . . and they can be. Little fingers and tools fit conveniently within the supposedly "protected" live socket. Probably the safest option is to use an oddball, but still locally available, AC plug-and-socket configuration that has the prongs perpendicular, slanted, or in some other configuration that makes it impossible to plug in to a standard 120-volt AC plug. These have lots of metal to metal contact, and will be rated for at least 15 amps continuously. Higher-amperage sockets are readily available if you need them. Although it is tempting to use standard inexpensive 120-volt AC outlets, we've seen this lead to disaster too often, and strongly advise against it. Murphy's Law finds enough ways to trip us up without leaving the welcome mat out. We know from personal experience that any savings are quickly lost when a wrong-voltage appliance is plugged in and destroyed.

Do NOT mix AC and DC wiring within a single electrical box! That's a big-time Code (and common sense) violation. It's fine to put your AC and DC outlets right next to each other—but they need to be in separate boxes.

The roots of the independent-energy movement grew from the automobile and recreational-vehicle industries, so plugs and outlets for low-voltage applications are usually based on that somewhat flimsy creation, the cigarette lighter socket. This is an unfortunate choice.

Wire Sizing Chart/Formula

This chart is useful for finding the correct wire size for any voltage, length, or amperage flow in any AC or DC circuit. For most DC circuits, particularly between the PV modules and the batteries, we try to keep the voltage drop to 3% or less. There's no sense using your expensive PV wattage to heat wires. You want that power in your batteries!

Note that this formula doesn't directly yield a wire gauge size, but rather a "VDI" number, which is then compared to the nearest number in the VDI column, and then read across to the wire gauge size column.

1. Calculate the Voltage Drop Index (VDI) using the following formula:

VDI = AMPS x FEET ÷
(% VOLT DROP x VOLTAGE)

Amps = Watts divided by volts

Feet = One-way wire distance

% Volt Drop = Percentage of voltage drop acceptable for this circuit (typically 2% to 5%)

2. Determine the appropriate wire size from the chart below.
   A. Take the VDI number you just calculated and find the nearest number in the VDI column, then read to the left for AWG wire gauge size.
   B. Be sure that your circuit amperage does not exceed the figure in the Ampacity column for that wire size. (This is not usually a problem in low-voltage circuits.)

Example: Your PV array consisting of four Siemens SP75 modules is 60 feet from your 12-volt battery. This is actual wiring distance, up pole mounts, around obstacles, etc. These modules are rated at 4.4 amps × 4 modules = 17.6 amps maximum. We'll shoot for a 3% voltage drop. So our formula looks like:

$$VDI = \frac{17.6 \times 60}{3[\%] \times 12[V]} = 29.3$$

Looking at our chart, a VDI of 29 means we'd better use #2 wire in copper, or #0 wire in aluminum. Hmmm. Pretty big wire.

What if this system was 24-volt? The modules would be wired in series, so each pair of modules would produce 4.4 amps. Two pairs × 4.4 amps = 8.8 amps max.

$$VDI = \frac{8.8 \times 60}{3[\%] \times 24[V]} = 7.3$$

Wow! What a difference! At 24-volt input you could wire your array with little ol' #8 copper wire.

WIRE SIZE	COPPER WIRE		ALUMINUM WIRE	
AWG	VDI	Ampacity	VDI	Ampacity
0000	99	260	62	205
000	78	225	49	175
00	62	195	39	150
0	49	170	31	135
2	31	130	20	100
4	20	95	12	75
6	12	75	•	•
8	8	55	•	•
10	5	30	•	•
12	3	20	•	•
14	2	15	•	•
16	1	•	•	•

Chart developed by John Davey and Windy Dankoff. Used with permission.

For wire and adapter products, see addendum on pages 558-559.

CHAPTER 4
Emergency Preparedness

MANY EVENTS DURING THE LAST FIVE YEARS—since this *Sourcebook* was last updated—have demonstrated dramatically that the world can be a hazardous place even in the best of times. The tragedy of the September 11, 2001, terrorist attacks—the electric power grid blackout of August 2003 affecting eight states and parts of Canada—the disastrous forest fire seasons of 2002 and 2003 in the western United States—these are only the most vivid demonstrations of the ways our society has become vulnerable to disruption of the various infrastructures that most of us depend on to live our daily lives. There seems to be little doubt that factors as diverse as growing population pressures, grinding poverty, increasing economic globalization, political conflict, climate change, and other natural disasters will continue to place enormous strains on the networks that supply us with food, water, power, and other necessities. We're likely to see bigger storms and hurricanes, stranger weather, overloaded utilities, more electrical failures, less available and much more costly oil, and more terrorist attacks in our lifetimes. And isn't it still about time for that next big California earthquake?

Short of living completely off the grid—several of them, really—most of us are likely to experience these kinds of disruptions and breakdowns in the foreseeable future. Won't you rest easier knowing that you've taken steps to protect or insulate yourself and your family from events beyond your control that threaten to undermine your ability to meet your basic needs? Gaiam Real Goods offers a variety of products and resources to get you up to speed on emergency preparedness.

> Won't you rest easier knowing that you've taken steps to protect or insulate yourself and your family from events beyond your control that threaten to undermine your ability to meet your basic needs?

Want Some Protection, or at Least a Little Backup?

The place to start, of course, is with electric power. For one thing, that's what we do best. But the key point is that, while not always apparent, many domestic systems, appliances, and operations that are not fueled by electricity are nevertheless dependent on it to function. When the grid goes down, you're going to lose a lot more than your lights and your computer.

The renewable energy industry has spent its maturing years learning how to provide highly reliable home energy systems under adverse and remote conditions. No utility power? No problem! No generator fuel? No problem! No easy access to the site? No problem! Gaiam Real Goods is ready to provide off-the-shelf solutions for short-term, long-term, or lifetime energy problems. We've collected and discussed a few of them in this chapter. Like individual lifestyles, energy needs and the necessary equipment to meet them are individual matters. If you don't see what you need, give our experienced technical staff a call. We're masters at putting together custom systems to meet individual needs, climates, and sites. Reliable power for your peace of mind is our business.

Steps to a Practical Emergency Power System

Following are some practical solutions to utility disruption. We've arranged them in financial order with the least expensive options first. Be aware that lowest-cost options will involve some sacrifices and won't be acceptable solutions for more than short time periods—emergencies, in other words. As back-up systems get more robust, they impose less restrictions, and are more pleasant to live with.

1. INSTALL A GOOD-QUALITY BACK-UP GENERATOR

A generator will be useful for short power outages, or for longer ones so long as you have fuel and you, and the neighbors, are willing to endure the noise. Higher-quality generators

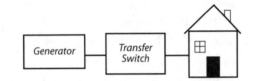

tend to be much quieter. Propane fuel is a good choice. It is delivered in bulk to your site (in all but the most remote locations), it doesn't deteriorate or evaporate with age, and it's piped to the generator, so you never need to handle it. Generators live longer on propane because it produces less carbon build-up. Your electrician will need to install a transfer switch between the utility line and the generator, so that your generator doesn't try to run the neighborhood, and doesn't threaten any utility workers.

Simple back-up generators use manual starting, and a manual transfer switch. When power fails somebody has to start the generator, then throw the switch. Fancier systems will do all this automatically. With any generator-based back-up system, even automated ones, there will be some dead time between when the power goes out, and when the back-up systems come online.

Most generator-based systems will require some electrical conservation. Don't plan on running big watt-sucker appliances like air conditioning, electric room or water heaters, or electric clothes dryers, unless you're willing to support a huge generator of 12,000 watts or larger. Generators run most efficiently at 50 to 80% of full load. A big generator is going to suck almost as much fuel to run a few lights as it is to

run all the lights and the air conditioning. Generators are not a case of "bigger is better."

2. ADD BATTERIES AND AN INVERTER/CHARGER

The generator only supplies power when it's running, and it isn't practical or economical to run it all the time. Running a load of just 200 watts is particularly expensive with a generator. They're happiest running at about 80% of full rated wattage, which keeps carbon build-up to a minimum and is the most efficient in terms of watts delivered per quantity of fuel consumed.

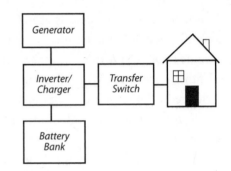

Adding batteries and an inverter/charger allows the best use of your generator. Whenever the generator runs, it will automatically charge the batteries. This means the generator is working harder, building up less carbon, and delivering more energy for the fuel consumed. Energy stored in the batteries can be used later to run lights, entertainment equipment, and smaller loads without starting the generator.

Many medium- to larger-size inverters come with built-in, automatic battery chargers. So long as the inverter senses that utility or generator power is available, that line power is passed right through. If outside AC power fails, the inverter will start using stored battery power automatically to supply any AC demands. When outside AC power returns, the batteries will be recharged automatically. For a power failure that lasts more than a couple of days, this means you'll only need to run the generator for a few hours per day, yet you'll have AC power full time.

Putting an inverter and battery pack into the system means that any AC power interruptions will be measured in seconds, or less, depending on the equipment chosen.

The renewable energy industry has spent its maturing years learning how to provide highly reliable home energy systems under adverse and remote conditions. No utility power? No problem! No generator fuel? No problem! No easy access to the site? No problem!

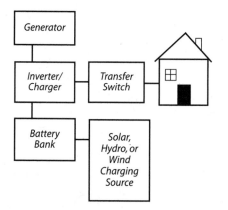

```
Generator
   |
Inverter/ ── Transfer ── [House]
Charger      Switch
   |
Battery ── Solar,
Bank       Hydro, or
           Wind
           Charging
           Source
```

3. ADD SOLAR, HYDRO, OR WIND POWER FOR BATTERY CHARGING

A generator has the lowest initial cost for power backup, but it's only cheap if you don't run it. Generators are expensive and noisy to operate, and fuel supplies have to be replenished. If you anticipate power outages lasting longer than a few weeks, then solar, hydro, or wind power will save you money in the long run, and deliver a reliable long-term energy supply.

Once you have the battery pack, it can accept recharging from any source. Electrons all taste the same to your battery. Whether they come from a belching, cheapo gas generator, or a clean, silent PV module, the effect is the same on the battery. Renewable energy sources have a higher initial cost, but very low costs over time for both your wallet and the environment.

Emergency and Back-up Power System Design

The average American home uses around 20 kWh of electricity per day. While it's certainly possible to put together an emergency power system to supply this level of use, most folks would find it prohibitively expensive. Most of us are willing to modify our lifestyle in the face of an emergency. Energy conservation is the cornerstone of reduced consumption, which will make getting by in case of disruptions that much more manageable.

What Do You Really Need in an Emergency?

Refrigeration, heat, drinking water, lights, cooking equipment, and communications are on the typical short list of what you really need in an emergency. Let's take a quick look at each of these necessities and see how they can best be supplied during an emergency.

REFRIGERATORS AND FREEZERS

Refrigeration is a major concern for most emergency or back-up systems. The fridge will probably be your major power consumer, typically using 3 to 5 kWh of energy per day. It's going to require a 1,500-watt generator, or a robust 1,000-watt inverter, at the very least, to start and run a home refrigerator or freezer. If your refrigerator is 1993 vintage or older, consider a new replacement. Refrigerators have made enormous advances in energy efficiency over the past few years. The average new 22-cubic foot fridge will use half the power of one produced before 1993. If you're shopping, watch those yellow EnergyGuide tags. They really level the playing field! All models are compared to the same standard, which happens to be a pessimistic 90°F ambient temperature. Most of us will see slightly lower operating costs, most of the time. Any fridge model sporting a tag for

under 500 kWh per year will serve you well and drop energy use to near 1.5 kWh per day. The very best mass-produced fridges will have ratings for 400 kWh/year or less. Top-and-bottom units have a slight energy advantage over side-by-side units.

If you expect power to be out for more than a few weeks, then a propane-powered fridge might be a better choice. Although small by usual American standards (about 7 to 8 cubic feet), they'll handle the real necessities for about 1.5 gallons of propane per week. Propane freezers are also available.

HEAT

If a natural disaster happens in winter, keeping warm suddenly may become overwhelmingly important, while keeping the food from spoiling is of no concern at all. Anyone with a passive solar house, a wood stove, and a supply of firewood will be sitting pretty when the next ice storm hits. Obviously, any heating solution that doesn't involve electricity will keep you warm more effectively following a natural disaster. Beyond wood stoves, other nonelectric solutions may include gas-fired wall heaters, or portable gas or kerosene heaters. Make sure you

Obviously, any heating solution that doesn't involve electricity will keep you warm more effectively following a natural disaster.

Disease carried by polluted drinking water does far more human harm than actual natural disasters.

If you have your own well or other private water supply that requires a pump for water delivery, you'll need some kind of power system that can run your water pump.

If you use compact fluorescent lamps instead of common incandescent lamps, you'll get four to five times as much light for your power expenditure.

get one with a standing pilot instead of electric ignition or you'll be out of luck. Also be aware that if your thermostat is electric, it won't work in an extended power outage and you'll need to use manual controls.

Wall heaters are preferable, because they're vented to the outside. Use extreme caution with portable heaters, or any other "ventless" heater that uses room oxygen and puts its waste products into the air you're breathing! Combustion waste products combined with depleted oxygen levels are a dangerous or even deadly combination. Neither brain damage nor freezing are acceptable choices. Those in earthquake country shouldn't count on natural gas supply to continue following a large earthquake.

If a central furnace or boiler is your only heat choice, you'll need some electricity to run it. If your generator- or inverter/battery-based emergency power system is already sized to run a fridge, it probably will be able to pick up the heating chores instead of the fridge during cold weather. (Put the food outside, Mother Nature will keep it cold.) Furnaces, which use an air blower to distribute heat, will generally use a bit more power than the average refrigerator. Boilers, which use a water pump or pumps to distribute heat, will generally use a bit less power than the fridge. Depending on the capabilities of your emergency power system, you might need to limit other uses while the heater is running.

Pellet stoves are a very poor choice for emergency heating. Although the fuel is compact and stores nicely, pellet stoves require a steady supply of power in order to operate. Pellet stoves have blowers for combustion air, other blowers for circulation of room air, and pellet feed motors. A better solution is a special stainless steel pellet basket that allows you to burn pellet fuel in any conventional wood stove. You'll find pellet baskets in chapter 5 under Appliances for Heating.

WATER

If you're hooked up to a city-supplied water system, you may have a problem that an emergency power system can't solve. Most cities use a water tower or other elevated storage to provide water pressure. They can continue to supply water for only a few days, or maybe just a few hours, when the power fails. Other than filling the bathtub, or draining the water heater tank, which can get you by for a few days, you may want to provide yourself with some emergency water storage. Water storage tanks are available from any farm supply store, and some larger

building and home supply stores. There are "utility" and "drinking water" grade tanks. Make sure you're buying the drinking water grade with an FDA-approved nonleaching coating.

If you have your own well or other private water supply that requires a pump for water delivery, you'll need some kind of power system that can run your water pump. AC water pumps require a large surge current to start. The size of the surge depends on pump type and horsepower. Submersible pumps require more starting power than surface-mounted pumps of similar horsepower.

For short-term backup, the easiest and cheapest solution is to use a generator to run the pump once or twice a day to fill a storage or large pressure tank. While the generator is running, you can also catch a little battery charging for later use to run more modest power needs that can be supplied by an inverter/battery system.

Longer-term solutions will require either a bigger inverter and battery bank to run the water pump on demand, or a smaller water pump that can be run more easily by solar or battery power. Our technical staff will by happy to discuss your situation and make the best recommendations for your needs and budget.

WATER QUALITY (IS IT SAFE TO DRINK?)

Natural disasters often leave municipal water sources polluted and unsafe to drink, sometimes for weeks. So even though you may have water, you'd better think twice before drinking it. Disease carried by polluted drinking water does far more human harm than actual natural disasters.

With just a little preparation, and at surprisingly low cost, you can be ready to treat your drinking water. And considering how truly miserable, long-lasting, and even life-threatening many water-borne diseases are, an ounce of prevention is worth many tons of cure. You can boil, which is time and energy intensive, you can treat with iodine, which is cheap and effective but distasteful, or you can filter, which is quick and effective with the right equipment. The "right equipment" usually means either ultrafine ceramic filtration, or a reverse osmosis system. We offer several models of both types in the product section that follows.

LIGHTING

If you use compact fluorescent lamps instead of common incandescent lamps, you'll get four to five times as much light for your power expenditure. Because power is valuable and expensive in an emergency or back-up situation, these efficient lights are a must. If CFLs happen to be part of your normal household lighting anyway, you'll enjoy some energy savings while waiting for The Big One. Since compact fluorescent lamps will outlast ten to thirteen normal incandescent lamps, and will use 75% less energy, they'll end up paying for themselves, and they'll put hundreds of dollars back in your pocket over their lifetime. Compact fluorescent lamps are a great investment, both for everyday living and for energy-efficient emergency backup.

Smaller, battery-powered, solar-recharged lanterns and flashlights can be extremely helpful during short-term outages, and can easily be carried to wherever light is needed. We offer several quality models that can be charged with their onboard solar cells, or simply have fresh batteries slipped in when time and conditions won't allow solar recharging.

COOKING

Cooking is one of the easier problems to tackle in a power failure. Let's start by looking at what you're using to cook and bake with now.

Electric Stoves and Ovens. These use massive amounts of electricity. It's not going to be practical to run your electric burners or oven with a back-up system, or even with a generator. Break out the camping gear. Camp stoves running on white gas or propane are a good alternative and widely available at modest cost. Charcoal barbecues can be used as well, but *only outdoors!* They produce carbon monoxide, which is odorless and deadly if allowed to accumulate indoors. Want a longer-term solution? Consider a gas stove.

Gas Stoves and Ovens. These can be fired by either propane or natural gas. Propane, which only depends on a small, locally installed tank, may be more dependable in a major emergency, particularly an earthquake. Major earthquakes break buried natural gas lines, but electrical outages don't affect natural gas systems. In either case, an electrical outage will put your modern spark-ignition burner lighters out of commission. Just use a match or camping-type stove igniter for the burners. Older stoves use a pilot light for ignition, which will work as normal with or without electricity.

Your gas oven probably won't work without electricity. Older stoves, using a standing pilot light, will be unaffected; all others will use either a "glow bulb" or a "spark igniter." Spark igniters, like those used on stovetop burners, will allow you to light the oven with a match. This is good. On the other hand, glow bulbs require 200 to 300 watts of electrical power all the time the oven is on, and will not allow match lighting. This is a problem. You can tell which type of igniter you have by opening the bottom broiler door, and then turning on the oven. You'll hear or see the spark igniter, and the unit will light up quickly. Glow bulbs take 30 seconds or longer to light up, and you'll probably see the glowing orange-red bulb once it gets warmed up.

Solar Ovens. If the weather allows, solar ovens do a great job at zero operational cost. They do require full sunlight, and are excellent for anything from a cup of rice, to a batch of muffins, to a pot full of stew. We offer several ovens, kits, and plans in the product section that follows.

COMMUNICATIONS

For simple communication needs, we offer several models of radios with wind-up and/or solar power sources. Then it gets more complicated.

Many folks now consider the Internet to be a basic mode of communication. Internet access requires a computer, and that means some minimum power use. If you have the choice, a laptop computer will use a fraction of the power that a desktop unit requires. The average desktop with 17" color monitor will use approximately 120 watts. The average laptop with color matrix screen only uses 20 to 30 watts. If you plan to spend more than a couple hours per day on the computer without grid support, then a good laptop is going to pay for itself in energy system savings. Laptops, and that other nearly ubiqitous electronic device, cell phones, require frequent recharging. These kinds of loads are easily manageable with a small generator back-up system or a renewable energy system. Most other personal communications devices, such as cell phones, CB radios, ham radios, and two-way radios, are designed for, or easily adapted to, 12-volt DC power use. This means the energy in your battery storage can be utilized directly. And if you want to be totally free of the telephone grid, DirecWay or StarBand two-way satellite devices provide faster-than-DSL internet service for net surfing.

It's not going to be practical to run your electric burners or oven with a back-up system, or even with a generator. Break out the camping gear!

If the weather allows, solar ovens do a great job at zero operational cost.

Conclusion

An ancient Chinese curse says, "May you live in interesting times." Like it or not, we have been thoroughly cursed with interesting times. We can make the best of it with some modest preparation, or we can sit back, do nothing, and wait to see what gets dealt to us, and how much it hurts. In the product section that follows, we've put together a few items and kits that can ease the pain a bit, or even a whole lot. Choose the level of protection that feels comfortable to you. After you're prepared, sit back feeling secure and smug, knowing that you're independent from the grid—at least for awhile. And if you're really ambitious, talk to our technicians about a full-time off-the-grid system.

EMERGENCY PREPAREDNESS PRODUCTS

Multifunction Survival Tool

Shovel, hammer, saw, hatchet, bottle opener, nail puller, wrench. Unscrew the calibrated compass and find matches, fishing hooks and line, nails, and more in a waterproof bag stashed in the handle. Comes with a snap-on carry case. 12" L x 4.5" W. China.

63-114   All-Purpose Survival Tool   **$20**

EMERGENCY LIGHTING PRODUCTS

Humphrey Propane Lights

Propane lighting is very bright. It is a good alternative if you don't have an electrical system. One propane lamp emits the equivalent of 50 watts of incandescent light, while burning only one quart of propane for 12 hours use. The Humphrey 9T contains a burner nose, a tie-on mantle, and a #4 Pyrex globe. The color is "pebble gray." The mantles seem to last around three months, and replacements are cheap. Note: Propane mantles emit low-level ionizing radiation due to their thorium content.

35-101	**Humphrey Propane Lamp (9T)**	**$59**
35-102	**Propane Mantle (5-pack)**	**$9.75**
35-103	**Replacement Globe**	**$13.95**

Flashlights

Create Light Without Batteries

No preparedness kit should be without the Dynamo. Continuous squeezing produces a steady beam of light. This is a reliable battery-free emergency light, with a shockproof and shatterproof lens, a spare bulb, a peg for attaching a cord, and a slide mechanism to lock the handle closed. Assorted gear colors. Canada.

37-374	**Hand-Cranked Dynamo Flashlight**	**$14**
	3 or More	**$12 each**

Solar Flashlight

Free sunlight fully recharges the built-in 600 mA nicad battery pack in about eight hours. No time for solar? Slip two AA batteries into the battery compartment, and you have light right now.  These optional AA batteries will be recharged by solar too. Includes three LED warning flashers on tail with focusing lens. Flashlight and warning flashers are switched separately, so one or both can be run. An adjustable holder for bikes or whatever is included, along with two spare bulbs and wrist strap. Run time on built-in battery alone is approximately 2.5 hours for the flashlight, or six days for the warning flasher. Optional NiMH batteries will increase run times by 300%; alkaline batteries by 250%. Waterproof and floats too! China.

37-295   Solar Flashlight w/Warning Flashers   **$25**

Forever Flashlight

This is an incredibly innovative flashlight that never wears out or runs down. Just shake the light to charge it up. Ten seconds of shaking delivers about two minutes of light. The internal magnetic generator turns shaking motion into electricity, which is stored in a large capacitor to run the 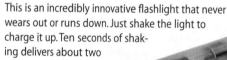 bright LED lamp. Clear construction lets you see the internal workings. This is one flashlight that will never let you down. Great for emergencies and car trunks. Taiwan.

06-0529 Forever Flashlight   **$29**

Solar Navigator Lantern

Whether outdoors at night or navigating a dark room, being able to switch on the light remotely can be an asset. Our multi-purpose Solar Navigator Lantern is a bril-

liant fluorescent light that can be turned on by remote control. Plus, the motion sensor can detect movement and switch on from as far away as 50 feet. Plugged in to an outlet, it also lights up automatically if the power is interrupted. Light head swivels 45° for variable lighting angles and has built-in solar module on back. Place in full direct sunlight for optimal battery charging capabilities. Comes with remote control on key chain, built-in battery, standard AC/DC adapter and convenient carrying handle. Each full charge lasts 7 hours. 14.5"H x 8.25"W x 2.5" diameter. China.

14-0218 Solar Navigator Lantern $99

EverLite Solar Spotlight

For camping, backyard, or emergency use, our high-tech solar spotlight is waterproof, safe in any climate, and runs over 24 hours per charge. Just three hours of full sun delivers 12 hours of ultra-bright LED run time from the built-in NiMH battery pack. Even 10 hours of overcast will deliver 12 hours of light. The bright, blue-white light has eight LEDs, delivering a total of 50 lumens. The lightweight lamp and base stows on the back of the solar panel along with the 15-foot cord. The lamp assembly can be unplugged for full portability. If the lamp arm is left in the ON position (up), our spotlight

will automatically turn on at dusk and off at dawn. Comes with a convenient nylon mesh bag that allows daytime charging right thru the bag. Options include plug-in chargers for 12- or 120-volt that allow quick recharging where the sun doesn't shine, and a nifty 12 vdc output cable to deliver power from the lamp base to your cellphone or other small 12v appliance. 10.25" x 6.6" x 2.5", weighs 2.7 lbs. China.

14-0299 EverLite Solar Spotlight with Carrying Case	$119
17-0290 EverLite 12V DC Charger	$12
17-0289 EverLite 120V AC Charger	$12
17-0291 EverLite 12V DC Output Adapter	$25

TerraLux LED Flashlight Conversion

Our astonishing LED conversion for common Mini Maglite® 2AA flashlights delivers two to three times as much bright blue-white light, runs two to four times longer on a set of batteries, and the lamp, with up to

100,000 hour life expectancy, will likely outlive you. Our kit supplies the MicroStar2 lamp and a custom reflector. You supply the MiniMag PowerPush™ technology keeps the bulb at full light output even with dropping battery voltage. When batteries are exhausted, it shifts to emergency moon mode for reduced, but continuing, light output. USA.

06-0524 MicroStar2 Flashlight Kit $25

The MicroStar2 kit is not made or endorsed by Mag Instruments, Inc.

Hi-Output LEDs Outshine All Others

The Princeton Tec Yukon HL headlamp delivers a major breakthrough in LED design. This hybrid features a 1-watt Luxeon emitter that delivers 25 lumens for brightness and distance that rivals the highest output incandescent, halogen, or xenon lamps. For closer work with a broad even beam spread, flip to the three hi-output 5mm LEDs. Both light sources deliver crisp white light that doesn't yellow as the batteries fade. Uses three AA batteries in rear pack. Runs approximately 44 hours on the big Luxeon LED, or 120 hours on the three LEDs. Impact-resistant, lightweight adjustable design fits helmets or bare heads. 8 oz. USA.

06-0538 Princeton Tec Yukon HL $59

The Flashlight of the Future

The Princeton Tec Impact XL is the best four-AA flashlight available. The powerful 1-watt Luxeon LED lamp delivers 25 lumens of crisp white light that never yellows as the batteries fade. It runs an incredible 70 hours on a set of batteries. The bulb lasts over 10,000 hours, and isn't bothered by anything less than a crushing impact. It is waterproof to 100 meters with secure rubber over a molded grip. This is the most reliable, longest-lasting flashlight you'll ever find. 6.4 oz. with batteries. USA.

06-0539 Princeton Tec Impact XL $39

WATER COLLECTION

Rainwater Collection for the Mechanically Challenged

Revised and expanded in 2003. Laugh your way to a home-built rainwater collection and storage system. This delightful paperback is not only the best book on the subject of rainwater collection we've ever found, it's

funny enough for recreational reading, and comprehensive enough to lead a rank amateur painlessly through the process. Technical information is presented in layman's terms, and accompanied with plenty of illustrations and witty cartoons. Topics include types of storage tanks, siting, how to collect and filter water, water purification, plumbing, sizing system components, freeze-proofing and wiring. Includes a resources list and a small catalog. 108 pages, softcover. USA.

**80-704 Rainwater Collection for
the Mechanically Challenged** **$20**

Rainwater Collection for The Mechanically Challenged Video

You say you're *really* mechanically challenged and want more than a few pictures? Here's your salvation. From the same irreverent, fun-loving crew that wrote the book above. See how all the pieces actually go together as they assemble a typical rainwater collection system, and discuss your options. This is as close as you can get to having someone else put your system together. 37 minutes and lots of laughs. USA.

21-0198 Rainwater Challenged Video **$19.95**

Rain Barrel and Diverter

Capture irrigation water at no cost with our virtually indestructible, 0.1875"-thick, recycled food-grade polyethylene rain barrel. Multiple barrels can be linked with an ordinary garden hose. Features overflow fitting, drain plug, screw-on cover and threaded spigot. Optional galvanized steel rain diverter (shown) fits any downspout (metal or plastic). When full, flip the diverter to a closed position to let downspout function as usual. 39"H x 24" diameter. 60-gallon capacity. 20 lbs. USA.

14-9201	**Rain Barrel and Diverter**	**$130**
14-0238	**Rain Barrel only**	**$114**
46209	**Diverter only**	**$22**

Allow 4–6 weeks for delivery. Can be shipped to customers in the contiguous U.S. only. Cannot be shipped to P.O. boxes. Sorry, express delivery not available.

Build Your Own Water Tank

By Donnie Schatzberg. An informative booklet that gives you all the details you need to build your own ferro-cement (iron-reinforced cement) water storage tank. Lots of drawings and illustrations. No special tools or skills are required. The information given in this book is accurate and easy to follow, with no loose ends. The author has considerable experience building these tanks, and has gotten all the "bugs" out. 58 pages, softcover. USA.

80-204 Build Your Own Water Tank **$14**

Kolaps-a-Tank

These handy and durable nylon tanks fold into a small package or expand into a very large storage tank. They are approved for drinking water, withstand temperatures to 140° F, and fit into the beds of several truck sizes. They will hold up under very rugged conditions, are self-supported, and can be tied down

with D-rings. All tanks have a 1.5" plastic valve with standard plumbing threads for input/output, and a 6-inch diameter by 2-foot long filler sleeve at the top for filling. Our most popular size is the 525-gallon model, which fits into a full-sized long-bed (5 x 8 ft.) pickup truck. USA.

47-401	**73 gal. Kolaps-a-Tank,** **40" x 50" x 12"**	**$311**
47-402	**275 gal. Kolaps-a-Tank,** **80" x 73" x 16"**	**$460**
47-403	**525 gal. Kolaps-a-Tank,** **65" x 98" x 18"**	**$558**
47-404	**800 gal. Kolaps-a-Tank,** **6´ x 10´ x 2´**	**$645**
47-405	**1,140 gal. Kolaps-a-Tank,** **7´ x 12´ x 2´**	**$889**
47-406	**1,340 gal. Kolaps-a-Tank,** **7' x 14' x 2'**	**$995**

WATER PURIFICATION

Seagull Undercounter Filter

Seagull water filtration products are widely used aboard international airliners. The unique dual-layer cartridge provides both carbon filtering for taste, odor, chlorine, and organic chemicals, plus an ultra-fine 0.1 micron ceramic element for giardia, cysts, bacteria, and parasites.

Seagull filters are manufactured in the United States from stainless steel and other high-grade components. The stainless housing with band clamp makes cartridge changes easy. The unique microfine structured matrix allows high flow rates of 1 gpm at standard pressures (30–40 psi). The replaceable cartridge should last anywhere from 9 to 15 months with ordinary use. This undercounter model is supplied with an ultra long-life ceramic disc chrome faucet. Although expensive initially, this is one of the best all-around filters available. This is the filter we put on our drinking fountains at the Solar Living Center. USA.

42650	Seagull Undercounter Filter	$429
42657	Replacement Cartridge	$90

Water Filter Crocks

Ceramic and carbon filtration reduces even E. coli by 99%. Pour in tap, well, or even rain water, and produce up to 6 gallon a day of clean and clear water—storing 2 gallon at

a time. Our crocks have a lead-free, non-toxic interior glaze that prevents mineral leaching, discoloration, and mold and mildew growth. Choose glazed sage or unglazed terra cotta. Sorry, not available in California. USA.

01-0391	Ceramic Water Filter Crock (17.5"H x 8" diameter, 20 lbs.)	$189
01-0401	Terra cotta Water Filter Crock 20"H x 10" diameter (17.5"H x 8" diameter, 20 lbs.)	$189
01-0092	Replacement Filter (replace yearly)	$59

The Katadyn Drip Filter

With no moving parts to break down, superior filtration, and a phenomenal filter life, there is simply no safer choice for potentially pathogen-contaminated water. There are no better filters than Katadyn for removing bacteria, parasites, and cysts. Three 0.2-micron ceramic filters process one gallon per hour. Clean filters by brushing the surface. Ideal for remote homes, RV, campsite, and home emergency use. Food-grade plastic canisters stack to 11" Dia. x 25" H. Weighs 10 lb. One-year manufacturer's warranty. Switzerland.

42-842	Katadyn Drip Filter	$289
42-843	Replacement Filter (needs 3)	$75 ea

WATER HEATING AND WASHING

4-Gallon Solar Shower

No more cold showers when you're camping. This low-tech invention uses solar energy to heat water for all your washing needs. The large 4-gallon capacity provides ample hot water for at least two hot showers. On a 70° day the Solar Shower will heat 60° water to 108° in only three hours. Great for camping, car trips, or emergency use. This has been one of our catalog's all-time best-sellers. Taiwan.

17-0169	Solar Shower	$15

James Washers

The James hand-washing machine is made of high-grade stainless steel with a galvanized lid. It uses a pendulum agitator that sweeps in an arc around the bottom of the tub and prevents clothes from lodging in the corner or floating on the surface. This ensures that hot suds are mixed thoroughly with the clothes. The James is sturdily built. The corners are electrically spot-welded. All moving parts slide on nylon surfaces, reducing wear. The faucet at the bottom permits easy drainage. Capacity is about 17 gallons. Wringer attachment pictured with the washer is available at an additional charge. USA.

63-411	James Washer	$449

Hand Wringer

The hand wringer will remove 90% of the water, while automatic washers remove only 45%. It has a rustproof, all-steel frame and a very strong handle. Hard maple bearings never need oil. Pressure is balanced over the entire length of the roller by a single adjustable screw. We've sold these wringers without a problem for over 17 years. USA.

63-412	Hand Wringer	$169

Super Wash Pressure Hand Washing Machine

Launders a 5-pound load of medium- to lightweight wearables quickly, effectively, and without a single watt of electricity. Portable, lightweight and easy to use—load up the barrel, add water and soap, screw on the pressure lid, and turn the handle. To rinse, drain, refill, and agitate for another 30 seconds. The only moving parts are two pivots; nothing to break down. Sturdy and durable. Assembled, the unit measures 19"H x 17.5"W x 15"D; limited manufacturer's warranty. USA.

**63860   Super Wash Pressure Hand
Washing Machine $49**

Fifty Feet of Drying Space

This is the most versatile clothes dryer we've seen yet. Don Reese and his family are still making them by hand, from 100% sustainably harvested New England pine and birch. Don's sturdy dryer can be positioned to create either a peaked top or a flat top ideal for sweaters. A generous 50 feet of drying space—and you can fold it up to carry back in the house (or to the next sunny spot) still loaded with laundry. Measures 58"H x 30"W x 33"D in peaked position; 46"H x 30"W x 40"D in flat-top position, and holds an average washer load. Folds to 4.5"D x 58"H x 30"W. USA.

10-8002 New England Clothes Dryer $59
Requires UPS ground delivery.

COOKING PRODUCTS

Emergency Candle Lasts 120 Hours

This candle will fit in your backpack, your car trunk, or on a shelf for emergencies. It's enclosed in a metal can so you can take it anywhere. Use it for lighting, heating, and cooking for up to 120 hours. The secret is its six long-burning movable wicks. Light just one for illumination, two or three for heating food. The specially formulated FDA-approved (food grade) paraffin is nontoxic. Includes six wicks, tweezers, and matches. USA.

63-452	**120-Hour Candle**	**$12.50**
	3 or More	**$9.95 each**

Hand-Cranked Blender

The human-powered Vortex blender is perfect for picnics, tailgate parties, camping, and for educating your children about energy. It features a 48-ounce (1.5-liter) graduated Lexan pitcher with an O-ring sealing top, a removable pour spout that's a 1-ounce shot glass, a brushed stainless steel finish, soft stable rubber feet, an ergonomic handle, two-speed operation, and a C-clamp for stable operation. The base fits inside the pitcher for easy packing. High speed operation is noisy, your camping neighbors will know you're making margaritas! Weighs 4.5 lbs. China.

63867   Hand-Cranked Blender $88

The World's Finest Solar Oven

Our Sun Oven is a great, portable solar cooker weighing only 21 pound. It's ruggedly built with a strong, insulated fiberglass case and tempered glass door. The reflector folds up and secures for easy portability on picnics. It's completely adjustable, and comes with a built-in thermometer. The interior oven dimensions are 14" W x 14" D x 9" H and temperatures range from 350° to 400°F. This is a very easy oven to use and it will cook most anything! After preheating, the Sun Oven will cook one cup of rice in 35 to 45 minutes. USA.

63-421   Sun Oven $249

Sport Solar Oven

We love the look on friends' faces when we present fresh bread and other home-baked treats while camping. So will you when you take our lightweight solar oven on your next trip. It concentrates the renewable heat of the sun to effortlessly (and deliciously!) roast meats, steam vegetables, bake breads and cookies, and prepare rice, soups, and stews. Ideal for backyard use, camping, boating, and picnicking. Requires only minimal sun aiming to cook most foods in two to four hours. Our complete kit includes two pots, oven thermometer, water pasteurization indicator, and recipe book. 12"H x 27"W x 16"D. 11 lbs.

01-0454 Sport Solar Oven   **$150**

Economy Solar Cooker

This solar cooker uses a foil-laminated foam reflector that folds flat to a 14" x 14" x 2" packet. It's waterproof, crushproof, and travel-friendly. This is a slow cooker that's good for larger stew or chili batches. It's not real fussy about aiming perfectly, so a bit of attention every couple of hours keeps it cooking happily. You provide your own 3- to 4-quart dark-colored pot. A high temperature baking bag (a turkey bag) is recommended to help hold the heat. Reaches 250 to 300°F. USA.

63-848   Economy Solar Cooker   **$24**

Cooking with the Sun
How to Build and Use Solar Cookers

By Beth and Dan Halacy. Solar cookers don't pollute, they use no wood or any other fuel resource, they operate for free, and they run when the power is out. And a simple and highly effective solar oven can be built for less than $20.

The first half of this book, generously filled with pictures and drawings, presents detailed instructions and plans for building a solar oven that will reach 400°F or a solar hot plate that will reach 600°F. The second half has 100 tested solar recipes. These simple-to-prepare dishes range from everyday Solar Stew and Texas Biscuits to exotica like Enchilada Casserole. 116 pages, softcover. USA.

80-187   Cooking with the Sun   **$9.95**

COMMUNICATIONS PRODUCTS

Solar-Dynamo Multi-Band Radio

The built-in, environmentally friendly NiMH battery is recharged from solar power, its hand-cranked dynamo, or your home electricity. Can also be supplemented by three AA batteries (not included). It picks up shortwave frequencies from 3.8 to 19.2 MHz, VHF television bands, AM/FM and National Weather Service bands. This radio packs in a ton of features for the price! One minute of hand-cranking can yield as much as 30 minutes of play time. AC/DC adapter and earplugs included. 4.375"H x 6.75"L x 2.25"W. China.

57-168   Solar Dynamo Multi-Band Radio   **$79**

Freeplay Solar-Powered Summit Radio

For a trip to the beach or a global trek, this stylish, compact radio works anywhere in the world! Covers four radio bands: FM, AM, SW, and LW for local and world radio. It's powered by a rechargeable NiMH battery pack. Charge with the built-in solar panel, with the built-in dynamo crank (30 seconds of cranking gives about 30 minutes of listening) or with the included AC/DC power adapters, which include world plug adapters. An indicator light shows optimum charging. Earphone socket included for private listening, or use the built-in speaker. Thirty station presets are available; ten each for AM and FM, five each for SW and LW, with precise digital tuning. Supplied with a shortwave antenna, worldwide travel plug adapter, an international shortwave guide and a very nice travel pouch. Weight: 1.3 lbs. Size: 6.8"W x 3.5"H x 3.1"D. Two-year manufacturer's warranty. Designed in UK, made in China.

53-0101 Freeplay Solar-Powered
       Summit Radio   **$99**

SideWinder™ Cell Phone Charger

The SideWinder is the world's smallest, lightest and most powerful portable cell phone charger ever made. At just 2.5 ounces, it's the ideal solution whenever you're away from traditional power sources. Two minutes of cranking delivers five to six minutes of talk time, or up to thirty minutes standby. Voltage output is regulated and is safe to use with any cell phone. Our universal kit has cords and connectors that fit more than 95% of phones, including Motorola V series, Nokia, Samsung/Kyocera, Ericsson, and Audiovox. A tough little zippered cordura bag keeps it all handy and contained.

SideWinder also has a bright-white LED emergency light that provides over five minutes of light, with only thirty seconds of cranking. 2.5" x 1.75" x 1.25". Jelly blue color. USA/China.

53-0102 SideWinder Cellphone Charger   **$25**
       2 for $23 each

Solar e-Power

The pocket-size, folding Solar e-Power charges most models of Nokia, Ericsson, Motorola, or Samsung cell phones. Power cord adapter included. It can also charge four AA or AAA batteries in the attached battery holder. Phone charging can be solar "direct" or energy stored in the batteries. The folding design keeps it small and protects the cells during transport. Delivers 250 mA @ 6.0 volt. 4.5" x 2.6" x 1.3". China.

50266   Solar e-Power   **$69**

BATTERY CHARGERS AND POWER SUPPLIES

iSun and Battpak

This 2.2-watt folding unit can be switched for 6- or 12- volt output, comes with two power cables and seven different DC plugs that cover more than 90% of cell phones, discmans, MP3 players, or GPS units. For more power, up to five iSun units can be plugged into each other with the included spine connector plug. Measures 7.25" x 4.25" (8.25" unfolded) x 1.25". Weighs 11 oz. Delivers 145mA @ 15.2v, or 290mA @ 7.6v. China.

The Battpak docks mechanically and electrically to the back of the iSun, or it can be used stand-alone with the included AC and DC charging cords. It will charge up to ten rechargeable AA or AAA size batteries. ON and FULL lights indicate charging activity. Two DC output plugs will deliver 1.5v to 12v depending on how many batteries are installed. Measures 7.5" x 2.9" x 1.25". Weighs 6.5 oz. China.

11613	**iSun Solar Charger**	**$80**
50267	**Battpak Charger**	**$30**
50268	**iSun with Battpak**	**$99**

AccuManager 20

Using a pulse-charging technology that rechargeable batteries just love, the AccuManager 20 charger will recharge your NiMH, nicad, or Accucell batteries faster, and better, than any other charger we've ever seen. You can mix different battery types, sizes, and charging times. After charging, each cell is stored at the peak of readiness with a gentle trickle charge, until you're ready for it. There are four bays that accept AAA, AA, C, or D cells, plus two bays that accept 9-volt cells. Charge times will run from 20 minutes to 12 hours depending on type and size. Supplied with both AC and DC charging cords and will run off standard 120vac, automotive 12 volts, or off a 10- to 15-watt solar panel with appropriate plug, such as our solar option sold below. 3-year manufacturer's warranty. Germany/China. For rechargeable batteries only.

50269   AccuManager 20 Battery Charger   **$69**
06-0384 Lightweight 10W Solar Option   **$149**

ALKALINE BATTERIES VERSUS RECHARGEABLES
What's the smart choice?

2,000 Alkaline Batteries @ $2.85 / 4 pack	=	4 Nimh Batteries $12.00
		1 Charger $29.00
		2,000 Charges @ 1¢ ea. $20.00
4,000 Ah for $1,425.00		4,000 Ah for $61.00

12-Volt Nicad Charger

This 12-volt charger will recharge AAA-, AA-, C-, or D-size NiMH or nicad batteries in 10 to 20 hours from your 12-volt power source. It will charge up to four batteries simultaneously. Batteries must be charged in pairs. Has a clever contact system on the positive end that will deliver different current flow depending on battery size. AAAs will receive 50 milliamps, AAs will receive 130 milliamps, Cs and Ds will receive 240 milliamps. It comes with a 12-volt cigarette lighter plug to go into any 12-volt socket. Make sure the cigarette lighter works with the key off when using it in your car. One or two charge cycles will not discharge the car's battery. China.

50214   12-Volt Nicad Charger $20

Hi-Speed Digital Charger

Our speedy, affordable battery charger comes with both AC and DC power cords, charges AA- or AAA-size NiMH or nicad batteries in pairs and stops charging automatically when batteries are full, then switches to gently trickle-charge your batteries 'til you're ready for them. Charges most batteries in 5 hours or less. LEDs show red when charging, green when finished, and flashes red/green if a battery is defective. Taiwan.

17-0281 Hi-Speed Digital Charger $29

Hi-Capacity NiMH Rechargeable Batteries

Nickel-metal hydride batteries just keep getting better and better. Our batteries excel at hard use and deliver steady voltage with no sag all the way down to 10% of charge. This makes them a better choice for heavy-discharge uses like digital cameras or remote-control toys, where you actually get to use more of the stored energy than any other battery type will deliver. There is no memory effect from short cycling and no toxic content. Because they self-discharge over 3 to 6 months, these cells aren't a good choice for emergency lights. China.

17-9275 4	AAA 4-pack (750mAh)	$12
17-9275 8	AAA 8-pack (750mAh)	$22
17-9276 4	AA 4-pack (2200mAh)	$12
17-9276 8	AA 8-pack (2200mAh)	$22
17-9175	C 2-pack (3500mAh)	$11
17-9176	D 2-pack (7000mAh)	$20
17-0284	9V single (160mAh)	$9
17-0285	9V 2-pack (160mAh)	$16

Quickcharger Battery Charger

For recharging nickel metal hydride (NiMH) and nickel cadmium (nicad) batteries, this sophisticated microprocessor-controlled charger will provide the fastest complete charge of any battery charger on the market without creating damaging heat. An automatic discharge circuit conditions each battery for full capacity and the longest life by running a controlled discharge at the beginning of the charge cycle. Works perfectly with the newer ultra-high-capacity NiMH cells. A graphic LCD display shows what's happening, and displays battery voltage. After fully recharging, it will hold the cells on a gentle trickle charge, so they always come out at the peak of performance. Accepts one to four AAA- through D- size cells. Batteries on each charging cycle should be the same size and chemistry. Runs on 12vdc, comes with an AC adapter. One-year manufacturer's warranty. China.

17-0184 Quickcharger Battery Charger $66

Mobile DC Power Adapter

Use a 12-volt source to run and recharge any battery-powered device using less than 12 volts DC. Our adapter delivers stable DC power at 3, 4.5, 6, 7.5, 9, or 12 volts at up to 2.0 amps. Voltage varies less than 0.5 volt in our tests. Perfect for your camera, DVD or MP3 player, camcorder, Game Boy®, cell phone, or whatever. The 6-foot fused input cord with LED power indicator plugs into any lighter socket, the 1-foot output cord has six adapter plugs. A storage compartment in the power adapter holds your spare plugs to prevent loss. Slide switch selects voltage. China.

53-0105 Mobile DC Power Adapter $24

12v DC PC Power Supplies

Run and recharge your PC from any 12-volt DC source. Available in 65- or 100-watt output with stabilized voltage adjustable from 15 to 21 volts. Output voltage varies less than 0.5 volt from zero to full load in our tests. External voltage adjustment is recessed for safety. Two-foot fused input cord with LED power indicator plugs into any lighter socket. Four-foot output cord has six different adapters to fit almost any PC laptop with a round power

socket. Will not recharge newer Macintosh products requiring 24-volt input. 65-watt unit has maximum 4.0 amp output. 100-watt unit has maximum 7.0 amp output. 3.8" x 2.3" x 1.5", weighs 8 oz. Taiwan.

53-0103   65-watt $59
53-0104   100-watt $75

POWER SUPPLY PACKAGES

Here is a handy way to carry a 12-volt power supply for laptops, cellphones, digital cameras, CD players, etc. Our power supply packages will accept solar, automotive, or 120vac charging, so no matter where you are, you should be able to rustle up enough power to keep all the equipment happy. Batteries are sealed, spill-proof, airline approved, and replaceable, with a three to four year life expectancy. LED indicators show battery state of charge. All have a one-year manufacuturer's warranty. China.

Power Depot

Our smallest power supply is 12-volt DC only with a 7Ah battery. It has a nice detachable 13-watt fluorescent work light that will run over 10 hours on one charge. Weighs 7.0 lbs. 8.25"H x 5.8"W x 3.25"D.

15856   Power Depot $59

Power Zone

Our medium-size power supply delivers AC or DC and

comes with a 12Ah battery. The built-in inverter delivers 150 watts AC. It has a single DC lighter socket and a DC charge input port. The pair of 6-watt fluorescent lights will run over 15 hours on one charge. The built-in AC charger stores its cord in a nice rear compartment, which also stores the DC adapter cord and the DC jumper cables. Weighs 17.6 lbs. 12"L x 9.5"H x 4.1"D.

15857   Power Zone $149

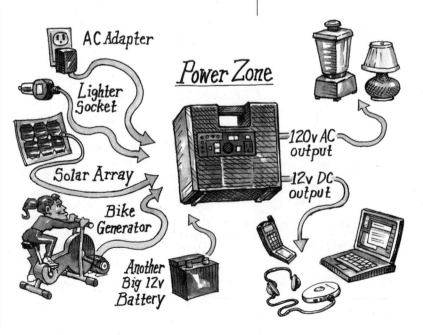

Power Depot 2000

Our larger power supply delivers AC or DC, features an 18Ah battery, a built-in 300-watt AC inverter with dual outlets, dual DC sockets, a built-in AC charger, and a small work light. Jumper cables with side storage pockets are included. Weighs 18.0 lb. 16"L x 14.5"H x 4.5"D.

15858 Power Depot 2000 $169

Sealed Replacement Batteries

Do not buy until your battery is ready for replacement! Batteries will die sitting on the shelf. We buy in small batches, or ship directly from the manufacturer to ensure fresh batteries.

**15200   12V/7Ah (6"L x 2.5"W x 3.9"H,
          5.5 lbs.)**   **$28**

**15202   12V/12Ah (6"L x 3.9"W x 3.9"H,
          8.7 lbs.)**   **$50**

**15209   12V/17.5Ah (7.2"L x 3"W x 6.6"H,
          12.8 lbs.)**   **$59**

Portable Power System

Our Portable Power Generator can supply lighting or run entertainment equipment at your occasional-use cabin or on your boat. Also ideal for emergency use and disaster relief, the system is housed in a rugged weathertight Rubbermaid box (21" x 15" x 12"). Already pre-wired are a 15-amp PV controller, a 150-watt Statpower Prowatt inverter for AC output, a pair of lighter plugs for DC output, an externally mounted DC voltmeter "fuel gauge," an input plug for PV module(s), and all the fusing to keep everything safe. All you need to add is the PV module(s) or other charging source of choice, and a 98Ah (or smaller) battery. All components except the battery and PV module(s) are pre-wired and ready to rock. The Portable Power Generator weighs 9 lbs. as delivered, or 81 lbs. with the maximum-capacity sealed battery listed below. One-year manufacturer's warranty. USA.

12107   Portable Power Generator $699
15214   Solar Gel Battery 12V, 98 Ah $189
PV module and battery sold separately; call for sizing.

XPower 1500 Powerpack

Our largest portable powerpack, the XPower 1500 delivers 120vac and 12vdc. It packs a large 60Ah sealed battery and a 1,500-watt inverter with dual outlets in a nice, rugged wheeled cart with a removable 38"-high waist handle for ease of movement. Includes a 5-amp battery charger for when AC power is available from a wall plug or generator, and a heavy-duty DC charging cord for lighter socket or a solar panel of your choice. Has a battery charge indicator. Weighs 60 lbs. Measures 14.8"H x 15.6"W x 12.3"D without handle. One-year manufacturer's warranty. Canada.

17-0283 XPower 1500 Powerpack $380

Human Power Generator

A small, dependable human-power generator—perfect for emergencies, power failures, marine and off-grid applications. Pedal or crank by hand to charge 12-volt batteries or run small appliances. As easy to operate as an ordinary bicycle. Use it to run back-up lighting, or ask your children to pedal enough power to watch their favorite television programs. Great exercise, energy saving, and experiential learning about power!

- Generate 50–125 watts (1/6 HP at 12 volts)
- Heavy-duty welded steel base and frame
- Rubber bumpers
- Permanent magnet DC generator
- Rugged one-piece crank arm
- Detachable pedals (hand cranks optional)
- Two-year warranty

17330   Human Power Generator $599
17338   Optional Hand Cranks $49

Bike Generator for Battery Charger

The human body can be highly efficient at delivering energy. Most adults can generate a steady 75 to 150 watts with spurts of double that power. Our bike generator provides a sturdy, adjustable stand that raises and clamps the rear axle of your bike, with an adjustable friction wheel, a flywheel for steady power, and a DC generator that can deliver over 500 watts. Even Lance Armstrong could find his comfort zone on this powerful generator. Many off-the-grid parents require their kids to power their TV watching with our bike generator. The

stand has large, heavy rubber feet for stability and quiet. The generator will deliver 12v or 24v power depending on the battery it's connected to (use higher gears with higher voltage). Charge your off-grid or RV household battery, or any portable battery pack. Bike not included. Stand made in Japan, generator made in USA.

47-0103 Bike Generator & Stand $589

SMALL PV PANELS

Deluxe Solar Charger

Our Deluxe Clip-on Solar Charger is a 250-milliamp, 12-volt solar module that keeps your boat, RV, tractor, golf cart, or motorcycle battery topped off, even when it's been sitting unused for extended periods of time. Its built-in, adjustable stand allows the panel to receive the maximum sunlight. Measures 10"W x 7.25"H x .5"D, and it comes with a 6-foot connecting cable and alligator clips. China.

17-0144 Deluxe Solar Charger   **$89**

13-Watt General Purpose Solar Module

Delivering a peak 13.5 watts, our single-crystal, tempered glass glazed module is useful for any 12-volt charging applications. Features include weathertight construction, a sturdy anodized aluminum frame with four adjustable mounting tabs, a standard 36 cells in series, built-in reverse-current diode and 7-foot wire pigtail output with color-coded clips. Good for trickle battery recharging on RVs, farm equipment, boats, and a perfect match with the Aquasolar 200 pump for backyard fountains. China.

Rated Watts: 13 watts @ 25°C
Rated Power: 17.5 volts, 710 mA
Open Circuit Volts: 18.0 volts @ 25°C
Short Circuit Amps: 750 mA @ 25°C
15.1"L x 14.4"W x 1.0"D
Construction: single crystal, tempered glass
Warranty: 90 days
Weight: 3.5 lbs./1.6 kg.

17-0170 13-Watt General Purpose
       Solar Module   **$139**

Lightweight Travel-Friendly Modules

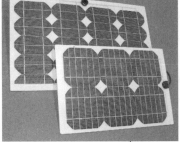

Featuring a lightweight "no-frame" design with fiberglass back, Tedlar cover, and grommet mounting holes in each corner, these unbreakable 36-cell modules are travel-friendly. A good choice for camping, expeditions, or just backyard battery charging. We use these versatile modules for a wide variety of small charging applications, such as laptop computers, digital cameras, and camp lighting. Comes standard with a blocking diode and a bare 6-foot output cord, or optionally with a preassembled female lighter socket on the 10W module. Mexico.

Rated Watts:	10 watts @ 25°C	20 watts @ 25°C
Rated Power:	16.8 volts, 580 mA	16.5 volts, 1.21 amps
L x W x D:	16.25" x 10.6" x .2"	21" x 12.75" x .2"
Construction:	single crystal, Tedlar coating	single crystal, Tedlar coating
Warranty:	3 years	3 years
Weight:	1.9 lb/.86 kg	3.25 lb/1.47 kg

41-0164 10W Lightweight Module   **$129**

06-0384 10W Lightweight Module
       with Lighter Socket   **$149**

41-0168 20W Lightweight Module   **$249**

PowerDock Power Station for Mobile Gadgets

The PowerDock has everything you need to run laptops, cell phones, GPSs, camp lights, and other modest-power gear in remote sites. This slick and sturdy power station has a zip-out, unbreakable 15-watt PV panel and 9.2 amp/hrs of sealed 12-volt batteries. Also included are a pair of fused lighter socket outlets, a power meter to show state of charge, and a water-resistant, canvas carrying case with five roomy compartments for converters and accessories. Will accommodate laptop computers up to 9" x 15" x 2". Want to double your charging power? Add the 15-watt Solar Boost. Plugs into one of the lighter sockets and folds into the PowerDock for travel. The quality construction is delightfully robust, with easy no-tools access to the batteries (lots of Velcro!). Has four batteries, which currently cost $8 each and typically need to be replaced in 2 to 4 years. Weighs 14.5 lbs, yet is comfortable to carry with the wide, well-padded shoulder strap. This product is quality all the way! Comes with an AC wall-watt charger for when the sun don't shine. One-year manufacturer's warranty, ten years on PV array. USA & Mexico.

53-0112 PowerDock   **$349**
41-0127 PowerDock 15W Boost   **$179**

CHAPTER 5

Energy Conservation

WE LIVE IN A DYNAMIC SOCIETY. Our economy is constantly growing. All economic activities require energy inputs, so economic growth demands increased energy consumption. To fuel this process we need to pump more oil out of the ground and build more power plants to generate more electricity. Right? Right????

Wrong. It is a well-documented fact that the cheapest, most cost-effective way to increase energy supplies under contemporary conditions is through simple acts of conservation. You know the drill: Raised on a steady diet of cheap fossil-fuel energy during the twentieth century, and especially since World War II, our society is enormously wasteful of energy. Not only are those fossil-fuel supplies finite and dwindling—most petroleum engineers agree that world oil production already has reached its absolute peak, or will do so within a couple of decades—but in the last ten or fifteen years we've come to understand and appreciate the dire ecological impact of fossil-fuel energy on the Earth's climate, in addition to their contribution to air and water pollution, habitat destruction, and other environmental ills. Energy conservation is the fundamental starting point for a sustainable future.

Oh yeah. Energy conservation saves us all money, too.

A couple of quick examples. We have calculated, using data from the California Energy Commission, that if every household in the state replaced four 100-watt incandescent light bulbs with four 27-watt compact fluorescent bulbs (the CF equivalent of a 100-watt light), burning on average for five hours a day, the state would save 22 gigawatt-hours per day (a gigawatt is 1,000 megawatts)—enough energy to *shut down* seventeen power plants. Similarly, if each of those households replaced one average-flow showerhead with a low-flow, energy-saving showerhead, California would save an additional 19.2 gWh per day—enough to shut down another fifteen power plants. Conservation is a very powerful tool.

So it's really a no-brainer. Conserving energy reduces greenhouse gas emissions, slows down the depletion of natural resources, decreases environmental pollution, takes strain off of the planet's inherent life-support systems, and takes a smaller bite out of your wallet. None of us can afford *not* to conserve energy. The further beauty of conservation is that regardless of what not-so-enlightened pundits and critics say, conservation does not mean sacrifice. Many European and Scandinavian societies enjoy a comparable standard of living to those of us in the United States, but they do it on a much tighter energy budget. As the authors of *Beyond the Limits* noted in 1992, "the North American economy could do everything it now does, with currently available technologies and at current or lower costs, using half as much energy," which would bring it to the efficiency levels of Western Europe and Japan. It's all about awareness, habits, and using energy efficiently. The tools, techniques, and technologies are readily available and well tested. We just have to start implementing them on a much wider basis.

It is a well-documented fact that the cheapest, most cost-effective way to increase energy supplies under contemporary conditions is through simple acts of conservation.

Energy conservation is the fundamental starting point for a sustainable future.

SOLAR LIVING SOURCEBOOK

Conserving energy reduces greenhouse gas emissions, slows down the depletion of natural resources, decreases environmental pollution, takes strain off of the planet's inherent life-support systems, and takes a smaller bite out of your wallet. None of us can afford *not* to conserve energy.

Energy conservation is especially important if you're interested in achieving some degree of self-sufficiency, independence from conventional utility networks, or off-the-grid living. The lower your household energy consumption, particularly for heating and electricity, the more easily renewable resources can contribute to meeting your needs with clean, reliable, sustainable energy. But regardless of your mode of living, the same few conservation principles apply to everyone: Build smaller rather than larger. Make sure your building envelope is tight. Use passive solar strategies to minimize your heating and cooling loads. Install low-flow showerheads and compact fluorescent lights. Buy the most efficient appliances possible for your needs and budget. Strive to use renewable energy resources.

Gaiam Real Goods supplies many types of products, appliances, systems, and informational resources to help families conserve energy and consume it efficiently. The material in this chapter focuses on domestic space heating and cooling, household appliances, and lighting. Energy conservation as applied to water heating and transportation is addressed in later chapters.

We'd like to mention one other idea. If you want to take the concept of energy conservation to its furthest reaches, consider the notion of *embodied energy*. Every commodity or product that you consume "embodies" all the energy it took to produce it and get it to you. From the big-picture perspective, then, you can also conserve by assessing the comparative energy required to produce and transport various goods that you choose to buy. If you live on the East Coast, for example, a locally grown organic tomato in season embodies much less energy than a California organic tomato that's calling out to your taste buds in January. The scenarios and possible comparisons are endless, of course, and it's not always easy to figure out how much energy is embodied in any given thing, or to measure the actual impact of one choice over another. Thinking about this issue will probably take you places you may not want to go, and drive you crazy in the process. Nonetheless, the concept of embodied energy has real environmental import. We suggest that cultivating the habit of thinking about embodied energy is a conscientious act of global citizenship in service to sustainability.

Living Well, But Inexpensively

The lower your household energy consumption, particularly for heating and electricity, the.more easily renewable resources can contribute to meeting your needs with clean, reliable, sustainable energy.

Imagine a pair of similar suburban homes on a quiet residential street. Both house a family of four living the American suburban lifestyle. The homes appear to be identical, yet one spends under $50 per month on utilities, and the other over $400. It can't be? Ah, but it can, and often is! This huge cost difference demonstrates the dramatic savings that careful building design, landscaping, and selection of energy-efficient appliances make possible without affecting a family's basic lifestyle.

Our example isn't based on some bizarre construction technique, or appliances that can be operated only by rocket scientists, but on simple, common-sense building enhancements and off-the-shelf appliances. Studies have shown that no investment pays as well as conservation. Banks, mutual funds, real estate investments . . . none of these options will bring the 100 to 300% returns that are achievable through simple, inexpensive conservation measures. Not even our own dearly beloved renew-

able energy systems will repay your investment as quickly as conservation.

At Gaiam Real Goods, we are experts in energy conservation. We have to be! After twenty-seven years designing renewable electrical systems for remote locations, we have learned to squeeze the maximum work out of every precious watt. Although our focus is on energy generation through solar modules or wind and hydroelectric generators, we've found we often have to backtrack a bit to basic building design, or retrofitting of existing buildings, before we start selecting appliances.

We'll tackle the broad, multifaceted subject of energy conservation by looking first at building design, followed by retrofitting an existing building, and then tips for selecting appliances. After that we'll delve more into space heating, cooling, energy-efficient appliances, and super-efficient lighting. Feel free to skip sections that do not apply to your current needs.

It All Starts with Good Design

The single most important factor that affects energy consumption in your house is design. All the intelligent appliance selection or retrofitting in the world won't keep an Atlanta, Georgia, family room with a four-by-six-foot skylight and a west-facing eight-foot sliding glass door from overheating every summer afternoon and consuming massive amounts of air conditioning energy. Intelligent solar design is the best place to start. Passive solar design works compatibly with your local climate conditions, using the seasonal sun angles at your latitude to create an interior environment that is warm in winter and cool in summer, without the addition of large amounts of energy for heating and cooling.

Passive solar buildings cost no more to design or build than "energy-hog" buildings, yet cost only a fraction as much to live in. You don't have to build a totally solar-powered house to take advantage of some passive solar savings.

Just a few simple measures, mostly invisible to your neighbors, can substantially improve your home's energy performance!

For an in-depth examination of passive solar design strategies, from the simple to the complex, see chapter 1, "Land and Shelter," pp. 16–22.

> You don't have to build a totally solar-powered house to take advantage of passive solar savings.

House orientation affects heating and cooling costs.

Retrofitting: Making the Best of What You've Got

If you're like many of us, you already have a house, have no intention of building a new one, and just want to make it as comfortable and economical as possible. By retrofitting your home with energy savings in mind, you can lower operating costs, boost efficiency, improve comfort levels, save money, and feel good about what you've accomplished.

WEATHERIZATION AND INSULATION

Increasing insulation levels and plugging air leaks are favorite retrofitting pastimes, which produce well-documented, rapid paybacks in comfort, and utility savings. Short of printing your own money, weatherization and insulation are the best bets for putting cash in your wallet—and they're a lot safer in the long run than counterfeiting.

Weatherization

Weatherization, the plugging and sealing of air leaks, can save 25 to 40% of heating and cooling bills. The average unweatherized house in the United States leaks air at a rate equivalent to a four-foot-square hole in the wall. Weatherization is the first place for the average home owner

to concentrate his or her efforts. You'll get the most benefit for the least effort and expense.

WHERE TO WEATHERIZE

■ The following suggestions are adapted from *Homemade Money* by Richard Heede and the Rocky Mountain Institute. Used with permission.

Here's a basic checklist to help you get started. Weatherization points are keyed in parentheses to the illustration on the next page.

1. In the attic

- Weatherstrip and insulate the attic access door (6).
- Seal around the outside of the chimney with metal flashing and high-temperature sealant such as flue caulk or muffler cement (3).
- Seal around plumbing vents, both in the attic floor and in the roof. Check roof flashings (where the plumbing vent pipes pass through the roof) for signs of water leakage while you're peering at the underside of the roof (8).
- Seal the top of interior walls in pre-1950s houses anywhere you can peer down into the wall cavity. Use strips of rigid insulation, and seal the edges with silicone caulk (7).

> Each year in the U.S. about $13 billion worth of energy, in the form of heated or cooled air—or about $150 per household —escapes through holes and cracks in residential buildings.
>
> —American Council for an Energy-Efficient Economy

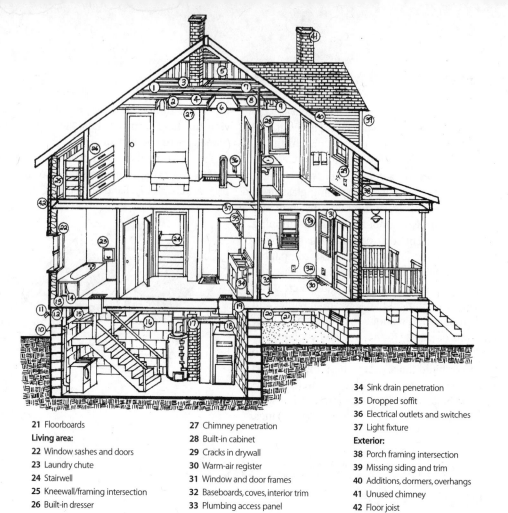

Attic:

1 Dropped ceiling
2 Recessed light
3 Chimney chase
4 Electric wires and box
5 Balloon wall
6 Attic entrance
7 Partition wall top plate
8 Plumbing vent chase
9 Exhaust fan

Basement and crawlspace:

10 Dryer vent
11 Plumbing/ utility penetrations
12 Sill plate
13 Rim joist
14 Bathtub drain penetration
15 Basement windows and doors
16 Block wall cavities
17 Water heater and furnace flues
18 Warm-air ducts
19 Plumbing chase
20 Basement/crawlspace framing

21 Floorboards

Living area:

22 Window sashes and doors
23 Laundry chute
24 Stairwell
25 Kneewall/framing intersection
26 Built-in dresser

27 Chimney penetration
28 Built-in cabinet
29 Cracks in drywall
30 Warm-air register
31 Window and door frames
32 Baseboards, coves, interior trim
33 Plumbing access panel

34 Sink drain penetration
35 Dropped soffit
36 Electrical outlets and switches
37 Light fixture

Exterior:

38 Porch framing intersection
39 Missing siding and trim
40 Additions, dormers, overhangs
41 Unused chimney
42 Floor joist

- Stuff fiberglass insulation around electrical wire penetrations at the top of interior walls and where wires enter ceiling fixtures. (But not around recessed light fixtures unless the fixtures are rated IC [for insulation contact]). Fluorescent fixtures usually are safe to insulate around; they don't produce a lot of waste heat. Incandescent fixtures should be upgraded to compact fluorescent lamps, but that's another story; see below, pp. 263–68) (2, 4).
- Seal all other holes between the heated space and the attic (9).

2. In the basement or crawlspace

- Seal and insulate around any accessible heating or A/C ducts. This applies to both the basement and attic (18).
- Seal any holes that allow air to rise from the basement or crawlspace directly into the living space above. Check around plumbing, chimney, and electrical penetrations (11, 14, 19, 21).
- Caulk around basement window frames (15).
- Seal holes in the foundation wall as well as gaps between the concrete foundation and the wood structure (at the sill plate and rim joist). Use caulk or foam sealant (12, 13).

3. Around windows and doors

- Replace broken glass and reputty loose panes. See the Window section, page 225, about upgrading to better windows, or retrofitting yours (22).
- Install new sash locks, or adjust existing ones on double-hung and slider windows (22).
- Caulk on the inside around window and door trim, sealing where the frame meets the wall and all other window woodwork joints (31).
- Weatherstrip exterior doors, including those to garages and porches (31).
- For windows that will be opened, use weatherstripping or temporary flexible rope caulk.

4. In living areas

- Install foam-rubber gaskets behind electrical outlet and switch trim plates on exterior walls (36).
- Use paintable or colored caulk around bath and kitchen cabinets on exterior walls (26, 35).
- Caulk any cracks where the floor meets exterior walls. Such cracks are often hidden behind the edge of the carpet (32).
- Got a fireplace? If you don't use it, plug the flue with an inflatable plug, or install a rigid insulation plug. If you do use it, make sure the

damper closes tightly when a fire isn't burning. See the section on Heating and Cooling (pp. 229–38) for more tips on fireplace efficiency (41).

5. On the exterior

- Caulk around all penetrations where electrical, telephone, cable, gas, dryer vents, and water lines enter the house. You may want to stuff some fiberglass insulation in the larger gaps first (9, 10, 11).
- Caulk around all sides of window and door frames to keep out the rain and reduce air infiltration.
- Check your dryer exhaust vent hood. If it's missing the flapper, or it doesn't close by itself, replace it with a tight-fitting model (10).
- Remove window air conditioners in winter; or at least cover them tightly, and make rigid insulation covers for the flimsy side panels.
- Caulk cracks in overhangs of cantilevered bays and chimney chases (27, 40). ■ ■

Insulation

Many existing homes are woefully underinsulated. Exterior wall cavities or underfloors are often completely ignored, and attic insulation levels are sometimes more in tune with the 1950s when heating oil was 12¢ per gallon than

R-Values of Loose-Fill Insulation

MATERIAL	R-VALUE PER INCH	USE
Fiberglass (low density)	2.2	Walls and Ceilings
Fiberglass (medium density)	2.6	Walls and Ceilings
Fiberglass (high density)	3.0	Walls and Ceilings
Cellulose (dry)	3.2	Walls and Ceilings
Wet-Spray Cellulose	3.5	Walls and Ceilings
Rock Wool	3.1	Walls and Ceilings
Cotton	3.2	Walls and Ceilings

R-Values of Rigid Foam and Liquid Foam Insulation

MATERIAL	R-VALUE PER INCH	USE
Expanded polystyrene	3.8 to 4.4	Foundations, walls, ceilings, and roofs
Extruded polystyrene	5	Same
Polyisocyanurate	6.5 to 8	Same
Roxul (mineral wool)*	4.3	Foundation exteriors
Icynene	3.6	Walls and ceilings
Air Krete	3.9	Walls and ceilings

* Rigid board insulation made from mineral wool.

—From Dan Chiras, *The Solar House: Passive Heating and Cooling*

Insulation

Since warm air rises, the best place to add insulation is in the attic. This will help keep your upper floors warm. Recommended insulation levels depend on where you live and the type of heating system you use. For most climates, a minimum of R-30 will do, but colder areas, such as the northern tier and mountain states, may need as much as R-49. Two useful booklets, BOE/CE-0180 (Aug. 1997) and DOE/GO-10097-431 (Sept. 1997), provide information about insulating materials and insulation levels needed in various locations by zip code. Write the U.S. Department of Energy, Office of Technical Information, P.O. Box 62, Oak Ridge, TN 37831.

There are four types of insulating materials: batts, rolls, loose fill, and rigid foam boards. Each is suitable for various parts of your house. Batts, designed to fit between the studs in the walls, or the joist in the ceilings, are usually made of fiberglass or rock wool. Rolls are also made of fiberglass, and can be laid on the attic floor. Loose-fill insulation, which is made of cellulose, fiberglass, or rock wool, is blown into attics and walls. Rigid board insulation, designed for confined spaces such as basements, foundations, and exterior walls, provides additional structural support. It is often used on the exterior of exposed cathedral ceilings.

the early twenty-first century, when oil prices are more than ten times higher, and promising to continue to escalate. Adequate insulation rewards you with a building that is warmer in the winter, cooler in the summer, and much less expensive to operate year-round.

Recommended insulation levels vary according to climate and geography. Your friendly, local building department can give you the locally mandated standards, but for most of North America you want a minimum of R-11 in floors, R-19 in walls, and R-30 in ceilings. These levels will be higher in more northern climes, and lower in more southern climes.

INSULATE-IT-YOURSELF?

Some retrofit insulation jobs, like hard-to-reach wall cavities, are best done by professionals who have the spiffy special tools and experience to do the job quickly, correctly, and with minimal disruption. There are insulation and weatherproofing companies that will inspect, advise, and give an estimate to bring your home up to an acceptable standard.

Easier-to-reach spaces, like attics, are improvable by the do-it-yourselfer with either fiberglass batt insulation, or with easier- and quicker-to-install blown-in cellulose insulation.

R-value is a measure of the resistance of a material to heat flow. The higher the number, the greater the resistance.

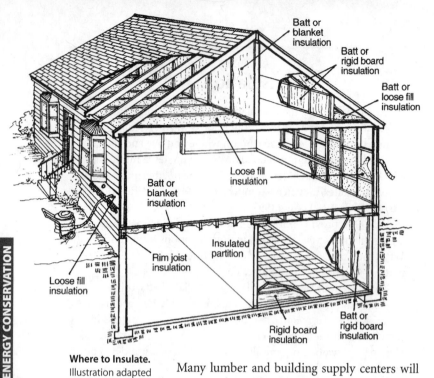

Where to Insulate.
Illustration adapted from *Reader's Digest* (1982), *Home Improvements Manual*, p. 360, Pleasantville, NY.

Many lumber and building supply centers will rent or loan the blowing equipment (and instructions!) when you buy the insulation. Some folks may choose to leave this potentially messy job to the pros.

Underfloor areas are usually easy to access, but working overhead with fiberglass insulation is no picnic. You might consider a radiant barrier product for underfloor use because of the ease of installation. A radiant barrier stapled on the bottom of floor joists performs equal to R-14 insulation in the winter. Performance will be less in the summer keeping heat out (R-8 equivalent), but this isn't a big consideration in most climates.

RADIANT BARRIERS—THE NEWEST WRINKLE IN INSULATION

Over the past few years, a new type of insulation called "radiant barrier" has become popular. Radiant barriers work differently from traditional "dead-air space" insulation. See the sidebar for a detailed explanation.

A radiant barrier will stop about 97% of radiant heat transfer. Multiple layers of barrier provide no cumulative advantage; the heat practically all stops at the first layer. Radiant barriers do not affect conducted or convected heat transfers, so they are usually best employed in conjunction with conventional dead-air-space types of insulation, such as fiberglass.

R-values can't be assigned to radiant barriers because they're only effective with the most common radiant transfer of heat. But for the sake of having a common understanding, the performance of a radiant barrier is often expressed in R-value equivalents. For instance, when heat is trying to transfer downwards, like from the underside of your roof into your attic in the summer, a radiant barrier tacked to the underside of your rafters will perform equivalent to R-14 insulation. When that same heat is trying to transfer back out of your attic at night, however, the radiant barrier only performs equivalent to R-8, since it has no effect on convection and convection transfer is in the upward direction. Therefore, conventional insulation is best for keeping warm in the winter when all the heat is trying to go up through convection. Radiant barrier material is most effective when it's reflecting heat back upwards (the direction of convection). So, two of the best applications of this product are under rafters to keep out unwanted summer heat, and under floors to keep desirable winter heat inside your house.

The independent, nonprofit Florida Solar Energy Center has tested radiant barrier products extensively and has concluded that installation on the underside of the roof deck will reduce summer air conditioning costs by 10 to 20%. And that's in a climate where humidity is more of a problem than temperature. This means rapid payback for a product as inexpensive, clean, and easy to install as a radiant barrier. FSEC also paid close attention to roof temperatures, as there's a logical concern about where all that reflected heat is going. They found that shingle temperatures increased by 2° to 10°F. To a roof shingle, that's beneath notice, and will have no effect on roof lifespan.

We only offer perforated radiant barriers now, due to the moisture barrier properties of unperforated barriers causing occasional problems when misapplied.

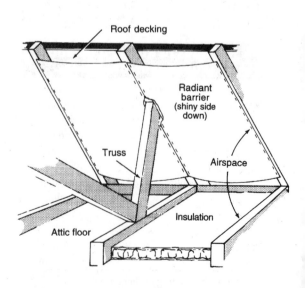

Typical radiant barrier attic installation.

How Radiant Barriers Work

Heat is transmitted three ways:

Conduction is heat flow through a solid object. A frying pan uses heat conduction.

Convection is heat flow in a liquid or gas. The warmer liquid or gas rises and the cooler liquid or gas comes down. Floor furnaces heat the air, which then rises in the room.

Radiation is the transmission of heat without the use of matter. It follows line of sight and you can feel it from the sun and a woodstove. Radiant heat flow is the most common method of heat transmission.

Radiant barriers use the reflective property of a thin aluminum film to stop (i.e., reflect) radiant heat flow. Gold works even better, but its use is limited mainly to spacecraft, for obvious reasons. A thin aluminum foil will reflect 97% of radiant heat. If the foil is heated, it will emit only 5% of the heat a dull black object of the same temperature would emit. For the foil to work as an effective radiant barrier, there must be an air space on at least one side of the foil. A vacuum is even better and is used in glass thermos bottles, where the double glass walls are aluminized and the space between them is evacuated. Without an air space, ordinary conduction takes place and there is no blocking effect.

A radiant barrier only blocks heat flow. It has no effect on conduction or convection. But then, so long as there's a small air space, the only practical way that heat can be transmitted across it is by radiant transfer.

See our product section for a selection of perforated radiant barriers in various widths (page 245).

AIR-TO-AIR HEAT EXCHANGERS (AKA HEAT-RECOVERY VENTILATORS)

Building and appliance technologies have made tremendous gains in the past twenty five years. Improvements in insulation, infiltration rates, and weatherstripping do much to keep our expensive warmed or cooled air inside the building. In fact, we've gotten so good at making tight buildings and reducing infiltration of outside air that indoor air pollution has become a problem. A whole new class of household appliances has been developed to deal with the byproducts of normal living. Air-to-air heat exchangers, also known as heat-recovery ventilators, take the moisture, radon, and chemically saturated indoor air and exchange it for outdoor air. They strip up to 75% of the heat from the outbound air in the winter, or the inbound air in summer, thereby increasing your home's energy efficiency. These heat exchangers use minimal power, and controls can be triggered by time, humidity, or usage. If you use gas for cooking, live in a moist climate, have radon or allergy problems, or simply live in a good, tight house, then air-to-air heat exchangers are worth considering.

The air-to-air heat exchanger.
A heat recovery ventilator's heat exchanger transfers 50–70% of the heat from the exhaust air to the intake air. Illustration courtesy of Montana Department of Natural Resources and Conservation.

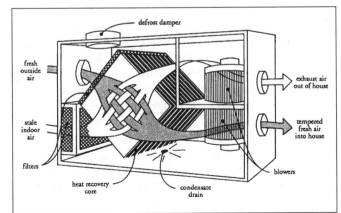

Windows

Window technology is one of the fastest developing fields of building technology. And it's about time! Windows are the weakest link in any building's thermal barrier. Until a few years ago, a window was basically a hole in the wall that let light in and heat out. The R-value of a single-pane window is a miserable 1. Nearly half of all residential windows in the United States only provide this negligible insulation value. An "energy-saving" thermopane window is worth only about R-2. When you consider that the lower-cost wall around the window is insulated to at least R-12, and probably to over R-16, a window needs a solid justification for being there. In cold climates, windows are responsible for up to 25% of a home's winter heat loss. In warmer climates, solar radiation, entering through improperly placed or shaded windows, can boost air-conditioning bills similarly.

Worth Retrofitting?

Major innovations in window construction have occurred in the past several years. The most advanced superwindows now boast R-values up to R-10. While superwindows only cost 20 to 50% more than conventional double-pane windows, the payback period for such an upgrade in an existing house is fifteen to twenty years due to installation costs. That's too long for most folks. But if you're planning to replace windows anyway due to remodeling or new construction, the new high-tech units are well worth the slightly higher initial cost, and will pay for themselves in just a few years.

Sneaky R-Values

When shopping for windows, be forewarned that until recently, most windows were rated on their "center of glass" R-value. Such ratings ignore the significant heat loss through the edge of the glass and the frame. Inexpensive aluminum-framed windows, which transmit large amounts of heat through the frame, benefit disproportionately from this dubious standard. A more meaningful measure is the newer "whole-unit" value. Make sure you're comparing apples to apples. Wood- or vinyl-framed units outperform similar metal-framed units, unless the metal frames are using "thermal break" construction, which insulates the outer frame from the inner frame. The *Consumer Guide to Home Energy Savings*, published by the American Council for an Energy-Efficient Economy, provides an in-depth overview of how to understand various window ratings.

Start with Low-cost Options

With windows, as with all household energy-saving measures, start with jobs that cost the least, and yield the most. A commonsense combination of three or four of the ideas in the next section will result in substantial savings on your heating/cooling bill, minimized drafts, more constant temperatures, and enhanced comfort, especially in areas near windows.

WINDOW SOLUTIONS FOR EFFICIENT HEATING

■ The following suggestions (cold- and warm-weather window solutions) are adapted from *Homemade Money* by Richard Heede and the Rocky Mountain Institute. Used with permission.

First, stop the wind from blowing in and around your windows and frames by caulking and weather stripping. After you've cut infiltration around the windows, the main challenge is to increase the insulating value of the window itself while continuing to admit solar radiation. Here are some suggestions for beefing up your existing windows in winter.

Install clear plastic barriers on the inside of windows

Such barriers work by creating an insulating dead-air space inside the window. After caulking, this is the least expensive temporary option to cut window heat loss. Such barriers can cut heat loss by 25 to 40%.

Repair and weatherize exterior storm windows

If you already own storm windows, just replace any broken glass, re-putty loose panes, install them each fall, and seal around the edges with rope caulk.

Add new exterior or interior storm windows

Storm windows are more expensive than temporary plastic options, but have the advantages of permanence, reusability, and better performance. Storm windows cost about $7.50 to $12.50 per square foot, and can reduce heat loss by 25 to 50%, depending on how well they seal around the edges. Exterior storm windows will increase the temperature of the inside window by as much as 30°F on a cold day, keeping you more comfortable.

Apply low-e films

Low-e films substantially reduce the amount of heat that passes through a window, with minimal effect on the amount of visible light passing through. The sidebar to the right explains how these films work. Don't use low-e films on, or built into, south-facing windows if you want solar gain! They can't tell the difference between winter and summer. See our product section for films that can be applied to existing windows. Note that low-e films will retard either heat loss or heat gain, depending on where they are applied.

Exotic infills

The other new technology commonly found in new windows is exotic infills. Instead of filling the space between panes with air, many windows are now available with argon or krypton gas infills that have lower conductivity than air, and boost R-values. Krypton has a higher R-value, but costs more. These inert gases occur naturally in the atmosphere, and are harmless even if the window breaks. (Krypton-filled windows are also safe from Superman breakage.) All this technology adds up to windows with R-values in the 6 to 10 range.

Low-E Films

The biggest news in window technology is "low-e" films, for low-emissivity. These thin metal coatings allow the shortwave radiation of solar energy to pass in, but block most of the long-wave thermal energy trying to get back out. A low-e treated window has less heat leakage in either direction. A low-e coating is virtually invisible from the inside, but most brands tend to give windows a semi-mirror appearance from the outside.

Low-e windows are available ready-made from the factory, where the thin plastic film with the metal coating is suspended between the glass panes; or, low-e films can be applied to existing windows.

We offer a do-it-yourself product that is applied permanently with soap and squeegee, or they can be applied professionally. These films offer the same heat reflecting performance as the factory-applied coatings, and are relatively modest in cost (especially compared to new windows!). They do require a bit of care in cleaning, as the plastic film can be scratched. Some utility companies offer rebates for after-market low-e films.

Install tight-fitting insulating shades

These shades incorporate layers of insulating material, a radiant barrier, and a moisture-resistant layer to help prevent condensation. Several designs are available. One popular favorite is Window Quilts. This quilted-looking material consists of several layers of spun polyester and radiant barriers with a cloth outer cover. Depending on style, they fold or roll down over your windows at night, providing a tight seal on all four sides, high R-value insulation, privacy, and soft quilted good looks. This allows your windows to have all the daytime advantages of daylighting and passive heat gain, while still enjoying the nighttime comfort of high R-values and no cold drafts. Because there is little standardization of window sizing, Window Quilts are generally custom cut to size, making them unsuitable for a mail-order retail operation like ours. You can order directly from the manufacturer, or they can direct you to a local retailer who can supply installation services. Contact them at 877-966-3678, or on the Web at www.1windowquilts.com.

Construct insulated pop-in panels or shutters

Rigid insulation can be cut to fit snugly into window openings, and a lightweight, decorative fabric can be glued to the inside. Pop-in panels aren't ideal, as they require storage whenever you want to look out the window, but they are cheap, simple, and highly effective. They are especially good for windows you wouldn't mind covering for the duration of the winter. Make sure they fit tightly so moisture doesn't enter the dead-air space and condense on the window.

Close your curtains or shades at night

The extra layers increase R-value, and you'll feel more comfortable not being exposed to the cold glass.

Open your curtains during the day

South-facing windows let in heat and light when the sun is shining. Removing outside screens for the winter on south windows can increase solar gain by 40%.

Clean solar-gain windows

Keep those south-facing windows clean for better light and a lot more free heat. Be sure to keep those same windows dirty in the summer. (Just kidding!)

WINDOW SOLUTIONS FOR EFFICIENT COOLING

The main source of heat gain through windows is solar gain—sunlight streaming in through single or dual glazing. Here are some tips for staying cool:

Install white window shades or miniblinds

Using shades or blinds is a simple, old-fashioned practice. Since our grandparents didn't have air conditioners, they knew how to keep the heat out. Miniblinds can reduce solar heat gain by 40 to 50%.

Close south- and west-facing curtains

Do this during the day for any window that lets in direct sunlight. Keep these windows closed too.

Install awnings

Awnings are another good, old-fashioned solution. Awnings work best on south-facing windows where there's insufficient roof overhang to provide shade. Canvas awnings are more expensive than shades, but they're more pleasing to the eye, they stop the heat on the outside of your building, and they don't obstruct the view.

Hang tightly woven screens or bamboo shades outside the window during the summer

Such shades will reduce your view, but they are inexpensive and stop 60%-80% of the sun's heat from getting to the window.

Plant trees or build a trellis

Deciduous (leaf-bearing) trees planted to the south or particularly to the west of your building provide valuable shade. One mature shade tree can provide as much cooling as five air conditioners (although they're a bit difficult to transplant at that stage, so the sooner you plant the better). Deciduous trees block summer sun, but drop their leaves to allow half or more of the winter sun's energy into your home to warm you on clear winter days.

Low-e films and exotic gas infills

Both low-e films and inert-gas infill windows will improve cooling efficiency as well as heating efficiency. See above, page 226, and the sidebar on page 227, for further details. ■■■

Landscaping

When retrofitting existing buildings, most folks don't think about the impact of landscaping on energy use. Appropriate landscaping on the west and south sides of your house provides valuable summertime shading that will reduce unwanted heating as much as 50%. Those are better results than we get from more expensive projects like window and insulation upgrades! If your landscaping is deciduous, losing its leaves in winter to let the warming winter sun through, you have the best of both worlds. Landscaping can also block cold winter winds that push through the little cracks and crevices of a typical house, and make outdoor patio and yard spaces more livable and inviting. For more information, see Home Cooling on page 240.

Conserving Electricity

Appliance selection is one area where we feel a greater sense of control. If you're renting or "financially challenged" (and who isn't?), you may not have control over your house design, window selection and orientation, or heating plant. But you can select the light bulb in your lamps and the showerhead in your bathroom. And you determine whether appliances get turned off (really off!) when not in use. Electricity generation is one of the largest contributors of greenhouse gas emissions, along with the internal combustion engine. So in addition to saving you money, conserving electricity helps you save the planet.

Many household appliances are important enough to justify entire chapters in the *Sourcebook*. Lighting, heating and cooling (with refrigeration), water pumping, water heating, composting toilets, and water purification are big topics that have significant impact on total home energy use. Please see the individual chapters or sections on the above products. Here we provide general information that applies to all appliances.

DON'T USE ELECTRICITY TO MAKE HEAT

Avoid products that use electricity to produce heat when there's any choice. Making heat from electricity is like using bottled water for your lawn. It gets the job done just fine, but it's terribly expensive and wasteful. Electric space heaters are not the only electrical appliances that produce heat; in fact, most of them do. This includes electric water heaters, electric ranges and ovens, hot plates and skillets, waffle irons, waterbed heaters, and the most common household electric heater . . . the incandescent light bulb. Most of these appliances can be replaced by other appliances that cost far less to operate. Incandescent bulbs are one of the most dramatic examples. Standard light bulbs only return 10% of the energy you feed them as visible light. The other 90% disappears as heat. Compact fluorescent lamps return better than 80% of their energy as visible light, and they last over 10 times as long per lamp.

If you have a choice about using gas or electricity for residential heating, water heating, or cooking, use gas, even if this means buying new appliances. For those lucky areas that have natural gas service, your monthly bills will be 50 to 60% lower than with electricity. Even in areas that have to bring in bottled propane, gas will be 30 to 40% cheaper than electric. The same goes for clothes drying. Better yet, use a zero-energy-cost clothesline, which has the side benefit of making your clothes last longer. Dryer lint is your clothes wearing out by tumbling.

PHANTOMS AND VAMPIRES!

Now what's that odd quote in the margin about televisions using power when "off"? Appliances that use power even when they're off create what are called phantom loads. Any device that uses a remote control is a phantom load because part of the circuitry must remain on in order to receive the "on" signal from the remote. For most TVs, this power use is 15 to 25 watts. VCRs are typically 5 to 10 watts. Together that'll cost you over $20 per year. Either plug these little watt-burners into a switched outlet, or use a switched plug strip so they really can be turned off when not in use.

The other increasingly common villain to watch out for are the little transformer cubes that are showing up on the end of many small appliance power cords. Officially, in the electric industry these are known as "vampires" because they constantly suck juice out of your system, even when there's no electrical demand at the appliance. Ever noticed how those little cubes are usually warm to the touch? That's wasted wattage being converted to heat. Most vampire cubes draw a few watts continuously. Vampires need to be unplugged when not in use, or plugged into switched outlets.

Go Low-Flow, Too

Showers typically account for 32% of home water use. A standard showerhead uses about 3 to 5 gallons of water per minute, so even a five-minute shower can consume 25 gallons. According to the U.S. Department of Energy, heating water is the second largest residential energy user. With a low-flow showerhead, energy use and costs for heating water for showers may drop as much as 50%. This is particularly important if you heat your water with electricity, which is the most expensive and energy-inefficient way. A recent study showed that changing to a low-flow showerhead saved 27¢ worth of water and 51¢ of electricity per day for a family of four. So, besides being good for the Earth, a low-flow showerhead will pay for itself in about two months!

WATER-SAVING SHOWERHEADS

As noted above, showers account for one-third of the average family's home water use, and heating water is the second-largest residential energy user. Our low-flow showerheads can easily cut shower water consumption by 50%. Add one of our instantaneous water heaters for even greater savings. See chapter 7 for details.

All the remote control televisions in the U.S., when turned to the "off" position, still use as much energy as the output of one Chernobyl-sized plant.

—Amory Lovins,
The Rocky Mountain Institute

House Heating and Cooling

Heating and cooling a home typically accounts for more than 40% of a family's energy bill (as much as two-thirds in colder regions), and costs on average well over $600 a year. That's a good bit of money, , and any of us could find more interesting ways to spend it. Fortunately, it isn't difficult to cut our heating and cooling bills dramatically with a judicious mixture of weatherization, additional insulation, landscaping, window upgrades, improved heating and cooling systems, and careful appliance selection.

Making improvements to your building envelope, including weatherization, insulation, window treatments, and landscaping for energy conservation, is covered in the preceding section. In this section we'll cover home heating and cooling systems, looking at sustainability issues, selecting the best options for your home, and improvements and fine tuning to keep those systems working efficiently.

Sustainability and Renewables

Whether you're concerned about costs, environmental impact, or both, it makes sense to consider home heating and cooling from a sustainability perspective. We know that the fossil fuels are going to run out eventually; we know that fossil-fuel prices are very likely to increase, probably dramatically, as supplies become used up during the twenty-first century; we know that our current dependence on oil from the Middle East, which will continue into the foreseeable future, carries various risks that portend the real possibility of price hikes sooner rather

If a house isn't resource-efficient, it isn't beautiful.

—Amory Lovins,
The Rocky Mountain Institute

than later; and we know that our enormous consumption of fossil fuels produces tremendous, potentially catastrophic, environmental damage. The solutions to all of these problems depend upon greater use of renewable energy resources: sun, wind, water, biomass, geothermal heat.

Home heating and cooling options can be more or less sustainable, depending on your choice of strategies, systems, and fuels. As discussed above and in chapter 1, maintaining a tight building envelope and incorporating passive solar strategies to whatever extent possible are the places to start.

Passive Solar Heat

Solar energy is the most environmentally friendly form of heating you can find. The amount of solar energy we take today in no way diminishes the amount we can take tomorrow, or tomorrow, or tomorrow. Free fuel is the cheapest fuel. Passive solar heating is nonpolluting and does not produce greenhouse gases. It uses no moving parts, just sunshine through south-facing insulated windows and thermal mass in the building structure to store the heat. (Those south of the equator please make the usual direction adjustments.)

A carefully designed passive solar building can rely on the sun for half or more of its heating needs in virtually any climate in the United States. Many successful designs have cut conventional heating loads by 80% or better. Some buildings, such as the Real Goods Solar Living Center in Hopland, California, or the Rocky Mountain Institute's headquarters high in the Rocky Mountains in Snowmass, Colorado, have

Sun path diagrams.
Passive solar heating is practicable in every climate. This simplified illustration shows the importance of a calculated roof overhang to allow solar heating in the winter but prevent unwanted solar heat gain in the summer. Adapted from an illustration by E SOURCE (1993), *Space Heating Technology Atlas.*

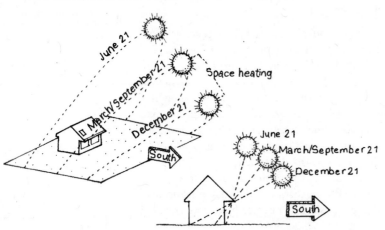

no need for a central heating system. RMI even grows banana trees indoors! By the same token, in hot climates, good design, shading, and passive cooling strategies can eliminate the need for mechanical cooling. Our Solar Living Center can survive over four weeks of daily high temperatures exceeding 100°F—and a steady flow of overheated visitors—while the interior temperature stays below 78°F without using air conditioning or mechanical cooling in any way.

Adding south-facing windows or a greenhouse is a way to retrofit your house to take advantage of free solar gain. See chapter 1, "Land and Shelter," for more detailed information and books about passive solar design.

ACTIVE SOLAR HEATING: BEST FOR HOT WATER

In many climates around the United States, flat-plate collector systems can provide enough solar heat to make a central heating system unnecessary, but in most regions it is prudent to add a small back-up system for cloudy periods. Active solar systems can be added at any time to supplement space- and water-heating needs. However, active solar space heating has a high initial cost, is complex, and has potential for high maintenance. We don't recommend it; use passive solar! The beauty of passive systems is in their simplicity and lack of moving parts. Remember the KISS rule: Keep It Simple, Stupid.

The best and most common use of active systems is for domestic water heating. Real Goods offers several solar hot-water systems for differing budgets. See our Water Heating chapter.

Buying a New Heating System

If your old heating system is about to die, you're probably in the market for a new one. Or, if you're currently spending over $1,000 per year on heating, it's likely that the $800 to $4,500 you'll spend on a more efficient heating system will reduce your bills enough to pay for your investment in several years. Your new system also will be more reliable, and it will increase the value of your house. As you shop, keep in mind the following factors that will save you money and increase your comfort:

If you've weatherized and insulated your home—and we'll remind you again that these are by far the most cost-effective things you can do—then you can downsize the furnace or boiler without compromising the capability of your heating system to keep you warm. An oversized heater will cost more to buy up front and more to run every year, and the frequent short-cycle on and off of an oversized system reduces efficiency. Ask your heating contractor to explain any sizing calculations and make sure he or she understands that you have a tight, well-insulated house, to verify that you don't get stuck with an oversized model (which the contractor gets to charge more money for).

Woodstoves, Masonry Stoves, and Pellet Stoves

After passive solar, heating with biomass is probably the next best environmentally friendly option. Greg Pahl, author of the award-winning book *Natural Home Heating: The Complete Guide to Renewable Energy Options*, explains why:

■ The following is adapted from *Natural Home Heating*, by Greg Pahl, Chelsea Green Publishing, used with permission.

Heating with wood does not contribute to global warming as long as it is part of a managed resource cycle that includes replanting trees. The carbon dioxide that is released when wood is burned is taken up by the replacement trees that grow in a sustainably managed forest. This cycle can be repeated indefinitely without increasing atmospheric carbon. The same rule applies to all "biomass" (wood, wood pellets, vegetation, grains, or agricultural waste) used as a fuel or energy source.

If you are considering heating with wood, find a reliable long-term source of firewood in your area. This may be difficult, especially in many urban locations, and could cause you to choose a different fuel source instead. Check local regulations too: Concerns about air pollution caused by smoky wood fires have prompted some municipalities and states to restrict the use of wood-burning appliances. You will also need a place to store your firewood. ■■

Wood heat is a mixed blessing. If harvested and used in a responsible manner, firewood can be a sustainable resource. But, to be honest, most wood gathering is not done sustainably. If you have your own woodlot, you can create a sustainable forest management plan, with the help of a professional forester, if necessary. If you buy wood, you can try to ascertain if your wood supplier cuts in a responsible manner, or buys logs from someone who does—and you can encourage him to do so. You can also participate in reforestation programs run by organizations

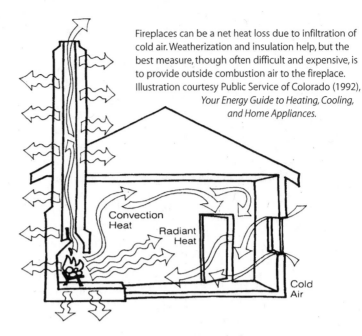

Fireplaces can be a net heat loss due to infiltration of cold air. Weatherization and insulation help, but the best measure, though often difficult and expensive, is to provide outside combustion air to the fireplace. Illustration courtesy Public Service of Colorado (1992), *Your Energy Guide to Heating, Cooling, and Home Appliances.*

such as American Forests; it costs very little to offset your annual fuel wood consumption by funding the planting of new tress.

Pollution is still a problem, though progress is being made. There are 27 million wood-burning stoves and fireplaces in the United States, the majority of them older pre-EPA designs, that contribute millions of tons of pollutants to the air we breathe. In Washington state, for example, wood heating contributes 90% of the particulates in two of the counties with the worst air pollution problems. Modern woodstoves have catalytic combustors and other features that boost their efficiencies into the 55 to 75% range while lowering emissions by two-thirds. Older woodstoves without air controls have efficiencies of only 20 to 30%. Wood burns at a higher temperature in a catalytic woodstove and the gases go through a secondary combustion at the catalyst, thereby minimizing pollution and creosote buildup.

If you're buying a new woodstove, make sure it's not too large for your heating needs. This will allow you to keep your stove stoked with the damper open, resulting in a cleaner and more efficient burn. Some stoves incorporate more cast iron or soapstone as thermal mass to moderate temperature swings and prolong fire life. If you live in a climate that requires heat continuously for several months, you definitely want the heaviest stove you can afford. Masonry heaters take the thermal mass idea one very comfortable step further.

MASONRY STOVES

Masonry stoves, also called Russian stoves or fireplaces, developed by a people who know

Masonry stoves are the best wood-burning solution for extreme climates that need twenty-four-hour heating.

heating, have long been popular in Europe. These large, free-standing masonry fireplaces work by circulating the heat and smoke from the combustion process through a long twisting labyrinth masonry chimney. The mass of the heater slowly warms up and radiates heat into the room for hours. Masonry stoves are best suited for extreme climates that need heat 24 hours per day. They are much more efficient than fireplaces. They have very low emissions because air flow is restricted only minimally during burning, giving a clean, high-temperature burn. How many woodstoves have seats built into them so that you can snuggle in? Russian fireplaces commonly do. As Mark Twain observed, "One firing is enough for the day . . . the heat produced is the same all day, instead of too hot and too cold by turns."

Masonry stoves are expensive initially, but will last for the life of the home, and are one of the cleanest, most satisfying ways to heat with wood. They're easiest to install during initial construction, and are best suited to well-insulated homes with open floor plans that allow the heat to radiate freely.

PELLET STOVES

If you can't burn wood but are looking for a similar way to ween yourself from oil, a pellet stove may be right for you. As renewable energy expert Greg Pahl observes, pellets have clear environmental benefits: "Pellets are a renewable resource, so burning pellets does not add any net carbon dioxide to the atmosphere. Because of the extremely hot combustion in most pellet-fired appliances, pellets burn cleanly with extremely low emissions and produce virtually no creosote. Compared to other fuels, burning 1 ton of pellets instead of heating with electricity will save 3,323 pounds of carbon emissions. You'll save 943 pounds of emissions per ton by replacing oil with pellets and 549 pounds per ton if you're replacing natural gas."

Pellet stoves have the advantage that the fuel pellets of compressed sawdust, cardboard, or agricultural waste are fed automatically into the stove by an electric auger. The feed rate is dialed up or down with a rheostat, so they'll happily run all night if you can afford the pellets. You fill the reservoir on the stove once every day or so. Pellet stoves are cleaner burning than wood-

To Burn or Not to Burn

I'm totally confused. The subject is wood, a material I have used to heat my home for the past 15 years. I know it's more work to burn wood, but I like the exercise. It makes more sense to burn calories splitting and hauling hardwood than running in front of a television set on a treadmill at the health spa. I enjoy the fresh air. I like the warmth. I like the flicker and glow of embers.

Sometimes, when I am stacking wood so that a summer's worth of sunshine can make it a cleaner, hotter fuel, I think of those supertankers carrying immense cargos of black gold from the ancient forests of the Middle East. I think of the wells that Saddam Hussein set afire in Kuwait, and the noxious roar of the jet engines of the warplanes we used to ensure our country's access to cheap oil.

My wood comes from hills that I can see. It's delivered by a guy named Paul, who cuts it with his chainsaw, then loads it in his one-ton pickup. We talk about the weather, the conditions in the woods, and the burning characteristics of different species. We finish by bantering about whether he's delivered a large cord, a medium, or a small. I can't imagine having the same conversation with the man who delivers oil or propane.

Unfortunately (and this is where I begin to get confused), there is another side to the wood-burning issue. Along comes evidence that the emissions from airtight stoves contain insidious carcinogens, and further indications that a home's internal environment can be more adversely affected by burning wood than by passive cigarette smoke. So now I burn my wood in a "clean" stove. But, my environmentally active friends say that the only good smoke is no smoke, and that even my supposedly

high-tech stove is smogging up their skies.

Theoretically, I should be able to go back to feeling good about wood, but life is never that simple. On the positive side of the ledger comes the information that wood-burning, in combination with responsible reforestation, actually helps the environment by reversing the greenhouse effect. The oxidation of biomass, whether on the forest floor or in your stove, releases the same amount of carbon into the atmosphere. It makes more sense for wood to be heating my home than contributing to the brown skies over Yellowstone. This is obviously a simplistic analysis, but wood should be a simple subject. In the meantime, I have to keep warm, so I'm choosing to burn wood. I keep coming back to the fact that humans have been burning oil for more than fifty years and wood for five million.

—Stephen Morris

stoves because combustion air is force-fed into the burning chamber. Unlimited access to oxygen and drier fuel than firewood mean a more efficient combustion. But the pellets—which look like rabbit food—may be slightly more expensive than firewood in most parts of the country, and simply unavailable in some areas. Pellets must be kept bone dry in order to auger and burn.

Pellet stoves have disadvantages, too. They have more mechanical parts than woodstoves, so they're subject to maintenance and breakdown, and they require electricity. No power—no fire. The fire will go out without the combustion blower running. Pellet stoves usually cannot be used with renewable electricity systems, unless you've got considerable extra power in the winter, like from a hydro system.

Now there is a way to burn pellets in a regular woodstove, too. See the product section for information about the innovative Prometheus pellet fuel basket.

Displace Oil with Biodiesel

If you're stuck with an oil-burning furnace or boiler, don't just sit there and grimace about all the greenhouse gases you're putting into the atmosphere—there is something you can do about it (besides making sure your house is tight and your system efficiency is high): biodiesel. The original Diesel engine exhibited at the Paris World's Fair in 1900 ran on vegetable oil, and the concept remains more than viable. Biodiesel is a renewable fuel that can be made easily from a variety of biomass feedstocks, even from recycled cooking oil. Biodiesel has been promoted primarily as a vehicle fuel, but it also works as a home heating fuel additive. Increasing numbers of people are burning B-100, or pure biodiesel, while others burn a blend of #2 heating oil mixed with 10 or 20% biodiesel; no conversion is required, and the extra maintenance needed is minimal. Biodiesel is currently more expensive than heating oil, though not prohibitively so, and the price is bound to become more competitive over time. Biodiesel manufacturing facilities are springing up everywhere and fuel distributors are looking into supplying it to consumers.

Any biodiesel you use to displace oil from your heating system lowers carbon dioxide emissions and reduces particulate pollution. Call a local fuel distributor, or your state energy office, for more information about obtaining biodiesel for heating purposes. See chapter 11, "Sustainable Transportation," for details on biodiesel as an alternative motor vehicle fuel.

Fuel Switching: Natural Gas Furnaces and Boilers

When available, natural gas is the lowest-cost heating fuel for most homes. If you have a choice, switching fuels may be cost effective, but only if this does not also require substantial changes to your home's existing heat distribution system. For example, replacing the old hot-water boiler with a high-efficiency forced-air furnace would require installation of all new forced-air ducts.

It is always cost effective to pay a little more up front in return for more efficiency. But it may not be cost effective to pay a lot more for the most efficient gas unit, since your reduced heating needs mean a longer payback. The differences between various models of gas furnaces and boilers can be significant, and it's worth your while to examine efficiency ratings before making a decision.

If you are currently using electricity for heating, switching to any other fuel will be cost-effective. See the next subject.

Electric Resistance Heating

Electric baseboard heating is by far the most expensive way to warm one's home; it's cheap to buy and install, but it costs two to three times as much to heat with electricity as with gas. The life-cycle cost of electric heating is, without exception, far higher than that of gas-fired furnaces and boilers, even taking into account the higher installation cost of the latter, and even in regions with exceptionally low electricity prices. Electric resistance is an energy-inefficient way to produce heat, and because grid-generated electric power is a key greenhouse gas culprit, this is a terrible environmental choice. (Electric heating may be excused if you have a plentiful, seasonal supply of renewably generated power, for example from a year-round micro-hydro system.)

If you must heat with electricity, weatherizing and insulating your home, along with installing a programmable thermostat, are especially lucrative options. If you also use air conditioning, and live in a mild winter climate, consider switching to a heat pump.

Geothermal Energy: Heat Pumps

■ (he following is excerpted from *Natural Home Heating,* by Greg Pahl, Chelsea Green Publishing, used with permission.

The concept behind geothermal heating is simple: The Earth is a huge heat-storage device. For millions of years, the Earth has been absorbing and storing solar energy in the air, water, and ground. This stored energy offers enormous potential to meet a major portion of our energy needs. The trick up to now, however, has been to figure out practical ways to harvest and use that energy. One of the most successful ways to accomplish that goal is to use a heat pump. Unlike most other home-heating devices, heat pumps are not based on combustion. Instead, heat pumps move heat from one location to another. ■ ■

While heat pumps are considered a renewable form of energy because they consume no fuel to generate heat, they do consume electricity in order to move heat around. Heat pumps therefore can be considered the most efficient form of electric heat.

There are three types of heat pumps: air-to-air heat pumps, water-source heat pumps, and ground-source heat pumps. Heat pumps collect heat from the air, water, or ground depending on type, concentrate the warmth, and distribute it through the home. Heat pumps typically deliver three times more energy in heat than they consume in electric power. Heat pumps also can be used to cool homes by reversing the process—collecting indoor heat and transferring it outside the building. Some newer heat pumps are designed to provide an inexpensive source of household hot water as well.

Not all heat pumps are created equal, but fortunately the Air Conditioning and Refrigeration Institute rates all heat pumps on the market. Look at the efficiency ratings and purchase a system designed for a colder climate. Some electric utilities offer rebates and other incentives to help finance the higher capital costs of these more efficient systems.

Air-to-air heat pumps are the most popular, the least expensive initially, and the most expensive to run in colder weather. Air-to-air units must rely on inefficient back-up heating mechanisms (usually electric resistance heat) when outside temperatures drop below a certain point. That point varies from model to model, so make sure you buy one that's designed for your climate—some models can cope with much colder weather before resorting to back-up heat. A heat pump running on back-up is no better than standard electric resistance heating. Some newer pumps have more efficient gas-fired back-up mechanisms, but if gas is available, why not use it directly?

Ground- or water-source heat pumps are one of the most efficient heating systems, and are substantially more cost-effective in colder climates. Because of the extensive burial of plastic pipe required, these systems are best installed during new construction. Tens of thousands of ground-source heat pumps have been installed in Canada, New England, and other frost-belt areas. See the annualized heating/cooling system cost chart on the next page for a comparison.

Fireplace Inserts

As noted by the authors of the *Consumer Guide to Home Energy Savings,* a fireplace is essentially part of a room's décor, not a source of heat. A customary fireplace installation generally will lose more heat than it provides. However, if outfitted with a well-sealed insert, a fireplace can provide some useful heat.

A fireplace insert fits inside the opening of a fireplace, operates like a woodstove, and offers improved heat performance. An insert is difficult to install because it must be lifted into the fireplace opening and the space around it must be covered with sheet metal and sealed with a cement grout to reduce air leaks. Such inserts are

Ground-source heat pump.
Ground-source heat pumps can be highly cost effective in new construction over the life cycle of heating and cooling equipment compared to more conventional options. Several designs are on the market; this illustration shows the Slinky™ design. Adapted from an illustration by E SOURCE (1993), *Space Heating Technology Atlas.*

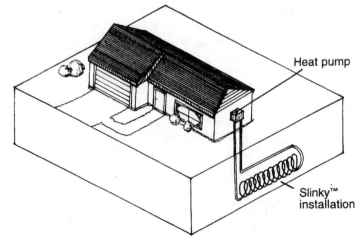

Heat pump

Slinky™ installation

only 30 to 50% efficient, which is a good improvement over a fireplace, but less efficient than a good freestanding woodstove, which is free to radiate heat off all sides, not just the front.

A fireplace can be retrofitted with an airtight woodstove installed in, or better yet in front of, the fireplace. This is the most efficient choice, since all six heat-emitting sides sit inside the living space, on the hearth. The stove's chimney is linked into the existing flue, and good installations line the chimney all the way to the top with stainless pipe to limit creosote buildup and provide easy flue cleaning. Simply shoving a few feet of woodstove pipe into the existing chimney is begging for a chimney fire of epic proportions. Any creosote in the smoke stream will condense on the large, cool chimney walls, providing a spectacular amount of high-temperature fuel in a few years. This has been well proven to be an excellent way to burn a house down—further demonstrations and experiments aren't necessary. Line your chimney when retrofitting a woodstove.

Gas Fireplaces

Gas fireplaces have combustion efficiencies up to 80%, whereas the initially less expensive gas logs are only 20 to 30% efficient. However, in case of gas leaks, many local codes require dampers to be welded open. This reduces their overall efficiency unless they have tight-fitting doors to cut down the amount of warmed interior air allowed to float out freely.

People Heaters

It often makes sense to use a gas or electric radiant or convection heater to warm only certain areas of the home, or, as in the case of radiant heaters, to keep individual people warm and comfortable. Radiant systems keep you and objects around you warm just like the sun warms your skin. This allows you to set the thermostat for the main heating system lower by 6° to 8°F.

Even if you heat with gas, electric convective or radiant spot heaters can save you money, depending on how the system is used. Radiant systems are designed and used more like task lights; you turn them on only when and where heat is needed, rather than heating the whole house. Most central heating systems give you little control in this respect, since they are designed to heat the entire home.

Heating Costs Comparison

While operating costs will vary by climate and region, and fuel and electricity prices differ across the country, the following chart represents estimated installation and operating costs for selected heating and cooling systems in a typical single-family house. It assumes that heat pumps will provide air conditioning in the summer as well as space heating in winter.

COMPARING HEATING FUEL COSTS

		In Dollars per million Btu
ELECTRICITY		
	Ground-source heat pump	$9.15
	Air-source heat pump	$13.70
	Baseboard resistance heater	$27.65
NATURAL GAS		
	Medium (90%) efficient central furnace	$14.85
PROPANE		
	Medium (85%) efficient central furnace	$21.25
FUEL OIL		
	Medium (85%) efficient central furnace	$16.80

Fuel costs are late 2004 national averages.

Electricity: $0.0934/kWh; natural gas: $13.78/thousand cubic feet; fuel oil: $2.00/gallon; propane: $1.65/gallon.

Fine Tuning Your Heating System

Let's assume, perhaps foolishly, that your home is already well insulated and weather tight. By weatherizing and insulating your home, you've reduced the amount of heat escaping in the most cost-effective way. It now makes sense to look at how efficiently your heating system produces and delivers the heat that you need. This section gives tips for making your existing heating system run more efficiently, whether it is powered by gas, electric, oil, wood, or solar energy. The following suggestions are either relatively inexpensive or free. All are highly cost effective.

Remember, if your existing system appears close to a natural death, or its maintenance costs are high, these fine-tuning tips may not be particularly cost effective. Modern heating systems have seen dramatic efficiency gains over the past few years, so replacing an old one could well save you money.

For safety reasons, adjustments, tune-ups, and modifications to your heating system itself are best done by a heating system professional. Home owners and renters can improve system performance by insulating ducts and pipes, cleaning registers, replacing filters, and installing programmable thermostats.

FURNACES AND BOILERS

Heating System Tune-Ups

Gas furnaces and boilers should be tuned every two years, while oil units should be tuned once a year. Your fuel supplier can recommend or provide a qualified technician. Expect to pay $60 to $150. It's money well spent. Also have the technician do a safety test to make sure the vent does not leak combustion products into the home. You can do a simple test by extinguishing a match a couple of inches from the spill-over vent: The smoke should be drawn up the chimney.

During a furnace tune-up, the technician should clean the furnace fan and its blades, correct the drive-belt tension, oil the fan and the motor bearings, clean or replace the filter (make sure you know how to perform this monthly routine maintenance), and help you seal ducts if necessary.

Efficiency Modifications

While the heating contractor is there to tune up your system, he or she may be able to recommend some modifications such as reducing the nozzle (oil) or orifice (gas) size, installing a new burner and motorized flue damper, or replacing the pilot light with an electronic spark ignition. If you have an older oil burner, installing a flame-retention burner head (which vaporizes the fuel and allows more complete combustion) typically will pay back your investment in two to five years.

Turn Off the Pilot Light during the Summer

This simple act will save you about $2 to $4 per month. Do this only if you can safely light it again yourself, so you don't have to pay someone to do it. Federal regulations now require that new natural gas–fired boilers and furnaces be equipped with electronic ignition, saving $30 to $40 per year in gas bills. Propane-fired units cannot be sold without a pilot light and cannot be retrofitted safely with a spark ignitor unless you install expensive propane-sniffing equipment. Propane is heavier than air, and any leak-ing propane can pool around the unit, creating an explosion potential when the unit tries to start. Natural gas is lighter than air.

Insulate the Supply and Return Pipes on Steam and Hot-Water Boilers

Use a high-temperature pipe insulation such as fiberglass wrap for steam pipes. The lower temperatures of hydronic or hot-water systems may allow you to use a foam insulation (make sure it's rated for at least 220°F).

Clean or Change Your Filter Monthly

Yes, filters are used for more than coffee and cars. A clogged furnace filter impedes air flow, makes the fan work harder, makes the furnace run longer, and cuts overall efficiency. Filters are designed for easy service, so the hardest part of this maneuver is turning into the hardware store parking lot with a note of the correct size in hand. While you're there, pick up several filters, since you'll want to replace the filter every month during the heating season. Most full-house air-conditioning systems use the same fan, ducts, and filter. So monthly changes during AC season are a must, too. They'll set you back a buck or two apiece. For five bucks, you can buy a reusable filter that will need washing or vacuuming every month, but will last for a year or two.

Seal and Insulate Air Ducts

Losses from leaky, uninsulated ducts—especially those in unheated attics and basements—can reduce the efficiency of your heating system by as much as 30%. Don't blow your expensive heated air into unheated spaces! Seal ducts thoroughly with mastic, caulking, or duct tape, and then insulate with fiberglass wrap.

RADIATORS AND HEAT REGISTERS

Vacuum the Cobwebs Out of Your Registers

Anything that impedes airflow makes the fan work harder, and the furnace run longer.

Reflect the Heat from Behind Your Radiator

You can make foil reflectors by taping aluminum foil to cardboard, or use radiant barrier material (left over from your attic or basement). Place them behind any radiator on an external wall with the shiny side facing the room. Foil-faced rigid insulation also works well.

Vacuum the Fins on Baseboard Heaters

Vacuuming up dust improves airflow and efficiency. Keeping furniture and drapes out of the way also improves airflow from the radiator.

Bleed the Air out of Hot-Water Radiators

Trapped air in radiators keeps them from filling with hot water and thus reduces their heating capacity. Doing this should also quiet down clanging radiators. You can buy a radiator key at the hardware store. Hold a cup or a pan under the valve as you slowly open it with the key. Close the valve when all the air has escaped and only water comes out. If it turns out that you have to do this more than once a month, have your system inspected by a heating contractor.

THERMOSTATS

Install a Programmable Thermostat

One-half of home owners already turn down their heat at night, saving themselves 6 to 16% of heating energy. A programmable or clock thermostat can do this for you automatically. In addition to lowering temperatures while you sleep, it also will raise temperatures again before you get out from under the covers. It not only sounds good, it's cost effective too. Such thermostats can be used to drop the house temperature during the day as well if the house is unoccupied. *Home Energy* magazine reports savings in excess of 20% with two eight-hour setbacks of 10°F each.

Electronic setback thermostats are readily available for $25 to $150. If you have a heat pump or central air conditioning, make sure the thermostat is designed for heat and A/C. Installation is simple, but make sure to turn off the power to the heating plant before you begin. If connecting wires makes you jittery, have an electrician or heating technician do it for you. It's well worth the investment.

HEAT PUMPS

Change or Clean Filters Once a Month

Dirty filters are the most common reason for heat pump failures.

Keep Leaves and Debris Away from the Outside Unit

Good airflow is essential to heat pump efficiency. Clear a 2-foot radius. This includes shrubbery and landscaping.

Listen Periodically to the Compressor

If the compressor cycles on for less than a minute, something is wrong. Have a technician adjust it immediately.

Dust Off the Indoor Coils of the Heat Exchanger at Least Once a Year

This dust-off should be part of the yearly tune-up if you pay a technician for this service.

Have the Compressor Tuned Up

Every year or two, a technician should check refrigerant charge, controls, and filters, and oil the blower motors.

WOODSTOVES

Use a Woodstove Only If You're Willing to Make Sure It Burns Cleanly

Woodstoves aren't like furnaces or refrigerators; you can't just turn them on and forget them. "Banking" a woodstove for an overnight burn only makes the wood smolder and emit lots of pollution. Sure, we've all done it, but we didn't know any better. Now we do. A well-insulated house will hold enough heat that you shouldn't need to keep the stove going all night. Heavier, masonry stoves are a good idea in colder climates. Once the large mass is warmed up, it emits slow, steady heat overnight without a smoky fire.

Thermostat Myths

Myth: Turning down your thermostat at night or when you're gone means you'll use more energy than you saved when you have to warm up the house again.

Fact: You always save by turning down your thermostat no matter how long you're gone. (The only exception is older electric heat pumps with electric resistance heaters for backup.)

Myth: The house warms up faster if you turn the thermostat up to 90° initially.

Fact: Heating systems have only two "speeds": ON or OFF. The house will warm up at the same rate no matter what temperature you set the thermostat, as long as it is high enough to be "ON." Setting it higher only wastes energy by overshooting the desired temperature.

Burn Well-Seasoned Wood and No Trash

Wood should be split at least six months before you burn it. Never burn garbage, plastics, plywood, or treated lumber.

Hotter Fires Are Cleaner

Another good case for masonry stoves, which burn hot, clean, short fires with no air restriction. The heat is soaked up by the masonry and emitted slowly over many hours.

If a Woodstove is Your Main Source of Heat, Use an EPA-Approved Model

EPA regulations now require woodstoves to emit less than 10% of the pollution that older models produced.

Check Your Chimney Regularly

Blue or gray smoke means your wood is not burning completely.

FIREPLACES

Fireplaces offer a good light show but are not an effective source of heat. In fact, fireplaces are usually a net heat loser because of the huge volume of warm interior air sucked up the chimney, which has to be replaced with cold air leaking into the house. Fireplaces average minus 10% to plus 10% efficiency. If you've done a good job of weatherizing your house and sealing up air leaks, you may lack adequate fresh air for combustion, or the chimney won't draw properly, filling the house with smoke. The two best things to do with a fireplace are don't use it and plug the flue; or install a modern woodstove and line the chimney.

Improve the Seal of the Flue Damper

To test the damper seal, close the flue, light a small piece of paper, and watch the smoke. If the smoke goes up the flue, there's an air leak. Seal around the damper assembly with refractory cement. (Don't seal the damper closed.) If the damper has warped from heat over the years, get a sheet-metal shop to make a new one, or consider a woodstove conversion.

Install Tight-Fitting Glass Doors

Controlling airflow improves combustion efficiency by 10 to 20%, and may help reduce air leakage up a poorly sealed damper.

Use a C-Shaped Tubular or Blower-Equipped Grate

Tube grates draw cooler air in the bottom and expel heated air out the top. Blower-equipped grates do the same job with a bit more strength.

Caulk Around the Fireplace and Hearth

Do this where they meet the structure of the house, using a butyl rubber caulk.

Use a Cast-Iron Fireback

Firebacks are available in a variety of patterns and sizes. Fireplace efficiency is improved by reflecting more radiant heat into the room.

Locate the Screen Slightly Away from the Opening

Moving the screen forward into the room a bit allows more heated air to flow over the top of the screen. While they do prevent sparks from flying into the room, fire screens also prevent as much as 30% of the heat from entering the room.

If You Never Use Your Fireplace, Put a Plug in the Flue to Stop Heat Loss

Seal the plug to the chimney walls with good-quality caulk and tell anyone who might build a fire (or leave a sign). Temporary plugs can be made with rigid board insulation, plywood with pipe insulation around the edges, or an inflatable plug.

Home Cooling Systems

Cooling your home, like heating it, can be accomplished in sensible, inexpensive, energy-efficient ways, or in the energy-voracious way that Americans have favored for the past several decades in the era of cheap fossil fuels: Install a big air conditioner, crank it up, and pay the bills. It's much better for your pocketbook, and for the planet, to pursue the common-sense approach: Reduce your cooling load, take advantage of passive solar cooling strategies, utilize low-energy methods like fans, and increase the efficiency of your existing air conditioner. Finally, if you determine that it's cost effective, you should explore the prospect of buying new, efficient cooling equipment.

Reduce the Cooling Load through Smart Design and Passive Solar Strategies

The best strategy for keeping a dwelling cool is to keep it from getting hot in the first place. This means preventing outside heat from getting inside, and reducing the amount of heat generated inside by inefficient appliances such as incandescent lights, unwrapped water heaters, or older refrigerators.

In a hot, humid climate, 25% of the cooling load is the result of infiltration of moisture, 25% is from outdoor heat penetrating windows, walls, and roof, 20% from unwanted solar gain through windows, and the remaining 30% from heat and moisture generated within the home. It makes a lot of economic sense to cut these loads before getting that new monster air conditioner. You'll be able to get by with a smaller, less-expensive unit that costs less to operate for the rest of its life. And by cutting your power consumption, you'll significantly reduce your personal emissions of greenhouse gases and other environmental pollutants.

The same kinds of smart design choices that reduce your heating load and make your home energy efficient also serve to reduce the cooling load. Weatherizing, insulating, radiant barriers, good-quality windows, environmental landscaping, effective ventilation, and roof whitening each has the capacity to reduce your cooling costs substantially, especially in combination. Funny how we keep hammering those points. These tricks for keeping outside heat from get-

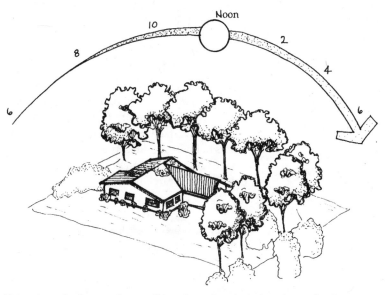

Mature trees shading your house will keep it much cooler by lowering roof and attic temperatures, blocking unwanted direct solar heat radiation from entering the windows, and providing a cooler microclimate outdoors. Adapted from an illustration by Saturn Resource Management, Helena, Montana.

ting inside should be considered, and implemented, before you move on to even low-energy methods that consume electricity—especially if you rely on renewable power systems.

Each of these methods is recapped briefly here. For more detailed discussions of these aspects of energy-efficient building design, see chapter 1.

WEATHERIZATION

Weatherization, usually done in northern climates to keep heat in, can also effectively keep heat and humidity out in the hot, humid climates or seasons. Cutting air infiltration by half, which is easily done, can cut air conditioning bills by 15% and save $50 to $100 per year in the average house in the southern United States. Much of the work done by an air conditioner in humid climates goes into removing moisture, which increases comfort. Reducing air infiltration is more important in humid climates than in dry ones.

INSULATION

Insulation slows heat transfer from outside to inside your home, or vice versa in the winter. Attic insulation levels should be R-19 or higher. If you also need to heat your house in the winter, insulation is even more important because of the greater temperature differential between inside and outside. In that case, you'll want to

> The best strategy for keeping a dwelling cool is to keep it from getting hot in the first place.

> Shading that blocks summer sun on the east, south, and west sides of your house, but not summer breezes, is one of the most effective ways to keep your home cooler.

Storing "coolth" works better if your house has thermal mass such as concrete, brick, tile, or adobe that will cool down at night and help prevent overheating during the day. These are the same things that make passive solar houses work better at heating in the winter.

have insulation levels of R-30 or higher. In hot climates, if you already have some significant amount of insulation, it will probably be more cost-effective to install a radiant barrier instead of adding more insulation.

RADIANT BARRIERS

Radiant barriers are thin metal films on a plastic or paper sheet to give them strength. They will reflect or stop approximately 97% of long-wave infrared heat radiation. Such barriers typically are stapled to the underside of attic rafters to lower summertime attic temperatures. Laying these barriers on the attic floor is also an option, but research indicates that the effective-

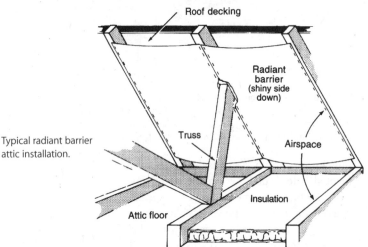

Typical radiant barrier attic installation.

ness of the barrier is reduced by dust build-up. Lowering attic temperatures—which easily can reach 160°F on a sunny day—reduces heat penetration into the living space below and is frequently a cost-effective measure to lower air-conditioning bills. A study by the Florida Solar Energy Center found that properly installed radiant barriers reduced cooling loads by 7 to 21%. The study also found that temperatures of the roofing materials were only increased by 2° to 10°F, which will have little or no impact on the life expectancy of a shingle. A radiant barrier will reduce, but not eliminate, high attic temperatures, so insulating to at least R-19 is still advisable to slow heat penetration. The cost for radiant barrier material is very moderate at approximately 20 to 25 cents per square foot. See our product section.

WINDOWS

For too long, a window has simply been a hole in the wall that lets light and heat through.

Solar-powered vent fans are a very cost-effective way to move out unwanted heat. By using solar power, they'll only run when the sun shines, and the brighter the sun, the faster they'll work.

Windows, even so-called "thermopane" (dual-glazed) windows, have very low R-values. Single-pane units have an R-value of 1. Thermopane windows have an R-value of 2. Both are pretty miserable.

Conventional windows conduct heat through the glass and frame, permit warm, moist air to leak in around the edges, and let in lots of unwanted heat in the form of solar radiation. South-facing windows with properly sized roof overhangs or awnings won't take in the high summer sun, but will accept the lower winter sun when it usually is wanted. East- and west-facing windows cannot be protected so simply. See the chapter on Land and Shelter for a complete discussion of window treatments, including newer high-tech windows with R-values up to 8 or 10.

LANDSCAPING FOR SHADE AND PASSIVE COOLING

You can take advantage of so-called "environmental landscaping" to provide shading and/or to channel cooling breezes through your home.

Shading that blocks summer sun on the east, south, and west sides of your house, but not summer breezes, is one of the most effective ways to keep your home cooler. Planting shade trees, particularly on the west and south sides of your house, can greatly increase comfort and coolness. (Remember that deciduous trees, which lose their leaves in winter but grow a thick canopy of leaves in the summer, should be used on the south side in colder climates to maximize solar gain in winter and minimize it in summer.) Awnings, porches, or trellises on those same sides of a building will reduce solar gain through the walls as well as through the windows. A home's inside temperature can rise as much as 20°F or more if the east and west windows and walls are not shaded.

Our own Dr. Doug greatly improved the comfort of his master bedroom by planting a jasmine-covered trellis over a west-facing wall. Without all those BTUs soaking through the wall four to eight hours after sunset, the bedroom stays 10° to 20°F cooler overnight than it used to. And the jasmine dramatically softens what used to be a glaring backyard patio in the afternoon, making both the outdoor and the indoor spaces more comfortable and pleasing.

Trees and shrubs planted in the proper configurations also can funnel and concentrate breezes, to increase airflow through and around a house. Increased airflow translates into increased cooling action, all thanks to Mother Nature.

MECHANICAL SHADING

In addition to trellises and awnings, various types of mechanical window-shading devices can be cost effective. Some are internal (such as drapes, curtains, and shades), some are external (such as louvers, shutters, and exterior roller shades). In the overall scheme of things, these devices are relatively inexpensive and they operate effectively to block sun and heat from entering your home. Explore the options and decide which ones will work best in various applications.

ROOF WHITENING

A white roof is an obvious solution. If you are replacing a roof, using a white or reflective roofing surface will reduce your heat load on the house. If you are not reroofing, coating the existing roof with a white elastomeric paint is a cost-effective measure in some climates. Be sure to get a specialty paint specifically designed for roof whitening: Normal exterior latex won't do.

The Florida Solar Energy Center reports savings from white roofs of 25 to 43% on air-conditioning bills in two poorly insulated test homes, and 10% savings in a Florida home with good R-25 roof insulation already in place. Costs are relatively high—the coatings cost 30 to 70 cents per square foot of roof surface area—and adding insulation or a radiant barrier may be a more cost-effective measure. Light-colored asphalt or aluminum shingles are not nearly as effective as white paint in lowering roof and attic temperatures.

PASSIVE NIGHTTIME COOLING

Opening windows at night to let cooler air inside is an effective cooling strategy in many states. This method is far more effective in hot, dry areas than in humid regions, however, because dry air tends to cool down more at night, and because less moisture is drawn into the house. The Florida Solar Energy Center found that when apartments in humid areas (and Florida certainly qualifies) opened their windows at night to let the cooler air in, the air conditioners had to work much harder the next day to remove the extra moisture that came along with the night air. Storing "coolth" works better if your house has thermal mass such as concrete, brick, tile, or adobe that will cool down at night and help prevent overheating during the day. These are the same things that make passive solar houses work better at heating in the winter.

Renewable Energy Systems and Home Cooling

Traditional air conditioners, even small room-sized units, and renewable energy systems are generally incompatible but it can be done. The equipment and knowledge is readily available. The problem is the sheer quantity of electricity required for air conditioning, and the long hours of continuous operation, augmented by the high initial cost of renewable energy systems. As a ballpark estimate, we have found that it will require about $3.50 worth of renewable energy "stuff" for every watt-hour needed.

A 1,500-watt room-size A/C unit running for six hours consumes 9,000 watt-hours, and needs over $30,000 worth of renewable energy system to support it.

ACTIVE SOLAR VENTILATION

Ventilating your attic is important for moisture control and keeping cool. Roofs can reach temperatures of 180°F, and attics can easily exceed 160°F. This heat eventually will soak through even the best insulation unless you get rid of it. Solar-powered vent fans are a very cost-effective way to move out this unwanted heat. By using solar power they'll only run when the sun shines, and the brighter the sun, the faster they'll work. We offer complete kits in two different sizes; they include fan, temperature switch—so it will only run when attic temperature is above 80°F—photovoltaic module(s), and mounting. See the product section.

WHOLE-HOUSE, CEILING, AND PORTABLE FANS

A whole-house fan is an effective means of cooling and is far less expensive to run than an

A whole-house fan can cool the house by bringing in cooler air—especially at night—and lowering attic temperatures. Adapted from an illustration by Illinois Department of Energy and Natural Resources (1987), *More for Your Money …Home Energy Savings.*

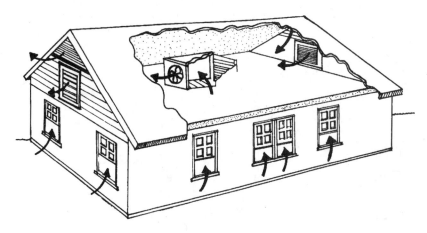

air conditioner. It can reduce indoor temperatures by 3° to 8°F, depending on the outside temperature. Through prudent use of a whole-house fan, you can cut air conditioner use by 15 to 55%. This is true particularly if you're lucky enough to live in a climate that reliably cools off in the evening. A high R-value house can be buttoned up during the heat of the day, heating up inside slowly like an insulated ice cube. In the evening when it's cooler outside than inside, a whole-house fan blows the hot air out of your house, while drawing cooler air through the windows. It should be centrally located so that it draws air from all rooms in the house. Be sure your attic has sufficient ventilation to get rid of the hot air. To ensure a safe installation, the fan must have an automatic shut-off in case of fire.

Ceiling, paddle, and portable fans produce air motion across your skin that increases evaporative cooling. A moderate breeze of one to two miles an hour can extend your comfort range by several degrees and will save energy by allowing you to set your air conditioner's thermostat higher or eliminate the need for air conditioning altogether. Less frequent use of air conditioning by setting the thermostat higher will cut cooling bills greatly. See our product section for more details.

EVAPORATIVE COOLING

If you live in a dry climate, an evaporative or "swamp cooler" is an excellent cooling choice, saving 50% of the initial cost and up to 80% of the operating cost of an air conditioner. The lower the relative humidity, the better they work. Evaporative coolers are not effective when your relative humidity is higher than about 40%.

PERSONAL COOLING

Cool drinks, the Real Goods Solar Hat Fan, a minimum of lightweight clothing, siestas, going for a swim, or sitting on a shady porch—all of these personal cooling methods work.

Cool drinks, the Real Goods Solar Hat Fan, a minimum of lightweight clothing, siestas, going for a swim, or sitting on a shady porch— all of these personal cooling methods work.

Personal cooling. Buying a block of ice isn't cost effective, but feeling cool is important. Cool drinks, small fans, less clothing, siestas, or sitting on a shady porch all work. Illustration courtesy of Saturn Resource Management, Helena, Montana.

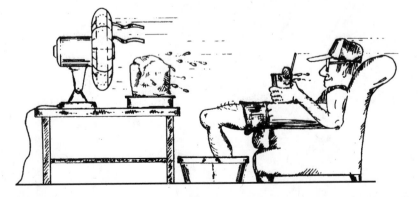

Improving Air Conditioner Efficiency

Once you've reduced heat penetration and taken advantage of passive solar and low-energy cooling methods, if you still have a need for additional cooling, it makes sense to improve the efficiency of your air conditioner.

CHECK DUCTS FOR LEAKS AND INSULATE

With central A/C systems, inspect the duct system for leaks and seal with duct tape or special duct mastic where necessary. The average home loses more than 20% of its expensive cooled or heated air before it gets to the register. It's well worth your time to find and seal these leaks. Just turn the fan on and feel along the ducts. Distribution joints are the most common leaky points.

If your ducts aren't already insulated in the attic or basement, insulate them with foil-faced fiberglass duct insulation. Batts can be fastened with plastic tie wraps, wire, metal tape, or glue-on pins. Your A/C technician can do this job or recommend someone who can, if you're not up for it.

HAVE YOUR HEAT PUMP OR CENTRAL A/C SERVICED REGULARLY

Regular maintenance will include having the refrigerant charge level checked—both under- and overcharging compromises performance—oiling motors and blowers, removing dirt buildup, cleaning filters and coils, and checking for duct leakage. Proper maintenance will ensure maximum efficiency as well as the equipment's longevity.

- Set your A/C to the recirculate option if your system has this choice—drawing hot and humid air from outside takes a lot more energy. Many newer systems only recirculate, so if you can't find this option it's probably already built in.
- Set your A/C thermostat at 78°F—or higher if you have ceiling fans. For each degree you raise the thermostat, you'll save 3 to 5% on cooling costs.
- Don't turn the A/C thermostat lower than the desired setting—cooling only happens at one speed. The house won't cool any faster, and you'll waste energy by overshooting.
- Turn off your A/C when you leave for more than an hour. It saves money.

- Close off unused rooms or, if you have central A/C, close the registers in those rooms and shut the doors.
- Install a programmable thermostat for your central A/C system to regulate cooling automatically. Such a thermostat can be programmed to ensure that your house is cool only when needed.
- Trim bushes and shrubs around outdoor condenser units so they have unimpeded airflow; a clear radius of 2 feet is adequate. Remove leaves and debris regularly.
- Provide shade for your room A/C or for the outside half of your central A/C if at all possible. This will increase the unit's efficiency by 5 to 10%.
- Remove your window-mounted A/C each fall. Their flimsy mounting panels and drafty cabinets offer little protection from winter winds and cold. If you are unable to remove the unit, at least close the vents and tightly cover the outside.
- Clean your A/C's air filter every month during cooling season. Normal dust buildup can reduce airflow by 1% per week.
- Clean the entire unit according to the manufacturer's instructions at least once a year. The coils and fins of the outside condenser units should be inspected regularly for dirt and debris that would reduce airflow. This is part of the yearly service if you're paying a technician for the service.

IF YOU LIVE IN A HUMID CLIMATE

An important part of what a conventional air conditioner does is removing moisture from the air. This makes the body's normal cooling—sweating—work better, and you feel cooler. Unfortunately, some of the highest-efficiency models don't dehumidify as well as less-efficient air conditioners. A/C units that are too big for their cooling load have this problem also. High humidity leads to indoor condensation problems and gives us a sweaty mid-August New York City feeling—minus the physical and aural assaults and the great food and nightlife. People tend to set their thermostats lower to compensate for the humidity, using even more energy. Here are some better solutions.

- Reduce the fan speed. This makes the coils run cooler and increases the amount of moisture that will condense in the A/C rather than inside your house.
- Set your A/C to recirculate if this is an option and you haven't already done so.

- Choose an A/C with a "sensible heat fraction" (SHF) less than 0.8. The lower the SHF, the better the dehumidification.
- Size your A/C carefully. Contractors should calculate dehumidification as well as cooling capacity. Many air conditioning systems are oversized by 50% or more. The old contractor's rule of thumb—one ton of cooling capacity per 400 square feet of living space—results in greatly oversized A/C units when reasonable weatherization, insulation, and shading measures have been implemented to reduce cooling load. One ton per 800 to 1,000 square feet may be a more reasonable rule of thumb.
- Investigate using a Dinh heat pipe/pump. These devices cool the air before it runs through the A/C evaporator, and warm it as it blows out, and this without energy input. They can help your A/C dehumidify several times more effectively.
- Explore "desiccant dehumidifiers," which, coupled with an efficient A/C, may save substantial energy. Desiccant dehumidifiers use a water-absorbing material to remove moisture from the air. This greatly reduces the amount of work the A/C has to do, thereby reducing your cooling bill.

Buying a New Air Conditioner

Air conditioners have become much more efficient over the last twenty years and top-rated models are 50 to 70% better than the current average. Federal appliance standards have eliminated the least-efficient models from the market, but builders and developers have little incentive to install a model that's more efficient than required, since they won't be paying the higher electric bills.

If you have an older model, it may be cost effective to replace it with a properly sized, efficient unit. The Office of Technology Assessment—a U.S. Congress research organization—estimates that buying the most-efficient room air conditioner, costing $70 more than a standard unit, has a payback of six and a half years, or less, if the additional features the better units include are taken into account.

AIR CONDITIONER EFFICIENCY

Several studies have found that most central air-conditioning systems are oversized by 50% or more. To avoid being sold an oversized model, or one that won't remove enough mois-

Air conditioners have become much more efficient over the last twenty years and top-rated models are 50 to 70% better than the current average.

Many contractors may simply use the "one ton of cooling capacity per 400 square feet of living space" rule of thumb. This will be far larger than you need, especially if you've effectively weatherized and insulated your home, and done some of the other measures we discuss here to reduce heat gain and the need for cooling.

Apply the weatherization, insulation, shading, and alternative cooling tips we have discussed, and your need for energy-intensive mechanical cooling will be eliminated in all but the most severe climates.

Real Goods sells a selection of DC-powered cooling appliances that can be run from batteries or directly from photovoltaic modules.

ture, ask your contractor to show you the sizing calculations and dehumidification specifications. Many contractors may simply use the "one ton of cooling capacity per 400 square feet of living space" rule of thumb. This will be far larger than you need, especially if you've effectively weatherized and insulated your home, and done some of the other measures we discuss here to reduce heat gain and the need for cooling.

One ton of cooling capacity—an archaic measurement equaling approximately the cooling capacity of one ton of ice—is 12,000 Btu per hour. Room air conditioners range in size from half a ton to one and a half tons, while typical central A/Cs range from two to four tons.

Desuperheaters

It's possible to use the waste heat removed from your house to heat your domestic water. In hot climates, where A/C is used more than five months per year, it makes economic sense to install a desuperheater to use the waste heat from the A/C to heat water. Some utilities give rebates to builders or homeowners who install such equipment.

Evaporative Coolers

If you live in a dry climate, an evaporative or "swamp" cooler can save 50% of the initial cost, and up to 80% of the operating cost, of normal A/C. An evaporative cooler works by evaporating water, drawing fresh outside air through wet, porous pads. This is an excellent choice for most areas in the Southwestern "sunbelt." The lower the relative humidity, the better they work. At relative humidity above 30%, performance will be marginal; above 40% humidity, forget it!

"Indirect" or "two-stage" models that yield cool, dry air rather than cool, moist air are also available, but the same low humidity climate restrictions apply.

Heat Pumps

If you live in a climate that requires heating in the winter and cooling in the summer, consider installing a heat pump in new construction, or if you are replacing the existing cooling systems. The ground- or water-coupled models are the most efficient; see the discussion on page 234, and the comparison chart included there. The air conditioning efficiency of a heat pump is equivalent to that of a typical air conditioner, as they're basically the same thing, but the heating mode will be much more efficient than the cooling mode, outperforming everything except natural gas-fired furnaces.

Appliances for Cooling

Since our core business is renewable energy, and the high energy consumption of air conditioning equipment makes renewable energy systems very expensive, we offer a minimal selection of appliances. Apply the weatherization, insulation, shading, and alternative cooling tips we have discussed, and your need for energy-intensive mechanical cooling will be eliminated in all but the most severe climates.

Real Goods sells a selection of DC-powered cooling appliances that can be run from batteries or directly from photovoltaic modules. These include small, highly efficient evaporative coolers, a selection of fans for attic venting or personal cooling, and some ceiling-mounted paddle fans.

INSULATION, RADIANT BARRIER, AND HEATING PRODUCTS

KeepCool Radiant Barrier

A radiant barrier is your best line of defense for keeping excess heat out of your attic in the summer and keeping your floors warm in the winter. In a barn, chicken

coop, or anywhere it's vital to keep farm animals and livestock warm in winter and cool in summer, a radiant barrier will provide invaluable protection. Stapled under the rafters or joists, it reflects 97% of the radiant heat. The nonprofit Florida Solar Energy Center has shown that an attic radiant barrier can reduce air conditioning costs by at least 20%. Performance in drier climates is even better. This perforated product won't become an unintentional moisture barrier. When the heat source is above the radiant barrier it performs equivalent to R-14 insulation. In order to function, it must have an air space on at least one side. It doesn't matter if it's the hot or cold side. Installation is a snap—just staple the radiant barrier to the bottom of rafters or joists. Perforated foil over a lightweight, tear-resistant paper/poly-fiber backing that passes UL-723 flame spread tests. Supplied in 48" wide rolls at 500 or 1,000 square feet. USA.

56-730	KeepCool Radiant Barrier	
	48" x 125'	$85
56-731	KeepCool Radiant Barrier	
	48" x 250'	$170

Window Tinting

Our do-it-yourself window-tinting solution reduces UV transmission by 97%. Equally important, window tinting cuts your utility demand and saves you money because it decreases heat penetration into your home by 60%. It applies easily to wet windows or Plexiglas, and static electricity—not adhesive—holds it in place. Lift it off instantly when the seasons change. Reduces glare and protects furniture, drapes, and carpeting from fading. Accommodate a variety of windows by choosing 24", 36", or 48" width. For multi-pane windows, apply to the outer pane. 15'L rolls. USA.

17-0177	Window Tinting	
	24"W x 15'L	$38
	36"W x 15'L	$48
	48"W x 15'L	$58
17-0178	Window Tinting Installation Kit (squeegee and knife)	$4

Draft Dodger

Classic weatherbeater locks out drafts so you stay warmer for less

It's a simple idea that's worked without fail for generations. Covered in hardy ThermaFleece, trimmed in satin piping, stuffed with insulating poly fibers, and weighted to stay right where you need it most, the Draft Dodger shuts out cold air to lower energy bills and keep you cozy. 4"H x 36"L. China.

17-0183	Draft Dodger	$20

HEATING PRODUCTS

Real Goods does not offer heat pumps or furnaces that require the services of a specialized technician for safe installation. We offer small, room-sized direct-vent gas heaters and wood-burning cookstoves.

SPACE HEATING

EMPIRE DIRECT-VENT HEATERS

Direct-venting makes these gas heaters safe and efficient for bedrooms, bathrooms, other closed rooms, or airtight homes. The entire combustion chamber is sealed off from inside air. Outside air is used for combustion, so there's no chance of combustion byproducts in your living space, no oxygen depletion, or wasting of already heated air. Most of these heaters must mount on an outside wall as intake and exhaust go out the back.

Empire MV-Series Heaters

(formerly Eco-Therm)
• Room-size heaters: 8,400, 14,000, and 20,000 Btu models
• Thermostat with modulating control
• Matchless piezo igniter with standing pilot
  • No electricity
  • Natural Gas or LP (propane)
  • 10-year warranty on combustion chamber, 1-year on all else

The slim, attractive, tan Euro-styled MV-series seems to float a couple inches off the wall with safe, pleasant radiant heat output. The thermostat modulates the gas flame for steady heat output with no fans or noise. Intake and exhaust are straight out the rear through a single double-wall pipe. Mounts to any wall 4.5" to 12" thick. Optional extensions to 18" wall thickness are available. Minimum clearances: floor 4.75", sides 4", back 0", top 36". AGA and CGA approved. Argentina, warranted by U.S. importer. Specify LP or Natural Gas.

Model #	MV-120	MV-130	MV-145
Max. (Btu/hr):	8,400	14,000	20,000
Min. (Btu/hr):	3,800	6,000	6,000
Width:	21.2"	27.2"	27.2"
Height:	24.4"	24.4"	24.4"
Depth:	6.75"	6.75"	6.75"
Weight:	33 lb.	42 lb.	55 lb.
Item #	54-0045	54-0046	54-0047
Price	$435	$545	$620

Empire DV-Series Heaters

• For medium to larger rooms: 25,000 and 35,000 Btu models
• Matchless piezo ignition with standing pilot
• Includes direct-vent kit and millivolt thermostat
• Optional automatic internal blower package
• Natural Gas or LP (propane) models available
• 10-year warranty on combustion chamber, 1-year on all else

Heating larger areas is convenient, economical, and safe with Empire Direct-Vent models. The millivolt thermostat requires no electricity. An optional automatic blower package delivers 75 cfm airflow to help distribute the heat more evenly for larger spaces. Outside vent cap included. The vinyl siding option offers extra heat protection. Beige color. Mounts to any outside wall 4.5" to 13" thick. Minimum clearances: floor 4", sides 6", back 0", top 48". AGA and CGA approved. USA. Specify LP or Natural Gas.

Model #	DV-25-SG	DV-35-SG
Input Btu/hr:	25,000	35,000
Width:	37.0"	37.0"
Height:	27.75"	27.75"
Depth:	11.5"	11.5"
Weight:	98 lb.	101 lb.
Item #	54-0048	54-0049
Price	$690	$750

54-0050 Automatic Blower Option	$125
54-0051 Vinyl Siding Vent Option	$30

Empire Console-type Heaters

A less-expensive alternative for larger spaces.

• Standard vented units (not direct-vent)
• For larger rooms, 65,000 Btu models
• Matchless piezo ignition with standing pilot
• Includes thermostat
• Automatic internal blower
• Natural Gas or LP (propane) models available
• 10-year warranty on combustion chamber, 1-year on all else

These larger standard vented units are available with either a closed front or with flame-view glass. A ceramic log option is available for the glass front model. Automatic blowers put warm air at floor level. 5" exhaust vent can run vertically or horizontally from heater back. Minimum clearances: floor 0", sides 6", back 2", top 55". AGA and CGA approved. USA. Specify LP or Natural Gas.

Model #	RH-65CB	RH-65B (w/glass)
Input Btu/hr:	65,000	65,000
Width:	34.0"	34.0"
Height:	29.6"	29.6"
Depth:	20.0"	20.0"
Weight:	142 lb.	142 lb.
Item #	54-0064	54-0063
Price	$850	$1,010

54-0065 Ceramic Log Option for RH-65B   **$159**

Prometheus Pellet Fuel Basket

Burn pellets in any woodstove

The Prometheus pellet fuel basket is a major breakthrough in wood-burning technology. Pellet fuel is a recycled product, produces far less creosote and ash than cordwood, is easier to handle, more readily available, and in many areas it now costs less. The only problem has been that, until now, it took a specialized stove to burn pellet fuel. The robust all-stainless steel Prometheus fuel basket makes pellet burning possible in any woodstove or fireplace. One-year manufacturer's warranty. USA.

Model Selection

For safe and reliable heating, it's important to choose the correct size Prometheus. Just multiply the inside width by the inside depth of your stove and compare your figure to the Minimum Hearth Area column on the chart. Your hearth area must be equal to or larger than this minimum area.

Item #	Model #	Woodstoves smallest hearth area allowed	Fireplaces min. opening width masonry	mftd.	Size (H x W x D)	Price
01-0299	6117	192 sq."	18"	21"	9⁵/₈" x 11⁵/₈" x 8¹/₄"	$199
01-0300	6147	244 sq."	21"	24"	9⁵/₈" x 14³/₄" x 8¹/₄"	$209
01-0301	6177	292 sq."	24"	26"	9⁵/₈" x 17⁵/₈" x 8¹/₄"	$229
01-0302	6227	377 sq."	29"	37¹/₂"	9⁵/₈" x 22³/₄" x 8¹/₄"	$249

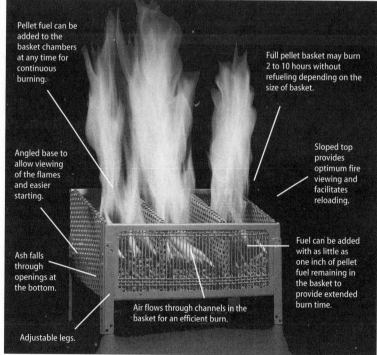

Pellet fuel can be added to the basket chambers at any time for continuous burning.

Full pellet basket may burn 2 to 10 hours without refueling depending on the size of basket.

Angled base to allow viewing of the flames and easier starting.

Sloped top provides optimum fire viewing and facilitates reloading.

Ash falls through openings at the bottom.

Fuel can be added with as little as one inch of pellet fuel remaining in the basket to provide extended burn time.

Air flows through channels in the basket for an efficient burn.

Adjustable legs.

Heartland Cookstoves

In an age when nothing seems to last, Heartland Cookstoves are a charming exception. Though modern refinements have been incorporated seamlessly into the design, the Cookstove still evokes time-honored traditions. Our black and white catalog can't do justice to the bright nickel trim, soft rounded corners, and porcelain enamel inlays in

seven optional colors. We'll be happy to send the factory color brochure.

Two woodburning models are available. The larger Oval features 50,000 Btu/hr output, a 35" x 26" cooking surface, and a large 2.4 cubic foot oven. The smaller

Sweetheart features 35,000 Btu/hr output, a 29" x 21" cooking surface, and a 1.7 cubic foot oven. Both feature a large firebox that takes 16" wood, and can be loaded from the front or top. A summer grate position directs the heat to the oven and cooktop, while limiting heating output. Flue size for both models is common 6", and is supplied locally.

Options include color porcelain, a stainless steel firebox water jacket for water heating, a 5-gallon copper water reservoir with spigot, a coal grate for burning coal rather than wood, and an outside combustion air kit.

Gas and electric models with all the traditional styling and color options are also available. Please call for options or a free color brochure.

61-113	**Oval Cookstove, Std. White**	**$4,615**
61-117	**Sweetheart Cookstove, Std. White**	**$3,790**
OPTIONS (same cost for both models):		
61-106	**Color Option (Almond)**	**$50**
61-107	**Color Option (Green, Blue, Teal, Black)**	**$199**
	(Red)	**$299**
61-119	**Stainless Steel Water Jacket (Oval)**	**$215**
61-122	**Stainless Steel Water Jacket (Sweetheart)**	**$199**
61-120	**Coal Grate**	**$450**
61-125	**Outside Air Kit, Oval**	**$79**
61-133	**Outside Air Kit, Sweetheart**	**$95**

Ecofans

Real Goods product of the decade

When you're heating with wood, a majority of your warmth goes up to the ceiling. The innovative Ecofan solves the problem, propelling much of that heat into your living space—increasing stove efficiency up to 30%. The hotter your stove gets, the faster it runs. Peak performance occurs

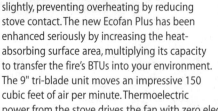

at surface temperatures of 400° to 700°F. This durable stove-top fan is very quiet and a temperature-sensitive, bimetal strip on the unit tilts it slightly, preventing overheating by reducing stove contact. The new Ecofan Plus has been enhanced seriously by increasing the heat-absorbing surface area, multiplying its capacity to transfer the fire's BTUs into your environment. The 9" tri-blade unit moves an impressive 150 cubic feet of air per minute. Thermoelectric power from the stove drives the fan with zero electricity needed. Simple and solid technology maximizes the warmth derived from precious timber resources. Made of anodized aluminum, fans won't rust or corrode. One-year manufacturer's warranty. Canada.

06-0146 Ecofan	**$109**
17-0180 Ecofan Plus	**$150**

Holy Smokes! Firestarter

Ignitable sticks set fires aglow without combustible chemicals or paper. Crafted from a combination of recycled church candles and wood fiber, Holy Smokes! Firestarters get a fire blazing in minutes. One-quarter stick of starter gets logs going. Habitat for Humanity receives a donation with your purchase. One package starts more than thirty fires. Scented. USA.

07-9294 Holy Smokes! Firestarter (set of 2)	**$16**

PEOPLE HEATING

Heated Foot Mat

Handy foot mat stops the cold where it starts, before it starts

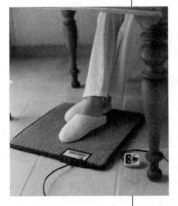

Cold extremities easily lead to head-to-toe chills. The cure for such discomfort often lies in highly localized warmth that stops the source of the sensation for full body relief. Using less energy than a light bulb, the carpeted Heated Foot Mat employs this efficient strategy. Under desks and tables, or next to sofas and chairs, two convenient remote-switched settings of radiant heat deliver gently soothing foot warmth that never burns and can't overload. On full power, unit uses 70 watts; low power uses 35 watts. 18"L x 19"W. USA.

13-0031 Heated Foot Mat **$99**

Car Cozy Heated Blanket

Plugs in 12V lighter socket

Everyone who travels in winter conditions should have one of these for emergencies. Soft, 100% polyester fleece travel blanket has an 8-foot cord, plugs into a 12-volt lighter socket for quick warmth whenever you need it. Draws 2.9 amps, about as much as your parking lights. You can stay warm without running the engine continuously. Available in Navy. China.

55339   Car Cozy Heated Blanket (41" x 57") $39

Organic Cotton Fireside Blanket

Our best-selling fireside blanket is now made of organic cotton

Densely woven from one of nature's purest fibers, the lightly textured Organic Cotton Fireside Blanket and Throw combine cashmere-like softness with easy-care comfort. Beautifully fashioned from 100% organic cotton that's free of dyes, chlorine, pesticides, and other chemical treatments typically used in the cultivation of conventional cotton, this blanket's pure plushness will simply delight your sense of touch. Finished with a whipstitched border. Machine washable. Portugal.

95-0902 Organic Cotton Fireside Blanket
Throw (51"W x 71"L)		$80
Twin (67"W x 90"L)		$115
Full/Queen (90"W x 90"L)		$150
King (108"W x 90"L)		$185

Books

Consumer Guide to Home Energy Savings, 8th ed.

Energy efficiency pays. Written by the nonprofit American Council for an Energy-Efficient Economy, the *Consumer Guide* has helped millions of folks do simple things around their homes that save hundreds of dollars every year. The new 8th edition has up-to-date lists of all the most energy-efficient appliances by brand name and model number, plus an abundance of tips, suggestions, and proven energy-saving advice for homeowners. 264 pages, soft-cover. USA.

21-0344 Consumer Guide to Home Energy
Savings, 8th Ed. **$9.95**

The Fuel Savers: A Kit of Solar Ideas for Your Home

What can you do right now, with your funky existing home and a minimum of cash, to save energy and live more comfortably? This book has a wealth of ideas from insulating curtains to solar hot water heaters. Each idea is examined for materials costs, fuel reduction, advantages, disadvantages, and cost effectiveness. A great step-by-step guide for improving the comfort and efficiency of any building. 83 pages, softcover. Updated in 2002. USA.

21-0357 The Fuel Savers **$15**

COOLING PRODUCTS

Air Cooler

Cool down your space inexpensively with a flow of energy-efficient chilled air. Our Air Cooler uses simple yet effective evaporative cooling technology, providing potent heat relief without ozone-depleting CFCs. Chill air as far away as 7 feet for about a dime a day. Use alone or to boost older A/C systems. Great for apartments, offices, and spaces needing quick cool-down. Includes two-speed fan, air filter, adjustable louvers, and easy-load large-capacity water reservoir with level indicator for 10 hours of continual cooling. Not recommended for humid climates. One-year replaceable wick. 12M"H x 12"L x 21H"W. UL listed. China.

01-0112	**Air Cooler**	**$149**
01-0427	**Replacement Wicks**	**$49**

Convair Prestige 900 Air Cooler

Evaporative air conditioners use the cooling power of water to keep your brain cells alert and happy on sweltering days. Portable so you don't have to use a window or make a hole in your wall. Three-speed fan control, oscillating action, and attractive pedestal stand. 15-liter (4-gallon) reservoir with built-in level indicator. Cools a room 12 to 15 square meters (120–150 square feet). The drier the air, the better it works (not recommended for humid climates). 40G"H (on stand) x 24G"W x 12"D. 120vac. 3,370 BTU/hr.

64511	**Convair Prestige 900 Air Cooler**	**$399**

TurboKool Evaporative Cooler for RVs

Using the proven, popular Recair design, TurboKool's spin-spray action washes out dust and pollen, delivering cool clean air through a standard 14" square opening. Streamlined styling and low-profile contour offers reduced wind resistance. Two-speed, 12-volt operation delivers 450 or 750 cubic feet per minute, drawing 2.1 or 4.6 amps. The non-organic industrial foam filter provides superior filtration and cooling without bacterial growth or unpleasant odors. The discharge grill has adjustable louvers. Weighs 16 lbs.; exterior dimensions are 35" x 22" x 10H". Like all evaporative coolers, TurboKool prefers low humidity. Ineffective when humidity exceeds 40%. USA.

53-0113	**TurboKool Evaporative**	**$699**

Solar-Powered Attic Vents

Venting your attic reduces cooling costs in summer, prevents moisture buildup in winter, and with these solar-powered fans, never costs a cent on your electric bill. Our solar-powered fan vents your attic using electricity from a built-in 10- or 20-watt PV module. The 12-volt, 12" diameter DC fan removes  air at 600 to 800 cfm with the 10-watt model, or at 1,200 to 1,400 cfm with the 20-watt model, completely venting a 1,000 square foot attic every few minutes. Airflow is highest when the sun is hottest, at midday. Shuts off automatically at night or on cloudy days. Select from flat or pitched-roof models. The pitched-roof model has a flat flange base that is designed to mount under shingles; the flat roof model features a crimped flange base that mounts onto a 19.5" x 16.75" x 8" framed curb. Three-year manufacturer's warranty. USA.

64-238	**10W Solar Attic Fan, Flat Roof**	**$399**
64-239	**10W Solar Attic Fan, Pitched Roof**	**$399**
14-0270	**20W Solar Attic Fan, Flat Roof**	**$595**
14-0271	**20W Solar Attic Fan, Pitched Roof**	**$595**

Solar-Powered Attic Fan Kits

Moving hot, stagnant air out of your attic, greenhouse, or barn is the least energy-intensive way to keep your space cool. Using solar power to do the work makes even more sense. The brighter the sun shines, the faster the fan runs. Our Solar Fan Kits give you all the major pieces. Kits include a high-efficiency DC fan, matched PV modules, mounting brackets, a thermoswitch to prevent running in cold weather, and instructions. You provide wire from PV modules to fan as needed.

The 12" Fan Kit has a 13-watt PV module and moves 600 to 800 CFM in full sun. The 16" Fan Kit uses a pair of 13-watt PV modules, and moves 800 to 1,000 CFM. Fan frames provide easy mounting and quiet operation. Modules are 15.1" x 14.4" single crystal with tempered glass and aluminum frame. Fans made in USA, three-year manufacturer's warranty. PV modules made in China, 90-day warranty.

64-479	**12" Attic Fan Kit**	**$249**
64-480	**16" Attic Fan Kit**	**$379**

12" and 16" DC Fans

These are the same fans used in our complete Attic Fan Kits. Both models use heavy-duty high-voltage DC motors that can be run with any 12- to 48-volt system. Frames have three mounting points. A nonadjustable thermostat switch that closes at 110°F and opens at 90°F (approximate temperature) is included. Input power and output in cubic feet per minute are given in the chart. Specs at 14 and 28 volts simulate PV-direct applications. Specs at 12 and 24 volts are battery-driven. Note that these specs are giving the fan all the power it wants. These fans will operate just fine on a 10-watt or larger PV module. Want your fan to spin faster? Feed it more watts! Three-year manufacturer's warranty. USA.

12" DC FAN

Voltage Input	Amps Draw	CFM Output
12 volts	1.17 amps	740
14 volts	1.47 amps	1,050
24 volts	3.20 amps	1,335
28 volts	4.20 amps	1,430

16" DC FAN

Voltage Input	Amps Draw	CFM Output
12 volts	1.47 amps	810
14 volts	1.84 amps	1,100
24 volts	4.20 amps	2,000
28 volts	5.30 amps	2,275

64-216	12" DC Fan	$149
64-217	16" DC Fan	$159

Oscillating 8" DC Table Fan

This 12-volt table fan features a weighted base with rubber feet, and 2-speed operation. Draws about 1.2 amps on low, and 2.0 amps on high. Oscillating feature can be switched on or off, just like a real fan! Taiwan.

64-241	12V, 8" Table Fan	
	$49	

Ultimate 12V Fan

Moves a lot of air for the size

This powerful fan plugs right into a cigarette lighter. 7" fan blade moves a lot of air, yet is soft and finger safe. Includes a suction cup and screw mount bases. Pivot the head up and down or swivel it sideways to aim the air in just the right direction. Perfect as a dash-

board fan or in a solar cabin or house. Easy on your batteries, draws just .59 amps. One-year warranty.

64514	12V Fan	$36
	2 or more for	$32 each

Compact Desk Fan

Powerful, quiet, battery-operated fan keeps air fresh and cool

This powerful, variable-speed fan will keep you cool and comfortable. The computerized speed control allows for a complete range of fan speeds—a better option over typical three-speed fans. The last setting is conveniently remembered when the fan is turned on and off again, and the tilt of the head is adjustable. Offering an impressive air flow, it runs over 300 hours on just one set of four "D" batteries (not included). Or, use the included 6-volt AC/DC adapter and plug in anywhere. Measures 5.75" W x 3.625" L x 8.75" H. One-year manufacturer's warranty. Canada.

64-255	Compact Desk Fan	$25

Thermofor Ventilator Opener

The Thermofor is a compact, heat-powered device that regulates window, skylight, or greenhouse ventilation according to temperature, using no electricity and requiring no wiring. You can set the temperature at which the window starts to open between 55° and 85°F. It will open any hinged window up to 15 pounds a full 15 inches, and can be fitted in multiples on long or heavy vents. These units are ideal for greenhouses, animal houses, solar collectors, cold frames, and skylights. United Kingdom.

64-302	Thermofor Solar Vent Opener	$59

Our Most-Efficient DC Ceiling Fan

For keeping comfortable with a 12- or 24-volt power system, you can't beat the efficiency and quiet operation of our 42" diameter DC ceiling fan. It comes with both a close flush-mount for flat ceilings, or a down rod ball-mount for sloped ceilings. The fan can be assembled with either four or five blades, showing either walnut or oak finish (five blades included). The fan body is a charcoal-gray ABS injection casting that can be painted to match or left stock. Reversible and speed-adjustable using the solid-state controls sold below. Close mount hangs down 9.5"; down rod needs 13".

Performance specs; at 12 volts: 60 rpm, 4.8 watts, 1,800 cfm; at 24 volts: 120 rpm, 19.2 watts, 4,000 cfm. Note that performance is …um…modest at 12 volts. 12-volters may be interested in the 12 to 24 voltage doubler speed control sold on the next page. Fan weight 11 lb. USA.

64-495	RCH DC Ceiling Fan	$199

Vari-Cyclone DC Ceiling Fan

The Vari-Cyclone utilizes revolutionary "Gossamer Wind" blade design technology that delivers 40% more air flow without any increase in power use. Intended for 12- or 24-volt off-grid DC power systems, this 60" diam-

eter fan can be assembled with either three or four blades. Finish of blades and body is white. Can be mounted as either a close flush mount for flat ceilings, or with a down rod ball mount for sloped ceilings. Hardware is included for both. Close mount hangs down 9.5", down rod takes 13". Fan becomes variable speed and reversing with addition of 12- or 24-volt solid state control listed below.

Performance specs (with 3-blade configuration): At 12 volt: 60 rpm, 7 watts, 2,500 cfm. At 24 volt: 120 rpm, 27 watts, 3,400 cfm. We recommend the voltage doubler adjustable fan speed control below for 12-volt use. Fan weight approximately 14 lbs. USA.

53-0100 Vari-Cyclone DC Fan $279

24V Variable Fan Speed Control with Reversing Switch

This solid-state DC speed control boasts 97% efficiency, smooth linear speed control from 0% to 100%, a reversing switch with a convenient center-off position, radio

frequency filtering, a silencing capacitor for quiet motor operation at low speeds, and it can even be connected to a standard millivolt thermostat for automatic operation. Whew! That seems to cover everything. Mounted on an ivory color single-gang electric box cover. Includes mounting screws and detailed, clearly illustrated installation instructions. 24V DC only. USA.

64-497 DC Fan Control w/Reverse 24V/1.5A $59

12V Variable Fan Speed Control With Reversing Switch

Identical looks, performance, and features with our 24V unit above, but made for 12-volt use. 2.0 amps, 12V DC only. USA.

64-504 DC Fan Control w/Reverse 12V/2.0A $59

Solar Hat Fan

A practical solar gadget that really cools you down on a hot day! It clamps on to your favorite hat (stiffer-type brims and baseball caps are recommended) and gets to work harvesting enough solar energy to directly power a working fan. Features a small photovoltaic panel with an adjustable mount. We sold these solar hat fans to a roofing company, whose workers can't be without them as they cool down from the 100°+ temperature. Measures 2.75" L x 2⅛"W x 2¼"H. China.

01-0343 Solar Fan $10
         2 or more $8 each

Voltage Doubler

Delivers 24v from a 12v source for DC ceiling fans. 2.0 amps output at 99% efficiency. Installed on an ivory face plate with on/off switch. Fits in a standard single-gang electric box. Accepts 8 to 16VDC input. Output voltage will be double the input. USA.

64-505 Voltage Doubler $59

Sun Tex Shadecloth and Super Solar Screen

The original Sun Tex Shadecloth blocks up to 80% of the sun's damaging glare, while still admitting plenty of natural light and air movement. The Super Solar Screen raises protection to 90%, with a finer mesh that provides for greater protection. A great way to create privacy, the cloths appear virtually opaque from outside, while allowing you to see out clearly. Made from heavy-duty coated polyester, they're mildew and fade resistant. Easy to install, easy to remove, they block the sun's glare through your windows, skylights, greenhouses, patios, and doors. These work best outside your windows. Stop the heat before it gets in your house! USA.

63-569	**Sun Tex 80% Shadecloth 36" x 84"**	**$18**
63-612	**Sun Tex 80% Shadecloth 48" x 84"**	**$24**
63-613	**Super Solar 90% Screen 36" x 84"**	**$22**
63-614	**Super Solar 90% Screen 48" x 84"**	**$28**

Energy-Efficient Appliances

Why Bother to Shop for Energy Efficiency?

Wasted energy translates into carbon dioxide production, air pollution, acid rain, and lots of money down the drain. The average American household spends more than $1,100 per year on appliances and heating and cooling equipment. You can easily shave off 50 to75% of this expense by making intelligent appliance choices.

For example, simply replacing a 20-year-old refrigerator with a new energy-efficient model will save you about $85 per year in reduced electric bills while saving 1,000 kWh of electricity, and reducing your home's CO_2 contribution by about a ton per year. Highly efficient appliances may be slightly more expensive to buy than comparable models with lower or average efficiencies. However, the extra first cost for a more efficient appliance is paid back through reduced energy bills long before the product wears out.

What Appliances Can I Buy from Real Goods?

Most appliance manufacturers prefer to sell through established appliance stores. Your mail-order choices are limited to more specialized or high-efficiency appliances. At present we offer several small washers, the full-size, highly efficient Staber horizontal-axis washing machine, the Spin-X centrifugal dryer, several varieties of incredibly efficient clotheslines, gas and DC refrigerators and freezers specifically for off-the-grid folks, high-efficiency tankless water heaters in both gas and electric, and the most problem-free solar water heaters available.

We want to enable our customers to make intelligent choices on their appliance purchases. Thanks to the nonprofit American Council for an Energy-Efficient Economy, we can offer the up-to-date eighth edition of the *Consumer Guide to Home Energy Savings*. This is a more-or-less annual listing of the most efficient refrigerators, clothes washers, dishwashers, tank-type water heaters, air conditioners, and home heating appliances available. Fortify yourself with this wonderful book before venturing into any major appliance purchase.

Refrigerators and Freezers

The refrigerator is likely to be the largest single power user in your home aside from air conditioning and water heating. Refrigerator efficiency has made enormous strides in the past decade, largely due to insistent prodding from the Feds with tightening energy standards; the most recent, upgraded standards went into effect in July 2001. An average new fridge with top-mounted freezer sold today uses under 500 kilowatt-hours per year, while the average model sold in 1973 used more than 1,800 kilowatt-hours per year. These are national average figures. The most efficient models available today are under 400 kilowatt-hours per year, and still dropping. The typical refrigerator has a life span of 15 to 20 years. The cost of running the thing over that time period will easily be two to three times the initial purchase price, so paying somewhat more initially for higher efficiency offers a solid payback. It may not be worth scrapping your 15-year-old clunker to buy a new energy-efficient model. But when it does quit, or it's time to upgrade, buy the most efficient model

> The refrigerator is likely to be the largest single power user in your home aside from air conditioning and water heating.

> Simply replacing a 20-year-old refrigerator with a new energy-efficient model will save you about $85 per year in reduced electric bills while saving 1,000 kWh of electricity, and reducing your home's CO_2 contribution by about a ton per year.

Refrigerators and CFCs

All refrigerators use some kind of heat transfer medium, or refrigerant. For conventional electric refrigerators, this has always been freon, a CFC chemical and prime bad-guy in the depletion of the ozone layer. Manufacturers all have switched to non-CFC refrigerants, and by now it should be nearly impossible to find a new fridge using CFC-based refrigerants in an appliance showroom.

What most folks probably didn't realize is that of the 2.5 pounds of CFCs in the typical 1980s fridge, 2 pounds are in the foam insulation. Until very recently, most manufacturers used CFCs as blowing agents for the foamed-in-place insulation. Most have now switched to non-CFC agents, but if you're shopping for new appliances, it's worth asking.

The CFCs in older refrigerators must be recovered before disposal. For the 0.5 pound in the cooling system, this is easily accomplished, and your local recycler, dump, or appliance repair shop should be able either to do it or recommend a service that will. To recover the 2.0 pounds in the foam insulation requires shredding and special equipment, which sadly only happens rarely.

Since 1999, all new fridges are completely CFC-free.

available. A great source for listings of highly efficient appliances is the Environmental Protection Agency's Energy Star® website, www.epa.gov/energystar/.

Be sure to have the freon removed professionally from your clunker before disposal. Most communities now have a shop or portable rig to supply this necessary service at reasonable cost.

Also note that snazzy features, such as side-by-side refrigerator-freezer units and through-the-door ice or cold water, increase energy consumption. If you can do without, it's cost effective to avoid these features.

REFRIGERATION AND RENEWABLE ENERGY SYSTEMS

When your energy comes from a renewable energy system, which has a high initial cost, the payback for most ultraefficient appliances is immediate and handsome. You will save the price of your appliances immediately in the lower operating costs of your power-generating system. This is particularly true for refrigerators.

Many owners of smaller, or intermittent-use renewable energy systems prefer gas-powered fridges and freezers, especially when the fuel is already on site anyway for water heating and cooking. Gas fridges have the clean 'n green advantage of using a mixture of ammonia and water, not freon, as a refrigerant. We offer several

gas-powered freezers and refrigerator/freezer combinations.

For many years, Sun Frost has held the enviable status as manufacturer of the "world's most efficient refrigerators," and that has produced a steady flow of renewable energy customers willing to pay the high initial cost in exchange for the savings in operating cost. Real Goods has been an unabashed supporter of this product for many years. But times are changing. The major appliance manufacturers have managed to implement huge energy savings in their mass-produced designs over the past several years. While few of the mass-produced units challenge Sun Frost for king-of-the-efficiency-hill, many are close enough, and less expensive enough, to deserve consideration. The difference between a $900 purchase and a $2,700 purchase will buy quite a lot of photovoltaic wattage. A number of mass-produced fridges have power consumption in the 1.0 to 1.5 kilowatt hours per day range, and can be supported on renewable energy systems. Any reasonably sized fridge with power use under 1.5 kilowatt hours per day is welcome in a renewable system; that's $45 to $50 per year on the yellow Energy Guide tag that every new mass-produced appliance wears in the showroom.

In Sun Frost's defense, we need to point out that conventional refrigerators run on 120 VAC, which must be supplied with an inverter, and will cost an extra 10% due to inverter efficiency. Also, if your inverter should fail, you could lose refrigeration until it gets repaired. Sun Frosts (if ordered as DC-powered units) will run on direct battery power, making them immune to inverter failures.

IMPROVING THE PERFORMANCE OF YOUR EXISTING FRIDGE

Here is a checklist of tips that will help any refrigerator do its job more easily and more efficiently.

- Cover liquids and wrap foods stored in the fridge. Uncovered foods release moisture (and get dried out), which makes the compressor work harder.
- Clean the door gasket and sealing surface on the fridge. Replace the gasket if damaged. You can check to see if you are getting a good seal by closing the refrigerator door on a dollar bill. If you can pull it out without resistance, replace the gasket. On new refrigerators with magnetic seals, put a flashlight inside the fridge some evening, turn off the room lights, and check for light leaking through the seal.

- Unplug the extra refrigerator or freezer in the garage. The electricity the fridge is using—typically $130 a year or more—costs you far more than the six-pack or two you've got stashed there. Take the door off, or disable the latch so kids can't possibly get stuck inside!
- Move your fridge out from the wall and vacuum its condenser coils at least once a year. Some models have the coils under the fridge. With clean coils, the waste heat is carried off faster, and the fridge runs shorter cycles. Leave a couple of inches of space between the coils and the wall for air circulation.
- Check to see if you have a power-saving switch or a summer-winter switch. Many refrigerators have a small heater (yes, a heater!) inside the walls to prevent condensation build-up on the fridge walls. If yours does, switch it to the power-saving (winter) mode.
- Defrost your fridge if significant frost has built up.
- Turn off your automatic ice maker. It's more energy efficient to make ice in ice trays.
- If you can, move the refrigerator away from any stove, dishwasher, or direct sunlight.
- Set your refrigerator's temperature between 38° and 42°F, and your freezer between 10° and 15°F. Use a real thermometer for this, as the temperature dial on the fridge doesn't tell real temperature.
- Keep cold air in. Open the refrigerator door as infrequently and briefly as possible. Know what you're looking for. Label frozen leftovers.

The EPA and Refrigerator Power Use Ratings

We're all familiar with the yellow energy-use tags on appliances now. These level the playing field, and make comparisons between competing models and brands simple. With some appliances, like refrigerators, the EPA takes a "worst case" approach. Fridges are all tested and rated in 90°F ambient temperatures. The test fridge is installed in a "hot box" that maintains a very stable 90° interior temperature. This results in higher energy use ratings than will be experienced in most settings. At the mid-70° temperatures that most of us prefer inside our houses, most fridges will actually use about 20 to 30% less energy than the EPA tag indicates.

- Keep the fridge full. An empty refrigerator cycles frequently without any mass to hold the cold. Beer makes excellent mass, and you probably always wanted a good excuse to put more of it in the fridge, but it tends to disappear. In all honesty, plain water in old milk jugs works just as well.

BUYING A NEW REFRIGERATOR

A new, more efficient refrigerator can save you $70 to $80 a year on electricity, and will completely pay for itself in nine years. It also frees you from the guilt of harboring a known ozone killer: Since 1999, all new fridges in the showroom are completely CFC-free.

Shop wisely by carefully reading the yellow EPA Energy Guide label found on all new appliances. Use it to compare models of similar size.

The American Council for an Energy-Efficient Economy is a nonprofit organization dedicated to advancing energy efficiency as a means of promoting economic development and environmental protection. Every year ACEEE publishes an extremely handy book called *Consumer Guide to Home Energy Savings*. This invaluable guide lists all the most efficient mass-produced appliances by manufacturer, model number, and energy use. We strongly recommend consulting this guide before venturing into any appliance showroom. It has an amazingly calming effect on overzealous sales personnel, and allows you to compare energy use against the top-of-the-class models. See the product section.

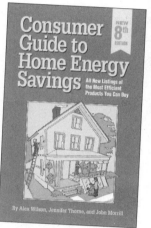

We strongly recommend consulting the *Consumer Guide to Home Energy Savings* before venturing into any appliance showroom. It has an amazingly calming effect on overzealous sales personnel.

Clothes Washers and Dryers

Laundry is another domestic activity that consumes considerable power. When it comes to appliances for washing and drying clothes, however, the energy-efficiency focus is decidedly lopsided. Much progress has been made recently on resource-efficient washing machines. Clothes dryers, on the other hand, are not required to display Energy Guide labels, so it is difficult to compare their energy usage and, apparently, manufacturers have little incentive to improve efficiency. The good news about drying, of course, is that it's easy to dry clothes using solar power—by hanging them up on a line or a rack, indoors or outdoors—thereby eliminating or greatly reducing the need to use a power-hungry machine.

Most of the energy use in washing clothes is for heating water. The new resource-efficient washing machines use less water than conventional models, which cuts their energy consumption by up to two-thirds. Some models are horizontal-axis front loaders, in which the clothes tumble through the water. Others are redesigned vertical-axis top loaders that spray the clothes or bounce them through the water rather than submerging them. Both types also use higher spin speeds to extract more water from the clean clothes, thereby reducing the energy required for drying (solar or otherwise).

The only energy-saving features available on dryers are sensors that shut off the machine when the clothes are dry. Most better-quality models on the market today have this feature. In addition to using a resource-efficient washer that spins more water out of your clean clothes, and taking advantage of free solar energy as much as possible, here are a few more tips for using a clothes dryer efficiently (courtesy of the American Council for an Energy-Efficient Economy).

- Separate clothes and dry similar types together.
- Dry two or more loads in a row, to take advantage of residual heat already in the machine.
- Clean the filter after each use. A clogged filter will reduce air flow.
- Dry full loads when possible. Drying small loads wastes energy because the dryer will heat up to the same extent regardless of the size of the load.
- Make sure the outdoor exhaust vent is clean and that the flapper opens and closes properly. If that flapper stays open, cold air will infiltrate into the house through the dryer, increasing your wintertime heating costs.

See the product section that follows for more information about the efficient washers and dryers Gaiam Real Goods sells.

ENERGY-EFFICIENT HOME APPLIANCES AND PRODUCTS

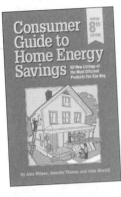

Consumer Guide to Home Energy Savings, 8th Ed.

Energy efficiency pays. Written by the nonprofit American Council for an Energy-Efficient Economy, the *Consumer Guide* has helped millions of folks do simple things around their homes that save hundreds of dollars every year. The new eighth edition has up-to-date lists of all the most energy-efficient appliances by brand name and model number, plus an abundance of tips, suggestions, and proven energy-saving advice for homeowners. 264 pages, softcover. USA.

**21-0344 Consumer Guide to Home
Energy, 8th Ed.** **$9.95**

ELECTRICALLY POWERED REFRIGERATION

Sun Frost Refrigerators & Freezers

For over fifteen years Sun Frost has been the first choice for off-the-grid homesteads. For folks who can run their fridge directly on 12 or 24 volt DC, they're still the most energy-efficient choice, thanks to a variety of design innovations that result in the best energy efficiency and quietest operation available. The smaller Sun Frosts (RF-12 and RF-16) are so efficient that they can be powered in most climates with as little as 150 watts of PV modules.

Sun Frost compressors are top-mounted for thermal efficiency. Without the usual plumbing and mechanical equipment on the bottom, a 13" stand is recommended to raise the bottom shelf level. Sun Frost's optional stand has two drawers. Standard models are off-white Navimar (a Formica-like laminate). Colors are available for $100, wood veneers are available for $150, unfinished.

In the listings below R stands for refrigerator, F for freezer, followed by the cubic footage. All models are 34.5" wide by 27.5" deep. Heights vary. Allow an additional 4" width if door will open against a wall. Doors will clear standard 24"-deep countertops. All models available in 12 or 24 volts DC or 120 VAC. Voltage is not field switchable; specify when ordering. Can be hinged on the right or left side (when facing the unit). Hinge side is not field switchable; specify when ordering. Sun Frosts are made to order, usually requires 4–6 weeks. Ships freight collect from Northern California. $60 crating charge included in price. USA.

62-135	Sun Frost RF-19 (DC)	$2,860
62-135-AC	Sun Frost RF-19, 120VAC	$2,705
62-125	Sun Frost R-19 (DC)	$2,599
62-125-AC	Sun Frost R-19, 120VAC	$2,520
62-115	Sun Frost F-19 (DC)	$2,915
62-115-AC	Sun Frost F-19, 120VAC	$2,759
62-134	Sun Frost RF-16 (DC)	$2,715
62-134-AC	Sun Frost RF-16, 120VAC	$2,560
62-133	Sun Frost RF-12	$2,029
62-133-AC	Sun Frost RF-12 120VAC	$1,949
62-122	Sun Frost R-10 (DC)	$1,640
62-122-AC	Sun Frost R-10 120VAC	$1,560
62-112	Sun Frost F-10 (DC)	$1,745
62-112-AC	Sun Frost F-10 120VAC	$1,665
62-131	Sun Frost RF-4 (DC)	$1,430
62-121	Sun Frost R-4 (DC)	$1,430
62-111	Sun Frost F-4 (DC)	$1,430
62-116	Sun Frost Stand, White	$259
62-117	Sun Frost Stand, Color	$299
62-118	Sun Frost Stand, Woodgrain	$325
62-120	Sun Frost Color Option	$100
62-119	Sun Frost Woodgrain Option	$150

Shipped freight collect from Northern California.

Chest-Style Off-Grid Sundanzer Refrigerators and Freezers

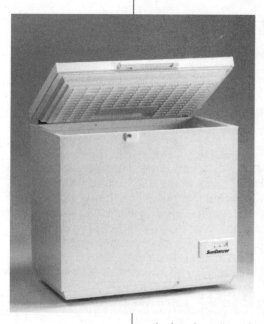

Sundanzer fridges and freezers provide the lowest energy use and save the most money we've ever seen for refrigeration—in fact, the Sundanzer often pays for itself in saved system expenses. Sorry AC users, these units are only for 12- or 24-volt DC use (the unique compressor accepts either voltage without modification). Sundanzer is available in 5.8- or 8.1-cubic-foot units, configured either as a fridge or a freezer. Features include thick 4.33" polyurethane insulation, a lockable lid with interior light, baskets for food organization, a patented low-frost system, automatic adjustable thermostat, ozone-safe R-134a refrigerant, powder-coated aluminum interior and galvanized steel exterior, and an easy-to-clean bottom drain. And did we mention the ultra-low power use that saves the remote home owner big dollars on support hardware? Sundanzer units are manufactured by one of the world's leading appliance manufacturers using stringent standards for quality and efficiency. We've heard nothing but rave reviews from folks who have purchased these coolers. See power use specs and sizing info below. Shipped by truck from Nevada. One-year manufacturer's warranty. USA.

Sundanzer	Fridge 5.8 cf	Fridge 8.1 cf	Freezer 5.8 cf	Freezer 8.1 cf
Watts/day @ 90°F	168	198	441	532
External width (All 26.2"D x 34.5"H)	36.8"	46.9"	36.8"	46.9"
Product weight	120 lbs.	140 lbs.	120 lbs.	140 lbs.
Item #	62581	62582	62583	62584
Price	$899	$999	$899	$999

Engel Portable Fridge/Freezers

The most reliable and efficient portable refrigerator/freezers available

Engel portable 12-volt coolers are extremely reliable, and are widely used in emergency vehicles. These models represent the latest, most efficient compressor technology out of Japan. They are robust, and are built to deliver many years of reliable operation. They are thermostat controlled, and can be either a freezer or a refrigerator depending on your needs. The power use is lower than any other portable cooler line thanks to high-efficiency patented "swing compressor" technology. This

is a great compressor design for portable coolers because it continues to function perfectly even when jostled and tilted, and has only one moving part. The larger models are ideal for sailboats or other power-challenged locations. All models have a 9' detachable power cord, and an attractive two-tone gray color scheme. 12-volt only. CFC-free. Japan.

Engel 15 has a 14 qt. capacity, weighs 23.5 lb. and measures 17" x 11" x 14.5" outside dimensions. One-year manufacturer's warranty.

62-575　Engel 15 Portable Cooler　$449
Engel 35 has a 34 qt. capacity, weighs 46.3 lb. and measures 25.5" x 14.3" x 16" outside dimensions. Two-year manufacturer's warranty.

62-576　Engel 35 Portable Cooler　$695
Engel 45 has a 43 qt. capacity, weighs 52.9 lb. and measures 25.5" x 14.3" x 20" outside dimensions. Two-year manufacturer's warranty.

62-577　Engel 45 Portable Cooler　$750
Engel 65 has a 64 qt. capacity, weighs 68.3 lb. and measures 31.1" x 19.3" x 17.4" outside dimensions. Two-year manufacturer's warranty.

62-578　Engel 65 Portable Cooler　$1,100

GAS-POWERED REFRIGERATION

Servel RGE 400 Propane Refrigerators

The state-of-the-art 7.7-cubic-foot RGE 400 model was introduced in the late 1990s. Average energy use per day is 0.27 gallon (1.1 lb.) propane, or 3.9 kWh electric. The body is white, the doors are hinged on the right as you face the unit (and are reversible on site), with four adjustable stainless shelves in the fridge and two in the freezer. Also includes two vegetable bins, egg and dairy racks, frozen juice rack, ice-cube trays, ice bucket, and battery-powered interior light. Base-mounted control panel with gas/electric selector, thermostat, piezo igniter, and flame safety valve. Operates on either propane or 120vac. 6.0 cu.ft. fridge, 1.7 cu.ft. freezer. 63.5"H x 23"W x 26.5"D, weighs 167 lbs. One-year manufacturer's warranty. Sweden/USA.

62-303   Servel RGE 400 Propane Fridge $1,295
Shipped freight collect from Ohio.

CRYSTAL COLD PROPANE REFRIGERATORS

Crystal Cold units are the largest and most fuel-efficient gas refrigerators available. Made in the American Midwest, these high-quality units were developed for the Amish with large families and high quality expectations. Styling is 1950s retro with heavy-duty casters for easy moving. Interiors are powder-coated white with front-mounted push-button igniter. Gas fridges need 10" to 12" of clearance above them to properly vent waste heat. The 15- and 18-cubic-foot models both use the same absorption cooling unit. In really hot, humid climates, the smaller 15-cubic-foot model will perform better. Natural gas models are available by special order; add $50 to cost. All Crystal Cold units ship by truck from Illinois.

18-Cubic-Foot Refrigerator/Freezer

This is the largest gas fridge on the market. Features a pair of vegetable crispers, reversible doors, and great retro styling. Burns approximately 1.5 lbs. (.35 gal.) of propane per day at 75°F ambient temperature. Refrigerator section is 13 cu.ft.; freezer is 4 cu.ft. Exterior size is 65"H x 28"W x 37"D. Choose propane or natural gas. Shipping weight is 300 lbs. USA.

62579   18 Cubic Foot Refrigerator/Freezer $2,499

15-Cubic-Foot Refrigerator/Freezer

The best unit for hot climates, this model has a single, deep vegetable crisper across the bottom. Doors are reversible on site. Refrigerator section is 10 cu.ft., the freezer is 3.3 cu.ft. Burns approximately 1.25 lbs. (.29 gal.) of propane per day at 75°F ambient temperature. Exterior size is 60.5"H x 28"W x 29.5"D. Choose propane or natural gas. Shipping weight is 275 lbs. USA.

62586   15 Cubic Foot Refrigerator/Freezer $2,430

15-Cubic-Foot Freezer

The largest, most dependable gas freezer on the market. Has three quick-freeze shelves and four door shelves. Built-in thermometer, magnetic door seals, and white exterior. Single door. 38"W x 26"D x 54"H. Propane only. One-year manufacturer's warranty. USA.

62574   15 Cubic Foot Freezer $2,680

Frostek Propane Freezer

A freezer that works without electricity or CFCs! The Canadian-produced Frostek is the largest chest-type propane-powered freezer on the market. Its 8.5-cubic-foot capacity offers reliable long-term food storage, and can accommodate an entire side of beef, or quarts upon quarts of frozen fruits and vegetables.

In carefully monitored laboratory tests, the powerful cooling unit maintained –8°F interior temperatures or less all the way up to an outside ambient temperature of 80°F. Even at 90°F, it still maintains 0°F inside. Fuel use is approximately 1.8 to 2.5 gallons per week, depending on ambient temperature (or 7.5 to 11 lb. if you buy propane by the pound). An intelligent exterior thermometer shows interior conditions

without opening the lid. Has easy-access bottom drain, piezo ignitor, thermostat on/off control, and runs silently.

The Frostek stands 38"H x 44.5"L x 31"D. Inside dimensions are 24" x 37" x 15.5". There are no awkward cooling fins jutting into the food storage area. Cabinet is rustproof steel with baked enamel finish. Color is gray and white. Approved by both the American Gas Association and Canadian Gas Association. Weight is 220 lb. Shipping weight is 275 lb. Call for availability; production quantities are limited. Canada.

62-308   Frostek Propane Freezer $1,999
Shipped freight collect from Ohio.

Staber Horizontal-Axis Washing Machine

The most advanced, efficient, and convenient washer available

The Staber System 2000 clothes washer combines top-loading convenience with the superior cleaning and low power use of horizontal-axis design. With the money you'll save on water, electricity, and laundry products, the Staber washer can actually pay for itself in about three years. The Staber is designed and built by a company that first spent twenty years rebuilding commercial washers. This washer is simple, durable, and if it ever needs it, extremely easy to fix.

The Staber is a unique horizontal-axis machine that loads from the top. This is more convenient for the user, and allows stronger support on both sides of the six-sided all stainless-steel drum. Horizontal-axis machines have proven to deliver superior cleaning with less power and water use due to the tumbling action. Compared to conventional vertical-axis washers, the Staber uses 66% less water, 50% less electricity, and 75% less detergent. The well-supported drum design allows much higher spin speeds, which shortens drying time too.

Other features that set the Staber apart include power use so low that a 1500-watt inverter will run it happily; no complex, power-robbing transmission; a self-cleaning, maintenance-free pump; rust and dent-proof ABS resin tops and lids with a 25-year warranty; easy front access to all parts, and a complete, fully illus-

trated service/owners manual for easy owner maintenance. Average power use per wash cycle is only 150 watt-hours. Average electric draw is 300 watts, peak is 550 watts. Water use: 6 gal. wash, 6 gal. first rinse, 6 gal. second rinse (6 gal. pre-wash).

Standard features include a large 18-pound capacity; variable water temperature; high/low water level selection; multiple wash cycles (normal, perm-press, delicate, pre-wash); an automatic dispensing system for detergent, pre-wash detergent, bleach, and softener; and a safety lid lock when running. EnergyGuide tagged at $7/year for natural gas water heater, or $26/year for electric water heater (hot water costs are included in washer comparisons). Measures 27"W x 26"D x 42"H. Manufacturer's warranty: 25 years on outer tub module, 10 years on tub and lid, 5 years on bearings & suspension, 1 year on all else. USA.

63-645   Staber HXW-2304 Washer $999
Shipped freight collect from Ohio.

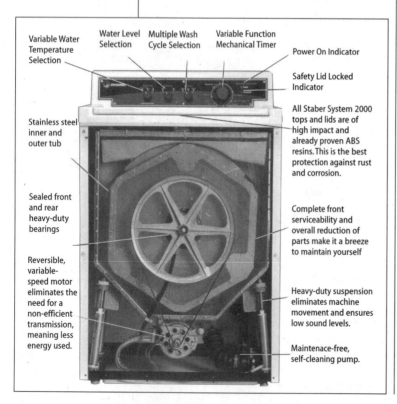

Variable Water Temperature Selection

Water Level Selection

Multiple Wash Cycle Selection

Variable Function Mechanical Timer

Power On Indicator

Safety Lid Locked Indicator

All Staber System 2000 tops and lids are of high impact and already proven ABS resins. This is the best protection against rust and corrosion.

Stainless steel inner and outer tub

Sealed front and rear heavy-duty bearings

Reversible, variable-speed motor eliminates the need for a non-efficient transmission, meaning less energy used.

Complete front serviceability and overall reduction of parts make it a breeze to maintain yourself

Heavy-duty suspension eliminates machine movement and ensures low sound levels.

Maintenace-free, self-cleaning pump.

Spin X Spin Dryer

Saves time, clothes, and money!

Clothes removed from your washer still contain considerable water. A tumble dryer will take up to an hour to remove that water, while the tumbling wears out your clothes, and the heat costs you money. Line drying takes even longer. In three minutes at 3,300 rpm, Spin X

removes 60% of the water with centrifugal force. Clothes come out barely damp, greatly reducing or even eliminating expensive dryer run time. A three-minute cycle in the Spin X costs fifteen seconds worth of electric dryer time.

The compact stainless steel Spin X dryer is only 13.5" diameter by 25.5" tall with chrome-plated top rim and extra-durable plastic lid. Its stainless steel drum has a 10-lb. load capacity. The one-hand safety catch/control switch is simple and obvious. Start up is smooth and gradual. An electronic timer limits run time to three minutes, and the drum is braked to a stop in less than three seconds when turned off. Plugs in to any 120VAC outlet, UL approved. Warranted one year by the U.S. importer. Will ship by UPS. Germany.

63-812   Spin X 10 lb. Stainless $395

Super Wash Pressure Hand Washing Machine

Launders a 5-pound load of medium- to lightweight wearables quickly, effectively, and without a single watt of electricity. Portable, lightweight, and easy to use—load up the barrel, add water and soap, screw on the pressure lid, and turn the handle. To rinse, drain, refill, and agitate for another 30 seconds. The only moving parts are two pivots; nothing to break down. Sturdy and durable. Assembled, the unit measures 19"H x 17.5"W x 15"D; limited manufacturer's warranty. USA.

63860   Super Wash Pressure Hand Washing Machine $49

Filtrol Septic System Protector

Polyester, nylon, and other synthetic fibers leave your washing machine as lint and don't break down as they flush through your septic system. These fibers clog up the soil of your leach field, causing your septic system to die prematurely and expensively. The simple Filtrol 200 is your solution. Installed on your washer drain and equipped with a reusable 200-micron filter bag, the Filtrol effectively removes most clothing lint before it gets to your drain. Simply dump the accumulated lint out of the filter bag periodically. Filter bags should be replaced every one to three years. Package includes clear filter housing, two 200-micron filter bags, sturdy wall mounting bracket, 6-foot washer drain hose extension, and hardware. Existing washer drain hose will fit 1" barbed fitting at bottom of Filtrol housing. Filter housing 7" diameter, 12" high. USA.

10-0043 Filtrol 200 Septic Protector Kit $179
10-0044 Filtrol 200 Repl. Bag $29

James Washer

A hand washer? That sounds like a lot of work. Not with the James Washer. Simply let the clothes soak in hot, soapy water, then just a few minutes of swishing the pendulum agitator back and forth does a normal load. Made of high-grade stainless steel with a galvanized lid and pine legs. Agitator swings on nylon bearings. Easy-draining bottom faucet, 17-gallon capacity. USA.

63-411   James Washer $449

Hand Wringer

The hand wringer will remove 90% of the water, while automatic washers remove only 45%. It has a rustproof, all-steel frame and a very strong handle. Hard maple bearings never need oil. Pressure is balanced over the entire length of the roller by a single adjustable screw. We've sold these wringers without a problem for over 17 years. Made by a Mennonite family in the USA.

63-412   Hand Wringer $169

Fifty Feet of Drying Space

This is the most versatile clothes dryer we've seen yet. Don Reese and his family are still making them by hand

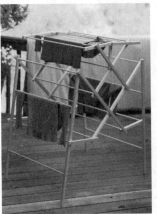

from 100% sustainably harvested New England pine and birch. Don's sturdy dryer can be positioned to create either a peaked top or a flat top ideal for sweaters. And you can fold it up to carry back in the house (or to the next sunny spot) still loaded with laundry.

Large size has 50 feet of space, measures 58"H x 30"W x 40"D in peaked position; 46"H x 30"W x 40"D in flat-top position, and holds an average washer load. Folds to 58"H x 30"W x 4.5"D. USA.

Small size has 30 feet of space, measures 41"H x 27"W x 40"D, and folds to 41"H x 27"W x 4"D. USA.

10-8002 Large New England Clothes Dryer $59
10-8003 Small New England Clothes Dryer $40
Requires UPS ground delivery.

Save Space and Conserve Energy with Fold-Out Clothes Dryer

This wooden drying rack is easy to install on any wall surface. Position unit on porch for solar drying, near back door for easy access, or in laundry room for indoor drying. It's a perfect drying rack for flowers, herbs, and

pasta, too! Space-saving dryer protrudes only 6" from wall when folded. Measures 12" x 27". Extends to 22" when in use. Constructed of sustainably

harvested sturdy Eastern white pine with durable birch hardwood dowels and four Shaker-style pegs. Comes fully assembled. Includes mounting hardware. USA.

10-0023 Wall Shelf Drying Rack $68
Requires UPS ground delivery.

Umbrella Clothes Dryer

Having beautifully fragrant clothes with the scent of the fresh outdoors is easier than ever! Our easy-to-assemble folding clothes dryer is a classic energy-efficient method for drying. It rotates easily, and has 164-feet of PVC-coated line with 4" spacing to allow easy hanging. The line holes in the arms are protected to prevent line

chafing. It can accommodate four to five loads of laundry, yet folds to only 4.5"W x 78.5"H. The diagonal width is an impressive 9'6" when open. The height conveniently adjusts from 4' to 6'. Polished aluminum 1.75" post, galvanized steel arms. Easy opening and closing for stowing between uses.

Optional indoor stand allows use in basement or garage. Cover for folded dryer prevents bird spots and UV degradation of line. Netherlands.

06-0229 Umbrella Clothes Dryer $159
10-0032 Indoor Stand $39
10-0034 Cover $16

Retractable Clothesline

Our Retractable Clothesline makes air drying laundry easy in any location. Extendable from any length up to 40 feet, the tangle-free line hooks to a bracket you secure at any desired distance and retracts into the impact-resistant cover for out-of-the-way storage. Supports a load's worth of sheets and towels while remaining taut. Easily installs to wood, metal, or drywall for convenient use either indoors or outdoors. Stainless steel springs won't rust in outdoor applications. Includes mounting hardware. Wipe with a damp cloth for cleaning. 7"H x 7"D x 6" diameter. USA.

10-0037 Retractable Clothesline $20

Super-Efficient Lighting

A revolution has taken place in commercial and residential lighting. Until recently, we were bound by Thomas Edison's ingenious discoveries. But that was more than a century ago, and Mr. Edison's magic now appears crude, even outdated. Mr. Edison was never concerned with how much energy his incandescent bulb consumed. Today, however, economic, environmental, and resource issues force us to look at energy consumption much more critically. Incandescent literally means "to give off light as a result of being heated." With standard light bulbs, typically only 10% of the energy consumed is converted into light, while 90% is given off as heat.

New technologies are available today that improve on both the reliability and the energy-consumption of Edison's invention. Although the initial price for these new, energy-efficient products is higher than that of standard light bulbs, their cost effectiveness is far superior in the long run. (See the discussion of whole-life cost comparison below.) When people understand just how much money and energy can be saved with compact fluorescents, they often replace regular bulbs that haven't even burned out yet.

Light output of a lamp is measured in lumens; energy input to a lamp is measured in watts. The efficiency of a lamp is expressed as lumens per watt. In the efficiency chart below, this number is listed for various lamps. High-pressure sodium and low-pressure sodium lights are primarily used for security and street lighting.

"Energy-Saving" Incandescent Lights?

Inside an incandescent light is a filament that gives off light when heated. The filament is delicate and eventually burns out. Some incandescents incorporate heavy-duty filaments or introduce special gases into the bulb to increase life. While this does not increase efficiency, it increases longevity by up to four times. Most manufacturers are now offering "energy-saving" incandescent bulbs. These are nothing more than lower-wattage bulbs that are marketed as having "essentially the same light output as . . ." In fact, General Electric lost a multimillion-dollar lawsuit relating to this very issue. A 52-watt incandescent bulb does not put out the same amount of light as a 60-watt incandescent bulb.

An "energy-saving" incandescent bulb will use less power, true, but it will also give you less light. This specious advertising claim is just a megacorporation excuse to avoid retooling by selling you a wimpier light bulb at a higher price. Don't be hoodwinked; you can do much better.

Quartz-Halogen Lights

Tungsten-halogen (or quartz) lamps are really just "turbocharged" incandescents. They are typically only 10 to 15% more efficient than standard incandescents; a step in the right direction, but nothing to write home about. Compared to standard incandescents, halogen fixtures produce a brighter, whiter light and are more energy efficient because they operate their tungsten filaments at higher temperatures than standard incandescent bulbs. In addition, unlike the standard incandescent light bulb, which loses approximately 25% of its light output before it burns out, a halogen light's output depreciates very little over its life, typically less than 10%.

To make these gains, lamp manufacturers enclose the tungsten filament inside a relatively small, quartz-glass envelope filled with halogen gas. During normal operation, the particles that evaporate from the filament combine with the halogen gas and are eventually redeposited back on the filament, minimizing bulb blackening and extending filament life. Where halogen lamps are used on dimmers, they need to be

Incandescent literally means "to give off light as a result of being heated." With standard light bulbs, typically only 10% of the energy consumed is converted into light, while 90% is given off as heat.

When people understand just how much money and energy can be saved with compact fluorescents, they often replace regular bulbs that haven't even burned out yet.

EFFICIENCY AND LIFETIME OF LAMPS

Lamp Type	Power	Lumens per watt	Lifetime in hours
Incandescent bulb	25w	8.4	2,500
Incandescent bulb	60w	14.0	1,000
Halogen bulb	50w	14.2	2,000
Incandescent bulb	100w	16.9	750
Halogen bulb	90w	17.6	2,000
Halogen floodlight	300w	19.8	2,000
G-16, compact fluorescent	16w	53.0	10,000
SLS 20, compact fluorescent	20w	60.0	10,000
Metal halide, floodlight	150w	60.0	10,000
D lamp, compact fluorescent	39w	71.0	10,000
T8, 4-ft fluorescent tube	32w	86.0	20,000
High-pressure sodium light	150w	96.0	24,000
Low-pressure sodium light	135w	142.0	16,000

The types of efficient lighting on the market include new-design incandescents, quartz-halogens, standard fluorescents, and compact fluorescents.

operated occasionally at full output to allow this regenerative process to take place.

Several styles of halogens are available for both 12-volt and 120-volt applications. The most popular low-voltage models for residential applications include the miniature multifaceted reflector lamp (product #33-108) and the halogen version of the conventional A-type incandescent lamp, which incorporates a small halogen light capsule within a protective outer glass globe. The reflector lamp offers a light source with very precise light-beam control, allowing you to distribute light exactly where you need it without wasteful spillover. A new family of 120-volt AC, sealed-beam reflector halogen lamps is now on the market to replace standard incandescent reflectors.

The higher operating temperatures used in halogen lamps produce a whiter light, which eliminates the yellow-reddish tinge associated with standard incandescents. This makes them an excellent light source for applications where good color rendition is important or fine-detail work is performed. Because tungsten-halogen lamps are relatively expensive compared to standard incandescents, they are best suited for applications where the optical precision possible with the compact reflector models can be effectively utilized. They are a favorite of high-rent retail stores.

Never touch the quartz-glass envelope of a halogen lamp with your bare hands. The natural oils in your skin will react with the quartz glass and cause it to fail prematurely. This applies to automotive halogen headlight bulbs as well. Because of this phenomenon, and for safety reasons, many manufacturers incorporate the halogen lamp capsule (generally about the size of a large flashlight bulb) within a larger outer globe.

Fluorescent Lights

Fluorescent lights are still trying to overcome a bad reputation. For many people the term "fluorescent" connotes a long tube emitting a blue-white light with an annoying flicker, a death-warmed-over color, and headaches. However, these limitations have been overcome completely with technological improvements.

The fluorescent tube, however it is shaped, contains a special gas at low pressure. When an arc is struck between the lamp's electrodes, electrons collide with atoms in the gas to produce ultraviolet light. This, in turn, excites the white phosphors (the white coating on the inside of the tube), which emit light at visible wave-

Clean Lights Save Energy

Building owners can save up to 10% of lighting costs simply by keeping light fixtures clean, says NALMCO, the National Association of Lighting Management Companies. Over the past three years, crews working on the U.S. EPA-funded project have collected scientific field measurements at offices, schools, health facilities, and retail stores to determine the benefits—improved lighting as well as reduced costs—of simple periodic cleaning of light fixtures. Recognizing that fixtures produce less light as they get dirty, current design standards compensate by adding more fixtures to maintain the desired light level. But if building owners committed to regular cleanings, designers could recommend fewer fixtures, reducing installation costs while trimming lighting energy use by at least 10%. For more information, call NALMCO at 515.243.2360.

lengths. The quality of the light that fluorescents produce depends largely on the blend of chemical ingredients used in making the phosphors; dozens of different phosphor blends are available. The most common and least expensive are "cool white" and "warm white." These provide a light with relatively poor color-rendering capabilities, making colors appear washed out, lacking luster and richness.

The tube may be long and straight, as with standard fluorescents, or there may be a series of smaller tubes, in a configuration that can be screwed into a common light fixture. This latter configuration is called a compact fluorescent light (or CF light).

Consumer awareness of and interest in compact fluorescents has exploded in recent years and the market for them is growing rapidly. A good variety of lamp wattages, sizes, shapes, and aesthetic packages is now readily accessible. Because this marketplace is growing and changing so quickly, the more-frequently published Gaiam Real Goods catalogs may feature lamp types not described here.

COLOR QUALITY

The color of the light emitted from a fluorescent bulb is determined by the phosphors that coat the inside surface. The terms "color temper-

LIGHT, COLOR RENDERING INDEX, AND COLOR TEMP

Type of Light	CRI	Deg. K
Incandescents	90–95	2,700
Cool white fluorescent	62	4,100
Warm white fluorescent	51	3,000
Compact fluorescent	82	2,700

ature" and "color rendering" are the technical terms used to describe light. Color temperature, or the color of the light that is emitted, is measured in degrees Kelvin (°K) on the absolute temperature scale, ranging from 9,000°K (which appears blue) down to 1,500°K (which appears orange-red). The color rendering index (CRI) of a lamp, on a scale of 0 to 100, rates the ability of a light to render an object's true color when compared to sunlight. A CRI of 100 is perfect, 0 is a cave. The accompanying table will demystify the quality-of-light debate.

Phosphor blends are available that not only render colors better, but also produce light more efficiently. Most notable of these are the fluorescent lamps using tri-stimulus phosphors, which have CRIs in the 80s. These incorporate relatively expensive phosphors with peak luminance in the blue, green, and red portions of the visible spectrum (those that the human eye is most sensitive to), and produce about 15% more visible light than standard phosphors. Wherever people spend much time around fluorescent lighting, specify lamps with higher (80+) CRI.

BALLAST COMPARISONS

All fluorescent lights require a ballast to operate, in addition to the bulb. The ballast regulates the voltage and current delivered to a fluorescent lamp, and is essential for proper lamp operation. The electrical input requirements vary for each type of compact fluorescent lamp, and so each type/wattage requires a ballast specifically designed to drive it. There are two types of ballasts that operate on AC: magnetic and electronic. The magnetic ballast, the standard since fluorescent lighting was first developed, uses electromagnetic technology. The electronic ballast, only recently developed, uses solid-state technology. All DC ballasts are electronic devices.

Magnetic ballasts can last up to 50,000 hours, and often incorporate replaceable bulbs. Magnetic ballasts flicker when starting and take a few seconds to get going. They also run the lamp

at 60 cycles per second, and some people are affected adversely by this flicker.

Electronic ballasts weigh less than magnetic ballasts, operate lamps at a higher frequency (30,000 cycles per second vs. 60 cycles), are silent, generate less heat, and are more energy efficient. However, electronic ballasts cost more, particularly DC units. Much-longer-lived electronic ballasts are possible, but costs go up dramatically.

Electronic ballasts last about 10,000 hours, the same as the bulb, and most do not have replaceable bulbs. Electronic ballasts start almost instantly with no flickering. They run the lamp at about 30,000 cycles per second. For the many people who suffer from the "60-cycle blues," this is the energy-efficient lamp of choice.

All of the compact fluorescent lamp assemblies and pre-wired ballasts reviewed here are equipped with standard medium, screw-in bases (like normal household incandescent lamps). The ballast portion, however, is wider than an incandescent light bulb just above the screw-in base. Fixtures having constricted necks or deeply recessed sockets may require a socket "extender" (to extend the lamp beyond the constrictions). These are readily available either in our catalogs or at most hardware stores.

Note: There is no difference between a fluorescent tube for 120 volts and one for 12 volts. Only the ballasts are different.

SAVINGS

The main justification for buying fluorescent lights is to save energy and money. Compact fluorescents provide opportunities for tremendous savings without any inconveniences. Simple payback calculations prove their cost effectiveness (see the Cost Comparison table). Fluorescent lights typically last ten to thirteen times longer and use one-fourth the energy of standard incandescent lights. The U.S. Congress's Office of Technology Assessment calls compact fluorescents the best investment in America today, with a 1.2-year payback.

The next question, of course, is, "How much do compact fluorescents save?" The more a light is used, the more energy and money you can save by replacing it with a compact fluorescent. Installing CFs in the fixtures that are used most and have high wattage saves the most. An excellent candidate, for example, is an all-night security light. The calculations in the accompanying table assume that the light is on for an average of six hours per day and power costs 10¢ per kWh (the approximate national average). The total

Fluorescent lights typically last ten to thirteen times longer and use one-fourth the energy of standard incandescent lights.

The U.S. Congress's Office of Technology Assessment calls compact fluorescents the best investment in America today, with a 1.2-year payback.

cost of operation is the cost of the bulb(s) plus the cost of the electricity used. The table shows that using the compact fluorescent saves about $40 and 540 kWh of electricity.

HEALTH EFFECTS

Migraine headaches, loss of concentration, and general irritation have all been blamed on fluorescent lights. These problems are caused not by the lights themselves, but by the way they operate. Common magnetic ballasts run the lamps at the same 60 cycles per second that is delivered by our electrical grid. This causes the lamps to flicker noticeably 120 times per second, every time the alternating current switches direction. Approximately one-third of the human population is sensitive to this flicker on a subliminal level.

The cure is to use electronic ballasts, which operate at around 30,000 cycles per second. This rapid cycling totally eliminates perceptible flicker and avoids the ensuing health complaints.

Fluorescents and Remote Energy Systems

Fluorescent lights are available for both AC and DC. Most people using an inverter choose AC lights because of the wider selection, a significant quality advantage, and lower price. Some older inverters may have problems running some magnetic-ballasted lights. If you have an older Heart inverter, buy one light to try it first.

COST COMPARISON*

	Philips SLS20, 20-Watt	Incandescent Bulb, 75-Watt
Cost of bulb	$19.95	50¢
Product life	4.5 years	167 days
Bulbs used in 4.5 years	1	10
Energy used annually	44 kWh	164 kWh
Energy used in 4.5 years	198 kWh	739 kWh
Total cost in 4.5 years	$39.75	$78.90
Savings in 4.5 years	$39.15	
Energy saved in 4.5 years	541 kWh	

*Computed at 6 hours of lamp use per day and electricity cost of 10¢ per kWh.

Most inverters operate compact fluorescent lamps satisfactorily. However, because all but a few specialized inverters produce an alternating current having a modified sine wave (versus a pure sinusoidal waveform), they will not drive compact fluorescents that use magnetic ballasts as efficiently or "cleanly" as possible, and some lamps may emit an annoying buzz. Electronic-ballasted compact fluorescents, on the other hand, are tolerant of the modified sine wave input, and will provide better performance, silently.

COMPACT FLUORESCENT APPLICATIONS

Due to the need for a ballast, a compact fluorescent light bulb is shaped differently from an Edison incandescent. This is the biggest obstacle in retrofitting light fixtures. Compact fluorescents are longer, heavier, and sometimes wider.

An Energy Tale, or Real Goods Walks Its Talk

In May of 1995 Real Goods headquarters moved into a new-to-us 10,000 square foot office building. Built in the early 1960s, the building had conventional 4-foot fixtures, each holding four 40-watt cool white tubes. The dreary, glaring "office standard." Light levels were much brighter than recommended when employees are using computers a lot. Plus, a number of our employees were having headache, energy level, and "attitude" problems after moving,

probably caused by the 60-cycle flicker from the old magnetic ballasts. Reducing overall light levels for less eyestrain was needed, as was giving employees a healthier working environment. And if we could save some money too, it wouldn't hurt.

Our solution was to retrofit the existing fixtures with electronic ballasts for no-flicker lighting, install specular aluminum reflectors, and convert from four cool white 40-watt T-12 lamps per fixture to two warm-colored 32-watt T-8 lamps. Power use per fixture was reduced by over 60%, but desktop light levels, because of the reflectors and more efficient lamps and ballasts, were reduced by

only 30% or less. The entire retrofit cost about $6,000 (not counting the $2,300 rebate from our local utility), yet our electrical savings alone are close to $5,000 per year. Plus our employees enjoy flicker-free, warm-colored, non-glaring light and greatly reduced EMF levels. The entire project paid for itself in less than a year! Improved working conditions came as a freebie!

Editor's Note: We realize that this story is now ten years old, but we felt it was so important that we're running it again in the twelfth edition *Sourcebook*. —JS

The ballast, the widest part, is located at the base, right above the screw-in adapter. Today, new "mini-spiral" CFs are available that fit in just about any fixture that currently uses a standard incandescent bulb. Not surprisingly, these CFs have quickly become the most popular model in the marketplace.

Compact fluorescents have many household applications—table lamps, floor lamps, recessed cans, desk lamps, bathroom vanities, and more. As manufacturers become attuned to this relatively new market, more light fixtures suited for compact fluorescents are becoming available. Gaiam Real Goods now offers a limited selection of fixtures; watch future catalogs for the latest offerings in this rapidly expanding field.

Table lamps are an excellent application for compact fluorescents. The metal harp that supports the lampshade may not be long or wide enough to accommodate these bulbs. We offer an inexpensive replacement harp that can solve this problem. Be aware also that heavier CFs may change a light but stable lamp into a top-heavy one.

Recessed can lights are limited by diameter, and sometimes cannot accept the wide ballast at the screw-in base. The base depth is adjustable on most recessed cans. There are a number of special bulbs available now just for recessed cans.

Desk lamps can be difficult to retrofit, but complete desk lights that incorporate compact fluorescents are readily available.

Hanging fixtures are one of the easiest applications for compact fluorescent lights. One of the best uses is directly over a kitchen or dining room table. Usually the shade is so wide that it won't get in the way. It may even be possible to use a Y-shaped two-socket adapter (available at hardware stores) and screw in two lights if more light is desired.

Track lighting is one of the most common forms of lighting today. When selecting your track system, choose a fixture that will not interfere with the ballasts located right above the screw-in base. It is best to use one of the reflector lamps, such as the SLS/R series.

Dimming

With limited exceptions, the current generation of compact fluorescent lights should never be dimmed. Using these lights on a dimmer switch may even pose a hazard. If you have a fixture with a dimmer switch that you wish to retrofit, you either can buy a CF lamp that specifically says that it is dimmable, or replace the switch. Replacing the switch is a simple, inexpensive procedure. Remember to turn off the power first!

Three-Way Sockets

Recently manufacturers have begun to make three-way compact fluorescent bulbs. While more efficient than conventional three-way bulbs, they are more expensive than one-way CFs and have a shorter lifespan.

Start-Up Time

The start-up time for compact fluorescent lamps varies. It is normal for most magnetic CFs to flicker for up to several seconds when first turned on while they attempt to strike an arc. Most electronically ballasted units start their lamps instantly. All fluorescent lamps start at a lower light output; depending on the ambient temperature, it may take anywhere from several seconds to several minutes for the lamp to come up to full brightness.

The very brief start-up time, which is only apparent in some of the fluorescent lights, is a small price to pay for the energy savings and your own peace of mind about doing less harm to our fragile environment.

COLD WEATHER/OUTDOOR APPLICATIONS

Fluorescent lighting systems are sensitive to temperature. Manufacturers rate the ability of lamps and ballasts to start and operate at various temperatures. Light output and system efficiency both fall off significantly when lamps are operated above or below the temperature range at which they were designed to operate. Most fluorescent lamps will reach full brightness at 50° to 70°F. In general, the lower the ambient temperature, the greater the difficulty in starting and attaining full brightness. Manufacturers are usually conservative with their temperature ratings. We have found that most lamps will start at temperatures 10° to 20°F lower than those stated by the manufacturer. Most compact fluorescent lamps are not designed for use in wet applications (for example, in showers or in open outdoor fixtures). In such environments, the lamp should be installed in a fixture rated for wet use.

SERVICE LIFE

All lamp-life estimates are based on a three-hour duty cycle, meaning that the lamps are tested by turning them on and off once every three hours until half of the test batch of lamps burn out. Turning lamps on and off more fre-

Generally, it is more cost effective to turn off standard fluorescent lights whenever you leave a room for more than 15 minutes. For compact fluorescents, the interval is 3 minutes, and you should turn incandescents off whenever you leave the room.

A typical compact fluorescent lamp will last as long as (or longer than) ten standard incandescent AC light bulbs or five standard incandescent AC floodlights, saving you the cost of numerous bulbs, in addition to a lot of electricity due to the greater efficiency.

quently decreases lamp life. Since the cost of operating a light is a combination of the cost of electricity used and the replacement cost, what is the optimum no-use period before turning off a light? Generally, it is more cost effective to turn off standard fluorescent lights whenever you leave a room for more than 15 minutes. For compact fluorescents, the interval is 3 minutes, and you should turn incandescents off whenever you leave the room. Since the rated lamp life represents an average, some lamps will last longer and some shorter. A typical compact fluorescent lamp will last as long as (or longer than) ten standard incandescent AC light bulbs or five standard incandescent AC floodlights, saving you the cost of numerous bulbs, in addition to a lot of electricity due to the greater efficiency. See the Cost Comparison table on page 266.

FULL-SPECTRUM FLUORESCENTS

Literally, "full-spectrum" refers to light that contains all the colors of the rainbow. As used by several manufacturers, "full-spectrum" refers to the similarity of their lamps' light (including ultraviolet light) to the midday sun. While we do believe that the closer an electric light source matches daylight the healthier it is, we have several practical reservations. The quality of light produced by "full-spectrum" lighting available today is very "cool" in color tone and in our opinion, not flattering to people's complexions or most interior environments. Also, the light intensity found indoors is roughly one hundred times less than that produced by sunlight. At these low levels, we question whether people receive the full benefits offered by full-spectrum lighting.

Ideally, we encourage you to get outdoors every day for a good dose of sunshine. Indoors, we think it is best to install lighting that is complimentary to you and the ambiance you desire to create. For most applications, people prefer a warmer-toned light source. Most of the compact fluorescent lamps we offer mimic the warm rosy color of 60-watt incandescent lamps, but we do offer a small selection of full-spectrum lamps for those who prefer them.

DC Lighting

When designing an independent power system, it is essential to use the most efficient appliances possible. Fluorescent lighting is four to five times more efficient than incandescent lighting. This means you can produce the same quantity of light with only 20 to 25% of the power use. When the power source is expensive, like PV modules, or noisy and expensive, like a generator, this becomes terribly important. Fluorescent lighting is quite simply the best and most efficient way to light your house.

Usually we recommend against using DC lights in remote home applications except in very small systems where there is no inverter. The greater selection, lower prices, and higher quality of AC products has made it hard to justify DC lights. DC wiring and fixture costs are significantly higher than with AC lighting, and lamp choices are limited. However, you might want to consider putting in a few DC lights for load minimization and emergency back-up. A small incandescent DC nightlight uses less than its AC counterpart plus inverter overhead, and a single AC device may not draw enough power to be detected by the inverter's load-seeking mode.

Since the introduction of efficient, high-quality, long-lasting inverters, the low-voltage DC appliance and lighting industries have almost disappeared except for RV equipment, which tends to be of lower quality. This is because most RV equipment is designed for intermittent use, where short life or lower quality isn't as noticeable.

Tube-type fluorescents, such as the Thin-Lite series, are great for kitchen counters and larger areas. These are RV units, with lower life expectancies compared to conventional AC units. Replacement ballasts for these units are available, and the fluorescent tubes are standard hardware-store stock. We also offer incandescent and quartz-halogen DC lights, but suggest caution in using these for alternative energy systems, where every watt is precious.

NATURAL LIGHTING AND LIGHTING PRODUCTS

SunPipe Skylights

Free sunlight is the best light

The SunPipe adds beautiful natural daylight to your house without high expense, heat gain or loss, bleaching of fabrics, or framing and drywall work. This is how skylights were always meant to be. Sunlight is collected above the roof and transferred thru a mirror-finish aluminum pipe to a diffusing lens in your ceiling. The Sun-Pipe transmits no direct sun rays, so there's no solar heat gain, bleaching or color fading. No cutting of joists and rafters, or framing and drywall work is required. Installation takes two to four hours on average, and can be done easily by any home handyman. The light is uniformly diffused by the interior lens, and the pipe is sealed top and bottom to prevent heat loss or condensation.

Two sizes are available. Approximately speaking, the SunPipe-9 is for up to 100 square feet of floor space, the SunPipe-13 covers up to 240 square feet. At midday the 13" SunPipe delivers more cool, free sunlight than a dozen 100-watt light bulbs. The 9" Sun Pipe deliver about half as much light.

The SunPipe Kits provide all the parts you'll need including templates and a complete installation guide. Order a pipe length that equals at least your ceiling to roof distance thru the attic, plus one-foot. The pipe has to extend up through the flashing. Add an extra 2-foot section if in doubt. Extra pipe length will telescope if needed. The standard galvanized flashing covers 0:12 to 6:12 roof pitches, the steep roof flashing covers 6:12 to 12:12 roof pitches and is a substitute for the standard flashing in the kit. The all-metal flashings are heavily galvanized to G-90 standard, and have a 60-year life expectancy. The optional elbows adjust from 0° to 45°. Ten-year manufacturer's warranty. USA.

SunPipe-9

17-0303	2-Foot Kit	$229
17-0304	4-Foot Kit	$270
17-0305	6-Foot Kit	$315
63-158	9" Pipe Extension 2-ft.	$42
63-823	9" Steep Flashing Option	$50
63-002	9" Elbow Option	$62

SunPipe-13

17-0306	2-Foot Kit	$299
17-0307	4-Foot Kit	$355
17-0308	6-Foot Kit	$410
63-181	13" Pipe Extension 2-ft.	$54
63-814	13" Steep Flashing Option	$99
63-004	13" Elbow Option	$74

AC LIGHTING

THE BULB	PRODUCT DESCRIPTION AND PRICE	SPECIAL FEATURES	CFL USAGE (WATTS)	INCAND. EQUIV. (WATTS)	LUMENS	MINIMUM OPERATING TEMP.	SIZE (HXW)	COLOR TEMP.[1]
	Sunwave Bulb 20W 31772 $25 23W 31773 $25 26W 31774 $25	Most advanced full-spectrum lighting in the world reduces eyestrain and increases visual acuity. Instant-starting electronic ballast. China.	20W 23W 26W	75W 100W 120W	1,300 1,600 2,000	14°F 14°F 14°F	5.7" x 2.2" 5.7" x 2.2" 6.2" x 2.5"	5,550K 5,550K 5,550K
	CF Spiral Bulb 11-0088 15W $8 20W $9 23W $10	Offers the same warm glow and performance as traditional incandescents. Available in warm glow (WG) or natural daylight (ND). China.	WG 15W ND 15W WG 20W ND 20W WG 23W ND 23W	60W 60W 75W 75W 100W 100W	880 880 1,150 1,150 1,350 1,350	14°F 14°F 14°F 14°F 14°F 14°F	5.2" x 2.1" 5.2" x 2.1" 5.2" x 2.1" 5.2" x 2.1" 5.2" x 2.1" 5.2" x 2.1"	2,700K 5,000K 2,700K 5,000K 2,700K 5,000K
	Mini CF Spiral Bulb 11-0087 Warm Glow 11W $8 Natural Daylight 11W $9	Use in areas where smaller bulbs or less light output is needed. Available in warm glow (WG) or natural daylight (ND). Hong Kong.	WG 11W ND 11W	50W 50W	550 550	14°F 14°F	4.6" x 1.8" 4.6" x 1.8"	2,700K 5,000K
	Natural Spectrum Reflector Floodlight 11-0137 $30	Natural spectrum technology in recessed and track lighting fixtures. A medium base fits standard sockets. Indoor use only. China.	15W	60W	650	-10°F	6" x 3.6"	6,400K
	Decorative Bulb 36616 $10	Looks great in open fixtures; standard base. Please note dimensions at right to ensure bulb fits your fixture. China.	5W	27W	250	-10°F	4.8" x 2"	2,700K
	Candelabra Bulb 36617 $10	For open fixtures. Smaller base designed specifically for candelabras. Please note dimensions at right to ensure bulb fits your fixture. China.	3W	15W	150	-4°F	4.8" x 1.5"	2,700K
	CF Mini Globe 11-0013 $18	Great for track lighting, vanities, and open fixtures. Available in warm glow or natural daylight.	ND 15W WG 15W	60W 60W	850 850	14°F 14°F	5.5" x 3.75" 5.5" x 3.75"	5,000K 2.700K

Chart is based on information provided by the manufacturer. All CFL bulbs have flicker-free electronic ballasts and a 10,000 hour life. CF=Compact Fluorescent.
[1]Color Temperature (CT)—Not to be confused with bulb temperature, this is a color spectrum rating; the higher the number, the bluer the light.

Open Indoor: Hanging pendant, ceiling lamps and wall fixtures

Shaded Lamps: Table and floor lamps

Enclosed Fixture: Ceiling and wall fixtures; exterior yard post lights, wall brackets, and hanging fixtures (check min. operating temp.). Life expectancy decreases in enclosed fixtures.

Recessed Can: Recessed ceiling downlights

Track Light: Ceiling and wall fixtures

Socket Extender

Sometimes your compact fluorescent lamps may have a neck that is too wide for your fixture. This socket extender increases the length of the base. The extender is also useful if a bulb sits too deeply inside a track lighting or can fixture, losing its lighting effectiveness. Extends lamp 1.75". USA.

11-0104 Socket Extender   **$2.95**

G A I A M   R E A L   G O O D S

FULL-SPECTRUM LIGHTING

Natural Spectrum Lamps

By simulating the pleasing brilliance of sunshine, the Verilux Desk and Floor Lamp filter out yellow tones to provide a natural, flicker-free light suitable for any task. Adjustable gooseneck design allows for easy placement of the light. Includes a compact fluorescent bulb that lasts 5,000 hours yet uses only 27W to provide the light of a regular 150W bulb. Replacement bulb has a 5,000-hour life. Specify graphite (black) or putty (gray). China.

11-0191	Floor Lamp (43"H x 10"W)	$139
11-0190	Desk Lamp (12"H x 10"W)	$89
11-0091	Replacement Bulb	$30

Natural Light Desk Lamp

Our sleek desk lamp uses less than 80% of the electricity needed by an equivalent incandescent bulb while producing a rich white light whose 5,100K temperature mimics natural light. Whether you use it for reading, paperwork, hobbies, or other tasks, you'll appreciate its buzz-free silence, its large illumination area, and the way its pure, flicker-free light eliminates eye strain. 16 watt. 13"H (level to desk) x 25"L x 6"D. China.

11-0182	Natural Light Desk Lamp	$49

Brushed Steel CFL Torchiere

A beautiful alternative to inefficient, dangerously hot halogen torchieres. Uses cool-burning, energy-efficient compact fluorescent bulbs. Our polished-satin nickel finish dimmable style creates bright light using only 58 watts of electricity (225W incandescent equivalent). Bulb included. 71"H x 13"W. Weighted base. UL listed. Assembled in China.

11-0183	Brushed Steel Torchiere	$99
11-0184	Replacement Bulb	$45

Happy Lite Deluxe Light Bath

Bathe in the healing power of 10,000 lux natural daylight spectrum light when you need the healing relief only sunlight can bring. The Verilux Happy Lite Deluxe Light Bath helps alleviate light deficiency symptoms, whether from rainy days, seasonal change, jet lag, or shift work. Compact and easy to carry; desk stand or wall-mount option. Flicker-free electronic ballast blocks harmful EMI and RF transmissions. Bulbs last 8,000 hours. 20"H x 12.5"W x 3.5"D. 4 lbs. China.

08-0330	Happy Lite Deluxe Light Bath	$189

Natural Daylight Fluorescent Tubes

Our Natural Spectrum Fluorescent Tubes offer a high light quality and are a low-energy solution for studios,

workbenches, and general use. Available in three sizes, the 20W (24" length, 800 lumens 94.5 CRI, 6,280°K) and 40W (48" length, 2,100 lumens 94.5 CRI, 6,280°K) sizes are T-12 diameter, about 1.25". The 32W (48" length, 2,900 lumens 85 CRI, 6,000°K) size is T-8 diameter, about 1". The T-8 bulbs are for use with electronic ballasts only and provide more light with less power. You must match your existing bulb size! All sizes offer an extra long life of 26,000 hours and feature the closest representation of natural light available. T-12 lamps, France. T-8 lamps, USA.

11-0052	40W (4-pack)	$50
11-0134	32W (4-pack)	$50
11-9001	20W (2-pack)	$25

DC LIGHTING

DC Compact Fluorescent Lamps

More choices and better quality! DC users at both 12 and 24 volts finally can have quality compact fluorescent lamps in a variety of sizes—and at reasonable prices. Our new German-engineered supplier manufactures in China and actually does a one-hour burn-in on

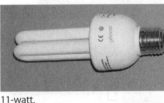

11-watt.

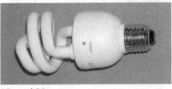

15- and 30-watt.

every lamp to eliminate possible failures. Available in 7-, 11-, 15-, and 30-watt sizes, which are approximately equivalent to 30-, 45-, 60- and 120-watt incandescent lamps. Warm lamps are 2,700K, cool lamps are 6,400K. Will operate at 14°F to 104°F, and at 11 to 15 volts (22 to 30 for 24V). Standard Edison medium base. Two-year manufacturer's warranty. Germany/China. Choose warm glow or natural daylight.

50-0104	**12V, 11W**	**$14**
50-0104	**12V, 15W**	**$19**
50-0104	**12V, 30W**	**$39**
50-0105	**24V, 15W**	**$39**

Solsum 12VDC Compact Fluorescent Lamps

Here's a truly compact, and even attractive DC fluorescent lamp. Using an instant-start solid-state electronic ballast, it will operate reliably from 10.5 to 14.5 volts, and uses a standard Edison screw base. The 11-watt lamp is equivalent to a 50-watt and measures 6.3" length, 2.25" diameter. Life expectancy is approximately 6,000 hours. Germany/China.

33-503 Solsum 11W, 12VDC Lamp $25

Thin-Lites

Thin-Lites are made by REC Specialties, Inc. Their 12-volt DC fluorescents are built to last, and are all UL listed. Easy to install, they have one-piece metal construction, non-yellowing acrylic lenses, and computer-grade rocker switches. A baked white enamel finish, along with attractive woodgrain trim, completes the long-lasting fixtures. Two-year warranty. All use easy-to-find, standard fluorescent tubes, powered by REC's highly efficient inverter ballast. Bulbs used in DC fixtures are exactly the same as those used in AC fixtures and can be purchased at any local hardware store—only the ballast is different. USA.

30-Watt 12-Volt DC Light
Uses two F15T8/CW fluorescent tubes (included).
• 18" x 5.5" x 1.375"
• 1.9 amps
• 1760 lumens

32-116 Thin-Lite #116, 30-Watt $48

15-WATT 12-VOLT DC LIGHT
Uses one F15T8/CW fluorescent tube (included).
• 18" x 4" x 1.375"
• 1.26 amps
• 800 lumens

32-115 Thin-Lite #115, 15-Watt $40

12-Volt Christmas Tree Lights

It just doesn't seem like Christmas without lights on the tree and these 12-volt lights will dazzle any old fir bush. These lights are actually great all year round for decorating porches, decks, your rolling art automobile, or your Harley. The light strand is 20 feet long and consists of 35 mini, colored, non-blinking lights (draw is 1.2 amps) with a 12-volt cigarette lighter socket on the end. Note: You can't connect them in series as you can with many 120-volt lights. Uses conventional 3-volt mini-lights, same as any 120-volt mini-light string, just wired for 12-volt input. Taiwan.

37-301 12-Volt Christmas Tree Lights $14.95
3 for $12.95 each

HALOGEN LIGHTING

Littlite High-Intensity Lamp

Our high-intensity 5-watt quartz halogen bulb concentrates light in a tightly controlled area, easily aimed exactly where you need it. Also great for any job involving accurate, close or detail work—drawing, computer, on a music stand, by a stereo system. The small hooded end won't get in your way, you have just the right amount of light, and you save energy (only 5 watts!). Includes 18" flexible gooseneck, mounting base, and fully adjustable dimmer. Weighted base converts into a movable free-standing lamp.

33514	**12V Model**	**$48**
33401	**Weighted Base**	**$18**
33109	**Replacement 5-watt bulb**	**$7**

Quartz Halogen Lamps Inside Frosted Globes

A small cottage industry in the Northeast manufactures these incredibly ingenious 12-volt light bulbs. On the outside they appear identical to an incandescent, but they have the increased efficiency and longevity of quartz halogens. Multiply the watts on halogens by 1.5 and you'll get an idea of their equivalent light output compared to an incandescent lamp. Available in 12- or 24-volt. USA.

33-102	Quartz Lamp, 10W/0.8amp	$18 or 2 for $30
33-103	Quartz Lamp, 20W/1.7amp	$18 or 2 for $30
33-104	Quartz Lamp, 35W/2.9amp	$18 or 2 for $30
33-105	Quartz Lamp, 50W/4.2amp	$18 or 2 for $30
33-106	Quartz Lamp, 20W/24V	$20
33-107	Quartz Lamp, 50W/24V	$20

12V Halogen Flood Lights

Shine light exactly where you want it with these 12V quartz halogen flood lights. They screw into a standard medium screw base fixture and, with 20 watts, provide a bright, focused light. Faceted dichronic reflector maximizes efficiency and gives a sharp clean light with excellent color rendition. Elegant brass base looks great.

33-513	20-Watt Halogen Flood, 12V	$24

LED LIGHTING

LED development as a light source is going the way of early computer development. Output doubles about every eighteen months while prices slowly drift downward. We're starting to see some high-quality, white light sources that may be appropriate for many off-the-grid home applications.

GOOD THINGS ABOUT LEDS ...

The white LEDs have a bright, slightly bluish light output that is extremely easy to read by. Black print just jumps off a white page with these lamps. The life expectancy of LEDs is exceptional, often exceeding 50,000 hours. They aren't bothered by vibration or modest impacts. Your flashlight won't die just because you dropped it. We're finally starting to see LED manufacturers listing the light output of their devices. Often listed in milli-candelas, 1,000 milli-candelas equals 1 candela, which is also equal to 1 lumen.

NOT SO GOOD THINGS ABOUT LEDS ...

Most LEDs are still fairly modest light sources, and they're expensive. LEDs tend to be highly directional in their output, they don't make good general illumination sources without serious diffusion help. Their lumens per watt ratio is improving, but currently is slightly less than compact fluorescent lamps. In late 2003, LEDs were averaging 25 to 30 lumens per watt, while CFLs averaged 40 to 60.

Brighter, More Affordable LED Lamps

LED lamps continue to evolve with higher light output, better looks, and lower costs. Our new DC lamps from Phocos feature a nice housing with standard screw-in base, and a life expectancy exceeding 50,000 hours. Available in blue/white with 9 high-output LEDs, or in warm-tone with 3 blue/white and 9 yellow LEDs—both draw only 1.2 watts. With no diffusers, the light output tends to be fairly focused with both these lights. The white version, which is also available in 24-volt, is particularly nice for reading. German-engineered, Chinese-manufactured. Two-year manufacturer's warranty.

50-0101	12V, 1.2W White Lamp	$38
50-0102	12V, 1.2W Warmtone Lamp	$32
50-0103	24V, 1.2W White Lamp	$38

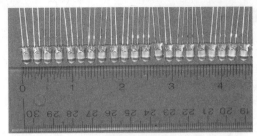

Hi-Output Individual White LEDs

Build your own low-voltage DC lights. Each 5mm LED delivers 8 candelas of blue/white light with 4 volts, 30 milliamps input. A candela is equal to a lumen. Wire three of these bulbs in series for 12-volt input. Sold in packs of 12 or 24. China.

| 32323 | White 8cd LED, 12-pack | $20 |
| 32324 | White 8cd LED, 24-pack | $35 |

SOLAR OUTDOOR LIGHTING

Solar Sensor Light

The Sensor Light eliminates the need to burn an outside light wastefully while you're away. The solar-charged detection circuit automatically turns on the light to welcome you when triggered by heat or motion at approximately 35 feet, then turns off after you leave the area.

This also can make it effective in deterring prowlers. The bright 20-watt quartz-halogen bulb will last over 100 hours and lights instantly at even the coldest temperatures. There are no timers and only one switch to set (off, charge only, on). An adjustable sensitivity control reduces false triggers by your cat, a raccoon, tiny UFOs, etc. The unit mounts easily to wall, fascia, soffit, or roof eaves. No wiring or electrician required. Has a 14-foot cord on solar module for best sun exposure. Extra battery capacity allows up to two weeks of operation without sun. Comes with mounting bracket hardware and bulb included. 7"H x 7"W x 7"D. Replaceable battery and bulb. One-year warranty. China.

Separate solar panel has 14' cord for optimal placement.

17-0134	Solar Sensor Light	$119
15-198	6V 4Ah Repl. Battery	$14.50
34-313	Solar Sensor Replacement Bulb	$14.50

A 12-Volt Powered Motion Sensor

This fully featured motion sensor runs on 12vdc, but the 10-amp contacts it controls are isolated, so you can run any AC or DC voltage up to 230v through them. Want to turn on a big spotlight that's powered by your inverter, but don't want to leave your inverter in "on" mode for hours and hours? Here's your answer. Sensor is adjustable for light level (works day or night), sensitivity (up to about 35' max), and lighting time (6 sec. to 12 min.). Standard 1/2" thread mount with optional-use weathertight terminal box, terminal strip, screws, and instructions. One-year manufacturer's warranty. Taiwan.

| 50-0109 | 12v Motion Sensor | $66 |

Durable and Powerful DC Floodlight

This tough polycarbonate outdoor floodlight is completely watertight with a UV-stabilized, nonyellowing lens. The 13-watt compact fluorescent lamp delivers 900 lumens of focused light, thanks to the ellipsoidal mirror reflector. Standard 1/2" male pipe thread mounting with lamp angle adjustable through 90°. This efficient floodlight is a great choice for automatic PV-powered sign lighting, and is included in our sign lighting kits. 12vdc only. USA.

| 32307 | Durable & Powerful DC Floodlight | $69 |

Coach-Style Yard Light

High-qualty black aluminum and glass make our new yard light durable enough for years of use, and the coach-style design makes it classy enough for the most refined landscaping. All-metal body construction means no rusting or degrading parts. Uses two bright LED lights that will last for decades. Black aluminum and glass and plastic solar panel housing. Two easily replaceable nicad AA batteries. Taiwan.

| 14-0317 | Coach-Style Yard Light | $42 |

Deluxe Stainless Yard Light

A wonderful accent to any yard, our lights are bright and easy to install without running any wires. Stainless steel and glass construction means they will last for decades without a second thought. Lights charge all day long, come on with a light sensor, and will last long into the night. Stainless with plastic solar panel housing. Two easily replaceable nicad AA batteries. Taiwan.

14-0318 Deluxe Stainless Yard Light   **$65**

FLASHLIGHTS

Dynamo AM/FM Flashlight

Solar flashlight/ AM/FM radio

No power, no batteries, no stores open late at night—no problem! The rechargeable nicad battery in this Solar Flashlight/AM/FM Radio retains its power from solar energy, incandescent light energy, hand-crank dynamo power, or with a 6-volt AC/DC adapter (not included). It can also run on three AA batteries (not included). Complete with siren and flashing signal, this is not only a quality sounding radio, but a great survival tool as well. Plus its bright, yellow color is easy to locate in an emergency. Measures 3.5″H x 8.5″L x 3.75″W. China.

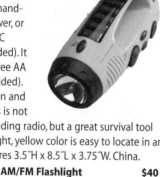

06-0403 Dynamo AM/FM Flashlight   **$40**

Solar Flashlight

Free sunlight fully recharges the built-in 600 mA nicad battery pack in about eight hours. No time for solar? Slip two AA batteries into the battery compartment, and you have light right now. These optional AA batteries will be recharged by solar too. Includes three LED warning flashers on the tail with focusing lens. Flashlight and warning flashers are switched separately, so one or both can be run. An adjustable holder for bikes or whatever is included, along with two spare bulbs and wrist strap. Run time on built-in battery alone is approximately 2.5 hours for the flashlight, or 6 days for the warning flasher. Optional NiMH batteries will increase run times by 300%, alkaline batteries by 250%. Waterproof and floats too! China.

17-0135 Solar Flashlight w/Warning Flashers   **$25**

Forever Flashlight

This is an incredibly innovative flashlight that never wears out or runs down. Just shake the light to charge it up. Ten seconds of shaking delivers about two minutes of light. The internal magnetic generator turns shaking motion into electricity, which is stored in a large capacitor to run the bright LED lamp. Clear construction lets you see the internal workings. This is one flashlight that will never let you down. Great for emergencies and car trunks. Taiwan.

06-0529 Forever Flashlight   **$29**

TerraLux LED Flashlight Conversion

Our astonishing LED conversion for common Mini Maglite® 2AA flashlights delivers two to three times as

much bright blue-white light, runs two to four times longer on a set of batteries, and the lamp, with up to 100,000 hour life expectancy, will outlive many flashlights. Our kit supplies the MicroStar2 lamp and a custom reflector. You supply the MiniMag PowerPush™ technology keeps the bulb at full light output even with dropping battery voltage. When batteries are exhausted, it shifts to emergency moon mode for reduced, but continuing, light output. USA.

06-0524 MicroStar2 Flashlight Kit   **$25**
The MicroStar2 kit is not made or endorsed by Mag Instruments, Inc.

Solar Navigator Lantern

Whether outdoors at night or navigating a dark room, being able to switch on the light remotely can be an asset. Our multi-purpose Solar Navigator Lantern is a brilliant fluorescent light that can be turned on by remote control. Plus, the motion sensor can detect movement and switch on from as far away as 50 feet. Plugged in to an outlet, it also lights up automatically if the power is interrupted. Light head swivels 45˚ for variable lighting angles and has built-in solar module on back. Place in full direct sunlight for optimal battery-charging capabilities. Comes with remote control on key chain, built-in battery, standard AC/DC adapter, and convenient carrying handle. Each full charge lasts 7 hours. 14.5"H x 8.25"W x 2.5" diameter. China.

14-0218 Solar Navigator Lantern   **$99**

EverLite Solar Spotlight

For camping, backyard, or emergency use, our high-tech solar spotlight is waterproof, safe in any climate, and runs over 24 hours per charge. Just three hours of full sun delivers 12 hours of ultra-bright LED run time from the built-in NiMH battery pack. Even 10 hours of overcast will deliver 12 hours of light. The bright, blue-white

light has eight LEDs, delivering a total of 50 lumens. The lightweight lamp and base stows on the back of the solar panel along with the 15-foot cord. The lamp assembly can be unplugged for full portability. If the lamp arm is left in the ON position (up), our spotlight will automatically turn on at dusk and off at dawn. Comes with a convenient nylon mesh bag that allows daytime charging right through the bag. Options include plug-in chargers for 12- or 120-volt that allow quick recharging where the sun doesn't shine, and a nifty 12 vdc output cable to deliver power from the lamp base to your cellphone or other small 12v appliance. 10.25" x 6.6" x 2.5", weighs 2.7 lbs. China.

14-0299 EverLite Solar Spotlight
         w/Carrying Case   **$119**
17-0290 EverLite 12V DC Charger   **$12**
17-0289 EverLite 120V AC Charger   **$12**
17-0291 EverLite 12V DC Output Adapter   **$25**

Hi-Output LEDs Outshine All Others

The Princeton Tec Yukon HL headlamp delivers a major breakthrough in LED design. This hybrid features a 1-watt Luxeon emitter that delivers 25 lumens for brightness and distance that rivals the highest output incandescent, halogen, or xenon lamps. For closer work with a broad even beam spread, flip to the three hi-output 5mm LEDs. Both light sources deliver crisp white light that doesn't yellow as the batteries fade. Uses three AA batteries in rear pack. Runs approximately 44 hours on the big Luxeon LED, or 120 hours on the three LEDs. Impact-resistant, lightweight adjustable design fits helmets or bare heads. 8 oz. USA.

06-0538 Princeton Tec Yukon HL   **$59**

The Flashlight of the Future

The Princeton Tec Impact XL is the best four-AA flashlight available. The powerful 1-watt Luxeon LED lamp delivers 25 lumens of crisp white light that never yellows as the batteries fade. It runs an incredible 70 hours on a set of batteries. The bulb lasts over 10,000 hours, and isn't bothered by anything less than a crushing impact. It is waterproof to 100 meters with secure rubber over a molded grip. This is the most reliable, longest-lasting flashlight you'll ever find. 6.4 oz. with batteries. USA.

06-0539 Princeton Tec Impact XL   **$39**

CHAPTER 6

Water Development

WATER IS THE SINGLE MOST IMPORTANT INGREDIENT in any homestead. Without a dependable water source, you can't call any place home for very long. An age-old saying goes, "You buy the water and the land comes for free." Gaiam Real Goods offers a wide variety of water development solutions for remote, and not so remote, homes. We specialize in products that are made specifically for solar-, battery-, wind-, or water-powered pumping. These pumps are designed for long hours of dependable duty in out-of-the-way places where the utility lines don't reach. For instance, ranchers are finding that small solar-powered submersible pumps are far cheaper and more dependable than the old wind pumps we're used to seeing dotting the Great Plains. Many state and national parks use our renewable-energy powered pumps for their backcountry campgrounds. The great majority of pumps we sell are solar- or battery-powered electric models, so we'll cover them first.

The initial cost of solar-generated electricity is high. Because of this high initial cost, most solar pumping equipment is scaled toward modest residential needs, rather than larger commercial or industrial needs. By wringing every watt of energy for all it's worth, we keep the start-up costs reasonable. Solar-powered pumps tend to be far more efficient than their conventional AC-powered cousins.

You don't miss your water till the well runs dry.

—Folk saying

The Three Components of Every Water System

Every rural water system has three easily identified basic components.

A Source. This can be a well, a spring, a pond, a creek, collected rainwater, or the big expensive tanker truck that hauls in a load every month.

A Storage Area. This is sometimes the same as the source, and sometimes it is an elevated or pressurized storage tank.

A Delivery System. This used to be as simple as the bucket at the end of a rope. But given a choice, most of us would prefer to have our water arrive under pressure from a faucet. Hauling water, because it's so heavy, and because we use such surprising amounts of it, gets old fast.

The first two components, source and storage, you need to produce locally; however, we can offer a few pointers based on experience and some of the better, and worse, stories we've heard.

Sources

WELLS

The single most common domestic water source is the well, which can be hand dug, driven, or drilled. Wells are less prone than springs, streams, or ponds to pick up surface contamination from animals, pesticides and herbicides, and the like. Hand-dug wells are usually about 3 feet or larger in diameter and rarely more than 40 feet deep. These are dangerous to build, they need regular cleaning, and it is dangerous to have them lying around unused on your property. Drilled or driven wells are standard practice now. The most common sizes for residential driven or drilled wells are 4 inches or 6 inches. Beware of the "do-it-yourself" well-drilling rigs, which rarely can insert pipe larger than 2 inches. This restricts your pump choices

Ranchers are finding that small solar-powered submersible pumps are far cheaper and more dependable than the old wind pumps we're used to seeing dotting the Great Plains.

APPROXIMATE DAILY WATER USE FOR HOME AND FARM

USAGE	GALLONS PER DAY
Home	
National average residential use	60–70 per person
As above, without flush toilet	25–40 per person
Drinking and cooking water only	5–10 per person
Lawn – Garden – Pool	
Lawn sprinkler, per 1,000 sq. ft., per sprinkling	600 (approx. 1 inch)
Garden sprinkler, per 1,000 sq. ft., per sprinkling	600 (approx. 1 inch)
Swimming pool maintenance, per 100 sq. ft. surface area	30 per day
Farm (maximum needs)	
Dairy cows	20 per head
Dry cows or heifers	15 per head
Calves	7 per head
Beef cattle, yearlings (90°F)	20 per head
Beef, brood cows	12 per head
Sheep or goats	2 per head
Horses or mules	12 per head
Swine, finishing	4 per head
Swine, nursing	6 per head
Chickens, laying hens (90°F)	9 per 100 birds
Chickens, broilers (90°F)	6 per 100 birds
Turkeys, broilers (90°F)	25 per 100 birds
Ducks	22 per 100 birds
Dairy sanitation – milk room and milking parlor	500 per day
Flushing barn floors	10 per 100 sq. ft.
Sanitary hog wallow	100 per day

Prior to developing a spring, you may want to give some consideration to any ecosystems it supports.

Ponds are one of the least expensive and most delightfully pleasing methods of supplying water, not only for your own needs, but also for the wildlife population that your pond will soon support.

to only the least-efficient jet-type pumps. A 4-inch casing is the smallest you want, that's the minimum for a submersible pump.

SPRINGS

Many folks with country property are lucky enough to have surface springs that can be developed. At the very least, springs need to be securely fenced to keep out wildlife. More commonly, springs are developed with either a backhoe and 2- or 3-foot concrete pipe sections sunk into the ground, or drilled and cased with one of the newer, lightweight horizontal drilling rigs. This helps to ensure that the water supply won't be contaminated by surface runoff or animals. To be respectful to your land and the critters that lived there before you came, prior to developing a spring you may want to give some consideration to any ecosystems it supports. If there are

no other springs nearby, try to ensure that some runoff will still continue after development.

PONDS, STREAMS, AND OTHER SURFACE WATER

Some water systems are as simple as tossing the pump intake into the existing lake or stream. Surface water usually is used only for livestock or agricultural needs—although for many homesteads with gardens and orchards, this can be 90% of your total water use. If you don't have a surface water source on your property, consider putting in a pond. Ponds are one of the least expensive and most delightfully pleasing methods of supplying water, not only for your own needs, but also for the wildlife population that your pond will soon support. They often are used to hold winter and spring runoff for summertime use. Don't drink or cook with surface water unless it has been treated or purified first (see chapter 8 for more help on this topic).

Storage

You might need to provide some means of water storage for any number of reasons. The most common are: To get through long dry periods, to provide pressure through elevated storage or pressurized air, to keep drinking water clean and uncontaminated, or to prevent freezing.

SURFACE STORAGE

Those with ponds, lakes, streams, or springs may not need any additional storage. Let the livestock find their own way to the water, or pump straight to whatever needs irrigation. However, many systems will need to pump water to an elevated storage site, or a large pressure tank, in order to develop pressure. You may need seasonal freeze protection, which means protected or underground storage.

POND LINERS

Custom-made polyvinyl liners are available for ponds or leaking tanks. Liners have revolutionized pond construction. They are designed for installation during pond construction and are then buried with 6 inches of dirt around the edges, practically guaranteeing a leak-free pond even when working with gravel, sand, or other problem soils. When buried, the life expectancy of these liners is 50 years-plus. They make reliable pond construction possible in locations that normally could not accommodate such inexpensive water storage methods.

TANKS

Covered tanks of one sort or another are the most common and longest-lasting storage solutions. The cover must be screened and tight enough to keep critters such as lizards, mice, and squirrels from drowning in your drinking water (always an exciting discovery). The most common tank materials are polypropylene, fiberglass, and concrete or ferro-cement.

Plastic Tanks. Both fiberglass and polypropylene tanks are commonly available. Your local farm supply store is usually a good source. All plastic tanks will suffer slightly from UV degradation in sunlight. Simply painting the outside of the tank will stop the UV degradation and will probably make the tank more aesthetically pleasing. If this tank is for your drinking water, make sure it's internally coated with an FDA-approved material for drinking water. Most farm supply stores will offer both "drinking water grade" and "utility grade" tanks. Utility grade tanks often use recycled plastic materials and probably will leach some polymers into the water, particularly when new. Some plastic tanks can be partially or completely buried for freeze protection. Ask before you buy if this is a consideration for you.

Liners make reliable pond construction possible in locations that normally could not accommodate such inexpensive water storage methods.

All plastic tanks will suffer slightly from UV degradation in sunlight. Simply painting the outside of the tank will stop the UV degradation and will probably make the tank more aesthetically pleasing.

A Mercifully Brief Glossary of Pump Jargon

Flow: The measure of a pump's capacity to move liquid volume. Given in gallons per hour (gph), gallons per minute (gpm), or for you worldly types who have escaped the shackles of archaic measurement, liters per minute (lpm).

Foot Valve: A check valve (one-way valve) with a strainer. Installed at the end of the pump intake line, it prevents loss of prime, and keeps large debris from entering the pump.

Friction Loss: The loss in pressure due to friction of the water moving through a pipe. As flow rate increases and pipe diameter decreases, friction loss can result in significant flow and head loss.

Head: Two common uses. 1) The pressure or effective height a pump is capable of raising water. 2) The height a pump is actually raising the water in a particular installation.

Lift: Same as Head. Contrary to the way this term sounds, pumps do not suck water, they push it.

Prime: A charge of water that fills the pump and the intake line, allowing pumping action to start. Centrifugal pumps will not self-prime. Positive displacement pumps will usually self-prime if they have a free discharge—no pressure on the output.

Suction Lift: The difference between the source water level and the pump. Theoretical limit is 33 feet; practical limit is 10 to 15 feet. Suction lift capability of a pump decreases 1 foot for every 1,000 feet above sea level.

Submersible Pump: A pump with a sealed motor assembly designed to be installed below the water surface. Most commonly used when the water level is more than 15 feet below the surface, or when the pump must be protected from freezing.

Surface Pump: Designed for pumping from surface water supplies such as springs, ponds, tanks, or shallow wells. The pump is mounted in a dry, weatherproof location less than 10 to 15 feet above of the water surface. Surface pumps cannot be submerged and be expected to survive.

Concrete Tanks. These are one of the best and longest-lasting storage solutions, but they are expensive and/or labor intensive initially. All but the smallest concrete tanks are built on site. They can be concrete block, ferro-cement, or monolithic block pours. Although monolithic pours require hiring a contractor with specialized forming equipment, these tanks are usually the most trouble free in the long run. Any concrete tank will need to be coated internally with a special sealer to be watertight. Concrete tanks can be buried for freeze protection, and to keep the water cool in hot climates. They can cost $0.40 to $0.60 per gallon, but in the long run they're well worth it.

Pressure Tanks. These are used in pumped systems to store pressurized water so that the pump doesn't have to start for every glass of water. They work by squeezing a captive volume of air, since water doesn't compress. The newer, better types use a diaphragm—sort of a big heavy-duty balloon—so that the water can't

> Pressure tanks are one of the few things in life where bigger really is better.

John Schaeffer's pond in Hopland, California.

absorb the air charge. This was a problem with the older-style pressure tanks. Pressure tanks are rated by their total volume. Draw-down volume, the amount of water that actually can be loaded into and withdrawn from the tank under ideal conditions, is typically about 40% of total volume. So a "20-gallon" pressure tank can really only deliver about 8 gallons before starting the

pump to refill. Pressure tanks are one of the few things in life where bigger really is better. With a larger pressure tank, your pump doesn't have to start as often, it will use less power and live longer, your water pressure will be more stable, and if power fails you'll have more pressurized water in storage to tide you over. Forty-gallon capacity is the minimum for residential use, and more is better.

Delivery Systems

A few lucky folks are able to collect and store their water high enough above the level of intended use (you need 45 vertical feet for 20 psi) so that the delivery system will simply be a pipe, and the weight of the water will supply the pressure free of charge. Most of us are going to need a pump, or two, to get the water up from underground, and/or to provide pressure. We'll cover electrically-driven solar and battery-powered pumping first, then water-powered and wind-powered pumps.

The standard rural utility-powered water delivery system consists of a submersible pump in the well delivering water into a pressure tank in some location that's safe from freezing. A pressure tank extends the time between pumping cycles by saving up some pressurized water for delivery later. This system usually solves any freezing problems by placing the pump deep inside the well, and the pressure tank indoors. The downside is that the pump must produce enough volume to keep up with any potential demand, or the pressure tank will be depleted and the pressure will drop dramatically. This requires a ⅓-hp pump minimally, and usually ½ hp or larger. Well drillers often will sell a much larger than necessary pump because it increases their profit and guarantees that no matter how many sprinklers you add in the future, you'll have sufficient water-delivery capacity. (And they don't have to pay your electric bill!) This is fine when you have large amounts of utility power available to meet heavy surge loads, but it's very costly to power with a renewable energy system because of the large equipment requirements. We try to work smarter, smaller, and use less-expensive resources to get the job done.

Solar-Powered Pumping

Where Efficiency is Everything

PV modules are expensive, and water is surprisingly heavy. These two facts dominate the solar-pumping industry. At 8.3 pounds per gallon, a lot of energy is needed to move water uphill. Anything we can do to wring a little more work out of every last watt of energy is going to make the system less expensive initially. Because of these economic realities, the solar-pumping industry tends to use the most efficient pumps available. For many applications that means a positive-displacement type of pump. This class of pumps prevents the possibility of the water slipping from high-pressure areas to lower-pressure areas inside the pump. Positive-displacement pumps also ensure that even when running very slowly—such as when powered by a PV module under partial light conditions—water still will be pumped. As a general rule, positive-displacement pumps manage four to five times the efficiency of centrifugal pumps, particularly when lifts over about 60 feet are involved. Several varieties of positive-displacement pumps are commonly available. Diaphragm pumps, rotary-vane pumps, piston pumps, and the newest darling, helical-rotor pumps, will all be found in our product pages.

Positive-displacement pumps have some disadvantages. They tend to be noisier, as the water is expelled in lots of little spurts. They usually pump smaller volumes of water, they must start under full load, most require periodic maintenance, and most won't tolerate running dry. These are reasons why this class of pumps isn't used more extensively in the AC-powered pumping industry.

Most AC-powered pumps are centrifugal types. This type of pump is preferred because of easy starting, low noise, smooth output, and minimal maintenance requirements. Centrifugal pumps are good for moving large volumes of water at relatively low pressure. As pressure rises, however, the water inside the centrifugal pump "slips" increasingly, until finally a pressure is reached at which no water is actually leaving the pump. This is 0% efficiency. Single-stage centrifugal pumps suffer at lifts over 60 feet. To manage higher lifts, as in a submersible well pump, multiple stages of centrifugal pump impellers are stacked up.

In the solar industry, centrifugal pumps are used for pool pumping, and for some circulation duties in hot-water systems. But in all applications where pressure exceeds 20 psi, you'll find us recommending the slightly noisier, occasional-maintenance-requiring, but vastly more efficient positive-displacement-type pumps. For instance, an AC submersible pump running at 7 to 10% efficiency is considered "good." The helical-rotor submersible pumps we promote run at close to 50% efficiency.

For Highest Efficiency, Run PV-Direct

We often design solar pumping systems to run PV-direct. That is, the pump is connected directly to the photovoltaic (PV) modules with no batteries involved in the system. The electrical-to-chemical conversion in a battery isn't 100% efficient. When we avoid batteries and deliver the energy directly to the pump, 20 to 25% more water gets pumped. This kind of system is ideal when the water is being pumped into a large storage tank, or is being used immediately for irrigation. It also saves the initial cost of the batteries, the maintenance and periodic replacement they require, plus the charge controllers and the fusing/safety equipment that batteries demand. PV-direct pumping systems, which are designed to run all day long, make the most of your PV investment, and help us get around the lower gallon-per-minute output of most positive-displacement pumps.

However (every silver lining has its cloud), we like to use one piece of modern technology on PV-direct systems that isn't often found on battery-powered systems. A linear current booster, or LCB for short, is a solid-state marvel that will help get a PV-direct pump running earlier in the morning, keep it running later in the evening, and sometimes make running at all a possibility on hazy or cloudy days. An LCB will convert excess PV voltage into extra amperage when the modules aren't producing quite enough current for the pump. The pump will run more slowly than if it had full power, but if a positive-displacement pump runs at all, it delivers water. LCBs will boost water delivery in most PV-direct systems by 20% or more, and we usually recommend a properly sized one with every system.

An AC submersible pump running at 7 to 10% efficiency is considered "good." The helical-rotor submersible pumps we promote run at close to 50% efficiency.

PV-direct pumping systems, which are designed to run all day long, make the most of your PV investment, and help us get around the lower gallon-per-minute output of most positive-displacement pumps.

Direct Current (DC) Motors for Variable Power

DC motors have the great advantage of accepting variable voltage input without distress. Common AC motors will overheat if supplied with low voltage. DC motors simply run slower when the voltage drops.

Pumps that are designed for solar use DC electric motors. PV modules produce DC electricity, and all battery types store DC power. DC motors have the great advantage of accepting variable voltage input without distress. Common AC motors will overheat if supplied with low voltage. DC motors simply run slower when the voltage drops. This makes them ideal partners for PV modules. Day and night, clouds and shadows; these all affect the PV output, and a DC motor simply "goes with the flow"!

Which Solar-Powered Pump Do You Want?

That depends on what you're doing with it, and what your climate is. We'll start with the most common and easiest choices, and work our way through to the less common.

PUMPING FROM A WELL

Do you have a well that's cased with a 4-inch or larger pipe, and a static water level that is no more than 750 feet below the surface? Perfect. We carry several brands of proven DC-powered submersible pumps with a range of prices, lift, and volume capabilities. The SHURflo Solar Sub is the lowest-cost system with lift up to 230 feet, and sufficient volume for most residential homesteads. The bigger helical-rotor type sub pumps like the Grundfos and ETA pumps are available in over a dozen models, with lifts up to 750 feet, or volume over 25 gpm, depending on the model. Performance, prices, and PV requirements are listed in the products section. The SHURflo Solar Sub is a diaphragm-type pump, and unlike almost any other submersible pump, it can tolerate running dry. The manufacturer says just don't let it run dry for more than a month or two! This feature makes this pump ideal for many low-output wells.

Complete submersible pumping systems— PV modules, mounting structure, LCB, and pump—range from $1,500 to $11,000 depending on lift and volume required. Options such as float switches that will automatically turn the pump on and off to keep a distant storage tank full are inexpensive and easy to add when using an LCB with remote control, as we recommend.

Because many solar pumps are designed to work all day at a slow but steady output, they won't keep up directly with average household

fixtures, like your typical AC sub pump. This often requires some adjustment in how your water supply system is put together. For household use, we usually recommend, in order of cost and desirability:

Option 1. Pumping into a storage tank at least 50 feet higher than the house, if terrain and climate allow.

Option 2. Pumping into a house-level storage tank if climate allows, and using a booster pump to supply household pressure.

Option 3. Pumping into a storage tank built into the basement for hard-freeze climates, and using a booster pump to supply household pressure.

Option 4. Using battery power from your household renewable energy system to run the submersible pump, and using a big pressure tank (80 gallons minimum).

Option 5. Using a conventional AC-powered submersible pump, large pressure tank(s), and your household renewable energy system with large inverter. There is some loss of efficiency in this set-up, but it's the standard way to get the job done in freezing climates, and your plumber won't have any problems understanding the system. We strongly recommend one of the helical-rotor-type Grundfos SQ Flex pumps for this option. Start-up surge will be kind to your inverter, output is 5 to 9 gpm, and power use is a quarter of conventional AC pumps.

Many folks, for a variety of reasons, already have an AC-powered submersible pump in their well when they come to us, but are real tired of having to run the generator to get water. For wells with 6-inch and larger casings, it's usually possible to install both the existing AC pump and a submersible DC pump. If your AC pump is 4 inches in diameter, the DC pump can be installed underneath it. The cabling, safety rope, and ½-inch poly delivery pipe from the DC submersible will slip around the side of the AC pump sitting above it. Just slide both pumps down the hole together. It's often comforting to have emergency back-up for those times when you need it, like when it's been cloudy for three weeks straight, or the fire is coming up the hill, and you want a lot of water fast!

PUMPING FROM A SPRING, POND, OR OTHER SURFACE SOURCE

Your choices for pumping from ground level are a bit more varied, depending on how high you need to lift the water, and how many gallons per minute you want. Surface-mounted pumps are not freeze tolerant. If you live in a freezing

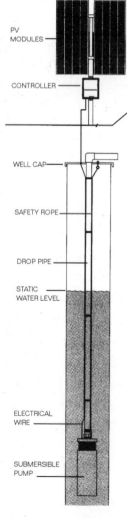

PV MODULES

CONTROLLER

WELL CAP

SAFETY ROPE

DROP PIPE

STATIC WATER LEVEL

ELECTRICAL WIRE

SUBMERSIBLE PUMP

A simple PV-direct solar pumping system

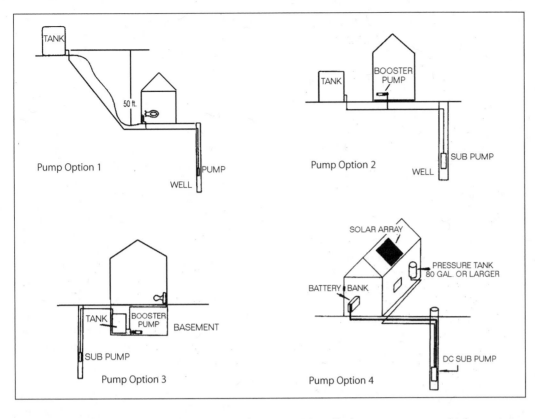

Pump Option 1 — TANK, 50 ft., PUMP, WELL

Pump Option 2 — TANK, BOOSTER PUMP, SUB PUMP, WELL

Pump Option 3 — TANK, BOOSTER PUMP, BASEMENT, SUB PUMP

Pump Option 4 — SOLAR ARRAY, PRESSURE TANK 80 GAL. OR LARGER, BATTERY BANK, DC SUB PUMP

Because many solar pumps are designed to work all day at a slow but steady output, they won't keep up directly with average household fixtures, like your typical AC sub pump.

climate, make sure that your installation can be (and is!) completely drained before freezing weather sets in. If you need to pump through the hard-freeze season, we recommend a submersible pump as described above.

Pumps Don't Suck! (They Push)

Pumps don't like to pull water up from a source. Or, put simply, Dr. Doug's #1 Principle of Pumping: Pumps don't suck, pumps push. In order to operate reliably, your surface-mounted pump must be installed as close to the source as practical. In no case should the pump be more than 10 feet above the water level. With some

positive-displacement pumps, higher suction lifts are possible, but not recommended. You are simply begging for trouble. If you can get the pump closer to the source, and still keep it dry and safe, do it! You'll be rewarded with more dependable service, longer pump life, more water delivery, and less power consumption.

Low-Cost Solutions

For modest lifts up to 50 or 60 feet and volumes of 1.5 to 3.0 gpm, we have found the SHURflo diaphragm-type pumps to be moderately priced and tolerant of abuses that would kill other pumps. They can tolerate silty water and sand without distress. They'll run dry for hours and hours without damage. But you get what you pay for. Life expectancy is usually two to five years depending on how hard, and how much, the pump is working. Repairs in the field are easy, and disassembly is obvious. We carry a full stock of repair parts, but replacement motors don't cost much less than a new pump. Diaphragm pumps will tolerate sand, algae, and debris without damage, but these may stick in the internal check valves and reduce or stop output, necessitating disassembly to clean out the debris. Who needs the hassle? Filter your intake!

Longer-Life Solutions

For higher lifts, or more volume, we often go to a pump type called a rotary-vane. Examples

Need Pump Help?

If you would like the help of our technical staff in selecting an appropriate pump, controller, power source, etc., we have a Solar Water Supply Questionnaire at the end of this chapter. It will give our staff the information we need in order to thoroughly and accurately recommend a water supply system for you. You can mail or fax your completed form. Please give us a daytime phone number if at all possible! Call us at 1-800-919-2400.

For wells with 6-inch and larger casings, it's usually possible to install both the existing AC pump and a submersible DC pump. It's often comforting to have emergency back-up for those times when you need it, like when it's been cloudy for three weeks straight, or the fire is coming up the hill, and you want a lot of water fast!

Any household pressure system requires a pressure tank. A 20-gallon tank is the minimum size we recommend for a small cabin; full-size houses usually have 40-gallon or larger tanks.

Diaphragm-type pump.

The smarter solar hot-water systems simply use a small PV panel wired directly to a DC pump.

Hot water circulation pump.

include the Slowpump and Flowlight Booster pumps. Rotary-vane pumps are capable of lifts up to 440 feet, and volumes up to 4.5 gpm, depending on the model. Of all the positive-displacement pumps, they are the quietest and smoothest. But they will not tolerate running dry, or abrasives of any kind in the water. It's very important to filter the input of these pumps with a 10-micron or finer filter in all applications. Rotary-vane pumps are very long-lived, but will eventually require a pump head replacement or a rebuild.

HOUSEHOLD WATER PRESSURIZATION

We promote two pumps that are commonly used to pressurize household water: The Chevy and Mercedes models, if you will.

The Chevy model. SHURflo's Medium-Flow pump is our best-selling pressure pump. It comes with a built-in 20 to 40 psi pressure switch. With a 2.5 to 3.0 gpm flow rate it will keep up with most household fixtures, garden hoses excluded. The diaphragm pump is reliable, easy to repair, but somewhat noisy, and has a limited life expectancy. We recommend 24-inch flexible plumbing connectors in a loop on both sides of this pump, and a pressure tank plumbed in as close as possible to absorb most of the buzz.

The Mercedes model. Flowlight's rotary-vane pump is our smoothest, quietest, largest-volume pressure pump. It delivers 3 to 4.5 gpm at full pressure, and is very long-lived, but quite expensive initially. This pump will keep up easily with garden hoses, sprinklers, and any other normal household use. Brushes are externally replaceable, and will last five to ten years. Pump life expectancy is fifteen to twenty years.

Any household pressure system requires a pressure tank. A 20-gallon tank is the minimum size we recommend for a small cabin; full-size houses usually have 40-gallon or larger tanks. Pressure tanks are big, bulky, and expensive to ship. Get one at your local hardware or building supply store.

SOLAR HOT-WATER CIRCULATION

Most of the older solar hot-water systems installed during the tax-credit heydays of the early 1980s used AC pumps with complex controllers and multiple temperature sensors at the collector, tank, plumbing, ambient air, etc. This kind of complexity allows too many opportunities for Murphy's Law.

The smarter solar hot-water systems simply use a small PV panel wired directly to a DC pump. When the sun shines a little bit, producing a small amount of heat, the pump runs slowly. When the sun shines bright and hot, producing lots of heat, the pump runs fast. Very simple, but absolutely perfect. System control is achieved with an absolute minimum of "stuff." We carry several hot-water circulation pumps for various-size systems. The best choice for most residential systems is the El-Sid pump. This is a solid-state, brushless DC circulation pump that was designed from scratch for PV-direct applications. The El-Sid pump comes in small, medium, and large sizes now, with open-discharge flow rates of 2, 3.3, and 6 gallons per minute. El-Sid pumps require a 5- to 30-watt PV module for drive, and life expectancy is three to four times longer than any other DC circulation pump. Volume and lift are sharply limited, however. These are circulation pumps, not lift pumps. They're meant to stir the fluid round and round in a closed system.

For solar hot-water systems with long, convoluted collection loops creating a lot of pipe friction, we can use multiple El-Sid pumps. If some amount of lift is involved, we have to look at more robust pumps like the Hartell or the SunCentric series, which will require substantially more PV power.

SWIMMING POOL CIRCULATION

Yes, it's possible to live off-the-grid and still enjoy luxuries like a swimming pool. In fact, pool systems dovetail nicely with household systems in many climates. Houses generally require a minimum of PV energy during the summer because of the long daylight hours, yet the maximum of energy is available. By switching a number of PV modules to pool pumping in the summer, then back to battery charging in the winter, you get better utilization of resources.

We offer several DC pool pumps. DC pumps run somewhat more efficiently than AC pumps, so a slightly smaller DC pump can do the same amount of work as a larger AC pump. We also strongly recommend using a low-back-pressure cartridge-type pool filter. Diatomaceous-earth filters are trouble. They have high back pressure, and will greatly slow circulation, or increase power use. See the product section for specs and prices. Please consult with our technical staff regarding pool filters, PV array sizing, and switching equipment.

Water-Powered Pumps

A few lucky folks have an excess supply of falling water available. This falling-water energy can be used to pump water. Both the High Lifter and ram-type water pumps use the energy of falling water to force a portion of that water up the hill to a storage tank.

RAM PUMPS

Ram pumps have been around for many decades, providing reliable water pumping at almost no cost. They are more commonly used in the eastern United States where modest falls and large flow rates are the norm, but they will work happily almost anyplace their minimum flow rate can be satisfied. Rams will work with a minimum of 1.5 feet to a maximum of about 20

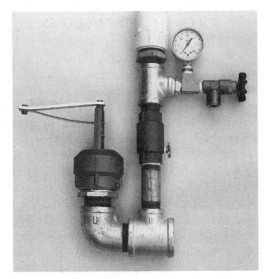

Ram pump.

feet of fall feeding the pump. Minimum flow rates depend on the pump size; see the product section for specs.

Here's how ram pumps work: A flow is started down the drive pipe and then shut off suddenly. The momentum of moving water slams to a stop, creating a pressure surge that sends a little squirt of water up the hill. How much of a squirt depends on the pump size, the amount of fall, and the amount of lift. Output charts accompany the pumps in the product section. Each ram needs to be tuned carefully for its particular site. Ram pumps are not self-starting. If they run short of water, they will stop pumping and simply dump incoming water, so don't buy too big. Rams make some noise. A lot less than a gasoline-powered pump, but the constant twenty-four-hours-a-day chunk-chunk-chunk is a consideration for some sites. Ram

pumps deliver less than 5% of the water that passes through them, and the discharge must be into an unpressurized storage tank or pond. But they work for free and have life expectancies measured in decades.

THE HIGH LIFTER PUMP

This pump is unique. It works by simple mechanical advantage. A large piston at low water pressure pushes a smaller piston at higher water pressure. High Lifters recover a much greater percentage of the available water than ram pumps do, but they generally require greater fall into the pump. This makes them better suited for more mountainous territory. They are available in two ratios, 4.5 to 1 and 9 to 1. Fall-to-lift ratios and waste-water-to-pumped-water ratios are also either 4.5 to 1 or 9 to 1. Note, however, that as the lift ratio gets closer to theoretical maximum, the pump is going to slow down and deliver fewer gallons per day. High Lifters are self-starting. If they run out of water, they will simply stop and wait, or slow down to match what water is available. This is a very handy trait for unattended or difficult to attend sites. The only serious disadvantage of the High Lifter is wear caused by abrasives. O-rings are used to seal the pistons against the cylinder walls. Any abrasives in the water wear out the O-rings quickly, so filtering of the intake is strongly encouraged. High Lifter pumps can be over-hauled fairly easily in the field, but the O-ring kit costs a lot more than filter cartridges. Output charts for this pump are included in the product section.

Wind-Powered Water Pumps

If you've been reading this far because you want to buy a nostalgic old-time jack-pump windmill, we're going to disappoint you. They are still made, but are very expensive, typically $7,500 and up. We don't sell them, and don't recommend them. They are quite a big deal to set up and install into the well, and require routine yearly service at the top of the tower. This is technology that largely has seen its day. Submersible pumps powered by PV panels are a much better choice for remote locations now.

There are a couple of ways to run a pump with wind power that have seen good success.

High Lifter pump.

Ram-type water pumps use the energy of falling water to force a portion of that water up the hill to a storage tank.

High Lifters recover a much greater percentage of the available water than ram pumps do, but they generally require greater fall into the pump.

The air is piped down the well and runs through a carefully engineered air injector. As it rises back up the supply tube, it carries slugs of water in between the bubbles.

Wind-powered pump.

COMPRESSED-AIR WATER PUMPS

Airlift pumps use compressed air. Three models are available depending on lift and volume needs. They all use a simple pole-mounted turbine that direct-drives an air compressor with a wind turbine. The air is piped down the well and runs through a carefully engineered air injector. As it rises back up the supply tube, it carries slugs of water in between the bubbles. The lift/submergence ratio of this pump is fairly critical. Lift is the vertical rise between the standing water level in the well and your tank. Approximately 30% of the lift is the recommended distance for the air injector to be submerged below the standing level. As lifts edge over 200 feet, the submergence ratio rises to a maximum of 50%. Too little submergence and the air will separate from the water; too much and the air will not lift the water, though considerable latitude exists between these performance extremes. This pump isn't bothered by running dry. Output depends on wind speed, naturally, but the largest model is capable of over 20 gpm at lower lifts, or can lift a maximum of 315 feet. The air compressor requires an oil change once a year. See the product section for more details and output specs.

WIND GENERATORS RUNNING SUBMERSIBLE PUMPS

The larger wind generator manufacturers, Bergey and Whisper, both offer options that allow the three-phase wind turbine output to power a three-phase submersible pump directly. These aren't residential-scale systems and are mostly used for large agricultural projects, or village pumping systems in less industrialized countries. They are moderately expensive initially. Contact our technical staff for more information on these options.

Freeze Protection

In most areas of the country, freezing is a major consideration when installing plumbing and water storage systems. For outside pipe runs, the general rule is to bury the plumbing below frost level. For large storage tanks burial may not be feasible, unless you go with concrete. In moderate climatic zones, simply burying the bottom of the tank a foot or two along with the input/output piping is sufficient. In some locations, due to climate or lack of soil depth, outdoor storage tanks simply aren't feasible. In these situations, you can use a smaller storage tank built into a corner of the basement with a separate pressure-

boosting pump, or you can pump directly into a large pressure-tank system.

Other Considerations and Common Questions

We hope that by this point you've zeroed in on a pump or pumps that seem to be applicable to your situation. If not, our technical staff will be happy to discuss your needs and recommend an appropriate pumping system—which may or may not be renewable-energy powered. Filling out the Solar Water Supply Questionnaire that follows will supply all the answers we're likely to need when we talk with you. At this point, another crop of questions usually appears, such as . . .

HOW FAR CAN I PUT THE MODULES FROM THE PUMP?

Often the best water source will be deep in a heavily wooded ravine. It's important that your PV modules have clear, shadow-free access to the sun for as many hours as practical. Even a fist-sized shadow will effectively turn off most PV modules. The hours of 10 A.M. to 2 P.M. are usually the minimum your modules want clear solar access, and if you can capture full sun from 9 A.M. to 3 P.M., that's more power for you. If the pump is small, running off one or two modules, then distances up to 200 feet can be handled economically. Longer distances are always possible, but consult with our tech staff or check out the wire-sizing formula on page 200 first, because longer distances require large (expensive!) wire. Many pumps routinely come as 24- or 48-volt units now, as the higher voltage makes long-distance transmission much easier. The Grundfos sub pump accepts DC input up to 300 volts. If you need to run more than 300 feet for sunshine, this may be just the ticket.

WHAT SIZE WIRE DO I NEED?

This depends on the distance, and the amount of power you are trying to move. See the Wire Sizing Chart on page 200 and formula that can properly size wire for any distance, at any voltage, AC or DC. Or, just give our tech staff a call, we do this kind of consultation all the time. Going down a well, 10-gauge copper submersible pump wire is the usual, although some of the larger sub pumps we're using now require 8-gauge, or even 6-gauge for very deep wells. If you need anything other than common 10-gauge wire, we'll let you know.

WHAT SIZE PIPE DO I NEED?

Many of the pumps we offer have modest flow rates of 4 gpm or less. At this rate, it's okay to use smaller pipe sizes such as ¾-inch or 1-inch for pumping delivery without increasing friction loss. However, that's for pumping delivery only. There's no reason you can't use the same pipe to take the water up the hill to the storage tank and also bring it back down to the house or garden. Pipes don't care which way the water is flowing through them. But if you do this, you'll want a larger pipe to avoid friction loss and pressure drop when the higher flow rate of the household or garden fixtures comes into play. We usually recommend at least 1¼-inch pipe for household and garden use, and 2-inch for fire hose lines. See the pipe friction loss charts on page 315 to be sure

WHAT SIZE PV MODULES DO I NEED?

A number of the pumps listed in the product section have accompanying performance and wattage tables. For instance, the rotary-vane Solar Slowpump model 2503 lifting 40 feet will deliver 2.5 gpm and requires 60 watts. Hey, what's to figure here? All PV modules are rated by how many watts they put out, right? So you just need a 60-watt module. Wrong. PV modules are rated under ideal laboratory conditions, not real life. If you want your pump to work on hot or humid days, then you'd better add 30% to the pump wattage. Heat reduces PV output, water vapor cuts available sunlight. So in our example we actually need 78 watts. Looking at available PV modules, we don't find exactly 78-watt modules, which means we buy a 75-watt module if this is a relatively temperate climate, or an 80- to 85-watt module if this is a hot (over 80°F) climate. The usual rule for sizing PV-direct arrays is to add about 20% to the pump wattage in a mild climate, or add 30% in a hot climate. And, of course, an LCB (linear current booster) of sufficient amperage capacity is practically standard equipment with any PV-direct system.

CAN I AUTOMATE THE SYSTEM?

Absolutely! Life's little drudgeries should be automated at every opportunity. The LCBs that we so strongly recommend with all PV-direct systems help us in this task. These units are all supplied with wiring for a remote control option. This allows you to install a float switch at the holding tank and a pair of tiny 18-gauge wires can be run as far as 5,000 feet back to the pump/controller/PV modules area. Float switches can be used either to pump up or pump

down a holding tank, and will automatically turn the system off when satisfied. The LCBs, float switches, and other pumping accessories are presented in the product section. LCBs can also be used on battery-powered systems when remote sensing and control would be handy.

Where Do We Go from Here?

If you haven't found all the answers to your remote water pumping needs yet, please give our experienced technical staff a call. They'll be happy to work with you selecting the most appropriate pump, power source, and accessories for your needs. We run into situations occasionally where renewable energy sources and pumps simply may not be the best choice, and we'll let you know if that's the case. For 95% of the remote pumping scenarios, there is a simple, cost-effective, long-lived renewable energy-powered solution, and we can help you develop it.

The usual rule for sizing PV-direct arrays is to add about 20% to the pump wattage in a mild climate, or add 30% in a hot climate.

Water Pumping Truths and Tips

1. Pumps prefer to **push** water, not **pull** it. In fact, most pumps are limited to 10 or 15 feet of lift on the suction side. Mother Nature has a theoretical suction lift of 33.9 feet at sea level, but only if the pump could produce a perfect vacuum. Suction lift drops 1 foot with every 1,000 feet rise in elevation. To put it simply, **Pumps Don't Suck, They Push.**

2. Water is heavy, 8.33 pounds per gallon. It can require tremendous amounts of energy to lift and move.

3. DC electric motors are generally more efficient than AC motors. If you have a choice, use a DC motor to pump your water. Not only can they be powered directly by solar modules, but your precious wattage will go further.

4. Positive-displacement pumps are far more efficient than centrifugal pumps. Most of our pumps are positive-displacement types. AC powered submersibles, jet pumps, and booster pumps are centrifugal types.

5. As much as possible, we try to avoid batteries in pumping systems. When energy is run into and out of a battery, 25% is lost. It's more efficient to take energy directly from your PV modules and feed it right into the pump. At the end of the day, you'll end up with 25% more water in the tank.

6. One pound per square inch (psi) of water pressure equals 2.31 feet of lift, a handy equation.

Solar Water Supply Questionnaire

To thoroughly and accurately recommend a water supply system, we need to know the following information about your system. Please fill out the form as completely as possible. Please give us a daytime phone number, too.

Name: _____

Address: _____

City: _____ State: _____ Zip: _____

email: _____ Phone: _____

DESCRIBE YOUR WATER SOURCE:

Depth of well: _____

Depth to standing water surface: _____

If level varies, how much? _____

Estimated yield of well (gallons per minute): _____

Well casing size (inside diameter): _____

Any problems? (silt, sand, corrosives, etc.)_____

_____

WATER REQUIREMENTS:

Is this a year-round home? _____ How many people full time?_____

Is house already plumbed? _____ Conventional flush toilets? _____

Residential gallons/day estimated: _____

Is gravity pressurization acceptable? _____

Hard freezing climate? _____

Irrigation gallons/day estimated: _____ Which months? _____

If you have a general budget in mind, how much? _____

Do you have a deadline for completion? _____

DESCRIBE YOUR SITE:

Elevation: _____

Distance from well to point of use: _____

Vertical rise or drop from top of well to point of use: _____

Can you install a storage tank higher than point of use? _____

    How much higher? _____ How far away? _____

Complex terrain or multiple usage? _____ (Please enclose map)

Do you have utility power available? _____ How far away? _____

Can well pump be connected to nearby home power system? _____

    How far? _____ Home power system power battery voltage: _____

Mail your completed questionnaire to:

Tech Staff/Water Supply, Real Goods, 13771 South Hwy. 101, Hopland, CA 95449.

Or fax to: 707-744-8771. Questions? Call us at 800-919-2400.

Dr. Doug's Homestead Plumbing Recommendations

(Learn from Dr. Doug's mistakes, so you can make your own creative new mistakes.)

TYPES OF PIPE

Metals

Black iron and galvanized iron—used for gas (propane or natural gas) plumbing, but little else. It is slow and difficult to cut and thread, and requires special, expensive tools. (Consider renting tools if needed.) A special plastic-coated iron pipe is used for buried gas lines. The insides of galvanized pipe corrode when used for water supply and eventually restrict water flow. Older houses often suffer from this affliction, which causes ultra-sensitivity of shower temperature and other exciting problems.

Cast iron—used to be the material of choice for waste plumbing, but ABS plastic thankfully has replaced it. Simple conversion adapters to go from cast iron to ABS are available if you find yourself having to repair an existing older system.

Copper—the most common material for water supply plumbing in new home construction. Most plumbing codes require copper for indoor work. Type M, the most common grade for interior work, comes in 20-foot rigid lengths. Houses typically use ¾-inch lines for feeders and ½-inch lines for fixtures. If gravity-fed water pressure will be lower than the 20- to 40-psi city standard, consider increasing your supply pipes one size. This cures many low-pressure woes by eliminating pressure loss within the house (and provides a shower that's nearly immune to temperature fluctuations). The thicker-walled flexible tubing L and K grades can be used legally in some localities, but they generally cost considerably more. Copper does not tolerate freezing at all! Copper is relatively quick and easy to work with, and you need only simple, inexpensive tools. Joints must be "sweat soldered," which takes about ten minutes to learn and is kinda fun afterward. Be sure to use the newer lead-free solders for water supply piping! Some states allow the use of flexible copper tubing for gas supply lines, in which case special compression-type fittings are used.

Bronze/Brass—beautiful stuff, but too expensive for common use. Pre-cut, pre-threaded nipples are used occasionally for dielectric water tank protection (5 inches of brass qualifies as protection under code). Use brass only where plumbing will be exposed and you want it to look nice.

Plastics

Poly (polyethylene)—this black, flexible pipe is used widely for drip and irrigation systems. Available in utility or domestic grades. Never use utility grade for drinking water! It's made with recycled plastics, and you don't know what will leach out. Poly is almost totally freezeproof. It's easy to work with (when warm) and requires common hand tools. It's sold in 100-foot or occasionally longer rolls. Barbed fittings and hose clamp fittings *will* leak (why do you suppose it's used in *drip* systems?). Do not use this cheap plumbing for any permanently pressurized installations, unless you can afford to throw water away. Poly also degrades fairly rapidly in sunlight. Two to three years is the usual unprotected lifespan. One good use for poly is with submersible well pumps where the flexibility makes for easy installations. Poly pipe is the usual choice for sub pumps unless they're pumping hundreds of feet of lift at very high pressure. Special 100- and 160-psi high-pressure versions of poly pipe are used for sub pumps.

PVC (polyvinyl chloride)—one of the wonders of chemistry that makes homesteading easy. This is the white plastic pipe that should be used for almost everything outside the house itself. It's easy to work with, and requires only common hand tools. (Although if you're doing a lot of it, a PVC cutter is a real time saver over the hacksaw.) PVC usually can survive mild freezing. Hard freezes will break joints, however, so use protection. It comes in 20-foot lengths. Sizes 1 inch and up generally are available with bell ends, saving you the expense and extra gluing of couplings. Use primer (the purple stuff), then glue (clear stuff) when assembling. PVC must be buried or protected. If exposed to sunlight, it will degrade slowly from the UV, but it will also grow algae inside the pipe! (Ask how I know....)

PVC grades: Schedule 40 is the standard PVC that is recommended for most uses. Some lighter-duty agricultural types are available. Class 125, class 160, and class 200 are graded by the psi rating of the pipe. Don't use these cheaper thin-wall classes if you're burying the pipe! Leaks from stones pushing through thinner walls are too hard to find and fix.

CPVC (chlorinated polyvinyl chloride)—a PVC formulation specifically made for hot water. The fittings and the pipe are a light tan color so you can tell the difference from ordinary PVC. Some counties and states allow the use of CPVC, most don't. It works just like PVC.

ABS (acrylonitrile butadiene styrene)—a black rigid pipe universally used for unpressurized waste and vent plumbing. It's easy to work with, requires common hand tools, and glues up quickly just like PVC. Make sure that the glue you use is rated for ABS. Some glues are universal and will do both PVC and ABS, others are specific. Buy a big can of ABS glue—these larger pipes use a lot more glue. Also, the bigger cans come with bigger swabs, which you'll need on larger pipe.

PB (polybutylene)—a flexible plastic pipe that is used occasionally for home plumbing, but mostly for radiant-floor heating systems. It requires special compression rings and tools for fittings. PB will survive considerable abuse during construction without springing leaks. It makes no reaction with concrete floors. PB is the best choice for radiant floors and has nearly infinite life expectancy when buried in concrete.

OTHER TIPS

Don't go small on supply piping. Long pipe runs from a supply tank up on the hill can produce substantial pressure drop. The typical garden needs 1¼-inch supply piping for watering and sprinklers. We recommend *at least* 1½-inch supply for most houses, and a 2-inch or larger line for a fire hose is often a very good idea.

Keep plumbing runs as short as possible. Try to arrange your house plan so that bathrooms, laundry room, and kitchen all fit back to back and can share a common plumbing wall. This saves thousands of dollars in materials costs, lots of time during construction, hours of time waiting for the hot water, and many gallons of expensive hot water.

Insulate *all* hot water pipes, even those inside interior walls before the walls are covered up. The savings and comfort over the life of the house quickly make up for the initial costs.

Store your glue cans upside down. This sounds wacky, but it works. I learned it from an old pro. Tighten the cap first, obviously. The glue seals any tiny air leaks, and keeps all the glue inside and on the cap threads from drying solid. Don't store your primer upside down, though; it can't seal the tiny leaks and will disappear.

WATER PUMPING PRODUCTS

BOOKS

Cottage Water Systems

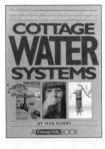

By Max Burns. An out-of-the-city guide to water sources, pumps, plumbing, water purification, and wastewater disposal, this lavishly illustrated book covers just about everything concerning water in the country. Each of the 12 chapters tackles a specific subject such as sources, pumps, plumbing how-to, water quality, treatment devices, septic systems, outhouses, alternative toilets, graywater, freeze protection, and a good bibliography for more info. This is the best illustrated, easiest to read, most complete guide to waterworks we've seen yet. 150 pages, softcover. USA.

80098   Cottage Water Systems $24.95

The Home Water Supply

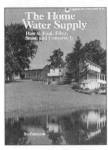

By Stu Campbell. Explains completely and in depth how to find, filter, store, and conserve one of our most precious commodities. A bit of history; finding the source; the mysteries of ponds, wells, and pumps; filtration and purification; and a good short discussion of the arcana of plumbing. After reading this book, you may still need professional help with pipes and pumps, but you will know what you're talking about. 236 pages, softcover. USA.

80205   The Home Water Supply $19

Build Your Own Water Tank

By Donnie Schatzberg. An informative booklet by one of our original customers that has been updated and greatly expanded with more text, drawings, and illustrations. It gives you all the details you need to build your own ferro-cement (iron-reinforced cement) water storage tank. No special tools or skills are required. The information given in this book is accurate and easy to follow, with no loose ends. The author has considerable experience building these tanks, and has gotten all the "bugs" out. 58 pages, softcover. USA.

80204   Build Your Own Water Tank $14

Rainwater Collection for the Mechanically-Challenged

By Suzy Banks with Richard Heinichen. Laugh your way to a home-built rainwater collection and storage system. This delightful paperback is not only the best book on the subject of rainwater collection we've ever found, it's funny enough for recreational reading, and comprehensive enough to lead a rank amateur painlessly through the process. Technical information is presented in layman's terms, and accompanied with plenty of illustrations and witty cartoons. Topics include types of storage tanks, siting, how to collect and filter water, water purification, plumbing, sizing system components, freeze-proofing, and wiring. Substantially revised and expanded in 2003. Includes a resources list and a small catalog. 108 pages, softcover. USA.

**80704   Rainwater Collection for
              the Mechanically-Challenged $20**

Rainwater for the Mechanically-Challenged Video

You say you're *really* mechanically challenged and want more than a few pictures? Here's your salvation. From the same irreverent, fun-loving crew that wrote the book. See how all the pieces actually go together as they assemble a typical rainwater collection system, and discuss your options. This is as close as you can get to having someone else put your system together. 37 minutes and lots of laughs. USA.

82568   Rainwater Challenged Video $19.95

A Great Water Pumping Video

Part of the Renewable Energy with the Experts series, this 59-minute video features solar-pumping pioneer Windy Dankoff, who has more than fifteen years experience in the field. Windy demonstrates practical answers to all the most common questions asked by folks facing the need for a solar-powered pump, and offers a number of tips to avoid common pitfalls. This is far and away the best video in this series, and offers some of the best advice and knowledge available for off-the-grid water pumping. This grizzled old technician/copyeditor even learned a few new tricks about submersible pump installations. Highly recommended! USA

80368   RE Experts Water Pumping Video $39

Build Your Own Ram Pump

Utilizing the simple physical laws of inertia, the hydraulic ram can pump water to a higher point using just the energy of falling water. The operation sequence is detailed with easy-to-understand drawings. Drive pipe calculations, use of a supply cistern, multiple supply pipes, and much more are explained in a clear, concise manner. The second half of the booklet is devoted to detailed plans and drawings for building your own 1- or 2-inch ram pump. Constructed out of commonly available cast iron and brass plumbing fittings, the finished ram pump will provide years of low-maintenance water pumping for a total cost of $50 to $75. No tapping, drilling, welding, special tools, or materials are needed. This pump design requires a minimum flow of 3 to 4 gallons per minute, and 3 to 5 feet of fall. It is capable of lifting as much as 200 feet with sufficient volume and fall into the pump. The final section of the booklet contains a setup and operation manual for the ram pump. Paperback, 25 pages. USA.

80501   All About Hydraulic Ram Pumps $10.95

WATER-CONSERVATION PRODUCTS

Water-Saving Showerheads

Showers typically account for 32% of home water use.

A standard showerhead uses about 3 to 5 gallons of water per minute, so even a five-minute shower can consume 25 gallons. According to the U.S. Department of Energy, heating water is the second largest residential energy user. With a low-flow showerhead, energy use and costs for heating water for showers may drop as much as 50%. Add one of our instantaneous water heaters for even greater savings. Our low-flow showerheads easily can cut shower water usage by 50%. A recent study showed that changing to a low-flow showerhead saved 27¢ worth of water and 51¢ of electricity per day for a family of four. So, besides being good for the Earth, a low-flow showerhead will pay for itself in about two months!

The Best Low-Flow Showerhead Ever

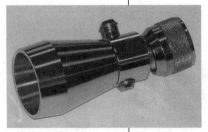

This is our longtime best-seller. Made in the USA of solid brass with a chrome-plate finish. Delivers a vigorous, well-controlled water spray pattern for a truly satisfying shower. Our favorite feature of this showerhead is that it's o-ringed and threaded together for easy cleaning of the stainless steel diffuser when the spray pattern gets erratic. This is a no-tools fix you can perform in the middle of a shower. Built-in soap-up valve. Maximum flow is 2.25 gpm at 80 psi (1.2 to 1.4 gpm is about average for most folks). This head cuts water use by 50 to 70% and can save a family of four up to $250 a year. Standard 1/2" pipe thread. Ten-year manufacturer's guarantee.

46104	**Best Low-Flow Showerhead**	**$12**
	2 or more	**$9.50 each**

Low-Flow Ultra Oxygenics Showerhead

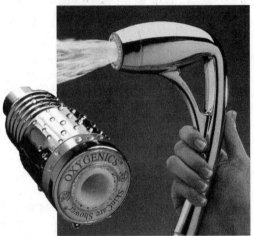

Ideal for homes with low water pressure or high energy or water costs, the patented technology inside our Ultra Oxygenics Showerhead pressurizes water to create a forcefully satisfying flow that belies the fact that up to 70% less water is being used—just 1.5 gallons per minute. To make your shower even better, internal mechanisms enrich shower water with up to 60% more oxygen to deep-clean skin, help fight free radicals, and improve circulation. Now available in a useful 2 gpm hand-held design that's great for cleaning the shower/tub and those hard-to-wash places. Not recommended for use with shower filters. USA.

02-0201	**Ultra**	**$30**
01-0452	**Hand-held**	**$55**

Earth Massage Showerhead

Enjoy a deluxe shower while only using 1.75 gallons per minute. Adjustable from fine spray to pulsating jet "turbo massage," this is the perfect showerhead for everyone. Three different spray patterns

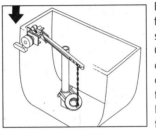

stay consistent in both low and high water pressures. Built-in "soak and soap" button reduces the flow rate to 1.0 gpm for soaping up and allows for the greatest amount of water savings possible. White. China.

46235   Earth Massage Showerhead $16

Wow Shower

Using powerful jet-pump action, our new Wow Shower filters and then recirculates warm water pooled in your tub back through the showerhead—saving water, energy and money while delivering a generous waterfall-like flow. This revolutionary system can save the average family of five up to 20,000 gallons of water per year! Pooled water is filtered constantly and refreshed with new water entering the system so you always feel clean. To return to conventional shower for a final rinse, simply depress diverter valve. Water flow delivers over 3 gallons per minute, but uses less than 1 gallon. Requires at least 35 psi of water pressure for operation. Simple five-step installation includes hardware. 72"H x 9"W x 9"D. 3.5 lbs. USA.

01-0445 Wow Shower $99

Revolutionary Water-Saving Toilet Kit

We're always looking for ways to conserve water, and this is one of the best: A complete multi-fixture kit that cuts the amount of water used by the typical bathroom in half. The center-piece is a revolutionary porcelain toilet whose patented design is seal- and flapper-free to eliminate leaks. The best flushing model of its kind, this #1 rated toilet comes with everything needed for easy

installation over both 10" and 12" rough-ins, including toilet seat, wax ring with sleeve, brass bolts, and stainless steel braided flex hose. With a large water surface to maintain a clean bowl, it uses just 1.6 gallons per flush. To complement this savings, our kit also includes a 2.0 gpm showerhead for a powerful shower without waste and a 1.0 gpm faucet aerator. We've even thrown in a kitchen aerator so you can extend your conservation efforts beyond the bathroom.

01-0453 Revolutionary Water-Saving Toilet Kit $185

Controllable Flusher

Convert your standard toilet into a low-flow toilet—without tools or a plumber—and save up to 35,000 gallons of water a year.

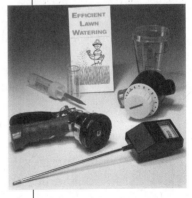

Because not all flushes need be the same, the dual-action Controllable Flusher controls the amount of water you use to flush waste matter. Push handle down for a conservative 1.5-gallon flush for liquid; lift for a powerful full flush for solids. Easily retrofits to standard front-flush toilets. 12.5"H x 5.5"W x 2"D. 5 oz. USA.

02-0205 Controllable Flusher $36

Touch Flow Swivel Spray Aerator

Our new kitchen faucet aerator has everything—swivel action for effective area cleaning, aerated jet, wide spray, and an easy touch flow restrictor. This kitchen faucet aerator not only makes your faucet more versatile and easier to use, but it will also save you money and water at only 1.5 gpm. USA.

09-0224 Touch Flow Swivel Spray Aerator $6
**   2 or more $ 5 each**

Outdoor Water Conservation Kit

Conserving water outdoors can be a challenge. It's often difficult to know how much to use, and it's hard to control how much you consume. Our kit contains tools that help you demystify the process. A rain gauge, drip gauge, and soil moisture meter tell you if watering is needed. If it is, a lawn sprinkler timer with 30-minute to 3-hour settings and a six-position low-flow hose nozzle reduce the possibility of waste. There's even a handy booklet that explains how to use these devices in tandem for maximum water savings.

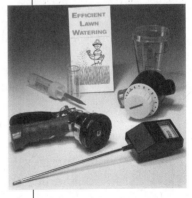

14-0289 Outdoor Water Conservation Kit $36

D'mand Water Saving System

Does your house make you wait, and wait, and wait for hot water at some faucets while wasting water down the drain? Here's the cure. With the push of a button, the D'mand system whisks hot water to the most remote faucet in seconds. The displaced cold water is pumped into the cold water plumbing, returning to the heater, so there's no waste. It shuts off automatically when hot water arrives and costs less than $1 per year

The Metlund® Hot Water D'MAND® System
Retrofit for Standard Plumbed Lines
©2002 ACT, Inc. Metlund Systems

D'mand S-02-PF-R

D'mand S-50-PF-R

D'mand S-69-PF-R

to operate. UL- and UPC-listed, the D'mand system is recognized by the Department of Energy as both a water- and power-saving device, and has a life expectancy exceeding fifteen years. It works with any type of water heater.

Our "no sweat" installation kit provides all the necessary plumbing bits and pieces, installs with common hand tools, and plugs into a standard outlet. Installed with a pushbutton at your furthest faucet, intervening faucets can be equipped with additional wireless pushbuttons to activate the system from up to 100 feet.

The only difference between kits is pump size. Longer plumbing runs need a larger pump to keep wait time under 20 to 30 seconds. Use S-50 model for runs up to 50', S-70 for runs up to 100', S-02 model for longer or commercial use. S-50 model has three-year manufacturer's warranty, other models are five-year. USA.

45-0100	S-50-PF-R Kit	$299
45-0101	S-70-PF-R Kit	$399
45-0102	S-02-PF-R Kit	$549
45-0103	Additional Transmitter	$20

> If each member of a family of four takes a five-minute shower each day, they will use more than 700 gallons of water every week—the equivalent of a three-year supply of drinking water for one person.
>
> —from *50 Simple Things You Can Do to Save the Earth*

STORAGE

Rain Barrel and Diverter

Capture irrigation water at no cost with our virtually indestructible, 0.1875" thick, recycled food-grade polyethylene rain barrel. Multiple barrels can be linked with an ordinary garden hose. Features overflow fitting, drain plug, screw-on cover, and threaded spigot. Optional galvanized steel rain diverter (shown) fits any downspout (metal or plastic). When full, flip the diverter to a closed position to let downspout function as usual. 39"H x 24" diameter. 60 gallon capacity. 20 lbs. USA.

14-9201	Rain Barrel and Diverter	$130
14-0238	Rain Barrel only	$114
46209	Diverter only	$22

Allow 4–6 weeks for delivery. Can be shipped to customers in the contiguous U.S. only. Cannot be shipped to P.O. boxes. Sorry, express delivery not available.

Kolaps-a-Tank

These handy and durable nylon tanks fold into a small package or expand into a very large storage tank. They are approved for drinking water, withstand temperatures to 140° F, and fit into the beds of several truck sizes. They will hold up under very rugged conditions,

are self-supported, and can be tied down with D-rings. All tanks have a 1.5" plastic valve with standard plumbing threads for input/output, and a 6" diameter by 2-foot long filler sleeve at the top for filling. Our most popular size is the 525-gallon model, which fits into a full-sized long-bed (5 x 8 ft.) pickup truck. USA.

47-401	73 gal. Kolaps-a-Tank, 40" x 50" x 12"	$311
47-402	275 gal. Kolaps-a-Tank, 80" x 73" x 16"	$460
47-403	525 gal. Kolaps-a-Tank, 65" x 98" x 18"	$558
47-404	800 gal. Kolaps-a-Tank, 6´ x 10´ x 2´	$645
47-405	1140 gal. Kolaps-a-Tank, 7´ x 12´ x 2´	$889
47-406	1340 gal. Kolaps-a-Tank, 7´ x 14´ x 2´	$995

Leak-Free Pond Liners
Heavy-duty gauges now available

A pond is a major investment, requiring lots of earth work, soil blending, and compaction. It's a gamble if it will hold water or not, or for how long. With a liner, earthwork can be kept to a minimum, and a leak-free pond can be established on any kind of soil. Our pond liners feature a very tough, UV-stabilized polyethylene that is highly resistant to punctures, tears, roots, or rodents. It's also nontoxic, nonleaching, and FDA/USDA approved for potable water supplies. Whether your plans call for a stocked aquaculture pond, an agricultural reservoir, or a garden feature, you need a liner that is absolutely leak-free to protect your investment of

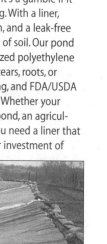

Easy-Shape PondGard Liner

PondGard membrane is friendly to do-it-yourselfers. This EPDM synthetic rubber is a thick 45-mil waterproofing membrane that's highly flexible and stable. It's far easier to work with in small spaces than our standard pond liner material at right, which tends to be quite stiff. PondGard is specially formulated to be safe for exposure to fish and plant life in decorative ponds. It can be shaped easily to fit the unique contours of any pond. Use it for pond lining, streams and fall features, erosion control, storage tanks, or wildlife ponds. PondGard has outstanding resistance to ultraviolet radiation (UV), ozone, and other environmental conditions. It requires little or no regular maintenance once installed. Available pre-cut in sizes listed. Ships via standard UPS. USA.

40-0074	**PondGard Liners 10' x 10'**	**$99**
	10' x 15'	$149
	15' x 15'	$225
	15' x 20'	$280
	15' x 25'	$375
	20' x 20'	$390

time and money. Our pond liners are custom-sized in any square or rectangular configuration. Heat welded factory seams are stronger than the surrounding material. Liners are "accordion-folded" for easy spreading on large sites. Four workers can handle up to 10,000 sq. ft.

Available in tough 20-mil thickness, in super heavy-duty 30-mil, or for really extreme sites, 40-mil. Weight per 1,000 sq. ft.: 20-mil = 100 lb., 30-mil = 150 lb., 40-mil = 200 lb.

100 sq. ft. minimum order. $15 handling charge on orders under 500 sq. ft. Free continental U.S. freight on orders over 2,000 sq. ft USA.

To figure size: Double the depth of your pond, add this to the width and length. Add an additional 2 to 5 feet for the buried apron. For instance, a 50' x 50' pond that's 5 feet deep needs a 62' x 62' liner at minimum.

47-212	**PondLiner Fab Tape 2" x 30'**	**$36**
47-227	**Handling Charge under 500 sq. ft.**	**$15**

POND LINER PRICE & COMPARISON CHART

Price per Square Foot

Item #	wgt.	100–200	201–600	601–2,000	2,001–5,000	5,001–8,000	8,001–12,000	12,001–20,000
47-203	20-mil	.56/sq. ft.	.50/sq. ft.	.44/sq. ft.	.42/sq. ft.	.31 sq. ft.	.28 sq. ft.	.26 sq. ft.
47-204	30-mil	.66/sq. ft.	.59/sq. ft.	.54/sq. ft.	.53/sq. ft.	.37 sq. ft.	.34 sq. ft.	.32 sq. ft.
47-211	40-mil	.72/sq. ft.	.69/sq. ft.	.66/sq. ft.	.65/sq. ft.	.46 sq. ft.	.43 sq. ft.	.41 sq. ft.

Pond liners are made to order, so please have dimensions ready when ordering. 100 sq. ft. minimum order. Please call for pricing on larger liners.

Solar-Powered Pond Aeration Kit

Aeration improves the aesthetics and overall health of ponds and lakes, reducing the chance of having unsightly algae blooms, the accumulation of sludge, and fish kills. To bring these benefits to your pond, you can either toss the water into the air, or you can toss the

air into the water. Since air weighs a lot less, your energy expenditure will be less and go further with an air pump—such as this 12-volt model that runs directly off a PV module or two. Our do-it-your-self kit supplies all the major pieces. The bottom-mounted diffuser delivers air as millions of tiny bubbles for better oxygenation and stirring of the pond to discourage stratification.

The Thomas 1/10 hp air pump delivers 1.4 cubic ft. per min. (CFM) at free discharge, or 0.8 CFM at 10 psi. It draws a maximum of 8.5 amps and can be driven to full output by a pair of 75- to 80-watt PV modules (sold separately below) with a linear current booster to help startup. The diffuser is properly sized to work with this pump. On site, you'll need to supply a 2.5" pipe for the pole-top mount, weather (and possibly noise) protection for the pump, wiring, small parts and weighed air line as needed by your site particulars. All products made in USA, except LCB made in Canada.

40-0072	Thomas 12V 1/10hp Air Pump	$275
40-0073	Air Diffuser	$240
41-0177	Sharp 80-watt PV Module	$395
42-0185	Uni-Rac 500040 Poletop Rack	$134
25-003	Solar Converters 10A LCB	$135

The weighted air diffuser is mounted in 12" PVC with a float. Includes ½" and ³⁄₈" air fittings.

FOUNTAIN PUMPS

A Solar Fountain Pump That Lasts

We've searched for years for a reliable PV-direct fountain pump. The extremely high-quality German-made submersible Aquasolar 200 finally meets our needs. This brushless, solid-state, sealed DC pump is designed to run at voltages from 12 to 22 volts; it will happily operate continuously at the 16- to 20-volt output of a 12V PV module. Using only 8 watts at the pump, it is rated for 12 liters per minute (over 3 gallons) at zero lift, or a trickle at the maximum lift of 1 meter (3.2 feet). This pump will run to full output with any 10- to 15-watt PV module. Output volume is externally adjustable. The pump housing acts as a large surface area intake screen. The outlet adapter will accept ½" flexible tubing, or with the supplied fitting, can accept ½" or ¾" threaded fittings. The housing, pump, and rotor can be disassembled easily for cleaning. Has a 15-foot cord with quick disconnect. Three-year manufacturer's warranty. Germany.

41914	Aquasolar 200 Fountain Pump	$179

Deluxe Fountain Kit: High-Volume Pump and Solar Module

Combines our 13-watt PV module with our best brushless, solid-state, sealed DC pump, the Aquasolar 200. Adjustable flow volume up to 3 gpm at zero lift (a trickle at 3-meter maximum lift); accepts ½" tubing; easy disassembly for cleaning; 15-foot cable, three-year manufacturer's warranty. Assembly instructions included. Germany/China. A $318 value if purchased separately.

41919	Deluxe Fountain Kit	$299

ATTWOOD SOLAR FOUNTAIN PUMPS

These submersible 12-volt pumps are designed as continuous-duty marine bilge pumps. We've found they make very impressive solar-powered fountain pumps. They won't lift very high, but they move an incredible

volume and use a tiny amount of power. The pump motor twist-locks onto the pump base to allow easy cleaning and removal of debris. They can withstand dry running for a limited time. Three-year manufacturer's warranty. USA.

Tech Note: In full-time PV-direct use, these pumps can be expected to last about one summer. Attwood is great about honoring the three-year warranty, so we recommend buying two pumps, and being prepared to rotate your stock.

Attwood V625 Pump

Model V625 is a good match with any 10- to 30-watt module.

- Amps: 1.3 @ 3 ft. head
- Max gpm: 7.5 @ 3 ft. head
- Max head: 5 ft.

41157	V625 Solar Fountain Pump	$30

Attwood V1250 Pump

Model V1250 is a good match with any 20- to 50-watt module.

- Amps: 2.9 @ 3 ft. head
- Max gpm: 15.5 @ 3 ft. head
- Max head: 8 ft.

41158	V1250 Solar Fountain Pump	$44

Pump and Panel for Solar Fountains

Use this nice little kit to create your own solar fountain, or to replace the failed pump on existing fountains. Our kit includes an unbreakable PV panel that measures 4.7" x 4.4" with a locking weatherproof electrical connector, 15 feet of flexible submersible cable, and a submersible DC pump that measures 2.5"L x 1.5"W x 2.0"H with a

$5/16$" tubing fitting and rubber suction-cup feet for quiet, stable operation. Under full sun it delivers $3/4$ gpm at zero lift, or a trickle at 24 inches of lift. The pump can be disassembled easily to clean the impeller. Motor brushes are not accessible or replaceable (expect about a two-year life). China.

40-0078 Sm. Solar Pump & Panel $49

CENTRIFUGAL PUMPS

EI-SID Hot Water Circulation Pumps

SID stands for Static Impeller Driver, the EI- is just for fun. These completely solid-state "motors" have no moving parts! There are no brushes, bearings, or seals to wear out. The rustproof bronze pump section is magnetically driven and is an existing off-the-shelf pump head that can be replaced easily if ever necessary. Life

EI-SID 20PV model   EI-SID 5 and 10 PV models

expectancy is many times longer than any other DC circulation pump.

PV-driven models are offered at 2.0, 3.3, and 6.0 gpm maximum flow rates. The battery-driven model delivers 3.3 gpm max at 12 or 24 volt. The 2.0 gpm PV pump requires a 5-watt PV module, the 3.3 gpm pump requires a 10- to 20-watt module, and the 6.0 gpm pump requires a 30-watt module. These are circulation pumps for closed-loop systems, and will not do any significant lift. See the specifications chart. Must be installed in a protected location (no rain exposure). Rated for temperatures up to 240°F, one-year manufacturer's warranty. USA.

41134	**EI-SID 5PV 2.0 gpm Pump**	**$215**
41133	**EI-SID 10PV 3.3 gpm Pump**	**$229**
40-0082	**EI-SID 20PV 6.0 gpm Pump**	**$305**
40-0083	**EI-SID 10B 12V 3.3 gpm Pump**	**$225**
40-0084	**EI-SID 10B 24V 3.3 gpm Pump**	**$275**

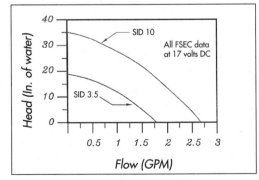

HARTELL HOT WATER CIRCULATION PUMPS

Hartell pumps feature magnetic drive with no rotating seals. Designed for solar hot water systems or other low-flow, modest-lift applications, they will operate directly off an 18- to 20-watt PV module. Brighter sun means faster pumping. Rated for temperatures up to 200°F, and pressure up to 150 psi. Brass pump bodies have $1/2$" MNPT fittings. We offer only Hartell's brushless pump model, with six-month manufacturer's warranty. USA.

Hartell MD-10-HEH

A brushless model that offers almost zero maintenance. Requires an 18- to 20-watt PV module. Measures 5.25" diameter by 9" length. Maximum volume, 5 gpm; maximum lift, 8 feet. See chart.

41529   Hartell Pump MD-10-HEH $410

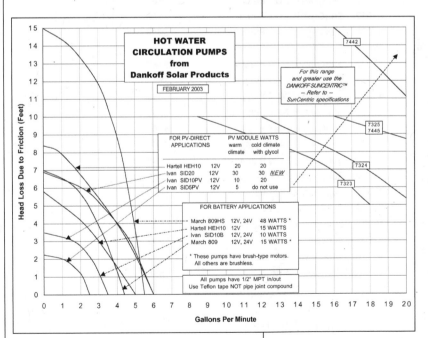

SunCentric Solar Surface Pumps

High-temperature option available

A reliable cast iron DC pump with one moving part. Good for swimming pools, agriculture, hydronic and solar heating, pond aeration, and more. Can do lifts up to 80 feet, or volumes up to 60 gpm depending on model; see performance charts. Maximum suction lift is 10 feet with a foot valve; centrifugal pumps will not self-prime. Mount the pump as close to the water source as possible, flooded inlet is best. Easy-starting on PV-direct systems; no electronic boost controls are needed. Both PV-direct and battery-powered models are available. Standard pumps use a glass-filled poycarbonate impeller with a temperature limit of 140°F. The $50 high-temperature option upgrades to a brass impeller with a 212°F limit. HT option reduces flow volume by 15%; power input is the same. Brush life is typically three to ten years. Two-year manufacturer's warranty. USA.

41010	SunCentric 7212 Pump	$705
41011	SunCentric 7213 Pump	$705
41012	SunCentric 7214 Pump	$705
41013	SunCentric 7321 Pump	$745
41014	SunCentric 7322 Pump	$745
41015	SunCentric 7323 Pump	$745
41016	SunCentric 7324 Pump	$705
41017	SunCentric 7325 Pump	$860
41018	SunCentric 7415 Pump	$860
41019	SunCentric 7423 Pump	$765
41020	SunCentric 7424 Pump	$765
41021	SunCentric 7425 Pump	$860
41022	SunCentric 7426 Pump	$855
41023	SunCentric 7442 Pump	$765
41024	SunCentric 7443 Pump	$765
41025	SunCentric 7444 Pump	$765
41026	SunCentric 7445 Pump	$860
41027	SunCentric 7446 Pump	$860
41028	SunCentric 7511 Pump	$810
41029	SunCentric 7521 Pump	$810
41030	SunCentric 7526 Pump	$905
41031	SunCentric 7622 Pump	$905
41032	SunCentric 7623 Pump	$905
40-0103	High Temp Option	$50

Replacement Parts For SunCentric Pumps

41033	SunCentric Brush Set	$45
41034	SunCentric 7xx1 to 4 Seal & Gasket Set	$18
41035	SunCentric 7xx5 or 6 Seal & Gasket Set	$18
40-0104	High Temp Gasket Set (specify model)	$38

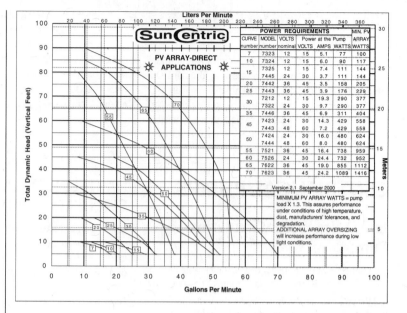

CURVE number	MODEL number	VOLTS nominal	VOLTS	AMPS	WATTS	MIN. PV ARRAY WATTS
7	7323	12	15	5.1	77	100
10	7324	12	15	6.0	90	117
15	7325	12	15	7.4	111	144
	7445	24	30	3.7	111	144
20	7442	36	45	3.5	158	205
25	7443	36	45	3.9	176	229
30	7212	12	15	19.3	290	377
	7322	24	30	9.7	290	377
35	7446	36	45	6.9	311	404
45	7423	24	30	14.3	429	558
	7443	48	60	7.2	429	558
50	7424	24	30	16.0	480	624
	7444	48	60	8.0	480	624
55	7521	36	45	16.4	738	959
60	7526	24	30	24.4	732	952
65	7622	36	45	19.0	855	1112
70	7623	36	45	24.2	1089	1416

Version 2.1 September 2000

MINIMUM PV ARRAY WATTS = pump load X 1.3. This assures performance under conditions of high temperature, dust, manufacturers' tolerances, and degradation. ADDITIONAL ARRAY OVERSIZING will increase performance during low light conditions.

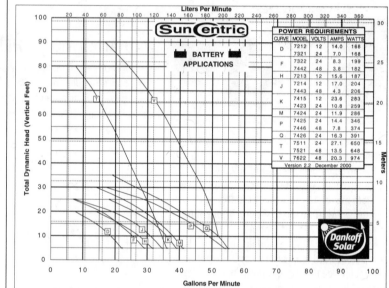

CURVE	MODEL	VOLTS	AMPS	WATTS
D	7212	12	14.0	168
	7321	24	7.0	168
F	7322	24	8.3	199
	7442	48	3.8	182
H	7213	12	15.6	187
J	7214	12	17.0	204
	7443	48	4.3	206
K	7415	12	23.6	283
	7423	24	10.8	259
M	7424	24	11.9	286
P	7425	24	14.4	346
	7446	48	7.8	374
Q	7426	24	16.3	391
T	7511	24	27.1	650
	7521	48	13.5	648
V	7622	48	20.3	974

Version 2.2 December 2000

Dankoff Solar

SunMotor M3 Solar-Driven Agricultural Pump

Great for large fountains and aeration

A professional, long-lived pump that floats mostly submerged in your pond. Delivers low-head, medium-volume water needs from surface water supplies. Just the thing to keep the stock from breaking down banks and

mucking up ponds. Common uses are livestock watering, small-scale irrigation, pond aeration, landscape ponds, and water transfer systems. The pump is magnetically coupled to eliminate any shaft seal and possibility of leakage. Also provides overload protection for the motor. This easy-starting centrifugal pump doesn't need a controller or booster to start. Fifty feet of submersible cable is already attached, so no watertight splice kit is needed. Motor brushes have approximately a two-year life expectancy, and are easily field replaceable with ordinary hand tools. An input screen prevents clogging by any foreign material. Anything that can pass through the screen is lifted easily by the pump. Kit includes pump, 50′ wire, 50′ ¼″ poly safety rope to hold pump in position, and 25′ flexible output hose. Depending on lift and volume required, it needs a single 80- to 115-watt PV module with mounting to complete a system. See output chart below. Canada.

41468	SunMotor M3 Pump	$995
41-0177	Sharp 80W module	$395
41-0124	Evergreen EC115 module	$585
42-0182	UniRac 500036 Poletop Mount	$121
	(fits either module above)	

M3 AGRICULTURAL PUMP PERFORMANCE CHART

Meters Lift (Feet)	80W Liters/Min. (GPM)	Approx. Daily Output Gallons	115W Liters/Min. (GPM)	Approx. Daily Output Gallons
1 (3.3)	22.7 (6.0)	1,500	31.8 (8.4)	2,100
2 (6.6)	21.5 (5.7)	1,400	28.6 (7.6)	1,900
3 (9.8)	20.0 (5.3)	1,300	25.4 (6.7)	1,700
4 (13.1)	16.7 (4.4)	1,200	22.0 (5.8)	1,450
5 (16.4)	13.6 (3.6)	1,100	18.3 (4.8)	1,250
6 (19.7)	11.4 (3.0)	900	13.9 (3.7)	950
7 (23.0)	7.6 (2.0)	600	9.9 (2.6)	650

Solar Pool-Pumping Systems

Pool pumping often can be a suburban home's largest single power use. Solar pool-pumping systems are moderately expensive initially, but run for nothing and have life expectancies measured in decades. Our solar-powered pool-pumping systems consist of a high-efficiency brushless motor coupled to a standard glass-filled polycarbonate pool pump head with a basket filter, clear cover, and 2″ inlet/outlet fittings. The pump is driven directly from a solar-electric array for simplicity and next-to-no maintenance. When the sun comes up, the pump runs.

Optional AC power converter provides back-up power from utility or generator power. Pump can draw power from solar and the converter simultaneously, with solar power as priority. Mounts sold separately. Manufacturers' warranties and countries of origin: 4 years on pump motor/control; China. 10 years on racks/mounts; USA. 20 years on PV modules; USA. Pump head: Germany.

Each system includes a pump assembly with controller plus BP Solar brand PV modules in 48-volt configuration. Choose a system wattage based on pool size and filter back pressure (see lower chart). Then choose your preferred mounting style for that size system.

40-0059	340–watt system	$5,075
40-0060	480–watt system	$5,890
40-0061	600–watt system	$6,980
48-0015	GFI Option (for roof mounted arrays)	$190
48-0016	AC Power Converter Option	$755

SOLAR ARRAY MOUNTING OPTIONS FOR POOL SYSTEMS

Mounting Type	340-watt System	480-watt System	600-watt System
Flush Rooftop	42-0050 $169	42-0051 $214	42-0052 $214
Fixed Pole-top	42-0053 $230	42-0054 $345	42-0055 $360
Tracking Pole-top	42-0056 $475	42-0057 $995	42-0058 $995

POOL PUMPING SYSTEMS PERFORMANCE

Solar Array Watts	Back pressure psi	Fixed Array Gallons per day	Fixed Array Max Pool size**	Tracking Array Gallons per day	Tracking Array Max Pool size**
340	3	17,600	25,100	26,600	38,000
480	3	23,400	33,500	35,400	50,600
600	3	27,500	39,300	39,600	56,600
340	6	11,100	15,900	16,900	24,200
480	6	17,200	24,500	26,000	37,200
600	6	21,100	30,200	31,700	45,300
340	8.5	6,400	9,200	9,800	14,000
480	8.5*	11,600	16,500	17,500	25,100
600	8.5	16,100	22,900	24,300	34,700
480	11	6,400	9,100	9,700	13,800
600	11	10,300	14,800	15,700	22,400

Daily performance based on solar irradiation of 6.0 peak sun hours per day, and 17% degradation of array output due to heat, dirt, and manufacturer's tolerances.

* 8 psi is typical back pressure for a system sized according to this selection table. Low friction (2″) piping and a large cartridge-type filter can reduce it further.

** Max. pool size is based on a turnover of 70% per day. For faster turnover choose a larger system if possible.

ROTARY VANE PUMPS

SOLAR SLOWPUMPS

Slowpumps are the original solar-powered pumps developed by Windy Dankoff in 1983. They set the standard for efficiency and reliability in solar water pumping where water demand is in the range of 50 to 3,000 gallons per day. The surface-mounted Slowpump is designed to draw water from shallow wells, springs, cisterns, tanks, ponds, rivers, or streams and to push it as high as 440 vertical feet for storage, pressurization, or irrigation.

These positive-displacement rotary-vane-type pumps feature forged brass bodies with carbon-graphite vanes held in a stainless steel rotor. No plastic!

Motor brushes are externally replaceable, and last 5 to 10 years. Pump life expectancy is 15 to 20 years. Rebuilt/exchange pump heads are available for under $150. Slowpumps are NSF approved for drinking water. A wide variety of pump and motor combinations are available for a variety of lifts and delivery volumes. Consult the performance charts.

Rotary-vane pumps must have absolutely clear water. They will not tolerate any abrasives. A 10-micron cartridge pre-filter is highly recommended for all installations. If your water is very dirty, improve the source or consider a diaphragm pump.

Fittings on 1300 and 1400 series pumps are ½-inch female, 2500 and 2600 series pumps have ¾-inch male fittings. Sizes and weights vary slightly with model but are approximately 6" x 16" and 16 lb. One-year manufacturer's warranty. USA.

¼-Horsepower PV Or Battery Powered Pumps, 12 Or 24 Volts

See chart. Specify voltage when ordering.

41102	Slowpump 1322 pump	$475
41104	Slowpump 1308 pump	$475
41106	Slowpump 1303 pump	$475
41108	Slowpump 2507 pump	$475

½-Horsepower Slowpumps

See chart. State voltage when ordering.

41931	Slowpump 1404 Pump	$695
41932	Slowpump 1403 Pump	$695
41933	Slowpump 2605 Pump	$695
41934	Slowpump 2607 Pump	$695

AC Version of any Slowpump Model

41110	Slowpump AC Option	Add $145

¼-HORSEPOWER 1300 & 2500 MODELS PV OR BATTERY POWER, 12V, 24V, OR 115VAC

Total Lift in Feet	1322 GPM	1322 Watts	1308 GPM	1308 Watts	1303 GPM	1303 Watts	2507 GPM	2507 Watts
20	0.51	27	1.25	30	2.50	48	4.00	57
40	0.51	32	1.25	48	2.50	60	3.95	78
60	0.51	36	1.20	54	2.40	78	3.90	102
80	0.49	40	1.20	60	2.30	93	3.90	120
100	0.49	45	1.20	66	2.30	105	3.85	144
120	0.48	50	1.20	70	2.25	121	3.80	165
140	0.47	56	1.20	75	2.20	138	3.65	195
160	0.47	62	1.20	84	2.20	153		
180	0.47	66	1.18	93	2.15	165		
200	0.45	74	1.16	101	2.15	180		
240	0.44	90	1.14	117	2.15	204		
280	0.41	102	1.12	135				
320	0.41	120	1.10	153				
360	0.41	134	1.05	171				
400	0.40	150	1.00	198				
440	0.39	168						

Actual performance may vary ±10% from specifications. Performance listed at 15V or 30V (PV-direct). Deduct 20% from flow and watts for battery. Watts listed are power use at pump.

½-HORSEPOWER 1400 & 2600 MODELS, 24 OR 48V BATTERY OR 36V PV DIRECT

Total Lift in Feet	1404 GPM	1404 Watts	1403 GPM	1403 Watts	2605 GPM	2605 Watts	2607 GPM	2607 Watts
160							4.30	283
180					3.35	280	4.25	305
200					3.33	296	4.20	338
240			2.55	266	3.30	331	4.05	396
280			2.50	302	3.25	373	4.00	444
320	1.66	255	2.50	338	3.20	410		
360	1.64	280	2.50	374	3.16	450		
400	1.62	312	2.50	406				
440	1.60	341	2.50	451				

Actual performance may vary ±10% from specifications. Watts listed are power use at pump.

Flowlight Booster Pumps

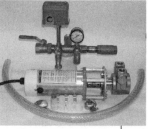

Sharing all the robust design features of the smaller Slowpumps, these larger rotary-vane pumps are designed specifically for household water pressurization. They use one-half to one-third as much energy as a similar capacity AC pressure pump, and eliminate the starting surges that are so hard on inverters. These pumps are quieter, smoother running, and far more durable than diaphragm-type pressure pumps. For full-time off-the-grid living, this is the pressure pump you want. Life expectancy is 15 to 20 years, and then they just need a simple pump head replacement.

Two models are available. Standard models feature higher flow rates, but are noisier, require 1" or larger intake plumbing, and have less than 10 feet of suction capacity. Low speed models are quieter, can accept ¾" intake plumbing and suction lifts greater than 10 feet, but have lower flow rates. Both models are available in 12 or 24 volts. Standard models also available in 48VDC and 115VAC.

BOOSTER PUMP PERFORMANCE CHART

	STANDARD MODEL				LOW SPEED MODEL			
PSI	30	40	50	65	30	40	50	65
GPM	4.5	4.3	4.3	4.1	3.4	3.3	3.1	2.7
Amps @24V	6.5	7.5	8.0	11.0	5.0	5.5	6.0	7.5
W/H per Gal	0.6	0.67	0.75	1.1	0.6	0.67	0.75	1.1

A pressure tank must be used with this, or any, pressure pump. Forty-gallon capacity is the minimum size recommended; larger is better. The Booster Pump Installation Kit and the Inline Sediment Filter, as shown with the pump picture, are strongly recommended. Flexible hose input and output hose assemblies and pressure relief valve are included with the pump. One year manufacturer's warranty. USA.

41141	Standard Booster Pump, 12V	$540
41142	Standard Booster Pump , 24V	$540
41000	Standard Booster Pump, 48V	$695
41147	Standard Booster Pump, 115VAC	$630
41145	Low Speed Booster Pump, 12V	$520
41146	Low Speed Booster Pump, 24V	$520

Booster Pump Installation Kit

This kit contains all the plumbing bits and pieces you'll need for installation of the Booster Pump. It includes a one-way check valve, a pressure switch, a pressure gauge, a drain valve, a shut-off valve, and the pressure tank tee all manifolded together. Shown with Booster Pump picture.

41143	Booster Pump Install Kit	$99
41137	Inline Sediment Filter	$49

Accessories for Slowpump Installation

Inline Sediment Filter

This is the recommended minimum filter for every Slowpump installation. This filter is designed for cold water only, and meets NSF standards for drinking water supplies. It features a sump head of fatigue-resistant Celcon plastic, an opaque body to inhibit algae growth if installed outdoors, and is equipped with a manually operated pressure-release button to simplify cartridge replacement. Rated for up to 125 psi and 100ºF, it is equipped with ¾-inch female pipe thread inlet and outlet fittings, and is rated for up to 6 gpm flow rates (with clean cartridge). It accepts a standard 10-inch cartridge with 1-inch center holes. Supplied with one 10-micron fiber cartridge installed. USA.

41137	Inline Sediment Filter	$49
42632	10 Micron Sediment Cartridge, 2 pack	$13

30" Intake Filter with Foot Valve

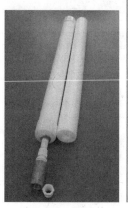

A great choice for any Slowpump with a surface water source. This large, triple-size 10-micron filter with no external cover can be lowered into wells, or floated on a pond. Large surface area means less frequent replacements. Equipped with a foot valve so you won't lose your pump prime. Included bushing allows ½ or ¾" pipe thread outlet. Comes with assembled filter and first replacement cartridge as pictured. USA.

40-0086	30" Intake Filter/Foot Valve	$70
40-0087	30" Filter Repl. 3-Pack	$44

Dry Run Shutdown Switch

Rotary-vane pumps will be damaged if allowed to run dry. This thermal switch clamps to the pump head of any new or older Slowpump. It will sense the temperature rise of dry running and shut off the pump before damage can occur. A manual reset is required to restart the pump. This switch will pay for itself many times over even if only used once during the lifetime of your Slowpump. USA.

41135	Dry Run Sw. for 1300 Series Pumps	$50
41144	Dry Run Sw. for all Larger Series Pumps	$50

Replacement Motor Brush Sets

The Slowpump brushes are easy to inspect and replace. Simply unscrew and withdraw the old brushes, then screw in the new ones if needed. Normal life expectancy is five to ten years. Sold in sets of two. USA.

41111	12V Slowpump Motor Brush Set	$20
41112	24V Slowpump Motor Brush Set	$20

Replacement Pump Heads

Slowpump pump heads typically enjoy a 15- to 20-year life expectancy if the water is clean, and they don't run dry. But hey, stuff happens, and some of the early Slowpumps have been out there for over 20 years now. Here's your replacement pump head, and it isn't going to hurt too bad. Specify your model when ordering. USA

41937	Replacement Pump Head for 1300 series	$142
41938	Replacement Pump Head for 2500 series	$193

DIAPHRAGM PUMPS

Guzzler Pumps

Human-powered pumps

These high-volume, low-lift pumps are driven by simple, dependable muscle power. Both hand pump and foot pump models are available. These are simple self-priming, diaphragm-type pumps with reinforced flapper valves for durable, dependable pumping. All models can deliver

about 12 feet of suction lift below the pump, and 12 feet of delivery head above the pump. Guzzlers make excellent emergency sump pumps, bilge pumps, or water delivery systems. The Model 400 series delivers up to 10 gpm, and uses 1" hose. Model 500 series delivers up to 15 gpm, and uses 1.5" hose. The Foot Pump models have a spring return stroke. USA.

41151	400 Hand Pump	**$55**
41152	500 Hand Pump	**$65**
41936	400 Foot Pump	**$133**
41937	500 Foot Pump	**$142**

Solar Surface Pump Kit

Our surface pump kit features one of our most popular pumps, the SHURflo Low Flow, along with enough PV power to lift up to 130 feet. This kit includes an Ever-

green 115-watt PV module, a pole-top mount for ease of installation, an LCB for best delivery under marginal sun conditions, and a screened foot valve for reliable operation. You supply a 10-foot length of 2.5"

steel pipe for the PV mount, plumbing and wiring as needed, and a surface water supply. Mount your pump as close to the water supply as practical, with a maximum of 10-feet suction lift. Pumps would much rather push. Kit pricing saves $47.

Delivers approximately 1.5 gpm, depending on lift and sunlight availability. In a typical six-hour summer day, it puts over 500 gallons into your garden or storage tank. All components U.S.-made, except LCB made in Canada.

41037	**Solar Surface Pump Kit**	**$889**

SHURFLO PUMPS

SHURflo produces a high-quality line of positive-displacement, diaphragm pumps for RV and remote household pressurization, and for lifting surface water to holding tanks. These pumps will self-prime with 10 feet of suction and a free discharge (no pressure on outlet). The three-chamber design runs quieter than other diaphragm pumps and will tolerate silty or dirty water or running dry with no damage. However, if your water is known to have sand or debris, it's best to use input filtering. Either SHURflo's small inline filter, or a standard 10" cartridge will save you trouble by preventing sand or other crud from lodging in the check valves. No damage, but you'll have to disassemble the pump to flush it out. Who needs the hassle? Put a filter on it!

Motors are slow-speed DC permanent-magnet types (except AC model) for long life and the most efficient performance. All pumps are rated at 100% duty cycle and may be run continuously. There is no shaft seal to fail; the ball bearing pump head is separate from the motor. Pumps are easily field repaired and rebuild kits are available for all models. All pumps have a built-in pressure switch to prevent over-pressurization. Some are adjustable through a limited range. Low flow model has 3/8-inch MIPT inlets and outlets. Medium and high flow models are 1/2-inch MIPT. All pumps carry a full one-year warranty. USA.

SHURflo Low Flow Pump

For modest water requirements or PV-direct applications. Available in 12 volts DC only. Delivers a maximum of 60 psi or 135 feet of head. This is an excellent surface water delivery pump used with the 7-amp PV Direct Pump Control (25002) and PV modules as necessary for your lift. (Add 30% to pump wattage to figure necessary PV wattage.) 3/8-inch MIPT inlet and outlet.

41450	**SHURflo Low Flow Pump**	**$109**

SHURflo Low Flow 12V		
PSI	GPM	Watts
0	1.75	37
10	1.66	41
20	1.57	50
30	1.48	59
40	1.38	67
50	1.30	76
60	1.23	86

SHURflo Medium Flow Pump

The Medium Flow is available in 12- or 24-volt DC; performance specs are similar for both (the 24V pump does slightly less volume). Maximum pressure is 40 psi or 90 feet of lift for both models. This model is a good choice for low-cost household pressurization; it will keep up with any single fixture. You must use a pressure tank with any pressurization system; obtain locally. It has a built-in 20- to 40-psi pressure switch, perfect for most households. Buzzy output can be reduced by looping 24-inch flexible connectors on both inlet and outlet. 1/2-inch MIPT inlet and outlet.

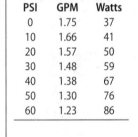

Medium Flow 12V			Medium Flow 24V		
PSI	GPM	Amps	PSI	GPM	Amps
0	3.30	5.1	0	3.00	2.5
10	3.02	5.6	10	2.87	3.2
20	2.82	7.0	20	2.64	3.8
30	2.65	8.4	30	2.41	4.3
40	2.48	9.6	40	2.20	4.7

41451	**SHURflo Medium Flow, 12 Volt**	**$149**
41452	**SHURflo Medium Flow, 24 Volt**	**$165**

SHURflo High Pressure Pump

SHURflo High Pressure 12V		
PSI	GPM	Amps
0	3.40	6.0
10	3.00	6.9
20	2.75	7.5
30	2.50	8.6
40	2.25	9.5

This model is higher pressure, 30- to 45-psi for household pressurization systems. For those who need slightly higher pressure than the standard Medium Flow pumps. Buzzy output can be reduced by looping 24-inch flexible connectors on both inlet and outlet. Available in 12 volts only. Maximum pressure is 45 psi or 104 feet of lift. 1/2-inch MIPT inlet and outlet.

41453   SHURflo High Pressure Pump $189

SHURflo AC Pump, 115 Volts

SHURflo 115v AC Pump		
PSI	GPM	Amps
0	3.05	59
10	2.85	65
20	2.60	82
30	2.40	96
40	2.24	108

For those who need to send power more than 150 to 200 feet away, this pump can save you big $$ on wire costs! The motor is thermally protected and under heavy continuous use will shut off automatically to prevent overheating. This should only happen at maximum psi after approximately 90 minutes of running. Pressure switch is adjustable from 25 to 45 psi. Maximum pressure is 45 psi. Weight 5.8 lb. 1/2-inch MIPT inlet and outlet.

41454   SHURflo 115 volt AC Pump $179

SHURflo Extreme High Pressure Pump

EXTREME PRESSURE 12V		
PSI	GPM	Amps
10	1.68	3.6
30	1.53	5.1
50	1.39	6.4
70	1.29	7.6
90	1.20	8.8
100	1.15	9.3

This is a special pump for very high pressures up to 100 psi. It can pump up to 230 feet of vertical lift. The built-in pressure switch is factory adjusted to turn on at 85 psi and off at 100 psi. Adjustment range is 80 to 100 psi. Not recommended for continuous operation. Has 3/8-inch female NPT inlet and outlet fittings. 12 Volt only. USA.

41357   SHURflo Extreme High Pressure, 12 Volt $109

ULTRAflo Pump and Accumulator Combo

Delivering over 5 gallons per minute, SHURflo offers this pair of high-output pumps pre-assembled with a stainless steel 2-gallon accumulator tank. This is a complete water pressure delivery system for 12- or 24-volt DC operation. The double-head diaphragm pumps deliver up to 5.6 gpm at 12v, or 6.5 gpm at 24v for multi-fixture operation. The built-in pressure switch is on at 35, off at 45 psi, with +/- 5 psi adjustment. Designed for marine and RV applications with limited space, these will perform better and live longer with the addition of a second larger pressure tank if you've got the space. 1/2" male pipe-thread outlet. 18.5"L x 14.5"H x 10"D, weighs 9 lb. One-year. mfanufacturer's warranty. USA.

40-0080   12v Pump/Accumulator $339
40-0081   24v Pump/Accumulator $345

12v Pump/Accumulator			24v Pump/Accumulator		
PSI	GPM	Amps	PSI	GPM	Amps
0	5.6	7.4	0	6.5	4.0
10	5.1	9.4	10	6.0	5.3
20	4.4	11.3	20	5.4	6.5
30	3.7	12.7	30	4.6	7.5
40	2.9	13.6	40	4.2	8.1

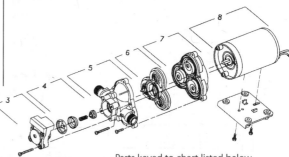

Parts keyed to chart listed below.

SHURFLO REPAIR AND REPLACEMENT PARTS

Pump Model	Press. Switch 3	Check Valve 4	Upper Housing 3, 4 & 5	Valve Kit 6	D&D Kit 7	Motor 8
Low Flow 12v 41450	41325 $20	41327 $10	n/a	41331 $20	41333 $37	41335 $100
Med. Flow 12v 41451	41326 $20	41326 $10	41329 $25	41332 $15	41334 $30	41335 $100
Med. Flow 24v 41452	41326 $20	41328 $10	41329 $25	41332 $15	41334 $30	41334 $119
High Press 12v 41453	n/a	41328 $10	41330 $50	41332 $15	41334 $30	n/a
115VAC 41454	41326 $20	41328 $10	41329 $25	41332 $15	41334 $30	41338 $169

SHURflo Inline Filter

For light-duty filtering, this stainless steel screen is all the highly tolerant SHURflo pumps need. It'll keep out anything big enough to cause trouble for the pump. Screws onto 1/2" male pump inlet, has 1/2" barbed fitting to water source. Disassemble for cleaning.

40-0088 SHURflo Inline Filter   **$9**

Solaram Pump

This is the highest yield, highest lift solar pump available. Using a large multi-piston industrial diaphragm pump and a permanent-magnet DC motor with positive gear belt drive, the Solaram series of pumps features models capable of lifts up to 960 feet, or over 9 gallons per minute. The performance chart shows delivery and watts required. If running PV-direct, the rated PV wattage must exceed the pump requirement by 25% or more. A pressure relief valve and flexible intake/outlet hoses are included.

In full-time daily use, yearly diaphragm replacement is recommended. Motor brushes typically last 10 years. The Diaphragm and Oil Kit provides a set of new diaphragms and the special nontoxic oil. The Long-Term Parts Kit contains three diaphragm and oil kits, plus a gearbelt and motor brush kit.

Inlet fittings are 1.5 inches, outlet is 1 inch. Dimensions are 28" long, 16" deep, 16.5" high. Weight is approximately 150 lb. (varies slightly with model) and is shipped in two parcels. Manufacturer's warranty is one year. USA.

Specify full 4-digit model number when ordering.

41480	**Solaram 8100 series Pump**	**$2,475**
41481	**Solaram 8200 series Pump**	**$2,475**
41482	**Solaram 8300 series Pump**	**$3,825**
41483	**Solaram 8400 series Pump**	**$3,825**
41484	**Diaphragm & Oil Kit**	**$197**
41485	**Long Term Parts Kit**	**$345**

Solaram™ Surface Pump Performance Chart

Model Numbers: First 2 digits ——
                 Second 2 digits ——

TOTAL LIFT		__ __ 21			__ __ 22			__ __ 23			__ __ 41			__ __ 42			__ __ 43			Model #
Feet	Meters	GPM	LPM	Watts	GPM	LPM	Watts	GPM	LPM	Watts	GPM	LPM	Watts	GPM	LPM	Watts	GPM	LPM	Watts	Volts
0-80	24	3.0	11.4	170	3.7	14.0	207	4.6	17.4	285	6.2	23.5	258	7.5	28.4	339	9.4	35.6	465	**81__**
120	37	2.9	11.0	197	3.7	14.0	238	4.5	17.1	319	6.0	22.7	305	7.3	27.7	396	9.1	34.5	539	**24V**
160	49	2.9	11.0	225	3.6	13.6	268	4.5	17.1	352	5.8	22.0	354	7.2	27.3	453	8.9	33.7	619	
200	61	2.9	11.0	247	3.6	13.6	296	4.5	17.1	388	5.7	21.6	400	7.1	26.9	513	8.9	33.7	693	
240	73	2.8	10.6	265	3.6	13.6	327	4.5	17.1	427	5.6	21.2	453	7.0	26.5	572	8.6	32.6	724	**82__**
280	85	2.8	10.6	286	3.6	13.6	356	4.4	16.7	466	5.5	20.8	499	6.9	26.2	628	8.4	31.8	801	**24V**
320	98	2.8	10.6	315	3.5	13.3	388	4.4	16.7	496	5.4	20.5	548	6.8	25.8	686	8.3	31.5	869	
360	110	2.8	10.6	342	3.5	13.3	416	4.4	16.7	536	5.4	20.5	592	6.6	25.0	733	8.2	31.1	927	
400	122	2.7	10.2	363	3.4	12.9	450	4.4	16.7	572	5.3	20.1	649	6.5	24.6	782	8.7	33.0	1122	**83__**
480	146	2.7	10.2	416	3.4	12.9	505	4.3	16.3	649	5.3	20.1	717	6.5	24.6	900	8.5	32.2	1265	**180V**
560	171	2.7	10.2	456	3.3	12.5	570	4.3	16.3	693	5.2	19.7	800	6.5	24.6	1045	8.4	31.8	1397	
640	195	2.7	10.2	502	3.3	12.5	623	4.2	15.9	774	5.1	19.3	893	6.5	24.6	1116	8.2	31.1	1540	**84__**
720	220	2.6	9.9	551	3.2	12.1	690	4.1	15.5	856	5.1	19.3	1031	6.4	24.3	1287	8.1	30.7	1683	**180V**
800	244	2.6	9.9	589	3.2	12.1	715	4.1	15.5	931	5.1	19.3	1114	6.4	24.3	1408	8.0	30.3	1815	
880	268	2.6	9.9	647	3.2	12.1	774	4.0	15.2	1082	5.1	19.3	1206	6.3	23.9	1529	8.0	30.3	1958	**85__**
960	293	2.6	9.9	705	3.1	11.7	838	4.0	15.2	1190	5.0	18.9	1289	6.1	23.1	1650	8.0	30.3	2145	**180V**

Performance may vary ± 10 %

PISTON PUMPS

Solar Force Piston Pump

Built like a tank, the cast iron Solar Force is an excellent choice for high-volume delivery of 4 to 9 gpm at lifts up to 220 feet. It can be used for water delivery to a higher storage tank, for irrigation, fire protection, or pressurization. It has a 25-foot suction lift at sea level. The durable cast iron and brass pump is designed to last for decades, and routine maintenance is only required every two to six years. A Linear Current Booster is needed for PV-direct drive. LCB sizing depends on model and voltage. Our tech staff will advise. Solar Force uses a non-slip gear belt drive on PV-direct models, and standard V-belt drive on battery models. Intake fitting is 1.25 inches, output is 1 inch. A pressure relief valve is included. Measures 22"W x 16"H by 13"D. Weighs approximately 80 lb. maximum, shipped in two or three parcels depending on model. Two-year manufacturer's warranty. USA.

The Solar Force Pump is available in three basic models:
Model 3010: available in 12 or 24 volts battery-driven only.
Model 3020: available in 12, 24, or 48 volts, battery or PV-direct, also 115VAC.
Model 3040: available in 12, 24, or 48 volts, battery or PV-direct, also 115VAC.
Please specify voltage (12, 24, 48, or 115) when ordering.

41254	**Model 3010 Solar Force (12 & 24V battery only)**	**$1,085**
41255	**Model 3020 Battery Powered**	**$1,350**
41256	**Model 3020 PV Powered**	**$1,525**
41257	**Model 3040 Battery Powered**	**$1,415**
41258	**Model 3040 PV Powered**	**$1,595**
41259	**Model 3020 or 3040 115VAC Powered**	**$1,350**

SOLAR FORCE ACCESSORIES AND PARTS

Heavy-Duty Pressure Switch

A DC-rated pressure switch for using the Solar Force as a pressure pump. Will handle the heavy starting surge. Good for 12-, 24-, or 48-volt systems.

41264	**Solar Force Pressure Switch**	**$75**

Surge Tank

This tank is included with the PV models. It helps absorb pressure pulsations when long piping runs are required between the pump and the tank. It keeps plumbing systems from being shaken apart.

41265	**Solar Force Pulsation Tank**	**$90**

Seal and Belt Kit

Contains all the repair parts needed for routine service required every two to six years. Spare gaskets, rod and valve seals, new drive belt, and two sets of piston seals are included. PV kits are higher priced due to gear belt drive instead of V-belt.

Specify pump model number when ordering.

41267	**Seal & Belt Kit for Battery Pumps**	**$45**
41268	**Seal & Belt Kit for PV Pumps**	**$75**

Long-Term Parts Kit

Everything included in the Seal & Belt Kit above, plus a second drive belt, a replacement cylinder sleeve, motor brushes, and two oil changes.

Specify pump model number when ordering.

41269	**Long-Term Parts Kit for Battery Pumps**	**$160**
41270	**Long-Term Parts Kit for PV Pumps**	**$210**

SOLAR FORCE PISTON PUMP PERFORMANCE

Vertical Lift Feet	Model 3110 GPM	Model 3110 Watts	Model 3020 GPM	Model 3020 Watts	Model 3040 GPM	Model 3040 Watts
20	5.9	77	5.2	110	9.3	168
40	5.6	104	5.2	132	9.3	207
60	5.3	123	5.1	154	9.2	252
80	5.1	152	5.1	182	9.2	286
100	5.1	171	5.0	202	9.1	322
120	4.9	200	5.0	224	9.1	364
140	4.9	226	5.0	252	9.1	403
160	This model 12 or 24v battery-driven only		4.9	269		
180			4.9	280		
200			4.8	308		
220			4.7	314		

SUBMERSIBLE PUMPS

Amazon Submersible

Delivers a maximum of 4 gpm at zero lift, or trickle at 30 feet of lift, this 1.1-pound pump has 13 feet of cable with battery clips, 1/2" barbed connections, and is 1.5" diameter. Draws 7 amps @ 12 volt maximum. Do not run dry or expect a long life. Intermittent use only. Great Britain.

41950	**Amazon 12-volt Pump**	**$87**
41963	**Amazon 24-volt Pump**	**$87**

Congo Submersible

Delivers a maximum of 7 gpm at zero lift, or a trickle at 30 feet of lift. Has 13' of cable with battery clips, 1/2" barbed connections, and is 1.6" diameter. Draws 8 amps @ 12 volt maximum. Do not run dry or expect a long life. Intermittent use only. Great Britain.

41956	**Congo 12-volt Pump**	**$120**

SHURflo Solar Submersible Pump

This small, efficient, submersible, diaphragm pump has been the standard for residential solar water pumping for over 10 years. For homesteads without large garden or orchard water needs, this is probably all the pump you need. The SHURflo Sub will deliver up to 230 feet of vertical lift and is ideal for deep well or freeze-prone applications. Remember, you're only lifting from the water surface, not from the pump. In a PV-direct application, it will yield from 300 to 1,000 gallons per day, depending on lift requirements and power supplied. Solar submersible pumps are best used for slow, steady water production into a holding tank, but may be used for direct pressurization applications as well. (See suggestions on this subject beginning on page 282.) Flow

SHURFLO SUBMERSIBLE PUMP PERFORMANCE CHART AT 24 VOLTS

Total Vertical Lift	Gallons Per Hour	Minimum PV Watts	Amps
20	117	58	1.5
40	114	65	1.7
60	109	78	2.1
80	106	89	2.4
100	103	99	2.6
120	101	104	2.8
140	99	115	3.1
180	93	135	3.6
200	91	141	3.8
230	82	155	4.1

rates range from .5 to 1.9 gpm depending on lift and power supplied. The pump can be powered from either 12- or 24-volt sources. Output is best with 24-volt panel-direct. Because it's a diaphragm-type, this pump is untroubled by running dry, so long as it doesn't last for more than two or three months. Yes, months! So this is a great choice for low-yielding wells. The best feature of this pump, other than proven reliability, is easy field serviceability. No special tools are needed. Experience has shown that the diaphragm needs replacement every four to five years, the motor has about a ten- to eleven-year life expectancy. Repair parts sold below.

The pump is 3.75 inches in diameter (will fit standard 4-inch well casings), weighs 6 pounds, uses 1/2-inch poly delivery pipe, making easy one-person installation possible. Maximum pump submersion is 100 feet. Requires flat-jacketed style of submersible cable for the watertight splice kit. Not always available locally, we offer it below. One-year warranty. USA.

In addition to the pump you will also need:
- PV modules and mounting appropriate for your lift (see chart)
- A controller for PV-direct systems (PV direct controller #25002 on page 310)
- 1/2-inch poly drop pipe or equivalent
- 10-gauge submersible pump cable (3-wire cable okay if flat style)
- Poly safety rope
- Sanitary well seal

Drop pipe, submersible cable, rope, and seal are all commonly available at most plumbing supplys. Watertight splice kit for flat cable is included.

(12-volt performance will be approximately 40% of figures listed in performance chart.)

41455	**SHURflo Solar Submersible**	**$695**
26540	**10-2 Jacketed Sub Pump Cable** (specify length)	**$1.00/ft.**

A Complete Solar Sub-Pump Kit

Our Sub-Pump Kit features the popular and durable Shurflo submersible pump. We've put it together with a pair of 80-watt modules for maximum output up to its maximum lift of 230 feet, a pole-top mount for ease of installation, and a current booster controller for good performance with marginal sun. Cost of components purchased separately is $1,725. Add the float switch for automatic shut-off when your tank fills. Add the special flat-jacketed sub pump cable as needed, if this type of sub pump cable is not available locally for you. Our pole-top mount requires nominal 2.5" Schedule 40 steel pipe, provided locally to save shipping. 10 feet is usually good, with 3 feet buried. You will also be providing wiring as needed, poly safety rope, 1/2" poly drop pipe, sanitary well seal, and hole in the ground (hopefully with water at the bottom). All components made in USA except controller made in Canada.

41128	**Shurflo Sub-Pump Kit**	**$1,695**
41638	**Float Switch "D"**	**$35**
26540	**10-2 Jacketed Sub Pump Cable** (specify length)	**$1.00/ft.**

SHURflo Solar Sub Service Parts

The drive and diaphragm on this hard-working little pump should be replaced at four to five year intervals. Do it *before* the diaphragm starts leaking, and you'll save yourself a $225 motor. If the lower canister (and motor) is full of water on disassembly, replace the motor. The carbon brushes soak up the water and soften. They'll only last a few days after you put it back in the well. Very frustrating doing repairs twice isn't it? O-rings should be replaced anytime you open the pump. We try very hard to keep all the common bits in stock for rush delivery. USA

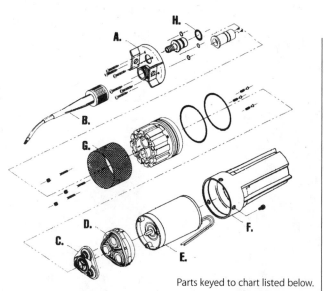

Parts keyed to chart listed below.

REPLACEMENT PARTS KIT LIST FOR SHURFLO SOLAR SUB PUMP

Complete Tool Kit	Lift Plate Kit A	Cable Plug Kit B	Valve Kit C	Drive & Diaphragm Kit D	Motor E	Canister F	Filter Screen G	Complete O-Ring Kit H
41-340	41-341	41-342	41-343	41-344	41-345	not	41-437	41-348
$20	$80	$125	$40	$100	$225	available	$30	$25

ETApump Mini Submersible

No-maintenance solar pumping

The helical-rotor ETApumps have proven themselves to be highly efficient, incredibly long-lasting, and almost completely maintenance-free. Now there's a smaller version of this terrific solar- or battery- powered pump for residential customers with less demanding water needs. ETApump Mini features a helical-rotor-type pump with a life expectancy equal to the best conventional AC sub pumps (15–20 years+). It has no brushes, cams, flapper valves, or plastic parts to wear out. All electronics are in the controller, at the surface, which has display lights for: system on, pump on, tank full, low water, overload, and for battery models, low battery. It runs 24-volt PV direct, or 24-volt battery-based using different models. Additionally there is a low-lift/high-volume model with lift to 65 feet, or a higher-lift/lower-volume model with lift to 165 feet, for a total of four models. Pumps can be installed in 4" or larger wells, tanks, streams, or ponds in any position, and at any submersion depth. A dry run switch with automatic 20-minute reset is included, and must be connected if there's a chance of running dry. The PV-direct model has an RPM limit adjustment for low-yielding wells.

ETApump Mini includes: 3.8" diameter pump, pump controller, low-water probe, and submersible splice kit. 1" NPT outlet. Requires 3-wire plus ground sub pump cable. Low-water probe requires additional 2-16ga. wires. Controller accepts optional reverse-logic float switch (closed contact shuts pump OFF). German-engineered, Chinese-manufactured, USA-assembled. Two-year manufacturer's warranty.

40-0068 ETApump Mini 04L-PV	**$1,510**	
40-0069 ETApump Mini 14L-PV	**$1,510**	
40-0070 ETApump Mini 04L-BAT	**$1,375**	
40-0071 ETApump Mini 04L-BAT	**$1,375**	

ETAPUMP MINI 24V PV-DIRECT

Vertical Lift	Model 04L-PV Peak GPM	Min. PV Watts	Model 14L-PV Peak GPM	Min. PV Watts
16	1.9	60	5.8	100
33	1.7	70	5.3	120
50	1.6	80	4.8	150
65	1.5	90	4.4	200
100	1.5	100	N/A	N/A
130	1.3	120	N/A	N/A
165	1.3	120	N/A	N/A

ETAPUMP MINI 24V BATTERY SYSTEM

Vertical Lift	Model 04L-BAT Peak GPM	Watts	Model 14L-BAT Peak GPM	Watts
16	1.5	24	4.6	38
33	1.4	29	4.4	55
50	1.3	34	4.0	74
65	1.2	38	3.7	91
100	1.2	48	N/A	N/A
130	1.1	58	N/A	N/A
165	0.9	65	N/A	N/A

SUBMERSIBLE ETAPUMP SYSTEMS

Complete packages with pump, controls, PV, and mounts

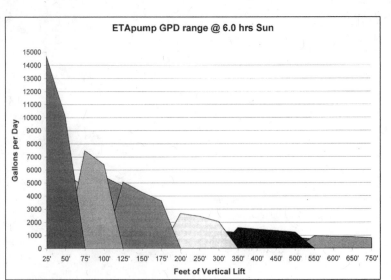

The ETApump is one of a new generation of "helical-rotor" pump types. Based on the ancient Archimedes screw, but now using a hardened stainless steel rotor in a durable flexible rubber sleeve, helical-rotor pumps feature life expectancies as long as the best traditional sub pumps (15–20 years+), with the best submersible pump efficiency we've seen from any pump type yet. Higher efficiency means more water for less PV dollars. ETApumps have all the electronic control parts above ground, and a brushless motor. No diaphragms or brushes to replace! This is probably the most trouble- and maintenance-free solar pumping system available.

ETApumps are sold as a complete system that includes pump, controller, solar modules, pole-top mounting rack, disconnect switch, submersible splice kit, low-water sensor, and a detailed, illustrated instruction manual. On site, you will supply a functioning well with 4" or larger casing, submersible pump cable (3-wire plus ground), drop pipe (1" or 1 1/4" depending on pump model), 1/4" poly or nylon safety rope, schedule 40 steel pipe for the pole-top mount, and small hardware and electrical parts as needed for your particular installation. The low-water sensor will shut off the pump to prevent dry run damage. Restart is automatic when water level recovers. Controllers are weatherproof, and will accept a float switch or other external on/off control with reverse logic. A closed float switch will turn off the pump .

ETApump models are also available for battery-based operation at 24 or 48 volts. These packages include a pump and the special battery-based controller only.

An AC Power Pack option is available for the ETA-pumps that will automatically use any solar energy available, then supplement it as needed with enough additional AC power to run the pump any time of day or night. Either a generator or utility power can be used. Can be set up to accept either 240 or 120vac input. PV input must be 48v nominal.

ETApump systems are sized for the lift and volume of water you need. There are *many* models, the performance chart printed above is only a partial listing. You must call or email our staff to size and price a system. ETApumps will lift up to 750 feet, and can deliver as much as 15,000 gallons per day (at 25' lift). Complete system prices run from $3,700 to $8,775, battery-based systems are $1,765 to $1,870. More lift and/or more volume costs more. ETApump delivery estimates are delightfully conservative and honest, they carry an excellent, industry-leading four-year warranty on the pump and controller when purchased as a complete system. German engineered, Chinese manufactured.

Complete ETApump PV-Powered
Systems   **$3,700 – $8,775**

ETApump Battery-Powered
Systems   **$1,765 – $1,870**

48-0016 ETApump AC Power Pak   **$755**

Please call for details & sizing assistance.

GRUNDFOS SQ FLEX SUBMERSIBLE PUMPS

High efficiency and adaptable to any power input

Featuring an incredible power adaptability, the submersible SQ Flex series comes from one of the most-experienced, highest-quality, and largest pump manufacturers in the world. It features a single motor that can be fitted with seven different pump heads, depending on your pumping requirements. The SQ Flex will happily accept any DC voltage from 30 to 300 volts, or any AC

voltage from 90 to 260 volts, 50 or 60hz, just flip a switch on the optional AC/DC control box to go back and forth. This makes the SQ Flex pump extremely easy to back up with a generator or utility power. When running on DC, the SQ Flex features electronic Maximum Power Point tracking to wring the best possible performance from your solar array. A low-water sensor is included and pre-wired to the pump input cable. Low-water shut-off and restart are automatic.

With a maximum of 900 watts input, the SQ Flex motor offers performance equivalent to a conventional ³/₄- to 1- horsepower AC pump. But this motor can be coupled to helical rotor pumps with efficiencies that conventional AC pumps can't even dream about. The three smaller pump heads are the new positive-displacement helical-rotor types. This is the newest pump type in the solar industry. They feature much greater efficiency than any other pump type, and don't wear out or require any routine maintenance. Water delivery ranges from over 24,000 gallons per day at a minimal 20 foot lift, to 1,000 gallons per day at the maximum lift of 525 feet. The four higher-volume pump heads are conventional centrifugal types.

An optional control box is required for switchable AC/DC operation, or for remote float level sensing. The nicely featured float control also features an LCD display of wattage consumption, operation indicator lights, on/off switch, and fault codes display.

Complete systems include pump, controller box(es), solar array, array wiring, and mounting structure. Complete system prices range from $3,100 to $10,900. Custom systems using 48-volt batteries, generators, or other power sources can be configured. One-year manufacturer's warranty. Made in Denmark and USA.

Because of the wide variety of lifts, volumes, and power sources, you need to call or email our tech staff to size and price a Grundfos SQ Flex pumping system.

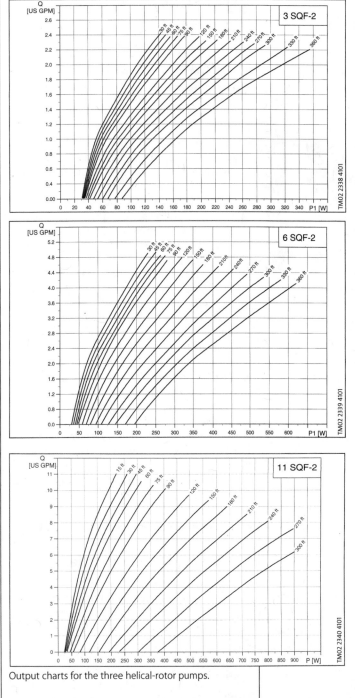

Output charts for the three helical-rotor pumps.

41980	Grundfos 3 SQF-2 Helical Rotor Pump	$1,385
41981	Grundfos 6 SQF-2 Helical Rotor Pump	$1,385
41982	Grundfos 11 SQF-2 Helical Rotor Pump	$1,385
41983	Grundfos 25 SQF-3 Centrifugal Pump	$1,385
41984	Grundfos 25 SQF-6 Centrifugal Pump	$1,385
41985	Grundfos 40 SQF-3 Centrifugal Pump	$1,385
41986	Grundfos 75 SQF-3 Centrifugal Pump	$1,385
25815	Grundfos IO100 std. on/off Control	$140
25816	Grundfos IO101 AC/DC sw. Control	$369
25817	Grundfos CU200 Float/Monitor Control	$299

SUBMERSIBLE PUMP OPTIONS

The Universal Solar Pump Controllers

Run ordinary AC pumps directly from PV power

From the same incredibly innovative company that produced GM's EV1 electric vehicle, and the human-powered plane that crossed the English Channel, comes an inverter/controller that will run any standard single- or three-phase AC submersible pump directly from PV power. The advanced USPC maintains a constant volts/Hz ratio that allows standard AC pumps to run at lower speeds without distress. This allows easy soft starts, and the efficient use of lower-than-peak PV outputs with off-the-shelf pumps. Minimum speed is owner-selectable to avoid no-flow conditions. The USPC's power electronics are 97% efficient, and will provide automatic shutdown in case of dry running, motor lock, or wiring problems. The USPC can drive any standard 50- or 60-Hz three-wire single-phase pump or three-phase induction motor that has a rated input voltage of 120, 208, or 230VAC. The model 2000 will run any 0.5 to 2.0 horsepower pump. The model 5000 is for pumps up to 5.0 horsepower, and a 10-horsepower model is available by special order. The enclosure is rugged outdoor-rated NEMA 3 steel. USPC complete systems, with PV array, controller, and pump run $5,000 to $50,000. Two-year manufacturer's warranty. UL certified. USA.

Please call our tech staff for help with PV sizing, which will be unique for your installation.

41913	**USPC-2000 Pump Controller**	**$2,320**
41923	**USPC-5000 Pump Controller**	**$4,300**

PV-Direct Pump Controllers

These current boosters from Solar Converters will start your pump earlier in the morning, keep it going longer in the afternoon, and give you pumping under lower light conditions when the pump would otherwise stall. Features include: Maximum Power Point Tracking, to pull the maximum wattage from your modules; a switchable 12- or 24-volt design in a single package; a float or remote switch input; a user-replaceable ATC-type fuse for protection; and a weatherproof box. Like all pump controllers, a closed float switch will turn the pump off, an open float switch turns the pump on.

Four amperage sizes are available in switchable 12/24 volt. Input and output voltages are selected by simply connecting or not connecting a pair of wires. Amperage ratings are surge power. Don't exceed 70% of amp rating under normal operation. The 7-amp model is the right choice for most pumping systems, and is the controller included in our submersible and surface pumping kits. 7- and 10-amp models are in a 4.5" L x

2.25" W x 2" D plastic box. 15-amp model is a 5.5" L x 2.9" W x 2.9" D metal box. 30-amp model is supplied in a NEMA 3R raintight 10" x 8" x 4" metal box. One-year manufacturer's warranty. Canada.

25002	**7A, 12/24V Pump Controller**	**$106**
25003	**10A, 12/24V Pump Controller**	**$135**
25004	**15A, 12/24V Pump Controller**	**$220**
25005	**30A, 12/24V Pump Controller**	**$395**

Float Switches

These fully encapsulated mechanical (no mercury!) float switches make life a little easier by providing automatic control of fluid level in the tank. You regulate the water level by merely lengthening or shortening the power cable and securing it with a band clamp or zip tie. Easily

installed, the switches are rated at 15 amps at 12 volts, or 13 amps at 120 volts. The "U" model will fill a tank (closed circuit when float is down, open circuit when up) and the "D" model will drain it (closed circuit when up, open circuit when down). When used with a PV-direct pump controller, the "U" and "D" functions are reversed: "D" will fill and "U" will empty.

Not sure if you want an "up" or "down" float switch? Confused by reverse-logic controllers? The Goof-Proof float switch is your salvation. This three-wire float switch can be connected either way. There's a common wire, an "up" wire, and a "down" wire. If it doesn't work right, just switch wires. 5-amp maximum current. Safe for domestic water. Two-year manufacturer's warranty. USA.

41637	**Float Switch "U"**	**$35**
41638	**Float Switch "D"**	**$35**
41036	**Goof-Proof Float Switch**	**$59**

Flat Two-Wire Sub Pump Cable

This flat, two-conductor, 10-gauge submersible cable is sometimes difficult to find locally. Works with the SHUR-flo Solar Sub splice kit. Cut to length; specify how many feet you need.

26-540	**10-2 Sub Pump Cable**	**$1.00/ft.**

WIND-POWERED PUMPS

Airlift Wind-Powered Water Pumps

Water pumping for windy sites

Airlift pumps use compressed air to lift water into a storage tank. Compressed air is delivered by a two- or four-cylinder air compressor driven by a wind turbine with tempered aluminum blades of 9- or 10-feet diameter, depending on the model. The variable pitch blades provide high-torque startup with winds of 4 to 6 mph, and reach optimum performance at 10 to 15 mph. In high wind conditions the turbine will automatically turn out of the wind, then return as wind speed decreases. Yearly maintenance is a simple oil change and intake filter cleaning. The turbine mounts on a 2.5" steel pipe

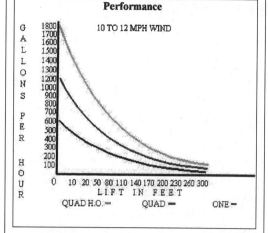

tower. The wind turbine can be hundreds of feet from the well. Air is routed to an air injection chamber submerged in the well. Air bubbles lift the water. Lifts up to 300 feet and 1 to 30 gpm are possible from well casings as small as 2". An air injection chamber and 200 feet of air line are included with each pump. There are no moving or wearing parts in the well. Pumped water can be moved up to 500 feet horizontally, but the delivery line must slope upwards at least slightly.

Airlift 1 will lift up to 300 feet maximum, and can deliver up to 10 gpm maximum. The Quad model is limited to 315 feet of lift, and can deliver up to 20 gpm. The Quad H.O. also lifts to 315 feet, but can deliver up to 30 gpm.

Submergence—how far the injection chamber is below water level—is critical for these simple pumps. Too little submergence and the air bubbles will separate from the water, too much submergence and the pump won't lift water. Lift is the vertical distance from the static water level to the storage tank. For lifts up to 150 feet, 70% submergence is recommended. So a 100-foot lift needs 70 feet of submergence. For lifts from 150 to 300 feet, 50% submergence is recommended.

For those lacking sufficient submergence depth, or needing to move long horizontal distances, a positive pump option is offered that will run with only 5 feet of submergence. The positive pump option will deliver approximately 20% less water than air injection. These air-powered pumps require a 10- to 20- gallon air pressure tank teed into the air line so the pump can always complete its cycle and stop in the "start-up" position. Positive pumps can lift up to 300 feet.

A 22-foot folding pipe tower kit is available, and highly recommended for ease of routine maintenance, or you can build your own. Tower is not included with pump. USA.

41713	Airlift 1	$2,495
41714	Airlift Quad	$3,295
43-0175	Airlift Quad H.O.	$3,895
42-0245	Airlift 22' Hinged Tower Kit	$1,295
41-715	Airlift Positive Pump 3.5" x 67"	$450
41-716	Airlift Positive Pump 4.5" x 48"	$430
41-717	Airlift Positive Pump 6.0" x 24"	$465

WATER-POWERED PUMPS

Ram Pumps

The ram pump works on a hydraulic principle using the liquid itself as a power source. The only moving parts are two valves. In operation, the water flows from a source down a "drive" pipe to the ram. Once each cycle, a valve slaps shut, causing the water in the drive pipe to stop suddenly. This causes a water-hammer effect and high pressure in a one-way valve leading to the "delivery" pipe of the ram, thus forcing a small amount of water up the pipe and into a holding tank. Rams will only pump into unpressurized storage. In essence, the ram uses the energy of a large amount of water falling a short distance to lift a small amount of water a large distance. The ram itself is a highly efficient device; however, only 2 to 10% of the liquid is recoverable. Ram pumps will work on as little as 2 gpm supply flow. The maximum head or vertical lift of a ram is about 500 feet.

SELECTING A RAM

Estimate amount of water available to operate the ram. This can be determined by the rate the source will fill a container. Make sure you've got more than enough water to satisfy the pump. If a ram runs short of water, it will stop pumping and simply dump all incoming water.

Estimate amount of fall available. The fall is the vertical distance between the surface of the water source and the selected ram site. Be sure the ram site has suitable drainage for the tailing water. Rams splash big-time when operating! Often a small stream can be dammed to provide the 1½ feet or more head required to operate the ram.

Estimate amount of lift required. This is the vertical distance between the ram and the water storage tank or use point. The storage tank can be located on a hill or stand above the use point to provide pressurized water. Forty or fifty feet water head will provide sufficient pressure for household or garden use.

Estimate amount of water required at the storage tank. This is the water needed for your use in gallons per day. As examples, a normal two- to three-person household uses 100 to 300 gallons per day, much less with conservation. A 20- by 100-foot garden uses about 50 gallons per day. When supplying potable water, purity of the source must be considered.

Using these estimates, the ram can be selected from the following performance charts. The ram installation will also require pouring a small concrete pad, a drive pipe five to ten times as long as the vertical fall, an inlet strainer, and a delivery pipe to the storage tank or use point. These can be obtained from your local hardware or plumbing supply store. Further questions regarding suitability and selection of a ram for your application will be promptly answered by our technical staff.

Aqua Environment Rams

We've sold these fine rams by Aqua Environment for over 15 years now with virtually no problems. Careful

attention to design has resulted in extremely reliable rams with the best efficiencies and lift-to-fall ratio available. Working component construction is of all bronze with O-ring seal valves. Air chamber is PVC pipe. The outlet gauge and valve permit easy startup. Each unit comes with complete installation and operating instructions.

Typical Performance and Specifications					
Vertical Fall (feet)	Vertical Lift (feet)	Pump Rate (Gallons/Day)			
		¾" Ram	1" Ram	1¼" Ram	1½" Ram
20	50	650	1350	2250	3200
20	100	325	670	1120	1600
20	200	150	320	530	750
10	50	300	650	1100	1600
10	100	150	320	530	750
10	150	100	220	340	460
5	30	200	430	690	960
5	50	100	220	340	460
5	100	40	90	150	210
1.5	30	40	80	130	190
1.5	50	20	40	70	100
1.5	100	6	12	18	25

Water Required to Operate Ram

¾" Ram - 2 gallons/minute	Maximum Fall - 25 feet
1" Ram - 4 gallons/minute	Minimum Fall - 1.5 feet
1¼" Ram - 6 gallons/minute	Maximum Lift - 250 feet
1½" Ram - 8 gallons/minute	

41811	Ram Pump, ¾"	$339
41812	Ram Pump, 1"	$339
41813	Ram Pump, 1¼"	$415
41814	Ram Pump, 1½"	$415

Folk Ram Pumps

The Folk is the most durable and most efficient ram pump available. It is solidly built of rustproof cast aluminum alloy and uses all stainless steel hardware. With minimal care, it probably will outlast your grandchildren. It is more efficient because it uses a diaphragm in the air chamber, preventing loss of air charge, and eliminating the need for a snifter valve in the intake that cuts efficiency.

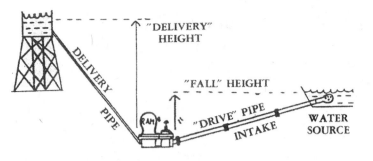

FOLK RAM PUMP CAPACITY		
Folk Model Drive Pipe Size	**Capacity In GPM Min.–Max.**	**Delivery Pipe**
1″	2–4	1″
1¼″	2–7	1″
1½″	3–15	1″
2″	6–30	1¼″
2½″	8–45	1¼″
3″	15–75	1¼″

Folk rams require a minimum of 2 gpm and 3 feet of drive pipe fall to operate. The largest units can accept up to 75 gpm, and a maximum of 50 feet of drive pipe fall. Delivery height can be up to fifteen times the drive fall, or a maximum of 500 feet lift. The drive pipe should be galvanized Schedule 40 steel, and the same size the entire length. Recommended drive pipe length is 24 feet for 3 feet of fall, with an additional 3 feet length for each additional 1 foot of fall. Be absolutely sure you have enough water to meet the minimum gpm requirements of the ram you choose. Rams will stop pumping and simply dump water if you can't supply the minimum gpm.

To estimate delivery volume, multiply fall in feet times supply in gpm, divide by delivery height in feet, then multiply by 0.61. Example: a ram with 5 gpm supply, a 20-foot drive pipe fall, and delivering to a tank 50 feet higher. 20ft x 5gpm ÷ 50ft x 0.61 = 2.0gpm x 0.61 = 1.22gpm, or 1,757 gallons per day delivery.

Warranted by manufacturer for one year. USA.

All Folk rams are drop-shipped via UPS from the manufacturer in Georgia.

41815	**Folk 1″ Ram Pump**	**$1,070**
41816	**Folk 1¼″ Ram Pump**	**$1,070**
41817	**Folk 1½″ Ram Pump**	**$1,070**
41818	**Folk 2″ Ram Pump**	**$1,330**
41819	**Folk 2½″ Ram Pump**	**$1,330**
41820	**Folk 3″ Ram Pump**	**$1,330**

High Lifter Pressure Intensifier Pump

The High Lifter Water Pump offers unique advantages for the rural user. Developed expressly for mountainous terrain and low summertime water flows, this water-powered pump delivers a much greater percentage of the input water than ram pumps can. This pump is available in either 4.5:1 or 9:1 ratios of lift to fall. The High Lifter is self-starting and self-regulating. If inlet water flow slows or stops the pump will slow or stop, but will self-start when flow starts again.

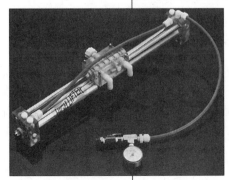

The High Lifter pump has many advantages over a ram (the only other water-powered pump). Instead of using a "water hammer" effect to lift water as a ram does, the High Lifter is a positive-displacement pump that uses pistons to create a hydraulic lever that converts a larger volume of low-pressure water into a smaller volume of high-pressure water. This means that the pump can operate over a broad range of flows and pressures with great mechanical efficiency. This efficiency means more recovered water. While water recov-

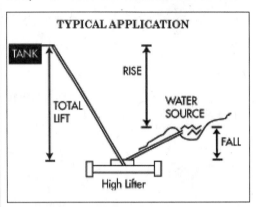

ery with a ram is normally about 5% or less, the High Lifter recovers 1 part in 4.5 or 1 part in 9 depending on ratio. In addition, unlike the ram pump, no "start-up tuning" or special drive lines are necessary. This pump is quiet, and will happily run unattended.

The High Lifter pressure intensifier pump is economical compared to gas and electric pumps, because no fuel is used and no extensive water source development is necessary. A kit to change the working ratio of either pump after purchase is available, as are maintenance kits. Maintenance consists of simply replacing a handful of O-rings. The cleaner your input water, the longer the O-rings last. Choose your model High Lifter pump from the specifications and High Lifter performance curves. One year parts and labor warranty from manufacturer. USA.

41801	**High Lifter Pump, 4.5:1 Ratio**	**$895**
41802	**High Lifter Pump, 9:1 Ratio**	**$895**
41803	**High Lifter Ratio Conversion Kit**	**$97**
41804	**High Lifter Rebuild Kit, 4.5:1 ratio**	**$65**
41805	**High Lifter Rebuild Kit, 9:1 ratio**	**$65**

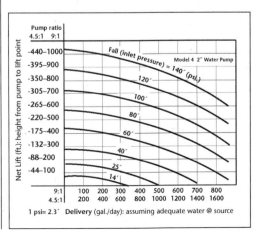

Pump Accessories

Floating Suction Filters

Get water from the cleanest point in your storage system —just below the water surface. Improves the quality of your rainwater and protects pumps from drawing up sediment without a need for pre-filtering. The floating ball lets the suction point rise and fall as well as preventing the filter from sinking to the bottom. Includes a float, fine filter housing with a non-return valve, and 1" barbed connection. Germany.

42908   **Float Suction Filter**   **$149**

Inline Sediment Filter

This filter is designed for cold water only, and meets NSF standards for drinking water supplies. It features a sump head of fatigue-resistant Celcon plastic, an opaque body to inhibit algae growth if installed outdoors, and is equipped with a manually operated pressure-release button to simplify cartridge replacement. Rated for up to 125 psi and 100ºF, it is equipped with 3/4-inch female pipe-thread inlet and outlet fittings, and is rated for up to 6 gpm flow rates (with clean cartridge). It accepts a standard 10-inch cartridge with 1-inch center holes. Supplied with one 10-micron fiber cartridge installed. USA.

41137   **Inline Sediment Filter**   **$49**
42632   **10 Micron Sediment Cartridge, 2 pack**   **$13**

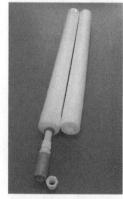

30" Intake Filter with Foot Valve

A great choice for any Slowpump with a surface water source. This large triple-size 10-micron filter with no external cover can be lowered into wells, or floated on a pond. Large surface area means less-frequent replacements. Equipped with a foot valve so you won't lose your pump prime. Included bushing allows 1/2 or 3/4" pipe-thread outlet. Comes with assembled filter and first replacement cartridge as pictured. USA.

40-0086   **30" Intake Filter/Foot Valve**   **$70**
40-0087   **30" Filter Repl. 3-Pack**   **$44**

Nominal Pipe Size vs. Actual Outside Diameter for Steel and Plastic Pipe

Nominal Size	Actual Size
1/2"	0.840"
3/4"	1.050"
1"	1.315"
1 1/4"	1.660"
1 1/2"	1.900"
2"	2.375"
2 1/2"	2.875"
3"	3.500"
3 1/2"	4.000"
4"	4.500"
5"	5.563"
6"	6.625"

Friction Loss Charts for Water Pumping

HOW TO USE PLUMBING FRICTION CHARTS

If you try to push too much water through too small a pipe, you're going to get pipe friction. Don't worry, your pipes won't catch fire. But it will make your pump work harder than it needs to, and it will reduce your available pressure at the outlets, so sprinklers and showers won't work very well. These charts can tell you if friction is going to be a problem. Here's how to use them:

PVC or black poly pipe? The rates vary, so first be sure you're looking at the chart for your type of supply pipe. Next, figure out how many gallons per minute you might need to move. For a normal house, 10 to 15 gpm is probably plenty. But gardens and hoses really add up. Give your-self about 5 gpm for each sprinkler or hose that might be running. Find your total (or something close to it) in the Flow GPM column. Read across to the column for your pipe diameter. This is how much pressure loss you'll suffer for every 100 feet of pipe. Smaller numbers are better.

Example: You need to pump or move 20 gpm through 500 feet of PVC between your storage tank and your house. Reading across, 1" pipe is obviously a problem. How about 1¼"? 9.7 psi times 5 (for your 500 feet) = 48.5 psi loss. Well, that won't work! With 1½" pipe, you'd lose 20 psi...still pretty bad. But with 2" pipe you'd only lose 4 psi ... ah! Happy garden sprinklers! Generally, you want to keep pressure losses under about 10 psi.

> If you try to push too much water through too small a pipe, you're going to get pipe friction. Don't worry, your pipes won't catch fire.

Friction Loss in PSI per 100 Feet of Scheduled 40 PVC Pipe

Flow GPM	Nominal Pipe Diameter in Inches							
	½"	¾"	1"	1-1/4"	1-1/2"	2"	3"	4"
1	3.3	0.5	0.1					
2	11.9	1.7	0.4	0.1				
3	25.3	3.5	0.9	0.3	0.1			
4	43.0	6.0	1.5	0.5	0.2	0.1		
5	65.0	9.0	2.2	0.7	0.3	0.1		
10		32.5	8.0	2.7	1.1	0.3		
15		68.9	17.0	5.7	2.4	0.6	0.1	
20			28.9	9.7	4.0	1.0	0.1	
30			61.2	20.6	8.5	2.1	0.3	0.1
40				35.1	14.5	3.6	0.5	0.1
50				53.1	21.8	5.4	0.7	0.2
60				74.4	30.6	7.5	1.0	0.3
70					40.7	10.0	1.4	0.3
80					52.1	12.8	1.8	0.4
90					64.8	16.0	2.2	0.5
100					78.7	19.4	2.7	0.7
150						41.1	5.7	1.4
200						69.9	9.7	2.4
250							14.7	3.6
300							20.6	5.1
400							35.0	8.6

Friction Loss in PSI per 100 Feet of Polyethylene (PE) SDR-Pressure Rated Pipe

Flow GPM	NOMINAL PIPE DIAMETER IN INCHES							
	0.5	0.75	1	1.25	1.5	2	2.5	3
1	0.49	0.12	0.04	0.01				
2	1.76	0.45	0.14	0.04	0.02			
3	3.73	0.95	0.29	0.08	0.04	0.01		
4	6.35	1.62	0.50	0.13	0.06	0.02		
5	9.60	2.44	0.76	0.20	0.09	0.03		
6	13.46	3.43	1.06	0.28	0.13	0.04	0.02	
7	17.91	4.56	1.41	0.37	0.18	0.05	0.02	
8	22.93	5.84	1.80	0.47	0.22	0.07	0.03	
9		7.26	2.24	0.59	0.28	0.08	0.03	
10		8.82	2.73	0.72	0.34	0.10	0.04	0.01
12		12.37	3.82	1.01	0.48	0.14	0.06	0.02
14		16.46	5.08	1.34	0.63	0.19	0.08	0.03
16			6.51	1.71	0.81	0.24	0.10	0.04
18			8.10	2.13	1.01	0.30	0.13	0.04
20			9.84	2.59	1.22	0.36	0.15	0.05
22			11.74	3.09	1.46	0.43	0.18	0.06
24			13.79	3.63	1.72	0.51	0.21	0.07
26			16.00	4.21	1.99	0.59	0.25	0.09
28				4.83	2.28	0.68	0.29	0.10
30				5.49	2.59	0.77	0.32	0.11
35				7.31	3.45	1.02	0.43	0.15
40				9.36	4.42	1.31	0.55	0.19
45				11.64	5.50	1.63	0.69	0.24
50				14.14	6.68	1.98	0.83	0.29
55					7.97	2.36	0.85	0.35
60					9.36	2.78	1.17	0.41
65					10.36	3.22	1.36	0.47
70					12.46	3.69	1.56	0.54
75					14.16	4.20	1.77	0.61
80						4.73	1.99	0.69
85						5.29	2.23	0.77
90						5.88	2.48	0.86
95						6.50	2.74	0.95
100						7.15	3.01	1.05
150						15.15	6.38	2.22
200							10.87	3.78
300								8.01

CHAPTER 7

Water Heating

MOST OF US TAKE HOT WATER at the turn of a tap for granted. It makes civilized life possible, and we get seriously annoyed by cold showers. Yet most people probably do not realize how much this convenience costs them. The average household spends an astonishing 20 to 40% of its energy budget on water heating. And for most folks, all those energy dollars are given to an appliance that has a life expectancy of only ten to fifteen years and throws away a steady 20% or more of the energy you feed it. Any improvements you provide in efficiency or longevity for your appliances reduces your environmental impact and leaves a better world for your children.

There are better, cheaper, and more durable ways to heat your home water supply than using an electric water heater. We'll briefly describe all the common water heater types, including solar water heaters, discussing the good and bad points of each. Then we'll offer suggestions for improvement, starting with the simplest solutions and working up to the more complex ones. Finally, we've developed a fairly accurate life-cycle cost comparison chart, so you can compare the real operation costs over the lifetime of the appliance. Points to consider for each heater type are initial cost, cost of operation, recovery time (how fast does it heat water?), ease of installation or ability to retrofit, life expectancy, and finally, life-cycle cost.

> The average household spends an astonishing 20 to 40% of its energy budget on water heating.

Common Water Heater Types

Storage or Tank-Type Water Heaters

Storage or tank-type water heaters are by far the most common kind of residential water heater used in North America. They typically range in size from 20 to 80 gallons, and can be fueled by electricity, natural gas, propane, or oil.

Storage heaters work by heating up water inside an insulated tank. Their good points are the modest initial cost, the ability to provide large amounts of hot water for a limited time, and the fact that they're well understood by plumbers and do-it-yourself home owners everywhere. Their bad points include constant standby losses, slow recovery times, and low life expectancy. Because heat always escapes through the walls of the tank (standby heat loss), energy is consumed even when no hot

water is being used. This wastes energy and raises operation costs. Most manufacturers are offering "high-efficiency" storage heaters now for a premium price. These just use more insulation to reduce standby losses slightly. Customer-installed water heater blankets do the same thing (although maybe not quite so well). Standby losses for gas and oil heaters are higher because the air in the internal flue passages is being warmed constantly, rising, and pulling in fresh cold air from the bottom. Although storage tanks can deliver large volumes of hot water rapidly, once they're exhausted, the limited energy input means a long recovery time. Who hasn't had to wait to get a hot shower, or suffered through a quick cold one, after someone else used up all the hot water?

Life expectancy for tank-type water heaters averages ten to fifteen years. Usually the tank rusts through at this age, and the entire appliance

> Because heat always escapes through the walls of the tank (standby heat loss), energy is consumed even when no hot water is being used.

An "add-on" heat pump water heater.

An "integral" heat pump water heater.

has to be replaced. Not a very efficient way to design relatively expensive appliances. Life expectancy can be prolonged significantly if the sacrificial anode rod is replaced at about five-year intervals. This metal rod is installed by the tank manufacturer. It keeps the tank from rusting, but only has about a five-year life span.

Storage water heaters are fairly cost effective if you use natural gas, which happens to be a real energy bargain. If you're heating with electricity or propane, then a storage water heater, though cheaper initially, is your highest-cost choice in the long run. See the Life-Cycle Comparison Chart on page 325.

Heat Pump Water Heaters

If you use electricity to heat water, heat pumps are three to five times more efficient than conventional tank-type resistive heaters. Heat pump water heaters use a compressor and refrigerant fluid to transfer heat from one place to another, like a refrigerator in reverse. Electricity is the only fuel choice. The heat source is air in the heat pump vicinity, although some better models can duct in warm air from the attic or outdoors. The warmer the air, the better: Heat pumps work best in warm climates where they don't have to work as hard to extract heat from the air. In these applications, the upper element of the conventional electric water heater usually remains active for back-up duty.

The good points of heat pump water heaters are that they use only 33 to 50% as much electricity as a conventional electric tank-type water heater, they will provide a small cooling benefit to the immediate area, and life expectancy is a good twenty years. The negatives are that heat pump heaters are quite expensive initially (average installed cost is approximately $1,200), and recovery rates are modest, variable, and can be fairly low if the pump doesn't have a warm environment from which to pull heat. Like all storage-tank systems, heat pumps have standby losses, and if installed in a heated room will rob heat from Peter (the furnace) to pay Paul (the water heater).

Heat pumps make better use of electricity because it's much more efficient to use electricity to *move* heat than to *create* it. Heat pump water heaters are available with built-in water tanks, called integral units, or as add-on units to existing water heaters. Add-on units may be a smarter investment, as the heat pump will probably outlive the storage tank. If you live in a warmer climate and use electricity to heat water,

a heat pump is your best choice. In fact, heat pump units stack up quite favorably against everything but natural gas if you can afford the initial purchase cost.

Heat pump water heaters are a bit complex to install, particularly the better units that take their heat from a remote site or use a separate tank. You will probably require a contractor to install one. Call your local heating and air-conditioning contractor for more details on availability in your area, cost, and installation estimates.

Tankless Coil Water Heaters

This type of heater is probably only found in older oil- or gas-fired boilers. A tankless coil heater operates directly off the house boiler; it does not have a storage tank, so every time there is a demand for hot water the boiler must run. This may be fine in the winter, when the boiler is usually hot from household heating chores anyway, but during the rest of the year it results in a lot of start-and-stop boiler operation. Tankless coil boilers may consume 3 Btu of fuel for every 1 Btu of hot water they deliver. This type of water-heating system is not recommended.

Indirect Water Heaters

Indirect water heaters also use the home heating system's boiler, but in a smarter setup. Hot water is stored in a separate insulated tank. Heat is transferred from the boiler using a small circulation pump and a heat exchanger. The separate insulated storage tank adds reserve capacity. This means the boiler doesn't have to turn on and off as frequently, which greatly improves fuel economy. When used with the new high-efficiency, gas-fired boilers and furnaces, indi-

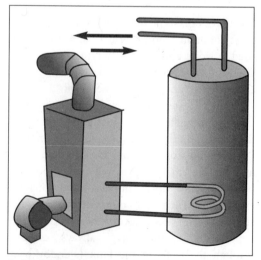

Indirect water heater.

rect water heaters are usually the least expensive long-term solution to provide hot water. They will add about $700 to the heating system cost. The life expectancy of the heat exchanger and circulation pump is an exceptional thirty years, although the storage tank may need replacement at fifteen to twenty years.

The big disadvantage of an indirect water-heating system is that it is an integral part of the household boiler/heating system that is best installed during new construction by your heating and cooling contractor. In addition, they suffer the usual standby losses of storage tanks.

Demand or Tankless Water Heaters

The tankless strategy is so obvious it's a no-brainer. Why keep 30 to 80 gallons of water at 120ºF all the time? Do you leave your car idling twenty-four hours a day just in case you need to run an errand in the middle of the night? Of course not. You start it when you need it. Our water heaters should work the same way, starting up only when we need hot water. Otherwise we heat up the water, it sits there in a big tank and loses heat, then we heat it up again, it sits there . . . and so on, ad infinitum. In many cases, this waste heat must even be removed from the home with expensive air conditioning! Due to a long honeymoon with cheap power, the United States is one of the few countries where people still use tank-type water heaters. Nearly everybody else in the world figured out the virtues of demand water heating a long time ago. Tank-type heaters use a minimum of 20% more energy than demand systems, due to heat loss. If yours is a small one- or two-person household, your heat losses are even greater, because the hot water spends more time sitting around, waiting to be used.

The advantages of demand water heaters are very low standby losses, lowest operation costs, unlimited amounts of hot water delivery, a very long life expectancy, and ease of installation for do-it-yourselfers and retrofitting. One disadvantage is that they cost a bit more initially. The other is that the unlimited amounts of hot water can't be taken too rapidly, because flow rates are limited to the heater's abilities, so hot water use may need to be coordinated at times. (People tend to be *very* sensitive and excitable about shower temperatures.)

HOW DEMAND WATER HEATERS WORK

Demand or tankless water heaters only go to work when someone turns on the hot water.

When water flow reaches a minimum flow rate, the gas flame or heater elements come on, heating the water as it passes through a radiator-like heat exchanger. Tankless heaters do not store any hot water for later use, but heat water only as demanded at the faucet. The minimum flow rates required for turn-on prevent any possibility of overheating at very low flow, and ensure that the unit turns off when the faucet is turned off. Minimum flow rates vary from model to model, but are generally about 0.5 to 0.75 gpm for household units. Other safety devices include the standard pressure/temperature relief valve that all water heaters in North America are required to carry, plus, tankless heaters use an additional overheat sensor or two on the heat exchanger.

The great advantage of tankless heaters is that you run out of hot water only when either the gas or the water runs out. On the other hand, tankless heaters will meter out the hot water at just so many gallons per minute. Excess water flow will result in lower temperature output. Some tankless heaters are limited to running just one fixture at a time. Larger household-sized heaters, such as the Aquastar 125 or the Takagi T-K2, can run multiple fixtures simultaneously. Showers are a touchy issue, so what complaints we hear usually revolve around showers and multiple-fixture uses. Tankless heaters are probably the best choice for smaller homesteads of three people or fewer, where hot water use can be coordinated easily. Larger homes with an intermittent use a long distance from the rest of the household hot-water plumbing, such as a master bedroom at the end of a long wing, are also good candidates for a tankless heater just to supply that isolated area.

See the More Details section below for a complete discussion of demand water heaters if this seems like a good choice for you.

CAN I GET AN ELECTRIC DEMAND WATER HEATER?

Yes you can, but first a warning or two. Electricity is easily the most expensive power source, unless you live in the Northwest where there are huge hydro projects. At average North American 2004 energy prices, natural gas is the least expensive choice for water heating, propane is about 60% more expensive, and electricity is about another 30% above that. It takes a shocking amount of electricity to heat water on the fly. Most households will need a 200-amp electric service at a minimum to support an instant electric water heater. That said, we've got a line of

Aquastar 125— A demand or tankless water heater.

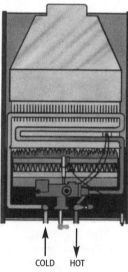

COLD   HOT

Basic demand water heater internal construction.

Do you leave your car idling twenty-four hours a day just in case you need to run an errand in the middle of the night? Of course not. Our water heaters should work the same way, starting up only when we need hot water.

The great advantage of tankless heaters is that you run out of hot water only when either the gas or the water runs out.

The typical solar water heater in North America will provide approximately two-thirds of a four-person family's hot water needs, saving you hundreds to thousands of dollars over its lifetime.

Standard "flat-plate" collector construction.

high-quality German electric water heaters with precise digital control and multiple temperature sensors that will do a great job with no standby losses.

Solar Water Heaters

The sun drops approximately 1,000 watts of free energy on every square meter of surface of the Earth's tropical and temperate zones at midday. Solar water heaters simply collect and store some of this free thermal energy. Although expensive initially, they can save large amounts of money over the long term. The typical solar water heater in North America will provide approximately two-thirds of a four-person family's hot water needs. As our Comparison Chart shows, a solar heater, working as a preheater before an electric- or propane-fueled back-up heater, will save you hundreds to thousands of dollars over its lifetime. If you're lucky enough to have low-cost natural gas as your energy source, then solar heaters have a marginal return on investment.

The advantages of solar water heaters are a no-cost/no-impact energy source (the amount you take today in no way diminishes the amount you can take tomorrow, and tomorrow, and tomorrow), the lowest operation cost, a retrofit-friendly technology, and a long life expectancy. The disadvantages are high initial cost, the potential of freezing with some less expensive models, and the need for a back-up heater in most climates.

TYPES OF SOLAR WATER HEATERS

Solar heaters are divided into two basic types: flat-plate collectors and batch collectors. These are discussed briefly below, and covered in more detail in our Best Choices section.

Flat-plate collectors are as simple as putting a black metal plate in the sun, then collecting the heat. Flat-plate collectors circulate a small amount of fluid through little passages in a sun-exposed, blackened copper plate, which is enclosed in an insulated, glass-covered box. Heat picked up by the black plate is transferred to the fluid, which carries it to the storage tank.

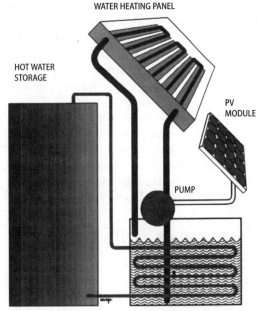

A flat-plate solar hot water system using an antifreeze solution in the collector loop with PV-powered pumping and a heat exchanger. This is a top-end "Lexus" solution.

Very simple homemade batch heater made from a water heater tank and an insulated box. Must be manually opened and closed daily. This is the low-end "bicycle" solution.

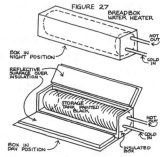

Flat-plate collectors don't store any energy. The metal is thin, and they typically hold less than a half-gallon of fluid. They have to move the heat to a storage tank for later use. Usually this will involve a pump and some control equipment.

Batch collectors are even simpler. Set a tankful of water out in the sun, then wait for it to get hot. Paint the tank black to help absorb heat, and put it into an insulated glass-covered box, and you've got a batch collector. Or for an even simpler setup, open and close your insulated box every day as in the illustration. The batch collector is plumbed in line before your back-up water heater. Any time hot water is used, the back-up heater gets preheated water from the solar collector, rather than stone-cold water.

If you think solar water heating is a good possibility for you, read more about it below.

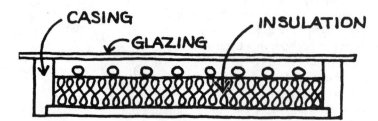

More Details for Your Best Water Heating Choices

Demand or Tankless Water Heaters

Tankless heaters are moderate cost initially, feature long (20 years or more) life expectancies, and next to solar, have the lowest operation costs, making them the best choice for most households. We covered operational basics above; what follows are more specifics of operation, and some installation tips.

THERMOSTATIC CONTROL

All the demand heater units we sell are thermostatically controlled. At lower flow rates the gas flame or electric element is reduced automatically to maintain a stable output temperature. The temperature is adjustable at the front panel from roughly 90°F to 140°F (110°F to 120°F is optimal for most uses). If the water flow rate increases, the heater will respond by increasing the heat input, up to the heater's Btu maximum. This makes a water heater that performs more like the storage tank heaters we're all used to. The system isn't perfect, because the sudden opening of a second tap still may cause the shower temperature to bobble. But this is also true of the plumbing in most homes with storage-tank heaters. With all the tankless heater models we sell, we recommend only running a single fixture, if that fixture is a shower. We're all touchy about our shower temperature, and this is the best way to ensure a steady temperature. The larger Takagi and Aquastar units have built-in flow restrictors, so that it is difficult to run more water through the unit than it can heat. If too many taps are opened simultaneously, the water pressure will fall and the water will simply run lukewarm, rather than ice-cold.

LIFE EXPECTANCY OF A TANKLESS WATER HEATER

Buying a tankless heaters is like buying a furnace. This appliance is going to give you decades of service. Demand heaters are designed so that all parts are repairable or replaceable. Aquastar warrants their heat exchanger for a full fifteen years. No model has any corrosible parts that touch water. The manufacturers have toll-free 800 service numbers backed up with full-time technicians and fully stocked parts warehouses. These heaters should last the rest of your home's life. Compare this to tank-type heaters, which have a ten- to fifteen -year life expectancy at best.

WHERE SHOULD YOU INSTALL A TANKLESS HEATER?

Just like tank-type heaters, the shorter the hot-water delivery pipe, the less energy is lost to warming up the plumbing. However, if you are willing to throw away the heat energy, and to wait for the hot water to reach the fixture, pipe length doesn't matter. A tankless heater does not need to be installed right at the fixture any more than a storage-type heater. If you are replacing an existing tank heater, it is probably most convenient to install the tankless unit in the same space. Your water and gas piping are already in the vicinity and will require only minor plumbing changes. If this is new construction, just pick the most central location. As with all gas appliances, the heater requires venting to the outside. In a retrofit installation the existing flue pipe will probably need to be upsized for the tankless heater. Do not reduce the flue size of the tankless unit to fit a smaller flue already in place. Do not install a tankless heater outside or in an unheated space, unless it never freezes in your climate. If the unit has a pilot light, it can be blown out easily if the unit is exposed to the wind.

SOLAR SYSTEM BACK-UP

As we mentioned above, the standard tankless units sense the outgoing water temperature and adjust the heat input accordingly to maintain a steady output temperature. As flow rate or

> Buying a tankless heater is like buying a furnace. This appliance is going to give you decades of service.

> A tankless heater does not need to be installed right at the fixture any more than a storage-type heater would.

Energy Conversions

Btu: British Thermal Unit, an archaic measurement, the energy required to raise one pound of water one degree Fahrenheit.

1 gal. liquid propane	= 91,500 Btu
1 gal. liquid propane	= 4 lb. (if you buy propane by the pound)
1 gal. liquid propane	= 36.3 cubic feet propane gas @ sea level
1 Therm natural gas	= 100,000 Btu
1 cubic foot natural gas	= 1,000 Btu
1 kW electricity	= 3,414.4 Btu/hr
1 horsepower	= 2,547 Btu/hr

incoming water temperature varies, the heat input will be modulated up or down to compensate. The standard Aquastar or Takagi units can only modulate heat input down to about 20,000 Btu, and are not particularly recommended for solar back-up. The gas-fired Aquastar "S" units, and all Stiebel-Eltron electric units, can modulate all the way down to zero Btu, and are highly recommended if there's any possibility you may use preheated water in the future. If the incoming water is already preheated to your selected output temperature, then the tankless heater will only come on briefly. If your preheated water is at 90°F, and you've got a 120°F output set, then the water heater will come on just enough to give 120°F output. We think that's pretty slick, and it gracefully slides us into the next good choice, solar water heaters.

Solar Water Heaters

THE BEST CHOICE FOR THE LONG RUN

Although solar water heaters are expensive initially, their long life expectancy and nearly zero cost of operation gives them the lowest life-

Tankless Water Heater Installation Tips

GAS PIPING

Most tankless heaters require a larger gas supply line than the tank-type heater you may be replacing. Just adapting an existing 1/2-inch supply line to 3/4-inch at the heater will choke the fuel supply and limit the water heater output. If the heater's installation manual says "3/4-inch supply line," that means all the way back to the gas pressure regulator.

Most of these heaters have a small safety override regulator on the gas inlet. This is a back-up for, and in addition to, the standard regulator at the tank or gas entrance. This secondary regulator is also the means for adjusting gas flow at higher altitudes (fully explained in the installation manual). The gas inlet is bottom center for all tankless heaters.

ELECTRIC SUPPLY

Even the smallest electric demand heater will require a dedicated heavy wiring circuit. This isn't something you can plug into an existing outlet, or tap off the existing water-heater circuit. Household-size units will require two or three 50-amp/240-volt circuits. So in a retrofit you'll be pulling new 6-gauge wire. Make sure you've got the space for breakers and the electrical capacity to run this much power.

PRESSURE/TEMPERATURE RELIEF VALVE

Unlike conventional tank-type heaters, there may not be a P/T valve port built into the heater. This important safety valve must be plumbed into your hot water outlet during installation. Simply tee into the hot water outlet.

VENTING

Tankless heaters must be vented to the outside. All models come with the draft diverter installed. Aquastar tankless heaters use conventional double-wall Type B vent pipe. This is the same type of vent used with conventional water and space heaters. The Aquastar FX model, and all Takagi heaters, use inexpensive 4-inch single-wall vent pipe, due to their power-vented exhaust. Vent pipe is not included, but is readily available at plumbing, building supply, or hardware stores. The vent piping used with most of these heaters is larger than tank-type vent piping. Do not adapt down to an existing vent size! Replace the vent and cut larger clearance holes as necessary if doing a retrofit. Venting may be run horizontally as much as 10 feet.

Maintain 1 inch of rise per foot of run. Aquastar recommends 10 feet of vertical vent at some point in the system to promote good flow.

INSTALLATION LOCATION

Don't install your tankless heater outside unless you are in a location where it never freezes. Freeze damage is the most common repair on these heaters. Don't tempt fate by trying to save a few bucks on vent pipe. If the heater is installed in a vacation cabin, put tee fittings and drain valves on both the hot and cold plumbing directly below the heater to ensure full drainage. The heat exchanger that works so well to get heat into the water works just as well to remove heat.

WARNING

It is illegal and dangerous to install a tankless water heater in an RV or trailer unless the heater is certified for RV service.

The following chart will help you plan your tankless heater installation before you buy. Aquastar heaters have zero clearance on the back for wall mounting. The Takagi heaters need 1" clearance on back, provided by the included wall-mount brackets.

Tankless Water Heater Clearance to Combustibles Chart

Heater	Btu/hr	Top	Bottom	Sides	Front
Aquastar 38	40,000	12"	12"	4"	4"
Aquastar 125B	117,000	12"	12"	4"	4"
Takagi T-K Jr	140,000	24"	–	R-2", L-12"	4"
Takagi T-K2	185,000	12"	–	R-2", L-6"	4"

cycle cost. Solar water heaters enjoyed a huge surge of interest following the Arab oil embargo during the 1970s and early 1980s, when they were supported by 50% federal tax credits. Unfortunately, those same tax credits encouraged a rash of sleazy door-to-door salesmen pushing less-than-the-best technology at astronomical prices. A lot of people got burned in the process, and the solar hot-water industry still hasn't recovered from the bad image this circus left behind.

All the solar heaters carried by Real Goods were designed after the Carter solar tax-credit days, and have benefited from lessons learned during this period. The solar heaters available now cost far less and are much more reliable. So the first lesson was quality. Don't buy junk. All these modern water heaters are high-quality units warranted for five to ten years by solid companies that are not going to disappear the day after the solar tax credits expire. These water heaters reasonably can be expected to last as long as your house. The second lesson was simplicity. The systems we sell don't use controllers, temperature sensors, or drain-down valves—all the complex hardware that experience showed us was most prone to breakdown. These heaters use simple passive designs with sunlight and heat as the controls.

FLAT-PLATE COLLECTORS

Flat-plate collectors are great at collecting solar energy, and moderately easy to construct, but almost always require a pump to circulate the heated fluid to storage, and a controller to tell the pump when to run. They're also prone to freezing easily, because of the many small passages and large surface area. The better systems use a freezeproof fluid, which then requires a heat exchanger between this fluid and the household water supply. This results in a moderately expensive system, but one that is absolutely freezeproof and has an excellent life expectancy in any climate. The Heliodyne systems we carry are examples of freezeproof flat-plate collector systems. This is a high-quality solar hot-water system that will work year-round, in any location no matter how cold, with no attention from the homeowner.

How about simplicity; in other words, how do we get rid of all the sensors and controllers that make the flat-plate collector work? Obviously you don't want the circulation pump(s) to run unless the collector is hotter than the storage tank. Most flat-plate collector systems accomplish this by using multiple temperature

Ask Dr. Doug

Is it possible to use one of the tankless heaters on my hot tub?

With some ingenuity and creativity, it has been done successfully. You need to bear in mind that these heaters are designed for installation into pressurized water systems. For safety reasons, they won't turn on until a certain minimum flow rate is achieved. The flow rate is sensed by a pressure differential between the cold inlet and the hot outlet. Tankless heater flow rates are 3 gallons per minute and less, which is a much lower flow rate than hot tubs. A hot tub system is very low pressure, but high volume. What usually works is to tee the heater inlet into the hot tub pump outlet, and the heater outlet into the pump inlet. This diverts a portion of the pump output through the heater, and the pressure differential across the pump usually is sufficient to keep the heater happy. Aquastar has a very inexpensive "recirc kit" for its heaters that lowers the pressure differential required. This kit is mandatory! In addition, you'll need an aquastat to regulate the temperature by turning the pump on and off (your tub may already have one if a heating system already was installed). You'll also need a willingness to experiment and a good dose of ingenuity.

sensors and a little controller "brain." Experience has shown that temperature sensors, wires, and splice connections leading to them, are the most trouble-prone parts of solar hot-water systems, followed closely by controllers. Our Heliodyne systems eliminate controllers, wiring splices,

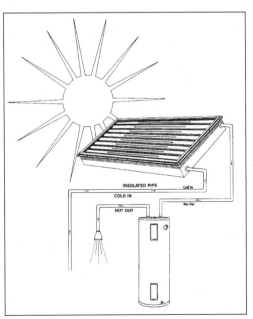

INSULATED PIPE — Cold In

COLD IN

HOT OUT — Hot Out

All the solar heaters carried by Real Goods were designed after the Carter solar-tax credit days, and have benefited from lessons learned during this period.

The ProgressivTube collector preheats the water before it reaches the hot water tank.

A homemade batch collector with glass glazing

Batch collectors are less expensive and less complex than flat-plate collectors, but are limited to use in climates without extended hard-freeze weather, or can be used for three seasons and drained during the winter.

and sensors by using a small PV module connected directly to the circulation pump. When the sun shines, the PV module starts producing electricity, and the pump starts running! The hotter and more direct the sunlight, the faster the pump runs. When the sun goes down, the pump stops working. Circulation speed is directly proportional to sunlight intensity. Simple elegance.

BATCH COLLECTORS

Batch heaters are designed to be simple, eliminating the pumps, controllers, sensors, and wiring at the start. The water isn't circulated from the collector to a storage tank. Batch collectors *are* the storage tank. This is simplicity itself. The primary disadvantage of this simple collector type is the potential for freezing. Batch collectors heat the domestic water directly, so they can't use an antifreeze fluid. Because of the

mass of 20 to 40 gallons of water inside the collector, the chance of freezing overnight in a cold snap is minimal, although the supply and return plumbing can be vulnerable. So, batch collectors are less expensive and less complex than flat-plate collectors, but are limited to use in climates without extended hard-freeze weather, or can be used for three seasons and drained during the winter. Hard freeze is defined as temperatures of 20°F or colder that last longer than a day. Three-season use is actually the most viable, cost-effective alternative for much of North America. Cloud cover in the winter will prevent collecting any meaningful amount of energy in much of the Northeast, and other portions of the continent. In a hard-freeze climate the batch collector is simply shut off and drained in the autumn, then refilled in the spring. The ProgressivTube collector that we carry is an example of a batch collector system.

Fuel Choices for Water Heating: Electric vs. Gas

Electricity is wonderfully useful for many tasks, but using it for larger-scale heating chores, such as household water and space heating, isn't a smart choice. While inexpensive and easy to install initially, electric tank heaters cost two to three times more to operate than gas-fired heaters. Over an average 13-year lifetime this amounts to a $3,000–$8,000 difference! See our Life-Cycle Cost Comparison Chart below for details of electric, gas, tankless, and solar water heating options. If you have gas or oil available, by all means use it! Over the lifetime of the heater you'll save many times the initial purchase and installation cost. In many rural homes where natural gas is not available, the higher life-cycle cost when using electricity may be a strong argument for bringing propane fuel on site, or for investing in one of the efficient heat pump water heaters. While propane costs about twice as much as natural gas, as the Comparison Chart shows, it's still significantly less expensive than electricity.

LIFE-CYCLE COST COMPARISON FOR WATER HEATERS AT 20 YEARS

Type	Cost[1]	Cost per Year[2]	Life Expectancy	20-year Cost[3]
STANDARD ELECTRIC TANK (R-value 14.5)	$435	@8¢kWh = $395 @10¢kWh = $493 @12¢kWh = $591	13 year	$8,770 $10,730 $12,690
HIGH-EFFICIENCY ELECTRIC TANK (R-value 25)	$525	@8¢kWh = $373 @10¢kWh = $467 @12¢kWh = $560	13 year	$8,510 $10,390 $12,250
ELECTRIC HEAT PUMP	$1,200	**@8¢kWh = $160** **@10¢kWh = $200** **@12¢kWh = $240**	20 year	**$4,400** **$5,200** **$6,000**
SOLAR W/ HI-E ELECTRIC TANK BACK-UP (solar covers 66%)	$2,500	@8¢kWh = $123 @10¢kWh = $154 @12¢kWh = $185	20 year	$4,960 $5,580 $6,200
STANDARD GAS-FIRED TANK (R-value 8)	$470	w/ NG = $244 w/ LP = $425	13 year	$5,820 $9,440
HIGH-EFFICIENCY GAS-FIRED TANK (R-value 16)	$570	w/ NG = $228 w/ LP = $398	13 year	$5,700 $9,100
Gas-fired Demand	$800	w/ NG = $205 w/ LP = $360	20 year	$5,800 $8,000
SOLAR W/ HI-E GAS TANK BACK-UP (solar covers 66%)	$2,545	**w/ NG = $75** w/ LP = $131	20 year	**$4,045** $5,165
SOLAR W/ GAS DEMAND BACK-UP (solar covers 66%)	$2,775	w/ NG = $68 **w/ LP = $119**	20 year	$4,135 **$5,155**

Figures in bold are lowest cost options over twenty years for electric, natural gas, and propane.

1. Approximate, includes $200 for installation.
2. Energy cost based on hot water needs for family of four. Gas costs: $0.96/therm for NG, $1.53/gal for LP (Feb. 2004 national averages).
3. Future operating costs neither discounted nor adjusted for infla- tion. Includes replacement cost for systems with less than twenty-year life expectancy.

Source: Adapted from American Council for an Energy-Efficient Economy life-cycle charts.

WATER HEATING PRODUCTS

TANKLESS WATER HEATERS

Aquastar Water Heaters

An Aquastar, besides being the most efficient gas water heater you can use, is probably the last water heater you'll ever purchase for your house. Unlike tank-type heaters, Aquastars are designed for decades of service. This is more like buying a furnace, it's going to be with you for a good long while. All parts that actually touch water are noncorrosive stainless steel, brass, or copper. Every part in the heater is repairable or replaceable for a lot less money than a new heater. There's a fifteen-year warranty on the heat exchanger, and two years on everything else.

Aquastars are controlled thermostatically to maintain your set water temperature regardless of flow rate, making them perform like a tank-type heater, but without the wasteful standby losses. Adjustment range is approximately 90° to 140°F. Most households will save immediately 15% of their gas use just by eliminating standby loss. For small households, where the water sits around waiting for use longer, the savings are even higher. In addition, the heat exchanger on tankless water heaters is much more efficient. According to the U.S. Department of Energy, tank-type gas heaters average about 75% efficiency when new (it drops with age and mineral accumulation). Aquastar heaters are rated at 80 to 87% efficiency, depending on model, and that changes very little with age. For safety, all models have redundant overheat protection, and a manual gas shut-off valve.

Aquastar was purchased by Robert Bosch Corp. in 1996, and all models were updated and re-engineered in the following years. The biggest changes are to the most popular, household-size, 125 model. Now with a more efficient heat exchanger, it burns 6.5% less fuel, but supplies the same amount of hot water. Activation flow rate is down from 0.75 to 0.5 gpm, making this unit less prone to turning off at low flow rates. Other new standard equipment includes a piezo ignitor, a pressure relief valve to comply with U.S. plumbing code, and a drain plug for easier, complete draining to prevent freeze damage. Cabinets have also been redesigned. A new electronic ignition model with no standing pilot light has been introduced, and is the best-selling model in the line. Electronic ignition saves about 10 to 15% of your gas use. There's also a power-vented model that can exhaust horizontally with less expensive single-wall pipe. New models have a "B," for Bosch, following the model number.

All Aquastar models except the power-vented one, use double-wall Type B gas vent. They recommend at least 10 feet of vertical vent at some point in the exhaust for proper gas scavenging. Horizontal runs up to about 11 feet are okay, but there still must be 10 feet of vertical vent at some point.

Aquastar is the only tankless water heater manufacturer offering a special model for preheated water. The "S" models are for installations that have (or may have in the future) solar preheated water. These models have greater adjustment range for the gas, and will modulate the gas burner all the way down to zero if needed to maintain set output temperature. This option won't affect normal operation with or without preheated water, but costs a whole lot more to add as a retrofit in the future.

The smallest model—38B—is intended for summer cabins or simple hand- and dish-washing. It will just run a low-flow shower so long as incoming water temperature is above 45°F. The model 125B series, in its several configurations, is our "standard" for multi-person households with normal suburban demands.

Aquastar heaters are manufactured in Portugal or France depending on model, and are warranted by the U.S. importer for fifteen years on the heat exchanger, and for two years on all other components. Support includes a toll-free 800 number for parts and service advice, with real technicians, who are backed up with a fully stocked parts warehouse. AGA approved. Specify LP (propane) or NG (natural gas) when ordering. Portugal or France.

Aquastar 38B

The new improved Aquastar 38B is the perfect water heater for a small cabin with a low-flow shower or anywhere the demand for hot water is low. Will run a shower so long as incoming water temperature is 45°F or above. The model 38 is equipped with a piezo igniter. Features an 80% efficiency rating. Specify propane (LP) or natural gas (NG).

- Btu Input: 40,000 Btu/hr.
- 45°F Temp. Rise: 1.3 gpm
- 55°F Temp. Rise: 1.0 gpm
- 65°F Temp. Rise: 0.9 gpm
- Minimum Flow: 0.6 gpm
- Vent Size: 4" Type B gas vent
- Water Connections: 1/2" NPT
- Min. Gas Supply Line: 1/2" NPT
- Min. Water Pressure: 15 psi
- Dimensions: 25.375" H x 10.625" W x 9.125" D
- Shipping Weight: 25 lb

| 45101LP | Aquastar 38B LP | $369 |
| 45101NG | Aquastar 38B NG | $369 |

Aquastar 125B

The new improved Aquastar 125B is the perfect residential heater. It's capable of delivering a constant supply of hot water at the temperature you select. The 125B can support a single shower or smaller multiple fixtures simultaneously. This model is the best choice for the average home. Appropriate for small-volume commercial installations also. All 125 models feature an excellent 82% efficiency rating.

Standard models will modulate the gas down to approximately 20,000 Btu; S models will modulate down to zero for preheated water. Get the S model if you have, or plan to have, a solar or woodstove water-heating system. S models otherwise operate the same as standard models. Both the standard 125B and the 125BS models have a standing pilot light with piezo ignitor.

Aquastar now offers a pilotless HX version of the 125. This model does away with the standing pilot light, using electronic ignition, and significantly boosts energy efficiency. Most households will save about 10% of their gas bill with this model. No electric plug or batteries are required; the ignition is powered by a tiny hydroelectric turbine that spins up when hot water is turned on. Trick German engineering at its best! All other specifications are identical. The HX model is (sadly) not available with the S solar option. You can't have everything.

The FX model combines electronic ignition with a built-in power vent exhaust using smaller, less expensive 4" single-wall exhaust pipe. Horizontal or vertical venting is okay with this unit. A high-quality stainless vent hood for wall vent is included. This is the one Aquastar unit that needs to plug into 120vac. It draws about 40 watts when venting, and 1- to 2 watts in standby. BTU input for this model is boosted slightly to 125,000.

- Btu/hr Input: 117,000 (125,000 for FX)
- 55°F Temp. Rise: 3.3 gpm
- 75°F Temp. Rise: 2.4 gpm
- 90°F Temp. Rise: 2.0 gpm
- Minimum Flow: 0.5 gpm
- Vent Size: 5" Type B gas vent (4" single-wall for FX)
- Water Connections: 1/2" NPT
- Min. Gas Supply Line: 3/4" NPT
- Min. Water Pressure: 15 psi
- Dimensions: 29.75" H x 18.25" W x 8.75" D
- Shipping Weight: 44 lb

Standard 125B Model with standing pilot.

| 45105 LP | Aquastar 125B Propane | $599 |
| 45105 NG | Aquastar 125B Natural Gas | $599 |

125B S Model for preheated water

| 45107 LP | Aquastar 125B S Propane | $699 |
| 45107 NG | Aquastar 125B S Natural Gas | $699 |

125B HX Model with pilotless electronic ignition

| 45114 LP | Aquastar 125B HX Propane | $699 |
| 45114 NG | Aquastar 125B HX Natural Gas | $699 |

Note: "S" option not available in pilotless models.

125B FX Model, pilotless with power venting

| 45003 LP | Aquastar 125B FX Propane | $929 |
| 45003 NG | Aquastar 125B FX Natural Gas | $929 |

Takagi T-KJr and T-K2 Tankless Water Heaters

The most powerful, efficient, and adaptable water heaters available. The gas-fired Takagi water heaters feature electronic ignition, powered ventilation, computer-modulated gas flame for steady output temperatures,

and the industry's best energy factor of 0.84. Both models can be used with solar preheated incoming water—only adding additional heat as needed to meet your set output temperature. All waterways are noncorrosive copper or brass, burners are stainless steel, and the cabinet is powder-coated for long life. Automatic freeze protection will protect down to 5°F with no wind (don't tempt fate). Water, gas, and electrical connections are on the lower left (as you face the heater). We supply wall mounting brackets and P/T relief valve. The optional remote control allows greater temperature control, and displays any self-diagnosing error codes if a problem develops. The stainless wall vent terminator is recommended for horizontal vents. The backflow prevention kit is required for hard-freezing climates. Vent can be run up to 35', vertically or horizontally with a

Model	T-KJr	T-K2
BTU Input Range:	NG 19,500–140,000	NG 20,000–185,000
	LP 17,000–140,000	LP 19,000–175,000
Max Flow @ 50ºF temp rise	4.5 gpm	6.2 gpm
Max Flow @ 70ºF temp rise	3.1 gpm	4.4 gpm
Max Flow @ 90ºF temp rise	2.4 gpm	3.4 gpm
Minimum Flow	0.6 gpm	0.6 gpm
	(0.75 gpm to light)	(0.75 gpm to light)
Vent Size	4" single wall	4" single wall
Water Connections	3/4" Male NPT	3/4" Male NPT
Minimum Gas Supply	3/4" (1" if over 25')	3/4" (1" if over 25')
Minimum Water Pressure	15 psi	15 psi
Electrical Supply	120v, 60hz, 100 watts	120v, 60hz, 100 watts
Weight	32 lb.	60 lb.
Dimensions H x W x D	20" x 14" x 6"	24.5" x 16.5" x 8.3"
Energy Factor	0.84	NG 0.81, LP 0.84

5' reduction for every elbow. Maximum of 3 elbows. AGA-approved. Manufacturer's warranted for seven years on heat exchanger, two years on all else. Japan.

45-0108 LP	**Takagi T-K Jr Propane**	**$849**
45-0108 NG	**Takagi T-K Jr Natural Gas**	**$849**
45554	**Takagi T-K2 Propane**	**$1,499**
45555	**Takagi T-K2 Natural Gas**	**$1,499**
45507	**Takagi Remote Control Option**	**$195**
45506	**Takagi Wall Vent Terminator Option**	**$100**
45533	**Takagi Exhaust Backflow Prevention Kit**	**$56**

Stiebel-Eltron Instant Electric Water Heaters

Quality heaters from Germany's largest manufacturer

When electricity is your choice for water heating, these high-quality tankless heaters will save you 15 to 30% by eliminating standing losses. We're offering four models, ranging from a small hand-washing heater to a large

DHC-E 20 & 30

DHC-E 10

full-scale house heater. All feature automatic anti-scald electronic controls that won't let the output temperature exceed 130°F. All but the smallest model have front panel digital temperature control. Simply dial the output temperature you need from 86°F to 125°F. Advanced microprocessor control with multiple temperature sensors, flow sensors, and safety thermal cut-offs assures the output temperature never deviates. Installation is straightforward and simple. All electrical and water connections are clearly labeled and easy access. Three-year manufacturer's warranty with North American distributor and parts warehouse. Germany.

DHC 3-1

Model	DHC 3-1	DHC-E 10	DHC-E 20	DHC-E 30
Recommended use:	Hand washing	Summer cabin	Modest house	Full-size house
Voltage	120v	240v	240v	240v
Amperage	25A	40A	80A	120A
Wattage	3kW	9.6kW	19.2kW	28.8kW
Min. Breaker Size	30A	50A	2x50A	3x50A
Wire Size AWG copper	10	8	2x8	3x8
Min. Flow to Activate, gpm	0.32	0.45	0.58	0.87
Max. Temp Increase @gpm/°F	0.32/65	0.75/87	2.25/58	2.25/87
	0.5/41	1.0/65	3.0/44	3.0/65
	0.75/27	1.5/44	4.5/29	4.5/44
Water Connections	1/2" NPT	1/2" NPT	3/4" NPT	3/4" NPT
Size H x W x D, inches	14.2 x 7.8 x 4.1	14.2 x 7.8 x 4.1	14.2 x 21.75 x 4.6	14.2 x 21.75 x 4.6
Weight, pounds	4.6	5.9	21.0	24.25
Item #:	**45-0109**	**45-0110**	**45-0111**	**45-0112**
Price:	**$190**	**$275**	**$495**	**$675**

OTHER WATER HEATING PRODUCTS

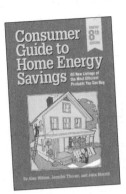

Consumer Guide to Home Energy Savings, 8th Ed.

Energy efficiency pays. Written by the nonprofit American Council for an Energy-Efficient Economy, the *Consumer Guide* has helped millions of folks do simple things around their homes that save hundreds of dollars every year. The new eighth edition has up-to-date lists of all the most energy-efficient appliances by brand name and model number, plus an abundance of tips, suggestions and proven energy-saving advice for home owners. 264 pages, softcover. USA.

**21-0344 Consumer Guide to
Home Energy Savings, 8th Ed.**   **$9.95**

D'mand Water Saving System

Does your house make you wait, and wait, and wait for hot water at some faucets while wasting water down the drain? Here's the cure. With the push of a button, the D'mand system whisks hot water to the most remote faucet in seconds. The displaced cold water is pumped into the cold water plumbing, returning to the heater, so there's no waste. It shuts off automatically when hot arrives and costs less than $1 per year to operate. UL- and UPC-listed, the D'mand system is recognized by the Department of Energy as both a water- and power-saving device, and has a life expectancy exceeding fifteen years. It works with any type of water heater.

Our "no sweat" installation kit provides all the necessary plumbing bits and pieces, installs with common hand tools, and plugs into a standard outlet. Installed with a pushbutton at your furthest faucet, intervening faucets can be equipped with additional wireless pushbuttons to activate the system from up to 100 feet.

The only difference between kits is pump size. Longer plumbing runs need a larger pump to keep wait time under 20 to 30 seconds. Use S-50 model for runs up to 50', S-70 for runs up to 100', S-02 model for longer or commercial use. S-50 model has three-year manufacturer's warranty, other models are five-year. USA.

45-0100 S-50-PF-R Kit	**$299**
45-0101 S-70-PF-R Kit	**$399**
45-0102 S-02-PF-R Kit	**$549**
45-0103 Additional Transmitter	**$20**

D'mand S-02-PF-R   D'mand S-50-PF-R   D'mand S-69-PF-R

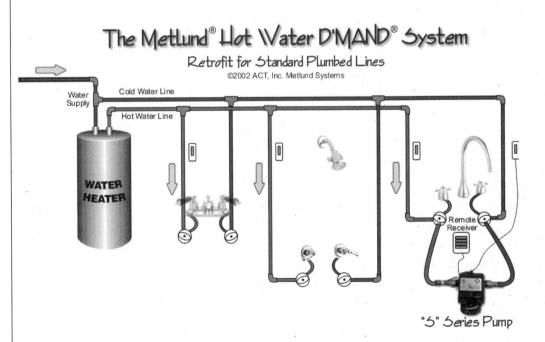

The Metlund® Hot Water D'MAND® System

Retrofit for Standard Plumbed Lines

©2002 ACT, Inc. Metlund Systems

"S" Series Pump

Snorkel Hot Tub Heaters

Snorkel stoves are woodburning hot tub heaters that bring the soothing, therapeutic benefits of the hot tub experience into the price range of the average person. Snorkel stoves are simple to install in wood tubs, easy to use, extremely efficient, and heat water quite rapidly. They may be used with or without conventional pumps, filters, and chemicals. The average tub with a 450-gallon capacity heats up at the rate of 30°F or more per hour. Once the tub reaches the 100°F range, a small fire will maintain a steaming, luxurious hot bath.

Stoves are made of heavy-duty, marine-grade plate aluminum that is powder-coated for the maximum in corrosion resistance. This material is very light, corrosion resistant, and strong. Aluminum is also a great conductor of heat, three times faster than steel. Stoves are supplied with secure mounting brackets and a sturdy protective fence.

Two stove models are offered, the full-size 120,000 Btu/hr Snorkel, or the smaller 60,000 Btu/hr Scuba. The Snorkel is recommended for 6- and 7-foot tubs. The Scuba is for smaller 5-foot tubs. These stoves are for wooden hot tubs only.

We also offer Snorkel's precision-milled, long-lasting Western Red Cedar hot tub kits that assemble in about four hours. A 5-foot tub holds two to four adults, a 6-

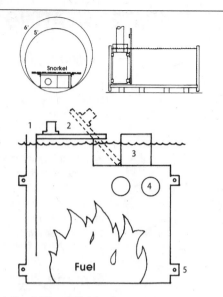

1. Special "Snorkel" air intake.
2. Sliding, cast aluminum, tilt-up door that services the fire box as well as regulates the air intake.
3. Sunken stack to increase efficiency.
4. Heat exchange tubes.
5. Mounting brackets for securing fence and stove.

foot tub holds four to five adults. The 7-foot tub is for large families or folks who entertain frequently. USA.

45400	**Scuba Stove**	**$599**
45401	**Snorkel Stove**	**$799**
45002	**Cedar Hot Tub 5´ x 3´**	**$1,410**
45207	**Cedar Hot Tub 6´ x 3´**	**$1,679**
45208	**Cedar Hot Tub 7´ x 3´**	**$2,210**
45504	**Drain Kit, PVC 1.5"**	**$45**
45501	**6´ Tub 3-Bench Kit**	**$349**

Shipped freight collect from Washington State.

SOLAR WATER HEATING

DOMESTIC WATER HEATERS

Solar Hot Water, Lessons Learned

Tom Lane probably has more hands-on practical experience from 1977 to 2002, and knowledge with all kinds of solar hot-water systems, than anyone on the planet. He is detailed, specific, and refreshingly honest about the good and bad points of commonly available solar hardware. Extensive use of pictures, drawings, and system schematics help make explanations clear. Beautifully organized with a detailed table of contents. Pick what you need. 120 pages, softcover. USA.

21-0336 Solar Hot Water, Lessons Learned **$29**

Four-Gallon Solar Shower

No more cold showers when you're camping. This incredible low-tech invention uses free solar energy to heat water for all your washing needs. Large 4-gallon capacity provides ample hot water for several hot showers. On a 70°F day, the Solar Shower will heat 60°F water to 108°F in only three hours for a tingling-hot shower. Great for camping, car trips, or emergency use. Taiwan.

17-0169 Solar Shower **$15**

ProgressivTube
Solar Water Heater

The best choice for mild climates

This passive batch-type heater is your most cost-effective water heater in climates that don't regularly freeze.* This is the simplest, yet most elegant and durable water heater we've seen. Plumbed inline ahead of your conventional water heater's cold inlet, it will preheat all incoming water, reducing your conventional heater's work load to near zero. Rated at approximately 30,000 Btu/day, the 4 ft x 8 ft collector is housed in a top-quality, bronze-finished aluminum box with stainless and aluminum hardware. There are no rusting components. The dual glazing has tempered low-iron solar glass on the outside, and non-yellowing or -degrading 96% transmittance Teflon film on the inside. High temperature, non-degrading phenolic foam board insulation is used on the sides (R-12.5), between internal tubes (R-12.5), and the bottom (R-16.7). The collector is composed of large 4-inch diameter selective-coated copper pipes aligned horizontally. They are connected in series, with cold water introduced at the bottom, and hot water taken off at the top. As the water warms, it stratifies and gradually works its

way toward the top. The outgoing hot water is never diluted by incoming cold water. There are no pumps, sensors, controls, or other moving parts that can fail. 30-, 40-, and 50-gallon models are produced by varying the number of collector pipes inside. Exterior size is 3 x 8 feet for the 30-gallon, and 4 x 8 feet for the 40 and 50 gallon models. BTUs collected per day under average North American sun conditions are 22,100 for the PT-30, 28,400 for the PT-40, and 28,700 for the PT-50, according to FSEC.

The collector and mounting system have been officially tested and approved for 180 mph winds, and 250 units installed in St. Croix were tested unofficially by the 200+ mph winds of Hurricane Hugo in 1989. Only six units suffered minor damage, and that was due to flying debris.

Use caution if roof-mounting a ProgressivTube collector. Filled, they weigh close to 600 lb. Rafters may need to be doubled up. Ground mounting is also a choice.

Both fixed (roof angle) and tilting mounts with 36 inches of cut-to-fit back leg tubes are offered. The two-way valve kit includes drain valves, a very important tempering valve, 150 psi pressure relief valve, and quality ball valves allowing you to choose between solar preheating or solar bypass. ProgressivTube has a great ten-year warranty. USA.

An overnight freeze won't threaten this heater, but three to four days of temps in the 20s would. Supply and return plumbing is much more vulnerable. Plumbing runs need to be short and well-insulated, and/or you need to be prepared and willing to drain the system seasonally.

45085	ProgressivTube PT-30	$1,180
45094	ProgressivTube PT-40	$1,465
45089	ProgressivTube PT-50	$1,565
45095	ProgressivTube Fixed Mounting	$75
45096	ProgressivTube Tilt Mounting	$99
45097	ProgressivTube 2-Way Valve Kit	$250

Shipped freight collect from Sarasota, FL.
Collector pricing includes $50 crating charge.

WATER HEATING

HELIODYNE SOLAR HOT WATER SYSTEMS

The best choice for hard-freezing climates

Freeze-Proof Heliodyne HX Solar Hot Water Systems

Heliodyne has survived in the solar hot-water business since 1976 because they offer the highest-quality equipment and the simplest designs. These systems use a separate anti-freeze-filled loop from the collector outside to the heat exchanger at your storage tank. They can't freeze and will continue collecting energy on cold sunny days. The PV-direct control system is as simple as sunrise. The brighter the sun, the faster the circulation pump runs. No temperature sensors, controllers, or wiring needed. The heat exchanger works with any off-the-shelf water heater tank, no expensive solar tanks needed.

Heliodyne kits provide: Gobi solar collectors as indicated, Flush Mount Kit, Blind Unions (supply bottom left, exit top right) Helix PV pumping station w/ 20-watt PV & DC pump, Heat Exchanger Assembly w/ 2 dial thermometers, check valve, filling ports, expansion tank, tank connections, pressure gauge, pressure relief valve, and flex tank connectors. You supply: ³/4" copper supply and return piping as needed and a 50- to 120-gallon water heater tank. All systems will need 1 to 2 gallons of non-toxic, FDA-approved anti-freeze (see below), which should be changed every five years.

Three systems are offered with FSEC ratings of 30,000, 46,000, or 60,000 BTU/day. The average North American person uses 10,000 to 15,000 BTU of hot water per day. Please call for more details. USA.

System pricing includes $75 crating charge. Shipped freight collect from San Francisco Bay area.

45508	Heliodyne HX13366 G PV System	$2,665
45509	Heliodyne HX1410 G PV System	$2,960
45532	Heliodyne HX23366 G PV System	$3,440

California Code Kit

Required in many states, and highly recommended for everyone, this kit provides an adjustable tempering valve that will automatically mix in cold water to prevent scalding. Also has an air vent for ease of system filling and bleeding. USA.

45511	California Code Kit	$95

Collection Anti-Freeze Fluid

A propylene-glycol solution with extra-strength corrosion inhibitors for high temperature use. Nontoxic, nonflammable, FDA-approved, safe for human contact. Every system will need 1 to 2 gallons, diluted 40 to 60%. Operating range is -50° to 325°F. Should be changed about every five years. USA.

45-515	Dyn-O-Flo, Single Gallon	$30
45100	Dyn-O-Flo, 4 Single Gallons in Cube	$98

Tilt Racking

(for HX systems only)

HX systems include flush mounts. Use this option if your collector(s) needs to be at an angle different from the roof. One kit per collector. USA.

45082	1 Tilt Racking Kit	$50
45082	2 Tilt Racking Kit	$59
45082	3 Tilt Racking Kit	$77

Solar Storage Tanks

Steel, glass-lined pressure vessels with R-16 insulation. 80-gallon tank is 24" diameter x 59"H. 120-gallon tank is 28" diameter x 62.5"H. USA.

45517	80-Gallon Storage Tank	$649
45531	120-Gallon Storage Tank	$990

Shipped freight collect from San Francisco Bay area.

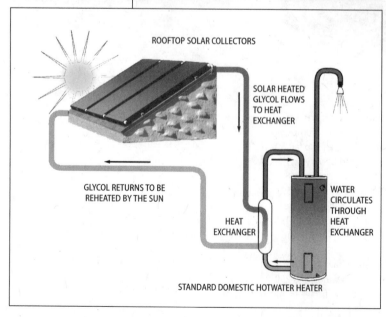

ROOFTOP SOLAR COLLECTORS

SOLAR HEATED GLYCOL FLOWS TO HEAT EXCHANGER

GLYCOL RETURNS TO BE REHEATED BY THE SUN

WATER CIRCULATES THROUGH HEAT EXCHANGER

HEAT EXCHANGER

STANDARD DOMESTIC HOTWATER HEATER

Heliodyne Solar Hot Tub Heaters

Our solar hot tub kit comes complete with a top-quality solar collector, flush-mount hardware, 120-volt AC controller, pump, temperature sensors, and air vent. You add the tub and plumbing to suit your site. These kits are designed to mount the collector on a nearby roof and connect to existing hot tub plumbing in such a way that all water can drain from the collector and pipes when not operating. Automatic operation is provided by differential temperature sensors and the controller. The system will only run if the tub needs heat, and there's heat to be had at the collector. Tub temperature is user adjustable, with an upper limit of 105°F. The pump is not self-priming, and must be installed below tub water level. It will lift water up to two stories.

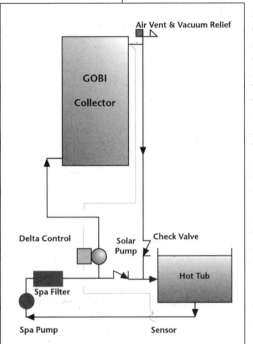

The smaller 3366 collector kit can deliver 18,000 to 30,000 Btu/day. With reasonable tub insulation, it will maintain daily tub temperature in North American climates, and will bring smaller to medium-size tubs up to temperature within a few days after a water change. The larger 4´ x 10´ kit can deliver 29,000 to 46,000 Btu/day. Use it for larger tubs or faster unassisted recovery times. USA.

| 45000 | Solar Hot Tub Kit 3366 | $1,380 |
| 45209 | Solar Hot Tub Kit 410 | $1,660 |

Shipped by truck from San Francisco Bay area.
System pricing includes $75 crating charge.

Heliodyne Gobi Solar Collectors

Heliodyne has produced the absolutely best solar collectors in the business since 1976. Absorbers feature 360° wrap-around fin-to-tube bonding that will never delaminate. Absorbers have a black chrome selective coating that is better at absorbing heat than radiating it off. Headers are extra-large 1-inch diameter, with wonderful solid brass O-ringed unions at all four corners for complete adaptability. Multiple collectors will couple up without any modification. The glazing is tempered 5/32" low-iron solar glass, with an antiglare finish. It's mounted in a full-perimeter rubber gasket. Frames are exceptionally strong bronze-anodized aluminum extrusions, with a mounting flange around the full perimeter. They are certified for survival at wind speeds up to 50 lb/sq.ft.; that's a bit over 100 mph. Each solar collector system will require a pair of blind unions to block off the unused header ends, and either flush mount or tilting rack mounts.

The 3366 collector is designed specifically to fit into trucks and standard shipping containers for reduced shipping costs. It measures 43.6" x 88.6", and is FSEC rated to deliver 21,000 to 26,000 Btu/day in domestic water-heating service.

The 408 collector is slightly larger at 48" x 97.6", and is FSEC rated to deliver 23,000 to 37,000 Btu/day in domestic water-heating service. Variations in collector ratings are due to low or medium temperature input to collector.

The 410 collector is the largest Heliodyne produces at 48" x 121.6". It is FSEC rated to deliver 29,000 to 46,000 Btu/day in domestic water-heating service. Because of its large size, this collector can be expensive to ship. The typical North American home requires about 10,000 to 15,000 Btu worth of hot water per person per day for comparison. USA.

45512	Heliodyne 3366 Gobi Collector	$689
45498	Heliodyne 408 Gobi Collector	$820
45500	Heliodyne 410 Gobi Collector	$969

Pricing includes $75 crating charge
Shipped from San Francisco Bay Area.

HELIODYNE OPTIONS

Solar Collection Fluid

A propylene-glycol solution with extra-strength corrosion inhibitors for high temperature use. Nontoxic, non-flammable, FDA-approved, safe for human contact. Every system will use 1 to 2 gallons, diluted 40 to 60%. Operating range is -50° to 325°F. Should be changed about every five years. USA.

| 45100 | Dyn-O-Flo, four one-gallon bottles | $98 |
| 45515 | Dyn-O-Flo, single one-gallon bottle | $30 |

Blind Unions

Every system needs a pair of these to block off the two unused unions after the plumbing is assembled. (The Complete PV-Powered Systems are supplied with these already.)

| 45495 | Heliodyne Blind Unions, 1 Pair | $20 |

Flush Mount Hardware

This hardware kit will mount one collector parallel to the roof, with an attractive stand-off clearance for code compliance and longest roof life. (The Complete PV-Powered Systems are already supplied with these.) See the Tilting Rack Options on the next page if your collector needs to be tilted in relation to the roof surface.

| 45496 | Heliodyne Flush Mount Kit | $49 |

Tilting Rack Options

Flush mounts are included with complete systems. If you need to tilt your collectors at something other than roof angle, here's the answer. Includes 42 inches of back leg tubing to allow cutting at needed collector angle.

45109	**Tilt Rack for 1 Collector**	**$96**
45110	**Tilt Rack for 2 Collectors**	**$153**
45111	**Tilt Rack for 3 Collectors**	**$219**

Nitroprene Pipe Insulation

Nitroprene is a new closed-cell, lightweight, hypoallergenic insulation material that's less expensive than conventional polyethylene foam, but superior in almost every way. It's highly UV resistant and tolerates from -104° to 230°F, making it the best available choice for solar hot-water systems. It's inherently self-extinguishing, highly flexible, will not support bacterial or fungal growth, and has a smooth clean surface texture. No CFCs, chlorine, or asbestos are used in manufacturing, and the material does not outgas. How is it not superior? Poly foam has an R-value of 4.0 per inch, Nitroprene is 3.6 per inch, but we're offering a much thicker product than the $3/8$" wall usually found in local hardware stores. Available in 5-packs of black pre-slit 6-foot lengths in $3/4$" and 1" thicknesses for $1/2$" and $3/4$" copper pipe sizes. USA.

06-0544	**$1/2$" pipe x $3/4$" wall x 6' long, 5-pack**	**$15**
06-0545	**$1/2$" pipe x 1" wall x 6' long, 5-pack**	**$24**
06-0546	**$3/4$" pipe x $3/4$" wall x 6' long, 5-pack**	**$19**
06-0547	**$3/4$" pipe x 1" wall x 6' long, 5-pack**	**$33**

CHAPTER 8

Water and Air Purification

IF YOU ARE CONCERNED about the degradation of our environment, you are probably also conscientious about the foods you eat. If you are conscientious about the foods you eat, then you may also want to pay attention to the quality of the air you breathe and the water you drink. Like food, air and water can carry into your body the contaminants that saturate the natural world. But you may not realize the extent to which your personal source of domestic water and the air inside your house may be polluted. This chapter examines the bad stuff that may be lurking in your air and water, the conditions that contribute to potentially unhealthful levels of these contaminants in your living space, and what you can do about it. At the end of the chapter, you'll find descriptions of our recommended air and water purification products and resources.

The Need for Water Purification

The Environmental Protection Agency (EPA) states that no matter where you live in the United States, some toxic substances are likely to be found in the groundwater. Indeed, the agency estimates that one in five Americans, supplied by one-quarter of the nation's drinking water systems, consume tap water that violates EPA safety standards under the Clean Water Act. Even some of the substances that are added to our drinking water to protect us, such as chlorine, can form toxic compounds—for example trihalomethanes, or THMs—and have been linked to certain cancers. The EPA has established enforceable standards for more than 100 contaminants. However, credible studies have identified well more than 2,000 contaminants in the nation's water supplies.

The most obvious solution to water pollution is a point-of-use water purification device. The tap is the end of the road for water consumed by our families, so this is the logical, and most efficient, place to focus water treatment. Different water purification technologies each have strengths and weaknesses, and are particularly effective against specific kinds of impurities or toxins. So the system you need depends first and foremost on the nature of the problem you have, which in turn requires testing and diagnosis.

We will describe these systems in detail after we discuss the contaminants that you might encounter. Most treatment systems are point-of-use and deal only with drinking and cooking water, which is less than 5% of typical home use. Full treatment of all household water is a very expensive undertaking, and is usually reserved for water sources with serious, health-threatening problems.

Contaminants in Water

Presumably your drinking water comes from a municipal system, a shallow dug well, or a deep, drilled well. If you drink bottled water, you've already taken steps to control what you consume. Bottled water drinkers should read on, too: Most good purification systems provide tasty potable water for far less money than the cost of regularly drinking bottled water. Each type of water supply is more or less vulnerable to different kinds of pollutants, because surface water (rivers, lakes, reservoirs), groundwater (underground aquifers), and treated water each are exposed to environmental contaminants in different ways. The hydrogeological characteristics of the area you live in, and localized activities and sites that create pollution, like factories,

Most good purification systems provide tasty potable water for far less money than the cost of drinking bottled water.

agricultural spraying, or a landfill, potentially will impact your water quality. And some impurities, such as lead and other metals, can be introduced by the piping that delivers the water to your tap.

Reliable and inexpensive tests are available to identify the biological and chemical contaminants that may be in your water. Here's what you may find.

Biological Impurities: Bacteria, Viruses, and Parasites

Microorganisms originating from human and animal feces, or other sources, can cause waterborne diseases. Approximately 4,000 cases of waterborne illness are reported each year in the United States. Additionally, many of the minor illnesses and gastrointestinal disorders that go unreported can be traced to organisms found in water supplies.

Biological impurities largely have been eliminated in municipal water systems with chlorine treatment. However, such treated water can still become biologically contaminated. Residual chlorine throughout the system may not be adequate, and therefore microorganisms can grow in stagnant water sitting in storage facilities or at the ends of pipes.

Water from private wells and small public systems is more vulnerable to biological contamination. These systems generally use untreated groundwater supplies, which could be polluted due to septic tank leakage or poor construction.

Organic Impurities: Tastes and Odors

If water has a disagreeable taste or odor, the likely cause is one or more organic substances,

ranging from decaying vegetation to algae to organic chemicals (organic chemicals are compounds containing carbon).

Inorganic Impurities: Dirt and Sediment, or Turbidity

Most water contains suspended particles of fine sand, clay, silt, and precipitated salts. This cloudiness or muddiness is called turbidity. Turbidity is unsightly, and it can be a source of food and lodging for bacteria. Turbidity can also interfere with effective disinfection and purification of water.

Total Dissolved Solids (TDS)

Total dissolved solids consist of rock and numerous other compounds from the Earth. The significance of TDS in water is a point of controversy among water purveyors, but here are some facts about the consequences of higher levels of TDS:

1. High TDS results in undesirable taste, which can be salty, bitter, or metallic.
2. Certain mineral salts may pose health hazards. The most problematic are nitrates, sodium, barium, copper sulfates, and fluoride.
3. High TDS interferes with the taste of foods and beverages.
4. High TDS makes ice cubes cloudy, softer, and faster melting.
5. High TDS causes scaling on showers, tubs, sinks, and inside pipes and water heaters.

Toxic Metals or Heavy Metals

The presence of toxic metals in drinking water is one of the greatest threats to human health. The major culprits include lead, arsenic, cadmium, mercury, and silver. Maximum limits for each of these metals are established by the EPA's Drinking Water Regulations.

Toxic Organic Chemicals

The most pressing and widespread water contamination problem results from the organic chemicals (those containing carbon) created by industry. The American Chemical Society lists more than four million distinct chemical compounds, most of which are synthetic (manmade) organic chemicals, and industry creates new ones every week. Production of these chemicals exceeds a billion pounds per year. Synthetic organic chemicals have been detected in many water supplies throughout the country. They get into the groundwater from improper disposal of industrial waste (including discharge into waterways), poorly designed and sited industrial lagoons, wastewater discharge from sewage treatment plants, unlined landfills, and chemical spills.

Studies since the mid-1970s have linked organic chemicals in drinking water to specific adverse health effects. However, only a fraction of these compounds have been tested for such effects. Well over three-quarters of the substances identified by the EPA as priority pollutants are synthetic organic chemicals.

Volatile Organic Compounds (VOCs)

Volatile organic compounds are very lightweight organic chemicals that easily evaporate into the air. VOCs are the most prevalent chemicals found in drinking water, and they comprise a large proportion of the substances regulated as priority pollutants by the EPA.

Chlorine

Chlorine is part of the solution and part of the problem. In 1974, scientists discovered that VOCs known as trihalomethanes (THMs) are formed when the chlorine added to water to kill bacteria and viruses reacts with other organic substances in the water. Chlorinated water has been linked to cancer, high blood pressure, and anemia.

The scientific research linking various synthetic organic chemicals to specific adverse health effects is not conclusive, and remains the subject of considerable debate. However, given the ubiquity of these pollutants and their known presence in water supplies, and given some demonstrated toxicity associations, it would be reasonable to assume that chronic exposure to high levels of synthetic organic chemicals in water could be harmful. If they're in your water, you want to get them out.

Pesticides and Herbicides

The increased use of pesticides and herbicides in American agriculture since World War II has had a profound effect on water quality. Rain and irrigation carry these deadly chemicals into groundwater, as well as into surface waters. These are poisonous, plain and simple.

Asbestos

Asbestos exists in water as microscopic suspended fibers. Its primary source is asbestos-cement pipe, which was commonly used after World War II for city water systems. It has been estimated that some 200,000 miles of this pipe are currently in use delivering drinking water. Because pipes wear as water courses through them, asbestos shows up with increasing frequency in municipal water supplies. It has been linked to gastrointestinal cancer.

Radionuclides

The Earth contains naturally occurring radioactive substances. Certain areas of the United States exhibit relatively high background levels of radioactivity, due to their geological characteristics. The three substances of concern to human health that show up in drinking water are uranium, radium, and radon (a gas). Various purification techniques can be effective at reducing the levels of radionuclides in water.

Testing Your Water

The first step in choosing a treatment method is to find out what contaminants are in your water. Different levels of testing are commercially available, including a comprehensive screening for nearly one hundred substances. See the product section that follows.

Even before you test your water, a simple comparison can help you figure out what level of treatment your domestic water supply may need to make you happy. Find some bottled water that you like. Note the normal levels of total dissolved solids (TDS), hardness, and pH, and also how high these levels range. If you don't find this information on the label, call the bottler and ask. Then get the same information about your tap water, which can be obtained by calling your municipal supplier. If your tap water has lower levels of these things than the bottled water, a good filter should satisfy your needs if you decide you want to treat it. If your tap water has higher levels than the bottled water you prefer, you may need more extensive purification.

If you're drinking water from a private well, you'll have to get it tested to know what's in it. Some common problems with well water, such as staining, sediment, hydrogen sulfide (rotten-egg smell), and excess iron or manganese, should be corrected before you purchase a point-of-use treatment system, because they will interfere with the effective operation of the system. These problems can be identified easily with a low-priced test; many state water-quality agencies will perform such tests for a nominal fee, or we offer some tests in our product section.

If you have any reason to worry that your water may be polluted, we strongly advise you to conduct a more comprehensive test.

Chlorinated water has been linked to cancer, high blood pressure, and anemia.

Certain areas of the United States exhibit relatively high background levels of radioactivity, due to their geological characteristics.

Water Choices: Bottled, Filtered, Purified

Many manufacturers claim that their purifiers produce clean water for just pennies a gallon; certainly these systems will pay for themselves in cost savings within a few years.

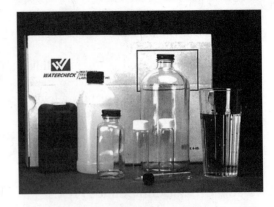

If you're not satisfied with your drinking water, you suspect it may be contaminated, or you've had it tested and you know it's unhealthy, you have two choices. You can buy bottled water, or you can install a point-of-use water treatment system in your home.

There are three kinds of bottled water: distilled, purified, and spring. Distillation evaporates the water, then recondenses it, thereby theoretically leaving all impurities behind (although some VOCs can pass right through this process with the water). Purified water is usually prepared by reverse osmosis, de-ionization, or a combination of both processes (see below for explanations). Spring water is usually acquired from a mountain spring or an artesian well, but it may be no more than processed tap water. Spring water generally will have higher total dissolved solids than purified water. Distilled water and purified water are better for batteries and steam irons because of their lower content of TDS.

But why pay for bottled water forever when treatment will be cheaper? Bottled water typically costs up to $2.00 per gallon or more. Many manufacturers claim that their purifiers produce clean water for just pennies a gallon; certainly these systems will pay for themselves in cost savings within a few years.

The two basic processes used to clean water are filtration and purification. The word "filter" usually refers to a mechanical filter, which strains the water, and/or a carbon filter system, which reduces certain impurities by chemically bonding them—especially chlorine, lead, and many organic molecules. Purification refers to a slower process, such as reverse osmosis, which greatly reduces dissolved solids, hardness, and certain other impurities, as well as many organics. Many systems combine both processes. Make sure that the claims of any filter or purifier you consider have been verified by independent testing.

Methods of Water Filtration and Purification

The most efficient and cost-effective solution for water purity is to treat only the water you plan to consume. A point-of-use water treatment system eliminates the middleman costs associated with bottled water, and can provide purified water for pennies per gallon. Devices for point-of-use water treatment are available in a variety of sizes, designs, and capabilities. Some systems only improve the water's taste and odor. Other systems reduce the various contaminants of health concern. Different systems work most effectively against certain contaminants. A system that utilizes more than one technology will protect you against a broader spectrum of biological pathogens and chemical impurities. Combining activated carbon filtration and reverse osmosis generally is considered the most complete and effective treatment.

When considering a treatment device, always pay careful attention to the independent documentation of the performance of the system for a broad range of contaminants. You should read the data sheets provided by the manufacturer carefully to verify its claims. Many companies are certified with the National Sanitation Foundation (NSF), whose circular logo appears on their data sheets.

The following is a brief analysis of the strengths and weaknesses of each option.

Mechanical Filtration

Mechanical filtration can be divided into two categories. Strainers are just fine mesh screens that generally remove only the largest particles present in water. Sediment filters (or prefilters) remove smaller particles such as suspended dirt, sand, rust, and scale—in other words, turbidity. When enough of this particulate matter has accumulated, the filter is discarded. Sediment filters greatly improve the clarity and appeal of the water. They also reduce the load on any more expensive filters downstream, extending their useful life. Sediment filters will not remove the smallest particles or biological pathogens.

Activated Carbon Filtration

Carbon adsorption is the most widely sold method for home water treatment because of its ability to improve water by removing disagreeable tastes and odors, including objectionable

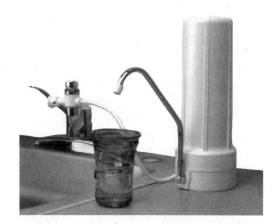

chlorine. Activated carbon filters are a very important piece of the purification process, although they are only one piece. Activated carbon effectively removes many chemicals and gases, and in some cases it can be effective against microorganisms. However, generally it will not affect total dissolved solids, hardness, or heavy metals. Only a few carbon filter systems have been certified for the removal of lead, asbestos, VOCs, cysts, and coliform.

Each type of carbon filter system—granular activated carbon and solid block carbon—has advantages and disadvantages (see below).

Activated carbon is created from a variety of carbon-based materials in a high-temperature process that creates a matrix of millions of microscopic pores and crevices. One pound of activated carbon provides anywhere from 60 to 150 acres of surface area. The pores trap microscopic particles and large organic molecules, while the activated surface areas cling to, or adsorb, small organic molecules.

The ability of an activated carbon filter to remove certain microorganisms and certain organic chemicals, especially pesticides, THMs (the chlorine by-product), trichloroethylene (TCE), and PCBs, depends upon several factors:

1. The type of carbon and the amount used.
2. The design of the filter and the rate of water flow (contact time).
3. How long the filter has been in use.
4. The types of impurities the filter has removed previously.
5. Water conditions (turbidity, temperature, etc.).

GRANULAR ACTIVATED CARBON

Any granular activated carbon filter has three inherent problems. First, it can provide a base

> A system that utilizes more than one technology will protect you against a broader spectrum of biological pathogens and chemical impurities.

> One pound of activated carbon provides anywhere from 60 to 150 acres of surface area. The pores trap microscopic particles and large organic molecules, while the activated surface areas cling to, or adsorb, small organic molecules.

for the growth of bacteria. When the carbon is fresh, practically all organic impurities (not organic chemicals) and even some bacteria are removed. Accumulated impurities, though, can become food for bacteria, enabling them to multiply within the filter. A high concentration of bacteria is considered by some people to be a health hazard.

Second, chemical recontamination of granular activated carbon filters can occur in a similar way. If the filter is used beyond the point at which it becomes saturated with the organic impurities it has adsorbed, the trapped organics can release from the surface and recontaminate the water, with even higher concentrations of impurities than in the untreated water. This saturation point is impossible to predict.

Third, granular carbon filters are susceptible to channeling. Because the carbon grains are held (relatively) loosely in a bed, open paths can result from the buildup of impurities in the filter and rapid water movement under pressure through the unit. In this situation, contact time between the carbon and the water is reduced, and filtration is less effective.

To maximize the effectiveness of a granular activated carbon filter and avoid the possibility of biological or chemical recontamination, it must be kept scrupulously clean. That generally means routine replacement of the filter element at six- to twelve-month intervals, depending on usage.

SOLID BLOCK CARBON

These filters are created by compressing very fine pulverized activated carbon with a binding medium and fusing the composite into a solid block. The intricate maze developed within the block ensures complete contact with organic impurities and, therefore, effective removal. Solid block carbon filters avoid the drawbacks of granular carbon filters.

Block filters can be fabricated with a porous structure fine enough to filter out coliform and other disease bacteria, pathogenic cysts such as giardia, and lighter-weight volatile organic compounds such as THMs. Block filters eliminate the problem of channeling. They are also dense enough to prevent the growth of bacteria within the filter.

Compressed carbon filters have two primary disadvantages compared with granular carbon filters. They have smaller capacity for a given size, because some of the adsorption surface is taken up by the inert binding agent, and because they tend to plug up with particulate matter.

Thus, block filters may need to be replaced more frequently. In addition, block filters are substantially more expensive than granular carbon filters.

LIMITATIONS OF CARBON FILTERS

To summarize, a properly designed carbon filter is capable of removing many toxic organic contaminants, but it will fall short of providing protection against a wide spectrum of impurities.

1. Carbon filters are not capable of removing excess total dissolved solids (TDS). To gloss over this deficiency, many manufacturers and sellers of these systems assert that minerals in drinking water are essential for good health. Such claims are debatable. However, some scientific evidence suggests that minerals associated with water hardness may have some preventive effect against cardiovascular disease.
2. Only a few systems have been certified for the removal of cysts, coliform, and other bacteria.
3. Carbon filters have no effect on harmful nitrates, or on high sodium or fluoride levels.
4. For both granular and block filters, the water must pass through the carbon slowly enough to ensure complete contact and filtration. An effective system therefore has to have an appropriate balance between a useful flow rate and adequate contact time.

Reverse Osmosis (Ultrafiltration)

Reverse osmosis (RO) is a water purification technology that utilizes normal household water pressure to force water through a selective semipermeable membrane that separates contaminants from the water. Treated water emerges from the other side of the membrane, and the accumulated impurities left behind are washed away. Sediment eventually builds up along the membrane and then it needs to be replaced.

Reverse osmosis is highly effective in removing several impurities from water: total dissolved solids (TDS), turbidity, asbestos, lead and other heavy metals, radium, and many dissolved organics. RO is less effective against other substances. The process will remove some pesticides (chlorinated ones and organophosphates, but not others), and most heavier-weight VOCs. However, RO is not effective at removing lighter-weight VOCs, such as THMs (the chlorine byproduct) and TCE (trichloroethylene), and certain pesticides. These compounds are either too small, too light, or of the wrong chemical structure to be screened out by an RO membrane.

To maximize the effectiveness of a granular activated carbon filter and avoid the possibility of biological or chemical recontamination, it must be kept scrupulously clean.

Reverse osmosis and activated carbon filtration are complementary processes. Combining them results in the most effective treatment against the broadest range of water impurities and contaminants. Many RO systems incorporate both a pre-filter of some sort, and an activated carbon post-filter.

RO systems have two major drawbacks. First, they waste a large amount of water. They'll use anywhere from 3 to 9 gallons of water per gallon of purified water produced. This could be a problem in areas where conservation is a concern, and it may be slightly expensive if you're paying for municipal water. On the other hand, this wastewater can be recovered or redirected for purposes other than drinking, such as watering the garden, washing the car, or the like. Second, reverse osmosis treats water slowly: It takes about three to four hours for a residential RO unit to produce one gallon of purified water. Treated water can be removed and stored for later use.

Other Treatment Processes

Two other water treatment processes are worth knowing about. Distillation is a process that creates clean water by evaporation and condensation. Distillation is effective against microorganisms, sediment, particulate matter, and heavy metals; it will not treat organic chemicals. Good distillers will have a carbon filter to remove organic chemicals.

Ultraviolet (UV) systems use UV light to kill microorganisms. These systems can be highly effective against bacteria and other organisms; however, they may not be effective against giardia and other cysts, so any UV system you buy should also include a 0.5-micron filter. Other than some moderately expensive solar-powered distillers on the market, any distiller or UV unit will require a power source.

Reverse osmosis and activated carbon filtration are complementary processes. Combining them results in the most effective treatment against the broadest range of water impurities and contaminants.

A Final Word on Water

Our best advice for those of you considering a drinking water treatment system consists of three simple points.

1. Test your water so you know precisely what impurities and contaminants you're dealing with.

2. Buy a system that is designed to treat the problems you have.

3. Carefully check the data sheets provided by the manufacturer to make sure that claims about what the system treats effectively have been verified by the National Sanitation Foundation (NSF) or another third party.

These actions will enable you to purchase a system that will most effectively meet your particular needs for safe drinking water.

WATER TREATMENT METHODS—GENERAL TREATMENT CAPABILITIES					
	GAC	Carbon Block	RO	Distillation	UV
Bacteria	no	maybe	no	yes	yes
Cysts	no	maybe	yes	yes	yes w/0.5 mic filter
Asbestos	no	maybe	yes	yes	
Heavy metals	lead only	maybe	yes	yes	
Turbidity	no	maybe	yes	yes	
TDS	no	no	yes	yes	
Chlorine	yes	yes	yes w/tfc	yes	
Chlorine by-products (THMs)	no	maybe	maybe	yes	
Organic matter	maybe	maybe	maybe	maybe	
Heavy organic chemicals	some	maybe	maybe	maybe	
Light organics (VOCs)	some	maybe	maybe	maybe	
Pesticides and herbicides	some	some	maybe	maybe	
Radium	no	no	yes	yes	
Radon	no	no	yes	yes	

WATER PURIFICATION PRODUCTS

NTL Water Check with Pesticide Option

This is the best analysis of your water available in this price range. Your water will be analyzed for 73 items,

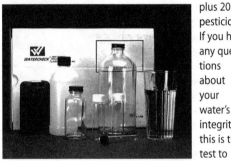

plus 20 pesticides. If you have any questions about your water's integrity, this is the test to give you peace of mind. The kit comes with five water sample bottles, a blue gel refrigerant pack (to keep bacterial samples cool for accurate test results), and easy-to-follow sampling instructions. You'll receive back a two-page report showing 93 contaminant levels, together with explanations of which contaminants, if any, are above allowed values. You'll also receive a follow-up letter with a personalized explanation of your test results, plus knowledgeable, unbiased advice on what action you should take if your drinking water contains contaminants above EPA-allowed levels. NTL, located outside of Detroit, must receive the package within 30 hours of sampling in order to provide accurate results. If you live in Alaska or Hawaii, you will need express shipping. Check with your shipper. This test is primarily for folks with private wells. Municipal water agencies are already doing this kind of testing; call them for results. USA.

42-003   NTL Check & Pesticide Option   **$175**

HOUSEHOLD POINT-OF-USE WATER FILTERS

Water Filter Crocks

Ceramic and carbon filtration reduces even E. coli by 99%. Pour in tap, well, or even rain water, and produce up to 6 gallons a day of clean and clear water—storing 2 gallons at a time. Our crocks have a lead-free, nontoxic interior glaze that prevents mineral leaching, discoloration and mold and mildew growth. Choose glazed sage or unglazed terra cotta. Sorry, not available in California. USA.

01-0391 Ceramic Water Filter Crock
(17.5"H x 8" diameter, 20 lbs.)   **$189**

01-0401 Terra cotta Water Filter Crock
20"H x 10" diameter
(17.5"H x 8" diameter, 20 lbs.)   **$189**

01-0092 Replacement Filter (replace yearly) **$59**

Paragon Water Filtration Systems

The beautiful choice for city water users

Prevent a host of the most common water problems as well as eliminate bad taste and odor. Using a titanium

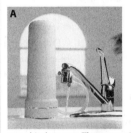

silicate/carbon block filter, these filtration systems remove virtually all hazardous contaminants commonly found in tap water.

The Basic Model (A) features extra-long tubing and a diverter system that sends filtered water out of your kitchen tap. The attractive Chrome Model (B) has a convenient spout. The white Deluxe Model (C) has its own spout and a microprocessor (2 AAA batteries included) that alerts you when it's time to change filters. The Under-the- Counter Model (D) easily mounts under your sink with the included bracket and attaches to your cold water line with quick-connect fittings. For cold water use. USA.

01-0144 Basic (12.5" x 4.25", 3 lbs.)   **$109**

01-0145 Chrome (12.5" x 5.75", 4 lbs)   **$159**

01-0146 Deluxe (12.5" x 4.25", 3 lbs.)   **$169**

01-0150 Under-the-Counter
(14.5" x 7", 3.5 lbs.)   **$139**

01-0224 Replacement Filter   **$55**

Maintenance-Free Water Filter

Get the health benefits of purified water without the fuss and expense of replacing filters. With a 5,000-gallon filtration capacity and cartridgeless design, our powerful Maintenance-free Water Filter never needs replacement filters and lasts about two to three years for a family of four. Uses carbon and KDF filtration plus three micro filters to reduce chlorine content by 95% and improve the taste and smell of your water. Attaches easily to faucet. 7"H x 3I" diameter. 3.25 lbs. USA.

01-0443 Maintenance-Free Water Filter   **$60**

Seagull Carbon/Ceramic Filters

Seagull water filtration products are used widely aboard international airliners. The unique dual-layer cartridge provides both carbon filtering for taste, odor, chlorine, and organic chemicals, plus an ultra-fine 0.1 micron ceramic element for giardia, cysts, bacteria, and parasites. Seagull filters are manufactured in the USA from stainless steel and other high-grade components. The stainless housing with band clamp makes cartridge changes easy. The unique microfine structured matrix allows high flow rates of 1 gpm at standard pressures

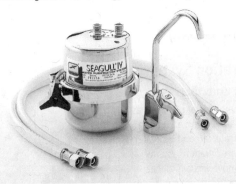

(30–40 psi). The replaceable cartridge should last anywhere from 9 to 15 months with ordinary use. This undercounter model is supplied with an ultra long-life ceramic disc chrome faucet. Although expensive initially, this is one of the best all-around filters available. This is the filter we put on our drinking fountains at the Solar Living Center. USA.

42-650	Seagull Undercounter Filter	$429
42-657	Seagull Replacement Cartridge	$99

O-SO PURE 3-Stage UV System

Good choice for less pristine private sources.

Triple treatment to remove most contaminants including

bacteria and viruses. Sediment filters remove sand, silt, and other particulates, then carbon second stage removes chlorine and other chemicals. Treated water is then exposed to high intensity ultraviolet to destroy bacteria and viruses. Filter elements and UV are all replaceable. Operate on 120 vac or 12 vdc—120-volt units include installation kit. Three-year warranty. These are residential units only.

42910	O-SO Pure 3-Stage UV POU System	$450
40-0067	O-SO Pure 3-Stage UV, 12V	$475
40-0064	12V Install Kit	$40
Replacement Filters		
42913	10" Sediment Filter	$8
42914	10" Carbon Filter	$29
42915	Replacement Bulb	$45

U.S. Pure Reverse Osmosis with Water Quality Monitor

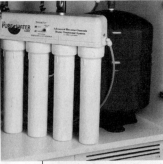

This advanced four-stage filtration system ends guesswork with a push-button water-quality monitor that tells you exactly when the membrane needs changing. This is a top-quality system, utilizing the best in filtration technology to purify your family's drinking water. Delivery is just under 5 gallons per day (depending on water pressure and temperature). Tested and certified to ANSI/NSF® 58 for specific contaminant claims and is California certified. Includes carbon pre- and post-filters, a sediment pre-filter, and membrane filter. Carbon and sediment filters last 6 to 12 months; service life for the membrane filter is two to five years. The 2-gallon storage tank and filters install under the sink, with a stainless steel lead-free faucet on the sink top for convenient access (system install time is about one hour). The reliable manufacturer insists on independent testing, top-quality materials, and outstanding performance. Two-year limited manufacturer's warranty. USA.

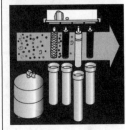

42-223	U.S. Pure Reverse Osmosis System	$399
42-224	Replacement Membrane	$75
42-225	Sediment Filter	$12.95
42-227	Carbon Pre/Post Filter (requires 2)	$14.95

Inline Sediment Filter

Sometimes your water isn't dirty enough to mess with fancy and expensive filtration systems and all you need is a simple filter. Our inline sediment filters accept standard 10-inch filters with 1-inch center holes. They are designed for cold-water lines only and meet National Sanitation Foundation (NSF) standards. Easily installed on any new or existing cold-water line (don't forget the shut-off valve), they feature a sump head of fatigue-resistant Celcon plastic. This head is equipped with a manually operated pressure release button to relieve internal pressure and simplify cartridge replacement. The housing is rated for 125 psi maximum and 100°F. It comes with a ³/4-inch FIPT inlet and outlet and measures 14 inches high by 4⁹/₁₆ inches in diameter. It accepts a 10-inch cartridge and comes with a 10-micron high-density fiber cartridge.

41-137	Inline Sediment Filter	$49
42-632	10 Micron Sediment Cartridge 2-pack	$13

Whole House Filtration System

Here's one of the most affordable ways we've seen to deliver healthy filtered water to every fixture in your home. Attached to your water supply at its entry point, this Whole House Filtration system uses two stages to provide pure water for drinking and bathing. An initial filtration preps your water by removing rust, dirt, debris, and other particles larger than 5 microns. Then a combination carbon and KDF media filter then targets chlorine, VOCs, pesticides, and other chemicals and heavy metals while removing odors and improving your water's taste. 100,000-gallon filters are easy to change and last at least a year before requiring replacement. Professional installation recommended. 18"H x 15"W x 8"D. Replacement cartridges are $19 for pre-filter, $65 for carbon/KDF available directly from the manufacturer. USA.

40-0077 Whole House Filtration System　　　**$499**

Safe Water Systems

One of the greatest health and environmental problems our world faces is providing safe drinking water. Pasteurizing water, raising it to a temperature of 174°F (79.4°C), has proven to be 99.999% effective for microbio-

The Sol*Saver can provide safe water for an entire village, school, or health clinic.

logically contaminated drinking water. It's how the dairy industry has sterilized milk for decades. Pasteurizing requires only one-eighth the energy of boiling.

The Sol*Saver and Wood*Saver pasteurizing systems can provide safe water for villages, schools, hospitals, health clinics, orphanages, or other such uses using either solar energy, wood fuel, or both. Two simple concepts create this revolutionary technology. The temperature of water standing in a glazed solar collector easily can exceed 200°F on a sunny day. A patented thermal outlet control valve opens only when the temperature exceeds the 174°F pasteurization point. This valve is rated for more than one million cycles. In the event of malfunction, no water will flow to the sanitary water storage tank (not included) but will be discharged through a safety dump port. The second great innovation is to run the now hot, sterile water through a heat exchanger, which preheats the water that is about to be pasteurized and saves much of the input energy! A supply tank just above the pasteurizer can provide simple gravity feed.

The Sol*Saver

Constructed of durable copper, aluminum, brass, and tempered glass for a long life, this unit includes an approximate 4-ft x 10-ft solar collector with mounts, high-efficiency heat exchanger, and fail-safe thermal control valve. Operation is

fully automatic, just add water and sun. Adjustable wrenches are the only tools needed for plumbing on site, no pipe soldering is required. An optional back-up solid-fuel burner can provide sterile water during prolonged cloudy weather. The solar collector can pasteurize up to 200 gallons per day. The back-up burner can produce more than 25 gallons per hour and will burn wood, coal, charcoal, or other solid fuels. Weight (crated): 521 pounds. Volume (crated): 63 cu. ft. USA.

42-190 Sol*Saver　　　**$3,500**
42-191 Sol*Saver w/ Burner　　　**$4,000**
Drop shipped freight collect from central California.

The Wood*Saver

Simply the solid wood burn chamber by itself, fail-safe thermal control valve, and heat exchanger. Weight (crated): 45 pounds. Volume (crated): 10 cu. ft. USA.

All Safe Water Systems have a life expectancy of fifteen years and are backed by a three-year warranty.

42-192 Wood*Saver　　　**$1,500**
Drop shipped freight collect from central California.

SHOWER AND BATH FILTERS

Bath Ball Faucet Filter

Now you can take a relaxing bath without chlorine (or without filling the tub with the showerhead). Our Bath Ball Faucet Filter is a plastic sphere filled with an ingenious technology that removes chlorine. Simply attach the sturdy vinyl strap under the tub faucet. Water flows into openings in the top and emerges 95% free of chlorine. You can notice the difference in the water quality with your very first bath. Lasts one year. Includes container to keep bath ball submerged when not in use. 4" diameter. USA.

01-1169	**Bath Ball Faucet Filter**	**$49**
01-0132	Replacement Filter	**$32**

Massage Showerhead

Our innovative Chlorine-Free Massage Showerhead has a hidden replaceable cartridge in the handle that filters 99% chlorine, hydrogen sulfide, iron oxides, trace lead, and odors from your water. Additionally, the patented filtration media Ph balances the water, inhibits bacterial growth, and reduces scale buildup on tiles. You notice softer skin and hair, and you'll feel safe from chlorine absorption and inhalation. Adjustable settings range from wide spray to pulse massage and the swivel design easily adjusts to different heights. Available in chrome or white finished heavy-duty ABS plastic. Handle and showerhead measure 10"L x 1.5" diameter. Hose is 72"L. USA.

02-0206	**Chrome Showerhead**	**$69**
02-0206	**White Showerhead**	**$59**
01-0143	Replacement Filters (2)	**$28**

High Output Shower Filter

Hot shower water exposes you to chlorine that can harm your hair, skin, eyes, and nasal membranes, and it creates chloroform, a toxic gas that's inhaled during washing. The special non-carbon filter inside the High Output Shower Filter removes 99% of chlorine and reduces lead, mold, mildew, iron deposits, sediments, and odors for healthier bathing. With an easy-open design for simple filter changes and filters that last up to a full year (depending on sediment levels), this is the most durable, highest capacity replaceable filter model made. Self-sealing threads prevent leaks. Use with your existing showerhead. 150 psi maximum. Choose white or chrome-plated. USA. Note: Showerhead not included.

01-0407	**High Output Shower Filter**	**$50**
01-0408	Replacement Filters	**$24**

Shower Clean Filter

Our Shower Clean filter removes odors, bacteria, organic waste, trihalomethanes, VOCs, and common carcinogens such as carbon tetrachloride, chloroform, benzene, and vinylidene chloride. Special design makes it height-adjustable. Three-filter system includes: a 5-micron sediment filter, granular activated carbon, and MicroCleen—a state-of-the-art bacteriostatic filter medium. Up to 3 gpm flow. Replace filter every 6 months. Showerhead not included.

01-0420	**Shower Clean**	**$99**
01-1421	Replacement Filter	**$15**

PORTABLE WATER FILTERS

The Katadyn Drip Filter

With no moving parts to break down, superior filtration, and a phenomenal filter life, there is simply no safer choice for potentially pathogen-contaminated water. There are no better filters than Katadyn for removing bacteria, parasites, and cysts. Three 0.2-micron ceramic filters process one gallon per hour. Clean filters by brushing the surface. Ideal for remote homes, RV, campsite, and home emergency use. Food-grade plastic canisters stack to 11" Dia. x 25" H. Weighs 10 lb. One-year manufacturer's warranty. Switzerland.

42-842	**Katadyn Drip Filter**	**$289**
42-843	**Replacement Filter**	**$75 ea.**

POOL AND SPA WATER PURIFICATION

Solar-Powered Pool Purifier

Slashes chlorine use

The Floatron kills algae, bacteria, and is a safer and economical alternative to the chemical marinade of most pools. It is solid-state, with no moving parts, no batteries, and is portable and cost  effective. The Floatron's solar panel generates a low-power electric current that energizes a specially alloyed electrode. The electrode delivers copper and silver ions into the water, which kill algae and bacteria cells on contact. The electrode element is consumed gradually, and needs replacement every one to three seasons, depending on pool size. The element is available directly from the manufacturer for about $50, and takes a couple of minutes to replace. The solar panel lasts for the life of your pool. No more chlorine allergies, red eyes, discolored hair, or bleached bathing suits! Floatron typically will reduce chemical expenses by an average of 80%. Effective for pools up to approximately 40,000 gallons. 2-yr. manufacturer's warranty. USA.

14-0172 Floatron   **$299**

Metal Foam Spa Purifier

Reduce chlorine or bromine usage by up to 90% with our safe, effective Metal Foam Spa Purifier. Our spa purifier destroys most of the same bacteria as standard chemical treatments, including E. coli and legionella pneumophilia. Once  a month remove the disc and backwash with a garden hose; replace disc twice annually. You'll still need to chlorinate lightly (small amount of bromine or bleach after use) to remove nonbacterial spa contaminants (sweat, skin products). Simply drop the disc into the skimmer or into the center hole of your spa's filter. Includes disc holder and two metal foam discs. USA.

42804   Metal Foam Spa Purifier   **$60**

Air Purification

Most of us are familiar with the idea that the water we drink at home might be polluted or contaminated. But did you realize that the air inside your home (or place of work) potentially could be even more hazardous to your health than the water?

The Problem with Indoor Air

The problem of indoor air quality has attracted attention only recently. One reason is that the sources of indoor air pollution are more mundane and subtle than the sources of outdoor air and water pollution. In addition, before highly insulated, tightly constructed buildings became common in the 1970s, most American homes were drafty enough that the build-up of harmful gases, particles, and biological irritants indoors was unlikely. However, the combination of tighter, well-insulated buildings that allow minimal infiltration of outside air, and the continual increase of synthetic products that we bring into our homes, has added up to a possible public health problem of major proportions.

The preceding remark is qualified because little is known with relative certainty about the unhealthful consequences of constant exposure to polluted indoor air. In any case, the EPA has concluded that many of us receive greater exposure to pollutants indoors than outdoors. EPA studies have found concentrations of a dozen common organic pollutants to be two to five times higher inside homes than outside, regardless of whether they were located in rural areas or industrialized areas. Furthermore, people who spend more of their time indoors and are thus exposed for longer periods to airborne contaminants are often the very people considered most susceptible to poor indoor air quality: very young children, the elderly, and the chronically ill.

SOURCES OF INDOOR AIR POLLUTION

You can probably identify some of the sources of indoor air pollution: new carpeting, adhesives, cigarette smoke, and maybe you've heard about dust mites. But many things that contribute potentially irritating or harmful substances to indoor air may not be obvious.

Indoor air contaminants typically fall into one or more of the following categories:

- Combustion products
- Volatile chemicals and mixtures
- Respirable particulates
- Respiratory products
- Biological agents
- Radionuclides (radon and its byproducts)
- Odors

What kinds of substances are we talking about? Let's start with chemicals and synthetics. Building materials and interior furnishings emit a broad array of gaseous volatile organic compounds (VOCs), especially when they are new. These materials include adhesives, carpeting, fabrics, vinyl floor tiles, some ceiling tiles, upholstery, vinyl wallpaper, particle board, drapery, caulking compounds, paints and stains, and solvents. In addition to producing emissions, many of these materials also act as "sponges," absorbing VOCs and other gases that can then be reintroduced into the air when the "sponge" is saturated. These kinds of building and furnishing materials also can be the source of respirable particulates, such as asbestos, fiberglass, and dusts.

Appliances, office equipment, and office supplies are another major source of VOCs and particulates. Among the culprits are leaky or unvented heating and cooking appliances; computers and visual display terminals; laser printers, copiers, and other devices that use chemical supplies; common items such as preprinted paper forms, rubber cement, and typewriter correction fluid; and routine cleaning and maintenance supplies, including carpet shampoos, detergents, floor waxes, furniture polishes, and room deodorizers.

Your own routine "cleaning and maintenance supplies" are not above suspicion, either. Many of the personal care and home cleaning products we use emit various chemicals (especially from the components that create fragrance), some of which may contribute to poor indoor air quality. Clothes returning from the dry cleaner may retain solvent residues, and studies have shown that people do breathe low levels of these fumes when they wear dry-cleaned clothing. Even worse, the pesticides we use to control roaches, termites, ants, fleas, wasps, and other insects are by definition toxic. Sometimes these poisons are tracked indoors on shoes and clothing. Do you have any of this stuff stored underneath the sink? One EPA study has suggested that up to 80% of most people's exposure to airborne pesticides occurs indoors.

Other human activities can also have a dramatic impact on indoor air quality. Tobacco smoke from cigarettes, cigars, and pipes is obviously an unhealthy pollutant. Residential heating and cooking activities may introduce carbon monoxide, carbon dioxide, nitrogen oxide and dioxide, and sulfur dioxide into the air. Candles and wood fires contribute CO and CO_2 as well as various particulates.

Many biological agents contribute to indoor pollution: We are surrounded by a sea of microorganisms. Bacteria, insects, molds, fungi, protozoans, viruses, plants, and pets generate a number of substances that contaminate the air. These range from whole organisms themselves to feces, dust, skin particles, pollens, spores, and even some toxins. Humans contribute too: We exhale microbes, and our sloughed skin is the primary source of food for the infamous dust mites. Furthermore, certain indoor environments, such as ducts and vents, or humid areas such as bathrooms and basements, provide ideal conditions for microbial growth.

Radon (and its decay products, known as "radon daughters") is one type of indoor pollutant that has received widespread notice. As noted above, indoor radon buildup can be a problem in certain parts of the United States where it is present in rocks and soil. A radioactive gas, radon has been implicated as a cause of lung cancer. Simple and inexpensive radon test kits are available at hardware stores. Contact your regional EPA office—you can look them up in the phone book—for more information about effective radon remediation strategies.

Odors are always present. Some are pleasant, others are unpleasant; some signal the presence of irritating or harmful substances, others do not. The perception of odors is highly subjective and varies from person to person. Most odors are harmless, more of a threat to state of mind than to health, but eliminating them from an indoor environment may be a high priority to keep people content and productive.

HEALTH EFFECTS OF INDOOR AIR POLLUTION

Before you become too alarmed about the health dangers that may be floating around in the air you breathe at home, remember the following proviso. As with most environmental pollutants, adverse health effects depend upon both the dose received and the duration of exposure. The combination of these factors, plus an individual's particular sensitivity to specific substances, will determine the potential health effect of the contaminant on that individual.

With that in mind, the health effects of indoor air pollution can be short term or long term, and can range from mildly irritating to severe, including respiratory illness, heart disease, and cancer. Most often the effects of such pollutants are acute, meaning they occur only in the presence of the substance; but in some cases, and perhaps in many cases, the effects may be cumulative over time and may result in a chronic condition or illness.

The clinical effects of indoor air pollution can take many forms. The most common clinical signs include eye irritation, sneezing or coughing, asthma attacks, ear-nose-throat infections, allergies, and migraine. More seriously, prolonged exposure to radon, tobacco smoke, and other carcinogens in the air can cause cancer. Most volatile organic compounds (VOCs) can be respiratory irritants, and many are toxic, although little applicable toxicity data exists about low levels of indoor VOCs. One VOC, formaldehyde, is now considered a probable human carcinogen. Needless to say, the best way to protect your health is to avoid or reduce your exposure to indoor air contaminants.

Nearly all observers agree that more research is needed to better understand which health effects occur after exposure to which indoor air pollutants, at what levels and for how long.

Improving Indoor Air Quality

There are three basic strategies for improving indoor air quality: source control, ventilation, and air cleaning.

Source control is the simplest, cheapest, and most obvious strategy for improving indoor air quality. Remove sources of air pollution from your home. Store paint, pesticides, and the like in a garage or other outbuilding. Read the labels on the products you bring into your home, and buy personal care, cleaning, and office products that contain fewer synthetic volatile compounds or other suspicious and potentially irritating substances.

Other sources can be modified to reduce their emissions. Pipes, ducts, and surfaces can be enclosed or sealed off. Leaky or inefficient appliances, such as a gas stove, can be fixed.

Regular house cleaning is effective: Sweeping, vacuuming, and dusting can keep a broad spec-

Building materials and interior furnishings emit a broad array of gaseous volatile organic compounds (VOCs), especially when they are new.

One EPA study has suggested that up to 80% of most people's exposure to airborne pesticides occurs indoors.

Source control is the simplest, cheapest, and most obvious strategy for improving indoor air quality.

Regular house cleaning is effective: Sweeping, vacuuming, and dusting can keep a broad spectrum of dust, insects, and allergens under control.

Unfortunately, indoor air cleaning or purification is more complicated than purifying water. Not many technologies are available for household use, most of them have limited effectiveness, and some of the more promising methods are highly controversial.

trum of dust, insects, and allergens under control. You should be conscientious about cleaning out ducts, and replacing furnace filters on a regular basis. Also, reduce excess moisture and humidity, which encourage the growth of microorganisms, by emptying the evaporation trays of your refrigerator and dehumidifier.

Ventilation is another obvious and effective way to reduce indoor concentrations of air pollutants. It can be as simple as opening windows and doors, operating a fan or air conditioner, and making sure your attic and crawl spaces are properly ventilated. Fans in the kitchen or bathroom that exhaust to the outdoors remove moisture and contaminants directly from those rooms. It is especially important to provide adequate ventilation during short-term activities that generate high levels of pollutants, such as painting, sanding, or cooking. Energy-efficient air-to-air heat exchangers that increase the flow of outdoor air into a home without undesirable infiltration have been available for several years. Such a system is especially valuable in a tight, weatherized house. See chapter 5, page 225, for more information.

Unfortunately, indoor air cleaning or purification is more complicated than purifying water. Not many technologies are available for household use, most of them have limited effectiveness, and some of the more promising methods are highly controversial. Air filtration is problematic because it's only really effective if you draw all of the indoor air through the filter. Even then, such filters only reduce particulate matter. Any furnace or whole-house air conditioning system has a filter. Some of them, such as electrostatic filters or HEPA (High Efficiency Particulate Air) filters, can reduce levels of even relatively small particulates significantly. Portable devices incorporating these technologies have also been developed, which can be used to clean the air in one room. Some of them have EnergyStar ratings (including the Blueair models that we sell), which means they'll consume less power than other models. Follow the manufacturer's instructions about how frequently to change the filter on your furnace or stand-alone air cleaner. However, be aware that air filters are not designed to deal with volatile organic compounds (VOCs) or other gaseous pollutants.

Ionization is a second method of cleaning particulates out of the air. A point-source device generates a stream of charged ions (via radio frequencies or some other method), which disperse through the air and meet up with oppo-

sitely charged particles. As tiny particles clump together, they eventually become heavy enough to settle out of the air, and you won't breathe them in. But like filtration, ionization is not effective against VOCs.

What About Ozone?

A somewhat controversial method of air purification uses ozone, in combination with ionization. Ozone (O_3) is a highly reactive and corrosive form of oxygen. You know about the ozone layer in the upper atmosphere, which prevents a lot of ultraviolet light from reaching the Earth's surface where it can damage living things. Ozone also exists naturally in small concentrations at ground level, where it has various effects. Ozone is a component of smog, produced by the reaction of sunlight upon auto exhaust and industrial pollution. Ozone is also produced by lightning, and is partially responsible for that fresh, clean smell you notice outside after a thunderstorm. Some studies suggest that slightly greater than natural outdoor levels of ozone can reduce respiratory, allergy, and headache problems, and possibly even enhance one's ability to concentrate. However, scientists agree that at slightly elevated levels in the air, ozone can be dangerous and unhealthful to people. In concentrations that exceed safe levels as determined by the EPA, ozone is considered an air pollutant.

Manufacturers of air cleaners that operate as ozone generators make strong claims about their products. They assert that these devices kill biological agents, such as molds and bacteria, that contribute to indoor air pollution; that the ozone penetrates into drapes and furniture, and filters down to the floor, into carpets and crevices, to oxidize and burn up absorbed contaminants and those that fall out of the air; and that the ozone breaks down potentially harmful or irritating gases, including many VOCs, into carbon dioxide, oxygen, water vapor, and other harmless substances. They claim that their machines accomplish all this with levels of ozone that are safe, and that ozone production can be monitored, either automatically or manually, to ensure that ozone concentration does not become excessive.

The EPA strongly disputes many of these claims and considers them misleading. The agency's position is that "'available' scientific evidence shows that at concentrations that do not exceed public health standards, ozone has little potential to remove indoor air contaminants

[and] does not effectively remove viruses, bacteria, mold, or other biological pollutants." While the EPA acknowledges that some evidence shows that ozone can combat VOCs, it points out that these reactions produce potentially irritating or harmful byproducts. In addition, the EPA is skeptical of manufacturer's claims that the amount of ozone produced by such devices can be controlled effectively so as not to exceed safe levels. The bottom line for the EPA is that "no agency of the federal government has approved ozone generation devices for use in occupied spaces. The same chemical properties that allow high concentrations of ozone to react with organic material outside the body give it the ability to react with similar organic material that makes up the body, and potentially cause harmful health consequences. When inhaled, ozone will damage the lungs. Relatively low amounts can cause chest pain, coughing, shortness of breath, and throat irritation." (This information was obtained in July 2004 from a publication posted on the EPA website, at www.epa.gov/iaq/pubs/ozonegen.html.)

The ozone issue is still being debated and at present, Gaiam Real Goods has limited the ozone purifiers in its product mix.

Resources

The best source of up-to-date information about indoor air pollution and residential air cleaning systems is the federal Environmental Protection Agency. The EPA maintains several useful telephone services, and distributes, free of charge, a variety of helpful publications. (You can view many of these publications online at www.epa.gov/iaq/pubs/) You can also request these publications from the regional EPA office in your area, which you can find in your local telephone book, under U.S. Government, Environmental Protection Agency. The following resources may be particularly valuable:

The Inside Story: A Guide to Indoor Air Quality, publication #402K93007.

Indoor Air Facts, No. 7: Residential Air Cleaners, publication #20A-4001.

Residential Air Cleaning Devices: A Summary of Available Information, publication #402K96001.

National Service Center for Environmental Publications and Information
800.490.9198
www.epa.gov/ncepihom/

Indoor Air Quality Information Clearinghouse
P.O. Box 37133
Washington, DC 20013-7133
800.438.4318
www.epa.gov/iaq/iaqxline.html

National Radon Hotline
800-SOS-RADON; information recording operates 24 hours a day.
www.epa.gov/radon/rnxlines.html

Manufacturers of air cleaners that operate as ozone generators claim that ozone kills biological agents that contribute to indoor air pollution; oxidizes and burns up contaminants absorbed by drapes and furniture; and breaks down potentially harmful or irritating gases, including many VOCs, all with levels of ozone that are safe and can be monitored. The EPA strongly disputes many of these claims and considers them misleading.

WATER AND AIR PURIFICATION

AIR PURIFICATION PRODUCTS

AIR FILTER COMPARISON CHART

Description	Item #	Max Sq Ft	# of speeds	Size (inches) (H x W x D)	Weight Lbs	Filter Type	CADR S / P / D	Warranty
Healthmate	01-8005	1,500	3	23 x 14.5 x 14.5	45	carbon/HEPA		5-year
Healthmate Junior	01-0163	500	3	16.5 x 11 x 11	18	carbon/HEPA		5-year
Vornado AQS-15	01-0355	224	3	14.6 x 9.6 x 13.8	13	carbon/HEPA	146/142/150	5-year
Roomaid HEPA Filter	01-0156	150	2	8.5 x 8.5 dia.	4	HEPA/VOC	not rated	5-year
Sun Pure Air Purifier	01-0137	2,000	4	18 x 21.5 x 8	23	carbon/HEPA/UV/ion	265/265/265	lifetime

CADR: Clean Air Delivery Rate, S/P/D: Smoke / Pollen / Dust (effective cubic feet per minute).
Life expectancy of any filter media is dependent on the concentration of the contaminants to which the system is exposed.
Chart is based on information provided by the manufacturer.

Sun Pure UV Air Filter

This unit covers all the bases, from allergens to airborne viruses. The six-stage filtering process combines gas absorption, ultraviolet technology, and ionization with the standard HEPA and carbon filters, and throws in a macro pre-filter for longer filter life—replace expensive filters every two years, instead of every six months! This is the system of choice for any environment where an active germ pool threatens family health, and it provides maximum relief for chemical sensitivities, asthma, and allergies in rooms as large as 50' x 40'. To keep energy expenditure at a minimum and convenience at a maximum, the Ultraviolet Air Purifier utilizes an infrared motion detector that "wakes up" the system when it's needed most (when you and your family are active and awake). When there is no movement in the room, or no dust in the air, it cycles down to the lowest setting to reduce operating costs. It also features a "sleep" mode that provides super-quiet operation while still purifying a room at 70% of normal capacity.

Step One: A treated pre-filter removes larger particles like dust (.5 microns and above) before they get to your expensive HEPA filter. The pre-filter also preserves the gas-absorption media from dust fouling.

Step Two: A specially formulated gas absorption media substantially reduces exhaust fumes, organic hydrocarbons, pesticides, formaldehyde, and other noxious gases emitted from household cleaners, solvents, chlorine, and paint. Critical for those with chemical sensitivities!

Step Three: Activated carbon (from coconuts) removes unpleasant household odors and industrial pollution. This filter is pleated to provide maximum surface area and a long filter life.

Step Four: The same HEPA filter used in professional health care settings, removing airborne contaminants as small as .3 microns (pollen, mold, fungal spores, dust mites, tobacco smoke, most bacteria).

Step Five: High-intensity ultraviolet light kills disease-causing viruses and bacteria on contact. This is what makes the system a "purifier," and not just a cleaner. A godsend for those with suppressed immune systems (Chronic Fatigue Syndrome, AIDS).

Step Six: An ionization chamber produces negative ions and activated oxygen, so indoor air tastes and smells like the real thing. Studies suggest that negative ions may even induce and enhance sleep for your resident insomniac.

The Ultraviolet Air Purifier measures 18" H x 21.5" W x 8" D, and boasts a limited lifetime manufacturer's warranty. Components are UL listed. Replacement set includes all filters and lamp needed for two years. USA.

01-0137 Sun Pure Air Purifier   **$599**
01-0138 Sun Pure Replacement Set
    (UV Lamp and Filters)   **$135**

Control panel closeup.

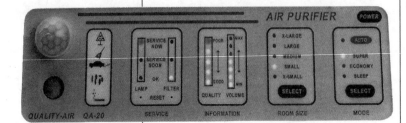

Healthmate HEPA Air Filter

These units clean rooms with minimal energy outlay, and relatively quietly. Because they adjust to a variety of room sizes, the Healthmates work only as hard as the room really needs, so power consumption is kept to a minimum. The larger

Healthmate Air Cleaner draws 115 watts on the highest setting (for rooms up to 10' x 15'); the Healthmate Junior draws 80 watts (for smaller rooms up to 7' x 10'). Both use a three-stage filtering system to clear the air effectively.

The HEPA filter traps airborne particles (dander, pollen, dust, mold spores, smoke) as small as .3 microns; the pre-filter traps larger particles before they reach the HEPA, extending the HEPA filter life (and saving replacement costs). The carbon/zeolite post-filter is primarily aesthetic, eliminating unpleasant household odors so your indoor air smells and tastes better. Features a 360-degree air intake, 3-speed adjustable air flow, casters for portability, and three color choices (black, white, or sandstone). The Healthmate Air Cleaner measures 23" H x 14.5" square, and weighs 45 pounds. The compact Junior model is only 16.5" H x 11" square, and weighs 18 pounds. Five-year manufacturer's warranty. UL listed. USA.

01-8005 Healthmate HEPA Air Filter $399
01-0163 Healthmate HEPA Junior $249
Allow 4–6 weeks for delivery. Can be shipped to customers in the contiguous U.S. only. Cannot be shipped to P.O. boxes.

Powerful Air Filter Delivers Massive Volumes Of Clean Air, Quietly

This is not an ordinary air filter. It's packed with features like the top exit air outlet, circulating clean air up and

out into the room, not across dirty floors. It comes with a simple, top-mounted, three-speed control and power light. It has a wide louver inlet grill, which can be removed easily for HEPA and filter replacement and cleaning. The true HEPA filter provides over 24 square feet of super-efficient filter action, removing a minimum of 99.97% of all particles. The PreSorb® pre-filter reduces odors and organic compounds including cooking odors, paint and solvent fumes, and musty bathroom odors. And with its small footprint, the Vornado doesn't take up a lot of floor space either; the flat back lets you put it against a wall or in a corner out of the way. Using an exclusive vortex action, all the air in

the room is circulated so all of the air in the room is cleaned. It measures 15" high x 13.5" wide x 10" deep and can filter a room as large as 16' x 14'. UL listed. Comes with three extra PreSorb® filters. USA.

01-0355 Vornado AQS-15 Air Filter $159
01-0356 PreSorb® Filters (set of 2) $15

Quiet HEPA Filtration for Small Spaces

The Roomaid HEPA Filter offers the benefits of larger units in a compact, ultra-quiet design. Small enough to fit comfortably on a desk, night table, or counter, the Roomaid makes so little noise it's perfect for bedrooms, nurseries or offices. Hidden within the Roomaid's unobtrusive package is a powerful three-stage filtration system that begins with a washable 1/8"-thick foam pre-filter, which eliminates large particulates such as dust and lint. Next, a replaceable HEPA filter removes 99.97% of particulates larger than 0.3 microns (pollen, bacteria, dander, smoke, and mold spores). Finally, a special activated-carbon VOC filter traps dangerous household gases from cleaners, paints, solvents, and other chemical substances. The two-speed unit uses just 4.5 watts, yet it filters a 10' x 10' room two times an hour. 7.5" H, 8.5" diameter, 5 lbs. Canada.

01-0156 Roomaid HEPA Filter $150
**01-0157 Replacement HEPA Filter
(lasts up to 5 yrs)** $89
**01-0158 Roomaid Filter Kit—3 VOC filters,
2 pre-filters (6–12 mo. ea.)** $22

Car Ionizer

Research shows that car interiors trap pollution and can often contain more contaminants than outside air. The Car Ionizer plugs into your car's lighter socket to release healthy negative ions that bond to odors and pollutants in the air and remove them, creating an oasis of clean air. With an output of 30,000 ions per cu. cm., this is the most compact and quiet model available. 4H"L. 6 oz. China.

01-0121 Car Ionizer $35
Note: Operation may interfere with lower frequency AM radio bands.

Deluxe Mini Personal Air Supply

Deliver a steady stream of pure, clean air to your breathing zone with the Deluxe Mini Air Supply. Operating with virtually silent, fanless operation, it uses corona-discharge technology to draw in and destroy particles such as smoke, dust, pollens, allergens, germs, and viruses. Producing 120 trillion ions per second, it has an airflow speed of 100 feet per minute—making it ideal for cars, airplanes, and even the outdoors. Includes a breakaway neck strap and belt clip. The lithium battery has a life span of 60 hours before it needs replacing. Pacemaker users should consult a physician before using. 2H"H x 1H"W x M"D., 1H oz. USA.

01-0440 Deluxe Mini Personal Air Supply $148

CHAPTER 9

Composting Toilets and Greywater Systems

Novel Solutions to an Age-Old Human Problem

Actually, all pollution is simply an unused resource. Garbage is the only raw material that we're too stupid to use.

—Arthur C. Clarke

A Short History of Waste Disposal

The way humans dispose of their biological waste is a bellwether of civilized society. Does this society protect its members from the horrors of dysentery and even worse diseases? Over the past century, sanitation in North America has evolved gradually from almost every home having an outhouse in the backyard, and rivers in major cities serving as open sewers, to modern flush-and-forget-it systems, where everything seems to simply disappear, and rivers are running very much cleaner. Making distasteful items disappear and rivers run clean is certainly going to win public approval, and we won't be so foolish as to suggest that something was mystically good in the old days. Outhouses smell bad in the summer, are too close to the house, and allow flies to spread filth and disease. In the winter they don't smell, but are too far away. However, in our rush to sanitize everything in sight, we end up throwing away a potentially valuable and money-saving resource (not to mention overdesigning expensive and energy-intensive disposal systems that still may pollute groundwater). Our conventional plumbing systems mix a few gallons of heavily polluted blackwater from our toilet with hundreds of gallons of very lightly polluted greywater from our sink, shower, tub, and washer. By going for the simplest possible short-term solution, we've taken a small problem and made it much larger, more difficult, and more expensive.

In this chapter, we're going to talk about what happens if we separate the really nasty stuff that goes down the toilet from the maybe just barely nasty stuff that goes down the shower, tub, or clothes-washer drain. Blackwater from toilets can and should be treated differently from grey-

water. Both are potential resources, and should not simply be thrown away. Greywater, with minimal or sometimes no treatment, can be used for landscape and garden watering. Human waste can be composted safely and odorlessly to kill the pathogens, then used as a highly beneficial compost.

Composting toilets can close the nutrient cycle, turning a dangerous waste product into safe compost, without smell, hassle, or fly problems. They are usually less expensive than conventional septic systems and they will reduce household water consumption by at least 25%. But like the venerable outhouse, composting toilets only deal with human excreta. Unlike a modern septic system, they won't provide greywater treatment. Greywater is covered later in the chapter.

> Blackwater from toilets can and should be treated differently from greywater. Both are potential resources, and shouldn't simply be thrown away.

A real two-story outhouse from the early 1900s.

What Is a Composting Toilet?

A composting toilet is a treatment system for toilet wastes that does not use a conventional septic system. Composting toilets were developed originally in Scandinavia, where thanks to recent (geologically speaking) glacier activity, almost no topsoils suitable for conventional septic systems exist. A composting toilet is basically a warm, well-ventilated container with a diverse community of aerobic microbes living inside that break down the waste materials. The process creates a dry, fluffy, odorless compost, similar to what you find in a well-maintained garden compost pile. Flowers and fruit trees love it, though we don't recommend using this compost on kitchen gardens, as some human pathogens possibly could survive the composting process.

The composting process in such a toilet does not smell. Rapid aerobic decomposition—active composting—which takes place in the presence of oxygen, is the opposite of the slow, smelly process that takes place in an outhouse, which works by anaerobic decomposition. Anaerobic microbes cannot survive in the presence of oxygen and the more energetic microbes that flourish in an oxygen-rich environment. If a composting toilet smells bad, it means something is wrong. Usually smells indicate pockets of anaerobic activity caused by inadequate mixing.

How Do Composting Toilets Work?

A composting toilet has three basic elements: a place to sit, a composting chamber, and an evaporation tray. Many models combine all three elements in a single enclosure, although some models have separate seating, with the composting chamber installed in the basement or under the house. In either case, the evaporation tray is positioned under the composting chamber to catch any liquids, and some sort of removable finishing drawer is supplied to carry off the finished compost material.

Ninety percent of what goes into a composting toilet is water. Compost piles need to be damp to work well, but many composting toilets suffer from too much water. Evaporation is the primary way a composting toilet gets rid of

excess water. If evaporation can't keep up, then many units have an overflow that is plumbed to the household greywater or septic system. Heat and air flowing through the unit assist the evaporation process. Every composting toilet has a vertical vent pipe to carry off moisture. Air flows across the drying trays, around and through the pile, then up the vent to the outside of the building. The low-grade heat produced by composting helps to provide sufficient updraft to carry vapor up the vent. However, like any passive vent with minimal heat, these are subject to downdrafts. Electric composters include vent fans and a small heating element as standard equipment. All manufacturers include, or offer, an optional vent fan with their nonelectric models that can be battery- or solar-driven. Every one to three days, the owner should add a cup of high-carbon-content bulking agent, such as peat moss, wood chips, or dry-popped popcorn. This helps soak up excess moisture, makes lots of little wicks to aid in evaporation, and creates air passages that prevent anaerobic pockets from forming.

The solids are treated with a diverse microbiological community—in other words, composted. Keeping this biological community happy and working hard requires warmth and plenty of oxygen, the same things needed for speedy evaporation. Smaller composters usually employ some kind of mixing or stirring mechanism to ensure adequate oxygen to all parts of the pile, and faster composting action. Com-

A modern BioLet composting toilet

A composting toilet is basically a warm, well-ventilated container with a diverse community of aerobic microbes living inside that break down the waste materials. The process creates a dry, fluffy, odorless compost, similar to what you find in a well-maintained garden compost pile.

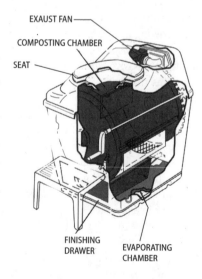

EXAUST FAN

COMPOSTING CHAMBER

SEAT

FINISHING DRAWER

EVAPORATING CHAMBER

A typical Sun-Mar composting toilet.

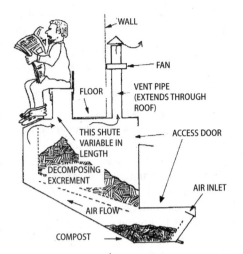

An early Multrum toilet—the first generation of composting toilets.

here, and give the pluses and minuses of each design.

First, some general comments. Smaller composters certainly cost less, but because the pile is smaller, they are more susceptible than larger models to all the problems that can plague any compost pile, such as liquid accumulation, insect infestations, low temperatures, and an unbalanced carbon/nitrogen ratio. Smaller composters require the user to take a more active role in the day-to-day maintenance of the unit. We have found that the smaller units with electric fans and thermostatically controlled heaters have far fewer problems than the totally nonelectric units. So be aware that the less-expensive composting toilets have hidden and long-range costs.

The solids are treated with a diverse microbiological community—in other words, composted. Keeping this biological community happy and working hard requires warmth and plenty of oxygen.

posting toilets work best at temperatures of 70°F or higher; at temperatures below 60°F the biological process slows to a crawl, and at temperatures below 50°F it comes to a stop. The composting action itself will provide some low-level heat, but not enough to keep the process going in a cold environment. It is okay to let a composting toilet freeze, like in a summer cottage over the winter, although it shouldn't be used when cold or frozen. Normal biological activity will resume when the temperature rises again.

The earliest composter designs, such as the Clivus Multrum system, use a lower-temperature decomposition process known as moldering, which takes place slowly over several years. These composters have air channels and fan-driven vents, but they lack supplemental heat or the capability of mixing and stirring. Therefore, these very large composters don't promote highly active composting. They have two other disadvantages. Liquids often have to be manually removed or pumped out of these units because the lower temperatures slow down the rate of evaporation. And, because temperatures don't get very high, there's a greater chance of pathogens or parasite eggs surviving the composting process. This type of large, slow composter is most effective in public access sites, and many are currently in successful use at state or national parks. Their large bulk lets them both absorb sudden surges of use, and weather long periods of disuse without upset.

Manufacturer Variations

Every compost toilet manufacturer will be delighted to tell you why their unit is the best. We'll attempt to take a more impartial attitude

Sun-Mar Composting Toilets

Sun-Mar is the largest and most experienced of the small residential-size composter manufacturers. Their composting toilets use a rotating drum design, like a clothes dryer, that allows the entire compost chamber to be turned and mixed easily—and remotely! This ensures that all parts of the pile get enough oxygen, and that no anaerobic pockets form. Routine once- or twice-a-week mixing also tumbles the newer material in with older material, so the microorganisms can get to work on recent additions more quickly. Periodically, as the drum approaches one-half to two-thirds full, a lock is manually released and the drum is rotated backwards. This action lets a portion of the composting material drop from the drum into a finishing drawer at the bottom of the composter. The next time the drum needs a portion moved along, the finishing drawer, which now contains a load of finished compost, is emptied on your orchard, ornamental plants, or selected parts of your garden, reinstalled, and a fresh load dumped in.

Sun-Mar has introduced a large Centrex 3000 model that has a composting drum twice as long as anything we've seen previously, with stacked drying trays. This design doubles the solids capacity, triples the drying tray area, and doubles the time that material can spend fully composting, alleviating many of the more common complaints we've heard about Sun-Mar units.

The main disadvantage of the Sun-Mar drum system is that fresh material can be included with the material dumped into the finishing drawer, and as a result, some pathogens may survive the composting process. Also, the limited size of most Sun-Mar composters means

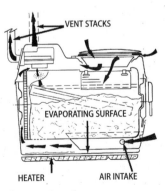

Some Sun-Mar models have the toilet separated from the composting chamber.

Cross-section of a typical Sun-Mar composter.

Carbon/Nitrogen Ratios

Proper composting requires a balance of carbon and nitrogen in the organic material being composted. Human excreta are not properly balanced as they are too high in nitrogen. They require a carbon material to be added for the encouragement of rapid and thorough microbial decomposition. In the mid 1800s the concept of balancing carbon and nitrogen was not known, and the high nitrogen content of humanure in dry toilets prevented the organic material from efficiently decomposing. The result was a foul, fly-attracting stench. It was thought that this problem could be alleviated by segregating urine from feces (which thereby reduced the nitrogen content of the fecal material) and dry toilets were devised to do just that. Today, the practice of segregating urine from feces is still widespread, even though the simple addition of a carbonaceous material to the feces/urine mix will balance the nitrogen of the material and render the segregation of urine unnecessary.

from *The Humanure Handbook,* 2nd Edition,
Joe Jenkins. Used with permission.

they work best for intermittent-use cabins or small households. With the exception of the large Centrex 3000 model, these composters can on occasion be overwhelmed by full-time use in larger households.

Sun-Mar produces both fully self-contained composting toilets and centralized units, in which the compost chamber is located outside the bathroom and connected to either an air-flush toilet, or an ultralow-flush toilet with standard 3-inch waste pipe. Electric models all feature thermostatically controlled heating elements to keep the microbe community happily active and assist with evaporation, and a small fan to ensure fresh airflow and negative air pressure inside the compost chamber. We strongly recommend the electric models for customers who have utility power available. For those with intermittent AC power, AC/DC models are available that take advantage of AC power when it's available but can operate adequately without it as well. Finally, Sun-Mar makes nonelectric models that don't need any power, although use should be limited to one or two people or a low-use cabin with these models.

BioLet Composting Toilets

BioLet composters use a composting chamber that sits upright like a bucket. Fresh material enters the top of the chamber, works its way down as composting progresses, and eventually is allowed to sift out the bottom of the chamber into the finishing drawer. Mixing and stirring rods rotate inside the drum to ensure oxygen supply, and to sift finished material out the bottom screen. It is not possible for fresh material to pass through the composter to the finishing tray in this design, although the mixing and stirring may not result in as complete aeration as with the tumbling-drum design.

BioLet produces two basic sizes, which, with various options, amount to four different models. The Standard is the smaller size, and is available in a nonelectric model with manual mixing, an electric model with manual mixing, and an electric model with automatic mixing. The 40% larger XL size comes in one electric model with automatic mixing. The BioLet heater warms the incoming air before it passes over the evaporation trays to speed evaporation. All models except the nonelectric require utility power, or a plentiful RE source such as a strong hydro system.

BioLet designs feature a number of innovative and user-friendly elements. A tasteful pair of clamshell doors immediately below the seat only open when weight is put on the seat. This means you don't have to look at your compost before sitting down. The heat level is adjustable, and the unit has a clear sight tube to see if liquid is accumulating (which means it's time to turn up the heat). All the fans, heaters, thermostat, and mechanical components are mounted on a single easy-access cassette in the top rear, so if repairs are ever needed, it's all easy to get to.

Disadvantages with these composters: They have a tendency to overly dry the compost, which leads it to clump up in big balls. This makes it really hard to stir. The shear pin or the nylon gears that drive the stirring rods are a common repair until folks learn to add water if

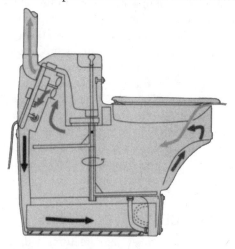

Cross-Section of a BioLet composter showing air flow.

clumping occurs. The tasteful clamshell doors under the seat mean that guys may need to be resourceful, or take it outside.

Carousel Composting Toilets

The Carousel is a very large composter for full-time residential use. It features a round carousel design with four separate composting chambers. Like most composters, it has drying trays under the composting chamber and a fan-assisted venting system to ensure no-smell operations. Each of the four chambers is filled in turn, which can take from two to six months; then a new chamber is rotated under the dry toilet.

This gives each of the four batches up to two years to finish composting without being disturbed. Batch composting has the advantage of not mixing new material with older, more advanced compost, allowing the natural ecological cascading of composting processes. This results in more complete composting, and greatly reduces the risk of surviving disease organisms. The Carousel can handle full-time use and large families or groups. Because it does not have a heating element it needs to be installed in a conditioned indoor space. But then with only the small vent fan to run, the power use will be very low. Its disadvantages are obvious: It's big, and it's expensive initially.

Will Your Health Department Love It As Much As You Do?

Probably not, but this depends greatly on the enlightenment quotient of your local health official. Some health departments will welcome composting toilets with open arms, and even encourage their use in some locales, while others will deny their very existence. We have often found that in lakeside summer cabin situations, health officials prefer to see composting toilets rather than pit privies or poorly working septic systems that leach into the lake. Several composters now carry the NSF (National Sanitation Foundation) seal of approval, which makes acceptance easier for local officials, but does not mandate it. It is really up to your local sanitarian. He can play God within his own district and there is nothing that you, Gaiam Real Goods, or the manufacturer can do about it. Our advice if you're trying to persuade your authorities? Be courteous. Health officials have to consider that

a composting toilet might suit you to a tee, and you'll take good care of it, but what if you sell the house? Will the next home owner be willing, or able, to take care of it, too?

Also, be aware that composting toilets only deal with toilet wastes, the blackwater. You still need a way to treat your greywater wastes—all the shower, tub, sink, and washer water. Unconventional greywater systems will face the same approval problems as composting toilets. Some localities happily will allow alternatives that use the greywater for landscape watering, but others may require a standard full-size septic system. In all cases, greywater systems must use subsurface disposal. Once the water goes down the drain, it can't see daylight, or the possibility of human contact, again. Greywater alternatives are covered in more detail below.

Composting Toilet Installation Tips

KEEP IT WARM

Compost piles like warmth, because it makes all the little microbes work faster. Thus, composting toilets work best in warm environments. If your installation is in a summer-use cabin, then a composting toilet is ideal. A Montana outdoor installation in wintertime isn't going to work at all. If you plan to use the composter year-round, then install it in a heated space on an insulated floor. The small electric heater inside the compost chamber will not keep it warm in an outdoor winter environment, but it will run your electric bill up about $25 a month trying. Talk to one of Real Goods' technicians if you have any doubts about proper model selection or installation.

PLUG IT IN

If you have utility power, by all means use one of the electrically assisted units; they have far fewer problems overall. If you have intermittent AC power from a generator or other source, use one of the AC/DC, or hybrid units. Anything that adds warmth or increases ventilation will help these units do their work. We strongly recommend adding the optional vent fans to non-electric units.

LET IT BREATHE

The compost chamber draws in fresh air and exhausts it to the outside. If the composter is inside a house with a woodstove and/or gas water heater that are also competing for inside

The Carousel Composter for full-time and larger family use.

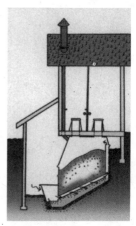

The CTS System for high-use public sites.

Some health departments will welcome composting toilets with open arms, and even encourage their use in some locales, while others will deny their very existence. Our advice if you're trying to persuade your authorities? Be courteous.

air, you may have lower air pressure inside the house than outside. This can pull cold air down the composter vent and into the house. Composters shouldn't smell bad, and something is wrong if they do, but you don't want to flavor your house with their gaseous by-products either. Do the smart thing: Give your woodstove outside air for combustion.

DON'T CONDENSE

Insulate the composter's vent stack when it passes through any unheated spaces. The warm, humid air passing up the stack will condense on cold vent walls, and moisture will run back down into the compost chamber. Most manufacturers include vent pipe and some insulation with their kits. Add more insulation if needed for your particular installation.

KEEP EVERYTHING AFLOAT

With central units such as the Centrex models, where the composting chamber is separate from the ultralow-flush toilet, the slope of any horizontal waste-pipe run is critical. The standard 3-inch ABS pipe needs to have $1/8$ to $1/4$ inch of drop per foot of horizontal run. More drop per foot than that allows the liquids to run off too quickly, leaving the solids high and dry.

With thanks to Joseph Jenkins and *The Humanure Handbook.*

Vertical runs are no problem, so if you want a toilet on the second floor, go ahead. Sun-Mar recommends horizontal runs of 18 feet maximum, but customers have successfully used runs of well over 20 feet. Play it safe and give yourself cleanout plugs at any elbow when assembling the ABS pipe. If you are using an air-flush unit, it must be installed directly over the composter inlet.

Greywater Systems

What Is Greywater?

Greywater is the mix of water, soap, and whatever we wash off at the shower, sink, dishwasher, or washing machine. Basically, it is any wastewater other than what comes from toilets. This is the majority of water used in most households. Most homes produce 20 to 40 gallons of greywater per person, per day. Greywater is also likely to contain bacteria, including small amounts of fecal coliform, protozoans, viruses, oil, grease, hair, food bits, and petroleum-based "whatevers."

WHY WOULD YOU WANT TO USE GREYWATER?

Utilizing greywater has many advantages. It greatly reduces the demands on your septic system, gives otherwise wasted nutrients to plants, reduces household water use, increases your awareness of natural cycles, provides very effective water purification, and probably reduces your use of energy and chemicals. How great is that? Greywater is used primarily for landscape irrigation, because soil is an excellent water purifier. By substituting for freshwater, greywater can make landscaping possible in dry areas. It may also keep your landscaping alive during drought periods. Sometimes a greywater system can make a house permit possible in sites that are unsuitable for a septic system.

WHEN WOULD GREYWATER BE A POOR CHOICE?

Greywater recycling isn't for everybody. High-rise apartment dwellers would be seriously challenged. But let's assume we're speaking to folks with at least a suburban plot around them. Even then, you may have insufficient yard and landscaping space, major bits of your drain plumbing may be encased in concrete, your climate may be too wet or too frozen, costs may outweigh benefits, or it might just be illegal where you live.

How Does a Greywater System Work?

Although the trace contaminants in greywater are potentially useful, even highly beneficial, for plants and landscaping, their presence demands

Most homes produce 20 to 40 gallons of greywater per person, per day.

At the simplest end of the spectrum, greywater can be hand carried to plants. At the most complex, it can be finely filtered and automatically distributed as needed through subsurface drip irrigation tubing.

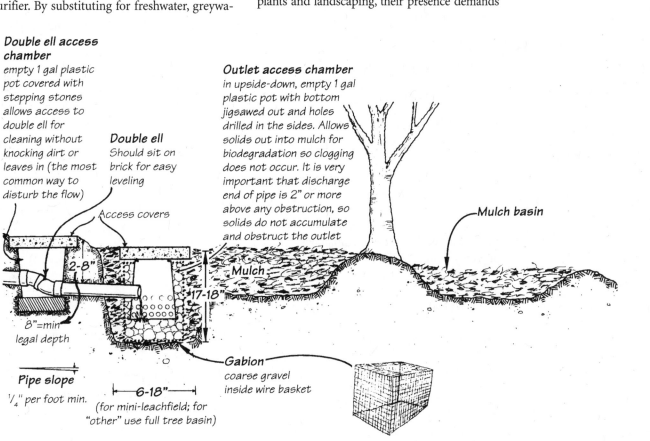

Double ell access chamber
empty 1 gal plastic pot covered with stepping stones allows access to double ell for cleaning without knocking dirt or leaves in (the most common way to disturb the flow)

Double ell
Should sit on brick for easy leveling

Access covers

Outlet access chamber
in upside-down, empty 1 gal plastic pot with bottom jigsawed out and holes drilled in the sides. Allows solids out into mulch for biodegradation so clogging does not occur. It is very important that discharge end of pipe is 2" or more above any obstruction, so solids do not accumulate and obstruct the outlet

Mulch

2-8"

17-18"

8"=min legal depth

Pipe slope
1/4" per foot min.

6-18"
(for mini-leachfield; for "other" use full tree basin)

Gabion
coarse gravel inside wire basket

Mulch basin

Generally, official attitudes are moving toward a more accepting and realistic stance, and authorities will often turn a blind eye toward greywater use.

Legal greywater systems tend toward engineering overkill, while many of the simple and economical methods that folks actually use are still technically illegal.

some modest caution in handling and disposal. How do you recycle it safely? All responsible greywater designs stem from two basic principles:

- Natural purification occurs while greywater passes slowly through healthy topsoil.
- No human contact should occur before purification.

Greywater system designs display great diversity. At the simplest end of the spectrum, greywater can be hand carried to plants. At the most complex, it can be finely filtered and automatically distributed as needed through subsurface drip irrigation tubing. Middle-of-the-road systems include connecting the washing machine to a hose that is moved from mulch basin to mulch basin, and branched drain systems, which take one flow and split it automatically into several permanent outlets.

Like other aspects of ecological design, greywater systems are highly site-specific. Each system needs to suit the site, climate, and perhaps the commitment level of the owner. The one general rule is that there are no general rules for greywater systems. The optimum system depends primarily on how much water you've got, how much irrigation you need, the soil permeability, your budget, and legal constraints. At www.greywater.net you can find a downloadable copy of the system selection chart from *Create an Oasis with Greywater,* which lists the attributes of twenty different greywater system design options.

Will Your Health Department Love Greywater as Much as You Do?

As with composting toilets, acceptance of greywater depends entirely on the folks at your local health department. Some will give you a choice about greywater; some won't. Some may allow greywater systems, so long as you also install a full-size, approved septic system. In most areas, the legality of greywater is ambiguous. Generally, official attitudes are moving toward a more accepting and realistic stance, and authorities will often turn a blind eye toward greywater use.

In the late 1970s, during a drought period, the state of California published a pamphlet that explained the illegality of greywater use, while showing detailed instructions of how to do it! Since then, California has become one of many places where you can install a greywater system legally, but with rules so ridiculous that less than 1% of greywater systems go through the permit-

ting process. Happily, there is a new trend toward reasonableness and realism in greywater regulation. Since 2001, the state of Arizona has allowed the installation of greywater systems of less than 300 gallons a day that meet a list of reasonable requirements without having to get a permit at all. A similar bill is currently pending in New Mexico. For up-to-date information, and much more info on presenting a greywater system to your friendly local health department, consult the Greywater Policy Center at www.greywater.net.

Recycling and greywater use may suit you perfectly, but someday you're likely to sell your perfect house. Will a greywater system suit the next owners as well? It's the people who will buy the house from you that the health department is thinking about. The best advice is simply to call your health department and ask. But ask as a hypothetical example. Health officials recognize the "hypothetical" question readily. This allows you to ask specific pointed questions, and them to answer fully and honestly, without anyone admitting that any crime or bending of the rules has occurred or is likely to occur.

How To Make a Greywater System

Legal greywater systems tend toward engineering overkill, while many of the simple and economical methods that folks actually use are still technically illegal. Start with the greywater guide series of booklets listed in the product section that follows. These provide the best overview of the subject and detail every type of greywater

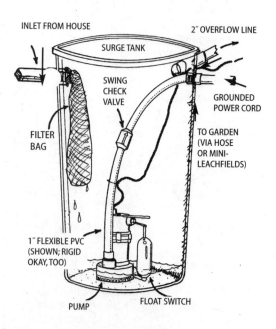

INLET FROM HOUSE

2" OVERFLOW LINE

SURGE TANK

SWING CHECK VALVE

GROUNDED POWER CORD

FILTER BAG

TO GARDEN (VIA HOSE OR MINI-LEACHFIELDS)

1" FLEXIBLE PVC (SHOWN; RIGID OKAY, TOO)

PUMP

FLOAT SWITCH

system, from the simple dishpan dump to fully automated designs. There are many possible ways to create a greywater system, and since the booklets cover them all with grace, style, a wealth of illustrations, and at minimal cost, we will not duplicate their work here. Aim for the best possible execution with the simplest possible system. During new home construction, it's fairly easy to separate the washer, tub, sink, and shower drains from the toilet drains. Retrofits are a bit more trouble, and in some cases impossible, like when the plumbing is encased in a concrete slab floor. Kitchen sinks are often excluded from greywater systems because of the high amount of oil, grease, and food particles. But this is a personal choice, and depends on how much maintenance you're willing to do on a regular basis.

GREYWATER PRODUCTS

BOOKS

THE GREYWATER GUIDE SERIES

Create an Oasis with Greywater—Revised 4th Edition

Choosing, building and using greywater systems

Describes how you can save water, save money, help the environment, and relieve strain on your septic tank or sewer by irrigating with reused wash water. Describes twenty kinds of greywater systems. The concise, readable format is long enough to cover everything you need to consider, and short enough to stay interesting. Plenty of charts and drawings cover health considerations, greywater sources, designs, what works and what doesn't, bio-compatible cleaners, maintenance, preserving soil quality, the list goes on.... 51 pages, softcover. USA.

82440   Create an Oasis with Greywater $14.95

Builder's Greywater Guide

A companion to Create an Oasis with Greywater

Will help you work within or around codes to successfully include greywater systems in new construction or remodeling. Includes reasons to install or not install a greywater system, flowcharts for choosing an appropriate system, dealing with inspectors, legal requirements checklist, design and maintenance tips, how the earth purifies water, and the complete text of the main U.S. greywater codes with English translations. 46 pages, softcover. USA.

80097   Builder's Greywater Guide $14.95

Branched Drain Greywater Systems

A companion to Create an Oasis with Greywater

Detailed plans and specific construction details for making a branched drain greywater system in any context. This ultra-simple, robust system is the best choice for over half of residential installations. It provides reliable, sanitary, and low-maintenance distribution of household greywater to downhill plants without filtration, pumping, or a surge tank. It is one of the few systems that is both practical and legal, and has been described as "the best greywater system since indoor plumbing." Note: the area to be irrigated has to be downhill from the house, and laying out the lines must be done with fanatical attention to proper slope. 52 pages, softcover. USA.

82494   Branched Greywater $14.95

Greywater Guide Booklet Set

Save over $5 by picking up all three Greywater booklets together.

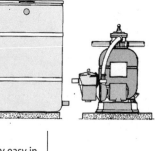

21-0360 Greywater Guide Booklet Set $39

GREYWATER SYSTEMS

Earthstar Greywater Systems

Automatic operation, sand filter with backwash cleaning. Simple, high quality, and easy maintenance. When greywater reaches a high level in the 55-gallon tank, a float switch starts a centrifugal pump that begins an irrigation cycle. A simple 5-minute backwash cleaning process every 2 months keeps everything in top shape. And you don't even have to get your hands wet. Comes complete with everything you need to install. Color-coded components make assembly easy in less than an hour. 12-gallon tank good for tight spaces. Lead time 8 to 10 weeks. USA.

44831	**Greywater System with 55 gallon tank**	**$1,199***
44832	**Greywater System with 12 gallon tank**	**$1,099***

Please allow 8-10 weeks for delivery; shipped directly from manufacturer.

Hard-to-Find Greywater Parts

Three-Way Valves. Installed at washers, kitchen sinks, or other drains that might not always be appropriate for greywater. Lets you choose to use your greywater system or your conventional septic system. Never needs lubrication. Guaranteed to never leak or break. Accepts two pipe sizes, one inside, one outside. CPVC material.

Double Ell with Cleanout. This specialized ell fitting with cleanout is required for the simple, low-maintenance branched-drain systems. See *Branched Drain Greywater Systems.* All other parts for these systems will be available locally. ABS material.

44833	**3-Way Valve 1.5 or 2"**	**$49**
44834	**3-Way Valve 2 or 2.5"**	**$49**
44835	**Double Ell w/Cleanout 1.5"**	**$18**
44836	**Double Ell w/Cleanout 2"**	**$24**

COMPOSTING TOILET PRODUCTS

BOOKS

The Humanure Handbook 2nd Ed.
A Guide to Composting Human Manure

Deemed "Most Likely to Save the Planet" by the Independent Publishers in 2000.

For those who wish to close the nutrient cycle, and don't mind getting a little more personal about it than our manufactured composting toilets require, here's the book for you. Provides basic and detailed information about the ways and means of recycling human excrement, without chemicals, technology, or environmental pollution. Includes detailed analysis of the potential dangers involved and how to overcome them. The author has been safely composting his family's humanure for the past twenty years, and with humor and intelligence has passed his education on to us. 302 pages, softcover. USA.

82308   The Humanure Handbook $19

The Composting Toilet Book

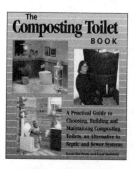

Even though off-the-grid homeowners have been using them for decades, convincing the local inspector to permit a composting toilet system in your conventional year-round home can be a frustrating experience. This book gives you the ammunition you need to get this water-saving, eco-friendly technology installed and permitted. This is the definitive handbook we've been wanting for at least ten years. It's packed with technical details of various permitable systems, state-by-state permitting information, maintenance and operation tips the manufacturers often do not provide, profiles of long-time composting toilet users, and a bonus section on greywater applications. While costs and restrictions for conventional systems have been rising, composting toilets have dropped in price and risen in quality. There's never been a better time to switch over to a composting toilet system, and there's never been a better resource on the topic than this one. 150 pages, softcover. USA.

80708   The Composting Toilet Book $29.95

COMPOSTING ACCESSORIES

Sun-Mar Home Composter
No-smell composting for kitchen waste

Sun-Mar has been handling tough composting jobs for more than twenty-five years. Their practical experience shows in this Home Composter for kitchen and garden waste. Designed for odorless indoor or outdoor use, the Home Composter employs Sun-Mar's proven mixing drum for easy and complete aeration. A well-turned compost pile is a happy, active, good-smelling compost pile, and the rotating drum on this composter makes it easy. Kitchen scraps, garden waste, peat moss, and a bit of water go into the composter upper chute. Material up to 1.5" chunks is acceptable. Smaller chunks compost faster. The gear-driven drum is rotated by turning a handle. A drum lock stops the drum in position to accept fresh material. As the compost tumbles and aerates, it moves along the main drum, and then back through the inner drum before dropping automatically into the finished compost drawer. To prevent saturation, any excess liquid drains through a screen into the liquid drawer to be used as compost tea.

Finished compost can be used for ornamental flower beds, fruit trees, or vegetable gardens. The compost tea can be used as liquid fertilizer.

Composting works best at temperatures over 60°F. Our busy little microbes pretty much go into hibernation at lower temperatures, so an indoor installation is recommended unless you live in a warm climate or only plan on seasonal use. Basements, heated garage, mudrooms, utility rooms, or laundry rooms are all good choices. The Home Composter is supplied with a 3-foot length of flexible dryer-type 3" vent with a through-the-wall exhaust fitting, and a low-wattage vent fan to ensure no-smells operation. The fan is 12-volt DC for safe operation, with a transformer cube that plugs into any conventional AC outlet. Compost starter and microbe mix are included to ensure a well-balanced biological community from initial start up. Canada.

Dimensions: 23" x 24" footprint, 33" height
Weight: 51 lb. (empty)
Shipping weight 70 lb.
Ships via: UPS directly from mfr.
Mfr.'s warranty: one year. Canada.

44814   Sun-Mar Home Composter $360

Compost Sure™ Starter Mulch

A mix of coarse peat moss and chopped hemp stalk. Designed to maintain porosity and retain oxygen while keeping compost moist, mixed, and well supplied with available carbon. 30 liter/8 gallon bag.

44253   Compost Sure™ Starter Mulch $22

Microbe Mix™

Special dried microbes and enzymes initiate and accelerate composting in all Sun-Mar composting toilets and systems. Packaged on a corn grit carrier; contains citronella to discourage insects. 500g (17.6 oz.) jar with scoop.

44839   Microbe Mix™ $18
2 for $32

Compost Quick™

Specially natural enzyme solution facilitates the action of aerobic bacteria. Also effective as a cleaner for bowl liners (self-contained toilets) or 1 pint toilets (central composting systems). 16 oz. with spray top.

44838   Compost Quick™ $20
2 for $36

COMPOSTING TOILETS

Sun-Mar Composting Toilets

Based in Canada, Sun-Mar has been in the composting toilet business longer than almost anyone. Sun-Mar designs are all based on a rotating drum, like a clothes dryer, that locks into the "use" position. Turning a crank on the outside of the composter rotates the drum, which ensures complete mixing and good aeration of the pile. A well-turned compost pile works faster and more effectively. Mixing should happen once or twice a week whenever the composter is in active use. When the drum is rotated in the normal mixing direction, the inlet flap swings shut, and all material stays in the drum. Periodically, when the drum approaches one-half to two-thirds full, a lock is released that allows the drum to rotate back-

Sun-Mar: Saving the Planet One Toilet at a Time

With toilets on all seven continents —including sites in Antarctica and Mount Everest base camp—the Sun-Mar company has single-handedly conserved more than 2 billion gallons of water.

That's why, since 1985, Real Goods has been proud to feature products by Sun-Mar, the world leader in consumer composting. This remarkable company was founded in the mid-1960s in Sweden with a mission to create waterless toilets off the traditional septic grid, using Mother Nature instead of hazardous chemicals to compost human waste into a healthy and usable product.

SMELLING LIKE A ROSE

As you might imagine, Sun-Mar's customer service team gets some interesting queries and comments— most notably, "what about the smell?" Answer: It's a non-issue with Sun-Mar composting toilets. Aerobic decomposition produces no sewer gases, and a partial vacuum is maintained by the venting system to ensure odor-free operation.

NEW OPTIONS WITH PATENTED SUN-MAR TECHNOLOGY

A company with an impressive track record of continual re-engineering and innovation, Sun-Mar will launch three new products in 2005 that feature its patented continuous-flow drum system. This ingenious system allows for a lot more compost to be produced in a much smaller space.

wards. This lets some compost drop into the finishing tray at the bottom. It finishes composting here until the next time the drum needs to dump some material.

The fully self-contained Sun-Mar composters, the Excel, Compact, and Space Saver models, are a good choice for summer camps, sites with intermittent use, sites without electricity, or full-time residences with winter heating. If you want your composter to work in the winter, it must be in a 65°F or warmer space.

The Centrex series puts the composting drum outside or in the basement for those who want a more traditional-looking toilet or toilets in the bathroom. Centrex models are available for a full range of uses from intermittent summer camps to full-time residential use. The entire Centrex line is NSF listed now, which will make it easier for local health officials to accept their use.

If you have utility electricity available at your site, we strongly recommend using one of the standard AC models with fan and heater. They compost faster, are more tolerant of variable uses, and are much less likely to suffer any performance problems. This electrical advice applies to all composting toilet brands!

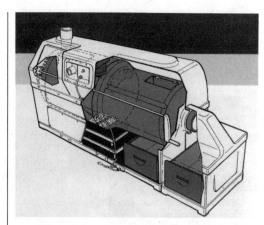

Sun-Mar Centrex 3000-Series

Our highest-capacity composting toilet

The NSF-listed Sun-Mar Centrex 3000-series is designed for heavy cottage use, or medium continuous residential use. The double-length composting drum in the 3000-series has 40% more capacity than any other Sun-Mar unit. This allows more time for composting, handles more volume or surge loads, and compost now cleverly drops into the finishing drawer automatically when the drum is turned. There are two finishing drawers to complete the composting process, and by moving a filled drawer from the end of the drum to the second station under the drum, the composting cycle can be extended further. The evaporating chamber for excess liquids now has two cascading trays above the composter bottom, greatly increasing available evaporation area. Electric models have a pair of 250-watt thermostatically controlled heaters to boost evaporation rates, and keep the compost chamber warm and friendly. A 1-inch overflow security drain hose is included with all models, which should be connected to the household greywater system.

The Centrex 3000 is available in a standard AC version, a non-electric version, or an AC/DC version. All versions are further available for use with either one-pint, ultralow-flush RV toilet(s), or an air-flush (dry) toilet. Toilets are purchased separately. Suggested capacities vary slightly with model, but generally, the 3000-series is good for five to eight adults in residential use, or seven to ten adults in vacation use.

Use a standard version if you have utility power. It comes with a 2-inch vent and 30-watt fan. The AC/DC version is for folks who run a generator more or less daily and has the 2-inch vent and fan, plus a second 4-inch vent with 2.4-watt DC fan. The non-electric version has the 4-inch vent with DC fan. Canada.

Dimensions: All versions 31.5" H x 69.5" W x 26.5" D. To remove finishing drawers, 52" of depth needed.
Weight: approximately 120 lb.
Shipping weight: 220 lb. (some models slightly less)
Ships via: Freight collect
Ships from: Buffalo, NY
Mfr.'s warranty: 3-yr. parts, 25-yr. body

44118	**Centrex 3000 Std**	$1,629
44119	**Centrex 3000 NE**	$1,469
44120	**Centrex 3000 AC/DC**	$1,729
44121	**Centrex 3000 Std A/F**	$1,679
44122	**Centrex 3000 A/F NE**	$1,519
44123	**Centrex 3000 A/F AC/DC**	$1,779
44815	**Sun-Mar 12v/3.4w 4" Stack Fan**	$50
44816	**Sun-Mar 24v/3.4w 4" Stack Fan**	$50

Shipped freight collect.

Sun-Mar Centrex 2000-Series

For medium to heavy seasonal use, or light residential use

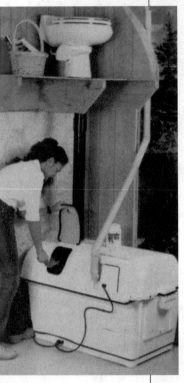

The NSF-listed Centrex 2000-series is probably the most popular of the central composting units. Good for medium to heavy seasonal use, or light residential use, the 2000-series is the refined result of over 25 years of composting experience. Available in a standard AC version, a non-electric version, or an AC/DC version. In addition, each version can be ordered for either a low-flush, RV-type toilet(s), or for a dry, air-flush toilet. Multiple low-flush toilets can serve a single central composter, but only a single air-flush toilet can be used with the A/F models. Toilet options are sold separately.

Standard AC units, and AC/DC units come with a 350-watt thermostatically controlled heater, a 2" vent, and a 30-watt fan. Non-electric units have no heater, and use a 4" vent. AC/DC units have both 2" and 4" vents. 1.4-watt DC vent fans are included with the NE and AC/DC air-flush composters only; they are optional with any other composter. A 1" overflow drain is included, and should be connected to the household greywater system with all composters. Suggested capacities vary slightly with model, but generally, the 2000-series is good for three to six adults in residential use, or five to eight adults in vacation use. Canada.

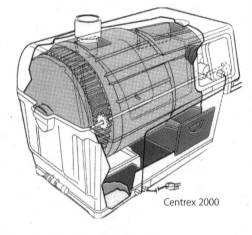

Centrex 2000

Dimensions: 27.5" H x 26.5" D x 44.75" W
(48.75" width to turn handle)
Weight: 100 lb.,
Shipping weight: 125 lb. (approx.; varies slightly with model)
Ships via: freight collect
Ships from: Buffalo, NY
Mfr.'s warranty: 3-yr. parts, 25-yr. body

44817	**Centrex 2000 Std.**	**$1,419**
44818	**Centrex 2000 N.E.**	**$1,259**
44819	**Centrex 2000 AC/DC**	**$1,519**
44820	**Centrex 2000 Std. A/F**	**$1,449**
44821	**Centrex 2000 N.E. A/F**	**$1,289**
44822	**Centrex 2000 AC/DC A/F**	**$1,549**
44803	**Sun-Mar 12v/1.4w 4" Stack Fan**	**$50**
44804	**Sun-Mar 24v/1.4w 4" Stack Fan**	**$50**

Sun-Mar Centrex 1000-Series

Our smallest central composter

The NSF-listed Centrex 1000 is designed for light to medium seasonal use. It is not recommended for continuous residential use. The Centrex 1000 has all controls, access, and venting on the front panel for ease of service. Front venting makes cottage installation much

easier, particularly when using the NE (non-electric) model, which requires a straight vent with no bends. The Centrex 1000 uses an RV toilet with an ultra-low flush of approximately 1 pint. Toilets are sold separately. The 1" overflow drain should be connected to the household greywater system.

Centrex 1000

Available in a standard AC model, an AC/DC model, and a non-electric model. The standard AC model, for use with utility AC power, has a single 2-inch vent stack with a 30-watt fan and thermostatically controlled 250-watt heater. The NE model has a single 4-inch vent stack, with no heater or fan. The optional 1.4-watt DC vent fan is recommended. The AC/DC model, for installations that run an AC generator several hours per day, has the heater, fan, and 2" vent just like the standard version, plus a 4-inch vent like the NE version. The 1.4-watt DC fan is optional, but recommended. Capacities vary slightly with models, but generally the 1000 is good for four to six people in intermittent weekend/vacation use. Canada.

Dimensions: 27.5" H x 26.25" D x 31.5" W (35.5" width to turn handle)
Weight: 50 lb.
Shipping weight: 100 lb. (approx. varies slightly with model)
Ships via: freight collect
Ships from: Buffalo, NY.
Mfr.'s warranty: 3-yr. parts, 25-yr. body.

44105	**Centrex 1000 Std.**	**$1,209**
44106	**Centrex 1000 N.E.**	**$999**
44107	**Centrex 1000 AC/DC**	**$1,309**
44803	**Sun-Mar 12v/1.4w 4" Stack Fan**	**$50**
44804	**Sun-Mar 24v/1.4w 4" Stack Fan**	**$50**

Toilets for Use with Centrex Composters

We offer the fine china Sealand, one-pint, RV-type toilet for use with the Centrex Series composters. These are flushed with a foot pedal, which opens the water and waste valves when pushed down. After flushing, a small amount of water is held in the bowl. These toilets will work just fine with even the lowest water pressure. The 510 Plus model has a standard-size toilet seat, connects to a standard 3" toilet floor flange, and requires a minimum of 10.5" clearance from the wall to the flange centerline. Available in white or bone color.

The Dry Toilet, for use only with Air Flush models, includes a bowl liner that can be removed for cleaning as required. Comes in white only. Add dry toilet pipe extensions as needed to reach the composting chamber.

44823	Sealand 510 Plus China Toilet specify white or bone	$249
44116	Sun-Mar White Dry Toilet	$249
44117	Dry Toilet 29" Pipe Extension	$99

All toilets shipped freight collect when ordered with a Sun-Mar composter; shipped UPS when ordered alone.

Sun-Mar Excel–AC

The fully self-contained, National Sanitation Foundation (NSF)-approved, Excel-AC is our most popular composter for installations with 120-volt electricity. It requires no water connections. Installed in a pro-

tected, moderate-temperature indoor location, this high-capacity unit can handle three to five people full-time, or even more intermittently, making it ideal for year-round or seasonal use. The Excel has a 25-watt fan that runs continuously, and a 250-watt thermostatically controlled heater. Normally the Excel will evaporate all liquids. A 1/2" emergency drain is fitted and should be connected if heavy use, or prolonged power outages are expected. Black bowl liner can be easily removed for cleaning. The drum crank handle is recessed. The

2" central vacuum-type vent pipe exits at the top rear of the unit, and can be installed invisibly through the wall.

The AC fan speed control is an option to install into the fan door cover. Use for composters that overdry the compost, or where slight fan noise is a customer concern. Canada.

Dimensions: 22.5" W x 32" H x 33" D. Requires 48"depth to remove finishing drawer.
Weight: 60 lb.,
Ship weight: 100 lb.
Ships via: freight collect
Ships from: Buffalo, NY
Mfr.'s warranty: 3-yr. parts, 25-yr. body

| 44102 | Sun-Mar Excel-AC Toilet | $1,209 |
| 44824 | AC Fan Speed Control | $40 |

Excel AC/DC Composting Toilet

The Excel AC/DC hybrid was co-developed by Real Goods and Sun-Mar specifically for Real Goods' off-the-grid customers who derive part of their power from generators. The AC/DC is identical to the Excel model, described above, except that it is fitted with an NE drain and with an additional 4-inch NE vent installed next to the Excel's 2-inch vent stack. The AC/DC provides the increased capacity of the Excel unit when a generator is run-

ning, but operates as a non-electric unit when 120-volt electricity is not available. For residential use, a 12-volt stack vent fan is included with this model, for use when the AC/DC is running in a non-electric mode. Stack fans are assembled inside a length of 4-inch vent pipe; they draw 1.4 watts. Canada.

Dimensions: 22.5" W x 32" H x 33" D
Weight: 60 lb.
Ship weight 110 lb.
Ships via: freight collect
Ships from: Buffalo, NY
Mfr.'s warranty: 3-yr. parts, 25-yr. body

44103	Sun-Mar Excel AC/DC Hybrid	$1,309
44803	Sun-Mar 12v/1.4w 4" Stack Fan	$50
44804	Sun-Mar 24v/1.4w 4" Stack Fan	$50
44805	Sun-Mar 110v AC 4" Stack Fan	$60

Sun-Mar Excel–N.E.

The Non-Electric is our most popular composting toilet. It's perfect for many of our customers living off-the-grid and not wanting to be dependent on their inverters. The N.E. uses the same body as the standard Excel, but with a larger 4-inch vent that exits straight out the top, and no fan or heater. The aeration and mixing action of the Bio-Drum, coupled with the help of a 4-inch vent pipe and the heat from the compost, creates a "chimney" effect that helps draw air through the system in a manner similar to that of a woodstove. In full-time residential use, an optional vent stack fan is recommended to improve capacity, aeration, and evaporation. A 1-inch drain is supplied, and must be connected to a greywater or other disposal system. Canada.

Dimensions: 22.5" W x 31" H x 33" D
Weight: 50 lb.
Ship weight: 95 lb.
Ships via: freight collect
Ships from: Buffalo, NY
Mfr.'s warranty: 3-yr. parts, 25-yr. body

44101	Sun-Mar Excel-NE Toilet	$999
44803	Sun-Mar 12v/1.4w 4" Stack Fan	$50
44804	Sun-Mar 24v/1.4w 4" Stack Fan	$50
44805	Sun-Mar 110v AC 4"Stack Fan	$60

Sun-Mar Compact

This is a scaled-down version of the Excel model. For those who don't need the larger capacity of the Excel and have access to 120-volt power, this is the recommended unit. The working components are the same as the Excel: a bio-drum for mixing and aeration, a thermostatically controlled base heater, and a small fan for positive air movement. What makes the Compact model different is a smaller, variable-diameter bio-drum, which allows a smaller overall size; an attractive rounded design; and most importantly, no more footrest! Also, the handle for rotating the bio-drum is now hinged and folds into the body when not being used. Recommended for one person in full-time residential use, or two to four people in intermittent cottage use. Canada.

Dimensions: 22" W x 27.5" H x 33" D
Installation length to remove drawer: 45"
Seat height: 21.5"
Weight: Shipping weight 90 lb., product weight 50 lb.
Ships via: freight collect
Ships from: Buffalo, NY
Vent & Drain: 2" vent pipe & fitting (supplied with unit), 1/2" emergency overload drain (optional hookup).
Electrical Power Requirements: 115-volt, 2.5-amp. 25-watt fan, 200-watt heater with replaceable thermostat. Approximate average demand: 125 watts.
Mfr.'s warranty: 3-yr. parts, 25-yr. body

| 44104 | Sun-Mar Compact | $1,109 |

Sun-Mar Space Saver Composting Toilet Family

Models for Homes, RVs, and Boats

The smallest model from Sun-Mar, the Space Saver is 19" wide by 22" deep, only slightly larger than a regular-sized toilet seat. Stands 28" high and uses a 3" vent stack. It will fit almost anywhere! This ultra-small model can handle two to three people in weekend use, or one person in residential use. Constructed of high-quality fiberglass and marine-grade stainless steel. The 1" overflow drain should be connected to household grey-water, or a mobile holding tank.

Three fairly different models are available in this family. The standard 120VAC Space Saver unit has a thermostatically controlled 120-watt heater and a 20-watt fan. The RV Ecolet unit has the same thermostatically controlled 120-watt AC heater for when AC power is available, plus a 12-volt, 4-watt stack fan. The Marine Ecolet unit is identical to the RV unit in heater and fan, but with a lower back that's sloped inward to accommodate installation against a sloping hull. It is U.S. Coast Guard certified. Both mobile units have heavy-duty mounting brackets, a gasketed finishing drawer, and higher air intakes to accommodate violent motion. A 12-volt, 120-watt heater with separate thermostat is optional for either mobile model. Canada.

Weight: 45 lb.
Ship weight: 80 lb.
Ships via: freight collect
Ships from: Buffalo, NY
Mfr.'s warranty: 3-yr. parts, 25-yr. body

44207	Space Saver 120VAC Toilet	$1,159
44208	Marine Ecolet	$1,159
44111	RV Ecolet	$1,109
44825	Optional 12v/120w Heater	$75

BIOLET COMPOSTING TOILETS

Made in Sweden, BioLet composting toilets are well-designed and some of the best-selling toilets in the world. They offer excellent performance and a compact design for homes that have utility power. Features include:

- Clamshell doors below the seat that hide the compost chamber until weight is put on the seat.
- A composting process that works with gravity. Fresh material enters the top; only the oldest, most completely composted material drops out the bottom, into the finishing tray.
- An adjustable thermostat to control liquid buildup. Won't use electricity unless actually needed.
- A mechanical cassette with all fans, heaters, and thermostats in the top rear makes any service quick, easy, and clean.

BioLet toilets are supplied with an installation kit consisting of 14 feet of 2-inch vent pipe; 39 inches of insulation; 39 inches of outer pipe jacket to cover insulation; vent cap; and a flexible roof flashing. All Biolet models are made in Sweden.

BioLet has done an exceptional job of packaging their products so they will ship via FedEx Ground at very reasonable rates. BioLets usually ship for half or less what other composter brands will cost to deliver.

BioLet produces two basic sizes, which, with various options, produces four different models. The Standard is the smaller size, and is available in a non-electric manual-mix model, an electric manual-mix model, and an electric automatic-mixing model. The 40% larger XL size comes in one electric model with automatic mixing. All models except the non-electric require utility power, or a plentiful RE source such as a strong hydro system.

BioLet XL Model

This is BioLet's top-of-the-line model with about 40% more composting capacity than the Deluxe or Standard models. The thermostatically controlled heating element is 305 watts. The compost-stirring mixer motor runs automatically for 30 seconds every time the lid is lifted and replaced. The quiet 25-watt fan runs continuously, the heating elements run as required by thermostat setting and room temperature. This model is for homes with utility power. The XL can handle four persons full-time, or six persons for intermittent vacation use. Sweden.

Dimensions: 26" H x 26" W x 32" D. Requires 26" x 54" of floor space for the finishing tray to slide out the front.
Weight: approx. 50 lb.
Ship weight: approx. 75 lb.
Ships via: FedEx Ground
Ships from: Massachusetts, $90 to continental U.S.
Mfr.'s warranty: 3-yr.

44109　BioLet XL Model　　　　　　$1,559

BioLet Deluxe Model

The Deluxe model, using BioLet's 40% smaller ABS plastic body, features the top-of-the-line automatic-stirring system, fans, and thermostatically controlled heaters like the XL model above. The heater for this smaller model is 250 watts, and will run as required by thermostat setting and room temperature. The quiet 25-watt fan runs continuously. The Deluxe can handle three persons full-time, or five persons for intermittent vacation use. This model is for homes with utility power. Sweden.

Dimensions: 26" H x 22" W x 29" D. Requires a footprint of 22" x 40" to allow the finishing tray to slide out.
Weight: approximately 50 lb.
Shipping weight: approx. 75 lb.
Ships via: FedEx Ground
Ships from: Massachusetts, $50 to continental U.S.
Mfr.'s warranty: 3 yr.

44124　BioLet Deluxe Model　　　　　$1,495

BioLet Standard Model

This model, BioLet's best-seller, has the 40% smaller ABS plastic body with a thermostatically controlled 250-watt heating element. Compost-stirring is done manually with the handle on top; instead of flushing you simply twirl the handle a few times. The mechanism is geared down ten to one, so little effort is needed. The quiet 25-watt fan runs continuously; the heater element runs as required by thermostat setting and room temperature. This model is for homes with utility power. The Standard model can handle three persons full-time, or five persons for intermittent vacation use. Sweden.

Dimensions: 27" H x 22" W x 29" D. Requires 22" x 40" of floor space for the finishing tray to slide out the front.
Weight: approx. 50 lb.
Ship weight: approx. 75 lb.
Ships via: FedEx Ground
Ships from: Massachusetts, $50 to continental U.S.
Mfr.'s warranty: 3-yr.

44110　BioLet Standard Model　　　　$1,390

BioLet Basic Non-Electric

The Basic Non-Electric model uses BioLet's smaller ABS body and the manual mixing system just like the Standard model. Unlike the Standard model, there is no heater, thermostat, or fan. An overflow fitting is supplied with this model, which needs to be routed to the home's greywater or septic systems. It needs a nice warm environment to live in, as there's no heater to help out the microbes. Temperatures below 70°F will slow it way down. In a friendly environment, this composter can handle up to three persons full-time. An optional 12-volt vent fan is available and recommended. Sweden.

Price for this mocel only includes FedEx Ground shipping anyplace in the continental U.S.

Dimensions: 26" H x 22" W x 29" D. Requires footprint of 22" x 40" to allow the finishing tray to slide out.
Weight: approx. 50 lb.
Ship weight: approx. 75 lb.
Ships via: Freight prepaid (within continental U.S.)
Ships from: Massachusetts
Mfr.'s warranty: 3-yr.

| 44125 | BioLet Basic Non-Electric Model | $995 |
| 44126 | Optional BioLet 12V Vent Fan | $60 |

BioLet Composting Toilet Mulch

BioLet strongly recommends using their own mulch for starting and ongoing maintenance, rather than commercial peat moss. We're increasingly finding that commercial peat moss has been ground almost to dust, and has very little structure or ability to open air channels. This leads to poor composting and material clumping.

BioLet's mulch is formulated specially for composting toilets. It has more tiny air channels, speeds composting by introducing oxygen, and resists clumping. Your composting toilet only needs a cup a day, or less, so it lasts a long time. 8-gallon bag, approximately 12 lbs. USA.

| 44827 | BioLet Composting Toilet Mulch | $39 |

CAROUSEL COMPOSTING TOILETS

The best choice for full-time or heavy use

Suitable for full-time residential use, the Carousel BioReactor system comes in two sizes to comfortably accommodate three to five people year-round, and as many as thirty-two people in a seasonal-use cabin. These systems use the superior batch composting method, resulting in more complete composting and greatly reducing the risk of disease organisms surviving the composting process. It features a round carousel design with four separate composting chambers, each of which is filled in turn. When one chamber is full, usually every two to six months, a new chamber is rotated under the dry toilet. This method has the distinct advantage of not mixing new material with older, more advanced compost; each of the four batches has from six months to a couple of years of undisturbed time to finish composting, allowing a more natural ecological cascading of the process.

Like most composters, it has drying trays under the compost chamber and a venting system with optional fan-assist to ensure no-smell operations. An excess liquid overflow is also provided. With the optional 120VAC heater and vent system, the Carousel is NSF-certified.

Carousel composters can use either a dry toilet with 8" connecting pipe, as sold below, or they can use an ultralow-flush RV toilet, such as the Sealand china toilet sold as a Sun-Mar accessory. Venting can be either passive or fan-assisted, and vent size is adaptable to comply with local codes. The complete and helpful Installation Manual details the many options available during installation.

Warranty is twenty-five years on fiberglass components, two years on all other components. USA.

Carousel Dry Toilet

Carousel Large Bio-Reactor

All Carousel models are 52" diameter, the Large is 51.2" high. With the heater option it handles 5+ people full-time, year-round, or 32 people seasonally.

| 44401 | Carousel Large Bio-Reactor w/heater & fan $4,335 |

Carousel Medium Bio-Reactor

Handles 3 people full-time, year-round, or 15 people seasonally. 52" diameter x 24.8" high.

| 44402 | Medium Carousel Bio-Reactor Only | $2,700 |

Carousel Options

| 44404 | Composting Toilet Starter Kit | $46 |
| 44403 | Dry Toilet with Liner | $317 |

The Storburn Incinerating Toilet

Let's face it, much as they suit our company philosophy of recycling everything, composting toilets just don't work everyplace you might need sanitary facilities. Say, an unheated cabin that sees use a few weekends every winter. Leave it to the Canadians to develop a cold-weather solution. The self-contained Storburn toilet uses no water, no electricity, no plumbing, no holding tank, has no moving parts beyond the lid and seat, and won't freeze. It uses propane or natural gas to incinerate wastes. It can be used forty to sixty times between incineration cycles, which take about 4.5 hours, and leave a sterile ash. Ash needs removal every other cycle. To reduce excitement, the burner cannot be activated while the unit is in use. A packet of anti-foam is added before lighting. The lid is locked, and the pilot light is started with the built-in ignitor. The burner shuts off automatically when finished. It has a written guarantee for no foul odors inside or outside. The cabinet is fiber-glass plastic, top deck is stainless steel with a heavy-duty seat and lid, storage chamber is cast nickel alloy for a long life expectancy. Maximum propane use for a capacity load will be about 10 pounds or 2.5 gallons. Smaller loads use slightly less gas, although partial loads cost almost as much due to preheating the combustion chamber. Chamber capacity is 3 gallons. Requires 17.75" x 31.25" footprint. Height is 53". Unit weighs 170 lb. One-year manufacturer's warranty. Canada.

Vent kit is purchased separately, and consists of 6 inches ID Type L vent pipe (don't use the cheaper Type B pipe). Vent kits consist of pipe, firestop and stack support, flashing, storm collar, and raincap. Mobile home kit is 82" total height; standard vent kit is 115" total height.

A packet of MK-1 Antifoam is required every burn cycle. Aerosol masking foam is optional. The foam helps ensure quick, clean, odorless operation. A sample of each is included.

Specify white (with white seat), or beige (with natural wood seat).
Choose propane or natural gas.
Shipped by truck from Ontario, Canada.

44807	**Storburn Model 60K, propane**	**$3,000**
44806	**Storburn Model 60K, natural gas**	**$2,780**
44808	**Storburn Std. Vent Kit**	**$300**
44809	**Storburn Mobile Vent Kit**	**$240**
44811	**Storburn MK-1 Antifoam, box of 24**	**$20**
44810	**Storburn Masking Foam, 4 Can Pack**	**$25**

CHAPTER 10

Off-the-Grid Living and Homestead Tools

MOST JOBS NEED TOOLS. The better the quality of your tools, the better your craftsmanship is likely to be. The job will also be easier and more enjoyable. Humans are a tool-using, tool-loving species—that's one of the primary traits that sets *Homo sapiens* apart from the rest of the animal kingdom. Good tools improve the quality of our lives and contribute to making them more fulfilling.

Tools for better lives and more efficient living are what this entire *Sourcebook* is about. We've managed to compartmentalize most of them neatly into chapters about solar power, water pumping, composting toilets, or the like, but we've always carried a selection of useful gizmos, such as our solar-powered hat fan, a motion-detecting deer chaser, or our push lawn mower, that defy nice, neat categorization. So here, in Off-the-Grid Living and Homestead Tools, you'll find our wild, diverse, and fun collection of things that defy being compartmentalized. Enjoy browsing!

WOOD-FIRED STOVES AND ACCESSORIES

Due to shipping costs, and concerns about safe installation, Gaiam Real Goods only offers a couple of very special wood burners. The Heartland stoves are nostalgic cookstoves, with all the charm of Grandma's house, and all the quality and refinement of over 100 years of production. They are also available in propane, natural gas, or electric versions, which are more popular than the wood-fired models.

The Snorkel and Scuba stoves are unique and durable wood-fired hot-tub heaters.

Heartland Cookstoves

In an age when nothing seems to last, Heartland Cookstoves are a charming exception. Though modern refinements have been incorporated seamlessly into the design, this cookstove still evokes time-honored traditions. Our black-and-white catalog can't do justice to the bright nickel trim, soft rounded corners, and porcelain enamel inlays in seven optional colors (but we'll be happy to send the factory color brochure).

Two wood-burning models are available. The larger Oval features 50,000 Btu/hr output, a 35" x 26" cooking surface, and a large 2.4 cu. ft. oven. The smaller Sweetheart features 35,000 Btu/hr output, a 29" x 21" cooking surface, and a 1.7 cu. ft. oven. Both feature a large firebox that takes 16" wood, and can be loaded from the front or top. A summer grate position directs the heat to the oven and cooktop, while limiting heating output. Flue size for both models is common 6", and is supplied locally. Options include color porcelain, a stainless steel firebox water jacket for water heating, a 5-gallon copper water reservoir with spigot, a coal grate for burning coal rather than wood, and an outside combustion air kit.

Gas and electric models with all the traditional styling and color options are also available. Please call for options or a free color brochure. Canada.

61113	**Oval Cookstove, Std. White**	**$4,615**
61117	**Sweetheart Cookstove, Std. White**	**$3,790**
OPTIONS (same cost for both models)		
61106	**Color Option (Almond)**	**$50**
61107	**Color Option (Green, Blue, Teal, Black)**	**$199**
61107	**Color Option (Red)**	**$299**
61119	**Stainless Steel Water Jacket**	**$215**
61120	**Coal Grate**	**$450**
61125	**Outside Air Kit**	**$79**

Snorkel Hot Tub Heaters

Snorkel stoves are wood-burning hot-tub heaters that bring the soothing, therapeutic benefits of the hot tub experience into the price range of the average person. Snorkel stoves are simple to install in wood tubs, easy to use, extremely efficient, and heat water quite rapidly. They may be used with or without conventional pumps,

filters, and chemicals. The average tub with a 450-gallon capacity heats up at the rate of 30°F or more per hour. Once the tub reaches the 100°F range, a small fire will maintain a steaming, luxurious hot bath.

Stoves are made of heavy-duty, marine-grade plate aluminum that is powder-coated for the maximum in corrosion resistance. This material is very lightweight, highly resistant to corrosion, and strong. Aluminum is also a great conductor of heat, three times faster than steel. Stoves are supplied with secure mounting brackets and a sturdy protective fence.

Two stove models are offered, the full-size 120,000 Btu/hr Snorkel, or the smaller 60,000 Btu/hr Scuba. The Snorkel is recommended for 6- and 7-foot tubs. The Scuba is for smaller 5-foot tubs. These stoves are for wooden hot tubs only.

We also offer Snorkel's precision-milled, long-lasting Western Red Cedar hot tub kits that assemble in about four hours. A 5-foot tub holds two to four adults, a 6-foot tub holds four to five adults. The 7-foot tub is for large families or folks who entertain frequently. USA.

45400	**Scuba Stove**	**$599**
45401	**Snorkel Stove**	**$799**
45002	**Cedar Hot Tub 5´ x 3´**	**$1,410**
47413	**Cedar Hot Tub 6´ x 3´**	**$1,679**
45208	**Cedar Hot Tub 7´ x 3´**	**$2,210**
45504	**Drain Kit, PVC 1.5"**	**$45**
45501	**6´ Tub 3-Bench Kit**	**$349**

Shipped freight collect from Washington State.

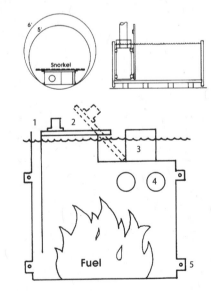

1. Special "Snorkel" air intake.
2. Sliding, cast aluminum, tilt-up door that services the fire box as well as regulates the air intake.
3. Sunken stack to increase efficiency.
4. Heat exchange tubes.
5. Mounting brackets for securing fence and stove.

HOMESTEAD TOOLS

Ecofans
Real Goods product of the decade

When you're heating with wood, a majority of your warmth goes up to the ceiling. The innovative Ecofan solves the problem, propelling much of that heat into your living space—increasing stove efficiency up to 30%. The hotter your stove gets, the faster it runs. Peak

performance occurs at surface temperatures of 400° to 700°F. This durable stove-top fan is very quiet and a temperature-sensitive, bimetal strip on the unit tilts it slightly, preventing overheating by reducing stove contact. The new Ecofan Plus has been enhanced seriously by increasing the heat-absorbing surface area, multiplying its capacity to transfer the fire's BTUs into your environment. The 9" tri-blade unit moves an impressive 150 cubic feet of air per minute. Thermoelectric power from the stove drives the fan with zero electricity needed. Simple and solid technology maximizes the warmth derived from precious timber resources. Made of anodized aluminum, fans won't rust or corrode. Canada.

06-0146 Ecofan   **$109**
17-0180 Ecofan Plus   **$150**

Holy Smokes! Firestarter

Ignitable sticks set fires aglow without combustible chemicals or paper. Crafted from a combination of recycled church candles and wood fiber, Holy Smokes! Firestarters get a fire blazing in minutes. One-quarter stick of starter gets logs going. Habitat for Humanity

receives a donation with your purchase. One package starts more than thirty fires. Scented. USA.

07-9294 Holy Smokes! Firestarter (set of 2)   **$16**

Prometheus Pellet Fuel Basket
Burn pellets in any woodstove

The Prometheus pellet fuel basket is a major breakthrough in wood-burning technology. Pellet fuel is a recycled product, produces far less creosote and ash than cordwood, is easier to handle, more readily available, and in many areas it now costs less. The only problem has been that, until now, it took a specialized stove to burn pellet fuel. The robust all-stainless steel Prometheus fuel basket makes pellet burning possible in any woodstove or fireplace. One-year manufacturer's warranty. USA.

Model selection

For safe and reliable heating, it's important to choose the correct size Prometheus. Just multiply the inside width by the inside depth of your stove and compare your figure to the Minimum Hearth Area column on the chart. Your hearth area must be equal to or larger than this minimum area.

Item #	Model #	Woodstoves smallest hearth area allowed	Fireplaces min. opening width masonry	Fireplaces min. opening width mftd.	Size (H x W x D)	Price
01-0299	6117	192 sq."	18"	21"	$9^5/8$" x $11^5/8$" x $8^1/4$"	$199
01-0300	6147	244 sq."	21"	24"	$9^5/8$" x $14^3/4$" x $8^1/4$"	$209
01-0301	6177	292 sq."	24"	26"	$9^5/8$" x $17^5/8$" x $8^1/4$"	$229
01-0302	6227	377 sq."	29"	$37^1/2$"	$9^5/8$" x $22^3/4$" x $8^1/4$"	$249

Pellet fuel can be added to the basket chambers at any time for continuous burning.

Full pellet basket may burn 2 to 10 hours without refueling depending on the size of basket.

Angled base to allow viewing of the flames and easier starting.

Sloped top provides optimum fire viewing and facilitates reloading.

Ash falls through openings at the bottom.

Fuel can be added with as little as one inch of pellet fuel remaining in the basket to provide extended burn time.

Air flows through channels in the basket for an efficient burn.

Adjustable legs.

YARD AND GARDEN TOOLS

Our Gaiam Real Goods color catalogs offer our widest selection of gardening tools and gizmos, but we have a few perennial favorites that deserve year-round space here in the *Sourcebook*. To order one of our periodic Gaiam Real Goods catalogs call 1.800.919.2400 or check out our website: www.realgoods.com.

Mighty Mule Gate Openers

These amazing gate openers are designed for home, ranch, or farm gates up to 16 feet in length or weighing up to 350 pounds per leaf. Designed for do-it-yourself installation with a video. Powered by a battery that's immune to power outages, it can be charged by the included AC transformer or by the optional solar panel with mounting arm. Closes automatically with an adjustable 0 to120 second delay. Kit includes opener arm, AC transformer, 12v battery, control box, receiver with 10' cable, mounting hardware, power cables, warning signs, and one entry transmitter. The dual gate kit includes a second opener arm and control cable.

Options include: Push-to-Open kit converts opener to swing out from property, outdoor digital keypad will accept up to 15 entry codes, an automatic gate lock (a MUST to actually lock the gate), and a buried gate opening sensor to automatically open the gate for outbound vehicles. One-year manufacutrer's warranty. UL-listed. USA.

63-126	Single Gate Opener	$595
14-0245	Dual Gate Opener	$925
14-0246	Solar Panel Kit	$175
63-144	Extra Transmitter	$33
14-0247	Key Ring Transmitter	$33
63-146	Digital Keypad	$60
63-147	Automatic Gate Lock	$199
64-500	Push-to-Open Kit	$18
14-0248	Buried Gate Opening Sensor	$209

Marie's Poison Oak Soap

This truly amazing product prevents poison oak or poison ivy outbreaks even after exposure. When used after contact, most people never get a rash at all. Using after a rash has started greatly accelerates healing. The mechanical action of the soap pulls poison oak and poison ivy oils off the skin,  while herbs with a natural antihistamine action help stop itching and redness. Other herbs promote healing of the skin. A completely natural blend of oils, herbs, extracts, clay, oatmeal and glycerin soap. No dyes, artificial ingredients or scents. The Oregon Highway Dept. buys over 2,000 bars every year for its employees. USA.

08-0346 Marie's Poison Oak Soap, 3 bars $14

Parmak Solar Fence Charger

 The 6-volt Parmak Fence Charger will operate for 21 days in total darkness and will charge up to 25 miles of fence. It includes a solar panel and a 6-volt sealed, leakproof, low-internal-resistance gel battery. It's made of 100% solid-state construction with no moving parts. Totally weatherproof, it has a full two-year warranty.

63128	Parmak Solar Fence Charger	$192
63138	Replacement Battery for Parmak, 6V	$39.95

Push Reel Lawn Mower

We've found a great manual lawn mower made by the oldest lawn mower manufacturer in the U.S. This mower is safe, lightweight, and very easy to push. It's perfect for small lawns and hard-to-cut landscaping. The reel mower provides a better cut than power  mowers, keeping lawns healthy and green, and it doesn't create harmful fumes, greenhouse gases, or noise pollution. The short grass clippings from the mower can be left on the lawn as natural fertilizer, or you can purchase the optional grass catcher and add the grass to your compost pile. Cutting width is 16 inches, cutting height is adjustable from 1½" to 2¼". It has 10-inch rubber wheels that adjust to three different heights, five blades, and a ball-bearing reel. USA.

14-0187	Lawn Mower	$129
14-0188	Grass Catcher	$34

Solar Mulching Mower

Envision power mowing your lawn without oil, gas, pull-cords, exhaust or engine roar ever again. Our Solar Powered Mulching Mower makes that dream come true with a quiet, thoroughly dependable Black & Decker electric mower that runs on the sun. It keeps lawns looking sharp with a 19" mulching blade that sends nutrient-laden cuttings back to your lawn for increased vitality. Quick-set height allows you to customize your cut. Includes a rear-bagger for mulching leaves. 3'H x 2'W x 5'L. 60 lbs. USA.

14-8008 Solar Mulching Mower $795
Allow 4 to 6 weeks for delivery. No express delivery available.

Put Earthworms to Work for a Healthier Garden

With our user-friendly composting system, a team of red worms digests kitchen wastes into worm castings—an organic, nutrient-rich garden amendment. The perforated stacking trays

separate the worms from their castings automatically, making it very easy for you to gather the worm castings left behind. The handy spigot allows you to capture the "worm tea"—a rich liquid amendment your plants will love. Sturdy, odorless, and pest resistant, this earth-friendly composting system is made from 100% post-consumer recycled plastic. Used indoors or out, it comes with complete instructions and 1 pound of worms. Measures 29"H x 20"D. Use the worm card to order extra or resupply in the spring. Can/Australia, worms/USA.

14-0176 Can-O-Worms $119
Requires UPS ground delivery.

14-0156 Worm Card (1 lb. worms) $37

Spinning Composter

This is the ideal composter for urban settings; it's compact enough to store on a balcony or patio, producing up to 7 cubic feet of fertile compost in as little as thirty days, without odors close neighbors might complain about. Meets all the requirements for sustainable living: It's manufactured in the United States of 50% post-consumer plastic, performs outstandingly, has a long useful life, and is totally recyclable. A liquid fertilizer reservoir is built into the base, designed to collect the nutritious liquid run-off for use on indoor plants, window boxes, or in the garden. It's an elegantly simple tumbler design that rolls with minimal effort on a stable base with eight built-in rollers. Secure latching door is hinged for filling, removes for emptying. Arrives assembled. 31" H x 26"L x 21"W. USA.

14-0020 Spinning Composter $129
Requires UPS ground delivery.

Our Most Efficient Composter

Today's enlightened gardener recognizes composting as a simple, efficient way to reduce waste and create a nutrient-rich amendment for plants. Made of 100% recycled plastic, this unobtrusive bin traps solar heat to accelerate the production of compost. Designed for convenience, it has a rodent-deterrent screen, adjustable air vents, an easy-access hinged lid for adding materials, and a sliding bottom door for compost removal. It snaps together in minutes—no tools are required. Includes a guide to successful backyard composting, and has a five-year manufacturer's warranty. 10.2-cubic-foot capacity, 40" H x 23"W. Canada.

14-9002 Garden Composter $70
Requires UPS ground delivery.

Turn and Aerate Compost More Easily

This perfectly simple compost turner is useful for any stationary compost pile, bin, or system, large or small, but it's especially well suited for bins with small top openings that make it difficult to manipulate a shovel or fork for the necessary turning and aerating of your compost. Just plunge the turner into the heart of the compost pile, then twist and pull. The folding wings at the tip open up to aerate and turn your compost. Won't rust or jam. Robust steel construction, with plastic handle grips. 36.5" long, handle is 7.75" across. Mexico.

14-0174 Compost Turner $20

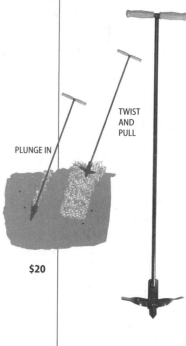

PLUNGE IN

TWIST AND PULL

Compost Activator

Starts compost fast, reduces odors, and deactivates weed seeds and pathogens. Combines high- and low-temp bacteria so it's effective in most climates. Instructions included. 2 lbs. USA.

14-0147 Compost Activator   **$18**

How to Grow More Vegetables

First published in 1974, this book has become the go-to guide for anyone trying to create or sustain a home garden. Author John Jevons shows how to use Alan Chadwick's biointensive techniques to reduce daily watering and general maintenance —from the beginning stages of setting up the garden beds to creating insect harmony. Includes charts and illustrations. 192 pages, softcover. USA.

21-0345 How to Grow More Vegetables   **$17.95**

The Perfect Sustainable Tree

In Asia they believe that planting a Paulownia tree attracts the good luck–bearing Phoenix. Paulownia trees can be the foundation for ecologically sustainable communities. Fast-growing deciduous trees, they provide shade, valuable hardwood, and create ideal microclimates for growing a wide variety of organic food crops. Imagine the difference if everyone planted a Paulownia tree. An eight-year study by the International Development Research Centre established that oxygen and relative humidity increased 10 to 15% as a result of these incredible trees. Their ability to grow and thrive in areas that have severe dust, smoke, and sulfur creates significantly reduced air pollution.

ORNAMENTAL BENEFITS
- Shade for home starting in just 16 to 18 months
- Beautiful edible lavender flowers
- Fully matured to 30 feet in three years
- Gentle scent like Jasmine

FOOD SOURCE BENEFITS
- Leaves are 18 to 20% protein with high nitrogen— excellent fodder or fertilizer
- Flowers are tasty and beautiful in salads
- Bees love the flowers—excellent source of organic honey

FORESTRY BENEFITS
- Grows 70 to 80' in ten to fifteen years
- Drought resistant, disease resistant
- Hardwood does not warp twist or crack, mahogany-like grain
- Lightweight with excellent weight/strength ratio
- Great for furniture, musical instruments, doors, moldings, poles, paper, and more
- Fire-resistant and high-water tolerant
- Regenerates from the stump after harvesting

AIR AND WASTE WATER BENEFITS
- Significantly reduces waterway contamination and odors
- Increases nitrogen uptake
- Absorbs dust, smoke, sulfur, and other chemicals

Kawakamii variety grows 30 feet high in three years; tolerates 0° to 130°F temperature range. Elongata grows 75 feet high in ten years; tolerates -5° to 130°F temperatures. Six- to twelve-inch seedling trees are priority-mailed early in the week with complete planting instructions. They will go into 1-gallon pots for a few months before permanent planting. Trees are guaranteed to thrive or will be replaced at no charge. USA.

81441	Kawakamii 30' high			$29 ea.
81442	Elongata 75' high			$29 ea.

No. of Trees	6+	12+	50+	100+
Price each	$19	$15	$12	$9

The World's Most Effective Deer Deterrent

Hook the Scarecrow up to your garden hose and stake it. The smart motion sensor detects an intruder (up to 35 feet away), and sends a full-pressure blast of water right at it. It switches off immediately, using a conservative two cups of water per discharge. You preset the detection area (protects a 1,000-sq.-ft. area) and sensitivity to prevent triggering by household pets. Sensor sensitivity is dampened automatically in windy conditions to avoid false triggers. Up to 1,000 discharges on a single 9-volt battery (not included). Reinforced nylon stake with sturdy step and hose flow-through. 24" High overall. Canada.

16-0023 Scarecrow Sprinkler   **$89**

We urge caution if planting these trees in the southeastern United States, where they may propagate without human help, and have been identified as an invasive species.

Bats Are the Ultimate House Guests

Mosquitoes don't have a chance around bats. Because they devour up to 600 night-flying insects in an hour and rarely bother humans, it makes sense to welcome bats to take up residence nearby. Made of Western Red Cedar sawmill trim, this handsome slatted bat conservatory provides shelter for approximately 40 bats. Measures 24.5" x 16" x 5.25"; attaches easily to your home, barn, or nearby tree. Put out the welcome mat for our native North American bats, and these fascinating creatures will reward your hospitality year after year. No more nasty mosquito repellents! Learn more about bats with Merlin Tuttle's 96-page book. USA.

16-0060 Bat Conservatory	$50
21-0306 *America's Neighborhood Bats*	$11

Solar Ant Charmer

The red imported fire ant was introduced accidentally in Mobile, Alabama, from South America in the 1930s. The species spread through the lower United States and continues to expand northward and westward into areas with mild climates and plentiful food. Fire ants attack animals, people, and even electrical devices. They cause failures in air-conditioning units and traffic control boxes. Fire ants kill livestock and also have been known to kill humans. The Solar Ant Charmer controls fire ants without poisons by eliminating a major part of the ant colony's defenses. It collects thousands of fire ants in just minutes. Fire ant poisons pollute the Earth and fire ant baits take up to six weeks to become effective. Use the Charmer over and over again for years without replacing any batteries—the solar cell does it all! USA.

63853 Solar Ant Charmer	$119

Solar Hat Fan

A practical solar gadget that really cools you down on a hot day! It clamps on to your favorite hat (stiffer-type brims and baseball caps are recommended) and gets to work harvesting enough solar energy to directly power a working fan. Features a small photovoltaic panel with an adjustable mount. We sold these solar hat fans to a roofing company, whose workers can't be without them as they cool down from the 100°+ temperature. Measures 2.75" L x 2⅛"W x 2¼"H. China.

01-0343 Solar Hat Fan	$10
2 or more	$8 each

Solar Mosquito Guard

We've received countless raves from satisfied Solar Mosquito Guard owners, all with the same theme: "These amazing things really work. We're mosquito-free, while our friends are swatting away." Our pocket-sized devices put out an audible, high-frequency wave that repels some mosquito species. No scientific evidence proves that mosquitoes can hear, yet they really seem to work! There is an on/off switch so you don't have to activate until the mosquitoes arrive. Battery fully recharges in three hours of sunlight. China.

90419	Solar Mosquito Guard	$10
	3–5	$9 each
	6 or more	$8 each

Natural Biological Mosquito Control Liquid

Use predation, not pesticides, to kill mosquitos in ponds and other sources of standing water. Each teaspoon of Mosquito Control treats up to 540 square feet of water for up to 14 days with non-toxic microbes that feed on mosquito larvae. Fish, plants, amphibians, and other aquatic life aren't affected, and its application is safe for areas used by people, pets, livestock, and wildlife. Mosquitos can't develop resistance to this completely organic method. 12 oz. container treats one medium pond. USA.

40-0076 Natural Biological Mosquito Control	$36

Medieval Bug Catcher

Much more attractive than those common plastic bug traps, our blown-glass Medieval Bug Catcher works with plain sugar water instead of potential toxins. Used since the Middle Ages because it works, the design is elegantly simple and aesthetically pleasing —and is made of 100% recycled bottle glass. It's about as humane as a pest control device can be—effective on wasps, hornets, yellow jackets, and mosquitoes. Designed to sit flat, not hang. 12"H x 7" diameter. Spain.

16-0057 Medieval Bug Catcher	$26

Recycled Tires Make Long-Lasting Buckets

Our buyers say these are the best buckets ever to bring compost to the garden or to get the car washed. It's the kind of tool that's irreplaceable when you need it, and having the right one on hand always makes your job easier. These long-lasting buckets are made of recycled tires and waste plastic, using a manufacturing process that is gentle to the Earth. Unlike metal buckets, they won't corrode or buckle, and the handsome galvanized steel handles hold up to the hardest knocks. To suit every need, we offer them in a great value set of three: 8-quart, 12-quart, and 18-quart. A product line that delivers where the rubber meets the road. USA.

06-9341 Tire Buckets (Set of 3)   **$30**

Recycled Tire Door Mats

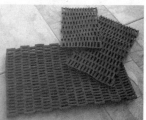

Made from sliced-up used tires, these mats will last a lifetime. Years of scuffing boots will only soften the surface and make them better mats. They clean easily as accumulated material simply falls through them when you shake them out. They work great as door mats, or as anti-fatigue mats in front of your tool bench or laundry folding table. Held together with heavy-duty galvanized steel wire. Three sizes available. USA.

18-0064 13.5" x 23"		**$17**
18-0065 18.5" x 28"		**$25**
18-0066 24" x 36"		**$45**

KITCHEN AND HOME TOOLS

This is a truly eclectic mix of products. Now we're really getting down to things that defy compartmentalization! The only common thread here is that these are all things you would use inside your home (probably).

SEVENTH GENERATION PRODUCTS

As the largest catalog retailer of Seventh Generation products, Gaiam has been providing safer solutions for healthier homes since 1989. With Seventh Generation's industry-leading nontoxic cleaners that use biodegradable and vegetable-based ingredients instead of synthetic chemicals—and household papers and trash bags that use recycled materials to save resources—we help you do your best for both your family and the environment.

Are you really making a difference? Yes! In the past two years alone, Gaiam customers have saved:

- 20,897 living trees, 53,953 cu. ft. of landfill, and 7,476,656 gallons of water by purchasing Seventh Generation recycled paper products!
- 328 barrels of petroleum and 2,626 lbs. of chlorine from entering the environment by purchasing Seventh Generation laundry products.

Dishwasher Detergent

While many commercial products contain phosphates and chlorine, these Automatic Dishwasher Detergents don't. The powder uses enzymes and non-chlorine bleach to get dishes clean, while the gel uses a unique system of polymers. 50 oz. A rinse aid is recommended with hard water. USA/Canada.

POWDER

28-3026 2 Boxes	**$8.95**
28-3024 4 Boxes	**$16.50**
28-3022 Case—12 Boxes	**$47.95**

GEL

28-3028 2 Bottles	**$9.95**
28-3029 Case—6 Bottles	**$26.95**

Dish Liquid

Mild on your hands and gentle on the environment, this Dish Liquid cuts grease and loosens caked-on food with the power of coconut oil, not petroleum. No artificial colors and lightly scented with natural fragrances. Not for use in automatic dishwashing machines. USA/Canada.

NATURAL CITRUS

28-3014 Two 28 oz. Bottles	**$5.95**
28-3015 Case—Twelve 28 oz. Bottles	**$31.95**

GREEN APPLE WITH ALOE VERA

28-9007 Two 28 oz. Bottles	**$5.95**
28-3005 Case—Twelve 28 oz. Bottles	**$31.95**

FREE & CLEAR DISH LIQUID (no fragrance)

28-3008 Two 28 oz. Bottles	**$5.95**
28-9034 Two 48 oz. Bottles	**$10.95**
28-9033 Case—Six 48 oz. Bottles	**$28.95**

Recycled Plastic Trash Bags

Sturdy and dependable Trash Bags are made from 100% recycled plastic with a minimum 30% post-consumer content, and require 25% less energy and 90% less water to produce than virgin plastic bags. The 13-gal. white bags are ideal for kitchen waste while the 2-ply 30-gal. bags fit snugly in 30 gal. containers. Use the 33-gal. bags for 32 -al. containers. USA.

13 GAL. TALL KITCHEN BAGS (30/box)

28-6006	3 Boxes	$9.50
28-6004	6 Boxes	$17.95
28-6002	Case—12 Boxes	$33.50

30 GAL. TRASH BAGS (20/box)

28-6016	3 Boxes	$9.50
28-6014	6 Boxes	$17.95
28-6012	Case—12 Boxes	$33.50

33 GAL. LARGE TRASH BAGS (15/box)

28-6026	3 Boxes	$9.50
28-6024	6 Boxes	$17.95
28-6022	Case—12 Boxes	$33.50
28-9002	Trash Bag Sampler (1 box ea. size)	$9.50

Drawstring Trash Bags

As an extra feature, we've added drawstrings to make closure a cinch—literally. Made of 65% recycled plastic with a minimum 20% post-consumer content. Choose 13-gal. Tall Kitchen or 30-gal. Drawstring. USA.

13-Gal. Drawstring Tall Kitchen Bags (20/box)

28-6008	3 boxes	$9.50
28-6009	6 boxes	$17.95
28-6010	Case—12 boxes	$33.50

30-Gal. Drawstring Trash Bags (10/box)

28-6018	3 boxes	$9.50
28-6019	6 boxes	$17.95
28-6020	Case—12 boxes	$33.50

Laundry Liquid

All-temperature Laundry Liquid uses powerful coconut-oil-based detergents to remove dirt, grime and stains effectively in cold or warm water. Fresh, natural citrus scent; convenient measuring cap. 33 loads per 100 oz. bottle; 16 per 50-oz. Canada.

28-2034	Two 50-oz. Bottles	$10.95
28-2024	Two 100-oz. Bottles	$18.50
28-2022	Case—Six 100-oz. Bottles	$49.95

Chlorine-Free Bleach

Chlorine-free bleach is safe for most colors; testing recommended for naturally dyed fabrics. 16 loads per 48-oz. bottle. Canada.

28-2044	Two 48 oz. Bottles	$7.50
28-2042	Case—Six 48 oz. Bottles	$19.95

Fabric Softener

Ultra-concentrated Fabric Softener uses renewable canola oil (rapeseed) to make laundry fluffy, downy soft, and static-free. And with 32 loads per 32-oz. bottle, it's also economical. Lavender scent. Biodegradable. Canada.

28-2068	2 Bottles	$8.95
28-2069	Case—6 Bottles	$25.50

Laundry Powder

Formulated with enzymes and chlorine-free bleach to tackle tough stains without harsh chemicals. Comes with a convenient measuring scoop. 42 loads per 112-oz. box. 18 loads per 48 oz. box. Canada.

28-2012	Two 48-oz. Boxes	$10.95
28-2004	Two 112-oz. Boxes	$21.95
28-2002	Case—Four 112-oz. Boxes	$38.95

Bathroom Tissue

Two-Ply Bathroom Tissue feels cushiony-soft and smooth against your skin. Each roll is made of 100% recycled paper with a minimum 80% post-consumer content and is safely bleached with sodium hydrosulfite. Septic system safe. 500 2-ply sheets. 58.6 sq. ft. USA. Safe for your septic system, 1-Ply Bathroom Tissue is made from 100% recycled fibers with a minimum 80% post-consumer content that are bleached with environmentally safe sodium hydrosulfite and hydrogen peroxide. With 1,000 sheets per roll, our 1-ply bathroom tissue lasts an extra long time. 116 sq. ft. Canada.

2-PLY BATHROOM TISSUE

28-1016	12 Rolls	$9.50
28-1014	Half Case—48 Rolls	$33.95
28-1012	Case—96 Rolls	$64.95

1-PLY BATHROOM TISSUE

28-1006	12 Rolls	$11.95
28-1004	Case—48 Rolls	$42.50

Facial Tissues

Facial Tissue made from 100% recycled paper with a minimum 20% post-consumer content and bleached with environmentally safe sodium hydrosulfite and hydrogen peroxide. USA.

100-COUNT (LEAF PRINT BOX)

28-1046	8 Boxes	$11.50
28-1044	Half Case—15 Boxes	$19.50
28-1042	Case—30 Boxes	$37.50

175-COUNT (FLORAL PRINT BOX)

28-1047	9 Boxes	$18.95
28-1048	Half Case—18 Boxes	$33.95
28-1049	Case—36 Boxes	$66.95

Brown Paper Towels

Brown towels are made from unbleached, undyed, 100% post-consumer "low-grade" recycled paper. Each roll packs more than twice the sheets of most brands, cutting down on packaging and fuel waste. 120 2-ply sheets per brown roll. 11" x 9". 123.75 sq. ft. USA. Brown.

28-1036	4 Rolls	$5.95
28-1034	8 Rolls	$11.95
28-1032	Case—15 Rolls	$20.95

Natural Napkins

The convenience of Paper Napkins with the lowest possible impact on the environment. Natural Napkins are made from an unbleached, dioxin-free 100% post-consumer mix of newspaper, corrugated boxes, and phone books. 1 ply brown. USA/Canada. (6" x 6H" folded, 500/pk)

28-1056	2 Packs	$8.50
28-1054	6 Packs	$23.95
28-1052	Case—12 Packs	$44.95

Baby Wipes

Lightly scented, extra-thick Baby Wipes are moistened with gentle aloe vera and water—not irritating alcohol. Just one does the job of two to three ordinary wipes. Bleached with hydrogen peroxide and chlorine dioxide, which are much less harmful than chlorine. 80 wipes per travel/refill pack or tub. USA.

28-7012	Tub With 1 Refill Pack	$7.95
28-7013	With 3 Refill Packs	$14.95

Toilet Cleaner

You don't need bleach or harsh chemicals for routine cleaning. Just squirt, brush, and flush. This vegetable-based Toilet Cleaner gently loosens stains like rust and hard water "scale" without polluting the environment. 32 oz. bottle. Canada.

28-4006	2 Bottles	$5.95
28-4004	Half Case—6 Bottles	$17
28-4002	Case—12 Bottles	$31.95

Pro-Ketch Mouse Trap

Protect your home and give yourself peace of mind with the humane trap professionals use. Made from tough galvanized steel, Pro-Ketch captures multiple mice without ever resetting. Placing bait inside entices mice to enter. As they walk in, the teeter-totter-style ramp lowers to allow mice in, but snaps back once they step off so they can't exit. Clear view lid allows easy monitoring. Simply take trap far from your home, open lid and let mice scurry out. USA.

16-0083 Pro-Ketch Mouse Trap	$17

BugZooka!

This battery-free wand generates ten times the suction power of motorized models to capture insect invaders humanely for outdoor release. Just push the bellows to compress some air, extend the telescoping arm up to 24", and push the trigger. Wasps, spiders, flies, and more are whisked inside a removable catch tube for transport outside. No unpleasant handling or squashing. No poisonous sprays. Kids can even get an educational close-up look at whatever you've caught before it's set free. Includes two clear catch tubes and wall mount. 38"L. 2 lbs. Taiwan/China.

16-0267 BugZooka!	$30

Orange Guard

Water-based Orange Guard relies on natural orange peel extract to kill insects on contact and repel others for weeks, including cockroaches, ants, and fleas. Certified for use in organic production. Great for both the garden and kitchen. Not hazardous to insect-eating birds or lizards. USA.

16-0041 Trigger Spray Bottle (32 oz.)	$12
Heavy-Duty Bottle with Sprayer(1 gal.)	$28

12V Blender Makes Off-the-Grid Margaritas

Make your margaritas without AC power. This robust blender comes with a cigarette lighter plug for 12-volt DC operation. Grind ice, make smoothies, etc., from your boat, RV, tailgate, or 12-volt cabin. The American-made Waring unit has a 15-foot-long fused cord and is built for serious heavy-duty use with a steel base. It comes with a five-year warranty on the motor (one year on the whole unit). It is rated for nonstop ice grinding, and will surge to around 12 amps starting up under a heavy load. More typical consumption is under 100 watts. USA.

63314   Waring 12-Volt Blender $145

Hand-Cranked Blender

The human-powered Vortex blender is perfect for picnics, tailgate parties, camping, and for educating your children

about energy. It features a 48 oz. (1.5 liter) graduated Lexan pitcher with an O-ring sealing top, a removable pour spout that's a 1-oz. shot glass, a brushed stainless steel finish, soft stable rubber feet, an ergonomic handle, two-speed operation, and a C-clamp for stable operation. The base fits inside the pitcher for easy packing. High-speed operation is noisy, your camping neighbors will know you're making margaritas! Weighs 4.5 lbs. China.

63867   Hand-Cranked Blender $88

Stove-Top Waffle Maker

Creates delicate, crisp waffles with extra deep pockets to capture your favorite toppings. And faster than an electric waffle iron! Long-lasting, heavy cast aluminum with a non-stick coating and cool bakelite handles. 16H" x 8H"x 1H", baking surface 7G" x 7G".

51647   Stove-Top Waffle Maker $59

Old-Fashioned Popcorn Popper

Bring a taste of nostalgia to evenings spent at home by making light, fluffy popcorn in a way the whole family can enjoy. Made from durable stainless steel, this Popper takes a cue from yesteryear and utilizes a wooden crank handle that you turn to keep the corn from sticking or burning. Works on electric or gas ranges, fireplaces and campfires. Includes a butter drip cup and window so you can watch the kernels burst. 8G"H x 16"W x 9"D. China.

09-0215 Old-Fashioned Popcorn Popper $49

Bag Dryer

Crafted from sustainably harvested birch and ash woods, our Counter Top Plastic Bag Dryer makes it easy to dry and reuse plastic bags. Includes a hanging hook. Folds for storage. 14"H x 7G"D. 6.3 oz. USA.

06-0007 Bag Dryer $19

Stainless Steel Dish Rack

Washing a few dishes by hand consumes far less energy than an automatic dishwasher. Now you can air dry your dishes in sleek, sustainable style and never buy a dish drainer again. Constructed from stainless steel, our heavy-duty Dish Rack won't chip or rust like chrome or split like wood versions. Includes detachable cutlery basket. Folds flat for out-of-the-way storage. 9"H x 24"W x 12"D. 5 lbs. China.

09-0197 Stainless Steel Dish Rack $38

Countertop Compost Bin

Put kitchen scraps conveniently out of sight—and significantly reduce odors—using our easy-to-clean compost bin before transporting to your garden compost. Its trim design fits handily under the sink or at the back of the counter. Built-in carbon-activated filters effectively prevent odors for up to six months. Made of durable plastic; lid snaps snugly to eliminate spills. 9.6 qt., 12"H x 9"W x 8"D. Green color. Canada.

09-0068 Compost Bin (9.6 qt.) $17
09-0079 Replacement Filter set of 3 $6

Kitchen Compost Crock

Our fully glazed ceramic crock looks beautiful and prevents stains and odor absorption. The 1.25-gallon interior holds a week's worth of scraps, and a lid with an activated carbon filter traps odors and lets air circulate to prevent rotting. The removable stainless steel handle makes carrying easy. Dishwasher safe. Replacement filters available in a set of 4. 13I"H x 7H"W x 7"D with lid. 6 lbs. China.

09-1051 Compost Crock	**$42**
09-9023 6 Replacement Filters	**$10**

Rotary Sweeper

Effectively cleaning floors and low-pile rugs without power, our Rotary Sweeper is a cleaning solution that exceeds sustainability standards. With a low profile, it easily reaches beneath furniture. Includes simple bottom-mount dustpan, exceptional design and craftsmanship. Lightweight and compact, it weighs only 5 lbs. 11" square x 37"H. Germany.

06-0482 Rotary Sweeper	**$70**

White Wizard

Completely safe for any surface or fabric from leather to upholstery, White Wizard is an odorless, biodegradable, nontoxic formula that virtually erases old or new stains like grease, blood, pet stains, and more, even on dry-clean-only items. Don't let a stain ruin anything. Just a dab of the wizard and it's gone! 10 oz. tub. USA.

06-0028 White Wizard	**$8.50**
06-0084 2 tubs	**$15.50**

Air Refreshers

Our nontoxic Air Refreshers contain volcanic minerals with a natural ionic charge that completely erase problem odors. Simply hang the packs up and walk away. A single pack deodorizes up to 600 sq. ft. of space. Placed in the air return of your central A/C or heating system, Refresh-A-House (not shown) keeps 1,000 sq. ft. naturally odor-free. Recharge in sunlight for a lifetime of use in any climate. USA.

06-0068 CLOS	**Closet Air Refresher (four 6 oz. and two 32 oz. packs)**	**$35**
06-0068 BAS	**Basement Air Refresher Set (red, 32 oz.)**	**$12**
06-0068 CAR	**Car Air Refresher (blue, 32 oz.)**	**$12**
06-0068 PET	**Pet Refresher (white, 32 oz.)**	**$12**
06-0082	**Refresh-A-House (light blue, 32 oz.)**	**$12**

All-Purpose Cleaner

Easily wipe off even the toughest grease and stains on countertops, walls and more with the All-Purpose Cleaner that uses renewable, vegetable-based surfactants —not toxic or petroleum-based chemicals. Fragrance-free, biodegradable, and nontoxic. Testing by an independent lab showed that it works as well as, or better than, leading national brands. 32 oz. trigger-spray bottle. USA.

28-0032 3 Bottles	**$12.50**
28-0033 Half Case — 6 Bottles	**$23.95**
28-0034 Case — 12 Bottles	**$44.95**

Glass and Surface Cleaner

Most traditional glass cleaners use solvents such as alcohol, glycol ethers, or ammonia, all respiratory irritants. Because this Glass and Surface Cleaner is fragrance-free and uses vegetable-based surfactants as the cleaning agent, it leaves behind no caustic fumes to inhale. All you'll notice after using this biodegradable, hypo-allergenic solution is a sparkling, streak-free shine. 32 oz. trigger-spray bottle. USA.

28-0002 3 Bottles	**$12.50**
28-0003 Half Case — 6 Bottles	**$23.95**
28-0004 Case — 12 Bottles	**$44.95**

Citra-Solv Natural Citrus Solvent

Citra-Solv is one of our favorite products. It will handle nearly all your cleaning needs. It dissolves grease, oil, tar, ink, gum, blood, fresh paint, and stains, to name just a few. It replaces carcinogenic solvents, such as lacquer and paint thinner, toxic drain cleaners, and caustic oven and grill cleaners. You won't believe how well it cleans your oven! You can also use Citra-Solv in place of soap scum removers, which use bleaching agents. Citra-Solv is composed of natural citrus extracts derived from the peels and pulp of oranges. It is highly concentrated and can be used on almost any fiber or surface in the home or workplace, except plastic. Citra-Solv is 100% biodegradable, and the packaging is 100% recyclable. USA.

06-0063 32oz Citra-Solv (quart)	**$17**
06-0063 1gal Citra-Solv (gallon)	**$49**

Lime-Eater

Lime-Eater eats away hard water stains, calcium deposits, soap scum and rust from sinks, tubs, tile and glass—even fiberglass and cultured marble—without hazardous solvents. Non-abrasive and Green Cross certified. Great for stainless steel and cooking gear, too! 21 oz. trigger-spray bottle. USA.

06-9015 Lime Eater 2 bottles   **$14**

Mildew Stain Away

The biodegradable formula used in Mildew Stain Away washes away mildew and its stains in the bathroom without chlorine or other toxic chemicals. Safe for any surface, even book covers. Just spray and rinse. Ideal for pre-treating shower curtains. Convenient 32 oz. trigger-spray bottle. USA.

06-0046 Mildew Stain Away   **$14**

Smells Be Gone

This product tackles heavy-duty odors like tobacco smoke or skunk spray; it is nontoxic, odorless, nonallergenic, and uses a deodorant system safe for a nursery or kitchen. USA.

06-0061 12oz Smells Be Gone 12 Oz.   **$13**
06-0061 24oz Smells Be Gone 24 Oz.   **$23**

Hardwood Floor Cleaning Kit

Used by professional floor refinishers, our nontoxic Hardwood Floor Cleaning Kit is safe for polyurethane finished and factory pre-finished wood floors. Also works on no-wax vinyl and ceramic tile. Includes a mop (not shown) with a 3-piece pressure-lock pole, 32 oz. cleaner, and an instant-on Velcro micro-fiber mop pad that lasts for up to 300 washings. USA.

06-0479	**Hardwood Floor Cleaning Kit**	**$34**
06-0470	**Hardwood Floor Cleaner**	
	32 oz. Trigger Spray Bottle	**$9**
	42 oz. Refill Bottle	**$11**
06-9481	**Replacement Mop Pad**	**$22**

ChemFree Toilet Cleaner

Keep your toilet clean and clear without toxins. First scrub your tank, then drop in ChemFree. Special "mineral magnets" kill bacteria on contact and prevent mineral stains, mildew, and fungus build-up. Lasts five years or 50,000 flushes. USA.

06-9029 Set of 2   **$16**

Septic Treatment

Imagine a septic treatment so effective you could significantly reduce costly, time-consuming tank pump-outs. Organica Cesspool/Septic Treatment uses an exclusive blend of naturally occurring microbes and enzymes that thrive in all septic conditions. With regular use, the treatment biodegrades septic wastes, preventing clogs in the system. Follow directions and simply pour down drain or toilet every couple of weeks to naturally break down organic wastes like grease, fat, paper, starch and protein. Biodegradable, non-corrosive, and safe for industrial and home use. Tub lasts 7 to 8 months. USA.

06-0224 Septic Treatment   **$16**

SOLAR COOKING

In addition to the big job of making worldwide weather, or the small local job of heating your domestic hot water or pool, the thermal energy from sunlight can be used to heat buildings (which we cover fully in chapter 1, Land and Shelter), and to cook food. Solar-powered cooking has enjoyed a huge surge in interest over the past two decades. In many developing countries, gathering fuel wood for cooking can occupy half the family members full time, and has resulted in barren moonscapes where every available scrap of fuel gets snatched up—including any reforestation efforts. Solar cookers are quickly taking the place of the family fire pit. A simple reflective box with a clear cover can easily attain 300° to 400°F temperatures on an average sunny day, making an excellent oven with no fuel cost, and no time spent collecting fuel. The nonprofit Solar Cookers International group provides cookers and simple cardboard and aluminum foil construction plans to developing countries. Some of our solar cooker product sales support their good efforts.

Solar cookers can be as simple as a cardboard box with aluminum foil reflectors and a dark pot with a clear turkey roasting bag wrapped over it, or they can be as fancy as our Sun Oven with insulated sides, self-leveling tray, and folding reflectors. Both will get the job done. Obviously, the Sun Oven will last longer, and has more user-friendly features, but it doesn't necessarily cook lunch any faster.

Cooking with the Sun: How to Build and Use Solar Cookers

Solar cookers don't pollute, use no wood or other fuel resources, operate for free, and still run when the power is out. A simple and highly effective solar oven can be built for less than $20 in materials.

The first half of this book, generously illustrated with pictures and drawings, presents detailed instructions and plans for building a solar oven that will reach 400°F or a solar hot plate that will reach 600°F. The second half has one hundred tested solar recipes. These simple-to-prepare dishes range from everyday Solar Stew and Texas Biscuits, to exotica like Enchilada Casserole. 116 pages, softcover. USA.

21-0273 Cooking with the Sun   **$9.95**

Sport Solar Oven

We love the look on friends' faces when we present fresh bread and other home-baked treats while camping. So will you when you take our lightweight solar oven on your next trip. It concentrates the renewable heat of the sun to effortlessly (and deliciously!) roast meats, steam vegetables, bake breads and cookies, and prepare rice, soups, and stews. Ideal for backyard use, camping, boating, and picnicking. Requires only minimal sun aiming to cook most foods in two to four hours. Our complete kit includes two pots, oven thermometer, water pasteurization indicator, and recipe book. 12"H x 27"W x 16"D. 11 lbs.

01-0454 Sport Solar Oven   **$150**

The Sun Oven with Thermometer

The Sun Oven is a great solar cooker. We've been using it for several years now, and love it. The Sun Oven is very portable (one piece) and weighs only 21 pounds. It is

ruggedly built with a strong, insulated fiberglass case, a tempered glass door, and it includes an oven thermometer. The reflector folds up and secures for easy portability on picnics, etc. It is completely adjustable, with a unique leveling device that keeps food level at all times. The interior oven dimensions are 14"W x 14"D x 9"H. It achieves temperatures from 360° to 400°F. This is a very easy-to-use oven and it will cook anything! After preheating, the Sun Oven will cook one cup of rice in 35 to 45 minutes. USA.

63421   Sun Oven   **$249**

Economy Solar Cooker

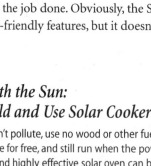

Our improved version of the Solar Cookit uses a closed-cell foam core, so it's rain-proof and crushproof. Folds down to a 14"x 14"x 2"-thick packet that takes up little space, and can even be used as a seat pad. With its foil-laminated reflective coating, this unit focuses available sunlight and slow-cooks like a crock-pot. You provide the dark-colored pot, up to 4 quarts. For faster cooking, particularly on windy days, cover the pot with a high-temperature turkey-baking bag to help hold in the heat. Highly portable, easily stored for camping or emergency use, and large enough to feed three to five people. USA.

63848   Economy Solar Cooker   **$24**

CHAPTER 11
Sustainable Transportation

IN A WAY, IT'S A SHAME that this chapter on Sustainable Transportation is so far back in the *Sourcebook*. It's near the end for a very practical reason: Gaiam Real Goods is not in the business of selling vehicles. We offer some excellent resources, but only a few relevant products. However, from the vantage points of energy consumption, greenhouse gas emissions, and long-term sustainable living on Earth, transportation is absolutely critical. Your personal transportation choices probably have a greater environmental impact than any other activity. It's hard to generalize because so many factors come into play, but it seems safe to say that for most Americans, at least a good quarter of our personal CO_2 emissions are attributable to how we get around, and for many of us that figure easily is closer to one-half.

And of course, our cultural addiction to fossil fuels creates problems beyond mere global warming, most notably air pollution and its detrimental effects on health, a large military establishment to protect our access to Middle East petroleum, and the resulting national insecurity and vulnerability to terrorism.

These days there's a lot of media buzz about alternative-fuel vehicles, the coming hydrogen economy, and more efficient SUVs, all just around the corner and coming to our collective rescue from climate change, high oil prices, and political terrorism. Meanwhile, our leaders lack the political will to mandate higher vehicle fuel-efficiency standards, our greenhouse gas emissions continue to rise, and the nation is mired in a war in Iraq whose purpose has always been at least as much about oil as about removing a dangerous dictator and inspiring a democratic uprising throughout the Arab world.

> Your personal transportation choices probably have a greater environmental impact than any other activity.

The problems associated with fossil-fuel consumption for transportation can be addressed in a number of ways. Perhaps the most painless (for everyone except Detroit's auto manufacturers, that is, and even they won't suffer too much) is to increase the fuel efficiency of our vehicle fleet. The United States managed to do this once before, something you young-uns might not be aware of, in the wake of the oil crisis of the 1970s. But due to corporate lobbying and government neglect, most of the fuel-efficiency gains of the past thirty years have been wiped out, as the popularity of energy-hog SUVs has soared. The most recent statistics available show that the average fuel economy of all new cars, passenger vans, SUVs, and pickups fell to 20.4 miles per gallon in 2002, the lowest level since 1980. Here's another interesting fact: According to Paul Roberts, author of *The End of Oil: On the Edge of a Perilous New World* (Houghton Mifflin, 2004), "By doubling the average fuel economy of cars and trucks to 40 miles per gallon (which existing hybrid technology could do), we could save 5 million barrels of oil a day by 2015—or more than twice our current imports from the Middle East." ("Energy Policy Should Look Beyond Oil," *Baltimore Sun*, July 13, 2004.) That would represent an enormous savings in CO_2 emissions, too. (5 million barrels × 55 gals./barrel × 365 days/year × 10 years × 23.8 lbs. CO_2/gal = **nearly 12 billion tons of CO_2!**) Switching to zero-emission vehicles that don't burn fossil fuels will make a big difference, and some day our economy might be powered

> Due to corporate lobbying and government neglect, most of the fuel-efficiency gains of the past thirty years have been wiped out, as the popularity of energy-hog SUVs has soared.

primarily by hydrogen that is itself produced without greenhouse gas emissions. But those days are a way off, and many of us feel considerable urgency to act now to begin withdrawing from our destructive fossil-fuel habit.

What can a mobility-loving, energy-conscientious individual do?

- Drive a vehicle that gets good gas mileage.
- Drive a vehicle that can burn biodiesel.
- Buy a hybrid gas-electric vehicle.
- Consider getting an electric vehicle.
- Use mass transit as much as you can.
- Take advantage of carpooling or car-sharing in communities where that system exists.
- Walk or ride your bike for a commuting change.

- Plant trees to offset your personal CO_2 emissions. Organizations such as American Forests (www.americanforests.org) and Trees For the Future (www.treesftf.org) have excellent reforestation programs in the United States and around the world—all you need to do is plunk down a little cash. The American Forests website even has an easy-to-use Personal Climate Change Calculator that tells you how many trees have to be planted to offset your annual contribution to global warming.

Let's take a look at the near and far horizons of vehicle development and see what choices are now on the market and likely to be available in the future.

Hybrids and Other Low-Emission Vehicles

At this writing, hybrid vehicles are all the rage. The idea is catching on with consumers and the auto manufacturers are having trouble keeping up with demand. This is good news all around, because gas-electric hybrids probably represent the simplest, most effective, most affordable short-term way to reduce fossil-fuel consumption in the transportation sector and the environmental impact of driving motor vehicles. In response to more stringent state and federal emissions standards, the manufacturers are also beginning to market Ultra Low-Emission and Super Ultra-Low-Emission gasoline vehicles (ULEVs and SULEVs) that feature redesigned, more efficient engines and more effective tailpipe emission controls.

A hybrid combines the best features of internal combustion and electric drive, while eliminating the worst parts of each. Hybrid vehicles have both an internal combustion engine and an electric motor, with a highly sophisticated automatic control system that chooses which one is running under what conditions. The electric motor is used at low speeds and for acceleration boost; the internal combustion motor delivers cruising speed and long range, but automatically shuts off at stop lights. Energy is captured during deceleration or braking by generating electricity to recharge the batteries.

Hybrid vehicles are available right now and getting better all the time. Several Real Goods employees at our Hopland Solar Living Center drive one to work every day. The 2004 Toyota Prius gets 60 MPG in the city and 51 MPG on the highway—a 15% improvement over the 2003 model. The 2004 Honda Civic hybrid gets 47 and 48 MPG, respectively. Even better, Honda's two-seater hybrid car, the Insight, rates at 57/56 MPG. These three models, all qualifying as SULEVS, have the highest "Green Scores" issued by the American Council for an Energy-Efficient Economy (except for a Civic model powered by compressed natural gas; these vehicles are available, but the fuel is hard to come by and currently more appropriate for fleets). These early hybrids continue to lean heavily on the internal combustion engine, because it's the technology we understand best. Future hybrids will rely more on electric storage and drive, and will increase fuel mileage greatly.

The other good news is that hybrids are affordable: The Prius, Civic, and Insight each cost about $20,000. Furthermore, Toyota recently announced that their hybrid sales have turned profitable, indicating that this technology is poised for longevity in the marketplace.

Hybrids are the technology that is most immediately accessible to manufacturers right now. Every major manufacturer has hybrid models in the works. The number of models that meet the toughest emissions standards nearly doubled between 2003 and 2004 (from 14 to 26). Ford has even just released a hybrid SUV, the Escape, that supposedly gets 36 MPG in city driving—but who knows if typical American SUV buyers will be drawn to the leaner and greener but smaller, lighter, and less powerful version?

Biodiesel and Other Biofuels

Like gas-electric hybrids, biodiesel is a readily available transportation choice that is more environmentally friendly than driving a conventional gasoline-powered car. The increasing public awareness and use of biodiesel actually represents a "back to the future" development. Believe it or not, the original engines invented by Rudolph Diesel in 1895 were designed to run on vegetable oil. That is the promise of biodiesel and other biofuels, such as ethanol, in the twenty-first century and beyond. These are clean-burning, relatively nonpolluting fuels made from renewable resources.

Biodiesel is a biodegradable, nontoxic fuel derived either from vegetable oils such as soybean or canola oil, or from recycled waste cooking oil. Biodiesel is made in a refinery process that removes glycerin from the oil. The process is not highly energy intensive, and the glycerin by-product can be sold for use in the manufacturing of soap and cosmetics. Biodiesel can be burned in compression-ignition (diesel) engines, as a pure fuel or blended with petroleum diesel, with little or no modifications to the vehicle.

Any use of biodiesel instead of petroleum diesel reduces emissions of pollutants and greenhouse gases. As demonstrated in Environmental Protection Agency tests, burning pure biodiesel reduces particulate emissions by 47%, polycyclic aromatic hydrocarbons by 75 to 85%, unburned hydrocarbons by 68%, carbon monoxide by 48%, and sulfur oxides and sulfates by nearly 100% compared to petroleum diesel. The only exception is nitrogen oxide emissions, which may increase slightly with biodiesel. The positive impact on climate change is even better. Not only

Gas-electric hybrids probably represent the simplest, most effective, most affordable short-term way to reduce fossil-fuel consumption in the transportation sector and the environmental impact of driving motor vehicles.

Alternative is the Norm at Real Goods

At our Hopland Solar Living Center, five of us live completely off the power grid, four more live with solar power on the grid, and many others use our renewable energy and energy efficiency products in our homes.

But how we get to work is often where we can make the biggest impact. Transportation alone consumes around 40% of the Earth's energy resources.

FROM BIKING TO BIODIESEL

That's why Real Goods team members are doing their part to minimize their footprint on the transportation grid. Technical experts Jason Sharpe, Jay Ruzicka, and Wes Kennedy, and Catalog Manager Bill Giebler all commute to work on their bicycles—up to 20 miles one way! And four other RG staffers drive diesel vehicles running on non-fossil-fuel based biodiesel fuel, sold at the Solar Living Center as "Grass-o-Lean." The Hopland store makes its own biodiesel using the FuelMeister kit and used oil from the local brew pub's fryer. And Real Goods President John Schaeffer and Technical Editor Doug Pratt have commuted in electric vehicles for over a decade, plugging into the solar-powered EV chargers at the Real Goods Solar Living Center.

ELECTRONIC COMMUTING

Of course, telecommuting is another smart eco-transportation option. As environmental visionary Amory Lovins puts it, "It's always more efficient to transport electrons than protoplasm." Jeff Oldham and several other RG solar techs telecommute regularly and save fuel while leaving their protoplasm at home.

Real Goods staffer Ben Dearnly filling up at the Solar Living biodiesel station.

Biodiesel is a readily available transportation choice that is more environmentally friendly than driving a conventional gasoline-powered car. Biodiesel can be burned in compression-ignition (diesel) engines, as a pure fuel or blended with petroleum diesel, with little or no modifications to the vehicle.

Not only does burning biodiesel reduce carbon dioxide emissions by 78%, but biodiesel (and other biofuels) is carbon-neutral: While combustion of any biofuel releases carbon dioxide into the atmosphere, growing the agricultural crop used in that fuel captures a similar amount of CO_2 through the process of photosynthesis.

does burning biodiesel reduce carbon dioxide emissions by 78%, but biodiesel (and other biofuels) is carbon-neutral: While combustion of any biofuel releases carbon dioxide into the atmosphere, growing the agricultural crop used in that fuel captures a similar amount of CO_2 through the process of photosynthesis. Some people today run B20, a blend of 20% biodiesel with 80% petroleum diesel. Burning B20 reduces tailpipe emissions by one-fifth of the percentages just noted. All the Volkswagen TDI (turbo diesel injection) models of the last several years run just fine on B100 or 100% biodiesel.

Biodiesel also has a positive energy balance. For every unit of energy consumed in the process of growing and manufacturing it, the resulting fuel contains anywhere from 2.5 to 3.2 units of energy. Fossil fuels have a negative energy balance, because extracting and refining them is highly energy intensive. While a given unit of biodiesel contains slightly less energy than the same unit of petroleum diesel, it has a slightly higher combustion efficiency, so using biodiesel in place of petroleum does not noticeably affect vehicle performance. And, oh yes, you can have a hybrid electric vehicle that uses a diesel motor instead of a conventional gasoline motor, and several major automakers are currently developing diesel hybrids.

While it's true that you can use biodiesel without making any engine modifications, it does have a solvent effect that may release accumulated deposits on tank and pipe walls, which may, in some vehicle models, initially clog filters and necessitate their replacement. Some experts also recommend replacing rubber hoses with ones made of synthetic materials in certain makes of vehicles. However, B20 does not appear to present a problem to those rubber components. Biodiesel also clouds and gels at higher temperatures than petroleum diesel, which could present a problem in colder climates. As with petroleum, however, winterizing agents added to biodiesel enable it to be used trouble-free to at least –10°F. During the winter of 2004, the road commissioner in chilly Norwich, Vermont, tested B20 in a town truck without a hitch.

Pure biodiesel currently costs about 60% more than petroleum diesel (aka dino-diesel). Running B20 therefore adds an extra 10% to your fuel cost, or approximately 15 cents per gallon. Many school districts and municipalities are able to achieve greater savings because they purchase their diesel in bulk. We all know that the price of oil is destined to rise drastically in our lifetime, and every price increase makes biodiesel that much more competitive in the marketplace. Here at the Solar Living Center we dispense B100—100% biodiesel—to numerous cars traveling the Northern California Highway 101 corridor. Six Real Goods employees currently drive biodiesel vehicles using B100. Our local supplier promises that with their new production plant, slated to be completed by the end of 2004, the price of 100% biodiesel will lower to around $2.50 per gallon, almost the same as "dino-diesel." Some small local biodiesel producers claim they can make biodiesel for prices well below those of "dino-diesel."

The list of cities, towns, and schools that are turning to biodiesel, out of concern for the health of schoolchildren and municipal employees as well as the global environment, continues to grow. Keene, New Hampshire, uses it in their truck fleet; Warwick, Rhode Island uses it in sixty school buses (and to heat school buildings); Deer Valley School District in Phoenix, Arizona, uses it in one hundred buses; and the states of Michigan and Minnesota have created incentive programs to encourage the switch. Companies such as L.L. Bean and Alcoa are experimenting with biodiesel, and even the federal government is getting into the act, testing biodiesel in its vehicle fleet in Yellowstone National Park.

Other biofuels may have even more promise than biodiesel produced from oil seeds or recycled cooking oil. Ethanol is familiar to many people, a fuel made primarily from corn today. It's controversial, because it's expensive to make, has a minimal positive energy balance (some claim it's actually negative), and is only commercially viable right now because of huge government subsidies. But ethanol can be produced more efficiently and sustainably from other biomass feedstocks, including paper mill sludge, perennial grasses such as switch grass, and rice or wheat straw. Scientists working to improve the production of ethanol from cellulose predict that foreseeable technological advances implemented on a large scale could lower the cost of ethanol to 50 cents a gallon. The potential environmental benefits are enormous.

And what would you think about biodiesel made from algae? That's right, I said algae. The National Renewable Energy Laboratory is sponsoring research on growing and processing algae for fuel. Algae are the most efficient organisms on the planet for consuming carbon dioxide and producing oil. According to NREL estimates, a 1,000-square-meter pond of algae can produce more than 2,000 gallons of oil per year. In comparison, the same crop area of high-yield canola can produce 50 gallons of oil. Even if only 10% of the algal oil can be extracted for use as fuel, it still outperforms the canola by a factor of 4. A half-million acres of algae ponds (that's equivalent to less than 1% of the fallow cropland in the U.S.) could produce something on the order of 10 billion gallons of oil annually, which could then be processed into biodiesel. Talk about renewable energy! Stay tuned.

Electric Vehicles

When we last updated this *Sourcebook,* in 2001, we devoted considerable attention to electric vehicles. At that time, we thought technological trends were heading in the direction of overcoming the limitations of pure electric vehicles. Just goes to show you, technological trends can be fickle, and prognostication is a dangerous sport. As of this writing, it appears that electric vehicles are more likely to become the domain of aficionados and purists rather than a leading-edge product finding favor with consumers.

An electric vehicle runs on an electric motor and contains a battery pack to store electrical energy. The great advantages of EVs are emissions and operating cost. EVs have no point-of-use emissions, and research has shown that even taking into account the power plants that generate the electricity, an EV releases 90% less emissions than a comparable gasoline-powered car. An EV can also be powered by solar energy, which of course would be the ideal scenario. EVs also are cheaper to operate than internal combustion vehicles. They don't need tune-ups, fuel and oil filters, oil changes, or mufflers, and because they have fewer moving parts, they require less maintenance. Research has demonstrated that overall life-cycle cost of operating an EV can be approximately 60% less than that of a comparable conventional vehicle.

It's the liabilities of electric vehicles that present the problem. The high cost of conversion ($5,000 or more for a typical passenger vehicle) and the limitations in range imposed by current battery technology (still only 50 or 60 miles before a recharge is needed) continue to conspire against EVs. What we wrote in 2001 remains true today: "We haven't reached the performance level of jumping in your EV at a moment's notice, driving 300 miles, and recharging it in ten minutes. In other words, the kind of vehicle performance we're used to."

In the EV's defense, according to the EPA, close to 80% of all North American auto trips involve a round trip of less than 20 miles. This is very comfortably within the range of even the most modest EVs. Therefore, EVs are far and away the best choice for shorter commutes and running errands in town. They do have an important role to play in our society's transportation mix—but to date they don't seem to offer a magic bullet. If you're interested in knowing more about the nuts and bolts of electric vehicle technology, see pages 440–48 of the eleventh edition *Solar Living Sourcebook.*

EVs have no point-of-use emissions, and research has shown that even taking into account the power plants that generate the electricity, an EV releases 90% less emissions than a comparable gasoline-powered car.

Fuel Cell and Hydrogen Vehicles

Fuel cells! The Hydrogen Economy! Now *there's* the magic bullet. Perhaps. The buzz about hydrogen intensifies all the time these days. Even our current administration, with their well-known ties to the oil industry, have endorsed the idea that hydrogen-powered fuel cell vehicles are a good idea. In his January 2003 State of the Union address President Bush announced a $1.2 billion government research initiative on fuel cells.

Hydrogen—the most common element in the universe—probably is the wave of the future. It may well become the primary storage medium of the twenty-first century, and that would be a good thing for people and the planet. However, we are persuaded by the analysts who argue that fuel cell and hydrogen vehicles have a long way to go before they're commercially viable, and therefore represent a long-term solution when more realistic short-term fixes are needed urgently now. A strategically targeted, well-funded crash hydrogen development program is advisable, but other steps can and should be taken immediately to reduce the greenhouse gases and other pollutants generated by the transportation sector of the economy.

Functionally, a fuel cell is close kin to a gas-powered generator. A fuel cell takes a hydrogen-rich fuel source and delivers electricity, heat, and

We are persuaded by the analysts who argue that fuel cell and hydrogen vehicles have a long way to go before they're commercially viable, and therefore represent a long-term solution when more realistic short-term fixes are needed urgently now.

some kind of exhaust through chemical conversion. The electrical output from the fuel cell is delivered to an electric motor, and you've got your basic fuel cell–powered vehicle. A modest battery pack sometimes is used to boost acceleration and help even out the electrical load on the fuel cell. Fuel cells are quiet, two to three times more efficient than an internal combustion engine, and produce less than half the emissions. In the ideal scenario, a fuel cell running on pure hydrogen will deliver only pure water vapor as exhaust. But on a practical level, hydrogen is found bound up with other elements to make water or hydrocarbons, so pure hydrogen is expensive. Most fuel cell vehicles under development use some kind of hydrocarbon fuel, such as natural gas, alcohol, or even gasoline, that run through an onboard "fuel processor" that boosts the hydrogen content. If you run something other than perfectly pure hydrogen into your fuel cell, you'll get something other than perfectly pure water vapor for exhaust. It still beats the pants off any kind of combustion process for cleanliness.

Hydrogen currently is considered a nonpetroleum fuel but it should be noted that it is not a source of energy. It is only a way of storing and transporting it. The Earth's gravity is not strong enough to hold free hydrogen, so it escaped into space billions of years ago. Hydrogen fuel must be "manufactured" by extracting it from fossil fuel or water. The energy available from hydrogen fuel is equal to the energy used to extract it, plus the energy from burning carbon in the process of extracting it from fossil fuel.

Electricity can be used to split water into hydrogen and oxygen, but this process, known as electrolysis, is expensive. A cheaper way to produce hydrogen, as long as fossil fuels are low-cost and abundant, is to re-form natural gas or spray steam on white-hot coals and make what's called "water gas" or "coal gas." The gas produced from fossil fuel using this method is 40% hydrogen and 50% carbon monoxide. Unfortunately, the carbon monoxide is highly poisonous. To render the carbon monoxide harmless, it can be burned, creating the greenhouse gas carbon dioxide. Coal gas is used in the Fisher-Tropsch process to make synthetic gasoline and alcohol.

Hydrogen is still far from an ideal automobile fuel. Even in its densest form (liquid), hydrogen has only one-third as much energy per liter as gasoline. If stored as compressed gas at 300 atmospheres (a more practical option), it delivers less than one-fifth the energy per volume of gasoline. Such low energy density means that

fuel storage would take up lots of room in a hydrogen-powered car—or, alternatively, a modest-sized fuel tank would severely restrict the vehicle's range between fill-ups.

Hydrogen will be made by whatever method is cheapest. In the short run, re-formed natural gas and electrolysis powered by electricity from nuclear plants seems to be Bush's plan. He has asked for $16 billion to subsidize the construction of six new nuclear power plants, ignoring the threat of radioactive waste and the $400 billion cost of the controversial nuclear waste dump in Yucca Mountain. Obviously, nuclear power should not be considered as a renewable fuel.

The Bush plan also will devote the bulk of federal fuel cell research funding to making hydrogen from fossil fuels rather than renewable sources. This means renewables will not be a part of the hydrogen economy and even if the rest of the world switches to hydrogen made from water, Americans will end up even more dependent on fossil fuels for generations.

This crucial difference between pure hydrogen and hydrocarbons points to the underlying problem with fuel cell technology: It's still in an early research-and-development stage. Joseph J. Romm, executive director of the Center for Energy and Climate Solutions, assistant secretary of energy for energy efficiency and renewable energy in the Clinton Administration, and author of the recent book *The Hype about Hydrogen: Fact and Fiction in the Race to Save the Climate* (Island Press, 2004), is a leading hydrogen skeptic. He points out that the current cost of producing hydrogen from natural gas is three times, and from water six to nine times, more than the price of gasoline. A February 2004 report from the National Academies' National Academy of Engineering and National Research Council concluded that in the best-case scenario, the transition to a hydrogen economy will take "many decades" and achieve insignificant reductions in oil imports and CO_2 emissions over the next twenty-five years. Another government report estimated that it will cost at least $500 billion to create and implement the infrastructure necessary to provide hydrogen fuel for just 40% of the cars on the road. Romm and others agree that hydrogen fuel cell vehicles cannot make a significant contribution to slowing climate change until after 2030. The problem is, global CO_2 emissions are projected to rise by more than 50% by 2030. It just doesn't make ecological sense to burn fossil fuels to produce hydrogen to run fuel cells.

Joseph Romm's argument is that running a fuel cell car on fossil fuel-produced hydrogen in 2020 has no life-cycle greenhouse gas advantage over the 2004 Prius hybrid running on gasoline.

Even in 2030, Romm notes, from the standpoint of greenhouse gas emissions it will be more cost-effective to use natural gas to displace coal in the generation of electricity than to use it to displace gasoline for transportation. Likewise, using renewable energy to generate electricity directly rather than to produce hydrogen for fuel cells has a more positive impact on climate change.

The bottom line for Romm: Most gas-electric hybrids available today are nearly as efficient as what is projected for fuel cell vehicles. His argument is that running a fuel cell car on fossil fuel–produced hydrogen in 2020 has no life-cycle greenhouse gas advantage over the 2004 Prius hybrid running on gasoline. If you're concerned about the environmental impact of your personal transportation choices, don't hold your breath waiting for fuel cell vehicles to save the day. Get a gas-electric or diesel-electric hybrid today, or make the switch to biodiesel.

The Hypercar

One other cutting-edge transportation concept is worth keeping an eye on. Amory Lovins of the Rocky Mountain Institute has been developing the Hypercar for the past decade. As explained on the RMI website: "A Hypercar® vehicle is designed to capture the synergies of: ultralight construction; low-drag design; hybrid-electric drive; and, efficient accessories to achieve 3- to 5-fold improvement in fuel economy, equal or better performance, safety, amenity and affordability, compared to today's vehicles. Rocky Mountain Institute's research has shown that the best (possibly, the only) way to achieve this is by building an aerodynamic vehicle body using advanced composite materials and powering it with an efficient hybrid-electric drivetrain." Lovins' idea is ultimately to power the Hypercar with a hydrogen fuel cell. A for-profit company now exists as a collaboration among RMI, automobile companies, and others, to bring the Hypercar to market in a commercially feasible way.

If anyone is capable of packaging together the most advanced strategies for vehicle design, materials, and fueling, and devising a product that can capture the attention of consumers at a reasonable price, these guys can. Keep your eyes and ears peeled for more news about the Hypercar. In the meantime, ride your bike, take the bus, and buy a hybrid or biodiesel car!

Sources of Further Information

Electric Auto Association (EAA)
P.O. Box 6661
Concord, CA 94514
www.eaaev.org
The national electric vehicle group since 1967. Has local chapters in many states. Yearly membership is $39, and includes an informative monthly newsletter with tech articles, industry news, and want ads. The website has links to practically everything and everybody EV related.

Electrifying Times
63600 Deschutes Market Road
Bend, OR 97701
www.electrifyingtimes.com
A commercial EV magazine that comes out three times yearly. Runs 60 pages or more. Much up-to-date information, and ads from suppliers and builders. Edited by Bruce Meland. $12/year.

Department of Energy
Office of Transportation Technologies
Hybrid Electric Vehicle Program
www.nrel.gov/vehiclesandfuels/hev/
The DOE's site contains a lot of good information. What hybrids are, how they work, where you can buy one.

Marshall Brain's How Stuff Works website
"How Hybrid Cars Work"
www.howstuffworks.com/hybrid-car.htm
The How Stuff Works website is great fun, highly informative, and has good interactive graphics, but you'll need to put up with some commercial content. Their hybrid site is very complete.

National Biodiesel Board
www.biodiesel.org

The Hypercar
www.hypercar.com

SUSTAINABLE TRANSPORTATION PRODUCTS

From the Fryer to the Fuel Tank
How to Make Cheap, Clean Fuel from Free Vegetable Oil

New updated and expanded version by Joshua and Kaia Tickell, who, with sponsorship from Real Goods, drove

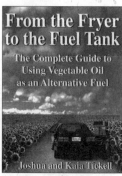

their unmodified Veggie Van over 10,000 miles around the United States during the summer of 1997, while stopping at fast-food restaurants for fuel fill-ups of used deep fat fryer oil. Vegetable oil provided 50% of their fuel. They towed a small portable fuel processing lab in a trailer behind their Winnebago. This inexpensive and environmentally friendly fuel source can be modified easily to burn in diesel engines by following the simple instructions. A complete introduction to diesel engines and potential fuel sources is included. Over 130 photographs, diagrams, charts, and tables. 162 pages, softcover. USA.

80961   From the Fryer to the Fuel Tank $24.95

Fuel from Water: Energy Independence with Hydrogen

By Michael Peavey. An in-depth and practical book on hydrogen fuel technology that details specific ways to generate, store, and safely use hydrogen. Hundreds of diagrams and illustrations make it easy to understand, very informative, and practical. If you're interested in using hydrogen in any way, you will find this book indispensable. 244 pages, softcover, USA.

80210   Fuel From Water $25

FuelMeister™ Biodiesel Processor

With a FuelMeister™ you don't have to be a plumber or mechanic to make your own biodiesel. This is not a hobby kit but a complete system engineered with quality materials. You get all the specialized equipment and supplies you need to start making biodiesel that same day. All you need to add is vegetable oil, methanol, electricity, and tap water. The whole setup fits comfortably in the corner of your garage. This is a "closed" system, so it won't smell, and you won't have to pour or stir any liquids. It will turn even heavily used cooking oil into clean burning, biodegradable, and smooth-running biodiesel. Whatever you own that runs on diesel fuel you can power with the biodiesel you produce with your own FuelMeister biodiesel production system.

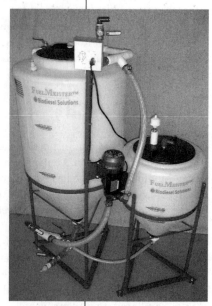

The FuelMeister produces approximately 40 gallons of biodiesel every forty-eight hours, with about one hour of operator time required per batch. Supplies typically cost 70 cents per gallon or less. Methanol is widely available as racing fuel; the

Optional oil pre-heat kit fits any 55-gallon barrel.

lye catalyst is available at any hardware store. A safe, manual methanol pump, and a precise digital scale for the lye is included, along with personal safety items (rubber gloves, safety glasses, etc.) and a titration kit for testing your raw oil to see precisely how much catalyst is required. This well-engineered package is designed for safety, quality, and ease of use with ball valves and see-through tanks and tubing.

The optional Oil Pre-Heater with thermostat accelerates production and improves yields in cold climates. It includes wrap-on insulation, heater, and tape to fit any standard 55-gallon barrel you provide. 40"H x 20"D. 120vac.

All components are UPS shippable. Requires about 62"H x 25" diameter when assembled. USA.

63-0015 FuelMeister Biodiesel Kit $2,999
63-0017 Oil Pre-Heat Kit $399

eGo Cycle: The electric scooter with moxie

Our employees road-tested this remarkable electric cycle on everything from commutes to grocery runs—and quickly declared it the little bike that could. The eGo has what it takes to get you where you need to go, at a running cost of just a penny a mile.

This ingenious mode of transport is built on a quasi-bicycle frame—making it affordable as well as light enough to transport you and a surprisingly sizable payload efficiently. A simple turn of a key and a quick throttle twist engage either of two electronically controlled operating modes ("Go Far" or "Go Fast"), letting you tailor your ride for either greater distance range or maximum speed. The eGo's regenerative braking system captures your momentum as you ride, recharging the battery and optimizing the DC motor's range. The built-in 5-amp charger plugs into any standard AC outlet for an 80% recharge in just 3 hours—and a full recharge from empty in only 4 to 6 hours.

The eGo's cargo features are both accommodating and versatile: Built-in rear rack holds briefcases, picnic baskets, and extra gear; optional folding rear baskets each hold a full bag of groceries; and optional front basket converts to shopping basket; optional cargo trailer attaches in a snap, folds flat for storage, can haul an additional 100 lbs., and keeps brake light visible. Available in four colors: Black, Red, Caribbean Sky (blue), and Wasabi (light green). 63"L x 40"H. 120 lbs. USA.

The Extended Range Battery Pack increases range by 50%—up to 40 miles! Includes padded case with tie down, 24-volt, 20-amp-hour battery, wiring harness, and instructions. Charges automatically when installed, comes off in seconds.

65-0037 eGo Cycle	**$1,099**
65-0038 Front Basket	**$29**
65-0039 Rear Folding Baskets	**$49**
65-0040 Cargo Trailer	**$150**
65-0041 Rain Cover	**$49**
65-0042 Extended Range Battery Pack	**$189**

Optional front basket removes and converts to a shopping basket.

Optional rear baskets fold flat.

Optional cargo trailer.

Optional rain cover.

Extended battery pak.

CHAPTER 12
Sustainable Living Library

"KNOWLEDGE IS POWER" the old saying goes, but at Real Goods knowledge is even more . . . it's our most important product.

In the *Solar Living Sourcebook* we provide products and technical information about solar power, but if that was all that was contained on these pages, this would be little more than a glorified catalog. The real value of this guidebook is that it is a hub that connects the reader to the entire world of solar living, an entire lifestyle that ranges from renewable energy to green building to the pond to the permaculture garden to sustainable transportation and all points in between. If we see our mission as moving the world to the paradigm shift of accepting sustainable living as our future mandate, then ideas will be just as important as equipment.

Even in the midst of an information revolution, the traditional book still reigns as the heavyweight champ. The book is amazingly efficient, fast, portable, durable, and even sensuous in terms of its ability to catalog, organize, and present data. The data accumulates into information, the information transforms into knowledge, and the knowledge into wisdom. No offense intended to the new technologies—they all have ways in which they shine, but none of them has come close to challenging the book's grip on the championship belt. It is for this reason that we have put significant energy into making the *Solar Living Sourcebook* a "book of books." The library we have assembled represents the end result of countless hours of research and selection. Moreover, you will find that each title, in turn, leads you into a new journey into the world of knowledge.

This is the first time we have collected our Sustainable Living Library in one place in the *Sourcebook*. You won't find such a comprehensive library anywhere, not on Amazon, not in all the Borders and Barnes & Nobles combined. If you come to our Solar Living Center in Hopland, however, please drop by and browse through the shelves, where our library comes alive. But if you can't, then do your browsing here. In doing so, you will be giving yourself the gift of power through the product of knowledge, that will ultimately result in wisdom.

LAND AND SHELTER

Superbia!

By Dan Chiras and Dave Wann. Practical ideas for creating more socially, economically, and environmentally sustainable neighborhoods. Discussions include transformation of a fictitious neighborhood that adopts the authors' thirty-one precepts. Examples from all over North America and the world provide evidence that citizens can create Superbia! Includes a remarkably comprehensive resource list. 229 pages, softcover. Canada.

21-0356 Superbia $19.95

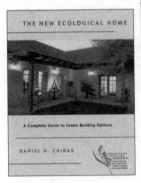

The New Ecological Home

Shelter, while crucial to our survival, has come at a great cost to the land. But, as author Dan Chiras shows in his follow-up book to *The Natural Home,* this doesn't have to be the case. With chapters ranging from green building materials and green power to landscaping, he covers every facet of today's homes. A good starting point for decreasing the environmental impact of new construction. 352 pages, softcover. USA.

21-0346 The New Ecological Home $35

The Not So Big House
A Blueprint for the Way We Really Live

By Sarah Susanka, with Kira Obolensky. This is one of the best, most inspiring home design books that's ever been written. Sarah's sense of what delights us, what makes us feel secure, and what simply works well for all humans is flawless. Is it a surprise that people gather in the kitchen? It shouldn't be; human beings are drawn to intimacy, and this book proposes housing ideas that serve our spiritual needs while protecting our pocketbook. Commonsense building ideas are combined with gorgeous home design and dozens of small simple touches that delight the senses. This is an inspiring book on building houses that make us feel safe, protected, and comfortably at home. 199 pages, softcover.

82-375 The Not So Big House $23

More Small Houses

In *More Small Houses,* you'll find that smaller is beautiful—and more intimate and affordable. Twenty homes, each less that 2,000 sq. ft., each a study in craft and efficiency, are used to explore space-saving design ideas. 159 pages, hardcover.

82-449 More Small Houses $24.95

Finding & Buying Your Place in the Country

By Les and Carol Scher. If you've never bought rural land before, you need this book before you start your search. This is the "bible" on buying rural property, now in its fifth edition. Even for those who've bought land, this well-organized book has plenty of cautions, evaluations, and checklists to  help you make informed development decisions. For practical dreamers, this book covers all the things you should consider before you buy rural land. 417 pages, softcover, USA.

82620 Finding & Buying Your Place in the Country, 5th Ed. $28

The Natural House
A Complete Guide to Healthy, Energy-Efficient Environmental Homes

By Daniel D. Chiras. Aimed at both builders and dreamers, this extraordinary book explores environmentally sound, natural, and sustainable building methods including rammed earth, straw bale, earthships, adobe, cob, and stone. Besides sanctuary to shelter our families, today we want something more—we want to live free of toxic materials. Chiras is an unabashed advocate of these new technologies, yet he openly addresses  the pitfalls of each, while exposing the principles that make them really work for natural, humane living. This book is devoted to analysis of the sustainable systems that make these new homes habitable. This is the short course in passive solar technology, green building material, and site selection, addressing more practical matters—such as scaling your dreams to fit your resources—as well. 469 pages, softcover.

82-474 The Natural House $35

Straw Bale Details

By Chris Magwood and Chris Walker. Nothing conveys a wealth of information like a good drawing by a professional architect, which is what this book is all about. It's the nitty-gritty of how to assemble a well-built, long-lasting straw-bale structure. All practical foundation, wall, door and window, and roofing options are illustrated, along with notes and possible modifications. Has a clever ring binding that lays flat for easy worksite reference. 68 pages, softcover. Canada.

21-0341 Straw Bale Details $32.95

Serious Straw Bale

Straw bale construction is a delightful form of natural, ecological building. But until the publication of *Serious Straw Bale: A Home Construction Guide for All Climates,* it

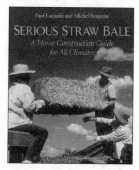

was assumed appropriate only for dry, arid climates. This book does not advocate universal use of straw bale, but provides practical solutions to the challenges brought on by harsh climates. This is the most comprehensive, up-to-date guide to building with bales ever published. Builders, designers, regulators, architects, and owner/builders will benefit from learning how bale building technique now includes climates and locales once thought impossible. 371 pages, softcover.

82-485 Serious Straw Bale $29.95

The Beauty Of Straw Bale Homes

By Athena and Bill Steen. This book is a visual treat, a gorgeous pictorial celebration of the tactile, timeless charm of straw bale dwellings. A diverse selection of building types is showcased— from personable and inviting smaller homes to elegant large homes and contemporary institutional buildings. The lavish photographs are accompanied by a brief description of each building's unique features. Interspersed throughout the book are insightful essays on key lessons the authors have learned in their many years as straw bale pioneers. 128 pages, softcover.

82-484 The Beauty of Straw Bale Homes $22.95

Buildings of Earth and Straw
Structural Design for Rammed Earth and Straw Bale

By Bruce King. Straw bale and rammed earth construction are enjoying a fantastic growth spurt in the United States and abroad. When interest turns to action, however, builders can encounter resistance from mainstream construction and lending communities unfamiliar with these materials. *Buildings of Earth and Straw* is written by structural engineer Bruce King, and provides technical data from an engineer's perspective. Information includes special construction requirements of earth and straw; design capabilities and limitations of these materials; and most importantly, the documentation of testing that building officials often require.

This book will be an invaluable design aid for structural engineers, and a source of insight and understanding for builders of straw bale or rammed earth. 190 pages, softcover with photos, illustrations, and appendices. USA.

80-041 Buildings Of Earth And Straw $25

Building with Earth
A Guide to "Dirt Cheap" and "Earth Beautiful"

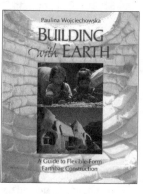

Building with Earth is the first comprehensive guide to the re-emergence of earthen architecture in North America. Even inexperienced builders can construct an essentially tree-free building, from foundation to curved roof, using recycled tubing or textile grain sacks. Featuring beautifully textured earth- and lime-based finish plasters for weather protection, "earthbag" buildings are being used for retreats, studios, and full-time homes in a wide variety of climates and conditions. This book tells (and shows in breathtaking detail) how to plan and build beautiful, energy-efficient earthen structures. Author Paulina Wojciechowska is founder of "Earth, Hands & Houses," a nonprofit that supports building projects that empower indigenous people to build shelters from locally available materials. 8" x 10", 200 pages, 180 illustrations and photos, softcover.

21-0323 Building with Earth $24.95

The Natural Plaster Book

This comprehensive step-by-step guide to choosing, mixing, and applying natural plasters takes the mystery out of using earth, lime, and gypsum plasters. Includes plaster recipes, tips on pigmentation, and common mistakes, and thorough insight on the tools and techniques for professional results. 252 pages, hardcover. USA.

21-0350 The Natural Plaster Book $29.95

Building with Structural Insulated Panels

By Michael Morley. Structural Insulated Panels (SIPs) are in the process of replacing the post-war norm of stick-framed, fiberglass-insulated houses and light commercial buildings. SIPs produce a structurally superior, better insulated, faster to erect, and more environmentally friendly house than

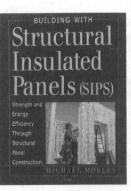

ever before possible. Experienced SIP builder Michael Morley covers choosing the right panels for the job, tools and equipment, wall and roof systems, and mechanical systems. This is an indispensable reference for anyone who wants to work with this innovative, field-tested construction system. 186 pages, hardcover, USA.

82621   Building with Structural Insulated Panels $35

Stone House

A great resource if you dream of building your own home. Topics range from traditional stonemasonry and slip forming to concrete-making and restoring recycled wooden doors and windows. Wonderfully illustrated with photographs and supported by an

online photo gallery. Includes a glossary and FAQ. 208 pages, softcover. USA.

21-0351 Stone House $33

The Hand-Sculpted House

Learn how to build your own home from a simple earthen mixture. Derived from the Old English word meaning "lump," cob has been a traditional building method for millennia. Authors Ianto Evans, Michael G. Smith, and Linda Smiley

teach you how to sculpt a cozy, energy-efficient home by hand without forms, cement, or machinery. Includes illustrations and photography. 384 pages, softcover. USA.

21-0331 The Hand-Sculpted House $35

Circle Houses

By David Pearson. The natural world is filled with circles, yet circular houses are unusual in industrialized societies. Among nomadic peoples, round-shaped homes have since time immemorial served to shelter families, echoing natural forms. A round home represents the universe in microcosm: the floor (Earth), the roof (sky), and the hole in the roof (sun). *Circle Houses* is a fascinating glimpse of tradition meeting timelessness, with stories of twenty-first century nomads, and basic instructions for designing and constructing your own tipi, yurt, or bent-frame tent. 95 pages, hardcover. USA.

82616   Circle Houses $17

Eco-Renovation Revisited

Eco-Renovation, published by Chelsea Green's U.K. affiliate, Green Books, is organized under four main headings: space, energy, health, and materials. The book guides you through each of these fields, and prioritizes choices according to the basic ecological principals of recycling, self-sufficiency, renewability, conservation, and efficiency. It includes a checklist for making an ecological assessment of your home, and detailed appendices that will enable you to locate "greener" products. *Eco-Renovation* will help direct your ecological thinking to a wide range of home improvements. It will show you how to reduce your heating bills substantially, choose the most energy-efficient appliances, select building materials and finishes that are ecologically benign, organize the space in your home to the maximum effect, and protect your family from toxic substances. By following the suggestions in this book, you can cut your domestic fuel bills by over 50 percent and reduce your household's carbon dioxide emissions by as much as 75%. 243 pages, softcover. United Kingdom.

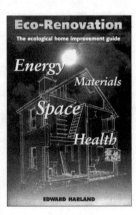

80-385 Eco-Renovation Revisited $16.95

Gaviotas

A Village to Reinvent the World

By Alan Weisman. Twenty-five years ago, the village of Gaviotas was established in one of the most brutal environments on Earth, the eastern savannas of war-torn, drug-ravaged Colombia. Read the gripping, heroic story of how Gaviotas came to sustain itself, agriculturally, economically, and artistically, eventually becoming the closest thing to a model human habitat since Eden. These incredibly resourceful villagers invented super-efficient pumps to tap previously inaccessible sources of water, solar kettles to sterilize drinking water, wind turbines to harness the mildest breezes—they even rediscovered a rain forest! To call this book inspiring is akin to calling the Amazon a winter creek. You won't be able to put it down. 240 pages, softcover. USA.

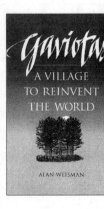

82-306   Gaviotas $14.95

Ecological Design

By Sim Van der Ryn and Stuart Cowan. This ground-breaking book profiles how the living world and humanity can be reunited by making ecology the basis for design. The synthesis of nature and technology, ecological design can be applied to create revolutionary forms of buildings, landscapes, and cities—as well as new technologies for a sustainable world. Some examples of innovations include sewage treatment plants that use constructed marshes to purify water and industrial "ecosystems" in which the waste from one process becomes fuel for the next. The first part of the book presents an overview of ecological design. The second part devotes chapters to five design principles fundamental to ecological design. A resource guide and annotated bibliography are included. 201 pages, softcover, 1996. USA.

80-512 Ecological Design   **$24**

Build a Fantasy House in the Trees

This eclectic and delightful collection of tree-dwellers and their dwellings will inspire even the diehard couch potato to grab a hammer and head for the backyard. Featuring real projects from all around the globe, this book depicts twenty distinctive tree dwellings, each one a masterwork of inventiveness. Lavishly illustrated with color photography, this book portrays not only the finished project, but also the construction process, giving enough information to complete a simple project. The extensive resource section offers places to visit, both real and virtual. David Pearson, noted author of *The Natural House Book*, has captured the combined spirits of innocence and inspiration in this awesome little book. 120 pages, hardcover.

82-614 Treehouses   **$16.95**

Architectural Resource Guide

From the Northern California Architects/Designers/Planners for Social Responsibility. Ever wondered where to find all those nontoxic, recycled, sustainable, or just plain better building materials? Well, here it is. This absolute gold mine of information has a huge wealth of products listed logically and intelligently following the standard architect's Uniform Specification System (i.e., concrete; thermal and moisture protection; windows and doors; finishes; etc.). Each section is preceded by a few pages explaining the pros and cons of particular materials or appliances. Each listing says what it is, gives the brand name, a brief explanation of the product or service, and access info, both national and NorCal local, for the manufacturer. Although written for the professional architect or builder, there's plenty of info for everyone here. Updated continuously, we order small quantities to ensure shipping the latest edition. 152 pages (approx.), softcover. USA.

80-395 Architectural Resource Guide   **$35**

Pocket Reference to the Physical World

Compiled by Thomas J. Glover. This amazing book, measuring just 3.2 by 5.14 inches, is like a set of encyclopedias or an Internet search engine in your shirt pocket! It's a comprehensive resource of everything from the physical properties of air to pipe-sizing information, and from calculating joist spacing to converting a broad range of weights and measures. A big chunk of knowledge in a very small package. No desktop is complete without one. 544 pages, softcover.

80-506 Real Goods Pocket Ref   **$12.95**

Home Work
Handbuilt Shelter

Building on the enormous success of the original *Shelter*, published in 1973, Lloyd Kahn continues his odyssey of finding and showing us the most magnificent and unusual hand-built houses in existence. Among the intriguing domiciles described in *Home Work* are a Japanese-style stilt house accessible only by a cable across a river; a stone house in a South African valley whose roof serves as a baboon trampoline; multi-level treehouses on the South China Sea; and a bottle house in the Nevada desert. Over 1,500 photos illustrate various innovative architectural styles and natural building materials that have gained popularity in the last two decades such as cob, papercrete, bamboo, adobe, strawbale, timber framing, and earthbags. Large 9 x 12 format with color and B&W photos and illustrations throughout. 256 pages. USA.

92-0262 Home Work   **$27**

Concrete Countertops

By Fu-Tung Cheng, master of design and construction. Concrete kitchen and bathroom countertops are currently very popular. Here at last is a complete, start-to-finish book on creating concrete countertops. Cheng takes you step-by-step through the process of making a concrete countertop—from building the mold and mixing and pouring concrete to curing, grinding, polishing, and installing the countertop.

You'll be inspired by the 350 color photographs that bring this exciting medium to life. And you'll discover

that the possibilities for creative expression with concrete are endless. You can color, polish, stamp, stain, and imbed objects in concrete. Countertops can be custom-formed, colored, and finished to look like marble, granite, glass, or sculpture— the look is limited only be the imagination. Throughout the book, Cheng offers valuable troubleshooting advice and useful tips on maintaining a concrete countertop. Large 9" x 11" format with the exquisite organization and layout that Taunton Press is famous for. 208 pages, softcover. USA.

21-0361 Concrete Countertops   **$30**

Rustic Retreats

By David and Jeanie Stiles. This delightful book delivers fantasy and fun for anyone in need of a rustic place to rest, relax, and retreat from the speed and stress of modern living. The Stiles provide more than twenty step-by-step plans for low-cost outdoor buildings, all richly illustrated with whimsical but detailed line draw-

ings. Among the projects are; a grape arbor, a hillside hut, a floating water gazebo, a log cabin, tree houses, a wigwam, a garden pavilion, a yurt, and a river raft. A bit of individual creativity can make any of the plans described and diagrammed into unique personal statements or fantasies. Many of the plans are so simple they can be built in a few hours or in less than a day, and most do not require high levels of carpentry or building skills. If you built forts as a kid—or wanted to—here's your chance to play at it again as an adult!

21-0362 Rustic Retreats   **$20**

Patterns of Home
The Ten Essentials of Enduring Design

By Max Jacobson, Murray Silverstein, and Barbara Winslow. Looking for inspiration and direction in the design or remodel of your home? Clearly written and profusely illustrated, *Patterns of Home* brings you the timeless lessons of residential design. The ten patterns described in the book—among them, "capturing light" and "the flow through rooms"—are drawn from hundreds of principles and presented with clarity by the authors, renowned architects who have designed homes together for more than thirty years. This book will jump-start the design process and make the difference between a home that satisfies material requirements and one that meets your personal needs of "home." *Patterns of Home* promises to become the "design bible" for homeowners and architects. Large 9" x 11" format. Lavishly illustrated in the rich Taunton Press style. 288 pages, hardcover. USA.

21-0363 Patterns of Home   **$35**

SUNSHINE TO ELECTRICITY

Power with Nature:
Solar and Wind Energy Demystified

Author Rex Ewing is right up to date, explaining all the latest offerings such as utility intertie systems and maximum power point tracking charge controls—all with extensive drawings, pictures, sidebars, and witticisms. You'll be enlightened by the wealth of practical, hands-on information for solar, wind, and micro-hydro power; plus common-sense explanations of the necessary components such as batteries, inverters, charge controllers, and more. 255 pages, softcover. USA.

21-0339 Power With Nature   **$24.95**

There Are No Electrons

Do you know how electricity works? Want to? Read this book and learn about volts, amps, watts, and resistance. So much fun to read, you won't even notice you're learning. And if you already "know" about all this, you'll enjoy it even more—just be prepared to rethink everything. 322 pages, hardcover.

82646   There Are No Electrons   **$16**

The Solar Design Studio v5.0

This CD-ROM contains a suite of eleven solar energy programs for professionals, involved educators, and interested individuals. These detailed simulations are practical and easy enough to be used by non-technical individuals.

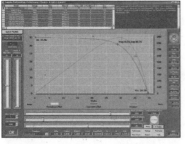

Programs will design stand-alone, grid-tied, or water-pumping PV systems, as well as solar hot water systems. Accessory programs deliver world climate info, sun path simulations, and output curves for all types of PV modules. Windows-based on CD-ROM. USA.

63-0018 The Solar Design Studio v5.0   **$149**

The Solar Electric House

By Steven Strong. The author has designed more than seventy-five PV systems. This fine book covers all aspects of PV, from the history and economics of solar power to the nuts and bolts of panels, balance of systems equipment, system sizing, installation, utility intertie, stand-alone PV systems, and wiring instruction. A great book for the beginner. 276 pages, softcover. USA.

80800   The Solar Electric House   **$21.95**

Photovoltaics: Design and Installation Manual

By Solar Energy International staff. Producing electricity from the sun using PV systems has become a major industry worldwide. Designing, installing and maintaining such systems requires knowledge and training. This

manual delivers the critical information to successfully design, install and maintain PV systems. The book contains an overview of photovoltaic electricity and a detailed description of PV system components, including PV modules, batteries, controllers and inverters. It also includes chapters on sizing photovoltaic systems, analyzing sites and installing PV systems, as well as detailed appendices on PV system maintenance, troubleshooting and solar insolation data for over 300 sites around the world. Topics covered also include: Stand-alone and PV/Generator hybrid system sizing; Utility-Interactive PV systems; Component specification; system costs and economics; Case studies and safety issues.

Those wanting to learn the skills of tapping the sun's energy can do so with confidence with this book in their library. 317 pages, softcover. Canada.

21-0366 PV: Design & Installation Manual   **$59**

Renewable Energy with the Experts

A professionally produced video series that covers basic technology introductions as well as hands-on applications. Each tape is presented by the leading expert in that field, and tackles a single subject in detail. Includes classroom explanations of how the technology works with performance and safety issues to be aware of. Then action moves into the field to review actual working systems, and finally assemble a real system on camera as much as possible. Other tapes in this series cover storage batteries with system sizing and water pumping. You'll find these tapes in appropriate chapters. This is the closest to hands-on you can get without getting dirt under your fingernails. All tapes are standard VHS format. USA.

80360   Set of Three Renewable Experts Tapes $100

Johnny Weiss has been designing and installing PV systems for over 15 years. He teaches hands-on classes at Solar Energy International in Colorado. 48 minutes.

80361   Renewable Experts Tape—PV   **$39**

Don Harris, the owner of Harris Turbines, practically invented the remote microturbine industry, and has manufactured over 1,000 in the past 15 years. 42 minutes.

80362   Renewable Experts Tape—Hydro   **$39**

Mick Sagrillo, the former owner of Lake Michigan Wind and Sun, has been in the wind industry for over 16 years with over 1,000 turbine installation/repairs under his belt. 63 minutes.

80363   Renewable Experts Tape—Wind   **$39**

Wind Power

Renewable Energy for Home, Farm and Business

By Paul Gipe. Wind energy technology has truly come of age—with better, more reliable machinery and a greater understanding of how and where wind power makes sense. In addition to expanded sections on gauging resources and turbine siting, this thoroughly updated edition contains a wealth of examples, case studies and human stories. Covers everything from small off-grid to large grid-tie, plus water pumping. This is the wind energy book for the industry. Everything you ever wanted to know about wind turbines. Has an abundance of pictures, charts, drawings, graphs and other eye candy, with extensive appendixes. 504 pages, softcover. USA.

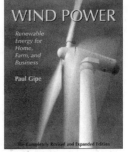

21-0343 Wind Power: Renewable Energy for Home, Farm, and Business   **$50**

Wind Energy Basics

A Guide to Small and Micro Wind Systems

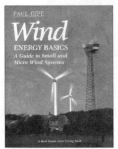

By Paul Gipe. A step-by-step manual for reaping the advantages of a windy ridgetop as painlessly as possible. Chapters include wind energy fundamentals, estimating performance, understanding turbine technology, off-the-grid applications, utility intertie, siting, buying, and installing, plus a thorough resource list with manufacturer contacts, and international wind energy associations. A Real Goods Solar Living Book. 222 pages, softcover. USA.

80928   Wind Energy Basics $19.95

Microhydro

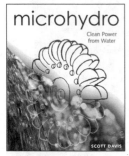

Hydroelectricity is the world's largest source of renewable energy. But until now there were no books that covered residential-scale hydropower, or micro-hydropower. This excellent guide covers the principles of micro-hydro, design and site considerations, equipment options, plus legal, environmental, and economic factors. Author Scott Davis covers all you need to know to assess your site, develop a workable system, and get the power transmitted to where you can use it. 157 pages, softcover. Canada.

21-0340 Microhydro $19.95

Battery Book for Your PV Home

This booklet by Fowler Electric Inc. gives concise information on lead-acid batteries. Topics covered include battery theory, maintenance, specific gravity, voltage, wiring, and equalizing. Very easy to read and provides the essential information to understand and get the most from your batteries. Highly recommended by Dr. Doug for every battery-powered household. 22 pages, softcover. USA.

80104   Battery Book for Your PV Home $8

Battery Care Kit

Our kit includes the informative *Battery Book for Your PV Home,* and a full-size graduated battery hydrometer. The *Battery Book* explains how batteries work and shows how to care for your batteries and get the best performance. Every battery-powered household needs a copy. The hydrometer is your most important tool for determining battery state of health. Detect and fix small problems before they require complete battery bank replacement.

Use Tip: The color-coded red, yellow, and green zones on the hydrometer are calibrated for automotive starting batteries: ignore them. Just look for differences among cells.

15754   Hydrometer/Book Kit $14

ENERGY CONSERVATION

Consumer Guide to Home Energy Savings, 8th Ed.

Energy efficiency pays. Written by the nonprofit American Council for an Energy-Efficient Economy, the *Consumer Guide* has helped millions of folks do simple things around their homes that save hundreds of dollars every year. The new eighth edition has up-to-date lists of all the most energy-efficient appliances by brand name and model number, plus an abundance of tips, suggestions and proven energy-saving advice for home owners. 264 pages, softcover. USA.

**21-0344 Consumer Guide to Home Energy
            Savings, 8th Ed. $9.95**

The Fuel Savers

A Kit of Solar Ideas for Your Home

What can you do right now, with your funky existing home and a minimum of cash, to save energy and live more comfortably? This book has a wealth of ideas from insulating curtains to solar hot water heaters. Each idea is examined for materials costs, fuel reduction, advantages, disadvantages, and cost-effectiveness. A great step-by-step guide for improving the comfort and efficiency of any building. 83 pages, softcover. Updated in 2002. USA.

21-0357 The Fuel Savers $15

WATER DEVELOPMENT

Solar Hot Water, Lessons Learned

Tom Lane probably has more hands-on practical experience from 1977 to 2002, and knowledge with all kinds of solar hot water systems, than anyone on the planet. He is detailed, specific, and refreshingly honest about the good and bad points of commonly available solar hardware. Extensive use of pictures, drawings, and system schematics help make explanations clear. Beautifully organized with a detailed table of contents. Pick what you need. 120 pages, softcover. USA.

21-0336 Solar Hot Water, Lessons Learned $29

The Sauna

By Rob Roy. A fascinating history of how and why our ancestors used saunas. This illustrated guide provides step-by-step instructions for building different types of saunas using traditional techniques. You'll also find complete advice on heating alternatives and a comprehensive list of sauna resources. 120 pages, softcover. USA.

82665   The Sauna $19.95

Cottage Water Systems

By Max Burns. An out-of-the-city guide to water sources, pumps, plumbing, water purification, and wastewater disposal, this lavishly illustrated book covers just about everything concerning water in the country. Each of the twelve chapters tackles a specific subject such as sources, pumps, plumbing how-to, water quality, treatment devices, septic systems, outhouses, alternative toilets, greywater, freeze protection, and a good bibliography for more info. This is the best illustrated, easiest to read, most complete guide to waterworks we've seen yet. 150 pages, softcover. USA.

80098   Cottage Water Systems $24.95

The Home Water Supply

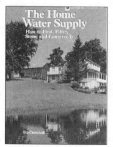

By Stu Campbell. Explains completely and in depth how to find, filter, store, and conserve one of our most precious commodities. A bit of history; finding the source; the mysteries of ponds, wells, and pumps; filtration and purification; and a good short discussion of the arcana of plumbing. After reading this book, you may still need professional help with pipes and pumps, but you will know what you're talking about. 236 pages, softcover. USA.

80205   The Home Water Supply $19

Rainwater for the Mechanically-Challenged Video

You say you're *really* mechanically challenged and want more than a few pictures? Here's your salvation. From the same irreverent, fun-loving crew that wrote the book below. See how all the pieces actually go together as they assemble a typical rainwater collection system, and discuss your options. This is as close as you can get to having someone else put your system together. 37 minutes and lots of laughs. USA.

82568   Rainwater Challenged Video $19

Rainwater Collection for the Mechanically-Challenged

Laugh your way to a home-built rainwater collection and storage system. This delightful paperback is not only the best book on the subject of rainwater collection we've ever found, it's funny enough for recreational reading, and comprehensive enough to lead a rank amateur painlessly through the process. Technical information is presented in layman's terms, and accompanied with plenty of illustrations and witty cartoons. Topics include types of storage tanks, siting, how to collect and filter water, water purification, plumbing, sizing system components, freeze-proofing and wiring. Substantially revised and expanded in 2003. Includes a resources list and a small catalog. 108 pages, softcover. USA.

**80704   Rainwater Collection for the
           Mechanically-Challenged $20**

Build Your Own Water Tank

By Donnie Schatzberg. An informative booklet by one of our original customers that has been updated and greatly expanded with more text, drawings, and illustrations. It gives you all the details you need to build your own ferro-cement (iron-reinforced cement) water storage tank. No special tools or skills are required. The information given in this book is accurate and easy to follow, with no loose ends. The author has considerable experience building these tanks, and has gotten all the "bugs" out. 58 pages, softcover. USA.

80204   Build Your Own Water Tank $14

A Great Water Pumping Video

Part of the Renewable Energy with the Experts series, this 59-minute video features solar-pumping pioneer Windy Dankoff, who has more than fifteen years experience in the field. Windy demonstrates practical answers to all the most common questions asked by folks facing the need for a solar-powered pump, and offers a number of tips to avoid common pitfalls. This is far and away the best video in this series, and offers some of the best advice and knowledge available for off-the-grid water pumping. This grizzled old technician/copyeditor even learned a few new tricks about submersible pump installations. Highly recommended! USA

80368   RE Experts Water Pumping Video $39

Build Your Own Ram Pump

Utilizing the simple physical laws of inertia, the hydraulic ram can pump water to a higher point using just the energy of falling water. The operation sequence is detailed with easy-to-understand drawings. Drive pipe calculations, use of a supply cistern, multiple supply pipes, and much more are explained in a clear, concise manner. The second half of the booklet is devoted to detailed plans and drawings for building your own 1- or 2-inch ram pump. Constructed out of commonly available cast iron and brass plumbing fittings, the finished ram pump will provide years of low-maintenance water pumping for a total cost of $50 to $75. No tapping, drilling, welding, special tools, or materials are needed. This pump design requires a minimum flow of 3 to 4 gallons per minute, and 3 to 5 feet of fall. It is capable of lifting as much as 200 feet with sufficient volume and fall into the pump. The final section of the booklet contains a setup and operation manual for the ram pump. 25 pages, softcover. USA.

80501   All About Hydraulic Ram Pumps $10.95

COMPOSTING TOILETS AND GREYWATER

THE GREYWATER GUIDE SERIES

Create an Oasis with Greywater— Revised 4th Edition

Choosing, building and using greywater systems

Describes how you can save water, save money, help the environment and relieve strain on your septic tank or sewer by irrigating with reused wash water. Describes twenty kinds of greywater systems. The concise, readable format is long enough to cover everything you need to consider, and short enough to stay interesting. Plenty of charts and drawings cover health considerations, greywater sources, designs, what works and what doesn't, bio-compatible cleaners, maintenance, preserving soil quality, the list goes on.... 51 pages, softcover. USA.

82440   Create an Oasis with Greywater $14.95

Builder's Greywater Guide

A companion to Create an Oasis with Greywater

Will help you work within or around codes to successfully include greywater systems in new construction or remodeling. Includes reasons to install or not install a greywater system, flowcharts for choosing an appropriate system, dealing with inspectors, legal requirements checklist, design and maintenance tips, how the earth purifies water, and the complete text of the main U.S. greywater codes with English translations. 46 pages, softcover. USA.

80097   Builder's Greywater Guide $14.95

Branched Drain Greywater Systems

A companion to Create an Oasis with Greywater

Detailed plans and specific construction details for making a branched drain greywater system in any context. This ultra-simple, robust system is the best choice for over half of residential installations. It provides reliable, sanitary, and low-maintenance distribution of household greywater to downhill plants without filtration, pumping, or a surge tank. It is one of the few systems that is both practical and legal, and has been described as "the best greywater system since indoor plumbing." Note: the area to be irrigated has to be downhill from the house, and laying out the lines must be done with fanatical attention to proper slope. 52 pages, softcover. USA.

82494   Branched Greywater $14.95

Greywater Guide Booklet Set

Save over $5 by picking up all three Greywater booklets together.

21-0360 Greywater Guide Booklet Set $39

The Humanure Handbook 2nd Ed.

A Guide to Composting Human Manure

Deemed "Most Likely to Save the Planet" by the Independent Publishers in 2000.

By Joseph Jenkins. For those who wish to close the nutrient cycle, and don't mind getting a little more personal about it than our manufactured composting toilets require, here's the book for you. Provides basic and detailed information about the ways and means of recycling human excrement, without chemicals, technology, or environmental pollution. Includes detailed analysis of the potential dangers involved and how to overcome them. The author has been composting his family's humanure safely for the past twenty years, and with humor and intelligence has passed his education on to us. 302 pages, paperback. USA.

82308   The Humanure Handbook $19

How to Shit in the Woods

By Katheleen Meyer. A humorous name for a useful book. Chock-full of practical tips that are useful in a variety of situations—and the great sense of humor makes learning fun. Written by a woman, it includes lots of extra information for women in the wild. Second edition covers health concerns, feedback from first edition and much more. 128 pages, softcover.

82640   How to Shit in the Woods $7

The Composting Toilet Book

Even though off-the-grid home owners have been using them for decades, convincing the local inspector to permit a composting toilet system in your conventional year-round home can be a frustrating experience. This book gives you the ammunition you need to get this water-saving, eco-friendly technology installed and permitted. This is the definitive handbook we've been

wanting for at least ten years. It's packed with technical details of various permittable systems, state-by-state permitting information, maintenance and operation tips the manufacturers often do not provide, profiles of long-time composting toilet users, and a bonus section on greywater applications. While costs and restrictions for conventional systems have been rising, composting toilets have dropped in price and risen in quality. There's never been a better time to switch over to a composting toilet system, and there's never been a better resource on the topic than this one. 150 pages, softcover. USA.

80708   The Composting Toilet Book $29.95

Septic Systems Owners Manual

This book, in simple terms, describes the conventional gravity-fed septic system, how it works, how it should be treated, how it should be maintained, and what to do if things go wrong. There is also basic information on the recent evolution in composting toilet systems, designs for simple graywater systems, and more. 163 pages, softcover.

82366   Septic Systems Owners Manual $14.95

OFF-THE-GRID LIVING AND HOMESTEAD TOOLS

How to Grow More Vegetables

First published in 1974, this book has become the go-to guide for anyone trying to create or sustain a home garden. Author John Jevons shows how to use Alan Chadwick's biointensive techniques to reduce daily watering and general maintenance—from the beginning stages of setting up the garden beds to creating insect harmony. Includes charts and illustrations. 192 pages, softcover. USA.

21-0345 How to Grow More Vegetables $17.95

Cooking with the Sun

How to Build and Use Solar Cookers

Solar cookers don't pollute, use no wood or other fuel resources, operate for free, and still run when the power is out. A simple and highly effective solar oven can be built for less than $20 in materials.

The first half of this book, generously filled with pictures and drawings, presents detailed instructions and plans for building a solar oven that will reach 400°F or a solar hot plate that will reach 600°F. The second half has one hundred tested solar recipes. These simple-to-prepare dishes range from everyday Solar Stew and Texas Biscuits, to exotica like Enchilada Casserole. 116 pages, softcover. USA.

21-0273 Cooking with the Sun $9.95

Build Your Own Earth Oven

By Kiko Denzer. Wonderfully illustrated and clearly written, this book covers everything from selecting materials and constructing the oven to baking and troubleshooting. The completed earth oven can cook anything you would typically cook inside, yet lends a rich, smoky flavor to your recipes. 132 pages, softcover.

21-0353 Build Your Own Earth Oven   **$14.95**

Let It Rot!

By Stu Campbell. This is the third edition of the classic guide to turning household waste into gardener's gold! Since 1975, *Let It Rot!* has helped countless gardeners recycle waste materials like household garbage, grass clippings, and ashes to create useful, soil-nourishing compost. This quietly humorous book provides an introduction to composting, covers reasons to compost; differing approaches; how decomposition works; various methods, ingredients, and containers; how to speed decomposition; and how to use the end result. Campbell is an experienced gardener, and the book goes into great detail, but the text remains clear and interesting. The simple black-and-white illustrations vary between decorative sketches and straightforward diagrams. 6" x 9" size, 153 pages, softcover. USA.

82459   Let It Rot!   **$13**

The Worm Book

By Loren Nancarrow and Janet Hogan Taylor. Got worms? This is the most comprehensive resource book on worm growing and worm composting available. It's also a sprightly and affectionate look at everyone's favorite invertebrates, and more fun than a barrel of worms! Includes plans for several types of composting bins, from basic to deluxe, directions for worm composting, essential earthworm biology, and everyone's favorite, worm recipes. Also resources for buying worms and ready-made composting systems. 6" x 9" format, 153 pages, softcover. USA.

21-0364 The Worm Book   **$12**

SUSTAINABLE TRANSPORTATION

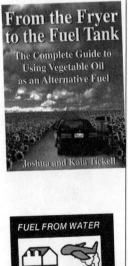

From the Fryer to the Fuel Tank

How to Make Cheap, Clean Fuel from Free Vegetable Oil

New updated and expanded version by Joshua and Kaia Tickell, who, with sponsorship from Real Goods, drove their unmodified Veggie Van over 10,000 miles around the United States during the summer of 1997, while stopping at fast food restaurants for fuel fill-ups of used deep fat fryer oil. Vegetable oil provided 50% of their fuel. They towed a small portable fuel processing lab in a trailer behind their Winnebago. This inexpensive and environmentally friendly fuel source can be modified easily to burn in diesel engines by following the simple instructions. A complete introduction to diesel engines and potential fuel sources is included. Over 130 photographs, diagrams, charts, and tables, 162 pages, softcover. USA.

80-961   From the Fryer to the Fuel Tank   **$24.95**

Fuel from Water

Energy Independence with Hydrogen

By Michael Peavey. An in-depth and practical book on hydrogen fuel technology that details specific ways to generate, store, and safely use hydrogen. Hundreds of diagrams and illustrations make it easy to understand, very informative, and practical. If you're interested in using hydrogen in any way, you will find this book indispensable. 244 pages, softcover. USA.

80-210   Fuel from Water   **$25**

Hydrogen: Hot Stuff, Cool Science

Rex A. Ewing shows us that science need not be dull, as he explains hydrogen energy and all the potential it holds for our society. Starting out with a section that takes us back to chemistry class, then delving into discovery and extraction of hydrogen, and ending with the potential hydrogen has to replace fossil fuels in everyday use, this book covers it all! Clean, renewable sources of energy to extract hydrogen, including wind, solar, and biomass, are explored, along with 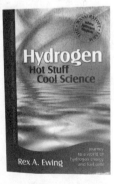 more traditional such as coal, natural gas, and nuclear power. Readers will discover why storage and transportation are such critical hydrogen issues. 256 pages, softcover. USA

92-0264 Hydrogen: Hot Stuff, Cool Science   **$24.95**

BOTTOM OF THE BARREL

What World Oil Peak Means for Our Way of Life

The pump says $7.98 a gallon. Your January heating bill reads $436. A pound of tomatoes rings up at $10.

It's not a scene from a science fiction novel. According to Richard Heinberg, author of *The Party's Over: Oil, War, and the Fate of Industrial Societies,* the world is about to start running out of cheap oil. And its loss—and rising cost—will permeate nearly every facet of our everyday lives.

Experts predict we'll reach the point of peak oil production sometime between 2006 and 2016, after which world oil supply will go into steep decline. And less oil, of course, means that we'll have to learn to live with less. Less transportation, less imported food and goods, and less of the everyday luxuries we've learned to take for granted. Oil is at the core of industrial growth, and its depletion will cause a dramatic change in life as we know it. "We're entering an era that is as different from the industrial era as the industrial era was to pre-industrialization," Heinberg says.

Over half of all oil-producing countries have passed their peak

It's no surprise that this is happening. Oil is a finite resource—one that petroleum geologist M. King Hubbert spent his life studying. In 1956, Hubbert concluded that when any given country had extracted half of its total oil, it would reach the point of peak production. That's because at about the halfway point, the cheap, easy oil already would have been extracted. What remained would be more costly and difficult to obtain. Thus, based on the law of diminishing returns, the rate of production would drop off. Hubbert predicted that U.S. oil production would peak in 1970 as it reached the halfway point of extraction. It did, and since then the same thing has happened in many other countries—out of 44 countries producing significant amounts of oil, 24 have already passed their peak.

"It's not a matter of some theoretical process that may or may not happen," Heinberg says of world peak oil production. "Everyone agrees it's going to happen. It's just a question of when."

Indeed, Heinberg is not alone. Matthew Simmons, CEO of Simmons and Co. International and energy advisor to President Bush, told a Swedish television station, "We need a wake up call. We need it desperately. We need basically a new form of energy. I don't know that there is one."

There's no magic elixir to replace oil

Oil as an energy source is the equivalent of "winning the lottery," Heinberg says. It delivers large amounts of energy for a small price (at least, up until now). Solar and wind technologies are certainly viable alternatives, but it will take immense resources—including oil—to build enough renewable energy systems to supply our current energy demand. To make the change without social chaos, we'll need to alter our energy habits drastically as we convert our infrastructure to sustainable power sources, he says.

Heinberg explores possible solutions in his new book *Powerdown: Options and Actions for a Post-Carbon World.* He has also started teaching

> With just 5% of the world's population, the U.S. consumes 25% of the world's oil.

students at New College of California about how to transition to a post-petroleum society. Among his recommendations is being as self-reliant as possible, by growing your own food and using less energy, for example. He also emphasizes sharing information, helping friends and neighbors understand the situation. "We've gotten ourselves into a real fix here," he says. "You have to make the best of a situation, no matter what it is."

Order The Party's Over *on page 412.*

OIL, POLITICS, AND GENERAL INTEREST

The Next Industrial Revolution

Working with nature's rules instead of continuing to flout them is what Bill McDonough and Michael Braungart believe is the next industrial revolution. Highlighting their work at the Environmental Studies center at Oberlin College and at locations around Europe, this film is truly encouraging—an inspirational video about how change can and is happening. Narrated by Susan Sarandon. VHS, 55 minutes.

21-0352 The Next Industrial Revolution **$24.95**

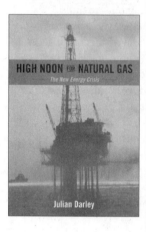

High Noon for Natural Gas— The New Energy Crisis

While much is discussed concerning petroleum supplies, the likelihood of difficulties in supply and distribution of natural gas is seldom mentioned. Julian Darley takes this issue head on—bringing the reader up to speed on the looming crisis with natural gas. While almost 20% of all electricity in the United States is supplied by natural gas, few of us consider the implications of this increasing reliance. *High Noon for Natural Gas* offers an in-depth analysis of several facets of this impending issue. A very important book for anyone with concerns about our nation's energy policy. 280 pages, softcover. USA.

265- **High Noon for Natural Gas** **$18**

The End of Suburbia

Oil Depletion and the Collapse of The American Dream

As we enter the twenty-first century, and the global demand for fossil fuels begins to outstrip supply, serious questions are beginning to emerge about the sustainability of the American way of life. With brutal honesty and a touch of irony, *The End of Suburbia* explores our prospects as the planet approaches a critical era—world oil peak and the inevitable decline of fossil fuels. A must see DVD for anyone—suburbanite or not—interested in the issue of world oil peak. DVD, 78 minutes.

90-0047 The End of Suburbia **$27.75**

The Party's Over

Oil, War and the Fate of Industrial Societies

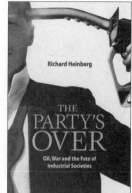

Without oil, what would you do? How would you travel? How would you eat? The world is about to change dramatically and forever as the global production of oil reaches its peak. Thereafter, even with a switch to alternative energy sources, industrial societies will have less energy available to do all the things essential to their survival. Author Richard Heinberg deals head-on with this imminent decline of cheap oil in *The Party's Over,* a riveting wake-up call that does for oil depletion what Rachel Carson's *Silent Spring* did for the issue of chemical pollution—raise to consciousness a previously ignored global problem of immense proportions. 288 pages, softcover. USA.

90-0267 The Party's Over **$17.95**

Powerdown

Options and Actions for a Post-carbon World

Now that we know that world oil peak is imminent, how can we avoid decades of war, economic collapse, and environmental catastrophe? Author Richard Heinberg's follow-up to *The Party's Over* explores realistic options available to industrial societies as oil and natural gas run out. Rejecting continued competition for resources and denial, it advocates cooperation, conservation, and sharing, discussing how the major players are likely to respond. *Powerdown* is crucial reading for everyone from policymakers to community members. 288 pages, softcover. USA.

92-0268 Powerdown **$16.95**

Appendix

Much of the Real Goods and Solar
Living Institute staff in 2003.

Real Goods' Mission Statement

Through our products, publications, and educational
demonstrations, Real Goods promotes and inspires
an environmentally healthy and sustainable future.

Part I: Who We Are and How We Got Here

Who Put the "Real" in Real Goods

As did many of his contemporaries in the 1960s and early 1970s, John Schaeffer, founder of Real Goods, experimented with an alternative lifestyle. After protracted exposure to nearly every strand of the lunatic fringe, he graduated from U.C. Berkeley in 1971 and moved to a commune called "Rainbow" outside of Boonville, California. There, in an isolated 290-acre mountain community, John pursued a picturesque life of enlightened self-sufficiency.

John Schaeffer, Real Goods founder.

ALAIN MCLAUGHLIN

Despite the idyllic surroundings, John soon found that certain key elements of life were missing. After several years reading bedtime stories to his children by the flickering light of a kerosene lamp, John began to squint. He grew tired of melted ice cream and luke-warm beer. He began to miss the creature comforts his family was lacking due to their "off-the-grid" lifestyle. He yearned for just a tiny amount of energy to strike a balance between the lifestyle he had grown up with and complete deprivation. In other words, John came to the realization that self-sufficiency was much more appealing as a concept than a reality.

Then he discovered 12-volt power. John hooked up an extra battery to his car that he charged while commuting to work, with just enough juice to power lights, a radio, and the occasional television broadcast. Despite his departure from a pure ascetic lifestyle, each and every time that *Saturday Night Live* aired, John's home became the most popular place on the commune. Eventually, when the tweve-hour community work days began to take their toll, John took a job as a computer operator in Ukiah, some thirty-five twisty miles from Boonville.

Once the word got out that John would be making the trek over the mountain to the "big city" daily, he became a one-man pick-up and delivery service, procuring the fertilizer, chicken wire, bone meal, tools, and supplies needed for the commune. As a conscientious and frugal person, John spent hours scrutinizing the hardware stores and home centers of Ukiah, searching for the best deals on the real goods needed for the communards' close-to-the-earth lifestyle. One day, while driving his VW bug back to the commune after a particularly vexing shopping trip, a thought occurred to John. "Wouldn't it be great," he mused, "if there was one store that sold all the products needed for independent living, and sold them at fair prices?" The idea of Real Goods was born. The company thrived opening up more retail stores and eventually morphed into a mail-order company.

Real Goods in the New Millennium

From its humble beginnings in 1978, Real Goods became a Real Business, with Real Employees serving Real Customers. In the 1990s Real Goods pioneered the "direct public offering" process, whereby it raised investment capital from its customers without the need for investment bankers or other financial middlemen. Before the Internet had really caught on, Real Goods was selling stock electronically and allowing its customers to print virtual stock certificates in the privacy of their own homes. The company now can lay claim to the title of the Oldest and Largest catalog firm devoted to the sale and service of renewable energy products in the world. Real Goods, now Gaiam Real Goods since its January 2001 merger with Gaiam, Inc., of Colorado, is still devoted to the same principles that guided its founding—quality products for fair prices, and customer service with courtesy and dignity.

Early on, John managed to turn his personal commitment to right livelihood into company policy, pioneering the concept of a socially conscious and environmentally responsible business. The company consistently has been honored and awarded for its ethical and environmental business standards. Plaudits include Corporate Conscience Awards (from the Council on Economic Priorities); inclusion in *Inc.* magazine's 1993 list of America's 500 Fastest-Growing Companies; three consecutive Robert Rodale Awards for Environmental Education; Northern California Small Business of the Year Winner for 1994; finalist for entrepreneur of the year two years running; and news coverage in *Time, Fortune, The Wall Street Journal, Mother Earth News,* numerous TV appearances, countless Japanese magazines, plus many thick scrapbooks full of press clippings.

Five Principles to Live By

Gaiam Real Goods is considered newsworthy not because our methods reflect the latest trends in corporate or business-school thinking, but because we unwittingly have helped to birth an astonishingly healthy "baby"—an ethical corporate culture based on environmental and social responsibility. Led by a certain naiveté and affection for simplicity, Gaiam Real Goods has discovered some simple principles that, by comparison to the "straight" business world, are wildly innovative. This has not been the work of commercial gurus or public relations mavens, but rather the result of realizing that business need not be so complicated that the average person cannot understand its workings. Our business is built around five simple principles.

PRINCIPLE #1: THIS IS A BUSINESS

A business is, first and foremost, a financial institution. You can have the most noble social mission on the planet, but if you can't maintain financial viability, you cease to exist. And so does your mission. The survival instinct is very strong at Gaiam Real Goods, and that reality governs many decisions. To say it another way, you can't be truly sustainable if your business isn't economically sustainable (read profitable). To ensure the continued flourishing of our mission, we pursue profitability through our catalogs, retail store, residential solar sales and installation business, internet presence, and our design and consulting business.

PRINCIPLE #2: KNOW YOUR STUFF

Knowledge seldom turns a profit, yet our social and environmental missions cannot be achieved without it. The independent lifestyle we advocate relies largely on technologies that often require a high degree of understanding, and a level of interaction that has been largely forgotten during our nation's half-century binge on cheap power. We aren't interested in selling things to people that they aren't well-informed enough about to live with comfortably and happily with. We want people to understand not only what we are selling and how it is used, but how a particular piece of hardware contributes to the larger goal of a sustainable lifestyle.

It goes against the grain of mainstream business to give anything away. Even a "loss leader" is designed to suck you into the store to buy other, higher-profit items. At Gaiam Real Goods, knowledge is our most important product, yet we give it away daily, through our catalogs (there's a lot in there to learn, even if you never buy a thing); through our Solar Living Center (free self-guided and group tours), now run by the nonprofit Solar Living Institute; through free workshops at our Hopland store, and through workshops we support through the SLI's annual SolFest renewable energy celebration. Our website acts as a launching pad for renewable energy research, leading you to fascinating information on sustainability topics of all kinds. We believe that as our collective knowledge of sustainability principles and renewable energy technology increases, the chance of achieving the Gaiam Real Goods mission increases, too.

PRINCIPLE #3: GIVE FOLKS A WAY TO GET INVOLVED

Just about all of our employees are also our customers and more than half of them live with solar—many live completely off-the-grid. Our Lifetime Membership program honors our community of customers with financial and educational benefits, and interns and volunteers are welcomed warmly by the SLI staff. Gaiam Real Goods is becoming a real community, acting in concert toward the common goal of a sustainable future. With the knowledge that the age of oil is likely soon coming to an end, it's comforting to know that we are all in this together.

PRINCIPLE #4: WALK THE WALK

At Gaiam Real Goods, we conduct our business in a way that is consistent with our social and environmental mission. We use the renewable energy systems we sell, and we sell what works. Our merchandising team makes absolutely sure the merchandise we sell performs as expected, is safe and nontoxic when used as directed, and is made from the highest quality sustainable materials. In 1990, we challenged our customers to help us rid the atmosphere of one billion pounds of CO_2 by the year 2000, and we achieved our goal three years ahead of schedule. Real Goods was recognized by being awarded the Rodale Award, as the business making the most positive contribution to the environment in America, for three years running. We don't just talk the talk at Gaiam Real Goods, we walk the walk. Come visit us at the Solar Living Center in Hopland, California, and see some of our innovative practices. Fill up your biodiesel vehicle on site, see water pumped

Water flows into the main pond at the Solar Living Center pumped by solar electricity.

JENNY MOON

from the sun, and learn from hydrogen fuel cell demonstrations.

PRINCIPLE #5: HAVE FUN!

We strongly support the best party of the year for 10,000 of our closest friends every summer in August on the grounds of our Solar Living Center in Hopland. We look forward to the Solar Living Institute's annual SolFest renewable energy celebration all year. If we've learned anything since 1978, it's that all work and no inspiration makes Jack and Jill a couple of burnt out zombies. SolFest is our little reminder to take care of ourselves with some good, clean fun, so we'll be rejuvenated and reinvested in the hard work of creating a sustainable future.

Part II: Living the Dream: The Real Goods Solar Living Center

Imagine a destination where ethical business is conducted daily amidst a diverse and bountiful landscape, where the gurgle of water flowing through its naturally revitalizing cycle heightens your perception of these ponds, these gardens, these living sculptures. You follow the sensuous curve of the hill and lazy meanders of the watercourse to a structure of sweeping beauty, where floor-to-ceiling windows and soaring architecture clearly proclaim this building's purpose—to take every advantage of the power of the sun throughout its seasonal phases. A few more steps and the spidery legs of a wind-generator come into view, and the top of a tree that looks as though it might be planted in a Cadillac. An awesome sense of place begins to reveal itself to you. Inside the building, sunlight and rainbows play across the walls and floors of a 5,000-square-foot showroom, and you begin to understand that all of this, even the offices and cash

The Ninth Annual SolFest Energy Festival from 1,000 feet in the air on August 21, 2004 (count the PV modules!).

registers, are powered by the energy of sun and wind. Welcome to the Solar Living Center in Hopland, California, the crowning achievement of the Gaiam Real Goods mission.

Our Solar Living Center began as the vision of Real Goods founder and president, John Schaeffer. His dream was to create an oasis of biodiversity, where the company could demonstrate the culture and technology of solar living, where the grounds and structures were designed to embody the sustainable living philosophy of Real Goods' catalogs and business. With the opening of the Solar Living Center in April 1996, John's vision is now a reality. As of mid-2004, 1.3 million people have made the two-hour trek north from San Francisco to visit the center, and have left this place with an overwhelming sense of inspiration and possibility.

In 1998, the Real Goods Solar Living Institute split off from its parent Real Goods Trading Corporation and became a legal 501(c)(3) nonprofit called the Solar Living Institute (SLI). Since then the SLI has nurtured and developed the 12-acre permaculture site that has flourished with fecundity.

Form and Function United: Designing for the Here and Now

If the "weird restrooms" sign doesn't grab them first, the 40,000 daily passersby on busy Highway 101 are bound to notice the striking appearance of the company showroom. This does not look like business as usual! The building design and the construction materials were selected with an eye toward merging efficiency of function, educational value, and stunning beauty.

The architect chosen to design the building was Sim Van der Ryn of the Ecological Design Institute of Sausalito, California. His associate, David Arkin, served as project architect, and Jeff Oldham of Real Goods managed the building of the project. Their creation is a tall and gracefully curving single-story building that is so adept in its capture of the varying hourly and seasonal angles of the sun that additional heat and light are virtually unnecessary. Wood-burning stoves provide back-up heating for the coldest winter mornings and solar-powered fluorescent lighting is available, but is rarely used. Through a combination of overhangs and manually controlled hemp awnings, excess insolation during the hot-weather months has been avoided. Solar-powered evaporative coolers provide a low-energy

alternative to air conditioning, and are also used to flush the building with cool night air, storing "coolth" in the six hundred tons of thermal mass of the building's walls, columns, and floor. Grape arbors and a central fountain with a "drip ring" for evaporative cooling are positioned along the southern exposure of the building to serve as a first line of defense against the many over-one-hundred-degree days that occur during the summer in this part of California.

Many of the materials used in the construction of the building were donated by companies and providers with a commitment similar to Real Goods'. As an example, the walls of the SLC were built with more than 600 rice straw bales donated by the California Rice Industries Association. Previously, rice straw has been disposed of by open burning, a practice that contributes to the production of carbon dioxide, the so-called "greenhouse gas" that is the leading cause of global warming. By using this agricultural by-product as a building material, everyone benefits. The farmers receive income for their straw bales, no carbon dioxide is produced, and the builder benefits from a low-cost, highly efficient building material that minimizes energy consumption.

At the SLC, visitors experience the practicality of applied solar power technology, including the generation of electricity and solar water pumping. The electrical system for the facility comprises nearly 150 kilowatts of photovoltaic power and 3 kilowatts of wind-generated power. Through an intertie with the Pacific Gas and Electric Company, the SLC sells the excess power it generates to the electric company, making the SLC 100% independent from the grid. Once again, like-minded companies have shared in the costs of developing the Solar Living Center as a demonstration site. Siemens Solar (now Shell Solar) donated more than 10 kilowatts of the latest state-of-the-art photovoltaic modules to the center, and periodically uses the SLC as a test site for new modules. Trace Engineering (now Xantrex) contributed four intertie inverters, which are on display behind the glass window of the SLC's "engine room" so that visitors can see the inner workings of the electrical system. More recently, Beacon Power donated a 5,000-watt intertie inverter to the electrical room.

In November 1999, a partnership between GPU, Astropower, Real Goods, and the Solar Living Institute installed a 132-kilowatt PV array on campus, one of the largest in power-hungry Northern California. This direct-intertie array delivers 163,000 kWh of power annually, enough to power fifty average California homes. On tours, either self-guided or with Solar Living Institute tour guides, visitors learn about the guiding principles of sustainable living, and are offered a chance to appreciate the beauty that lies in the details of the project. The site also provides a wonderful space for presentations by guest speakers and special events, and serves as the main campus and classroom for the workshop series staged by the Solar Living Institute, a nonprofit dedicated to education and inspiration toward sustainable living.

The Natural World Reclaimed: The Grounds and Gardens

Learning potential is intrinsic to the award-winning landscape, designed by Chris and Stephanie Tebbutt of Land and Place. For this project, the design of the grounds, gardens, and waterworks was the first phase of construction and contributed much to establishing the character of the site. This is a radically different approach than most commercial building projects, where the landscaping appears to be a cosmetic afterthought. At the SLC, the gardens are a synthesis of the practical and the profound. Most of the plantings produce edible and/or useful crops, and the vegetation is utilized to maximize the site's energy efficiency while portraying the dramatic aspects of the solar year. Plantings and natural stone markers follow the lines of sunrise and sunset for each equinox and solstice, emanating from a sundial at the exact center of the oasis. More sundials and unique solar calendars scattered throughout the site encourage visitors to establish a feeling for the relationship between this specific location and the sun. Those of us who work and play here daily have discovered an almost organic connection with the seasonal shifts of the solar year and the natural rhythms of the Earth.

The gardens themselves follow the sun's journey through the seasons, with zones planted to represent the ecosystems of different latitudes. Woodland, Wetland, Grassland and Dryland zones are manifested through plantings moving from north to south, with the availability of water the definitive element. Trees are planted to indicate the four cardinal directions. The fruit garden, perennial beds, herbs, and grasses reflect the abundance and fertility of a home-based garden economy. Visitors discover aesthetic statements in design and landscape tucked into nooks and crannies all over the grounds, and unexpected simple pleasures, too, like shallow water channels for cooling aching feet, and perfect hidden spots for picnics and conversation.

Unique to these gardens are the "Living Structures," which reveal their architectural nature according to the turn of the seasons. Through annual pruning, plants are coaxed into various dynamic forms, such as a willow dome,

Duncan Peak looms high over the gorgeous permaculture gardens at the Solar Living Center.

a hops tipi, and a pyramid of timber bamboo. These living structures grow, quite literally, out of the garden itself. Visitors unaccustomed to the heat of a Hopland summer find relief inside the "agave cooling tower," where the turn of the dial releases a gentle mist into the welcome shade of vines and agave plants. By the time visitors leave the center, they've begun to understand the subtle humor of the "memorial car grove," where the rusting hulks of '50s and '60s "gas hog" cars have been turned into planter boxes for trees. These "grow-through" cars make a fascinating juxtaposition to the famous Northern California "drive-through" redwood trees!

A Place to Play

In case this all sounds awfully serious, it should be pointed out that the SLC is a wonderful place to play! Upon entering the showroom, one is greeted by a delightful rainbow spectrum cre-

RIGHT: Annual Earth Day for Kids features hands-on cob building. BELOW: Grow through cars instead of drive through trees grace the Solar Living Center showing nature taking back the 1950's gas guzzling monstrosities.

ated by a large prism mounted in the roof of the building. Visitors need not understand on a conscious level that this rainbow functions throughout the year as a "solar calendar," or that the prism's bright hues mark the daily "solar noon" - this is a deeper learning that ignites the place where inspiration happens, not a raw scientific dissertation. Outside, interactive games and play areas tempt the young at heart to forget about the theory and enjoy the pleasure of pure exploration. A six-station bicycle "generator" allows riders to see how much energy humans are capable of generating compared to the average American home's energy needs. It doesn't take long for riders to really feel how much energy is required to create the tiniest bit of energy; almost everyone takes the time to compare the aching results of muscle power to the ease with

APPENDIX

which the same amount of energy is harvested from the sun with a solar panel.

The hands-down favorite for kids is the sand and water area. A solar-powered pump provides a water source that can then be channeled, diverted, dammed, and flooded through whatever sandy topography emerges from the maker's imagination. A shadow across the solar panel stops the flow of water, and it doesn't take long for kids to become immersed in starting and stopping the flow at will. Without even realizing it, these little scientists are learning about engineering, hydrology, erosion, and renewable energy theory!

A Global Warming displays shows a map of the greater San Francisco Bay Area as it sits today and what the water level would look like with a 1-meter rise in the ocean's level. If current levels of carbon dioxide production continue, a 1-meter rise is expected by the mid- to late twenty-first century. The display shows the Sacramento valley and much of the oceanside and bayside towns inundated.

A hydrogen fuel cell display shows visitors how solar power electrolyzes water into hydrogen and oxygen and how those two elements are fed back into a fuel cell to power a small fan. These are only a very few of the dozens of hands-on interactive displays available free to the public at the Solar Living Center.

Where to Find the Solar Living Center

The Real Goods Solar Living Center is located 94 miles north of San Francisco on Highway 101 and is open every day except Thanksgiving and Christmas. There is no admission charge for regularly scheduled or self-guided tours and picnicking is strongly encouraged.

Customized group tours for students, architects, gardeners, or others with special interests are available on a fee basis by advanced reservation, through the nonprofit Solar Living Institute. The Institute also offers a variety of structured learning opportunities, including intensive, hands-on, one-day to one-week seminars on a variety of renewable energy, sustainable living, and permaculture gardening topics. Please call the Institute at 707.744.2017 for more information, or visit their website at www.solarliving.org.

A Place in the Sun

By John Schaeffer. A fascinating behind-the-scenes account of the creation of the Gaiam Real Goods Solar Living Center, in which a barren, polluted site was transformed into a masterpiece of environmental design. This inspirational book outlines how the principles of sustainability were incorporated into every facet of design, from the self-regulating straw bale showroom to the wind turbine and solar panels that power the facility. Each chapter was written by different members of the collaborative design/construction team, with dozens of photos, illustrations, and diagrams, and an eight-page section (printed on bamboo paper!) of beautiful color photos. Anyone interested in design, architecture, or the environment will be delighted with this lively resource. This is the rare book that entertains while it teaches and inspires. 200 pages, softcover, USA.

80-044   A Place in the Sun $24.95

Part III: Spreading the Word: The Solar Living Institute

In April 1998, the Institute for Solar Living separated from Real Goods Trading Corporation to become a legal nonprofit 501(c)(3). By severing financial ties to a for-profit corporation, the Institute is now free to develop its educational mission without the constraint of profitability, and to focus its efforts solely on environmental education.

Since Real Goods' merger with Gaiam, Inc., in January 2001, the Solar Living Institute (SLI) has received several generous donations from Gaiam, which have enabled it to greatly expand its interactive displays and programs and heighten its impact on environmental education.

The SLI's Four-Fold Mission

The Institute's mission is to teach people of all ages how to live more sustainably on the Earth, to teach interdependence between people and the environment, to replace fossil fuels with renewable energy, and to honor biodiversity in all its forms. To further its mission, the Institute is focusing its resources on three endeavors: Sustainable Living Workshops, the Solar Living Center's Interactive Displays, Exhibits, and Educational Tours, and the SolFest Energy Fair, Educational Celebrations, and EarthDay for Kids events.

#1: Sustainable Living Workshops

One of the Institute's short-term goals is the continual expansion of the Sustainable Living Workshop series, offered at the Solar Living Center in Hopland, California, and other locations. Now in its fourteenth season with over 2,000 students annually, Institute workshops provide an ideal opportunity to learn hands-on skills, meet people of similar interests, work beside them, and build the family up close and hands-on. The main campus (the SLC) is not only an immensely inspirational setting, but a living, breathing model for sustainable development, restorative permaculture landscaping, and renewable technologies. Part park and part museum, the SLC is an invaluable resource for the Institute, the center of testing and evaluation for the Gaiam Real Goods enterprise, and a

JENNY MOON

SLI workshop students wire a PV array in one of dozens of hands-on solar training workshops.

unique opportunity for students to see sustainable living technologies at work. Almost all Institute workshops relate directly to technologies employed at the Solar Living Center.

Institute sessions are a fascinating blend of technological independence, rural simplicity, and community building. The secret to the success of the Institute's workshops, as near as we can tell, consists of three ingredients perfectly combined: the wonderfully real SLI students and other Solar Living Institute family members who come from far and wide, the allure and promise of living independently with energy and sustenance harvested under independently produced power, and the practical, logical skills and ideas offered by the Institute's teachers and staff. One graduate expressed it best, "I can't decide if I've just had the best short vacation of my life, or the best learning experience! Could it be both?"

These intensive one-day, two-day, and five-day workshops offer a full survey of the terrain of a sustainable lifestyle and a thorough exploration of the energy-related aspects of the specific topic. At the end of each session, graduates have become informed consumers, able to ask the right questions and to understand the answers. They report that the experience inspires them to go further forward along the path toward independent living. Those who arrive with some technical skill leave qualified to put together an independent energy system, to begin the design of an energy efficient home made of alternative and sustainable building

JENNY MOON

SLI teaches hundreds of local electricians how to wire commercial solar systems.

mation hungry students. Most classes take place at the Solar Living Center. Classes are one-day, two-day, or week-long and most cost between $95 and $135 per day. Accreditation is available for many classes from various educational institutions. Workshops are scheduled year round but mostly between March and November. In 2004, the SLI instituted "Teleclasses" where anyone on the planet with a telephone can tune in for an hour to learn a variety of subjects for minimal cost.

TYPICAL CLASSES OFFERED AT THE SOLAR LIVING INSTITUTE'S WORKSHOP SERIES

Here is a representative list of workshops scheduled for the 2004 workshop season. Because the Institute is strongly committed to expanding its workshop series, this list will continue to develop. Potential students are encouraged to check with the Institute for an updated list (www.solarliving.org) and to let them know which additional topics you might be interested in pursuing.

materials, to know what to look for in a piece of property, or to create balance and harmony in their own garden. Energy is the heart of the workshop series: rational, comfortable, lasting systems for homes and businesses on and off the electric grid. Most courses emphasize direct hands-on experience. The mission of each Institute workshop is to share excitement, sense of purpose, and knowledge with others interested in living sensibly and lightly on the Earth.

Throughout all sessions, give and take between instructors and attendees is a topic in and of itself. Students and faculty learn to know each other, and value the efforts at independence each has contemplated, or already begun to bring to their lifestyles. If the instructors don't have an answer, often a student attending the workshop will. Product demonstrations, cost-benefit analyses, and a tour of applied technologies at the SLC provide students with a practical grasp on the tools and techniques available for taking their energy lives in hand. Classes focus on hands-on instruction of popular topics on energy and sustainability. The Institute's expert faculty has decades of experience and a strong passion for passing along its expertise to infor-

On-Grid Solar
Green Building and Strawbale Construction
Gardening—the Works
Freedom from Petroleum Fuel Weekend
All about Strawbale Construction
Oil and Natural Gas: Signs and Impacts of Peak and Decline
Powerdown: Rational Responses to Energy Resource Depletion
Show Me the Money—California Solar Rebates
Hydrogen 101
Introduction to Permaculture
Soil Building—Permaculture Style
Solar Electric 101—Introduction to Photovoltaics
On-Grid Photovoltaic Design and Installation
5-Day Hands-On On-Grid Photovoltaic Design and Installation
Off-Grid Photovoltaic Design and Installation
Advanced Off-Grid Photovoltaic Design and Installation
Solar for Electricians
Commercial Applications for PV on Public and Private Facilities
Women in Solar
Carpentry for Women
Find Your Dream Job in Solar
Renewable Energy 5-Day Intensive
Conservation and Efficiency
Wind Power 101
Solar Hot Water

Intern Shauna Dudley masters PV wiring at an SLI workshop.

JENNY MOON

Biodiesel—Fuel from Vegetables

Driving on Veggie Oil

Hydrogen Energy—the Emerging Fuel in America

5-Day Hands-on Strawbale Construction Project

5-Day Natural Building Intensive

Ecological Design

Cob and Natural Building Materials

Building with Stone

Building with Bamboo

Hybrid Adobe

Fundamentals of Permaculture and Aquaculture

Wastewater/Greywater Systems

Biointensive Gardening, Soil Management, and Composting

Permaculture and Edible Landscaping

Heirloom Seed Saving and Medicinal Herb Basics

5-Day Hands-on Permaculture Garden Design and Installation

Rural Water Development

Alcohol Can Be a Gas

FEEDBACK FROM INSTITUTE GRADUATES

■ "The Institute has benefited greatly from the willingness of Institute attendees to provide feedback on the benefits and potential improvements that could be made."

■ "It's the helpful hints and learning from mistakes that others have made that brought me to your workshop—in addition to wanting to meet the people I will someday do substantial business with. So, less time with generalities and more specifics, hands-on or applied theory with real life examples/pictures/equipment of success and failures. That is why your people are different—real experience, not hypotheticals!"

■ "I needed to get beyond the reading stage and become more familiar with the subjects. I feel we got a very well-thought-out and balanced presentation on many subjects, and I am very appreciative that you folks are taking the lead and setting a good example in presenting the information so lovingly and enthusiastically. Thank you!"

■ "My wife is leading me toward a better lifestyle. I arrived as a skeptic and am leaving with the feeling this is possible for me."

■ "I am so excited to get home and get started that I can't sit still. One night I was sizing my system all night. Last night I dreamt I had five dif-

Solar Electric 101 students get a little hands-on wiring experience at the SLI's Sustainable Living Workshop program.

ferent PV systems laid out around the top of a green meadow. All night I was choosing the best one—a combination—for my needs. Jeff and Ross and Nancy were there talking to other people then answering all my questions!"

■ "An honest possibility even for a conservative Republican that voted twice for Reagan. It is not as difficult as others want you to believe."

■ "Good spiritual company, great food, wonderful setting, excellent information."

■ "You guys are great people, and your knowledge and experience is invaluable to the rest of us—that is why we came to you. Please concentrate on conveying what is unique to you and your experience. The rest we can get from the books you sell."

WHERE TO FIND THE INSTITUTE: UPDATED SCHEDULES AND REGISTRATION

To register for classes, or for more information on Institute sessions, content, or availability, call 707.744.2017, write the Solar Living Institute at P.O. Box 836, Hopland, CA 95449, or check out the website at www.solarliving.org. An Institute brochure and schedule is available upon request.

#2: Interactive Displays and Curriculum

The Solar Living Institute is continually expanding and enriching the Solar Living Center's interactive and educational displays. The staff won't rest until the SLC is recognized nationwide as a

Doug Livingston demonstrates the hydrogen fuel cell display at the annual Earth Day for Kids event.

renewables curriculum materials for schools, and promoting the SLC as a destination for school groups. The SLI has also partnered with a curriculum developer and will be distributing K-12 renewable energy and sustainable living education to teachers through the United States.

#3: SolFest—Annual Summer Energy Festival and Educational Fair

The Solar Living Institute's premier annual event is the SolFest energy fair, scheduled each year in August. The first ever SolFest took place in June 1996, celebrating the Grand Opening of the Real Goods Solar Living Center in Hopland, California. The gala three-day event was attended by 10,000 people, inaugurating the Solar Living Center as the premier destination for those interested in learning about renewable energy and other sustainable living technologies. The educational theme of the event, in concert with the uniquely beautiful setting of the Solar Living Center, has inspired thousands of visitors. World-class speakers, unique entertainment, educational workshops, exhibitor booths with a renewable energy orientation, along with a parade and display of electric vehicles, all helped to create a lively and very successful event.

major learning campus and educational tour destination. As a nonprofit organization, the SLI seeks sources of funding in keeping with its educational goals. For example, the Institute has applied for grant monies to fund additional interactive exhibits on hydro power, solar water heating, photovoltaics, wind power, and hydrogen fuel cells at the SLC. The SLI has engaged a professional designer (whose client list includes the Monterey Bay Aquarium and a smattering of other Northern California museums) to create more engaging, effective exhibits and interactive displays on the site. The SLI is also in the process of designing

Nearly 200 solar-powered vendor booths line up for Solfest 2004.

BARBARA BOURNE

BARBARA BOURNE

Over the years, SolFest has featured incredible speakers, including Amory Lovins of the Rocky Mountain Institute, Wes Jackson of the Land Institute, Ralph Nader, Jim Hightower, Ben Cohen of Ben and Jerry's Homemade, Julia Butterfly-Hill, Helen Caldicott, Amy Goodman (Democracy Now), Darryl Hannah, and many others. World-class entertainers who have graced the SolFest stage include Bruce Cockburn, Michelle Shocked, Spearhead, Michael Franti, David Grisman, Charlie Hunter, and more. SolFest also features over 50 ongoing educational workshops exploring topics like solar energy, straw bale construction, building with bamboo, biodiesel, electric vehicles, eco-design, socially responsible investing, unconventional financing, climate change, industrial hemp, community nature centers, the Headwaters Forest, solar cooking, utility deregulation, and much, much more.

RIGHT: The Solar Living Institute Intern graduating class of 2004. BELOW: Bruce Cockburn serenades thousands at SolFest 2004.

SolFest provides a unique opportunity for renewable energy and sustainable living aficionados to gather and swap stories, and to have a great time for two days in the sun while learning and playing. Call the SLI at 707.744.2017 or visit the website (www.solarliving.org) to find the exact date and bill of entertainment—the tenth annual SolFest will occur on August 20–21, 2005 and promises to be the best ever.

LEFT: Daryl Hannah preaches the biodiesel gospel at SolFest 2004.
ABOVE: Ten thousand kindred spirits attend SolFest every year.

The Institute's Partnership Program

It takes a lot of energy to manifest the SLI's vision, and not the kind that is measured in kilowatts. The Institute extends a warm welcome to any volunteers willing to give their time and energy to promoting sustainable living through inspirational environmental education. You don't need to live in California to help; volun-

LEFT: Interns prepare mud and straw for the new "eco-shanty." CENTER: One of many brotherhoods of solar interns at the Solar Living Institute. RIGHT: Interns raise a straw bale building at the Solar Living Center.

teering could take the form of consultation, publicity, physical labor, day-to-day support, or . . . you tell them where your talents are and how you'd like to help. The Institute already has a well-established internship program for college students studying organic and bio-dynamic gardening, renewable energy of any kind, and general sustainability education. SLI volunteers and interns are honored and appreciated for their contributions, and come away with a lasting sense of achievement. The SLI has an extensive internship program with students of all ages coming to the Solar Living Center from all over the world to live, work, learn, and experience the joys of sustainable living. Interns come for twelve-week stints and consistently depart with the remark that their internships have been one of their life's peak experiences. For more information on the SLI's internship program check out the website at www.solarliving.org.

The other kind of energy needed is the green kind. The SLI asks each of you seriously to consider partnering with them in building the premier renewable energy education facility on the planet. They have applied for grants and are seeking large-scale donations from like-minded businesses and individuals, and expect to find some support. However, it has always been grassroots mom-and-pop generosity that got any nonprofit ball rolling. As little as $35 per year (for a Basic Partnership) will help expand the Sustainable Living Workshop Series, build more educational exhibits at the Solar Living Center, and help reach schoolchildren with environmental educational programs.

Partnership includes concrete benefits, like the occasional free magazine subscription. Donors of larger amounts receive a free copy of *A Place in the Sun* (the story of the evolution of the Solar Living Center), and a personalized tour of the Solar Living Center. But the real benefits of Institute Partnership will never be counted in quantifiable units. Becoming a Partner in the Solar Living Institute means joining a growing community of individuals whose vision and dedication have the potential to effect truly far-reaching change for our planet. To become a Partner in the Institute, to make a donation, or to volunteer, please contact the SLI at 707.744.2017 or email sli@solarliving.org.

Solar Living Institute interns manage the organic farm and harvest and market the produce.

Part IV: Getting Down to Business

Remember our first principle? We never forget it. When the profitability principle is met, we are free to pursue our educational mission effectively. We achieve profitability through our catalogs, retail store, design and consulting group, and our website (www.realgoods.com). We support the principles of knowledge, involvement, credibility (and even fun!) through a variety of innovative programs, from educational opportunities to networking. When our business goals and principles work in tandem with our educational goals and principles, we sleep easier at night. These are the elements and programs which, taken together, make our business real.

Catalogs

The Gaiam Real Goods catalogs include products and information geared to folks who have been thinking a lot about bringing elements of sustainable living into their lives. Our focus is on high-impact, user-friendly merchandise like compact fluorescent lighting, air and water purification systems, energy-saving household implements, and mainstream products that introduce the concepts behind renewable energy without being dauntingly technical. We also include lots of information on sustainable living, environmental responsibility, and books, books, books. The full-color Gaiam Real Goods

catalog is a lot like the Gideon's Bible in a hotel drawer. We don't expect it to produce any wholesale conversions, but we hope it steers a few people in the right direction. The Real Goods catalog also has evolved into our technical publication, for individuals who have made a serious commitment to reducing their impact on the planet, and need the tools to get on with it. Along with renewable energy system components (like PV modules, wind generators, hydroelectric pumps, cables, inverters, and controllers), our catalog offers off-the-grid and DC appliances, energy-saving climate control products, solar ovens, air and water purification systems, a smattering of green building materials, and even more books.

Since our merger with Gaiam in 2001, our customers now have access to the Gaiam Harmony catalog, which includes a huge eco-selection of products for the home and outdoors and the Gaiam Living Arts catalog, which includes great products for yoga and "mind-body-fitness." You can order either of these fine catalogs at 1.800.919.2400 or through www.gaiam.com.

Retail Store

As a catalog and Internet business, Gaiam Real Goods can reach almost every nook and cranny in America. Even so, there will always be people

Real Goods' flagship retail store at the Solar Living Center—the largest strawbale store in the world.

APPENDIX

who want to "kick the tires" before making a
purchase. Our Hopland Solar Living Center
flagship retail store is second to none. Not only
do customers have an opportunity to visit 12
gorgeous acres of permaculture gardens and
enjoy dozens of interactive displays, but they get
to see all the products in this *Sourcebook* and our
other catalogs come to life. Our retail store offers
an opportunity to snag curious passers-by and
show them how sustainable living can enhance
their lives.

Real Goods Design and Consulting Group

In response to the growing demand for large-
scale commercial renewable energy systems, we
launched a dynamic new division in early 1999:
the Real Goods Design and Consulting Group.
The D&C Group was established specifically to
work on large-scale commercial renewables
projects, particularly in the arena of Eco
Tourism. Our experienced team of renewables
experts offers a full spectrum of services for
established and proposed resorts, eco-lodges,
remote facilities, and villages. Our D&C Group
has brought high-efficiency and renewable
energy resources to thousands of clients over the
last twenty years. Customers are rewarded with
lower operating costs, high reliability, and
diminished impact on the world's vanishing
resources. We have consulted with clients rang-

ing from the Kingdom of Tonga and the Vatican
to the White House and a lone mountaineer.
Each receives the same qualified, individual
service.

Services range from site assessment and
planning to architectural design, specializing in
renewable energy system design (solar, wind,
and hydroelectric), supply, installation, and
training. Our D&C Group will facilitate and
expedite the development of site infrastructure
to minimize your need for energy, water, and
other resources. We accomplish this through
methodical assessment of the proposed site and
its environment. We consult with you to develop
the best design, fulfillment, and installation
strategies. Our goal is to help you create an
infrastructure that is economically, environmentally, and culturally sustainable.

We have assembled some of the world's leading experts in renewable energy, architecture,
community planning, and sustainability into an
effective, productive team capable of implementing these concepts worldwide. Every member of our D&C Group has a strong and sincere
commitment to the environment, local people,
and cultures. Our client list includes:

- The White House
- U.S. National Park Service
- U.S. Department of the Interior
- U.S. Department of Defense
- U.S. Department of Energy
- World Health Organization
- Government of Brazil
- Jet Propulsion Laboratory
- Disney
- AT&T
- Pacific Bell
- Los Angeles Department of Water and
  Power
- Kingdom of Tonga
- Karuk Tribe of California
- The Essene Way
- Women's Front of Norway
- Sierra Club
- The Vatican
- Public Citizen
- National Resource Defense Council
- Discovery Channel
- Pan American Health Organization
- Rocky Mountain Institute
- NASA
- City and County of San Francisco
- Greenpeace
- The Nature Conservancy
- U.S. Agency for International Development
- PG&E

- Sony
- U.S. Forest Service
- Lakota Sioux
- Maho Bay Camps / Harmony / Concordia
- University of California
- CBS
- U.S. Virgin Islands Office of Energy
- Tassajara Zen Center . . . and many more.

If you have a facility where utility power is unavailable or of poor quality, consider our resources:

- Renewable Energy System Design (solar, wind, hydro)
- Complete Installation Services, On-Site Training, and Turnkey Systems
- Large-Scale Uninterruptible Back-up Power including Disaster Preparedness
- Utility Intertie with Renewable Energy
- Power Quality Enhancement
- Site Planning and Engineering
- Solar Architecture
- Passive and Active Design
- Green Building Techniques
- Whole Systems Integration
- High-Efficiency Appliances
- Biological Waste Treatment Systems
- Water Quality and Management
- Earth Sensitive Landscaping

The Real Goods Design and Consulting team will help you realize your vision. Contact us at 303.222.3814 or 707.744.1816 or visit our website www.solardevelopment.com.

Real Goods Technical Staff

The Real Goods Technical staff are experts at designing, procuring, and installing residential renewable energy systems—both off-the-grid and line intertie systems to your utility. You don't want your project to be a test bed for unproved technology or experimental system design. You want information, products, and service that are of the highest quality. You want to leave the details to a company with the capability to do the job right the first time. We welcome the opportunity to design and plan entire systems. Here's how the Real Goods technical services work:

1. To assess your needs, capabilities, limitations, and working budget, we ask you to complete a specially created worksheet (see page 431). The information required includes a list of your energy needs, an inventory of desired appliances, site information, and potential for hydroelectric and wind development. Note that we only need this detailed energy use information for off-the-grid systems.

2. A member of our technical staff will determine your wattage requirements, and design an appropriate system with you. Or, for intertie systems, we will determine either how much you want to spend or how much utility power you wish to offset, and then design an appropriate system.

3. At this point, we will begin tracking the time we spend in helping you plan the details of your installation. Your personal tech rep will work with you on an unlimited time basis until your system has been completely designed and refined. He/she will order parts and talk you through assembly, assuring you that you get precisely what you need. He will also consult and work with your licensed contractor, if need be. The first hour is free and part of our service. Beyond this initial consultation, time will be billed in ten-minute intervals at the rate of $75 per hour. If the recommended parts and equipment are purchased from Gaiam Real Goods, however, this time will be provided at no charge.

Real Goods has been providing clean, reliable, renewable energy to people all over the planet for over twenty-six years and has provided solar energy systems for over 60,000 homes and businesses in that time. With the creation of our separate commercial division, our renewables technicians now have a tighter focus on smaller, noncommercial residential systems; this specialization means we are better prepared than ever to design a utility intertie or independent home system that meets your unique needs.

You will leave your Real Goods technical consultation with the information you need to make informed decisions on which technologies are best suited for your particular application. You'll learn how to utilize proven technologies and techniques to reap the best environmental advantages for the lowest possible cost. And we won't leave you hanging once your system is installed; our technicians will help you with troubleshooting, ongoing maintenance, and future upgrades. The Real Goods commitment to custom design, proven technology, and ongoing support has resulted in more than two decades of exceptionally high customer satisfaction. Real Goods specializes in residential and commercial systems of 100 watts to 100 kilowatts output, including:

- Utility Intertie with Renewable Energy
- Off-the-Grid Renewable Energy Systems Design and Installation (solar, wind, hydro)

- Large-Scale Uninterruptible Back-up Power
- High-Efficiency Appliances
- Biological Waste Treatment Systems
- Power Quality Enhancement
- Whole Systems Integration
- Water Quality and Management

Real Goods technical services and sales for residential and commercial renewable energy systems are available by phone 800.919.2400, from 7:30 A.M. to 6 P.M., Pacific Time, Monday through Friday, and 7:30 A.M. to 4 P.M. on Saturday. We endeavor to answer technical email within 48 hours, and snail mail (USPS) within one week.

Cyberspace Is a Tree-Free World: The Real Goods Website

The biggest conundrum we've ever faced as a company has been trying to live our environmental mission while being in the catalog business, which by definition survives by consuming trees. We have been, and will continue to be, as environmentally responsible as possible in a print media business. Our print catalogs use maximum post-consumer content recycled paper and soy-based inks, and we mail only to customers who have the highest potential for buying from us. Still you just can't beat a virtual catalog for environmental responsibility. In fact, our website uses only post-consumer recycled electrons of the finest quality. Since we first established an Internet presence in 1995, our online business has grown by leaps and bounds. One unexpected benefit of our "tree free catalog" is the almost limitless amount of space available. Our website has evolved from a simple online catalog to a compendium of renewables resources. You'll find links to fascinating sustainability and renewable energy sites, and you can find out just about anything about our company. It's quite possible that virtual catalogs like ours will reduce (or even eliminate) the more wasteful aspects of the mail-order business in the not-too-distant future.

The free Real Goods Solar Times email newsletter is published more or less every month. It includes news on renewable energy, the environment, the latest happenings at Gaiam Real Goods, and even some great products. To subscribe, visit our home page and sign up: www.realgoods.com.

Real Goods Wedding Registry

Do your dreams of nuptial bliss lean toward the practical, rather than the extravagant? Does your vision of unwrapping wedding gifts reveal inverters and compact fluorescent light bulbs, rather than sterling silver tea service and a ten-slice toaster? Why not register with Real Goods, so that well-wishing friends can honor you with the gifts that won't end up in the closet. Is this a joke? No way. A number of young married couples have embarked upon their dream of living lightly on the planet with a running start, thanks to the Gaiam Real Goods Wedding Registry. Then there are those reluctant folks who need a little nudge in the direction of sustainability, and those who prefer simply to let their friends and family choose for themselves. We'll happily send a gift certificate in any amount to the person of your choice, along with our latest edition of the Gaiam Real Goods catalog.

SYSTEM SIZING WORKSHEET

AC device	Device watts	×	Hours of daily use	×	Days of use per week	÷	7	=	Average watt-hours per day
		×		×		÷	7	=	
		×		×		÷	7	=	
		×		×		÷	7	=	
		×		×		÷	7	=	
		×		×		÷	7	=	
		×		×		÷	7	=	
		×		×		÷	7	=	
		×		×		÷	7	=	
		×		×		÷	7	=	
		×		×		÷	7	=	
		×		×		÷	7	=	
		×		×		÷	7	=	
		×		×		÷	7	=	
		×		×		÷	7	=	
		×		×		÷	7	=	
		×		×		÷	7	=	
		×		×		÷	7	=	
		×		×		÷	7	=	
		×		×		÷	7	=	
		×		×		÷	7	=	
		×		×		÷	7	=	
		×		×		÷	7	=	

1   Total AC watt-hours/day

2   × 1.1 = Total corrected DC watt-hours/day

DC device	Device watts	×	Hours of daily use	×	Days of use per week	÷	7	=	Average watt-hours per day
		×		×		÷	7	=	
		×		×		÷	7	=	
		×		×		÷	7	=	
		×		×		÷	7	=	
		×		×		÷	7	=	
		×		×		÷	7	=	
		×		×		÷	7	=	
		×		×		÷	7	=	

3   Total DC watt-hours/day

SYSTEM SIZING WORKSHEET

3 (from previous page)	Total DC watt-hours/day	
4	Total corrected DC watt-hours/day from Line 2 +	
5	Total household DC watt-hours/day =	
6	System nominal voltage (usually 12 or 24) ÷	
7	Total DC amp-hours/day =	
8	Battery losses, wiring losses, safety factor × 1.2	
9	Total daily amp-hour requirement =	
10	Estimated design insolation (hours per day of sun, see map on p. 434) ÷	
11	Total PV array current in amps =	
12	Select a photovoltaic module for your system	
13	Module rated power amps ÷	
14	Number of modules required in parallel =	
15	System nominal voltage (from line 6 above)	
16	Module nominal voltage (usually 12) ÷	
17	Number of modules required in series =	
18	Number of modules required in parallel (from Line 14 above) ×	
19	Total modules required =	

BATTERY SIZING

20	Total daily amp-hour requirement (from line 9)	
21	Reserve time in days ×	
22	Percent of useable battery capacity ÷	
23	Minimum battery capacity in amp-hours =	
24	Select a battery for your system, enter amp-hour capacity ÷	
25	Number of batteries in parallel =	
26	System nominal voltage (from line 6)	
27	Voltage of your chosen battery (usually 6 or 12) ÷	
28	Number of batteries in series =	
29	Number of batteries in parallel (from line 25 above) ×	
30	Total number of batteries required	

Typical Wattage Requirements for Common Appliances

Use the manufacturer's specifications if possible, but be careful of nameplate ratings that are the highest possible electrical draw for that appliance. Beware of appliances that have a "standby" mode and are really "on" twenty-four hours a day. If you can't find a rating, call us for advice (800.919.2400).

DESCRIPTION	WATTS
Refrigeration (*Refrigerators only are listed in watt-hours per day*):	
5–10-yr.-old 22-cu. ft. auto defrost	3,000
New 22-cu. ft. auto defrost (with Energy Star rating of 400 kwh/yr)	1,100
12-cu. ft. Sun Frost refrigerator	600
5–10-yr.-old standard freezer	3,500
Sundanzer 8-cu. ft. freezer	530
Dishwasher:	
cool dry	700
hot dry	1,450
Trash compactor	1,500
Can opener (electric)	100
Microwave (.5 cu. ft.)	900
Microwave (.8 to 1.5 cu. ft.)	1,500
Exhaust hood	144
Coffeemaker	1,200
Food processor	400
Toaster (2-slice)	1,200
Coffee grinder	100
Blender	350
Food dehydrator	600
Mixer	120
Range, small burner	1,250
Range, large burner	2,100
Water Pumping:	
AC Jet Pump (⅓ hp), 300 gal per hour, 20' well depth, 30 psi	750

DESCRIPTION	WATTS
AC submersible pump (½ hp), 40' well depth, 30 psi	1,000
DC pump for house pressure system (typical use is 1–2 hours per day)	60
DC submersible pump (typical use is 6 hours per day)	50
Shop:	
Worm drive 7¼" saw	1,800
AC table saw, 10"	1,800
AC grinder, ½ hp	1,080
Hand drill, ⅜"	400
Hand drill, ½"	600
Entertainment/Telephones:	
TV (27-inch color)	170
TV (19-inch color)	80
TV (12-inch black & white)	16
Video games (not incl. TV)	20
Satellite system, VCR	30
DVD/CD player	30
AC-powered stereo (avg. volume)	55
AC stereo, home theater	500
DC-powered stereo (avg. volume)	15
CB (receiving)	10
Cellular telephone (on standby)	5
Cordless telephone (on standby)	5
Electric piano	30
Guitar amplifier (avg. volume)	40
(Jimi Hendrix volume)	8,500
General Household:	
Typical fluorescent light (60W equivalent)	15
Incandescent lights	(as indicated on bulb)
Electric clock	4
Clock radio	5
Electric blanket	400

DESCRIPTION	WATTS
Iron (electric)	1,200
Clothes washer (vertical axis)	900
Clothes washer (horizontal axis)	250
Dryer (gas)	500
Dryer (electric)	5,750
Vacuum cleaner, average	900
Central vacuum	1,500
Furnace fan:	
¼ hp	600
⅓ hp	700
½ hp	875
Garage door opener: ¼ hp	550
Alarm/security system	6
Air conditioner: per 1 ton or 10,000 BTU/hr capacity	1,500
Office/Den:	
Computer	55
17" color monitor	100
17" LCD "flat screen" monitor	45
Laptop computer	25
Ink jet printer	35
Dot matrix printer	200
Laser printer	900
Fax machine (plain paper):	
standby	5
printing	50
Electric typewriter	200
Adding machine	8
Electric pencil sharpener	60
Hygiene:	
Hair dryer	1,500
Waterpik	90
Whirlpool bath	750
Hair curler	750
Electric toothbrush: (charging stand)	6

Solar Insolation Maps

The maps below show the sun-hours per day for the U.S. Charts courtesy of D.O.E.

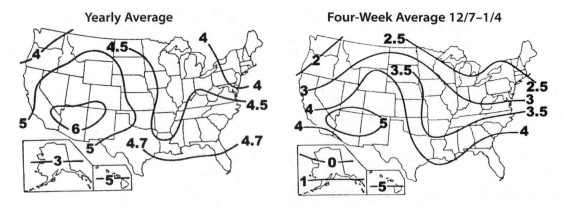

Yearly Average · Four-Week Average 12/7–1/4

Solar Insolation by U.S. City

This chart shows solar insolation in kilowatt-hours per square meter per day in many U.S. locations. For simplicity, we call this figure "Sun Hours/Day." To find average sun hours per day in your area (last column), check local weather data, look at the maps above or find a city in the table that has similar weather to your location. If you want year-round autonomy, use the lowest of the two figures. If you want only 100% autonomy in summer, use the higher figure.

State	City	High	Low	Avg
AK,	Fairbanks	5.87	2.12	3.99
AK,	Matanuska	5.24	1.74	3.55
AL,	Montgomery	4.69	3.37	4.23
AR,	Bethel	6.29	2.37	3.81
AR,	Little Rock	5.29	3.88	4.69
AZ,	Tucson	7.42	6.01	6.57
AZ,	Page	7.3	5.65	6.36
AZ,	Phoenix	7.13	5.78	6.58
CA,	Santa Maria	6.52	5.42	5.94
CA,	Riverside	6.35	5.35	5.87
CA,	Davis	6.09	3.31	5.1
CA,	Fresno	6.19	3.42	5.38
CA,	Los Angeles	6.14	5.03	5.62
CA,	Soda Springs	6.47	4.4	5.6
CA,	La Jolla	5.24	4.29	4.77
CA,	Inyokern	8.7	6.87	7.66
CO,	Grandby	7.47	5.15	5.69
CO,	Grand Lake	5.86	3.56	5.08
CO,	Grand Junction	6.34	5.23	5.85
CO,	Boulder	5.72	4.44	4.87
DC,	Washington	4.69	3.37	4.23
FL,	Apalachicola	5.98	4.92	5.49
FL,	Belie Is.	5.31	4.58	4.99
FL,	Miami	6.26	5.05	5.62
FL,	Gainsville	5.81	4.71	5.27
FL,	Tampa	6.16	5.26	5.67
GA,	Atlanta	5.16	4.09	4.74
GA,	Griffin	5.41	4.26	4.99
HI,	Honolulu	6.71	5.59	6.02
IA,	Ames	4.8	3.73	4.4
ID,	Boise	5.83	3.33	4.92
ID,	Twin Falls	5.42	3.42	4.7
IL,	Chicago	4.08	1.47	3.14
IN,	Indianapolis	5.02	2.55	4.21

State	City	High	Low	Avg
KS,	Manhattan	5.08	3.62	4.57
KS,	Dodge City	6.5	4.2	5.6
KY,	Lexington	5.97	3.6	4.94
LA,	Lake Charles	5.73	4.29	4.93
LA,	New Orleans	5.71	3.63	4.92
LA,	Shreveport	4.99	3.87	4.63
MA,	E. Wareham	4.48	3.06	3.99
MA,	Boston	4.27	2.99	3.84
MA,	Blue Hill	4.38	3.33	4.05
MA,	Natick	4.62	3.09	4.1
MA,	Lynn	4.6	2.33	3.79
MD,	Silver Hill	4.71	3.84	4.47
ME,	Caribou	5.62	2.57	4.19
ME,	Portland	5.23	3.56	4.51
MI,	Sault Ste. Marie	4.83	2.33	4.2
MI,	E. Lansing	4.71	2.7	4.0
MN,	St. Cloud	5.43	3.53	4.53
MO,	Columbia	5.5	3.97	4.73
MO,	St. Louis	4.87	3.24	4.38
MS,	Meridian	4.86	3.64	4.43
MT,	Glasgow	5.97	4.09	5.15
MT,	Great Falls	5.7	3.66	4.93
MT,	Summit	5.17	2.36	3.99
NM,	Albuquerque	7.16	6.21	6.77
NB,	Lincoln	5.4	4.38	4.79
NB,	N. Omaha	5.28	4.26	4.9
NC,	Cape Hatteras	5.81	4.69	5.31
NC,	Greensboro	5.05	4	4.71
ND,	Bismark	5.48	3.97	5.01
NJ,	Sea Brook	4.76	3.2	4.21
NV,	Las Vegas	7.13	5.84	6.41
NV,	Ely	6.48	5.49	5.98
NY,	Binghampton	3.93	1.62	3.16
NY,	Ithaca	4.57	2.29	3.79

State	City	High	Low	Avg
NY,	Schenectady	3.92	2.53	3.55
NY,	Rochester	4.22	1.58	3.31
NY,	New York City	4.97	3.03	4.08
OH,	Columbus	5.26	2.66	4.15
OH,	Cleveland	4.79	2.69	3.94
OK,	Stillwater	5.52	4.22	4.99
OK,	Oklahoma City	6.26	4.98	5.59
OR,	Astoria	4.76	1.99	3.72
OR,	Corvallis	5.71	1.9	4.03
OR,	Medford	5.84	2.02	4.51
PA,	Pittsburg	4.19	1.45	3.28
PA,	State College	4.44	2.79	3.91
RI,	Newport	4.69	3.58	4.23
SC,	Charleston	5.72	4.23	5.06
SD,	Rapid City	5.91	4.56	5.23
TN,	Nashville	5.2	3.14	4.45
TN,	Oak Ridge	5.06	3.22	4.37
TX,	San Antonio	5.88	4.65	5.3
TX,	Brownsville	5.49	4.42	4.92
TX,	El Paso	7.42	5.87	6.72
TX,	Midland	6.33	5.23	5.83
TX,	Fort Worth	6	4.8	5.43
UT,	Salt Lake City	6.09	3.78	5.26
UT,	Flaming Gorge	6.63	5.48	5.83
VA,	Richmond	4.5	3.37	4.13
WA,	Seattle	4.83	1.6	3.57
WA,	Richland	6.13	2.01	4.44
WA,	Pullman	6.07	2.9	4.73
WA,	Spokane	5.53	1.16	4.48
WA,	Prosser	6.21	3.06	5.03
WI,	Madison	4.85	3.28	4.29
WV,	Charleston	4.12	2.47	3.65
WY,	Lander	6.81	5.5	6.06

Magnetic Declinations in the United States

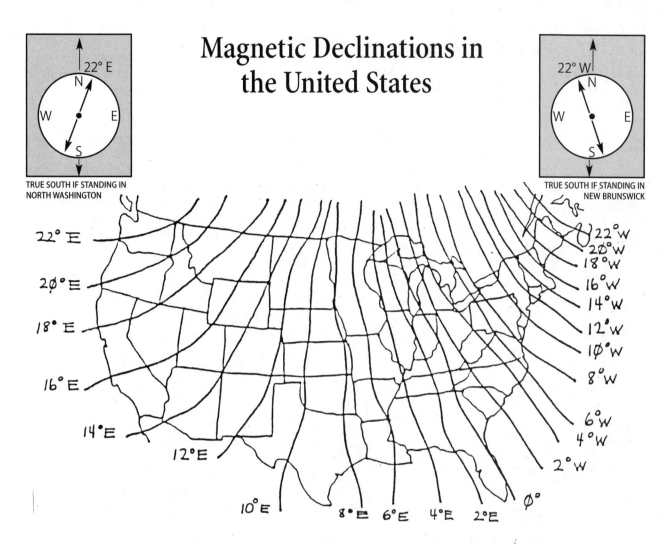

TRUE SOUTH IF STANDING IN NORTH WASHINGTON
22° E

TRUE SOUTH IF STANDING IN NEW BRUNSWICK
22° W

22° E
20° E
18° E
16° E
14° E
12° E
10° E
8° E 6° E 4° E 2° E

22° W
20° W
18° W
16° W
14° W
12° W
10° W
8° W
6° W
4° W
2° W
∅°

Figure indicates correction of compass reading to find true north. For example, in Washington state when your compass reads 22°E, it is pointing due north.

MAXIMUM NUMBER OF CONDUCTORS FOR A GIVEN CONDUIT SIZE

Conduit size		½"	¾"	1"	1¼"	1½"	2"
	#12	10	18	29	51	70	114
	#10	6	11	18	32	44	73
	#8	3	5	9	16	22	36
	#6	1	4	6	11	15	26
Conductor size	#4	1	2	4	7	9	16
	#2	1	1	3	5	7	11
	#1		1	1	3	5	8
	#1/0		1	1	3	4	7
	#2/0		1	1	2	3	6
	#3/0		1	1	1	3	5
	#4/0		1	1	1	2	4

Battery Wiring Diagrams

The following diagrams show how 2-, 6- and 12-volt batteries are connected for 12-, 24- and 48-volt operation.

—⌇— Fuse symbol (always use appropriate fusing at your battery.)

Wiring Basics

We answer a lot of basic wiring questions over the phone, which we're always happy to do, but there's nothing like a picture or two to make things apparent.

Battery Wiring

The batteries for your energy system may be supplied as 2-volt, 6-volt, or 12-volt cells. Your system voltage is probably 12, 24, or 48 volts. You'll need to series wire enough batteries to reach your system voltage, then parallel wire to another series group as needed to boost amperage capacity. See our drawings for correct series wiring. Paralleled groups are shown in dotted outline.

PV Module Wiring

PV modules are almost universally produced as nominal 12-volt modules. For smaller 12-volt systems this is fine. Most larger residential systems are configured for 24- or 48-volt input now.

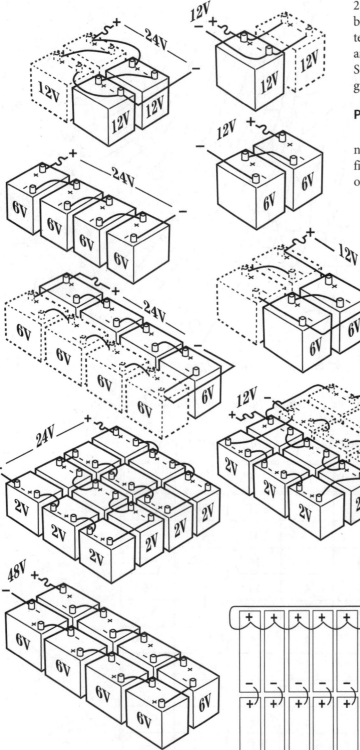

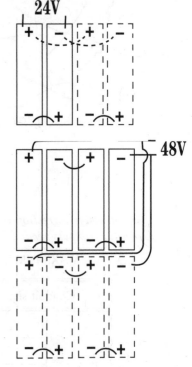

Longevity/Safety Tip for Wiring Larger PV Arrays

If you have a large PV array that produces close to, or over, 20 amps, multiple power take-off leads are a good idea. They may prevent toasted terminal boxes. Instead of taking only a single pair of positive and negative leads off some point on the array, take off one pair at one end, and another pair off the opposite end. Then join them back together at the array-mounted junction box where you're going to the larger wire needed for transmission. This divides up the routes that outgoing power can take, and eases the load on any single PV junction box.

Wire Sizing Chart/Formula

This chart is useful for finding the correct wire size for any voltage, length, or amperage flow in any AC or DC circuit. For most DC circuits, particularly between the PV modules and the batteries, we try to keep the voltage drop to 3% or less. There's no sense using your expensive PV wattage to heat wires. You want that power in your batteries!

Note that this formula doesn't directly yield a wire gauge size, but rather a "VDI" number, which is then compared to the nearest number in the VDI column, and then read across to the wire gauge size column.

1. Calculate the Voltage Drop Index (VDI) using the following formula:

VDI = AMPS x FEET ÷
    (% VOLT DROP × VOLTAGE)
Amps = Watts divided by volts
Feet = One-way wire distance
% Volt Drop = Percentage of voltage drop acceptable for this circuit (typically 2 to 5%)

2. Determine the appropriate wire size from the chart below.
   A. Take the VDI number you just calculated and find the nearest number in the VDI column, then read to the left for AWG wire gauge size.
   B. Be sure that your circuit amperage does not exceed the figure in the Ampacity column for that wire size. (This is not usually a problem in low-voltage circuits.)

Example: Your PV array consisting of four Siemens SP75 modules is 60 feet from your 12-volt battery. This is actual wiring distance, up pole mounts, around obstacles, etc. These mod-ules are rated at 4.4 amps × 4 modules = 17.6 amps maximum. We'll shoot for a 3% voltage drop. So our formula looks like:

$$VDI = \frac{17.6 \times 60}{3[\%] \times 12[V]} = 29.3$$

Looking at our chart, a VDI of 29 means we'd better use #2 wire in copper, or #0 wire in aluminum. Hmmm. Pretty big wire.

What if this system was 24-volt? The modules would be wired in series, so each pair of modules would produce 4.4 amps. Two pairs × 4.4 amps = 8.8 amps max.

$$VDI = \frac{8.8 \times 60}{3[\%] \times 24[V]} = 7.3$$

Wow! What a difference! At 24-volt input you could wire your array with little ol' #8 copper wire.

WIRE SIZE	COPPER WIRE		ALUMINUM WIRE	
AWG	VDI	Ampacity	VDI	Ampacity
0000	99	260	62	205
000	78	225	49	175
00	62	195	39	150
0	49	170	31	135
2	31	130	20	100
4	20	95	12	75
6	12	75	•	•
8	8	55	•	•
10	5	30	•	•
12	3	20	•	•
14	2	15	•	•
16	1	•	•	•

Chart developed by John Davey and Windy Dankoff. Used by permission.

Friction Loss Charts for Water Pumping

HOW TO USE PLUMBING FRICTION CHARTS

If you try to push too much water through too small a pipe, you're going to get pipe friction. Don't worry, your pipes won't catch fire. But it will make your pump work harder than it needs to, and it will reduce your available pressure at the outlets, so sprinklers and showers won't work very well. These charts can tell you if friction is going to be a problem. Here's how to use them:

PVC or black poly pipe? The rates vary, so first be sure you're looking at the chart for your type of supply pipe. Next, figure out how many gallons per minute you might need to move. For a normal house, 10 to 15 gpm is probably plenty. But gardens and hoses really add up. Give your-self about 5 gpm for each sprinkler or hose that might be running. Find your total (or something close to it) in the Flow GPM column. Read across to the column for your pipe diameter. This is how much pressure loss you'll suffer for every 100 feet of pipe. Smaller numbers are better.

Example: You need to pump or move 20 gpm through 500 feet of PVC between your storage tank and your house. Reading across, 1" pipe is obviously a problem. How about 1¼"? 9.7 psi times 5 (for your 500 feet) = 48.5 psi loss. Well, that won't work! With 1½" pipe, you'd lose 20 psi ... still pretty bad. But with 2" pipe you'd only lose 4 psi ... ah! Happy garden sprinklers! Generally, you want to keep pressure losses under about 10 psi.

Friction Loss in PSI per 100 Feet of Scheduled 40 PVC Pipe

Flow GPM	Nominal Pipe Diameter in Inches							
	½"	¾"	1"	1-1/4"	1-1/2"	2"	3"	4"
1	3.3	0.5	0.1					
2	11.9	1.7	0.4	0.1				
3	25.3	3.5	0.9	0.3	0.1			
4	43.0	6.0	1.5	0.5	0.2	0.1		
5	65.0	9.0	2.2	0.7	0.3	0.1		
10		32.5	8.0	2.7	1.1	0.3		
15		68.9	17.0	5.7	2.4	0.6	0.1	
20			28.9	9.7	4.0	1.0	0.1	
30			61.2	20.6	8.5	2.1	0.3	0.1
40				35.1	14.5	3.6	0.5	0.1
50				53.1	21.8	5.4	0.7	0.2
60				74.4	30.6	7.5	1.0	0.3
70					40.7	10.0	1.4	0.3
80					52.1	12.8	1.8	0.4
90					64.8	16.0	2.2	0.5
100					78.7	19.4	2.7	0.7
150						41.1	5.7	1.4
200						69.9	9.7	2.4
250							14.7	3.6
300							20.6	5.1
400							35.0	8.6

Friction Loss in PSI per 100 Feet of Polyethylene (PE) SDR-Pressure Rated Pipe
Pressure loss from friction in psi per 100 feet of pipe.

Flow GPM	NOMINAL PIPE DIAMETER IN INCHES							
	0.5	0.75	1	1.25	1.5	2	2.5	3
1	0.49	0.12	0.04	0.01				
2	1.76	0.45	0.14	0.04	0.02			
3	3.73	0.95	0.29	0.08	0.04	0.01		
4	6.35	1.62	0.50	0.13	0.06	0.02		
5	9.60	2.44	0.76	0.20	0.09	0.03		
6	13.46	3.43	1.06	0.28	0.13	0.04	0.02	
7	17.91	4.56	1.41	0.37	0.18	0.05	0.02	
8	22.93	5.84	1.80	0.47	0.22	0.07	0.03	
9		7.26	2.24	0.59	0.28	0.08	0.03	
10		8.82	2.73	0.72	0.34	0.10	0.04	0.01
12		12.37	3.82	1.01	0.48	0.14	0.06	0.02
14		16.46	5.08	1.34	0.63	0.19	0.08	0.03
16			6.51	1.71	0.81	0.24	0.10	0.04
18			8.10	2.13	1.01	0.30	0.13	0.04
20			9.84	2.59	1.22	0.36	0.15	0.05
22			11.74	3.09	1.46	0.43	0.18	0.06
24			13.79	3.63	1.72	0.51	0.21	0.07
26			16.00	4.21	1.99	0.59	0.25	0.09
28				4.83	2.28	0.68	0.29	0.10
30				5.49	2.59	0.77	0.32	0.11
35				7.31	3.45	1.02	0.43	0.15
40				9.36	4.42	1.31	0.55	0.19
45				11.64	5.50	1.63	0.69	0.24
50				14.14	6.68	1.98	0.83	0.29
55					7.97	2.36	0.85	0.35
60					9.36	2.78	1.17	0.41
65					10.36	3.22	1.36	0.47
70					12.46	3.69	1.56	0.54
75					14.16	4.20	1.77	0.61
80						4.73	1.99	0.69
85						5.29	2.23	0.77
90						5.88	2.48	0.86
95						6.50	2.74	0.95
100						7.15	3.01	1.05
150						15.15	6.38	2.22
200							10.87	3.78
300								8.01

Nominal Pipe Size versus Actual Outside Diameter for Steel and Plastic Pipe

Nominal Size	Actual Size
1/2"	0.840"
3/4"	1.050"
1"	1.315"
1 1/4"	1.660"
1 1/2"	1.900"
2"	2.375"
2 1/2"	2.875"
3"	3.500"
3 1/2"	4.000"
4"	4.500"
5"	5.563"
6"	6.625"

Temperature Conversions

°C = Degrees Celsius. 1 degree is 1/100 of the difference between the temperature of melting ice and boiling water.

°F = Degrees Fahrenheit. 1 degree is 1/180 of difference between the temperature of melting ice and boiling water.

TEMPERATURE CONVERSION CHART

°C	°F	°C	°F	°C	°F	°C	°F
200	392	140	284	80	176	15	59
195	383	135	275	75	167	10	50
190	374	130	266	70	158	5	41
185	365	125	257	65	149	0	32
180	356	120	248	60	140	−5	23
175	347	115	239	50	122	−10	14
170	338	110	230	45	113	−15	5
165	329	105	221	40	104	−20	−4
160	320	100	212	35	95	−25	−13
155	311	95	203	30	86	−30	−22
150	302	90	194	25	77	−35	−31
145	293	85	185	20	68	−40	−40

The Real Goods Resource List

Although this *Sourcebook* is a complete source of renewable energy and environmental products, we can't be everything to everyone. Here is our current list of other trusted resources for information and products. We've selected these organizations carefully, and since we have worked directly with many of them, we are giving you the benefit of our experience. Still, we'll offer the standard disclaimer that Gaiam Real Goods does not necessarily endorse all the actions of each group listed, nor are we responsible for what they say or do.

This list can never be complete. We apologize for resources we may have overlooked and for contact information that may have changed since this twelfth edition of the *Sourcebook* went to press. We welcome your suggestions. Please mail a brief description of the organization and access info to Gaiam Real Goods, Attention: Resource List.

American BioEnergy Association
314 Massachusetts Ave. NE
Suite 200
Washington, DC 20002
Website: www.biomass.org
The leading voice in the United States for the bioenergy industry. Works to build support through tax incentives, increased biomass research and development budgets, regulations, and other policy initiatives.

American Council for an Energy-Efficient Economy (ACEEE)
1001 Connecticut Avenue, Suite 801
Washington, DC 20036
Website: www.aceee.org
Email: info@aceee.org
Phone (Research and Conferences):
202-429-8873
Publications: 202-429-0063
Publishes books, papers, yearly guides, and comparisons of appliances and vehicles based on energy efficiency. Their website is an excellent source of efficient appliance info.

The American Hydrogen Association
1789 West 7th Avenue
Mesa, AZ 85202
Website: www.clean-air.org/
Email: answerguy@clean-air.org
Phone: 480-827-7915
A nonprofit organization that promotes the use of hydrogen for fuel and energy storage. Publishes Hydrogen Today, *a bimonthly newsletter.*

American Society of Landscape Architects (ASLA)
636 Eye Street, NW
Washington, DC 20001-3736
Website: www.asla.org
Phone: 202-898-2444
Fax: 202-898-1185
This professional organization advocates on public policy issues such as livable communities, surface transportation, the environment, historic preservation, and small business affairs.

American Solar Energy Society (ASES)
2400 Central Avenue, Suite A
Boulder, CO 80301
Website: www.ases.org/
Email: ases@ases.org
Phone: 303-443-3130
Fax: 303-443-3212
ASES is the United States Section of the International Solar Energy Society, a national organization dedicated to advancing the use of solar energy for the benefit of U.S. citizens and the global environment. Publishers of the bimonthly Solar Today *magazine for members, and sponsors of the yearly National Tour of Solar Homes.*

The American Wind Energy Association.
122 C Street NW, Suite 380
Washington, DC 20001
Website: www.awea.org/
Email: windmail@awea.org
Phone: 202-383-2500
Fax: 202-383-2505
A national trade association that represents wind-power plant developers, wind turbine manufacturers, utilities, consultants, insurers, financiers, researchers, and others involved in the wind industry. These folks primarily work with utility-level wind systems. Not a good source for residential info.

Battery Recycling (for Nicads)
Rechargeable Battery Recycling Corp.
1000 Parkwood Circle, Suite 450
Atlanta, GA 30339
Website: www.rbrc.org
Email: consumer@rbrc.com
Phone: 678-419-9990
Fax: 678-419-9986
A nonprofit public service organization to promote the recycling of nicad batteries. Just type your zip code in the website and get a list of local stores that will accept your old nicads for recycling.

BuildingGreen, Inc.
122 Birge St., Suite 30
Brattleboro, VT 05301
Website: www.buildinggreen.com
Email: info@buildinggreen.com
Phone: 802-257-7300
A leading source of objective information on green building and renewable energy. Publishes the journal Environmental Building News *and the* GreenSpec Directory *of green building products.*

California Energy Commission
1516 Ninth Street
Sacramento, CA 95814-5512
Website: www.energy.ca.gov
Email: Phone: 800-555-7794 (inside Calif.); 916-654-4058 (outside Calif.)
The State of California's primary energy policy and planning agency. Strongly supports energy efficiency and small-scale utility intertie projects. A

good source of honest intertie information.

CalStart
Northern California Office:
1160 Brickyard Cove, Suite 101
Point Richmond, California 94801
Phone: 510-307-8700
Southern California Office:
2181 E. Foothill Boulevard
Pasadena, CA 91107
Phone: 626-744-5600
Email: calstart@calstart.org
Website: www.calstart.org
The latest information on electric, natural gas, and hybrid electric vehicles, and other intelligent transportation technologies.

Center for Energy and Climate Solutions/Cool Companies
Website: www.cool-companies.org
CECS promotes clean and efficient energy technologies as a money-saving tool for reducing greenhouse gas emissions and other pollutants. Its website contains good information, including more skeptical views about a hydrogen-based economy.

Center for Renewable Energy and Sustainable Technology (CREST)
See Renewable Energy Policy Project

Chelsea Green Publishing
P.O. Box 428
85 North Main Street, Suite 120
White River Junction, VT 05001
Website: www.chelseagreen.com
Email: seaton@chelseagreen.com
Phone: 802-295-6300
Fax: 802-295-6444
One of the world's premier publishers of books on renewable energy, natural building, and sustainable living.

Electric Auto Association
P.O. Box 6661
Concord, CA 94514
Website: www.eaaev.org
Email: contact@eaaev.org
The national electric vehicle association with local chapters in most states and Canada. Dues are $39/year with an excellent monthly newsletter.

The website has links to practically everything in the EV biz.

Energy Efficiency and Renewable Energy Clearinghouse (EREC)
Mail Stop EE-1
Department of Energy
Washington, DC 20585
Website: www.eere.energy.gov/
Email: EEREMailbox@EE.DOE.Gov
Phone: 877-337-3463
The best source of information about renewable energy technologies and energy efficiency. These folks seem to be able to find good information about anything energy related. One of the best overall websites about energy.

Energy Information Administration
1000 Independence Ave. S.W.
Washington, DC 20585
Website: www.eia.doe.gov
Email: infoctr@eia.doe.gov
Phone: 202-586-8800
Official energy statistics from the U.S. government. Gas, oil, electricity, you name it, there's more supply, pricing, and use info here than you can shake a stick at.

Energy Star®
Environmental Protection Agency (EPA)
Climate Protection Partnerships Division
ENERGY STAR Programs Hotline & Distribution (MS-6202J)
1200 Pennsylvania Ave NW
Washington, DC 20460
Website: www.energystar.gov/
Email: info@energystar.gov
Phone: 888-STAR-YES (888-782-7937)
Good, up-to-date listings of the most efficient appliances, lights, windows, home and office electronics, and more.

Environmental Protection Agency (EPA)
Ariel Rios Building
1200 Pennsylvania Avenue N.W.
Washington, DC 20460
Website: www.epa.gov
Email: public-access@epamail.epa.gov
Phone: 202-272-0167

Florida Solar Energy Center (FSEC)
1679 Clearlake Road
Cocoa, FL 32922
Website: www.fsec.ucf.edu
Email: webmaster@fsec.ucf.edu
Phone: 321-638-1000
Fax: 321-638-1010
Provides highly regarded independent third-party testing and certification of solar hot-water systems and other solar or energy-efficiency goods.

Gas Appliance Manufacturers Association
2107 Wilson Boulevard, Suite 600
Arlington, VA 22201
Website: www.gamanet.org
Email: information@gamanet.org
Phone: 703-525-7060
Fax: 703-525-6790

Geothermal Resources Council
P.O. Box 1350
2001 Second Street, Suite 5
Davis, CA 95617-1350
Website: www.geothermal.org
Email: grclib@geothermal.org
Phone: 530-758-2360
Fax: 530-758-2839
A nonprofit organization dedicated to geothermal research and development.

The Green Power Network
Net Metering and Green Power information
Website:
www.eere.energy.gov/greenpower/
Information on the electric power industry's green power efforts. Includes up-to-date info on green power availability and pricing. Also the best state-by-state net metering details on the Net.

The Heartwood School for the Homebuilding Crafts
Johnson Hill Road
Washington, MA 01223
Website: www.heartwoodschool.com
Email: info@heartwoodschool.com
Phone: 413-623-6677
Fax: 413-623-0277
Offers a variety of hands-on building workshops. Specializes primarily in timber frames.

Home Power **Magazine**
P.O. Box 520
Ashland, OR 97520
Website: www.homepower.com
Email: hp@homepower.com
Phone: 800-707-6585
Fax: 541-512-0343
The journal of the renewable energy industry. Published bimonthly. U.S. subscription is $22.50/yr.

The Last Straw Online
PO Box 22706
Lincoln, NE 68542-2706
Website: www.thelaststraw.org
Email: thelaststraw@thelaststraw.org
Phone: 402-483-5135
Fax: 402-483-5161
An excellent hub site with links to many other straw bale and sustainable building sites. This is also the online home of The Last Straw, *an excellent straw bale quarterly journal($28/yr.).*

National Association of Home Builders (NAHB)
1201 15th Street, NW
Washington, DC 20005
Website: www.nahb.com
Email: info@nahb.com
Phone: 800-368-5242

National Association of State Energy Officials (NASEO)
1414 Prince Street, Suite 200
Alexandria, VA 22314
Website: www.naseo.org
Email: information@naseo.org
Phone: 703-299-8800
Fax: 703-299-6208
An excellent portal site with access info for all state energy offices.

National Biodiesel Board
3337a Emerald Lane
PO Box 104898
Jefferson City, MO 65110-4898
Website: www@biodiesel.org
Email: info@biodiesel.org
Phone: 800-841-5849
Fax: 573-635-7913
This advocacy organization has an excellent website brimming with all things biodiesel.

National Center for Appropriate Technology
P.O. Box 3838
3040 Continental Drive
Butte, MT 59702
Phone: 800-275-6228
Website: www.ncat.org
Email: info@ncat.org
A nonprofit organization and information clearinghouse for sustainable energy, low-income energy, resource-efficient housing, and sustainable agriculture.

National Renewable Energies Laboratory (NREL)
1617 Cole Boulevard
Golden, CO 80401-3393
Website: www.nrel.gov
Email: webmaster@nrel.gov
Phone: 303-275-3000
The nation's leading center for renewable energy research. Tests products, conducts experiments, and provides information on renewable energy.

New Society Publishers
P.O. Box 189
Gabriola Island, BC
Canada, V0R 1X0
Website: www.newsociety.com
Email: info@newsociety.com
Phone: 250-247-9737
Fax: 250-247-7471
Leading publisher on books for renewable energy, alternative fuels, and sustainable living. Distributor of this twelfth edition Solar Living Sourcebook.

Renewable Energy Policy Project and Center for Renewable Energy and Sustainable Technology (CREST)
1612 K St NW, Suite 202
Washington DC 20006
Phone: 202-293-2898
Fax: 202-293-5857
Website: http://solstice.crest.org
Email: info2@repp.org
An excellent online resource for sustainable energy information. Provides Internet services, software, databases, resource lists, and discussion lists on a wide variety of topics.

Rocky Mountain Institute
1739 Snowmass Creek Road
Snowmass, CO 81654-9199
Website: www.rmi.org
Email: outreach@rmi.org
Phone: 970-927-3851
A terrific source of books, papers, and research on renewable energy, energy-efficient building design and components, and sustainable transportation.

Solar Cookers International
1919 21st St. #101
Sacramento, CA 95814
Website: http://solarcooking.org/
Email: info@solarcookers.org
Phone: 916-455-4499
Fax: 916-455-4498
A nonprofit organization that promotes and distributes simple solar cookers in developing countries to relieve the strain of firewood collection. Newsletter and small products catalog available.

Solar Electric Light Fund
1612 K Street NW, Suite 402
Washington, DC 20006
Phone: 202-234-7265
Email: solarlight@self.org
Website: www.self.org/
A nonprofit organization promoting PV rural electrification in developing countries.

Solar Energy Industries Association (SEIA)
1616 H Street NW, Suite #800
Washington, DC 20006
Website: www.seia.org
Email: info@seia.org
Phone: 202-628-7745
Fax: 202-628-7779

Solar Energy Info for Planet Earth
Website:
http://eosweb.larc.nasa.gov/sse/
Complete solar energy data for anyplace on the planet! Thanks to NASA's Earth Science Enterprise program, there's more solar data here than even the Gaiam Real Goods techies can find a use for. Just point to your location on a world map.

Solar Energy International
P.O. Box 715
76 S. 2nd St.
Carbondale, CO 81623
Website: www.solarenergy.org
Email: sei@solarenergy.org
Phone: 970-963-8855
Fax: 970-963-8866
Offers hands-on and online classes on renewable technology.

Solar Living Institute
13771 So. Highway 101
Hopland, CA 95449
www.solarliving.org
E-mail: sli@solarliving.org
Phone: 707-744-2017
Fax: 707-744-1682
A nonprofit educational organization that offers world-class workshops on renewable energy, sustainable living, green building, and permaculture. Sponsors annual SolFest energy fair. Sign up for electronic newsletter: www.solarliving.org.

Sustainable Architecture, Building and Culture
P.O. Box 30085
Santa Barbara, CA 93130
Website: www.SustainableABC.com
Email: royprince@sustainableabc.com
Phone: 805-898-0079
Fax: 805-898-9199
A unique compendium of links and content oriented to the global community of ecological and natural building proponents.

Sustainable Buildings Industry Council
1112 16th Street NW, Suite 240
Washington, DC 20036
Website: www.sbicouncil.org
Email: sbic@sbicouncil.org
Phone: 202-628-7400
This trade organization has a nice website with lots of information about workshops, publications, and links to solar energy organizations. Especially good for professionals, but useful for home owners, too.

The Union of Concerned Scientists
2 Brattle Square
Cambridge, MA 02238-9105
Website: www.ucsusa.org
Email: usc@ucsusa.org
Phone: 617-547-5552
Fax: 617-864-9405
Organization of scientists and citizens concerned with the impact of advanced technology on society. Programs focus on energy policy, climate change, and national security.

Wind Energy Maps for U.S.
Website:
http://rredc.nrel.gov/wind/pubs/atlas
The complete Wind Energy Resource Atlas for the United States in downloadable format. Offers averages, regions, states, seasonal; more slicing and dicing than you can imagine. Includes Alaska, Hawaii, Puerto Rico, and Virgin Islands.

Worldwatch Institute
1776 Massachusetts Avenue NW
Washington, DC 20036-1904
Website: www.worldwatch.org
Email: worldwatch@worldwatch.org
Phone: 202-452-1999
Fax: 202-296-7365
A nonprofit public policy research organization dedicated to informing policymakers and the public about emerging global problems and trends. Offers a magazine and publishes a variety of books.

State Energy Offices Access Information

Here is all the straight, up-to-date information on utility intertie, and on any state programs that might help pay for a renewable energy system. All states have provided mail, and phone access. Many also provide fax, web, and email access. We've listed everything available as of press time. The ◇ symbol denotes a state that allows net-metering (as of early 2004). In addition, the states of Colorado, Florida, Kentucky, Illinois, Idaho, and Arizona have *some* utilities that support net metering without a state-wide law.

Also check www.dsireusa.org for the most up-to-date information on rebate and other state programs.

Alabama Energy Office
Department of Economic and Community Affairs
401 Adams Avenue, P.O. Box 5690
Montgomery, Alabama 36103-5690
Phone: (334) 242-5292
Fax: (334) 242-0552
Web Site: www.adeca.alabama.gov/ columns.aspx?m=4&id=19&id2=106

Alaska Energy Authority
Alaska Industrial Development and Export Authority
813 W. Northern Lights Boulevard
Anchorage, AK 9950
Phone: (907) 269-4625
Fax: (907) 269-3044
Web: www.aidea.org/aea.htm

American Samoa Energy Office
Territorial Energy Office
American Samoa Government
Samoa Energy House, Tauna
Pago Pago, American Samoa 96799
Phone: (684) 699-1101
Fax: (684) 699-2835

◇ Arizona Energy Office
Arizona Department of Commerce
1700 West Washington, Suite 220
Phoenix, AZ 85007
Phone: (602) 771-1100
Fax: (602) 771-1203
Email: energy@ep.state.az.us
Web: www.azcommerce.com /Energy/

◇ Arkansas Energy Office
Arkansas Industrial Development Commission
One Capitol Mall
Little Rock, Arkansas 72201
Phone: (501) 682-7377
Fax: (501) 682-2703
Web Site: www.1800arkansas.com/ Energy/

◇ California Energy Office
California Energy Commission
1516 Ninth Street, MS #32
Sacramento, California 95814
Phone: (916) 654-4287
Fax: (916) 654-4420
Email: mediaoffice@energy.state.ca.us
Web: www.energy.ca.gov

◇ Colorado Energy Office
Governor's Office of Energy Management and Conservation
225 East 16th Avenue, Suite 650
Denver, Colorado 80203
Phone: 303-894-2383
Fax: 303-894-2388
Email: oemc@state.co.us
Web: www.state.co.us/oemc

◇ Connecticut Energy Office
Policy Development and Planning— Energy
Connecticut Office of Policy and Management
450 Capitol Ave MS#52Enr
PO Box 341441
Hartford, CT 06134-1441
Phone: (860) 418-6374
Fax: (860) 418-6495
Email: john.mengacci@po.state.ct.us
Web: www.opm.state.ct.us/pdpd2/ energy/enserv.htm

◇ Delaware Energy Office
Department of Natural Resources and Environmental Control
146 South Governors Avenue
Dover, DE 19901
Phone: (302) 739-1530
Fax: (302) 739-1527
Web Site: www.delaware-energy .com

◇ District of Columbia Energy Office
2000 14th Street, NW, Suite 300E
Washington, DC 20009
Phone: (202) 673-6718
Fax: (202) 673-6725
Email: clintondc@aol.com
Web: www.dcenergy.org

Florida Energy Office
Florida Department of Environmental Protection
3800 Commonwealth Boulevard MS-19
Tallahassee, FL 32399-3000
Phone: (850) 245-2940
Fax: (850) 245-2947
Email: alexander.mack@dep.state.fl.us
Web: www.dep.state.fl.us /energy/

◇ Georgia Energy Office
Division of Energy Resources
Georgia Environmental Facilities Authority
100 Peachtree Street, NW, Suite 2090
Atlanta, Georgia 30303-1911
Phone: (404) 656-5176
Fax: (404) 656-7970
Web: www.gefa.org/energy_program .html

Guam Energy Office
Pacific Energy Resources Center
548 N. Marine Drive
Tamuning, Guam 96911
Phone: (671) 646-4361
Fax: (671) 649-1215
Email: energy@ns.gov.gu

✧ Hawaii Energy Office
Energy Resources and Technology
Division
Department of Business, Economic
Development and Tourism
235 South Beretania Street, Room 502
P.O. Box 2359
Honolulu, Hawaii 96804-2359
Phone: (808) 587-3807
Fax: (808) 586-2536
Email: gmishina@dbedt.hawaii.gov
Web: www.hawaii.gov/dbedt/ert/
energy.html

✧ Idaho Energy Office
Energy Division
Idaho Department of Water Resources
1301 N. Orchard Street
Boise, Idaho 83706
Phone: (208) 327-7900
Fax: (208) 327-7866
Email: energyguys@idwr.state.id.us
Web Site: www.idwr.state.id.us/
energy/

✧ Illinois Energy Office
Energy & Recycling Bureau
Illinois Department of Commerce
and Economic Opportunity
620 East Adams
Springfield, IL 62701
Phone: (217) 782-7500
Fax: (217) 785-2618
Web: www.illinoisbiz.biz/ho_recycling
_energy.html

✧ Indiana Energy Office
Energy Policy Division
Indiana Department of Commerce
One North Capitol, Suite 700
Indianapolis, Indiana 46204-2288
Phone: (317) 232-8939
Fax: (317) 232-8995
Web: www.state.in.us/doc/ energy/

✧ Iowa Energy Office
Energy Bureau
Energy and Geological Resources
Division
Iowa Department of Natural
Resources
Wallace State Office Building
East 9th & Grand Avenue
Des Moines, Iowa 50319

Phone: (515) 281-8681
Fax: (515) 281-6794
Web: www.state.ia.us/dnr/energy

✧ Kansas Energy Office
Energy Programs Section
Kansas Corporation Commission
1500 SW Arrowhead Road
Topeka, Kansas 66604-4027
Phone: (785) 271-3349
Fax: (785) 271-3268
Email: j.ploger@kcc.state.ks.us
Web: www.kcc.state.ks.us/energy

Kentucky Energy Office
Kentucky Division of Energy
663 Teton Trail
Frankfort, Kentucky 40601
Phone: (502) 564-7192
Fax: (502) 564-7484
Email: john.davies@ky.gov
Web Site: www.energy.ky.gov

Louisiana Energy Office
Technology Assessment Division
Department of Natural Resources
P.O. Box 44156
617 North Third Street
Baton Rouge, LA 70802
Phone: (225) 342-1399
Fax: (225) 342-1397
Email: tangular@dnr.state.la.us
Web: www.dnr.state.la.us/SEC/
EXECDIV/TECHASMT/

✧ Maine Energy Office
Division of Energy Programs
Maine Public Utilities Commission
State House Station No. 18
Augusta, ME 04333-0018
Phone: (207) 624-7495
Fax: (207) 287-8461
Web: http://bundlemeup.org/

✧ Maryland Energy Office
Maryland Energy Administration
1623 Forest Drive, Suite 300
Annapolis, MD 21403
Phone: (410) 260-7511
Fax: (410) 974-2875
Email: mea@energy.state.md.us
Web: www.energy.state.md.us

✧ Massachusetts Energy Office
Division of Energy Resources
Department of Economic
Development
100 Cambridge Street, Suite 1020
Boston, MA 02114
Phone: (617) 727-4732
Fax: (617) 727-0030
Email: energy@state.ma.us
Web: www.magnet.state.ma.us/doer

Michigan Energy Office
Michigan Public Service Commission
Michigan Department of Consumer
and Industry Services
P.O. Box 30221
Lansing, MI 48909
Phone: (517) 241-6180
Fax: (517) 241-6101
Web: www.michigan.gov/mpsc

✧ Minnesota Energy Office
Energy Division
Minnesota Department of Commerce
85 7th Place East, Suite 500
St. Paul, MN 55101-2198
Phone: (651) 297-2545
Fax: (651) 282-2568
Web: www.commerce.state .mn.us

Mississippi Energy Office
Energy Division
Mississippi Development Authority
P.O. Box 849
510 George Street, Suite 300
Jackson, MS 39205-0850
Phone: 601) 359-6600
Fax: (601) 359-6642
Email: energydiv@mississippi.org
Web: www.mississippi.org/
programs/energy/energy_overview
.htm

Missouri Energy Office
Energy Center
Department of Natural Resources
P.O. Box 176
169A E. Elm Street
Jefferson City, MO 65102
Phone: (573) 751-4000
Fax: (573) 751-6860
Web: www.dnr.state.mo.us/energy/
homeec.htm

❖ Montana Energy Office
Department of Environmental
Quality
P.O. Box 200901
1100 North Last Chance Gulch Room
401-H
Helena, MT 59620-0901
Phone: (406) 841-5240
Fax: (406) 841-5222
Web Site: www.deq.state.mt.us/
energy/

Nebraska Energy Office
P.O. Box 95085
1111 "O" Street, Suite 223
Lincoln, Nebraska 68509-5085
Phone: (402) 471-2867
Fax: (402) 471-3064
Email: energy@mail.state.ne.us
Web: www.nol.org/home/NEO/

❖ Nevada Energy Office
Nevada State Office of Energy
727 Fairview Drive, Suite F
Carson City, Nevada 89701
Phone: (775) 687-5975
Fax: (775) 687-4914
Email: dmcneil@dbi.state.nv.us
Web: http://energy.state.nv.us

❖ New Hampshire Energy Office
Governor's Office of Energy &
Community Services
State of New Hampshire
57 Regional Drive
Concord, New Hampshire 03301
Phone: (603) 271-2155
Fax: (603) 271-2615
Web: www.state.nh.us/governor/
energycomm/index.html

❖ New Jersey Energy Office
Office of Clean Energy
New Jersey Board of Public Utilities
Two Gateway Center
Trenton, NJ 07102
Phone: (609) 777-3335
Fax: (609) 777-3330
Email: energy@bpu.state.nj.us
Web: www.bpu.state.nj.us

❖ New Mexico Energy Office
Energy Conservation and Manage-
ment Division
New Mexico Energy, Minerals and
Natural Resources Department
1220 S. St. Francis Drive
P.O. Box 6429
Santa Fe, NM 87505
Phone: (505) 476-3310
Fax: (505) 476-3322
Web: www.emnrd.state.nm. us/ecmd/

❖ New York Energy Office
New York State Energy Research and
Development Authority
17 Columbia Circle
Albany, NY 12203
Phone: (518) 862-1090
Fax: (518) 862-1091
Web: www.nyserda.org

North Carolina Energy Office
State Energy Office
North Carolina Department of
Administration
1340 Mail Service Center
Raleigh, NC 27699
Phone: (919) 733-2230
Fax: (919) 733-2953
Web: www.energync.net

❖ North Dakota Energy Office
Division of Community Services
North Dakota Department of
Commerce
P.O. Box 2057
1600 East Century Avenue, Suite 2
Bismarck, ND 58502-2057
Phone: (701) 328-5300
Fax: (701) 328-2308
Web: www.state.nd.us/dcs/Energy/
default.html

**North Mariana Islands Energy
Office**
Energy Division
Commonwealth of the Northern
Mariana Islands
P.O. Box 340
Saipan, NMI 96950
Phone: (670) 664-4480/1/2
Fax: (670) 664-4483

❖ Ohio Energy Office
Office of Energy Efficiency
Ohio Department of Development
77 South High Street, 26th Floor
P.O. Box 1001
Columbus, OH 43216-1001
Phone: (614) 466-6797
Fax: (614) 466-1864
Web: www.odod.state.oh.us /cdd/oee/

❖ Oklahoma Energy Office
Division of Community Affairs and
Development
Oklahoma Department of Commerce
P.O. Box 26980
6601 North Broadway
Oklahoma City, Oklahoma 73126
Phone: (405) 815-6552
Fax: (405) 815-5344
Web: www.odoc.state.ok.us/

❖ Oregon Energy Office
625 Marion Street, N.E.
Salem, Oregon 97310
Phone: (503) 378-4040
Fax: (503) 373-7806
Email: energy.in.internet@state.or.us
Web: www.energy.state.or.us

❖ Pennsylvania Energy Office
Office of Pollution and Compliance
Assistance
Department of Environmental
Protection
P.O. Box 2063
400 Market Street, RCSOB
Harrisburg, Pennsylvania 17105
Phone: (717) 783-0542
Fax: (717) 783-2703
Web: www.paenergy.state.pa.us/

Puerto Rico Energy Office
Energy Affairs Administration
P.O. Box 9066600, Puerta de Tierra
San Juan, Puerto Rico 00906-6600
Phone: (787) 724-8777
Fax: (787) 721-3089
Email: quintanaj@caribe.net

❖ Rhode Island Energy Office
One Capital Hill, 2nd Floor
Providence, Rhode Island 02908
Phone: (401) 222-3370
Fax: (401) 222-1260
Web: www.riseo.state.ri.us/

South Carolina Energy Office
1201 Main Street, Suite 600
Columbia, South Carolina 29201
Phone: (803) 737-8030
Fax: (803) 737-9846
Web: www.state.sc.us/energy/

South Dakota Energy Office
Governor's Office of Economic
Development
711 E. Wells Avenue
Pierre, South Dakota 57501-3369
Phone: (605) 773-5032
Fax: (605) 773-3256
Web Site: www.sdgreatprofits .com/

Tennessee Energy Office
Department of Economics &
Community Development
Tennessee Tower
312 8th Avenue, North, 9th Floor
Nashville, Tennessee 37243
Phone: (615) 741-2994
Fax: (615) 741-5070
Web: www.state.tn.us/ecd/energy.htm

✧ Texas Energy Office
State Energy Conservation Office
Texas Comptroller of Public Accounts
111 E. 17th Street, 11th Floor
Austin, TX 78701
Phone: (512) 463-1931
Fax: (512) 475-2569
Web: www.seco.cpa.state.tx.us/

✧ Utah Energy Office
Utah Energy Office
1594 West North Temple
Suite 3610
P.O. Box 146480
Salt Lake City, UT 84114-6480
Phone: (801) 538-5428
Fax: (801) 538-4795
Email: dbeaudoin@utah.gov
Web: www.energy.utah.gov

✧ Vermont Energy Office
Energy Efficiency Division
Vermont Department of Public
Service
112 State Street, Drawer 20
Montpelier, Vermont 05620-2601
Phone: (802) 828-2811
Fax: (802) 828-2342
Email: vtdps@psd.state.vt.us
Web: www.state.vt.us/psd/ee/ee.htm

✧ Virginia Energy Office
Division of Energy
Virginia Department of Mines, Minerals & Energy
202 North Ninth Street, 8th Floor
Richmond, Virginia 23219
Phone: (804) 692-3200
Fax: (804) 692-3238
Web: www.mme.state.va.us/de

Virgin Islands Energy Office
Department of Planning and Natural
Resources
45 Mars Hill
Frederiksted, St. Croix, USVI 00840
Phone: (340) 773-1082
Fax: (340) 772-2133
Email: energydir@viaccess.net
Web Site: www.vienergy.org/

✧ Washington Energy Office
Washington State Energy Program
925 Plum Street, SE, Bldg #4
P.O. Box 43165
Olympia, WA 98504-3165
Phone: (360) 956-2000
Fax: (360) 956-2217
Web: www.energy.wsu.edu

West Virginia Energy Office
Energy Efficiency Program
West Virginia Development Office
Building 6, Room 645
State Capitol Complex
Charleston, West Virginia 25305
Phone: (304) 558-0350
Fax: (304) 558-0362
Web: www.wvdo.org/community
/eep.htm

✧ Wisconsin Energy Office
Division of Energy and
Public Benefits
Department of Administration
101 E. Wilson Street, 6th Floor
P.O. Box 7868
Madison, Wisconsin 53707-7868
Phone: (608) 266-8234
Fax: (608) 267-6931
Web: www.doa.state.wi.us /energy/

✧ Wyoming Energy Office
Minerals, Energy & Transportation
Wyoming Business Council
214 West 15th Street
Cheyenne, WY 82002
Phone: (307) 777-2800
Fax: (307) 777-2837
Email: jnunle@missc.state.wy.us
Web: www.wyomingbusiness.org/
minerals/

Glossary

A

AC: alternating current, electricity that changes voltage periodically, typically sixty times a second (or fifty in Europe). This kind of electricity is easier to move.

activated stand life: the period of time, at a specified temperature, that a battery can be left stored in the charged condition before its capacity fails.

active solar: any solar scheme employing pumps and controls that use power while harvesting solar energy.

A-frame: a building that looks like the capital letter A in cross-section.

air lock: two doors with space between, like a mud room, to keep the weather outside.

alternating current: AC electricity that changes voltage periodically, typically sixty times a second.

alternative energy: "voodoo" energy not purchased from a power company, usually coming from photovoltaic, micro-hydro, or wind.

ambient: the prevailing temperature, usually outdoors.

amorphous silicon: a type of PV cell manufactured without a crystalline structure. Compare with single-crystal and multi- (or poly-) crystalline silicon.

ampere: an instantaneous measure of the flow of electric current; abbreviated and more commonly spoken of as an "amp."

amp-hour: a one-ampere flow of electrical current for one hour; a measure of electrical quantity; two 60-watt 120-volt bulbs burning for one hour consume one amp-hour.

angle of incidence: the angle at which a ray of light (usually sunlight) strikes a planar surface (usually of a PV module). Angles of incidence close to perpendicularity (90°) are desirable.

anode: the positive electrode in an electrochemical cell (battery) toward which current flows; the earth ground in a cathodic protection system.

antifreeze: a chemical, usually liquid and often toxic, that keeps things from freezing.

array: an orderly collection, usually of photovoltaic modules connected electrically and mechanically secure; array current: the amperage produced by an array in full sun.

avoided cost: the amount utilities must pay for independently produced power; in theory, this was to be the whole cost, including capital share to produce peak demand power, but over the years supply-side weaseling redefined it to be something more like the cost of the fuel the utility avoided burning.

azimuth: horizontal angle measured clockwise from true north; the equator is at 90°.

B

backup: a secondary source of energy to pick up the slack when the primary source is inadequate. In alternatively powered homes, fossil fuel generators are often used as "backups" when extra power is required to run power tools or when the primary sources— sun, wind, water—are not providing sufficient energy.

balance of system: (BOS) equipment that controls the flow of electricity during generation and storage baseline: a statistical term for a starting point; the "before" in a before-and-after energy conservation analysis.

baseload: the smallest amount of electricity required to keep utility customers operating at the time of lowest demand; a utility's minimum load.

battery: a collection of cells that store electrical energy; each cell converts chemical energy into electricity or vice versa, and is interconnected with other cells to form a battery for storing useful quantities of electricity.

battery capacity: the total number of ampere-hours that can be withdrawn from a fully charged battery, usually over a standard period.

battery cycle life: the number of cycles that a battery can sustain before failing.

berm: earth mounded in an artificial hill.

bioregion: an area, usually fairly large, with generally homogeneous flora and fauna.

biosphere: the thin layer of water, soil, and air that supports all known life on Earth.

blackwater: what gets flushed down the toilet.

blocking diode: a diode that prevents loss of energy to an inactive PV array (rarely used with modern charge controllers).

boneyard: a peculiar location at Gaiam Real Goods where "experienced" products may be had for ridiculously low prices.

Btu: British thermal unit, the amount of heat required to raise the temperature of 1 pound of water 1 degree Fahrenheit. 3,411 Btus equals one kilowatt-hour.

bus bar: the point where all energy sources and loads connect to each other; often a metal bar with connections on it.

bus bar cost: the average cost of electricity delivered to the customer's distribution point.

buy-back agreement or contract: an agreement between the utility and a customer that any excess electricity generated by the customer will be bought back for an agreed-upon price.

C

cathode: the negative electrode in an electrochemical cell.

cell: a unit for storing or harvesting energy. In a battery, a cell is a single chemical storage unit consisting of electrodes and electrolyte, typically producing 1.5 volts; several cells are usually arranged inside a single container called a battery. Flashlight batteries are really flashlight cells. A photovoltaic cell is a single assembly of doped silicon and electrical contacts that allow it to take advantage of the photovoltaic effect, typically producing .5 volts; several PV cells are usually connected together and packaged as a module.

CF: compact fluorescent, a modern form of lightbulb using an integral ballast.

CFCs: chlorinated fluorocarbons, an industrial solvent and material widely used until implicated as a cause of ozone depletion in the atmosphere.

charge controller: device for managing the charging rate and state of charge of a battery bank.

controller terminology:

adjustable set point: allows adjustment of voltage disconnect levels.

high-voltage disconnect: the battery voltage at which the charge controller disconnects the batteries from the array to prevent overcharging.

low-voltage disconnect: the voltage at which the controller disconnects the batteries to prevent overdischarging.

low-voltage warning: a buzzer or light that indicates low battery voltage.

maximum power tracking: a circuit that maintains array voltage for maximal current.

multistage controller: a unit that allows multilevel control of battery charging or loading.

reverse current protection: prevents current flow from battery to array.

single-stage controller: a unit with only one level of control for charging or load control.

temperature compensation: a circuit that adjusts setpoints to ambient temperature in order to optimize charge.

charge rate: the rate at which a battery is recharged, expressed as a ratio of battery capacity to charging current flow.

clear-cutting: a forestry practice, cutting all trees in a relatively large plot.

cloud enhancement: the increase in sunlight due to direct rays plus refracted or reflected sunlight from partial cloud cover.

compact fluorescent: a modern energy-efficient form of lightbulb using an integral ballast.

compost: the process by which organic materials break down, or the materials in the process of being broken down.

concentrator: mirror or lens-like additions to a PV array that focus sunlight on smaller cells; a very promising way to improve PV yield.

conductance: a material's ability to allow electricity to flow through it; gold has very high conductance.

conversion efficiency: the ratio of energy input to energy output across the conversion boundary. For example, batteries typically are able to store and provide 90% of the charging energy applied, and are said to have a 90% energy efficiency.

cookie-cutter houses: houses all alike and all-in-a-row. Daly City, south of San Francisco, is a particularly depressing example. Runs in direct contradiction to our third guiding principle, "encourage diversity."

core/coil-ballasted: the materials-rich device required to drive some fluorescent lights; usually contains radioactive Americium (see **electronic ballasts**).

cost effectiveness: an economic measure of the worthiness of an investment; if an innovative solution costs less than a conventional alternative, it is more cost effective.

cross section: a "view" or drawing of a slice through a structure.

cross-ventilation: an arrangement of openings allowing air to pass through a structure.

crystalline silicon: the material from which most photovoltaic cells are made; in a single crystal cell, the entire cell is a slice of a single crystal of silicon, while a multicrystalline cell is cut from a block of smaller (centimeter-sized) crystals. The larger the crystal, the more exacting and expensive the manufacturing process.

cut-in: the condition at which a control connects its device.

cutoff voltage: in a charge controller, the voltage at which the array is disconnected from the battery to prevent overcharging.

cut-out: the condition at which a control interrupts the action.

cycle: in a battery, from a state of complete charge through discharge and recharge back to a fully charged state.

D

days of autonomy: the length of time (in days) that a system's storage can supply normal requirements without replenishment from a source; also called days of storage.

DC: direct current, the complement of AC, or alternating current, presents one unvarying voltage to a load.

deep cycle: a battery type manufactured to sustain many cycles of deep discharge that are in excess of 50% of total capacity.

degree-days: a term used to calculate heating and cooling loads; the sum, taken over an average year, of the lowest (for heating) or highest (for cooling) ambient daily temperatures. Example: if the target is 68° and the ambient on a given day is 58°, this would account for ten degree-days.

depth of discharge: the percent of rated capacity that has been withdrawn from the battery; also called DOD.

design month: the month in which the combination of insolation and loading require the maximum array output.

diode: an electrical component that permits current to pass in only one direction.

direct current: the complement of AC, or alternating current, presents one unvarying voltage to a load.

discharge: electrical term for withdrawing energy from a storage system.

discharge rate: the rate at which current is withdrawn from a battery expressed as a ratio to the battery's capacity; also known as C rate.

disconnect: a switch or other control used to complete or interrupt an electrical flow between components.

doping, dopant: small, minutely controlled amounts of specific chemicals introduced into the semiconductor matrix to control the density and probabilistic movement of free electrons.

downhole: a piece of equipment, usually a pump, that is lowered down the hole (the well or shaft) to do its work.

drip irrigation: a technique that precisely delivers measured amounts of water through small tubes; an exceedingly efficient way to water plants.

dry cell: a cell with captive electrolyte.

duty cycle: the ratio between active time and total time; used to describe the operating regime of an appliance in an electrical system.

duty rating: the amount of time that an appliance can be run at its full rated output before failure can be expected.

E

earthship: a rammed-earth structure based on tires filled with tamped earth; the term was coined by Michael Reynolds.

Eco Desk: an ecology information resource maintained by many ecologically-minded companies.

edison base: a bulb base designed by (a) Enrico Fermi, (b) John Schaeffer, (c) Thomas Edison, (d) none of the above. The familiar standard residential light bulb base.

efficiency: a mathematical measure of actual as a percentage of the theoretical best. See **conversion efficiency.**

electrolyte: the chemical medium, usually liquid, in a battery that conveys charge between the positive and negative electrodes; in a lead-acid battery, the electrodes are lead and the electrolyte is acid.

electromagnetic radiation: EMR, the invisible field around an electric device. Not much is known about the effects of EMR, but it makes many of us nervous.

electronic ballasts: an improvement over core/coil ballasts, used to drive compact fluorescent lamps; contains no radioactivity.

embodied: of energy, meaning literally the amount of energy required to produce an object in its present form. Example: an inflated balloon's embodied energy includes the energy required to blow it up.

EMR: electromagnetic radiation, the invisible field around an electric device. Not much is known about the effects of EMR, but it makes many of us nervous.

energy density: the ratio of stored energy to storage volume or weight.

energy efficient: one of the best ways to use energy to accomplish a task; for example, heating with electricity is never energy efficient, while lighting with compact fluorescents is.

equalizing: periodic overcharging of batteries to make sure that all cells are reaching a good state of charge.

externalities: considerations, often subtle or remote, that should be accounted for when evaluating a process or product, but usually are not. For example, externalities for a power plant may include downwind particulate fallout and acid rain, damage to lifeforms in the cooling water intake and effluent streams, and many other factors.

F

fail-safe: a system designed in such a way that it will always fail into a safe condition.

feng shui: an Asian system of placement that pays special attention to wind, water, and the cardinal directions.

ferro-cement: a construction technique; an armature of iron contained in a cement body, often a wall, slab, or tank.

fill factor: of a photovoltaic module's I-V (current/voltage) curve, this number expresses the product of the open circuit voltage and the short-circuit current, and is therefore a measure of the "squareness" of the I-V curve's shape.

firebox: the structure within which combustion takes place.

fixed-tilt array: a PV array set in a fixed position.

flat-plate array: a PV array consisting of nonconcentrating modules.

float charge: a charge applied to a battery equal to or slightly larger than the battery's natural tendency to self-discharge.

FNC: fiber-nickel-cadmium, a new battery technology.

frequency: of a wave, the number of peaks in a period. For example, alternating current presents sixty peaks per second, so its frequency is sixty hertz. Hertz is the standard unit for frequency when the period in question is one second.

G

gassifier: a heating device which burns so hotly that the fuel sublimes directly from its solid to its gaseous state and burns very cleanly.

gassing: when a battery is charged, gasses are often given out; also called outgassing.

golf cart batteries: industrial batteries tolerant of deep cycling, often used in mobile vehicles.

gotcha!s: an unexpected outcome or effect, or the points at which, no matter how hard you wriggle, you can't escape.

gravity-fed: water storage far enough above the point of use (usually 50 feet) so that the weight of the water provides sufficient pressure.

greywater: all other household effluents besides blackwater (toilet water); greywater may be reused with much less processing than blackwater.

grid: a utility term for the network of transmission lines that distribute electricity from a variety of sources across a large area.

grid-connected system: a house, office, or other electrical system that can draw its energy from the grid; although usually grid-power-consumers, grid-connected systems can provide power to the grid.

groundwater: as distinct from water pumped up from the depths, groundwater is run off from precipitation, agriculture, or other sources.

H

heat exchanger: device that passes heat from one substance to another; in a solar hot water heater, for example, the heat exchanger takes heat harvested by a fluid circulating through the solar panel and transfers it to domestic hot water.

high-tech glass: window constructions made of two sheets of glass, sometimes treated with a metallic deposition, sealed together hermetically, with the cavity filled by an inert gas and, often, a further plastic membrane. High-tech glass can have an R-value as high as 10.

homeschooling: educating children at home instead of entrusting them to public or private schools; a growing

trend quite often linked to off-the-grid powered homes.

homestead: the house and surrounding lands.

homesteaders: people who consciously and intentionally develop their homestead.

hot tub: a quasi-religious object in California; a large bathtub for several people at once; an energy hog of serious proportions.

house current: in the United States, 117 volts root mean square of alternating current, plus or minus 7 volts; nominally 110-volt power; what comes out of most wall outlets.

HVAC: heating, ventilation, and air conditioning; space conditioning.

hydro turbine: a device that converts a stream of water into rotational energy.

hydrometer: tool used to measure the specific gravity of a liquid.

hydronic: contraction of hydro and electronic, usually applied to radiant in-floor heating systems and their associated sensors and pumps.

hysteresis: the lag between cause and effect, between stimulus and response.

I

incandescent bulb: a light source that produces light by heating a filament until it emits photons—quite an energy-intensive task.

incident solar radiation: or insolation, the amount of sunlight falling on a place.

indigenous plantings: gardening with plants native to the bioregion.

inductive transformer/rectifier: the little transformer device that powers many household appliances; an "energy criminal" that takes an unreasonably large amount of alternating electricity (house current) and converts it into a much smaller amount of current with different properties; for example, much lower-voltage direct current.

infiltration: air, at ambient temperature, blowing through cracks and

holes in a house wall and spoiling the space conditioning.

infrared: light just outside the visible spectrum, usually associated with heat radiation.

infrastructure: a buzz word for the underpinnings of civilization; roads, water mains, power and phone lines, fire suppression, ambulance, education, and governmental services are all infrastructure. *Infra* is Latin for "beneath." In a more technical sense, the repair infrastructure is local existence of repair personnel and parts for a given technology.

insolation: a word coined from incident solar radiation, the amount of sunlight falling on a place.

insulation: a material that keeps energy from crossing from one place to another. On electrical wire, it is the plastic or rubber that covers the conductor. In a building, insulation makes the walls, floor, and roof more resistant to the outside (ambient) temperature.

Integrated Resource Planning: an effort by the utility industry to consider all resources and requirements in order to produce electricity as efficiently as possible.

interconnect: to connect two systems, often an independent power producer and the grid; see also **intertie.**

interface: the point where two different flows or energies interact; for example, a power system's interface with the human world is manifested as meters, which show system status, and controls, with which that status can be manipulated.

internal combustion engines: gasoline engines, typically in automobiles, small stand-alone devices like chainsaws, lawnmowers, and generators.

intertie: the electrical connection between an independent power producer—for example, a PV-powered household—and the utility's distribution lines, in such a way that each can supply or draw from the other.

inverter: the electrical device that changes direct current into alternating current.

irradiance: the instantaneous solar radiation incident on a surface; usually expressed in Langleys (a small amount) or in kilowatts per square meter. The definition of "one sun" of irradiance is one kilowatt per square meter.

irreverence: the measure, difficult to quantify, of the seriousness of Gaiam Real Goods techs when talking with utility suits.

IRP: See **Integrated Resource Planning.**

I-V curve: a plot of current against voltage to show the operating characteristics of a photovoltaic cell, module, or array.

K

kilowatt: one thousand watts, a measure of instantaneous work. Ten 100-watt bulbs require a kilowatt of energy to light up.

kilowatt-hour: the standard measure of household electrical energy use. If the ten bulbs left unfrugally burning in the preceding example are on for an hour, they consume 1 kilowatt-hour of electricity.

L

landfill: another word for dump.

Langley: the unit of solar irradiance; 1 gram-calorie per square centimeter.

lead-acid: the standard type of battery for use in home energy systems and automobiles.

LED: light-emitting diode. A very efficient source of electrical lighting, typically lasting 50,000 to 100,000 hours.

life: You expect an answer here? Well, when speaking of electrical systems, this term is used to quantify the time the system can be expected to function at or above a specified performance level.

life-cycle cost: the estimated cost of owning and operating a system over its useful life.

line extensions: what the power company does to bring their power lines to the consumer.

line-tied system: an electrical system connected to the power lines, usually having domestic power generating capacity and the ability to draw power from the grid or return power to the grid, depending on load and generator status.

load: an electrical device, or the amount of energy consumed by such a device.

load circuit: the wiring that provides the path for the current that powers the load.

load current: expressed in amps, the current required by the device to operate.

low-emissivity: applied to high-tech windows, meaning that infrared or heat energy will not pass back out through the glass.

low-flush: a toilet using a smaller amount (usually about 6 quarts) of water to accomplish its function.

low pressure: usually of water, meaning that the head, or pressurization, is relatively small.

low-voltage: usually another term for 12- or 24-volt direct current.

M

maintenance-free battery: a battery to which water cannot be added to maintain electrolyte volume. All batteries require routine inspection and maintenance.

maximum power point: the point at which a power conditioner continuously controls PV source voltage in order to hold it at its maximum output current.

ME: mechanical engineer; the engineers who usually work with heating and cooling, elevators, and the other mechanical devices in a large building.

meteorological: pertaining to weather; meteorology is the study of weather.

micro-climate: the climate in a small area, sometimes as small as a garden or the interior of a house. Climate is distinct from weather in that it speaks for trends taken over a period of at least a year, while weather describes immediate conditions.

micro-hydro: small hydro (falling water) generation.

millennia: one thousand years.

milliamp: one thousandth of an ampere.

module: a manufactured panel of photovoltaic cells. A module typically houses 36 cells in an aluminum frame covered with glass or acrylic, organizes their wiring, and provides a junction box for connection between itself, other modules in the array, and the system.

N

naturopathic: a form of medicine devoted to natural remedies and procedures.

net metering: a desirable form of buy-back agreement in which the line-tied house's electric meter turns in the utility's favor when grid power is being drawn, and in the system owner's favor when the house generation exceeds its needs and electricity is flowing into the grid. At the end of the payment period, when the meter is read, the system owner pays (or is paid by) the utility depending on the net metering.

NEC: the National Electrical Code, guidelines for all types of electrical installations including (since 1984) PV systems.

nicad: slang for nickel-cadmium, a form of chemical storage often used in rechargeable batteries.

nominal voltage: the terminal voltage of a cell or battery discharging at a specified rate and at a specified temperature; in normal systems, this is usually a multiple of twelve.

nontoxic: having no known poisonous qualities.

normal operating cell temperature: defined as the standard operating temperature of a PV module at 800 W/m°, 20° C ambient, 1 meter per second wind speed; used to estimate the nominal operating temperature of a module in its working environment.

N-type silicon: silicon doped (containing impurities), which gives the lattice a net negative charge, repelling electrons.

O

off-peak energy: electricity during the baseload period, which is usually cheaper. Utilities often must keep generators turning, and are eager to find users during these periods, and so sell off-peak energy for less.

off-peak kilowatt: a kilowatt hour of off-peak energy.

off-the-grid: not connected to the power lines, energy self-sufficient.

ohm: the basic unit of electrical resistance; I=RV, or Current (amperes) equals Resistance (ohms) times Voltage (volts).

on-line: connected to the system, ready for work.

on-the-grid: where most of America lives and works, connected to a continent-spanning web of electrical distribution lines.

open-circuit voltage: the maximum voltage measurable at the terminals of a photovoltaic cell, module, or array with no load applied.

operating point: the current and voltage that a module or array produces under load, as determined by the I-V curve.

order of magnitude: multiplied or divided by ten. One hundred is an order of magnitude smaller than one thousand, and an order of magnitude larger than ten.

orientation: placement with respect to the cardinal directions north, east, south, and west; azimuth is the measure of orientation.

outgas: of any material, the production of gasses; batteries outgas during charging; new synthetic rugs outgas

when struck by sunlight, or when warm, or whenever they feel like it.

overcharge: forcing current into a fully charged battery; a bad idea except during equalization.

overcurrent: too much current for the wiring; overcurrent protection, in the form of fuses and circuit breakers, guards against this.

owner-builder: one of the few printable things building inspectors call people who build their own homes.

P

panel: any flat modular structure; solar panels may collect solar energy by many means; a number of photovoltaic modules may be assembled into a panel using a mechanical frame, but this should more properly be called an array or subarray.

parallel: connecting the like poles to like in an electrical circuit, so plus connects to plus and minus to minus; this arrangement increases current without affecting voltage.

particulates: particles that are so small that they persist in suspension in air or water.

passive solar: a shelter that maintains a comfortable inside temperature simply by accepting, storing, and preserving the heat from sunlight.

passively heated: a shelter that has its space heated by the sun without using any other energy.

patch-cutting: clear-cutting (cutting all trees) on a small scale usually less than an acre.

pathetic fallacy: attributing human motivations to inanimate objects or lower animals.

payback: the time it takes to recoup the cost of improved technology as compared to the conventional solution. Payback on a compact fluorescent bulb (as compared to an incandescent bulb) may take a year or two, but over the whole life of the CF the savings probably will exceed the original cost of the bulb, and payback will take place several times over.

peak demand: the largest amount of electricity demanded by a utility's customers; typically, peak demand happens in early afternoon on the hottest weekday of the year.

peak kilowatt: a kilowatt hour of electricity taken during peak demand, usually the most expensive electricity money can buy.

peak load: the same as peak demand but on a smaller scale, the maximum load demanded of a single system.

peak power current: the amperage produced by a photovoltaic module operating at the "knee" of its I-V curve.

peak sun hours: the equivalent number of hours per day when solar irradiance averages one sun (1 kW/m°). "Six peak sun hours" means that the energy received during total daylight hours equals the energy that would have been received if the sun had shone for six hours at a rate of 1000 watts per square meter.

peak watt: the manufacturer's measure of the best possible output of a module under ideal laboratory conditions.

Pelton wheel: a special turbine, designed by someone named Pelton, for converting flowing water into rotational energy.

periodic table of elements: a chart showing the chemical elements organized by the number of protons in their nuclei and the number of electrons in their outer, or valence, band.

PG&E: Pacific Gas and Electric, the local and sometimes beloved utility for much of Northern California.

phantom loads: "energy criminals" that are on even when you turn them off: instant-on TVs, microwaves with clocks; symptomatic of impatience and our sloppy preference for immediacy over efficiency.

photon: the theoretical particle used to explain light.

photophobic: fear of light (or preference for darkness), usually used of insects and animals. The opposite, phototropic, means light-seeking.

photovoltaic cell: the proper name for a device manufactured to pump electricity when light falls on it.

photovoltaics: PVs or modules that utilize the photovoltaic effect to generate useable amounts of electricity.

photovoltaic system: the modules, controls, storage, and other components that constitute a stand-alone solar energy system.

plates: the thin pieces of metal or other material used to collect electrical energy in a battery.

plug-loads: the appliances and other devices plugged into a power system.

plutonium: a particularly nasty radioactive material used in nuclear generation of electricity. One atom is enough to kill you.

pn-junction: the plane within a photovoltaic cell where the positively and negatively doped silicon layers meet.

pocket plate: a plate for a battery in which active materials are held in a perforated metal pocket on a support strip.

pollution: any dumping of toxic or unpleasant materials into air or water.

polyurethane: a long-chain carbon molecule, a good basis for sealants, paints, and plastics.

power: kinetic, or moving energy, actually performing work; in an electrical system, power is measured in watts.

power-conditioning equipment: electrical devices that change electrical forms (an inverter is an example) or assure that the electricity is of the correct form and reliability for the equipment consuming it; a surge protector is another example.

power density: ratio of a battery's rated power available to its volume (in liters) or weight (in kilograms).

PUC: Public Utilities Commission; many states call it something else, but this is the agency responsible for regulating utility rates and practices.

PURPA: this 1978 legislation, the Public Utility Regulatory Policy Act, requires utilities to purchase power

from anyone at the utility's avoided cost.

PVs: photovoltaic modules.

R

radioactive material: a substance which, left to itself, sheds tiny, highly energetic pieces that put anyone nearby at great risk. Plutonium is one of these. Radioactive materials remain active indefinitely, but the time over which they are active is measured in terms of half-life, the time it takes them to become half as active as they are now; plutonium's half-life is a little over 22,000 years.

ram pump: a water-pumping machine that uses a water-hammer effect (based on the inertia of flowing water) to lift water.

rated battery capacity: manufacturer's term indicating the maximum energy that can be withdrawn from a battery at a standard specified rate.

rated module current: manufacturer's indication of module current under standard laboratory test conditions.

renewable energy: an energy source that renews itself without effort; fossil fuels, once consumed, are gone forever, while solar energy is renewable in that the sun we harvest today has no effect on the sun we can harvest tomorrow.

renewables: shorthand term for renewable energy or materials sources.

resistance: the ability of a substance to resist electrical flow; in electricity, resistance is measured in ohms.

retrofit: to install new equipment in a structure that was not originally designed for it. For example, we may retrofit a lamp with a compact fluorescent bulb, but the new bulb's shape may not fit well with the lamp's design.

romex: an electrician's term for common two-conductor-with-ground wire, the kind houses are wired with.

root mean square: RMS, the effective voltage of alternating current,

usually about 70 % (the square root of two over two) of the peak voltage. House current typically has an RMS of 117 volts and a peak voltage of 167 volts.

RPM: rotations per minute.

R-value: Resistance value, used specifically of materials used for insulating structures. Fiberglass insulation three inches thick has an R-value of 13.

S

seasonal depth of discharge: an adjustment for long-term seasonal battery discharge resulting in a smaller array and a battery bank matched to low-insolation season needs.

secondary battery: a battery that can be repeatedly discharged and fully recharged.

self-discharge: the tendency of a battery to lose its charge through internal chemical activity.

semiconductor: the chief ingredient in a photovoltaic cell, a normal insulating substance that conducts electricity under certain circumstances.

series connection: wiring devices with alternating poles, plus to minus, plus to minus; this arrangement increases voltage (potential) without increasing current.

set-back thermostat: combines a clock and a thermostat so that a zone (like a bedroom) may be kept comfortable only when in use.

setpoint: electrical condition, usually voltage, at which controls are adjusted to change their action.

shallow-cycle battery: like an automotive battery, designed to be kept nearly fully charged; such batteries perform poorly when discharged by more than 25% of their capacity.

shelf life: the period of time a device can be expected to be stored and still perform to specifications.

short circuit: an electrical path that connects opposite sides of a source without any appreciable load, thereby allowing maximum (and possibly disastrous) current flow.

short circuit current: current produced by a photovoltaic cell, module, or array when its output terminals are connected to each other, or "short-circuited."

showerhead: in common usage, a device for wasting energy by using too much hot water; in the Gaiam Real Goods home, low-flow showerheads prevent this undesirable result.

silicon: one of the most abundant elements on the planet, commonly found as sand and used to make photovoltaic cells.

single-crystal silicon: silicon carefully melted and grown in large boules, then sliced and treated to become the most efficient photovoltaic cells.

slow-blow: a fuse that tolerates a degree of overcurrent momentarily; a good choice for motors and other devices that require initial power surges to get rolling.

slow paced: a description of the life of a Gaiam Real Goods employee . . . not!

solar aperture: the opening to the south of a site (in the northern hemisphere) across which the sun passes; trees, mountains, and buildings may narrow the aperture, which also changes with the season.

solar cell: see **photovoltaic cell.**

solar fraction: the fraction of electricity that may be reasonable harvested from sun falling on a site. The solar fraction will be less in a foggy or cloudy site, or one with a narrower solar aperture, than an open, sunny site.

solar hot water heating: direct or indirect use of heat taken from the sun to heat domestic hot water.

solar oven: simply a box with a glass front and, optionally, reflectors and reflector coated walls, which heats up in the sun sufficiently to cook food.

solar panels: any kind of flat devices placed in the sun to harvest solar energy.

solar resource: the amount of insolation a site receives, normally expressed in kilowatt-hours per square meter per day.

specific gravity: the relative density of a substance compared to water (for liquids and solids) or air (for gases) Water is defined as 1.0; a fully charged sulfuric acid electrolyte might be as dense as 1.30, or 30 percent denser than water. Specific gravity is measured with a hydrometer.

stand-alone: a system, sometimes a home, that requires no imported energy.

stand-by: a device kept for times when the primary device is unable to perform; a stand-by generator is the same as a back-up generator.

starved electrolyte cell: a battery cell containing little or no free fluid electrolyte.

state of charge: the real amount of energy stored in a battery as a percentage of its total rated capacity.

state-of-the-art: a term beloved by technoids to express that this is the hottest thing since sliced bread.

stratification: in a battery, when the electrolyte acid concentration varies in layers from top to bottom; seldom a problem in vehicle batteries, due to vehicular motion and vibration, this can be a problem with static batteries, and can be corrected by periodic equalization.

stepwise: a little at a time, incrementally.

subarray: part of an array, usually photovoltaic, wired to be controlled separately.

sulfating: formation of lead-sulfate crystals on the plates of a lead-acid battery, which can cause permanent damage to the battery.

superinsulated: using as much insulation as possible, usually R-50 and above.

surge capacity: the ability of a battery or inverter to sustain a short current surge in excess of rated capacity in order to start a device that requires an initial surge current.

sustainable: material or energy sources that, if managed carefully, will provide at current levels indefinitely. A theoretical example: redwood is sustainable if it is harvested sparingly (large takings and exportation to Japan not allowed) and if every tree taken is replaced with another redwood. Sustainability can be, and usually is, abused for profit by playing it like a shell game; by planting, for example, a fast-growing fir in place of the harvested redwood.

system availability: the probability or percentage of time that an energy storage system will be able to meet load demand fully.

T

temperature compensation: an allowance made by a charge controller to match battery charging to battery temperature.

temperature correction: applied to derive true storage capacity using battery nameplate capacity and temperature; batteries are rated at 20°C.

therm: a quantity of natural gas, 100 cubic feet; roughly 100,000 Btus of potential heat.

thermal mass: solid, usually masonry volumes inside a structure that absorb heat, then radiate it slowly when the surrounding air falls below their temperature.

thermoelectric: producing heat using electricity; a bad idea.

thermography: photography of heat loss, usually with a special video camera sensitive to the far end of the infrared spectrum.

thermosiphon: a circulation system that takes advantage of the fact that warmer substances rise. By placing the solar collector of a solar hot-water system below the tank, thermosiphoning takes care of circulating the hot water, and pumping is not required.

thin-film module: an inexpensive way of manufacturing photovoltaic modules; thin-film modules typically are less efficient than single-crystal or multi-crystal devices; also called amorphous silicon modules.

tilt angle: measures the angle of a panel from the horizontal.

tracker: a device that follows the sun and keeps the panel perpendicular to it.

transformer: a simple electrical device that changes the voltage of alternating current; most transformers are inductive, which means they set up a field around themselves, which is often a costly thing to do.

transparent energy system: a system that looks and acts like a conventional grid-connected home system, but is independent.

trickle charge: a small current intended to maintain an inactive battery in a fully charged condition.

troubleshoot: a form of recreation not unlike riding to hounds in which the technician attempts to find, catch, and eliminate the trouble in a system.

tungsten filament: the small coil in a lightbulb that glows hotly and brightly when electricity passes through it.

turbine: a vaned wheel over which a rapidly moving liquid or gas is passed, causing the wheel to spin; a device for converting flow to rotational energy.

turnkey: the jail warden; more commonly in our context, a system that is ready for the owner-occupant from the first time he or she turns the key in the front-door lock.

TV: a device for wasting time and scrambling the brain.

two-by-fours: standard building members, originally 2 by 4 inches, now 1.5 inches by 3.5inches; often referred to as "sticks".

U

ultrafilter: of water, to remove all particulates and impurities down to the submicron range, about the size of giardia, and larger viruses.

uninterruptible power supply: an energy system providing ultrareliable power; essential for computers, aircraft guidance, medical, and other systems; also known as a UPS.

V

Varistor: a voltage-dependent variable resistor, normally used to protect sensitive equipment from spikes (like lightning strike) by diverting the energy to ground.

VCR: videocassette recorder, a device for making TVs slightly more responsive and useful.

VDTs: video display terminals, like televisions and computer screens.

vented cell: a battery cell designed with a vent mechanism for expelling gasses during charging.

volt: measure of electrical potential: 110-volt house electricity has more potential to do work than an equal flow of 12-volt electricity.

voltage drop: lost potential due to wire resistance over distance.

W

watt: the standard unit of electrical power; 1 ampere of current flowing with 1 volt of potential; 746 watts make one horsepower.

watt-hours: 1 watt for 1 hour. A 15-watt compact fluorescent consumes 15 of these in 60 minutes.

waveform: the characteristic trace of voltage over time of an alternating current when viewed on an oscilloscope; typically a smooth sine wave, although primitive inverters supply square or modified square waveforms.

wet shelf life: the time that an electrolyte-filled battery can remain unused in the charged condition before dropping below its nominal performance level.

wheatgrass: a singularly delicious potation made by squeezing young wheat sprouts; said to promote purity in the digestive tract.

whole-life cost analysis: an economic procedure for evaluating all the costs of an activity from cradle to grave, that is, from extraction or culture through manufacture and use, then back to the natural state; a very difficult thing to accomplish with great accuracy, but a very instructive reckoning nonetheless.

wind-chill: a factor calculated based on temperature and wind speed that expresses the fact that a given ambient temperature feels colder when the wind is blowing.

wind spinners: fond name for wind machines, devices that turn wind into usable energy.

Photovoltaic Power Systems and the National Electric Code

Suggested Practices

John Wiles
Southwest Technology Development Institute
New Mexico State University
PO Box 30001, MSC 3 SOLAR
Corner of Research Drive and Sam Steel Way
Las Cruces, NM 88003

ABSTRACT

This suggested practices manual examines the requirements of the *2002 National Electrical Code (NEC)* as they apply to photovoltaic (PV) power systems. The design requirements for the balance-of-systems components in a PV system are addressed, including conductor selection and sizing, overcurrent protection device rating and location, and disconnect rating and location. PV array, battery, charge controller, and inverter sizing and selection are not covered, as these items are the responsibility of the system designer, and they in turn determine the items in this manual. Stand-alone, hybrid, and utility-interactive PV systems are all covered. References are made to applicable sections of the *NEC*.

National Electrical Code® and *NEC* ® are registered trademarks of the National Fire Protection Association, Inc., Quincy, Massachusetts 02269

Acknowledgments

Numerous people throughout the photovoltaic industry and the electrical inspector community reviewed the earlier editions of this manual and provided comments that are incorporated in this edition. The author sends heartfelt thanks to all who have supported the efforts to improve and expand the document. With inputs from all involved, each succeeding edition is progressively more detailed and complete. Ward Bower at Sandia National Laboratories deserves special thanks for his dedicated and continuing support and review of this document. Ronald Donaghe, Southwest Technology Development Institute, performed document editing and layout.

Technical Comments to:

John C. Wiles
Southwest Technology Development Institute
New Mexico State University
P.O. Box 30001/MSC 3 SOLAR
Corner of Research Drive and Sam Steel Way
Las Cruces, New Mexico 88003-0001

Phone: 505-646-6105 FAX 505-646-3841
E-mail: jwiles@nmsu.edu

Copies may be downloaded from the PV and Codes links on the SWTDI/NMSU web site at
http://www.nmsu.edu/~tdi

Purpose

This document is intended to contribute to the widespread installation of safe, reliable PV systems that meet the requirements of the *National Electrical Code*.

Disclaimer

This guide provides information on how the *2002 National Electrical Code* (*NEC*) applies to photovoltaic systems. The guide is not intended to supplant or replace the *NEC*; it paraphrases the *NEC* where it pertains to photovoltaic systems and should be used with the full text of the *NEC*. Users of this guide should be thoroughly familiar with the *NEC* and know the engineering principles and hazards associated with electrical and photovoltaic power systems. The information in this guide is the best available at the time of publication and is believed to be technically accurate. Application of this information and results obtained are the responsibility of the user.

In most locations, all electrical wiring including photovoltaic power systems must be accomplished by, or under the supervision of, a licensed electrician and then inspected by a designated local authority. Some municipalities have additional codes that supplement or replace the *NEC*. The local inspector has the final say on what is acceptable. In some areas, compliance with the *NEC* is not required.

National Fire Protection Association (NFPA) Statement

The *National Electrical Code* including the 2005 *National Electrical Code* is published and updated every three years by the National Fire Protection Association (NFPA), Batterymarch Park, Quincy, Massachusetts 02269. The *National Electrical Code* and the term *NEC* are registered trademarks of the National Fire Protection Association and may not be used without their permission. Copies of the current edition of the *National Electrical Code* and the *National Electrical Code Handbook* are available from the NFPA at the above address, most electrical supply distributors, and many bookstores.

Table of Contents

List of Figures

NEC SUGGESTED PRACTICES

APPLICABLE ARTICLES in the 2002 NATIONAL ELECTRICAL CODE

Although most portions of the *National Electrical Code* apply to all electrical power systems, including photovoltaic power systems, those listed below are of particular significance.

Article	Contents
90	Introduction
100	Definitions
110	Requirements
200	Grounded Conductors
210	Branch Circuits
240	Overcurrent Protection
250	Grounding
300	Wiring Methods
310	Conductors for General Wiring
334	Nonmetallic Sheathed Cable
336	Power and Control Tray Cable: Type TC
338	Service-Entrance Cable
340	Underground Feeders
352	Rigid Nonmetallic Conduit
356	Liquidtight Flexible Metal Conduit and Liquidtight Flexible Nonmetallic Conduit
366	Auxiliary Gutters
400	Flexible Cords and Cables
408	Switchboards and Panelboards
445	Generators
480	Storage Batteries
490	Equipment, Over 600 Volts
690	Solar Photovoltaic Systems
705	Interconnected Electric Power Production Sources
720	Circuits and Equipment Operating at Less Than 50 Volts
Ch 9, Table 8	Conductor Properties
Annex C	Conduit Fill Tables

SUGGESTED PRACTICES

OBJECTIVE
- Safe, Reliable, Durable Photovoltaic Power Systems

- Knowledgeable Manufacturers, Designers, Dealers, Installers, Consumers, and Inspectors

METHOD
- Widespread Dissemination of These Suggested Practices and Knowledge of the NEC

- Technical Interchange Among Interested Parties

Introduction

The National Fire Protection Association has acted as sponsor of the *National Electrical Code (NEC)* since 1911. The original Code document was developed in 1897. With few exceptions, electrical power systems installed in the United States in the 20th and 21st centuries have had to comply with the *NEC*. This compliance requirement applies to most permanent installations of photovoltaic (PV) power systems. In 1984, Article 690 Solar Photovoltaic Systems, which addresses safety requirements for the installation of PV systems, was added to the Code. This article has been updated and expanded in each edition of the *NEC* since 1984.

Many of the PV systems in use and being installed today may not be in compliance with the *NEC* and other local codes. There are several contributing factors to this situation:

FACTORS THAT HAVE REDUCED LOCAL AND NEC COMPLIANCE

- The PV industry has a strong "grass roots," do-it-yourself faction that is not fully aware of the dangers associated with low-voltage and high-voltage, direct-current (dc) and alternating-current (ac) electrical power systems.
- Electricians and electrical inspectors have not had significant experience with direct-current portions of the Code or PV power systems.
- The electrical equipment industries do not advertise or widely distribute equipment suitable for dc use that meets *NEC* requirements.
- Popular publications present information to the public that implies that PV systems are easily installed, modified, and maintained by untrained personnel.

- Photovoltaic equipment manufacturers have, in some cases, been unable to afford the costs associated with testing and listing by approved testing laboratories like Underwriters Laboratories (UL), Canadian Standards Association (CSA) or ETL, Inc.
- Photovoltaic installers and dealers in many cases have not had significant training or experience installing ac residential and/or commercial power systems.

Some PV installers in the United States are licensed electricians or use licensed electrical contractors and are familiar with all sections of the *NEC*. These installer/contractors are trained to install safe and more reliable PV systems that meet the *NEC* and minimize the hazards associated with electrical power systems. On the other hand, some PV installations have numerous defects that typically stem from unfamiliarity with electrical power system codes or unfamiliarity with dc currents and power systems. These installations often do not meet the requirements of the *NEC*. Some of the more prominent problems are listed below.

OBSERVED PV INSTALLATION PROBLEMS

- Improper ampacity of conductors
- Improper types of conductors
- Improper or unsafe wiring methods
- Lack of or improper overcurrent protection on conductors
- Inadequate number and placement of disconnects
- Improper application of listed equipment
- No, or underrated, short-circuit or overcurrent protection on battery circuits
- Use of non-listed components when listed components are available

- Improper system grounding
- Lack of, or improper, equipment grounding
- Use of underrated hardware or components
- Use of ac components (fuses and switches) in dc applications

The *NEC* generally applies to any PV power system, regardless of size or location. A single, small PV module may not present a significant hazard, and a small system in a remote location may present few safety hazards because people are seldom in the area. On the other hand, two or three modules connected to a battery can be lethal if not installed and operated properly. A single deep-cycle storage battery (6 volts, 220 amp-hours) can discharge about 8,000 amps into a terminal-to-terminal short-circuit. Systems operate with voltages ranging from 12 volts to 600 volts or higher and can present shock hazards. Short circuits, even on lower voltage systems, present fire and equipment hazards. Storage batteries can be dangerous; hydrogen gas and acid residue from lead-acid batteries, although not *NEC*-specific, need to be dealt with safely.

The problems are compounded because, unlike with ac systems, there are few listed components that can be easily "plugged" together to result in a safe PV system. The available PV hardware does not have mating inputs or outputs, and the knowledge and understanding of "what works with what" is not second nature to the installer. The dc PV "cookbook" of knowledge does not yet exist.

Methods of Achieving Objectives

To meet the objective of safe, reliable, durable photovoltaic power systems, the following suggestions are offered:

- Dealer-installers of PV systems should become familiar with the *NEC* methods of and requirements for wiring residential and commercial ac power systems.
- All PV installations should be permitted and inspected, where required, by the local inspection authority in the same manner as other equivalent electrical systems.
- Photovoltaic equipment manufacturers should build equipment to meet UL or other recognized standards and have equipment tested and listed.
- Listed subcomponents should be used in field-assembled equipment where formal testing and listing is not possible.
- Electrical equipment manufacturers should produce, distribute, and advertise, listed, reasonably priced, dc-rated components.
- Electrical inspectors should become familiar with dc and PV systems.
- The PV industry should educate the public, modify advertising, and encourage all installers to comply with the *NEC*.
- Existing PV installations should be upgraded to comply with the *NEC* and other minimum safety standards.

Scope and Purpose of the *National Electrical Code*

Some local inspection authorities use regional electrical codes, but most jurisdictions use the *National Electrical Code*–sometimes with slight modifications. The *NEC* states that adherence to the recommendations made will reduce the hazards associated with electrical installations. The *NEC* also says these recommendations may not lead to improvements in efficiency, convenience, or adequacy for good service or future expansion of electrical use [90.1]. (Numbers in brackets refer to sections in the 2002 *NEC*.)

The *National Electrical Code* addresses nearly all PV power installations, even those with voltages of less than 50 volts [720]. It covers stand-alone and utility-interactive systems. It covers billboards, other remote applications, floating buildings, and recreational vehicles (RV) [90.2(A), 690]. The Code deals with any PV system that has external wiring or electrical components that must be assembled and connected in the field and that is accessible to the untrained and unqualified person.

There are some exceptions. The *National Electrical Code* does not cover PV installations in automobiles, railway cars, boats, or on utility company properties used for power generation [90.2(B)]. It also does not cover micro-power systems used in watches, calculators, or self-contained electronic equipment that have no external electrical wiring or contacts.

Article 690, Solar Photovoltaic Systems of the *NEC* specifically deals with PV systems, but many other sections of the *NEC* contain requirements for any electrical system including PV systems [90.2, 720]. When there is a conflict between Article 690 of the *NEC* and any other article, Article 690 takes precedence [690.3].

The *NEC* suggests (in some cases requires), and most inspection officials require, that equipment identified, listed, labeled, or tested by an approved testing laboratory be used when available [90.7, 100, 110-3]. The three most commonly encountered national testing organizations commonly acceptable to most jurisdictions are the *Underwriters Laboratories* (UL), *Canadian Standards Association* (CSA) and *ETL Testing Laboratories, Inc.* (ETL). *Underwriters Laboratories* and UL are registered trademarks of Underwriters Laboratories Inc. ETL is a registered trademark of ETL Testing Laboratories, Inc. CSA is a registered trademark of the Canadian Standards Association.

Most building and electrical inspectors expect to see a listing mark (UL, CSA, ETL) on electrical products used in electrical systems in the United States. This listing requirement presents a problem for some in the PV industry, because low production rates may not justify the costs of testing and listing by UL or other laboratory. Some manufacturers claim their product specifications exceed those required by the testing organizations, but inspectors readily admit to not having the expertise, time, or funding to validate these unsubstantiated claims.

This Guide

The recommended installation practices contained in this guide progress from the photovoltaic modules to the electrical outlets (in a stand-alone system) or to the utility interconnection (in a utility-interactive system). For each component, *NEC* requirements are addressed, with the appropriate Code sections referenced in brackets. A sentence, phrase, or paragraph followed by a *NEC* reference refers to a requirement established by the *NEC*. The words "will," "shall," or "must" also refer to *NEC* requirements. Suggestions based on field experience with PV systems are worded as such and will use the word "should." The recommendations apply to the use of listed products. The word "Code" in this document refers to the *2002 NEC*. In some places references will also be made to Article 690 from the *2005 NEC*.

In recent times, monetary incentives have resulted in large numbers of utility-interactive PV systems being installed. While most of these systems are purely grid-tied, many have batteries included to provide energy during blackouts, and some even include generators. With these added features, there are many similarities between the code requirements for utility-interactive systems and stand-alone systems. In this suggested practices manual, the code requirements are addressed at the component level and at the interconnection level between components. Where unique requirements apply, they are addressed as they relate to a particular system. Appendices provide additional details.

Appendix A provides a limited list of sources for dc-rated and identified, or listed, products, and references to the products are made as they are discussed.

Other appendices address details and issues associated with implementing the *NEC* in PV installations. Examples are included.

Longevity, Materials, and Safety

Although PV modules are warranted for power output for periods from 10-25 years, they can be expected to deliver dangerous amounts of energy (voltage and current) for periods of 40 to 50 years and longer. The warning on the back of PV modules is worth reading and heeding. See Figure 1. Each and every designer and installer of PV systems should strive to make the installation as durable and as safe as possible. The *NEC* provides only minimal safety requirements and general guidance on materials, and does not fully address the durability issues associated with installing electrical systems that must last for 50 years or longer. The PV module environment is harsh with temperatures ranging from –50°C to +85°C, very dry to monsoon moisture conditions, long-term ultraviolet exposure, and high mechanical loading from winds and ice. The use of materials tested and listed for outdoor exposure in the outdoor sections of the system is an absolute safe-practices requirement. Exceeding Code minimums for materials and installation practices is encouraged to ensure PV array and system longevity.

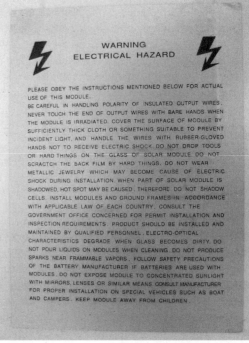

Figure 1. Warning Label

Testing and Approval

The *NEC* suggests (and in some cases requires), and many inspectors require that listed devices be used throughout a PV system. A *listed* device by UL or other approved testing laboratory is tested against an appropriate UL standard. A *recognized* device is tested by UL or other approved testing laboratory to standards established by the device manufacturer. In most cases, the requirements established by the manufacturer are less rigorous than those established by UL. Few inspectors will accept recognized devices, particularly where they are required for overcurrent protection. *Recognized* devices are generally intended for use within a factory assembly or equipment that will be further *listed* in its entirety.

Photovoltaic Modules

Numerous PV module manufacturers offer listed modules. In some cases (building integrated or architectural structures), unlisted PV modules have been installed, but these installations should have been approved by the local authority having jurisdiction (electrical inspector).

Module Marking

INFORMATION SUPPLIED BY MANUFACTURER

Certain electrical information **must** appear on each module. The information on the factory-installed label shall include the following items [690.51]:
- Polarity of output terminals or leads
- Maximum series fuse for module protection
- Rated open-circuit voltage
- Rated operating voltage
- Rated operating current
- Rated short-circuit current
- Rated maximum power
- Maximum permissible system voltage [690.51]

Figure 2 shows a typical label that appears on the back of a module.

Although not required by the *NEC*, the temperature rating of the module terminals and conductors are given to determine the temperature rating of the insulation of the conductors and how the ampacity of those conductors **must** be derated for temperature [110.14(C)]. While module terminals are usually rated for 90°C, most other terminals throughout the PV system will have terminals rated only for 60°C or 75°C. These terminal temperatures may significantly affect conductor ampacity.

Note: Other critical information, such as mechanical installation instructions, grounding requirements, tolerances of indicated values of I_{sc}, V_{oc} and P_{max}, and statements on artificially concentrated sunlight are contained in the installation and assembly instructions for the module.

Methods of connecting wiring to the modules vary from manufacturer to manufacturer. A number of manufacturers make modules with 48-inch lengths of interconnection cables permanently connected to the modules. There are no junction boxes for connection of conduit. The *NEC* does not require conduit, but local jurisdictions, particularly in commercial installations, may require conduit. The Code requires that strain relief be provided for connecting wires. If the module has a closed weatherproof junction box, strain relief and moisture-tight clamps should be used in any knockouts provided for field wiring. Where the weather-resistant gaskets are a part of the junction box, the manufacturer's instructions **must** be followed to ensure proper strain relief and weatherproofing [110.3(B), UL *Standard 1703*]. Figure 2 shows various types of strain relief clamps. The one on the left is a basic cable clamp for interior use with nonmetallic-sheathed cable (Romex®) that cannot be used for module wiring. The clamps in the center (Heyco) and on the right (T&B) are watertight and can be used with either single or multiconductor cable—depending on the insert.

Wiring Module Interconnections

Copper conductors are recommended for almost all photovoltaic system wiring [110.5]. Copper conductors have lower voltage drops and better resistance to corrosion than other types of comparably sized conductor materials. Aluminum or copper-clad aluminum wires can be used in certain applications, but the use of such cables is not recommended—particularly in dwellings.

Figure 2. Label on Typical PV Module

SHARP
SOLAR MODULE
NT-S5E1U

c(UL)us LISTED
2P09
PHOTOVOLTAIC MODULE
E160673

THE ELECTRICAL CHARACTERISTICS ARE WITHIN ±10 PERCENT OF THE INDICATED VALUES OF Isc, Voc, AND PMAX UNDER STANDARD TEST CONDITIONS (IRRADIANCE OF 1000W/m², AM1.5 SPECTRUM AND CELL TEMPERATURE OF 25°C)

MAXIMUM POWER	(PMAX)	185.0 W
OPEN-CIRCUIT VOLTAGE	(VOC)	44.9 V
SHORT-CIRCUIT CURRENT	(ISC)	5.75 A
MAXIMUM POWER VOLTAGE	(VPMAX)	36.21 V
MAXIMUM POWER CURRENT	(IPMAX)	5.11 A
MAXIMUM SYSTEM VOLTAGE		600 V
FUSE RATING		10 A
FIRE RATING		CLASS C
FIELD WIRING		COPPER ONLY 14 AWG MIN. INSULATED FOR 90°C MIN.
SERIAL No.		034090273

SHARP CORPORATION
282-1 HAJIKAMI, SHINJO-CHO, KITAKATSURAGI-GUN, NARA, 639-2198, JAPAN

Figure 3. Strain Reliefs

All wire sizes presented in this guide refer to copper conductors.

The *NEC* requires 12 AWG (American Wire Gage) or larger conductors to be used with systems under 50 volts [720.4]. Article 690 ampacity calculations yielding a smaller conductor size might override Article 720 considerations, but some inspectors are using the Article 720 requirement for dc circuits [690.3]. The Code has little information for conductor sizes smaller than 14 AWG, but Section 690.31(D) provides some guidance. Many listed PV modules are furnished with attached 14 AWG conductors.

Single-conductor, Type UF (Underground Feeder—Identified (marked) as Sunlight Resistant), Type SE (Service Entrance), or Type USE/USE-2 (Underground Service Entrance) cables are permitted for module interconnect wiring [690.31(B)]. Type UF cable **must** be marked "Sunlight Resistant" when exposed outdoors as it does not have the inherent sunlight resistance found in SE and USE conductors [UL Marking Guide for Wire and Cable]. Unfortunately, single-conductor, stranded, UF sunlight-resistant cable is not readily available and may have only a 60°C temperature rating. This 60°C-rated insulation is not suitable for long-term exposure to direct sunlight at temperatures likely to occur near PV modules. Such wire has shown signs of deterioration after four years of exposure. Temperatures exceeding 60°C normally occur in the vicinity of the modules; therefore, conductors with 60°C insulation cannot be used. Stranded wire is suggested to ease servicing of the modules after installation and for durability [690.34].

The widely available Underground Service Entrance Cable (USE-2) is suggested as the best cable to use for module interconnects. When manufactured to the UL Standards, it has a 90°C temperature rating and is sunlight resistant even though not commonly marked as such. The "-2" marking indicates a wet-rated 90°C insulation, the preferred rating. Additional markings indicating XLP or XLPE (cross-linked polyethylene) and RHW-2 (90°C insulation when wet) ensure that the highest quality cable is being used [Tables 310.13, 16, and 17]. An additional marking (not required) of "Sunlight Resistant" indicates that the cable has passed an extended UV exposure test over that normally required by USE-2. USE-2 is acceptable to most electrical inspectors. The RHH and RHW-2 designations frequently found on USE-2 cable allow its use in conduit inside buildings. USE or USE-2 cables, without the other markings, do not have the fire-retardant additives that SE and RHW/RHW-2 cables have and cannot be used inside buildings.

If a more flexible, two-conductor cable is needed, electrical tray cable (Type TC) is available but **must** be supported in a specific manner as outlined in the *NEC* [336 and 392]. TC is sunlight resistant and is generally marked as such. Although sometimes used (improperly) for module interconnections, SO, SOJ, and similar flexible, portable cables and cordage may not be sunlight resistant and are not approved for fixed (non-portable) installations [400.7, 8].

The temperature derated ampacity of conductors at any point **must** generally be at least 156% of the module (or array of parallel-connected modules) rated short-circuit current at that point [690.8(A), (B)]. See later sections of this manual for details on ampacity calculations.

Tracking Modules

Where there are moving parts of an array, such as a flat-plate tracker or concentrating modules, the *NEC* does allow the use of flexible cords and cables [400.7(A), 690.31(C)]. When these types of cables are used, they should be selected for extra-hard usage with full outdoor ratings (marked "WA" or "W" on the cable). They should not be used in conduit. Temperature derating information is provided by Table 690.31C. A temperature correction factor in the range of 0.33 to 0.58 should be used for flexible cables used as module interconnects.

Trackers in PV systems operate at relatively slow angular rates and with limited motion. Normal stranded wire (exposed USE-2 or THWN-2 inside flexible conduit) has demonstrated good performance without deterioration due to flexing.

Another possibility is the use of extra flexible (400+ strands) building cable type USE-RHH-RHW or THW. This cable is available from the major wire distributors (Appendix A). However, it should be noted that few mechanical terminals (screw or setscrew types) are listed for use with other than the normal Class B or C stranded cables (7, 19 or 37 strands). Cable types, such as THW or RHW that are not sunlight resistant, should be installed in flexible liquidtight conduit.

Terminals

Module junction boxes have various types of terminals inside junction boxes or permanently-connected leads (with and without connectors). The instructions furnished with each module will state the acceptable size and type of wires for use with the terminals. Ampacity calculations will dictate the minimum conductor sizes allowed. Some modules may require the use of crimp-on terminals when stranded conductors are used. The use of a crimp-on (compression) terminal is usually required when fine stranded conductors are being used with mechanical terminals (setscrew or screw fasteners) unless the terminal is marked for use with fine stranded cables. Very few, if any, are marked for use with fine stranded conductors.

Light-duty crimping tools designed for crimping smaller wires used in electronic components usually do

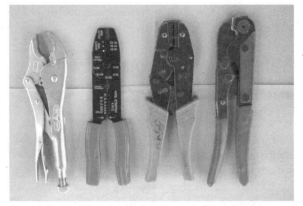

Figure 4. Terminal Crimping Tools—Two on Left Unlisted, Two on Right Listed

Figure 5. Crimp Terminals/Lugs-All Listed, but Not All Suitable for All Applications

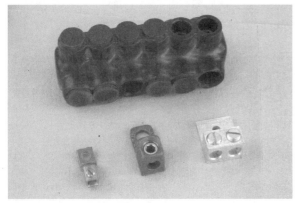

Figure 6. Listed Mechanical Terminals

not provide sufficient force to make long-lasting crimps on connectors for PV installations even though they may be sized for 12-10 AWG. Insulated terminals crimped with these light-duty crimping tools frequently develop high-resistance connections in a short time and may even fail as the wire pulls out of the terminal. It is strongly suggested that only listed or device specific, heavy-duty industrial-type crimping tools be used for PV system wiring where crimp-on terminals are required. Figure 4 shows four styles of crimping tools. On the far left is a common handyman locking pliers that *should not* be used for electrical connections. On the left center is a stripper/crimper used for electronics work that will crimp only insulated terminals. These two

types of crimping tools are frequently used to crimp terminals on PV systems; however, since they are not listed devices, their use is discouraged. The two crimping tools on the right are listed, heavy-duty industrial designs with ratcheting jaws and interchangeable dies that will provide the highest quality connections. They are usually available from electrical supply houses.

Figure 5 shows some examples of insulated and uninsulated terminals. In general, uninsulated terminals are preferred (with insulation applied later if required), but the heavier, more reliable listed electrical terminals, not unlisted electronic or automotive grades, are required. Again, an electrical supply house rather than an electronic or automotive parts store is the place to find the required items. Terminals are listed only when installed using the instructions supplied with the terminals and when used with the related crimping tool (usually manufactured or specified by the manufacturer of the terminals). If the junction box provides mechanical pressure terminals, it is not necessary to use crimped terminals unless fine stranded conductors are used.

Figure 6 shows a few mechanical terminals. The screws and setscrews used in these devices usually indicate that they are not listed for use with fine stranded, flexible conductors, but are intended for use only with the normal 7 or 19 strand conductors. Any terminal block used **must** be listed as suitable for use with "field-installed wiring [110.3(B)]."

Transition Wiring

Because of the relatively higher cost of USE-2 and TC cables and wire, they are usually connected to less expensive cable at the first junction box leading to an interior location. In many cases, a PV combiner as shown in Figure 7 is used to make the transition from the single conductor module wiring to one of the standard wiring methods. All PV system wiring **must** be made using one of the specific installation/materials methods included in the *NEC* [690.31, Chapter 3]. Single-conductor, exposed wiring is not permitted except for module wiring or with special permission [Chapter 3]. The most common methods used for PV systems are individual conductors in electrical metallic tubing (EMT) [358], rigid nonmetallic conduit (RNC) [352], or liquidtight flexible nonmetallic conduit (LFNC) [356].

Where individual conductors are used in conduit installed in outdoor, sunlit locations, they should be conductors with at least 90°C insulation such as RHW-2, THW-2, THWN-2 or XHHW-2. Conduits installed in exposed locations are considered to be installed in wet locations [100-Locations (wet, damp, dry)]. These conduits may have water trapped in low spots and therefore only conductors with wet ratings are acceptable in conduits that are located in exposed or buried locations.

Figure 7. PV Combiner with Circuit Breakers. INSET: Combiner with cover in place.

The conduit can be either thick-wall (rigid, galvanized-steel, RGS, or intermediate, metal-conduit, IMC) or thin-wall electrical metallic tubing (EMT) [358], and if rigid nonmetallic conduit is used, electrical (gray) PVC (Schedule 40 or Schedule 80) rather than plumbing (white) PVC tubing **must** be used [352].

Two-conductor (with ground) UF cable (a jacketed or sheathed cable) or tray cable (type TC) that is marked sunlight resistant is sometimes used between the module interconnect wiring and the PV disconnect device.

Interior exposed cable runs can also be made with sheathed, multi-conductor cable types such as NM, NMB, and UF. The cable should not be subjected to physical abuse. If abuse is possible, physical protection **must** be provided [300.4, 334.15(B), 340.12]. *Exposed*, single-conductor cable (commonly used improperly between batteries and inverters) **shall not** be used—except as module interconnect conductors [300.3(A)]. Battery-to-inverter cables are normally single-conductor cables installed in conduit.

PV conductors **must not** be routed through attics unless they are installed in a metallic raceway between the point of first penetration of the building structure and the first dc disconnect [690.14, 690.31(E)]. Attic temperatures will be at higher-than-outdoor temperatures due to solar heating, and the ampacity of the conductors will have to be derated for these elevated temperatures. However, due to the PV disconnect location requirements established by *NEC* Section 690.14, conductors routed through attics are becoming less frequent.

Module Connectors

Module connectors that are concealed at the time of installation **must** be able to resist the environment, be polarized, and be able to handle the short-circuit current. They **shall** also be of a latching design with the terminals guarded. The equipment-grounding member, if used, **shall** make first and break last [690.32, 33]. UL *Standard 1703* also requires that the connectors for pos-

itive and negative conductors should not be interchangeable.

Module Connection Access

All junction boxes and other locations where module wiring connections are made **shall** be accessible. Removable modules and stranded wiring may allow accessibility [690.34]. The modules should not be permanently fixed (welded) to mounting frames, and solid wire that could break when modules are moved to service the junction boxes should be used sparingly. Open spaces behind the modules would allow access to the junction boxes.

Splices

All splices (other than the connectors mentioned above) **must** be made in approved junction boxes with an approved splicing method [300.15]. Conductors **must** be twisted firmly to make a good electrical and mechanical connection, then brazed, welded, or soldered, and then taped [110.14(B)]. Mechanical splicing devices such as split-bolt connectors or terminal strips are also acceptable. Crimped splicing connections may also be made if listed splicing devices and listed, heavy-duty crimping tools are used. Splices in the module conductors where made of jacketed two-conductor UF or TC cable when located outside **must** be protected in rainproof junction boxes such as NEMA type 3R [300.15]. Cable clamps **must** also be used [300.15(C)]. Figure 8 shows some common splicing devices. Many of the "power blocks" (on the left) are only "Recognized" by UL for use inside factory-assembled, listed devices. These "Recognized" devices are not suitable for installation or assembly in the field.

Splices can be exposed in exposed single-conductor USE-2 cables and may be made by soldering and covering the splice with appropriate heat shrink tubing listed for outdoor use containing sealant. The electrical and mechanical properties of the spliced conductor and the insulation around the splice **must** equal or exceed the unspliced conductor. Inline mechanical crimped splices may be used when listed for the application and

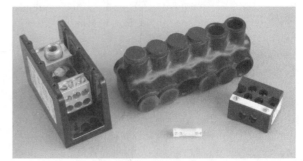

Figure 8. Common Splicing Devices

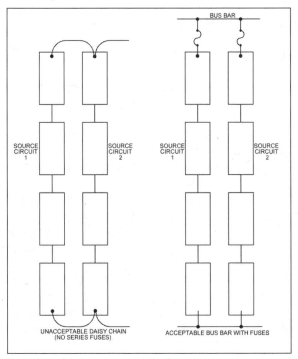

Figure 9. Module Interconnect Methods

when used in rainproof enclosures. However, few are listed for use with any type of conductor other than the normal Class B stranded wires (7 and 19 strands). Fuse blocks, fused disconnects, and circuit breakers frequently have these mechanical pressure terminals.

Twist-on wire connectors (approved for splicing wires), when listed for the environment (dry, damp, wet, or direct burial), are acceptable splicing devices. Unless specifically marked for ac only, they may be used on either ac or dc circuits. In most cases, they **must** be used inside enclosures, except when used in direct-burial applications [110.3(B), 310.15].

Where several modules are connected in series and parallel, a terminal block or bus bar arrangement **must** be used so that one source circuit can be disconnected without disconnecting the grounded (on grounded systems) conductor of other source circuits [690.4(C)]. On grounded systems, this indicates that the popular "Daisy Chain" method of connecting modules may not always be acceptable, because removing one module in the chain may disconnect the grounded conductor for all of those modules in other parallel chains or source circuits. This becomes more critical on larger systems where paralleled sets of long series strings of modules are used. Figure 9 shows unacceptable and acceptable methods. The required module-protective fuse or other overcurrent device is usually required on each module (12-volt systems) or string of modules.

installed with appropriately rated insulation listed for outdoor applications.

Properly used box-type mechanical terminal connectors (Figures 6 and 8) give high reliability. If used, they should be listed for at least damp conditions even

Conductor Color Codes

The *NEC* established color codes for electrical power systems many years before either the automobile or electronics industries had standardized color codes. PV systems are being installed in an arena covered by the *NEC* and, therefore, **must** comply with *NEC* standards that apply to both ac and dc power systems. In a system where one conductor is grounded, the insulation on all grounded conductors **must** be white, gray or have three white stripes or be any color except green if marked with white plastic tape or paint at each termination (marking allowed only on conductors larger than 6 AWG). Conductors used for module frame grounding and other exposed metal equipment grounding **must** be bare (no insulation) or have green or green with yellow-striped insulation or identification [200.6, 7; 210.5; 250.119]. Any insulated equipment-grounding conductor used to ground PV module frames must be an outdoor-rated conductor such as USE-2.

The *NEC* requirements specify that the grounded conductor be white. In most PV-powered systems that are grounded, the grounded conductor is the negative

conductor. Telephone systems that use positive grounds require special circuits when powered by PV systems that have negative grounds. In older PV systems where the array is center tapped, the center tap **must** be grounded [690.41], and this becomes the white conductor. There is no *NEC* requirement designating the color of the ungrounded conductor, but the convention in ac power wiring is that the first two ungrounded conductors are colored black and red. This suggests that in two-wire, negative-grounded PV systems, the positive conductor could be red or any color with a red marking except green or white, and the negative grounded conductor **must** be white. In a three-wire, center-tapped system, the positive conductor could be red, the grounded, center tap conductor **must** be white, and the negative conductor could be black.

The *NEC* allows grounded PV array conductors, such as non-white USE/USE-2, UF or SE that are smaller than 6 AWG, to be marked with a white marker [200.6(A)(2)].

PV Array Ground-Fault Protection

Article 690.5 of the *NEC* requires a ground-fault detection, interruption, and array disconnect (GFPD) device for fire protection if the PV arrays are mounted on roofs of dwellings. Ground-mounted arrays are not required to have this device. Several external devices or devices built into utility-interactive inverters are available that meet this requirement. These particular devices generally require that the system grounding electrode conductor be routed through or connected to the device. These devices include the following code-required functions:

- Ground-fault detection
- Ground-fault current interruption
- Array disconnect/inverter shutdown
- Ground-fault indication

Ground-fault detection, interruption, and indication devices might, depending on the particular design, accomplish the following actions automatically:

- Sense ground-fault currents exceeding a specified value
- Interrupt the fault currents
- Open the circuit between the array and the load

- Indicate the presence of the ground fault

Ground-fault devices have been developed for both grid-tied inverters (Figure 11) and stand-alone systems (Figure12), and others are under development. See Appendix H for more details.

The 1999 *NEC* added a Section 690.6(D) *permitting* (not requiring) the use of a device (undefined) on the ac branch circuit being fed by an ac PV module to detect ground-faults in the ac wiring. There are no commercially available devices as of mid 2004 that meet this permissive requirement. Standard 5-milliamp anti-shock receptacle GFCIs or 30-milliamp equipment protection circuit breakers should not be used for this application. The receptacle GFCIs interrupt both the hot (ungrounded) and neutral (grounded) conductor, and the equipment protection circuit breaker may be destroyed when backfed.

The 2005 NEC will allow ungrounded PV arrays and the requirements for ground-fault protection will differ slightly from these requirements for grounded systems. See Appendix L.

PV Array Installation and Service

Article 690.18 requires that a mechanism be provided to allow safe installation or servicing of portions of the array or the entire array. The term "disable" has several meanings, and the *NEC* is not clear on what is intended. The *NEC* Handbook does elaborate. "Disable" can be defined several ways:

- Prevent the PV system from producing any output
- Reduce the output voltage to zero
- Reduce the output current to zero
- Divide the array into non-hazardous segments

The output could be measured either at the PV source terminals or at the load terminals.

Fire fighters are reluctant to fight a fire in a high-voltage battery room because there is no way to turn off a battery bank unless the electrolyte can somehow be removed. In a similar manner, the only way a PV system can have zero output at the array terminals is by preventing light from illuminating the modules. The output voltage may be reduced to zero by shorting the PV module or array terminals. When this is done, short-circuit current will flow through the shorting conductor which, in a properly wired system, does no harm. The output

current may be reduced to zero by disconnecting the PV array from the rest of the system. The PV disconnect switch would accomplish this action, but open-circuit voltages would still be present on the array wiring and in the disconnect box. In a large system, 100 amps of short-circuit current (with a shorted array) can be as difficult to handle as an open-circuit voltage of 600 volts.

During PV module installations, the individual PV modules can be covered to disable them. For a system in use, the PV disconnect switch is opened during maintenance, and the array is either short circuited or left open circuited depending on the circumstances. In practical terms, for a large array, some provision (switch or bolted connection) should be made to disconnect portions of the array from other sections for servicing. As individual modules or sets of modules are serviced, they may be covered and/or isolated and shorted to reduce the potential for electrical shock. Aside from measuring short-circuit current, there is little that can be serviced on a module or array when it is shorted. The circuit is usually open circuited for repairs.

The code requirement that the PV source and output conductors be kept outside the building until the readily accessible disconnect is reached indicate that these conductors are to be treated in a manner similar to ac service entrance conductors [690.14]. First response personnel are less likely to cut these energized cables since they are on the outside of the building.

Even in dim light conditions (clouds, dawn, dusk) when sunlight is not directly illuminating the PV module or PV array, voltages near the open-circuit value will appear on PV source and output circuit wiring. Distributed leakage paths caused by dirt and moisture will ground-reference, supposedly ungrounded, disconnected conductors, and they may be energized with respect to ground posing a safety hazard.

Grounding

Definitions

The subject of grounding is one of the most complex issues in electrical installations. Definitions from Articles 100 and 250 of the *NEC* will help to clarify the situation when grounding requirements are discussed.

Grounded: Connected to the earth or to a metallic conductor or surface that serves as earth.

Grounded Conductor: (white or gray or three white stripes) A system conductor that normally carries current and is intentionally grounded. In PV systems, one conductor (normally the negative) of a two-conductor system or the center-tapped conductor of a bipolar system is grounded.

Equipment Grounding Conductor: (bare, green, or green with yellow stripe) A conductor not normally carrying current used to connect the exposed metal portions of equipment that might be accidentally energized to the grounding electrode system or the grounded conductor.

Grounding Electrode Conductor: A conductor not normally carrying current used to connect the grounded conductor to the grounding electrode or grounding electrode system.

Grounding Electrode: The conducting element in contact with the earth (e.g., a ground rod, a concrete-encased conductor, grounded building steel, and others).

Grounding—System

For a two-wire PV system over 50 volts (125% of open-circuit PV-output voltage), one dc current-carrying conductor **shall** be grounded. In a three-wire system, the neutral or center tap of the dc system **shall** be grounded [690.41]. These requirements apply to both stand-alone and grid-tied systems. Such system grounding will enhance personnel safety and minimize the effects of lightning and other induced surges on equipment. In addition, the grounding of all PV systems (even 12-volt systems) will reduce radio frequency noise from dc-operated fluorescent lights and inverters.

SIZE OF DC GROUNDING ELECTRODE CONDUCTOR

Section 250.166 of the *NEC* addresses the size of the dc grounding electrode conductor (GEC). Many PV systems can use a 6 AWG GEC if that is the *only connection* to the grounding electrode [250.166(C)] and that grounding electrode is a rod, pipe, or plate electrode. In some cases (a very small system with circuit conductors less than 8 AWG), an 8 AWG GEC may be used and should be installed in conduit for physical protection. Many inspectors will allow a 6 AWG GEC to be used without additional physical protection. Other grounding electrodes will require different sizes of grounding electrode conductors. In a few cases, the direct-current system-grounding electrode conductor **shall not** be smaller than 8 AWG or the largest conductor supplied by the system [250.166(B)]. If the conductors between the battery and inverter are 4/0 AWG (for example) then the grounding-electrode conductor from the negative conductor (assuming that this is the grounded conductor) to the grounding electrode may be required to be as large as 4/0 AWG. However, in most PV installations, a smaller GEC (usually 6 AWG) will be allowed if it is connected only to a rod, pipe, or plate electrode [250.166(C)].

If the grounding electrode were a concrete-encased conductor, then a 4 AWG GEC would be required. [250.66(B), 250.166(D)]

POINT OF CONNECTION

In stand-alone systems, primarily, the system grounding electrode conductor for the direct-current portion of a PV system **shall** be connected to the PV-output circuits [690.42] at a single point. When this connection (the dc bonding point) is made close to the modules, added protection from surges is afforded. However, real-world considerations affect this connection point.

In stand-alone PV systems, the charge controller may be considered a part of the PV-output circuit, and the

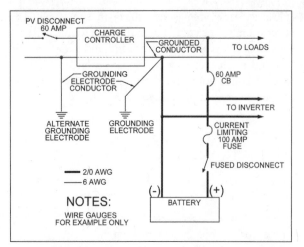

Figure 10. Typical System: Possible Grounding Conductor Locations

point-of-connection of the grounding electrode conductor could be before or after the charge controller. However, this grounding conductor may be a very large conductor (e.g., 4/0 AWG) while the conductors to and from the charge controller may be 10 AWG or smaller. Connecting the 4/0 AWG grounding conductor on the array side of the charge controller, while providing some degree of enhanced surge suppression from lightning induced surges, may not meet the full intent of the grounding requirements. Connecting the grounding conductor to the system on the battery side of the charge controller at a point where the system conductors are the largest size will provide better system grounding at the expense of less lightning protection. Since the *NEC* allows smaller grounding electrode conductors in many circumstances, either grounding conductor point of connection may be acceptable [250.166]. Figure 10 shows two possible locations for the grounding electrode conductor.

The *NEC* does not specifically define where the PV-output circuits end. Circuits from the battery toward the load are definitely load circuits. Since the heaviest conductors are from the battery to the inverter, and either end of these conductors is at the same potential, then

Figure 11. Utility-Interactive Inverter with Internal DC Bonding Point, GFPD, and Connection Point for Grounding Electrode Conductor

either end could be considered a point for connecting the grounding conductor. The negative dc input to the inverter is connected to the metal case in some unlisted stand-alone inverter designs, but this is not an appropriate place to connect the grounding electrode conductor and other equipment-grounding conductors, since this circuit is a dc-branch circuit and not a PV-output circuit. Connection of the grounding electrode conductor to or near the negative battery terminal would avoid the "large-wire/small-wire" problem outlined above.

However, the presence of ground-fault protection devices [690.5] may dictate that this bonding point be made in the ground-fault device or inside the inverter. Many utility-interactive inverters have internal ground-fault protection devices that dictate the connection point for the dc grounding electrode conductor. Figure 11 shows a utility-interactive inverter installation where the grounding electrode conductor is connected to a point inside the inverter and the inverter furnishes the bond to the grounded conductor.

It is imperative that there be no more than one ground connection to the dc grounded conductor of a PV system. Failure to limit the connections to one (1) will allow objectionable currents to flow in uninsulated equipment-grounding conductors and will create unexpected ground faults in the grounded conductor [250.6]. The ground-fault protection systems will sense these extra connections as ground faults and may not function correctly. There are exceptions to this rule when PV arrays, generators, or loads are some distance from the main loads [250.32].

UNUSUAL GROUNDING SITUATIONS

Some unlisted stand-alone inverter designs use the entire chassis as part of the negative circuit. Also, the same situation exists in certain radios—automobile and shortwave. These designs will not pass the current UL standards for consumer electrical equipment or PV systems and will probably require modification in the future since they do not provide electrical isolation between the exterior metal surfaces and the current-carrying conductors. They also create the very real potential for multiple grounded-conductor connections to ground.

Since the case of these non-listed inverters and other non-listed products is connected to the negative conductor and that case **must** be grounded as part of the equipment ground described below, the user has no choice whether or not the *system* is to be grounded [250 VI]. The system is grounded even if the voltage is less than 50 volts and the point of system ground is the negative input terminal on the inverter. It is strongly suggested that these unlisted inverters not be used and, in fact, to use them or any unlisted component may result in the inspector not accepting the system.

Some telephone systems ground the positive conductor, and this may cause problems for PV-powered telephone systems with negative grounds. An isolated-ground, dc-to-dc converter may be used to power subsystems that have different grounding polarities from the main system. In the ac realm, an isolation transformer will serve the same purpose.

In larger utility-tied systems and some stand-alone systems, high impedance grounding systems or other methods that accomplish equivalent system protection and that use equipment listed and identified for the use might be used in lieu of, or in addition to, the required hard ground [690.41]. The discussion and design of these systems are beyond the scope of this guide. Grounding of grid-tied systems will be discussed later in this manual.

CHARGE CONTROLLERS—SYSTEM GROUNDING

In a grounded system, it is important that the charge controller does not have electronic devices or relays in the grounded conductor. Charge controllers listed to the current edition of UL *Standard 1741* meet this requirement. Relays or transistors in the grounded conductor create a situation where the grounded conductor is not at ground potential at times when the charge controller is operating. This condition violates provisions of the *NEC* that require all conductors identified as grounded conductors always be at the same potential (i.e. grounded). A shunt in the grounded conductor is equivalent to a wire, if properly sized, but the user of such a charge controller runs the risk of having the shunt bypassed when inadvertent grounds occur in the system. The best charge controller design has only a straight-through conductor between the input and output terminals for the grounded current-carrying conductor (usually the negative conductor).

UNGROUNDED SYSTEMS

Section 690.35 of the 2005 *NEC*, will permit (not require) ungrounded PV systems when a number of conditions are met. These conditions are intended to make ungrounded PV installations in the United States as safe as equivalent ungrounded PV systems in Europe. Given the 100+-year history of grounded electrical systems, the U.S. PV industry and the electricians and inspectors may not have the experience, knowledge, and infrastructure to properly and safely install and inspect ungrounded PV systems. The *NEC* requirements were developed to bring the US PV industry in line with the rest of the world by adopting some of the European techniques and experience for installing ungrounded systems. They include:

Overcurrent protection and disconnects on all circuit conductors
Ground-fault protection on all systems

Jacketed or sheathed multiconductor cables or raceways
Additional warning labels
Inverters listed specifically for this use

Equipment Grounding

All non-current-carrying exposed metal parts of junction boxes, equipment, and appliances in the entire electrical system that may be accidentally energized **shall** be grounded [690.43; 250 VI; 720.10]. All PV systems, regardless of voltage, **must** have an equipment-grounding system for exposed metal surfaces (e.g., module frames and inverter cases) [690.43]. The equipment-grounding conductor **shall** be sized as required by Article 690.45 or 250.122. Generally, this will mean an equipment-grounding conductor (in other than PV source and output circuits) based on the size of the overcurrent device protecting the ac or dc circuit conductors. Table 250.122 in the *NEC* gives the sizes. For example, if the inverter-to-battery conductors are protected by a 400-amp fuse or circuit breaker, then at least a 3 AWG conductor **must** be used for the equipment ground for that circuit [Table 250.122]. If the current-carrying conductors have been oversized to reduce voltage drop, then the size of the equipment-grounding conductor **must** also be proportionately adjusted [250.122(B)].

In the PV source and output circuits, the equipment grounding conductors should generally be sized to carry at least 125% of the short-circuit currents from the PV circuits (not including backfeed currents from other sources) at that point in the circuit. They should not be less than 14 AWG to afford some degree of mechanical strength—particularly when they are installed between modules in free air. Where the circuit conductors are oversized for voltage drop, the equipment-grounding conductor shall be proportionately oversized in accordance with 250.122(B) except where there are no overcurrent devices protecting the circuit as allowed by 690.9 EX. [690.45]. See Appendix G for additional details on grounding PV modules.

If exposed, single-conductor cables are run along and adjacent to metal racks, then these racks may be subject to being energized and should be grounded. Installations using conductors in conduits may not seem to require grounded racks since the module grounding and the conduit grounding (if metallic conduits were used) would provide the code-required protection. However, modules have been known to shatter and conductive elements come into contact with the racks, therefore the racks should also be grounded. Frequently, module racks are grounded to provide additional protection against lightning.

Inverter AC Outputs

The inverter output (120 or 240 volts) **must** be connected to the ac distribution system in a manner that does not create parallel paths for currents flowing in grounded conductors [250.6]. The *NEC* requires that both the green or bare equipment-grounding conductor and the white ac neutral conductor be grounded, and this is normally accomplished by the ac distribution equipment or load center and not in the inverter. The Code also requires that current not normally flow in the equipment-grounding conductors. If the stand-alone inverter has ac grounding receptacles as outputs, the equipment-grounding and neutral conductors are most likely connected to the chassis and, hence, to chassis ground inside the inverter. This configuration allows plug-in devices to be used safely. However, if the outlets on the inverter are plug and cord connected (not allowed) to an ac load center used as a distribution device, then problems can occur.

The ac load center usually has the grounded neutral and equipment-grounding conductors connected to the same bus bar. This bus bar is also connected to the enclosure and has a grounding electrode conductor connected to a grounding electrode. Parallel current paths are created with neutral currents flowing in the equipment-grounding conductors when the inverter also has the neutral bonded to the equipment-grounding conductor. This problem can be avoided (where stand-alone inverters with internal bonding are used) by using a load center with an isolated/insulated neutral bus bar that is separated from the equipment-grounding bus bar.

Inverters with hard-wired outputs may or may not have internal bonding connections. Most listed stand-alone inverters and all utility-interactive inverters do not have an internal neutral-to-ground bond. Some stand-alone inverters with ground-fault circuit interrupters (GFCIs) for ac outputs **must** be connected in a manner that allows proper functioning of the GFCI [110.3(B)]. A case-by-case analysis will be required.

PV Inverters Create Separately Derived Systems

PV systems will generally have dc circuits and ac circuits and both **must** be properly grounded [250, 690 V]. Although the *NEC* has parts of Article 250 that deal with the grounding of ac systems and parts that deal with the proper grounding of dc systems, it does not specifically deal with systems that have both ac and dc components.

In Article 100 of the *NEC*, the definition of "Separately Derived Systems" includes PV systems, and in most cases this is correct. Most, but not all, PV systems (both stand-alone systems and utility-interactive systems) employ an inverter that converts the dc from the

PV modules to ac that is used to feed loads or the utility grid. These inverters use a transformer that isolates the dc side of the system from the ac side. The grounded dc circuit conductor is not directly connected to the grounded ac circuit conductor. Although the normal definition of separately derived systems applies only to ac systems with transformers, in fact, the isolation between ac and dc circuits in PV inverters makes many PV systems also separately derived systems.

AC GROUNDING

As in any separately derived system, both parts **must** be properly grounded [250.30]. There is usually no internal bond between the ac grounded circuit conductor and the grounding system inside either stand-alone or utility-interactive inverters. Both of these PV systems rely on the neutral-to-ground main bonding jumper in the service equipment (utility-interactive systems) or the bonding jumper in the first load center (stand-alone systems) for grounding the ac side of the system.

DC GROUNDING

The dc side of the system **must** also be grounded when the system voltage (open-circuit PV voltage times a temperature-dependent constant) is above 50 volts. See *NEC* Section 690.41 for more details. *NEC* Table 690.7 gives the temperature-dependent constant, and the application of this constant usually indicates that PV systems with a nominal voltage of 24-volts or greater **must** have the dc side grounded. Only infrequently are 12-volt dc systems found that do not have one of the dc circuit conductors grounded, and even those systems **must** have an equipment-grounding system [690.43]. Most of the 12-48 volt balance-of-systems PV equipment is designed to be used only with a grounded system. See *NEC* Section 690.43. Nearly all utility-interactive PV systems operate with a nominal voltage of 48 volts or higher so they **must** have one of the dc circuit conductors grounded [690.41].

Properly grounding the dc side of a PV system is somewhat complicated by Section 690.5 of the *NEC* that

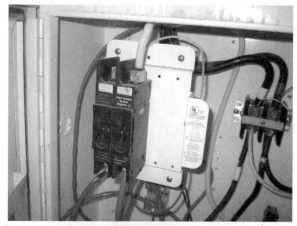

Figure 12. External Ground-Fault Protection Device

requires a ground-fault protection device (GFPD) on some PV systems. Many utility-interactive inverters have an internal GFPD (Figure 11). Inverters (both stand-alone and utility-interactive) that are used in systems with PV modules mounted on the roofs of dwellings that do not have the internal GFPD **must** have an external GFPD installed in the system [690.5]. See Figure 12. In nearly all cases, these GFPDs (either inside the inverter or externally mounted) actually make the grounded circuit conductor-to-ground bond.

For systems employing a GFPD, there should be no external bonding conductor, and to add one to these systems would bypass the GFPD and render it inoperative.

In most dc systems, the negative conductor is the grounded conductor.

A dc bond inside the inverter with a GFPD or a dc bond in a GFPD external to the inverter establishes the need for, and connection location of, a dc grounding electrode conductor. Some inverters with an internal GFPD have a terminal designated for connecting the usual 8 AWG to 4 AWG grounding electrode conductor. Other inverters lack this connection. Some inverter manufacturers provide a field-installed lug kit for this connection that has been evaluated by their listing agency. PV systems with an externally installed GFPD will have an appropriate connection place (and instructions) for the grounding electrode conductor.

PV systems that do not have PV modules mounted on the roofs of dwellings are not required to have the GFPD that is required in Section 690.5, but many inverters in those systems will have it anyway. In those systems not requiring or having a GFPD, the dc bonding jumper may be installed at any single point on the PV output circuits, and this is where the dc grounding electrode conductor should be connected.

Backup Generators

Backup ac generators used for battery charging pose problems similar to using inverters and load centers. Many of these smaller generators usually have ac outlets that may have the neutral and grounding conductors bonded to the generator frame. When the generator is connected to the system through a load center to a stand- alone inverter with battery charger, or to an external battery charger, parallel ground paths are likely. These problems need to be addressed on a case-by-case basis. A stand-alone PV system, in any operating mode (inverting or battery charging), **must not** have currents in the equipment-grounding conductors [250.6].

In some cases, manual or automated transfer switches **must** be used that switch both the grounded neutral conductor as well as the ungrounded circuit conductor [250.6]. In some cases, this neutral switching can eliminate the double bonding points.

Utility-interactive PV systems with batteries and possibly backup generators may have similar or more complex grounding and bonding issues.

Suggested AC Grounding

Auxiliary ac generators and inverters should be hardwired to the ac-load center. Neither should have an internal bond between the neutral and grounding conductors. Neither should have any receptacle outlets that can be used when the generator or inverter is operated when disconnected from the load center. The single bond between the neutral and ground should be made in the system ac load center. If receptacle outlets are desired on the generator or the inverter, they should be ground-fault-circuit-interrupting devices (GFCI).

Section 250.32 of the *NEC* presents alternate methods of achieving a safe grounding system in a limited number of installations where the various parts of the system (generator, PV modules, dc load center, and inverter) are remotely located from each other.

Grounding Electrode

The dc system grounding electrode **shall** be common with, or bonded to, the ac grounding electrode (if any) [690.47, 250 III]. The dc system grounded conductor and the equipment-grounding conductors **shall** be tied to the same grounding electrode or grounding electrode system. The conductors are usually first connected by a main dc bonding jumper and then a grounding electrode conductor is run from the bonding point to the grounding electrode. Even if the PV *system* is ungrounded (optional at less than 50 volts [typically 125% of V_{oc}]), equipment-grounding conductors **must** be used and **must** be connected to a grounding electrode [250.110]. Metal water pipes and other metallic structures as well as concrete encased electrodes are to be used in some circumstances [250.50]. When a manufactured grounding electrode is used, it **shall** be a corrosion resistant rod, a minimum of 5/8 inch (16mm) in diameter (1/2 inch (13mm) if stainless steel)) with at least 8 feet (2.4m) driven into the soil at an angle no greater than 45 degrees from the vertical [250.52]. Listed connectors **must** be used to connect the grounding electrode conductor to the ground rod [110.3(B)].

A bare-metal well casing makes a good grounding electrode. It should be part of a grounding electrode system. The central pipe to the well should not be used for grounding, because it is sometimes removed for servicing.

For maximum protection against lightning-induced surges, it is suggested that a grounding electrode **system** be used with at least two grounding electrodes. One electrode would be the main-system grounding electrode as described above. The other would be a supple-

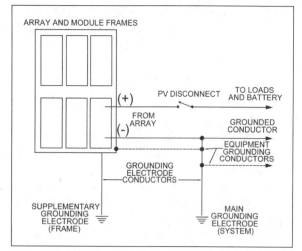

ARRAY AND MODULE FRAMES

Figure 13. Example Grounding Electrode System

mentary grounding electrode located as close to the PV array as practical. The module frames and array frames would be connected directly to this grounding electrode

to provide as short a path as possible for lightning-induced surges to reach the earth. This electrode is usually not bonded to the main system grounding electrode [250.54]. This supplementary ground rod is an auxiliary to the module frame grounding that is required to be connected with an equipment-grounding conductor connected to the main grounding electrode as discussed in the section on Equipment Grounding, above.

Do not connect the negative current-carrying conductor to the grounding electrode, to the equipment-grounding conductor, or to the module or array frame at the modules. *There should be one and only one point in the system where the dc grounding electrode conductor is attached to the dc system grounded conductor.* See Figure 13 for clarification. The wire sizes shown are for illustration only and will vary depending on system size. Chapter 3 of the *NEC* specifies the ampacity of various types and sizes of conductors. As is common throughout the *NEC*, there are exceptions to this guidance. See *NEC* Section 250.32(B).

Conductor Ampacity

NEC Tables 310.16 and 310.17 give the ampacity (current-carrying capacity in amps) of various sized conductors at temperatures of 30°C (86°F). There are several adjustments that normally **must** be made to these ampacity numbers before a conductor size can be selected [310.15].

The installation method **must** be considered. Are the conductors in free air [Table 310.17] or are they bundled together or placed in conduit [Table 310.16]?

What is the ambient air temperature, if not 30°C (86°F)?

How many current-carrying conductors are grouped together?

These adjustments are made using factors presented in Chapter 3 of the *NEC*.

Additionally, most conductors used in electrical power systems are restricted from operating on a continuous basis at more than 80% of their rated ampacity [210.19, 215.2, 690.8]. This 80% factor also applies to overcurrent devices and switchgear unless listed for operation at 100% of rating [210.20(A)]. PV conductors are also restricted by this factor (0.8=1/1.25) [690.8(B)].

Conductors carrying PV module currents are further restricted by an additional derating factor of 80% because of the manner in which PV modules generate electrical energy in response to sunlight and because the noon-time intensity of the sunlight may exceed the standard test condition value of 1000 W/m^2 [690.8(A)]. Also, nearby reflective surfaces (sand, snow, and water)

may enhance the solar intensity on the module and increase its output.

It should be noted that these ampacity adjustment factors may be applied to the basic conductor ampacities (e.g., multiply them by 0.80) or they may be applied to the anticipated current in the circuit (e.g., multiply the current by 1.25, the reciprocal of 0.8).

Photovoltaic modules are limited in their ability to deliver current. The short-circuit current capability of a module is nominally 10 to 15% higher than the operating current. Normal, daily values of solar irradiance may exceed the standard test condition of 1000W/m^2. These increased currents are considered by using the 1.25 adjustment in the ampacity calculations. Another design requirement for PV systems is that the conductors connected to PV modules or in contact with the back of PV modules may operate at temperatures as high as 75-80°C when the modules are mounted close to a structure, there are no winds, and the ambient temperatures are high. Temperatures in module junction boxes frequently occur within this range. This will require that the ampacity of the conductors be derated or corrected with factors given in *NEC* Table 310.16 or 310.17. For example, a 10 AWG USE-2/RHW-2 single-conductor cable used for module interconnections in conduit has a 90°C insulation and an ampacity of 40 amps in an ambient temperature of 26-30°C. When it is used in ambient temperatures of 61-70°C, the ampacity of this cable is reduced to 23.2 amps.

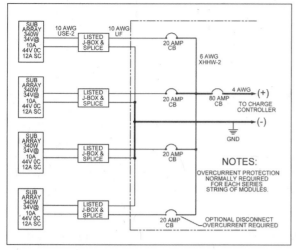

Figure 14. Typical Array Conductor Overcurrent Protection (with Optional Subarray Disconnects)

Figure 15. Listed Branch-Circuit Rated Breakers-Three on left are DC rated

Figure 16. Recognized (left) and listed (right) DC Circuit Breakers

that conductor [240.1]. Figure 14 shows an example of array conductor overcurrent protection for a medium-size array broken into subarrays. The cable sizes and types shown are examples only. The actual sizes will depend on the ampacity needed.

Either fuses or circuit breakers are acceptable for overcurrent devices provided they are rated for their intended uses—i.e., they have dc ratings when used in dc circuits, the ampacity is correct, and they can interrupt the necessary currents when short circuits occur

[240]. Figure 15 shows typical branch-circuit-rated, dc-rated, listed circuit breakers. The *NEC* allows the use of less-robust listed supplementary-type overcurrent devices only for PV source circuit protection [690.9(C)]. See Figures 16 and 17.

Some overcurrent devices rated at less than 100 amps may have terminals that are rated for use with 60°C conductors unless marked for use with 75°C conductors. The ampacity calculations of the connected cables may have to be adjusted. See Appendix I for the details of how the ratings of overcurrent devices are calculated.

Branch Circuits

DC branch circuits in stand-alone systems start at the battery and go to the receptacles supplying the dc loads or to the dc loads that are hard wired, such as inverters. In direct-connected systems (no battery), the PV output circuits go to the power controller or master dc power switch and a branch circuit goes from this location to the load. In utility-intertie systems, the circuit between the inverter and the ac-load center may be considered a feeder or possibly a branch circuit.

Fuses used to protect dc or ac branch (load) and feeder circuits **must** be listed for that use. They **must** also be of different sizes and markings for each amperage and voltage group to prevent unintentional interchange [240 VI]. These particular requirements eliminate the use of glass fuses and plastic automotive fuses as branch-circuit overcurrent devices because they are neither tested nor rated for this application. DC-rated fuses that meet the requirements of the *NEC* are becoming more available. Figure 17 shows *listed*, dc-rated, time-delay fuses on the right that are acceptable for branch circuit use, which would include the battery fuse. The cut-away fuse shows the complexity of the mechanisms required to interrupt dc currents. Acceptable dc-rated, *listed* fast-acting *supplementary* fuses are shown on the left and can be used in the PV source cir-

Figure 17. Listed Supplementary (two on left) and Branch Circuit (right) Fuses

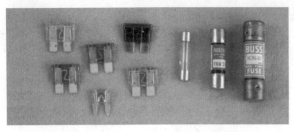

Figure 18. Unlisted, Unacceptable Automotive Fuses (left) and Listed, Unacceptable AC Fuses (right)

cuits. The fuses shown are made by Littelfuse (Appendix A) and Bussmann. Ferraz and others also have listed dc ratings on the types of fuses that are needed in PV systems.

Automotive fuses have no dc rating by the fuse industry or the testing laboratories and **should not be used in PV systems**. When rated by the manufacturer, they have only a 32-volt maximum rating, which is less than the open-circuit voltage from a 24-volt PV array. Furthermore, these fuses have no rating for interrupt current, nor are they generally marked with all of the information required for branch-circuit fuses. They are not considered supplementary fuses under the UL listing or component recognition programs. Figure 18 shows unacceptable automotive fuses on the left and unacceptable (for dc applications) ac fuses on the right. Unfortunately, even the listed ac fuses are intended for ac use and frequently have no dc ratings.

Circuit breakers also have specific requirements when used in branch circuits, but they are generally available with the needed dc ratings [240 VII].

To provide maximum protection and performance (lowest voltage drop) on branch circuits (particularly on 12 and 24-volt systems), the ampacity of the conductors might be increased, but the rating of the overcurrent devices protecting that cable should be as low as possible consistent with load currents. A general formula for cable ampacity and overcurrent device rating is 100% of the noncontinuous loads and 125% of the continuous loads anticipated [215.2]. Normally only worst-case continuous currents are used for ampacity and overcurrent calculations in PV systems. See Appendix I for the details of selecting appropriate overcurrent devices.

Amperes Interrupting Rating (AIR)—Short-Circuit Conditions

Overcurrent devices—both fuses and circuit breakers—are required to be able to safely open circuits with short-circuit currents flowing in them. Since PV arrays are inherently current limited, high short-circuit currents from the PV array are normally not a problem when the

conductors are sized as outlined above. In stand-alone systems with storage batteries, however, the short-circuit condition is very severe. A single 220 amp-hour, 6-volt, deep-discharge, lead-acid battery may produce short-circuit currents as high as 8,000 amps for a fraction of a second and as much as 6,000 amps for a few seconds in a direct terminal-to-terminal short circuit. Such high currents can generate high temperatures and magnetic forces that can cause an underrated overcurrent device to burn or blow apart. Two paralleled batteries could generate nearly twice as much current, and larger capacity batteries would be able to deliver proportionately more current under a short-circuit condition. In dc systems, particularly stand-alone systems with batteries, the interrupt capability of every overcurrent device is important. This interrupt capability or interrupt rating is specified as Amperes Interrupting Rating (AIR) and sometimes Amperes Interrupting Capability (AIC).

Some dc-rated, listed, branch circuit breakers that can be used in PV systems have an interrupt rating of 5,000 amps at 48 volts dc. However, Heinemann and AirPax make numerous circuit breakers with interrupt ratings of 25,000 amps at voltages from 65 to 125 volts (Appendix A). Some dc-rated, listed supplementary circuit breakers have an AIR of only 3,000 amps. Many listed, dc-rated class-type fuses have an AIR of up to 20,000 amps.

Fuses or circuit breakers **shall never be** paralleled or ganged to increase current-carrying capability unless done so by the manufacturer and listed for such use [240.8].

Since PV systems may have transients—lightning and motor starting as well as others—inverse-time circuit breakers (the standard type) or time-delay fuses should be used in most cases. In circuits where no transients are anticipated, fast-acting fuses can be used. They should be used if relays and other switchgear in dc systems are to be protected. Time-delay fuses that can also respond very quickly to short-circuit currents may also be used for system protection.

Fusing of PV Source Circuits

The *NEC* allows supplementary overcurrent devices (fuses and circuit breakers) to be used in PV source circuits [690.9(C)]. (See Figure 17.) A supplementary overcurrent device is one that is designed for use inside a piece of listed equipment. These devices supplement the main branch-circuit overcurrent device and do not have to comply with all of the requirements of fully rated branch overcurrent devices. They **shall**, however, be dc rated, listed, and able to handle the short-circuit currents they may be subjected to [690.9(D)]. Unfortunately, many supplementary fuses are not dc rated, and if they are, the interrupt rating (when available) is usually less than 5,000 amps. A mitigating factor is that the location of supplementary fuses in PV source circuits and PV output circuits places them at some electrical distance from potentially high short-circuit currents from the battery. At this location, the available short-circuit currents may be within their interrupt rating. The use of ac-only-rated supplementary fuses **is not** allowed for the dc circuits of PV systems [110.3(B)].

Current-Limiting Fuses— Stand-Alone Systems

A current-limiting fuse **must** be used in each ungrounded conductor from the battery where the down-stream overcurrent devices or switchgear have interrupt ratings less than the available short-circuit currents [690.71(C), 240.2, 110.9]. This fuse will limit the current that a battery bank can supply to a short circuit and should reduce the short-circuit currents to levels that are within the capabilities of downstream equipment [690.71(C)]. These fuses are available with dc ratings of 125, 300, and 600 volts dc, currents of 0.1 to 600 amps, and a dc interrupt of 20,000 amps. They are classified as RK5 or RK1 current-limiting fuses and should be mounted in Class-R rejecting fuse holders or dc-rated, fused disconnects. Class J or T fuses with dc ratings might also be used. For reasons mentioned previously, time-delay fuses should be specified, although some designers are getting good results with Class T fast-acting fuses.

One of these fuses and the associated disconnect switch should be used in **each** bank of batteries with a paralleled amp-hour capacity up to 1,000 amp-hours. Many 12, 24 and 48-volt battery banks are connected without overcurrent devices in each string of batteries and these have proved durable over the years. However, as batteries age and load conditions change, string current becomes unbalanced and the fuse in each string may help to prevent total battery bank failures under normal and fault conditions. On battery systems with higher than 48 volts nominal rating, the use of a disconnect and overcurrent device in each string of cells is necessary to prevent system failures that could result in fires and explosions and to allow for proper servicing [690.71].

Batteries with single-cell amp-hour capacities higher than 1,000 amp-hours will require special design considerations, because these batteries may be able to generate short-circuit currents in excess of the 20,000 AIR rating of the current-limiting fuses. When calculating the available short-circuit currents at a particular point in the circuit, the resistances of all connections, terminals, wire, fuse holders, circuit breakers, and switches to that point need to be considered. These resistances serve to reduce the magnitude of the available short-circuit currents at any particular point. The suggestion of one fuse per 1,000 amp-hours of battery size is only a general estimate, and the calculations are site specific. The listed branch-circuit fuses shown in Figure 17 are current limiting.

In lieu of current-limiting fuses, circuit breakers with high interrupt ratings may be used throughout the system for all overcurrent devices. These circuit breakers are not current limiting, even with the high interrupt rating, so they cannot be used to protect other types of fuses or circuit breakers. An appropriate use would be in the conductor between the battery bank and the inverter. This single device would minimize voltage drop and provide the necessary disconnect and overcurrent features. When high interrupt rating circuit breakers are used throughout a PV system, there is *NO* requirement for a current-limiting fuse, since each circuit breaker is capable of interrupting the short-circuit currents that may be impressed upon it.

Current-Limiting Fuses— Utility-interactive Systems

Normal electrical installation practice requires that utility service entrance equipment have fault-current protection devices that can interrupt the available short-circuit currents [110.9]. This requirement applies to the utility side of any power conditioning system in a PV installation. If the service is capable of delivering fault currents in excess of the interrupt rating of the overcurrent devices used to connect the inverter to the system, then current-limiting overcurrent devices **must** be used [110.9]. In utility-interactive systems that are connected to the line side of the service disconnect, particular attention should be paid to the amount of available short-circuit current from the utility feeder.

However, many utility-interactive PV systems make the utility connection through a back-fed circuit breaker in an existing load center and the existing load center is designed to handle the available short-circuit currents. No additional current limiting is required. If a new service entrance is added for the output of the PV system, then the service entrance equipment **must** have the appropriate ratings [690.64]. See Appendix C for additional details.

Fuse Servicing

Whenever a fuse is used as an overcurrent device and is accessible to unqualified persons, it **must** be installed in such a manner that all power can be removed from both ends of the fuse for servicing. It is not sufficient to reduce the current to zero before changing the fuse. There **must** be no voltage present on either end of the fuse prior to service. This may require the addition of switches on both sides of the fuse location—a complication that increases the voltage drop and reduces the reliability of the system [690.16]. Because of this requirement, the use of a fusible pullout-style disconnect, "finger-safe" fuse holder, or circuit breaker is recommended.

Disconnecting Means

There are many considerations in configuring the disconnect switches for a PV system. The *National Electrical Code* deals with safety first and other requirements last—if at all. The PV designer should also consider equipment damage from over voltage, performance options, equipment limitations, and cost.

A photovoltaic system is a power generation system, and a specific minimum number of disconnects are necessary to deal with that power. Untrained personnel will be operating the systems; therefore, the disconnect system **must** be designed to provide safe, reliable, and understandable operation [690 III].

Disconnects may range from nonexistent in a self-contained PV-powered light for a sidewalk to those found in the space-shuttle-like control room in a large, multi-megawatt, utility-tied PV power station. Generally, local inspectors will not require disconnects on totally enclosed, self-contained PV systems like a PV-powered, solar, hot-water circulating system. This would be particularly true if the entire assembly were listed as a unit and there were no external contacts or user serviceable parts. However, the situation changes as the complexity of the device increases and separate modules, inverters, batteries, and charge controllers having external connections are wired together and possibly operated and serviced by unqualified personnel.

Photovoltaic Array Disconnects

Article 690 requires all current-carrying conductors from the PV power source or other power source to have *disconnect* provisions. This provision includes the grounded conductor, if any [690 III]. *Ungrounded* conductors **must** have a switch or circuit breaker disconnect [690.13, 15, 17]. *Grounded* conductors which normally remain connected at all times, may have a bolted disconnect (terminal or lug) that can be used for service operations and for meeting the *NEC* requirements. Disconnect switches **must not** open grounded conductors [690.13]. Grounded conductors of faulted source circuits in roof-mounted dc PV arrays on dwellings are allowed to be automatically interrupted as part of ground-fault protection requirements in 690.5. [690.13]

Optionally ungrounded 12-volt and some 24-volt PV systems require an overcurrent device in both of the ungrounded conductors of each circuit. Since an equipment-grounding system is required on all systems, grounding the system and using overcurrent devices only in the remaining ungrounded conductors may reduce costs.

In an ungrounded 12-volt PV system (as allowed by [690.41]), both positive and negative conductors **must** be switched, since both are ungrounded. Since all systems **must** have an equipment-grounding system, costs may be reduced and performance improved by grounding 12-volt systems and using one-pole disconnects on the remaining ungrounded conductor.

Ungrounded systems operating at higher voltages, as will be allowed by the *2005 NEC* in 690.35, will also require switched disconnects and overcurrent protection in all of the circuit conductors since both the positive and negative circuit conductors will be ungrounded. See Appendix L for additional discussions of ungrounded PV systems.

PV Disconnect Location

Let us first consider the ac utility service to the typical residence. Either an overhead or an underground feeder will deliver the power. Before this service feeder gets into the house, it usually first goes through a billing kilowatt-hour meter and then the service entrance disconnect. In many jurisdictions, the local code allows the main disconnect to be immediately inside the home at the point of first penetration by the conductors of the building as allowed by the *National Electrical Code* (*NEC*) See *NEC* Section 230. In other locations, and the number is increasing, the service entrance disconnect **must** be located on the outside of the house with the load center sans disconnect inside the house [local codes]. In all cases this disconnect **must** be "readily accessible," which means it **must** not be in locked compartments, no ladders are required to access it, and no building material **must** be removed to get to it [690.14, 100-readily accessible]. These requirements were established many years ago to allow fire response personnel to quickly and safely shut off power to a building on fire that might require the firefighters to enter and cut holes in walls, ceilings and roofs. In life threatening situations, time is of the essence.

The *NEC* in Section 690.14 requires that the main PV disconnect be in a similar location. It therefore **must** be in a readily accessible location (no bathrooms, no attics—unless served by a permanent fixed stairs) at the point of first penetration of the dc PV source or output conductors. As in the ac service entrance disconnect, this PV disconnect may be located immediately inside the point of first penetration of the conductors. If the attic is reached by fixed stairs (not pull down), then the disconnect might be mounted in that location. Disconnects in bathrooms are not allowed. Other readily accessible rooms are acceptable as long as there are no locked doors.

Although commonly done in the past, many inspectors are not allowing PV conductors from the roof-mounted PV array to penetrate the attic and be run through the walls to the first floor or the basement where the main PV disconnect is located. These "always energized" conductors pose hazards to fire response personnel and possibly a fire hazard since they are in locations where potential short circuits might start fires.

The *2005 NEC* will allow an inside circuit installation provided it meets certain additional requirements. If the conductors are installed in a metal conduit or raceway, they will be permitted (not required) to be routed inside the house to the dc disconnect located at some distance from the point of first penetration. The disconnect will still have to be readily accessible, but this allowance, if adopted, will permit more design and installation flexibility. The metal conduit/raceway provides for added fire protection (does not burn), mechanical protection (difficult to accidentally cut), and ground-fault detection (in the event there is an internal ground fault).

Equipment Disconnects

Each piece of equipment in the PV system **shall** have disconnect switches to disconnect it from all sources of power. The disconnects **shall** be circuit breakers or switches and **shall** comply with all of the provisions of Section 690.17. DC-rated switches are expensive; therefore, the ready availability of moderately priced dc-rated circuit breakers with ratings up to 125 volts and 110 amps would seem to encourage their use in all 12-, 24-, and 48-volt systems. When properly located and used within their approved ratings, circuit breakers can serve as both the disconnect and overcurrent device. In simple stand-alone systems, one switch or circuit breaker disconnecting the PV array and another disconnecting the battery may be all that is required.

In larger utility-interactive systems, there may be several string disconnect switches, sub array disconnects, main PV disconnects for each inverter, ac output disconnects for each inverter and a complete system ac disconnect (sometimes operating at 12 kV).

A 2,000-watt inverter on a 12-volt system can draw more than 235 amps at full load. A 250kW utility-interactive inverter may have a PV dc input disconnect that carries 800 amps at 300 volts or more. Disconnect switches **must** be rated to carry the current and have

appropriate voltage and interrupt ratings [110.3(B)]. Again, a dc-rated, listed circuit breaker may prove less costly and more compact than a switch and fuse with the same ratings; at least in systems operating up to a nominal voltage of 48 volts.

Battery Disconnect

When the battery is disconnected from the stand-alone system, either manually or through the action of a fuse or circuit breaker, care should be taken that the PV system not be allowed to remain connected to the load. Depending on the design of the charge controller, small loads may allow the PV array voltage to increase from the normal battery charging levels to the open-circuit

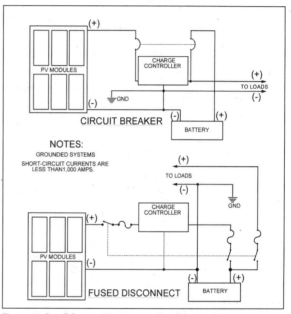

Figure 19. Small System Disconnects. Small System Disconnects

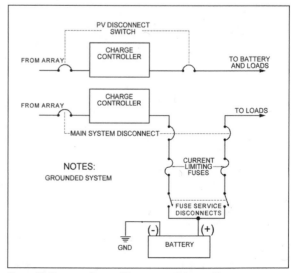

Figure 20. Separate Battery Disconnects. Separate Battery Disconnects

voltage, which will shorten dc lamp life and possibly damage electronic components.

This potential problem can be alleviated somewhat by using ganged multi-pole circuit breakers or ganged fused disconnects as shown in Figure 20. This figure shows two ways of making the connection. Of course, fuses in a ganged unit may operate independently, which may still create a problem. Separate circuits, including disconnects and fuses between the charge controller and the battery and the battery and the load, as shown in Figure 19, may be used if it is desired to operate the loads without the PV array being connected. If the design requires that the entire system be shut down with a minimum number of switch actions, the switches and circuit breakers could be ganged multi-pole units.

Charge Controller Disconnects

Some unlisted charge controllers are fussy about the sequence in which they are connected and disconnected from the system. These charge controllers do not respond well to being connected to the PV array and not being connected to the battery. The sensed battery voltage (or lack thereof) would tend to rapidly cycle between the array open-circuit voltage and zero as the controller tried to regulate the nonexistent charge process. This problem will be particularly acute in self-

contained charge controllers with no external battery sensing. The use of charge controllers listed to UL *Standard 1741* will minimize this problem. In this case, such a listed charge controller has been designed to operate properly with all of the overcurrent protection and disconnects required by the *NEC*.

Again, the multi-pole switch or circuit breaker can be used to disconnect not only the battery from the charge controller, but the charge controller from the array. Probably the safest method for self-contained charge controllers is to have the PV disconnect switch disconnect both the input and the output of the charge controller from the system. Larger systems with separate charge control electronics and switching elements will require a case-by-case analysis—at least until the controller manufacturers standardize their products. Figure 21 shows two methods of disconnecting the charge controller.

UNGrounded Systems

Systems that do not have one of the current-carrying conductors grounded **must** have disconnects *and* overcurrent devices in all of the ungrounded conductors [240.20, 690.13]. This means two-pole devices for the PV, battery, and inverter disconnects and overcurrent devices. The additional cost is considerable. See Appendix L for more information.

Multiple Power Sources

When multiple sources of power are involved, the disconnect switches **shall** be grouped and identified [230.72, 690.13(C)(5)]. No more than six motions of the hand will be required to operate all of the disconnect switches required to remove all power from the system [230.71]. These power sources include PV output, the battery system, any generator, and any other source of power. Multi-pole disconnects or handle ties should be used to keep the number of motions of the hand to six or fewer.

Article 230 in the NEC allows each structure to have more that one source of supply. The sources might be a utility connection and a PV system. The disconnects of these two sources of supply do not have to be grouped [230.2, 230.71]. However placards are required showing where all of the disconnects are located [230.70, 690.54, 705.10].

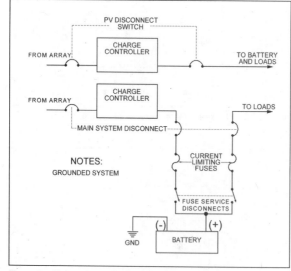

Figure 21. Charge Controller Disconnects

Panelboards, Enclosures, and Boxes

Disconnect and overcurrent devices **shall** be mounted in listed enclosures, panelboards, or boxes [240 III]. Wiring between these enclosures **must** use a *NEC*-approved method [110.8]. Appropriate cable clamps, strain-relief methods, or conduit **shall** be used. All openings not used **shall** be closed with the same or similar material to that of the enclosure [110.12(A)]. Metal enclosures **must** be bonded to the equipment-grounding conductor [408.20]. The use of wood or other combustible materials is discouraged. Conductors from *different* systems such as utility power, gas generator, hydro, or wind **shall not** be placed in the same enclosure, box, conduit, etc., as PV source conductors unless the enclosure is partitioned [690.4(B)]. This requirement stems from the need to keep "always live" PV source conductors separate from those that can be turned off. The ac outputs of a specific PV system may be routed in the same conduit or raceway as the dc PV source conductors from the *same* system providing that all conductors meet the insulation requirements of 300.3(C)(1).

When designing a PV distribution system or panel board, a listed NEMA type box and listed disconnect devices and overcurrent devices should be used. The requirements for the internal configuration of these devices are established by *NEC* Articles 110, 408, portions of article 690 as well as other articles in the code and **must** be followed. Dead front-panelboards with no exposed current-carrying conductors, terminals, or contacts are generally required [408.18]. Underwriters Laboratories also establishes the standards for the internal construction of panelboards and enclosures. The use of a listed commercial product designed for use in PV systems is encouraged

Batteries

In general, *NEC* Articles 480 and 690 VIII should be followed for installations having storage batteries. Battery storage in PV systems poses several safety hazards:

- Hydrogen gas generation from charging batteries
- High short-circuit current
- Acid or caustic electrolyte
- Electric shock potential

Hydrogen Gas

When flooded, non-sealed, lead-acid batteries are charged at high rates, or when the terminal voltage reaches 2.3 - 2.4 volts per cell, the batteries produce hydrogen gas. Even sealed batteries may vent hydrogen gas under certain conditions. This gas, if confined and not properly vented, poses an explosive hazard. The amount of gas generated is a function of the battery temperature, the voltage, the charging current, and the battery-bank size. Hydrogen is a light, small-molecule gas that is easily dissipated and is very difficult to contain. Small battery banks (i.e., up to 20, 220-amp-hour, 6-volt batteries) placed in a large room or a well-ventilated (drafty) area may not pose a significant hazard. Larger numbers of batteries in smaller or tightly enclosed areas require venting. Venting manifolds attached to each cell and routed to an exterior location are not recommended because flames in one section of the manifold may be easily transmitted to other areas in the system. The instructions provided by the battery manufacturer should be followed.

Closed battery boxes with single vents to outside-the-house air may pose problems unless carefully designed. Wind may force hydrogen back down the vent.

A catalytic recombiner cap (Hydrocap® Appendix A) may be attached to each cell of a flooded, lead-acid battery to recombine some of the evolved hydrogen and oxygen to produce water. If these combiner caps are used, they will require occasional maintenance. It is rarely necessary to use power venting. Flame arrestors are required by *NEC* Section 480.9, and battery manufacturers can provide special vent caps with flame-arresting properties when the local authority requires them.

Certain charge controllers are designed to minimize the generation of hydrogen gas, but lead-acid batteries need some overcharging to fully charge the cells. This produces gassing that should be dissipated.

In *no case* should charge controllers, switches, relays, or other devices capable of producing an electric spark be mounted in a battery enclosure or directly over a battery bank. Care needs to be exercised when routing conduit from a sealed battery box to a disconnect. Hydrogen gas may travel in the conduit to the arcing contacts of the switch. It is suggested that any conduit openings in battery boxes be made below the tops of the batteries, since hydrogen rises to the top of the enclosure as it displaces the air.

Battery Rooms and Containers

Battery systems are capable of generating thousands of amps of current when shorted. A short circuit in a conductor not protected by overcurrent devices can melt wrenches or other tools, battery terminals and cables, and spray molten metal around the room. Exposed battery terminals and cable connections **must** be protected. Live parts of batteries **must** be guarded [690.71]. This generally means that the batteries should be accessible only to a qualified person. A locked room, battery box, or other container and some method to prevent access by the untrained person should reduce the hazards from short circuits and electric shock. The danger may be reduced if insulating caps or tape are placed on each terminal and an insulated wrench is used for servicing. Note that with protective caps, corrosion may go unnoticed on the terminals. The *NEC* requires certain spacing around battery enclosures and boxes and other equipment to allow for unrestricted servicing—generally about three feet [110.26]. Batteries should not be installed in living areas, nor should they be installed below any enclosures, panelboards, or load centers [110.26].

One of the more suitable, readily available battery containers is the lockable, heavy-duty black polyethylene toolbox. Such a box can hold up to four L-16 size batteries and is easily cut for ventilation holes in the lid and for conduit entrances.

NEC Section 690.71(D) prohibits the use of conductive cases for flooded, lead-acid batteries operating above 48-volts nominal. Racks for these batteries may have no conductive parts within than 6" (150 mm) of the tops of the cases. These requirements were established to minimize the probability of high-voltage ground faults developing in the dust and electrolyte film that develops on these vented batteries during normal operation.

Acid or Caustic Electrolyte

A thin film of electrolyte can accumulate on the tops of the battery and on nearby surfaces. This material can cause flesh burns. It is also a conductor and, in high-voltage battery banks, poses a shock hazard, as well as a potential ground-fault path. The film of electrolyte should be removed periodically with an appropriate neutralizing solution. For lead-acid batteries, a dilute solution of baking soda and water works well. Commercial neutralizers are available at auto-supply stores.

Charge controllers are available that minimize the dispersion of the electrolyte and water usage because they minimize battery gassing. They do this by keeping the battery voltage from climbing into the *vigorous* gassing region where the high volume of gas causes electrolyte to mist out of the cells. A moderate amount of gassing is necessary for proper battery charging and destratification of the electrolyte in flooded cells.

Battery servicing hazards can be minimized by using protective clothing including facemasks, gloves, and rubber aprons. Self-contained eyewash stations and neutralizing solution are good precautionary additions to any battery room. Water should be used to wash acid or alkaline electrolyte from the skin and eyes.

Anti-corrosion sprays and greases are available from automotive and battery supply stores and they generally reduce the need to service the battery bank. Hydrocap® Vents also reduce the need for servicing by reducing the need for watering.

Electric Shock Potential

Storage batteries in dwellings **must** operate at less than 50 volts (48-volt nominal battery bank) unless live parts are protected during routine servicing [690.71(B)(1)]. It is recommended that live parts of any battery bank should be guarded [690.71(B)(2)].

Battery and Other Large Cables

Battery cables, even though they can be 2/0 AWG and larger, **must** be a standard building-wire type conductor [Chapter 3]. Welding and automobile "battery" cables (listed and non-listed) are not allowed. Flexible, highly-stranded, building-wire type cables (USE/RHW and THW) are available for this use. Flexible cables, identified in Section 400 of the *NEC* are permitted (not required) from the battery terminals to a nearby junction box and between battery cells. These cables shall be listed for hard service use and moisture resistance [690.74]. As is the case with flexible PV module interconnecting cables, it is rarely necessary to use anything other than the normal building wire types of cables identified in Chapter 3 of the *NEC*. Also the types of terminals that can be used with these flexible cables are limited. In general, the manufacturer's data should be consulted or the terminal or lug should be marked indicating compatibility with the fine stranded cables. The few lugs that are compatible are made of solid copper, have a flared entry section and look somewhat like the three lugs on the far right in Figure 5.

Generators

Other electrical power generators such as wind, hydro, and gasoline/propane/diesel **must** comply with the requirements of the *NEC*. These requirements are specified in the following *NEC* articles:

Article 230   Services
Article 250   Grounding
Article 445   Generators
Article 700   Emergency Systems
Article 701   Legally Required Standby Systems
Article 702   Optional Standby Systems
Article 705   Interconnected Power Production Sources

When multiple sources of ac power are to be connected to the PV system, they **must** be connected with an appropriately rated and listed transfer switch [702.6]. AC generators frequently are rated to supply larger amounts of power than that supplied by the PV/battery/inverter. The transfer switches (external to the inverter or a relay built into listed inverters) **must** be able to safely accommodate either power source [110.3(B)].

Grounding, both equipment and system, needs to be carefully considered when a generator is connected to an existing system. There **must** be no currents flowing in the equipment-grounding conductor under any normal operating mode of the system [250.6]. Bonds (connections) between the ac grounded conductor (neutral)

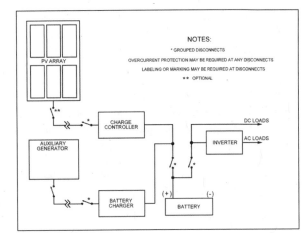

Figure 22. Disconnects for Remotely Located Power Sources. Disconnects for Remotely Located Power Sources

and the grounded frame in generators are common and have caused circulating, unwanted currents.

The circuit breakers or fuses that are built into the generator are usually not sufficient to provide *NEC*-required protection for the conductors from the generator to the PV system. An external (branch circuit rated) overcurrent device (and possibly a disconnect) **must** be mounted close to the generator [240.21]. The conductors from the generator to this overcurrent device **must** have an ampacity of not less than 115% of the nameplate current rating of the generator [445.12]. Figure 22 shows a typical one-line diagram for a system with an auxiliary backup generator.

Charge Controllers

A charge controller or self-regulating system **shall** be used in a stand-alone system with battery storage. The mechanism for adjusting state of charge **shall** be accessible only to qualified persons [690.72].

There are several charge controllers on the market that have been tested and listed to UL standards by recognized testing organizations.

Surface mounting of unlisted charge controllers with external terminals readily accessible to the unqualified person will not be accepted by the inspection authority. Dead-front panels with no exposed contacts are generally required for safety. Figure 23 shows a typical charge controller and remote display panel. It is a listed device, has no exposed terminals, is ready for installation with conduit, and has no readily-accessible user adjustments.

Electrically, listed charge controllers are designed with a "straight" conductor between the negative input and output terminals. A shunt is sometimes placed in that conductor. This design will allow the controller to

be used in a grounded system with the grounded conductor running through the controller. The installation manual of the charge controller **must** be reviewed to ensure proper system grounding [110.3(B)].

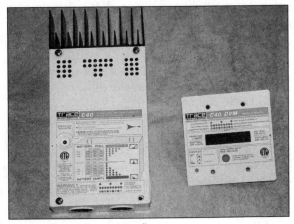

Figure 23. Typical Charge Controller

NEC SUGGESTED PRACTICES

Inverters

Inverters can have stand-alone, utility-interactive, or combined capabilities.

The ac output wiring is not significantly different from the ac wiring in residential and commercial construction, and the same general requirements of the Code apply. In the case of utility-interactive systems and combined systems, ac power may flow through circuits in both directions. This two-way current flow will normally require overcurrent devices at both ends of the circuit.

The dc input wiring associated with stand-alone or hybrid inverters is the same as the wiring described for batteries. Most of the same rules apply; however, the calculation of the dc input current needs special consideration since the *NEC* does not take into consideration some of the finer points required to achieve the utmost in reliability. Appendix F discusses these special requirements in greater detail.

The dc input wiring associated with utility-interactive inverters is similar, in most cases, to the wiring in PV source and output circuits.

Inverters with combined capabilities will have both types of dc wiring: connections to the batteries and connections to the PV modules.

Stand-Alone Distribution Systems

The *National Electrical Code* has evolved to accommodate supplies of relatively cheap energy. As the Code was expanded to include other power systems such as PV, many sections were not modified to reflect the recent push toward more efficient use of electricity in the home. Stand-alone PV systems *may* be required to have dc services with 60- to 100-amp capacities to meet the Code [230.79]. DC receptacles for appliances and lighting circuits, where used, *may* have to be as numerous as their ac counterparts [220, 422]. In a small one- to four-module system on a remote cabin where no utility extensions or local grids are possible, these requirements may be excessive, since the power source may be able to supply only a few hundred watts of power.

Changes in the *1999 NEC* in Section 690.10 clarified some of the code requirements for stand-alone PV systems.

The local inspection authority has the final say on what is, or is not, required and what is, or is not, safe. Reasoned conversations may result in a liberal interpretation of the Code. For a new dwelling, it seems appropriate to install a complete ac electrical system as required by the *NEC*. This will meet the requirements of the inspection authority, the mortgage company, and the insurance industry. Then the PV system and its dc distribution system can be added. If an inverter is used, it can be connected to the ac service entrance. *NEC* Section 690.10 elaborates on these requirements and allowances. DC branch circuits and outlets can be added where needed, and everyone will be happy. If or when grid power becomes available, it can be integrated into the system with minimum difficulty. If the building is sold at a later date, it will comply with the *NEC* if it has to be inspected. The use of a listed dc power center will facilitate the installation and the inspection.

Square D has received a direct current (dc), UL listing for its standard QO residential **branch** circuit breakers. They can be used up to 48 volts (125% PV open-circuit voltage) and 70 amps dc. This limits their use to a 12-volt nominal system and a few 24-volt systems in hot climates [Table 690.7]. The AIR is 5,000 amps, so a current-limiting fuse (RK5 or RK1 type) **must** be used when they are connected on a battery system [690.71(C)]. The Square D QOM **main** breakers (used at the top of the load center) **do not** have this listing, so the dc load center based on Square D QO circuit breakers should be obtained with main lugs and no main breakers (Appendix A).

In a small 12-volt PV system (less than 5000 amps of available short-circuit current), a two-pole Square D QO breaker could be used as the PV disconnect (one pole) and the battery disconnect (one pole). Alternatively, a fused disconnect or fusible pullout could be used in this configuration. This would give a little more flexibility since the fuses can have different current ratings. Figure 19 shows both systems with only a single branch circuit.

In a system with several dc branch circuits, the Square D QO load center can be used. A standard, off-the-shelf Square D QO residential load center without a main breaker can be used for a dc distribution panel in 12-volt dc systems and a very few hot-climate 24-volt systems. The main disconnect would have to be a "back fed" QO breaker, and it would have to be connected in one of the normal branch circuit locations. Back-fed circuit breakers **must** be identified for such use [690.64(B)(5)] and clamped [408.16(F)]. See Appendix C for additional details. Since the load center has two separate circuits (one for each line), the bus bars will have to be tied together in order to use the entire load center. Figure 24 illustrates this use of the Square D load center.

Another possibility is to use one of the line circuits to combine separate PV source circuits, then go out of the load center through a breaker acting as the PV discon-nect switch to the charge controller. Finally, the conductors would have to be routed back to the other line circuit in the load center for branch-circuit distribution. Several options exist in using one and two-pole breakers for disconnects. Figure 25 presents an example.

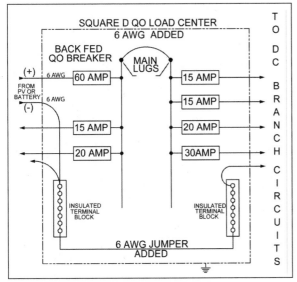

Figure 24. 12-Volt DC Load Center

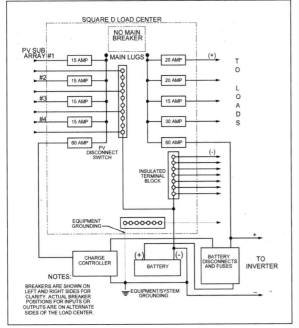

Figure 25. 12-Volt DC Combining Box and Load Center

Interior DC Wiring and Receptacles

Any dc interior wiring used in PV systems **must** comply with the *NEC* [300]. Nonmetallic sheathed cable (type NM - "Romex") may be used, and it **must** be installed in the same manner as cable for ac branch circuits [334, 690.31(A)]. The bare grounding conductor in such a cable **must not** be used to carry current and cannot be used as a common negative conductor for combination 12/24-volt systems [334.108]. Exposed, single-conductor cables are not permitted—they **must** be installed in conduit [300.3(A)]. Conductors in the same current (i.e., positive and negative battery conductors and equipment-grounding conductors) **must** be installed in the same conduit or cable to prevent increased circuit inductances that would pose additional electrical stresses on disconnect and overcurrent devices [300.3(B)].

The code allows the equipment-grounding conductors for dc circuits *only* to be run apart from the current-carrying conductors [250.134(B) EX2]. However, separating the equipment-grounding conductor from the circuit conductors may increase any fault-circuit time constant and impair the operation of overcurrent devices. The effects of transient pulses are also enhanced when equipment-grounding conductors are separate. It is suggested that dc equipment-grounding conductors be run in the same conduit or cable as the dc circuit conductors.

The receptacles used for dc **must** be different from those used for any other service in the system [406.3(F)]. The receptacles should have a rating of not less than 15 amps and **must** be of the three-prong grounding type [406.2(B), 406.3(A)]. Numerous different styles of listed receptacles are available that meet this requirement. These requirements can be met in most locations by using the three-conductor 15-, 20-, or 30-amp 240-volt NEMA style 6-15, 6-20, 6-30 receptacles

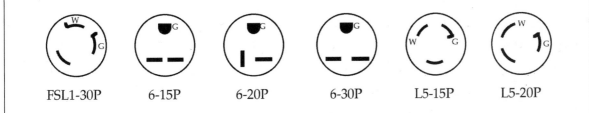

| FSL1-30P | 6-15P | 6-20P | 6-30P | L5-15P | L5-20P |

Figure 26. NEMA Plug Configurations

for the 12-volt dc outlets. If 24-volt dc is also used, the NEMA 125-volt locking connectors, style L5-15 or L5-20, are commonly available. The NEMA FSL-1 is a locking 30-amp 28-volt dc connector, but its availability is limited. Figure 26 shows some of the available configurations. Cigarette lighter sockets and plugs frequently found on "PV" and "RV" appliances *do not* meet the requirements of the *National Electrical Code* and should not be used.

It is not permissible to use the third or grounding conductor of a three-conductor plug or receptacle to carry common negative return currents on a combined 12/24-volt system. This terminal **must** be used for equipment grounding and may not carry current except in fault conditions [406.9(C)].

A 30-amp fuse or circuit breaker protecting a branch circuit (with 10 AWG conductors) **must** use receptacles rated at 30 amps. Receptacles rated at 15 and 20 amps **must not** be used on this 30-amp circuit [Table 210.21(B)(3)].

Smoke Detectors

Many building codes require that smoke and fire detectors be wired directly into the ac power wiring of the dwelling. With a system that has no inverter, two solutions might be offered to the inspector. The first is to use the 9-volt or other primary-cell, battery-powered detector. The second is to use a voltage regulator to drop the PV system voltage to the 9-volt or other level required by the detector.

The regulator should be able to withstand the PV open-circuit voltage and supply the current required by the detector alarm. Building such a device should only be attempted by the well-qualified individual.

On inverter systems, the detector on some units may trigger the inverter into an "on" state, unnecessarily wasting power. In other units, the alarm may not draw enough current to turn the inverter on and thereby produce a reduced volume alarm or, in some cases, no alarm at all. Small, dedicated inverters might be used, but this would waste power and decrease reliability when dc detectors are available.

Most building codes require detectors to be connected to the power line and have a battery backup. Units satisfying this requirement might also be powered by dc from the PV system battery and by a primary cell.

Ground-Fault Circuit Interrupters

Some ac ground-fault circuit interrupters (GFCI) do not operate reliably on the output of some non-sine-wave inverters. If the GFCI does not function when tested, it should be verified that the neutral (white-grounded) conductor in the system is solidly grounded

and bonded to the equipment-grounding (green or bare) conductor and both are connected to ground in the required manner. If this bond is present and does not result in the GFCI testing properly, other options are possible. Changing the brand of GFCI may rectify the solution. A direct measurement of an intentional ground fault may indicate that slightly more than the 5 milliamp internal test current is required to trip the GFCI. The inspector may accept this. Some modified square wave inverters will work with a ferro-resonant transformer to produce a waveform more satisfactory for use with GFCIs, but the no-load power consumption may be high enough to warrant a manual demand switch. A sine-wave inverter should be used to power those circuits requiring GFCI protection. Since sine-wave stand-alone inverters are becoming the norm, the problems of using GFCIs (and smoke detectors) with non sine-wave inverters are diminishing.

Interior Switches

Switches rated for *ac only* **shall not** be used in dc circuits [404.14(A)]. AC-DC general-use "snap" switches are available by special order from most electrical supply houses, and they are similar in appearance to normal "quiet switches" [404.14(B)].

NOTE: There have been some failures of dc-rated snap switches when used as PV array and battery disconnect switches. If these switches are used on 12- and 24-volt systems and are not activated frequently, the contacts may build up oxidation or corrosion and not function properly. Periodically (recommend monthly) activating the switches under load will keep the contacts clean.

Multiwire Branch Circuits

Stand-alone PV and PV/Hybrid systems are frequently connected to a building/structure/house that has been previously completely wired for 120/240-volts ac and has a standard service entrance and load center.

These structures may employ one or more circuits that the *National Electrical Code* (*NEC*) defines as a multiwire branch circuit. See Section 100 in the *NEC*, "Branch Circuit, Multiwire." These circuits take a three-conductor plus ground feeder from the 120/240-volt load center and run it some distance to a location in the structure where two separate 120-volt branch circuits are split out. Each branch circuit uses one of the 120-volt hot, ungrounded conductors from the 120/240-volt feeder and the common neutral conductor. See Figure 27.

In a utility-connected system or a stand-alone system with a 120/240-volt stacked pair of inverters, where the 120/240-volt power consists of two 120-volt lines that are 180 degrees out of phase, the currents in the common neutral in the multiwire branch circuit are limited to the difference currents from any unbalanced load. If the loads on each of the separate branch circuits were equal, then the currents in the common neutral would be zero.

A neutral conductor overload may arise when a single 120-volt inverter is tied to both of the hot input conductors on the 120/240-volt load center as shown in Figure 27. This is a common practice for stand-alone PV homes. At this point the two hot 120-volt conductors are being delivered voltage from the single 120-volt inverter and that voltage is in phase on both conductors. In multiwire branch circuits, the return currents from each of the separate branch circuits in the common neutral *add* together. A sketch of the multiwire branch circuit is presented in Figure 27.

Each branch circuit is protected by a circuit breaker in the ungrounded conductor in the load center. The neutral conductor is usually the same size as the ungrounded conductors and can be overloaded with the in-phase return currents. The circuit breakers will pass current up to the ampacity of the protected conductors, but when both branch circuits are loaded at more than 50%, the unprotected, common neutral conductor is *overloaded and may be carrying up to twice its rated currents.*

A definite fire and safety hazard exists. All existing stand-alone PV installations using single inverters tied to both ungrounded conductors at the service entrance should be examined for multiwire branch circuits.

The *NEC* requires that multiwire branch circuits *in some, but not all, cases* use a two-pole circuit breaker so that both circuits are dead at the same time under fault conditions and for servicing. This two-pole, side-by-side circuit breaker rated at 15 or 20 amps may be one indication that multiwire branch circuits have been used. Common handle circuit breakers rated at 30 amps and higher are usually dedicated to 240-volt circuits for ranges, hot water heaters, dryers, and the like and the conductors are usually 8 AWG and larger. The Code requires that there **must** be no 240-volt outlets in a structure fed by a single 120-volt inverter [690.10].

Examination of the wiring in the load center may show a three-wire cable (14 or 12 AWG red, black, and white conductors) with bare ground leaving the load center. This may be connected to a multiwire branch circuit. The circuit breakers connected to this cable and the

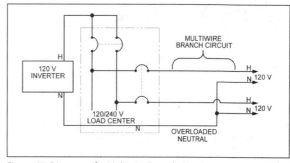

Figure 27. Diagram of a Multiwire Branch Circuit

outputs of this cable should be traced to determine the presence or absence of a multiwire branch circuit.

The following options are suggested for dealing with this situation:

- *Disconnect or rewire the multiwire circuits as separate circuits ("home runs") from the load center.*
- *Connect both "hot" (ungrounded) conductors of the multiwire branch circuit to a single circuit breaker rated for the ampacity of the neutral conductor.* Note: This may violate local code limitations on the number of outlets per branch circuit.
- *Install a transformer to provide a 120/240-volt output from a 120-volt inverter.*
- *Install a stacked pair of inverters to provide 120/240V ac.*

In systems where multiwire branch circuits are used with stacked (120/240-volt) inverters, a sign should be placed near the inverters warning that single inverter use (one inverter removed for repair and the system is rewired to accommodate all branch circuits) may cause overloaded circuits. The maximum current from the single inverter should be limited to the ampacity of the common neutral conductor.

In all systems (multiwire or not), the neutral busbar of the load center **must** be rated at a higher current than the output of the inverter [690.10(C)]. In other words, do not connect an inverter with a 33-amp output to a load center rated at 20 or 30 amps.

Additional information is found in the *NEC* in sections 100, 210.4, 240.20(B), and 300.13(B), and in the *NEC* Handbook. Section 690.10(C) provides requirements and allowances on connecting a single inverter to a code-compliant ac wiring system.

AC PV Modules

An AC PV module is a photovoltaic device that has an alternating current output (usually 120 volts at 60 Hz in the U.S.). The AC PV module is listed (by UL and other listing agencies) as a unified device and is actually a standard dc PV module with an attached (non-removable) utility-interactive inverter. The ac output is only available when the ac PV module is connected to a utility grid circuit where there is a stable 120 volts at 60 Hz present. With no utility power, there will be NO energy flow from the ac PV modules.

A number of ac PV modules may be connected on the same circuit (according to ampacity limitations), but that circuit **must** be dedicated to the ac PV module(s) and **must** terminate in a dedicated circuit breaker [690.6].

There are no external dc circuits in the ac PV module and none of the dc code requirements apply. Unlisted combinations of small listed inverters mated to listed dc PV modules do not qualify as an ac PV module and will have to have all code-required dc switchgear, overcurrent, and ground-fault equipment added.

AC PV modules **shall** be marked with the following:
- Nominal AC Voltage
- Nominal AC Frequency
- Maximum AC Power
- Maximum AC Current
- Maximum Overcurrent Device Rating for AC Module Protection [690.52]

System Labels and Warnings

Photovoltaic Power Source

A permanent label **shall** be applied near the PV disconnect switch that contains the following information: [690.53]
- Operating Current (System maximum-power current)
- Operating Voltage (System maximum-power voltage)
- Maximum System Voltage
- Short-Circuit Current

This data will allow the inspector to verify proper conductor ampacity and overcurrent device rating. It will also allow the user to compare system performance with the specifications.

Multiple Power Systems

Systems with multiple sources of power such as PV, gas generator, wind, hydro, etc., **shall** have a permanent plaque or directory showing the interconnections [705.10]. Diagrams are not required, but may be useful and should be placed near the system disconnects.

Interactive System Point of Iinterconnection

All interactive system(s) points of interconnection with other sources **shall** be marked at an accessible location at the disconnecting means as a power source with the maximum ac output operating current and the operating ac voltage [690.54].

Switch or Circuit Breaker

If a switch or circuit breaker has all of the terminals energized when in the open position, a label should be placed near it indicating: [690.17]

WARNING - ELECTRIC SHOCK HAZARD - DO NOT TOUCH TERMINALS. TERMINALS ON BOTH THE LINE AND LOAD SIDES MAY BE ENERGIZED IN THE OPEN POSITION

General

Each piece of equipment that might be opened by unqualified persons should be marked with warning signs. In some cases, a listed product is required to have similar warnings:

WARNING - ELECTRIC SHOCK HAZARD - DANGEROUS VOLTAGES AND CURRENTS - NO USER SERVICEABLE PARTS INSIDE - CONTACT QUALIFIED SERVICE PERSONNEL FOR ASSISTANCE

Each battery container, box, or room should also have warning signs to encourage safety for both qualified and unqualified people:

WARNING - ELECTRIC SHOCK HAZARD - DANGEROUS VOLTAGES AND CURRENTS - EXPLOSIVE GAS - NO SPARKS OR FLAMES - NO SMOKING - ACID BURNS - WEAR PROTECTIVE CLOTHING WHEN SERVICING

Inspections

Involving the inspector as early as possible in the planning stages of the system will begin a process that should provide the best chance of installing a safe, durable system. The following steps are suggested.

- Establish a working relationship with a local electrical contractor or electrician to determine the requirements for permits and inspections.
- Contact the inspector and review the system plans. Solicit advice and suggestions from the inspector.
- Obtain the necessary permits.
- Involve the inspector in the design and installation process. Provide information as needed. Have one-line diagrams and complete descriptions of all equipment available.

Insurance

Most insurance companies are not familiar with photovoltaic power systems. They are, however, willing to add the cost of the system to the homeowner's policy if they understand the additional liability risk. A system description may be required. Evidence that the array is firmly attached to the roof or ground is usually necessary. The system **must** usually be permitted and inspected if those requirements exist for other electrical power systems in the locale [Local Codes].

Some companies will not insure homes that are not grid connected because there is no source of power for a high-volume water pump for fighting fires. In these instances, it may be necessary to install a fire-fighting system and water supply that meets their requirements. A high-volume dc pump and a pond might suffice.

As with the electrical inspector, education and a full system description emphasizing the safety features and code compliance will go a long way toward obtaining appropriate insurance.

APPENDIX A: Sources of Equipment Meeting the Requirements of The *National Electrical Code*

A number of PV distributors and dealers stock the equipment needed to meet the *NEC* requirements. Some sources are presented here for specialized equipment. This list is not intended to be all-inclusive or to promote any of the products.

Conductors

Standard multiconductor cable such as 10-2 with ground Nonmetallic Sheathed Cable (NM and NMC), Underground Feeder (UF), Service Entrance (SE), Underground Service Entrance (USE and USE-2), larger sizes (8 AWG) single-conductor cable, uninsulated grounding conductors, and numerous styles of building wire such as THHN can be obtained from electrical supply distributors and building supply stores. See NEC Table 310-13 for cable types and characteristics.

Flexible, fine-stranded cables should not be used with terminals or lugs that have a setscrew or screw mechanical attachment. These terminals and lugs (also found on circuit breakers, fuse holders, and PV equipment) are not generally listed for use with other than normal 7, 19, and 37 stranded conductors. Appendix K presents additional details.

The highest quality, most durable USE-2 cable will also have RHW-2, and 600V markings and be made of cross-linked polyethylene (marked XLP or XLPE). Flexible USE, RHW, and THW cables in large sizes (1/0 - 250 kcmil) and stranded 8-, 10-, and 12-AWG USE single conductor cable can be obtained from electrical supply houses and wire distributors. The following short list provides information on a cable distributor and manufacturer.

> Anixter Bros.
> 2201 Main Street
> Evanston, Illinois 60202
> 800-323-8166 for the nearest distributor
> 847-677-2600

> Cobra Wire and Cable, Inc.
> PO Box 790
> 2930 Turnpike Drive
> Hatboro, PA 19040
> 215-674-8773

DC-Rated Fuses

DC-rated 15, 20, 30 amp and higher rated fuses can be used for dc branch-circuit overcurrent protection depending on conductor ampacity and load. Larger sizes (100 amp and up) are used for current-limiting and overcurrent protection on battery outputs. DC rated, listed fuses are manufactured by the following companies, among others:

> Bussmann
> P.O. Box 14460
> St. Louis, MO 63178-4460
> 314-527-3877
> 314-527-1270 (Technical Questions)

> Gould/Ferraz Inc.
> 374 Merrimac Street
> Newburyport, MA 01950
> 508-462-6662

> Littelfuse
> Power Fuse Division
> 800 E. Northwest Highway
> Des Plaines, Illinois 60016
> (708) 824-1188
> 800-TEC FUSE (Technical Questions)
> 800-227-0029 (Customer Service)

The following fuses may be used for battery circuit and dc branch circuit overcurrent protection and current limiting applications. If transients are anticipated in PV circuits, these fuses can also be used in those locations.

Fuse Description	Size	Manufacturer	Mfg #
125-volt dc, RK5 Time delay, current-limiting	0.1-600 amp	Bussmann	FRN-R
125-volt dc, RK5 Time delay, current-limiting	0.1-600 amp	Littelfuse	FLNR
300-volt dc, RK5 Time delay, current-limiting fuse	0.1-600-amp	Bussmann	FRS-R
300-volt dc, RK5 Time delay, current-limiting fuse	0.1-600 amp	Gould	TRS-R
300-volt dc, RK5 Time delay, current-limiting fuse	0.1-600 amp	Littelfuse	FLSR
600-volt dc, RK5 Time delay, current-limiting fuse	0.1-600 amp,	Littelfuse	IDSR
600-volt dc, RK5 Time delay, current-limiting fuse	70-600 amp	Gould	TRS70R-600R

The following fuses should be used for PV source-circuit protection if problems are not anticipated with transients. They may also be used inside control panels to protect relays and other equipment.

Fuse Description	Size	Manufacturer	Mfg #
Fast-acting, midget fuse	0.1-30 amp	Gould	ATM
Fast-acting, midget fuse	0.1-30 amp	Littelfuse	KLK-D

Encloures and Junction Boxes

Indoor and outdoor (rainproof) general-purpose enclosures and junction boxes are available at most electrical supply houses. These devices usually have knock-outs for cable entrances, and the distributor will stock the necessary bushings and/or cable clamps. Interior component mounting panels are available for some enclosures, as are enclosures with hinged doors. If used outdoors, all enclosures, clamps, and accessories **must** be listed for outdoor use [110.3(B)]. For visual access to the interior, NEMA 4X enclosures are available that are made of clear, transparent plastic.

Hydrocaps

Hydrocap® Vents are available from Hydrocap Corp. and some PV distributors on a custom-manufactured basis.

Hydrocap
975 NW 95 St.
Miami, FL 33150
305-696-2504

APPENDIX B: PV Module Operating Characteristics Drive *NEC* Requirements

Introduction

As the photovoltaic (PV) power industry moves into a mainstream position in the generation of electrical power, some people question the seemingly conservative requirements established by Underwriters Laboratories (UL) and the *National Electrical Code* (*NEC*) for system and installation safety. This short discourse will address those concerns and highlight the unique characteristics of PV systems that dictate the requirements.

The *National Electrical Code* (*NEC*) is written with the requirement that all equipment and installations are approved for safety by the authority having jurisdiction (AHJ) to enforce the *NEC* requirements in a particular location. The AHJ readily admits to not having the resources to verify the safety of the required equipment and relies exclusively on the testing and listing of the equipment by independent testing laboratories such as Underwriters Laboratories (UL). The AHJ also relies on the installation requirements for field wiring specified in the *NEC* to ensure safe installations and use of the listed equipment.

The standards published by UL and the material in the *NEC* are closely harmonized by engineers and technicians throughout the electrical equipment industry, the electrical construction trades, the national laboratories, the scientific community, and the electrical inspector associations. The UL Standards are technical in nature with very specific requirements on the construction and testing of equipment for safety. They in turn are coordinated with the construction standards published by the National Electrical Manufacturers Association (NEMA). The *NEC*, however, is deliberately written in a manner to allow uniform application by electricians, electrical contractors, and electrical inspectors in the field.

The use of listed equipment (by UL or other nationally recognized testing laboratory) ensures that the equipment meets well-established safety standards. The application of the requirements in the *NEC* ensures that the listed equipment is properly connected with field wiring and is installed in a manner that will result in an essentially hazard-free system. The use of listed equipment and installing that equipment according to the requirements in the *NEC* will contribute greatly not only to safety, but also the durability, performance, and longevity of the system.

Unspecified Details

The *NEC* does not present many highly detailed technical specifications. For example, the term "rated output" is used in several cases with respect to PV equipment. The conditions under which the rating is determined are not specified. The definitions of the rating conditions (such as Standard Test Conditions (STC) for PV modules) are made in the UL Standards that establish the rated output. This procedure is appropriate because of the *NEC* level of writing and the lack of appropriate test equipment available to the *NEC* user or inspector.

NEC Requirements Based on Module Performance

VOLTAGE

Section 690.7 of the *NEC* establishes a temperature-dependent voltage correction factor that is to be applied to the rated (at STC) open-circuit voltage (V_{oc}) in order to establish the system voltage. This factor on the open-circuit voltage is needed because, as the operating temperature of the module decreases, V_{oc} increases. The rated V_{oc} is measured at a temperature of 25°C and while the normal operating temperature is 40-50°C when ambient temperatures are around 20°C, there is nothing to prevent sub-zero ambient temperatures from yielding operating temperatures significantly below the 25°C standard test condition.

A typical crystalline silicon module will have a voltage coefficient of -0.38%/°C. A system with a rated open-circuit voltage of 595 volts at 25°C might be exposed to ambient temperatures of -30°C. This voltage (595V) could be handled by the common 600-volt rated conductors and switchgear. At dawn and dusk conditions, the module will be at the ambient temperature of -30°C, will not experience any significant solar heating, and can generate open-circuit voltages of 719 volts (595 x (1 + (25 - 30) x -0.0038)). This voltage substantially exceeds the capability of 600-volt rated conductors, fuses, switchgear, and other equipment. High wind speeds can also cause modules to operate at or near ambient temperatures, even in the presence of moderate levels of sunlight. The very real possibility of this type of condition substantiates the *NEC* requirement for the temperature dependent factor on the rated open-circuit voltage.

Thin-film PV technologies may have other voltage-temperature relationships, and the manufacturers of modules employing such technologies should be consulted for the appropriate data.

CURRENT

NEC Section 690.8(A) requires that the rated (at STC) short-circuit current of the PV module be multiplied by 125% before any other factors, such as continuous current and conduit fill factors, are applied. This factor is to provide a safe margin for wire sizes and overcurrent devices when the irradiance exceeds the standard 1000 W/m^2. Depending on season, local weather conditions, and atmospheric dust and humidity, irradiance exceeds 1000 W/m^2 every day around solar noon. The time can be as long as four hours with irradiance values that approach 1200 W/m^2, again depending on the aforementioned conditions and the type of tracking system being used. These daily irradiance values can increase short-circuit currents 20% over the 1000 W/m^2 value. Since these increased currents can be present for three hours or more, they are considered continuous currents. By multiplying the short-circuit current by 125%, the PV output currents are adjusted in a manner that puts them on the same basis as other continuous currents in the *NEC*

Enhanced irradiance due to reflective surfaces such as sand, snow, or white roofs, and even nearby bodies of water can increase short-circuit currents by substantial amounts and for significant periods of time. Reflections from cumulus clouds also can increase irradiance by as much as 50%. These transient factors are not considered continuous and are not addressed by either UL or the *NEC*

Another factor that needs to be addressed is that PV modules typically operate at 30-40°C above the ambient temperatures when not exposed to cooling breezes. In crystalline silicon PV modules, the short-circuit current increases as the temperature increases. A typical factor might be 0.1%/°C. If the module operating temperature was 60°C (35°C over the STC of 25°C), the short-circuit current would be 3.5% greater than the rated value. PV modules have been measured operating over 75°C. The combination of increased operating temperatures, irradiances over 1000 W/m^2 around solar noon, and the possibility of enhanced irradiance provide additional justification for the *NEC* requirement [690.8(A)] of 125% on the rated short-circuit current.

Additional *NEC* Requirements

The *NEC* requires that the continuous current of any circuit be multiplied by 125% before calculating the ampacity of any cable or the rating of any overcurrent device used in these circuits [690.8(B) and 240]. This factor is in addition to the required 125% discussed above and is needed to ensure that overcurrent devices and conductors are not operated above 80% of rating.

Since short-circuit currents in excess of the rated value are possible from the discussion of the Section 690.8(A) requirements above, and these currents are independent of the requirements established by Section 690.8(B), the *NEC* dictates that both factors will be used at the same time. This yields a multiplier on short-circuit current of 1.56 (125% x 125%).

The *NEC* also requires that the ampacity of conductors be derated for the operating temperature of the conductor. This is a requirement because the ampacity of cables is given for cables operating in an ambient temperature of 30°C. In PV systems, cables are operated in an outdoor environment and should be subjected at least to a temperature derating due to an ambient temperature of 40°C to 45°C. PV modules operate at high temperatures and, in some installations, may be over 75°C. Concentrating modules operate at even higher temperatures. The temperatures in module junction boxes approach these temperatures. Conductors in free air that lie against the back of these modules are also exposed to these temperatures. These high temperatures require that the ampacity of cables be derated by factors of 0.33 to 0.58 depending on cable type, installation method (free air or conduit), and the temperature rating of the insulation [310.16, 310.17].

Cables in conduit where the conduit is exposed to the direct rays of the sun are also exposed to elevated operating temperatures.

Cables with insulation rated at 60°C have no ampacity at all when operated in environments with ambient temperatures over 55°C. This precludes their use in most PV systems. Cables with 75°C insulation have no ampacity when operated in ambient temperatures above 70°C. Because PV modules may operate at temperatures in the 45-75°C range, it is strongly suggested that only cables with an insulation rated at 90°C be used.

Summary

The conditions under which PV modules operate (high and low ambient temperatures, high and low winds, high and low levels of sunlight) and the electrical characteristics of those modules dictate that all of the requirements in the *NEC* be fully considered and applied.

There appears to be little question that the temperature-dependent correction factor on voltage is necessary in any location where the ambient temperature drops below 25°C. Even though the PV system can provide little current under open-circuit voltage conditions, these high voltages can damage electronic equipment and stress conductors and other equipment by exceeding their voltage breakdown ratings.

In ambient temperatures from 25 to 40°C and above, module short-circuit currents are increased at the same time conductors are being subjected to higher operating temperatures. Irradiance values over the standard rating condition may occur every day. Therefore the *NEC* requirements for adjusting the short-circuit current are necessary to ensure a safe and long-lived system.

APPENDIX C: Utility-Interactive Systems

Utility-interactive (grid-connected) systems present some unique challenges for the PV designer and installer in meeting the NEC.

Inverters

Utility-interactive inverters that connected to the utility grid should meet the requirements established by UL *Standard 1741* and be so listed. Some of the larger inverters cannot have both the dc PV circuits and the ac output circuits grounded as required by code without causing operational and functional problems. These units require an external ac isolation transformer. Newer versions of these inverters may have solutions for this problem, and the *2005 NEC* will allow ungrounded PV systems as are used in Europe.

UTILITY CONNECTION

NEC part 690 VII and section 690.64 provide some detailed requirements for connecting the utility-interactive inverter to the utility. Most are relatively clear. However, 690.64(B)(2) needs elaboration. Consider the diagram of a backfed commercial load center shown in Figure C1.

In this figure, a 400-amp load center has a 400-amp main breaker (a common arrangement where the main breaker is sized the same as the load center rating). The maximum continuous loads on the load center, in a properly designed system, should not exceed 320 amps (80% of the main breaker rating). Although the sum of the rating of the circuit breakers *supplying loads* connected to the panel will usually significantly exceed the rating of the panel, the actual loads should be less than 320 amps. Otherwise, the main breaker would trip on the overloads, thereby protecting the load center and the feeder.

A utility-interactive PV system is connected to this panel through a 100-amp backfed circuit breaker installed as shown at or near the top of the load center. As long as the loads on the system do not exceed 320 amps, no problems exist as far as safety.

However, at some later date, the loads on the load center may increase. This may be due to added circuits, which should be installed by an electrician or by just increasing the existing plug loads. For example, office modules with outlets may be added with high desktop publishing loads. If the extra loads are present only during the daytime, the main breaker will not trip since the PV system will be picking up the excess loads. However, the bus bar in the load center at point B will be carrying more than its 400-amp rating. Up to 400 amps can be supplied to the load center through the main breaker

and up to 100 amps can be supplied through the backfed PV breaker. This current of up to 500 amps will cause excess heating of the bus bar. It may cause nuisance tripping of breakers in the load center and may also result in premature failure of the load center or the circuit breakers. No circuit breakers will be overloaded, none will trip, and no one will be alerted to the problem. In any event, the load center is being used in a manner for which it was not designed. In fact, NEC requirements generally dictate that the load center bus bars will not be required to handle more than 320 amps on a continuous basis.

Using the requirement of 690.64(B)(2) ensures that the currents being supplied by the PV system (as limited by the backfed breaker) plus the currents supplied by the utility (as limited by the main circuit breaker) will not exceed the rating of the load center. In commercial installations, each feeder panel subject to backfed PV currents **must** meet the requirements of 690.64(B)(2).

In commercial installations, the requirements of 690.64(B)(2) may be met in several ways.

1. If the existing load center is fully loaded (i.e. 320 amps on a 400-amp load center), then the load center may be replaced with a larger unit (i.e. 600 amps) with a smaller main breaker (i.e. 400 amps). This will leave 200 amps of capacity for backfeeding PV.

2. When an analysis of the electrical system reveals that several load centers (feeder panels) in the building need to be replaced/upgraded to handle backfed PV currents, it is usually easier to install a second service entrance on the building. In this case the service entrance conductors between the utility meter and the primary service disconnect can be tapped and

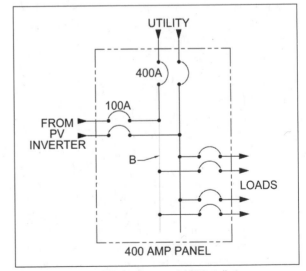

Figure C-1. 400 Amp Panel – Commercial PV Installation

routed to a dedicated disconnect which serves as a second service entrance just for the PV system.

PV Source-Circuit Conductors

Some older grid-tied inverters operate with PV arrays that are center tapped and have cold-temperature open-circuit voltages of ±325 volts and above. The system voltage of 650 volts or greater exceeds the insulation rating of the commonly available 600-volt insulated conductors. Each disconnect and overcurrent device and the insulation of the wiring **must** have a voltage rating exceeding the system voltage rating [110.3(B)]. Type G and W cables are available with the higher voltage ratings, but are flexible cords and do not meet *NEC* requirements for fixed installations. Cables suitable for *NEC* installations requiring insulation greater than 600 volts are available (Appendix A).

Other older inverters have been designed to operate on systems with open-circuit voltages exceeding ±540 volts requiring conductors with 2000-volt or higher insulation. See Appendix D for a full discussion of this area.

Overcurrent Devices

When UL tests and lists fuses for dc operation, the voltage rating is frequently one-half the ac voltage rating. This results in a 600-volt ac fuse rated for 300-volt dc. Fuses with high enough dc ratings for grid systems operating at ±300 volts or 600 volts to ground (600-volt system voltage) need to be carefully selected. There are a number of listed, dc-rated 600-volt fuses available. See Appendix A.

Backfed Circuit Breakers, The *National Electrical Code* and UL Standards

UTILITY-INTERACTIVE PV SYSTEMS

1. Section 690.64(B)(5) of the *National Electrical Code (NEC)* requires that backfed circuit breakers be identified for the use.

   Underwriters Laboratories (UL) standards indicate that any circuit breaker that is not marked "Line" and "Load" is identified as suitable for backfeeding. Most circuit breakers used in residential and commercial load centers are not marked "Line" and "Load" and are suitable for backfeeding.

   It should be noted that when the ac output of a utility-interactive inverter is connected to a circuit breaker, the current/power flow through the breaker is indeed backward. The closed breaker, connected in only one of the current-carrying conductors, is not affected by which way the ac power or current is flowing. However, when a fault occurs in this circuit, it will be grid current flowing through the breaker in the forward direction toward a fault in the inverter side that causes the breaker to trip. There is no reverse current or backfeeding of current in this breaker under fault conditions.

2. Section 408.16(F) of the *NEC* requires that "plug-on" backfed circuit breakers be clamped to the load center.

   This is certainly a valid requirement when the circuit breaker under discussion is a backfed main breaker. It would also be valid when the backfed breaker was connected to a voltage source such as a rotating generator. In both cases, pulling the circuit breaker from the load center bus bars could result in a energized surface (the plug-on contact) exposed on the breaker—either at grid voltage or voltage from the output of the generator. However, if an unqualified person has access the exposed panel, there is at least as great a hazard from the exposed bus bars and main lugs as from the possibly energized breaker contact.

   This requirement also originated in the days of exposed industrial panel boards that did not have dead fronts and where the plug-on breakers were easily accessed and pulled off without much thought.

3. In PV systems, where the backfed breaker is being fed by the output of a utility-interactive PV inverter identified and listed for such use, the situation changes substantially.

   IEEE Standards 929 and 1547 and UL Standard 1741 require that utility interactive inverters cease exporting power within 0.1 seconds upon loss of ac utility voltage (voltage below 50% of nominal). This means that when a backfed breaker from a PV utility-interactive inverter is pulled off of a load center bus bar, the breaker essentially becomes completely de-energized in a fraction of a second; probably before it is moved more that a small fraction of an inch away from the bus bars. There is no electric shock hazard from either terminal on the circuit breaker after it has been disconnected from the bus bar.

   Furthermore all currently available load centers, both residential and commercial, have dead front covers that are fastened with one to four or more screws. This very effectively clamps all internal circuit breakers to the bus bars. Although not explicitly stated, there is an implicit "rule" in the code that a tool **must** be used to gain access to energized circuits. This applies nearly universally from the screw-cover on an ac receptacle outlet or ac wall switch to the screw covers on terminal boxes and termination boxes for transformers, motors and other equipment.

   If the unqualified or qualified person gains access by removing the clamping front cover on a load center with backfed breakers, the exposed main lugs and the exposed bus bars pose greater immediate shock hazards than the backfed breaker which has not yet even been unplugged.

SUMMARY

There appears to be no safety hazard (either shock or fire) that would require backfed circuit breakers connected to the output of utility-interactive PV inverters to be clamped to the load center bus bars.

Disconnects

Many utility-interactive inverters operate at dc PV voltages in the 250-600 volt range and these voltages preclude the use of a circuit breaker as a dc disconnect. In most cases, a fused or unfused safety switch is used as a disconnect. These safety switches normally require that two of the three poles be wired in series to achieve the 600-volt dc rating and these two poles are then used to open the ungrounded positive PV conductor. This requirements dictate that one switch be used for each inverter or for each string of PV modules.

In the smaller inverters (up to about 3.5-4 kW), the dc currents are in the 10-15 amp range. Square D has obtained a special listing on their three-pole, 600-volt fused (H361) and unfused (HU361) Heavy Duty Safety Switches that allows them to be used on PV systems with only one switch pole per string of PV modules or one switch pole per inverter where the maximum currents are less than about 18 amps (rated short-circuit currents less than 11 amps).

Blocking Diodes

The *NEC* does not require blocking diodes. The language of the code simply allows their use, which is rapidly declining. The use of the required overcurrent device in each series string of modules provides the necessary reverse-current protection.

Blocking diodes are not overcurrent devices. They block reverse currents in direct-current circuits and help to control circulating ground-fault currents if used in both ends of high-voltage strings. Lightning induced surges are tough on diodes. If isolated case diodes are used, at least 3500 volts of insulation is provided between the active elements and the normally grounded heat sink. Choosing a peak reverse voltage as high as is available but at least twice the PV open-circuit voltage will result in longer diode life. Substantial amounts of surge suppression will also improve diode longevity.

Blocking diodes may not be substituted for the UL-*1703* requirement for module protective fuses in each series-connected string of modules.

Surge Suppression

Surge suppression is covered only lightly in the *NEC* because it affects system performance more than safety. Surges are a utility problem at the transmission line level in ac systems [280]. PV arrays mounted in the open, on the tops of buildings, act like lightning rods. The PV designer and installer should provide appropriate means to deal with lightning-induced surges coming into the system.

Array frame grounding conductors should be routed directly to supplementary ground rods located as near as possible to the arrays [250.54].

Metal conduit will add inductance to the array-to-building conductors and slow down any induced surges as well as provide some electromagnetic shielding.

Metal oxide varistors (MOV) commonly used as surge suppression devices on electronic equipment have several deficiencies. They draw a small amount of current continually. The clamping voltage lowers as they age and may reach the open-circuit voltage of the system. When they fail, they fail in the shorted mode, heat up, and frequently explode or catch fire. In many installations, the MOVs are protected with fast acting fuses to prevent further damage when they fail, but this may limit their effectiveness as surge suppression devices. Other electronic devices are becoming available that do not change performance characteristics as they age or are subjected to surges.

Several companies specialize in lightning protection equipment, but much of it is for ac systems. Electronic product directories, such as the *Electronic Engineers Master Catalog* should be consulted.

APPENDIX D: Cable and Device Ratings at High Voltages

There is a concern in designing PV systems that have system open-circuit voltages above 600 volts. The concern has two main issues—device ratings and *NEC* limitations.

Equipment Ratings

Some discontinued, out of production, utility-intertie inverters operate with a grounded, bipolar (three-wire) PV array. In a bipolar PV system, where each of the monopoles is operated in the 220-235-volt peak-power range, the open-circuit voltage can be anywhere from 290 to 380 volts, depending on the module characteristics such as fill-factor. Such a bipolar system can be described as a 350/700-volt system (for example) in the same manner that a 120/240-volt ac system is described. This method of describing the system voltage is consistent throughout the electrical codes used not only in residential and commercial power systems, but also in utility practice.

In all systems, the voltage ratings of the cable, switchgear, and overcurrent devices are based on the higher number of the pair (i.e., 700 volts in a 350/700-volt system). That is why 250-volt switchgear and overcurrent devices are used in 120/240-volt ac systems and 600-volt switchgear is used in systems such as the 277/480-volt ac system. Note that it is not the voltage to ground, but the higher line-to-line voltage that defines the equipment voltage requirements.

The *National Electrical Code* (*NEC*) defines a nominal voltage for ac systems (120, 240, etc.) and acknowledges that some variation can be expected around that nominal voltage. Such a variation around a nominal voltage is not considered in dc PV systems, and the *NEC* requires that a temperature-related connection factor on the open-circuit array voltage **must** be used [690.7(A)]. The open-circuit voltage is defined at Standard Test Conditions (STC) because of the relationship between the UL Standards and the way the *NEC* is written. The *NEC* Handbook elaborates on the definition of "circuit voltage," but this definition may not apply to current-limited dc systems. Section 690.7(A) of the *NEC* requires that the voltage used for establishing dc circuit requirements in PV systems be the computed open-circuit voltage for crystalline PV technologies. In new thin-film PV technologies, open-circuit voltages are determined from manufacturers' specifications for temperature coefficients.

The *2002 NEC* specifically defines the PV system voltage as the product of a temperature-dependent factor (that may reach 1.25 at –40°C) and the STC open-circuit voltage [690.7]. The systems voltage is also defined as the highest voltage between any two wires in a 3-wire (bipolar) PV system [690.2].

The comparison to ac systems can be carried too far; there are differences. For example, the typical wall switch in a 120/240-volt ac residential or commercial system is rated at only 120 volts, but such a switch in a 120/240-volt dc PV system would have to be rated at 240 volts. The inherent differences between a dc current source (PV modules) and a voltage source (ac grid) bear on this issue. Even the definitions of circuit voltage in the *NEC* and *NEC* Handbook refer to ac and dc systems, but do not take into account the design of the balance of systems required in current-limited PV systems. In a PV system, all wiring, disconnects, and overcurrent devices have current ratings that exceed the short-circuit currents by at least 25%. In the case of bolted faults or ground faults involving currents from the PV array, the overcurrent devices do not trip because they are rated to withstand continuous operation at levels above the fault levels. In an ac system, bolted faults and ground faults generally cause the overcurrent devices to trip or blow removing the source of voltage from the fault. Therefore, the faults that pose high-voltage problems in PV, dc systems cause the voltage to be removed in ac, grid-supply systems. For these reasons, a switch rated at 120 volts can be used in an ac system with voltages up to 240 volts, but in a dc, PV system, the switch would have to be rated at 240 volts.

Another consideration that we are dealing with is the analogy of dc supply circuit and ac load circuits. An analysis of ac supply circuits would be similar to dc supply circuits.

Underwriters Laboratories (UL) Standard 1703 requires that manufacturers of modules listed to the standard include, in the installation instructions, a statement that the open-circuit voltage should be multiplied by 125% (crystalline cells), further increasing the voltage requirement of the balance-of-systems (BOS) equipment. This requirement has been in the *NEC* Section 690.7 as a temperature-dependent constant since the 1999 edition of the Code.

Current PV modules that are listed to the UL *Standard 1703* are listed with a maximum system voltage of 600 volts. A few are listed to 1000 volts to meet European standards. Engineers caution all installers, factory and otherwise, to not exceed this voltage. This restriction is not modified by the fact that the modules undergo high-pot tests at higher voltages.

Although not explicitly stated by the *NEC*, it is evident that the intent of the Code and the UL Standards is

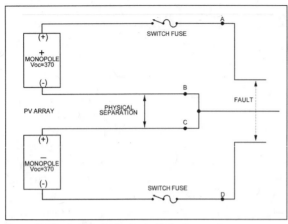

Figure D-1. Typical Bipolar System with Fault

that all cable, switches, fuses, circuit breakers, and modules in a PV system be rated for the maximum system voltage. This is clarified in the 1999 *NEC* [690.7(A)].

While reducing the potential for line-to-line faults, the practice of wiring each monopole (one of two electrical source circuits) in a separate conduit to the inverter does not eliminate the problem. Consider the bipolar system presented in Figure D-1 with a bolted fault (or deliberate short) from the negative to the positive array conductor at the input of the inverter. With the switches closed, array short-circuit current flows, and neither fuse opens.

Now consider what happens in any of the following cases.

1. A switch is opened
2. A fuse opens
3. A wire comes loose in a module junction box
4. An intercell connection opens or develops high resistance

5. A conductor fails at any point

In any of these cases, the entire array voltage (740 volts) stresses the device where the circuit opens. This voltage (somewhere between zero at short-circuit and the array open-circuit voltage) will appear at the device or cable. As the device starts to fail, the current through it goes from I_{sc} to zero as the voltage across the device goes from zero to V_{oc}. This process is very conducive to sustained arcs and heating damage.

Separating the monopoles does not avoid the high-voltage stress on any component, but it does help to minimize the potential for some faults. There are other possibilities for faults that will also place the same total voltage on various components in the system. An improperly installed grounding conductor coupled with a module ground fault could result in similar problems.

Section 690.5 of the *NEC* requires a ground-fault device on PV systems that are installed on the roofs of dwellings. This device, used for fire protection, **must** detect the fault, interrupt the fault current, indicate the fault, and disconnect the array.

Some large (100 kW) utility-interactive PV systems like the one at Juana Diaz, Puerto Rico have inverters that, when shut down, crowbar the array. The array remains crowbarred until the ac power is shut off and creates a similar fault to the one pictured in Figure D-1.

NEC Voltage Limitation

The second issue associated with this concern is that the *NEC* in Section 690.7(C) only allows PV installations up to 600 volts in one and two-family dwellings. Inverter and system design issues may favor higher system voltage levels.

Voltage Remedies

System designers can select inverters with lower operating and open-circuit voltages. Utility-intertie inverters are available with dc input voltages as low as 24 volts. The system designer also can work with the manufacturers of higher voltage inverters to reduce the number of modules in each series string to the point where the cold-temperature open-circuit voltage is less than 600 volts. The peak-power voltage would also be lowered. Transformers may be needed to raise the inverter ac output voltage to the required level. All utility-interactive inverters listed in the US operate with PV arrays that have open-circuit voltages of less than 600 volts.

Cable manufacturers produce UL-Listed, cross-linked polyethylene, single-conductor cable. It is marked USE-2/RHW-2, Sunlight Resistant and is rated at 2000 volts. This cable could be used for module interconnections in conduit after all of the other *NEC* requirements are met for installations above 600 volts.

Several manufacturers issue factory certified rating on their three-pole disconnects to allow higher voltage, non-load break operation with series-connected poles. The *NEC* will require an acceptable method of obtaining non-load break operation.

Some OEM circuit breaker manufacturers will factory certify series-connected poles on their circuit breakers. Units have been used at 750 volts and 100 amps with 10,000 amps of interrupt rating. Higher voltages may be available.

High-voltage industrial fuses are available, but dc ratings are unknown at this time.

Individual 600-volt terminal blocks can be used with the proper spacing for higher voltages.

Module manufacturers can have their modules listed for higher system voltages. Most are currently limited to 600 volts.

Power diodes may be connected across each monopole. When a bolted line-to-line fault occurs, one of the diodes will be forward biased when a switch or fuse opens, thereby preventing the voltage from one monopole from adding to that of the other monopole. The diodes are mounted across points A-B and C-D in Figure D-1. Each diode should be rated for at least the system open-circuit voltage and the full short-circuit current from one monopole. Since diodes are not listed as over-voltage protection devices, this solution is not recognized in the *NEC*.

The *NEC* allows PV installations over 600 volts in non-residential applications, which will cover the voltage range being used in most current designs.

It should be noted that there are numerous requirements throughout the *NEC* that apply specifically to installations over 600 volts:

- All equipment **must** be listed for the maximum voltage.
- Clearance distances and mechanisms for achieving that clearance are significantly more stringent as voltages increase above 600 volts.

Section 690.7(E) allows specially configured and listed inverters to be used in a system where the voltages are measured line-to-ground rather than line.

APPENDIX E: Example Systems

The systems described in this appendix and the calculations shown are presented as examples only. The calculations for conductor sizes and the ratings of overcurrent devices are based on the requirements of the *National Electrical Code* (*NEC*) and on UL *Standard 1703* which provides instructions for the installation of UL-Listed PV modules. Local codes and site-specific variations in irradiance, temperature, and module mounting, as well as other installation particularities, dictate that these examples should not be used without further refinement. Tables 310.16 and 310.17 from the *NEC* provide the ampacity data and temperature derating factors.

Cable Sizing and Overcurrent Protection

The procedure presented below for cable sizing and overcurrent protection of that cable is based on *NEC* requirements in Sections 690.9, 690.8, 110.14(C), 210.20(A), 215.2, 215.3, 220.10, 240.3(B), and 240.6(A). See Appendix I for a slightly different method of making ampacity calculations based on the same requirements.

1. **Circuit Current.** For circuits carrying currents from PV modules, multiply the short-circuit current by 125% and use this value for all further calculations. For PV circuits in the following examples, this is called the CONTINUOUS CURRENT calculation. In the *2002 NEC*, this requirement has been included in Section 690.8, but also remains in UL 1703. This multiplier should not be applied twice. For dc and ac inverter circuits in PV systems, use the rated continuous currents. AC and dc load circuits should follow the requirements of Sections 210, 220, and 215.

2. **Overcurrent Device Rating.** The overcurrent device **must** be rated at 125% of the current determined in Step 1. This is to prevent overcurrent devices from being operated at more than 80% of rating. This calculation, in the following examples, is called the 80% OPERATION.

3. **Cable Sizing.** Cables **shall** have a 30°C ampacity of 125% of the current determined in Step 1 to ensure proper operation of connected overcurrent devices. There are no additional deratings applied with this calculation.

4. **Cable Derating.** Based on the determination of Step 3 and the location of the cable (raceway or free-air), a cable size and insulation temperature rating (60, 75, or 90°C) are selected from the *NEC* Ampacity Tables 310.16 or 310.17. Use the 75°C cable ampacities to get the size, then use the ampacity from the 90°C col-

umn—if needed—for the deratings. This cable is then derated for temperature, conduit fill, and other requirements. The resulting derated ampacity **must** be greater than the value found in Step 1. If not greater, then a larger cable size or higher insulation temperature **must** be selected. The current in Step 3 is not used at this point to preclude over sizing the cables.

5. **Ampacity vs. Overcurrent Device.** The derated ampacity of the cable selected in Step 4, **must** be equal to or greater than the overcurrent device rating determined in Step 2 [240.4]. If the derated ampacity of the cable is less than the rating of the overcurrent device, then a larger cable **must** be selected. The next larger standard size overcurrent device may be used if the derated cable ampacity falls between the standard overcurrent device sizes found in *NEC* Section 240.6.

Note: This step may result in a larger conductor size than that determined in Step 4.

---

6. **Device Terminal Compatibility.** Since most overcurrent devices have terminals rated for use with 75°C (or 60°C) cables, compatibility **must** be verified [110.3(B)]. If a 90°C-insulated cable was selected in the above process, the 30°C ampacity of the same size cable with a 75°C (or 60°C) insulation **must** be greater than or equal to the current found in Step 1 [110.14(C)]. This ensures that the cable will operate at temperatures below the temperature rating of the terminals of the overcurrent device. If the overcurrent device is located in an area with ambient temperature higher than 30°C, then the 75°C (or 60°C) ampacity **must** also be derated [110.3(B)].

Here is an example of how the procedure is used:

The task is to size and protect two PV source circuits in conduit, each with an $I_{sc} = 40$ amps. Four current-carrying conductors are in the conduit and are operating in a 45°C ambient temperature.

Step 1: 1.25 x 40 = 50 amps. (continuous current)

Step 2: The required fuse (with 75°C terminals) is 1.25 x 50 = 62.5 amps. The next standard fuse size is 70 amps. (ensures operation below 80% of rating)

Step 3: Same calculation as Step 2. Cable ampacity without deratings must be 62.5 amps.

Step 4: From Table 310.16, cables with 75°C insulation: A 6 AWG conductor at 65 amps is needed. This meets Step 3 requirements. At least a 6 AWG XHHW-2 cable with 90°C insulation and a 30°C ampacity of 75 amps should be installed. Conduit fill derating is

0.8 and temperature derating is 0.87. Derated ampacity is 52.2 amps (75 x 0.8 x 0.87). This is greater than the required 50 amps in Step 1 and meets the requirement.

Step 5: It is acceptable to protect a cable with a derated ampacity of 52.2 amps with a 60-amp overcurrent device since this is the next larger standard size. However, this circuit requires at least a 62.5 amp device (Step 2). Therefore, the conductor **must** be increased to a 4 AWG conductor with a derated ampacity of 66 amps (95 x 0.87 x 0.8). A 70-amp fuse or circuit breaker is acceptable to protect this cable since it is the next larger standard size.

Step 6. The ampacity of a 4 AWG cable with 75°C insulation (because the fuse has 75°C terminals) operating at 45°C is 70 amps (85 x .82), and is higher than the calculated continuous circuit current of 50 amps found in Step 1. Using the 75°C column in Table 310.16 or 310.17 for starting Step 4 usually ensures that this check will be passed.

Examples

EXAMPLE 1 Direct-Connected Water Pumping System

Array Size: 4, 12-volt, 60-watt modules; I_{sc} = 3.8 amps, V_{oc} = 21.1 volts
Load: 12-volt, 10-amp motor

Description

The modules are mounted on a tracker and connected in parallel. The modules are wired as shown in Figure E-1 with 10 AWG USE-2 single-conductor cable. A loop is placed in the cable to allow for tracker motion without straining the rather stiff building cable. The USE-2 cable is run to a disconnect switch in an enclosure mounted on the pole. From this disconnect enclosure, 8 AWG XHHW-2 cable in electrical nonmetallic conduit is routed to the wellhead. The conduit is buried 18 inches deep. The 8 AWG cable is used to minimize voltage drop.

The *NEC* requires the disconnect switch. Because the PV modules are current limited and all conductors have an ampacity greater than the maximum output of the PV modules, no overcurrent device is required, although some inspectors might require it and it might serve to provide some degree of lightning protection. A dc-rated disconnect switch or a dc-rated fused disconnect **must** be used. Since the system is ungrounded, a two-pole switch **must** be used. All module frames, the pole, the disconnect enclosure, and the pump housing **must** be grounded with equipment-grounding conductors, whether the system is grounded or not.

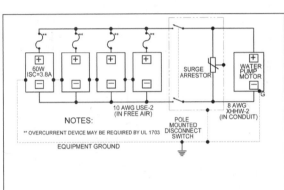

Figure E-1. Direct Connected System

The fuses shown connected to each module are required to protect the module from reverse currents from all sources. In this system, the only sources of potential reverse currents for an individual module are the modules connected in parallel. Those other three (out of four modules) could source 3 x 3.8 x 1.25=14.25 amps of current into a fault in a single module. If the module series protective fuse were 15 amps or less, these fuses would not be required; the potential currents could not damage any module. Of course, the conductors to each module should also

Calculations

The array short-circuit current is 15.2 amps (4 x 3.8).

CONTINUOUS CURRENT: 1.25 x 15.2.= 19 amps (Step 1)

No fuse, no Step 2

80% OPERATION: 1.25 x 19 = 23.75 amps (Step 3)

The ampacity of 10 AWG USE-2 at 30°C is 55 amps.

The ampacity at 61-70°C is 31.9 amps (0.58 x 55) which is more than the 19 amp requirement. (Step 4)

The equipment grounding conductors should be 10 AWG (typically 1.25 I_{sc} with a 14 AWG minimum).

The minimum voltage rating of all components is 26 volts (1.25 x 21.1).

EXAMPLE 2 Water Pumping System with Current Booster

Array Size: 10, 12-volt, 53-watt modules; I_{sc} = 3.4 amps, V_{oc} = 21.7 volts
Current Booster Output: 90 amps
Load: 12-volt, 40-amp motor

Description

This system has a current booster before the water pump and has more modules than in Example 1. Initially, 8 AWG USE-2 cable was chosen for the array connections, but a smaller cable was chosen to attach to the module terminals. As the calculations below show, the array was split into two subarrays. There is potential for malfunction in the current booster, but it does not seem possible that excess current can be fed back into the array wiring, since there is no other source of energy in the system. Therefore, these conductors do not need overcurrent devices if they are sized for the entire array current. If smaller conductors are used, then overcurrent devices will be needed.

However, there are now 10 modules in parallel connected via the two 30-amp circuit breakers. The potential reverse current from 9 modules would be 9 x 3.4 x 1.25 = 38.25 amps. This is well in excess of the ability of the module to handle reverse currents (possibly as low as 10 amps), so a fuse or circuit breaker must be used in series with each module. The minimum value would be 1.56 x 3.4 amps = 5.3 amps and the next higher standard value is 6 amps. These fuses would normally be contained in a PV combiner mounted in the shade behind the PV array.

Since the array is broken into two subarrays, the maximum short-circuit current available in either sub-

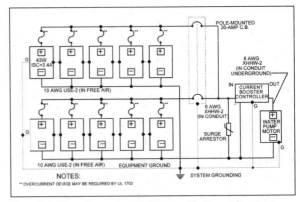

Figure E-2. Direct-Connected PV System with Current Booster

array wiring is equal to the subarray short-circuit current under fault conditions plus any current coming back through one of the 30-amp breakers form the other subarray. Overcurrent devices are needed to protect the subarray conductors under these conditions.

A grounded system is selected, and only single-pole disconnects are required. Equipment grounding and system grounding conductors are shown in Figure E-2

If the current booster output conductors are sized to carry the maximum current (3-hour) of the booster, then overcurrent devices are not necessary, but again, some inspectors may require them.

Calculations

The entire array short-circuit current is 34 amps (10 x 3.4).

CONTINUOUS CURRENT: 1.25 x 34 = 42.5 amps

80% OPERATION: 1.25 x 42.5 = 53.1 amps

The ampacity of 6 AWG USE-2 cable at 30°C in conduit is 75 amps.

The ampacity at 45°C (maximum ambient air temperature) is 65.5 amps (0.87 x 75), which is greater than the 42.5 amp requirement; so a single array could have been used. However, the array is split into two subarrays for serviceability. Each is wired with 10 AWG USE-2 conductors.

The subarray short-circuit current is 17 amps (5 x 3.4).

CONTINUOUS CURRENT: 1.25 x 17 = 21.3 amps

80% OPERATION: 1.25 x 21.25 = 26.6 amps

The ampacity of 10 AWG USE-2 at 30°C in free air is 55 amps.

The ampacity at 61-70°C (module operating temperature) is 31.9 amps (0.58 x 55), which is more than the 21.3 amp requirement. Since this cable is to be connected to an overcurrent device with terminals rated at 60°C or 75°C, the ampacity of the cable **must** be evaluated with 60°C or 75°C insulation. Overcurrent devices rated at 100 amps or less may have terminals rated at only 60°C. The ampacity of 10 AWG 75°C cable operating at 30°C is 35 amps, which is more than the 21.3 amps requirement. Therefore, there are no problems with the terminals on a 75°C overcurrent device.

Thirty-amp circuit breakers are used to protect the 10 AWG subarray conductors. The required rating is 1.25 x 21.25 = 26.6 amps, and the next largest size is 30 amps. Note: The maximum allowed overcurrent device for a 10AWG conductor is 30 amps [240.4(D)].

The current booster maximum current is 90 amps.

The current booster average long-term (3-hours or longer) current is 40 amps (continuous current).

80% OPERATION: 1.25 x 40 = 50 amps

The ampacity of 8 AWG XHHW-2 at 30°C in conduit is 55 amps.

The ampacity does not need temperature correction since the conduit is buried in the ground. The ampacity requirements are met, but the cable size may not meet the overcurrent device connection requirements when an overcurrent device is used.

The 6 AWG conductors are connected to the output of the circuit breakers, and there is a possibility that heating of the breaker may occur. It is therefore good practice to make the calculation for terminal overheating. The ampacity of a 6 AWG conductor evaluated with 75°C insulation (the rated maximum temperature of the terminals on the overcurrent device) is 65 amps, which is greater than the 27-amp requirement. This means that the overcurrent device will not be subjected to overheating when the 6 AWG conductor carries 27 amps.

All equipment-grounding conductors should be 10 AWG. The grounding electrode conductor should be 8 AWG or larger.

Minimum voltage rating of all components: 1.25 x 21.7 = 27 volts

EXAMPLE 3 Stand-Alone Lighting System

Array Size: 4, 12-volt, 64-watt modules; I_{sc} = 4.0 amps, V_{oc} = 21.3 volts

Batteries: 200-amp-hours at 24 volts

Load: 60 watts at 24 volts

Description

The modules are mounted at the top of a 20-foot pole with the metal-halide lamp. The modules are connected in series and parallel to achieve the 24-volt system rating. The lamp, with an electronic ballast and timer/controller, draws 60 watts at 24 volts. The batteries, disconnect switches, charge controller, and overcurrent devices are mounted in a box at the bottom of the pole. The system is grounded as shown in Figure E-3.

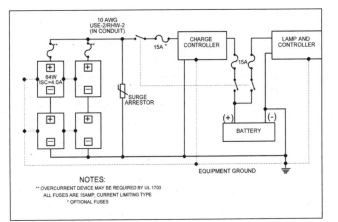

Figure E-3. Stand-Alone Lighting System

Calculations:

The array short-circuit current is 8 amps (2 x 4).

CONTINUOUS CURRENT: 1.25 x 8 = 10 amps

80% OPERATION: 1.25 x 10 = 12.5 amps

Load Current: 60/24 = 2.5 amps (continuous)

80% OPERATION: 1.25 x 2.5 = 3.1 amps

Cable size 10 AWG USE-2/RHW-2 is selected for module interconnections and is placed in conduit at the modules and then run down the inside of the pole.

The modules operate at 61-70°C, which requires that the module cables be temperature derated. Cable 10 AWG USE-2/RHW-2 has an ampacity of 40 amps at 30°C in conduit. The derating factor is 0.58. The temperature-derated ampacity is 23.2 amps (40 x 0.58), which exceeds the 10-amp requirement. Checking the cable with a 75°C insulation, the ampacity at the fuse end is 35 amps, which exceeds the 10-amp requirement. This cable can be protected by a 15-amp fuse or circuit breaker (125% of 10 is 12.5). An overcurrent device rated at 100 amps or less may only have terminals rated for 60°C, not the 75°C used in this example. Lower temperature calculations may be necessary.

The same USE-2/RHW-2, 10 AWG cable is selected for all other system wiring, because it has the necessary ampacity for each circuit.

A three-pole fused disconnect is selected to provide the PV and load disconnect functions and the necessary overcurrent protection. The fuse selected is a RK-5 type, providing current limiting in the battery circuits. A pullout fuse holder with either Class RK-5 or Class T fuses could also be used for a more compact installation. Fifteen amp fuses are selected to provide overcurrent protection for the 10 AWG cables. They are used in the load circuit and will not blow on any starting surges drawn by the lamp or controller. The 15-amp fuse before the charge controller could be eliminated since that circuit is protected by the fuse on the battery side of the charge controller. The disconnect switch at this location is required.

One of the two strings of PV modules could be subjected to reverse currents from the other string (1.25 x 4 = 5 amps) plus 15 amps from the battery through the 15-amp fuse. If this 20-amp potential backfed current exceeds the module series fuse requirement, then the string fuses and a PV combiner must be added to the system.

The equipment-grounding conductors should be 10 AWG conductors. An 8 AWG (minimum) conductor would be needed to the ground rod.

The dc voltage ratings for all components used in this system should be at least 53 volts (2 x 21.3 x 1.25).

EXAMPLE 4 Remote Cabin DC-Only System

Array Size: 6, 12-volt, 75-watt modules; I_{sc} = 4.8 amps, V_{oc} = 22 volts

Batteries: 700 amp hours at 12 volts

Load: 75 watts peak at 12-volts dc

Description

The modules are mounted on a rack on a hill behind the house. Nonmetallic conduit is used to run the cables from the junction box to the control panel. A control panel is mounted on the back porch, and the batteries are in an insulated box under the porch. All the loads are dc with a peak-combined power of 75 watts at 12 volts due, primarily, to a pressure pump on the gravity-fed water supply. The battery bank consists of four 350-amp-hour, 6-volt, deep-cycle batteries wired in series and parallel. Figure E-4 shows the system schematic.

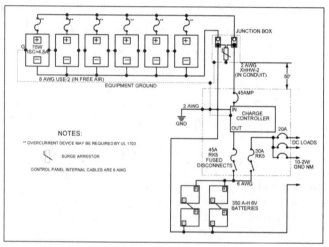

Figure E-4. Remote Cabin DC-Only System

Calculations

The array short-circuit current is 28.8 amps (6 x 4.8).

CONTINUOUS CURRENT: 1.25 x 28.8 = 36 amps

80% OPERATION: 1.25 x 36.0 = 45 amps

The module interconnect wiring and the wiring to a rack-mounted junction box will operate at 65°C. If USE-2 cable with 90°C insulation is chosen, then the temperature derating factor will be 0.58. The required ampacity of the cable at 30°C is 62 amps (36/0.58), which can be handled by 8 AWG cable with an ampacity of 80 amps in free air at 30°C. Conversely, the ampacity of the 8 AWG cable is 46.4 amps (80 x 0.58) at 65°C which exceeds the 36 amp requirement.

A PV combiner with a fuse for each module will be required because the available potential short-circuit current from these six modules in parallel plus the 45-amps from the PV disconnect circuit breaker will far exceed the maximum reverse current rating of a module.

From the rack-mounted junction box to the control panel, the conductors will be in conduit and exposed to 40°C temperatures. If XHHW-2 cable with a 90°C insulation is selected, the temperature derating factor is

0.91. The required ampacity of the cable at 30°C would be 36/0.91 = 39.6 amps in conduit. Cable size 8 AWG has an ampacity of 55 amps at 30°C in conduit. Conversely, the 8 AWG conductor has an ampacity of 50 amps (55 x 0.91) at 40°C in conduit that exceeds the 39.6 amp requirement at this temperature.

The 8 AWG cable, evaluated with a 75°C insulation, has an ampacity at 30°C of 50 amps, which is greater than the 36 amps that might flow through it on a daily basis.

The array is mounted 200 feet from the house, and the round trip cable length is 400 feet. A calculation of the voltage drop in 400 feet of 8 AWG cable operating at 36 amps (125% I_{sc}) is 0.778 ohms per 1000 feet x 400 / 1000 x 36 = 11.2 volts. This represents an excessive voltage drop on a 12-volt system, and the batteries cannot be effectively charged. Conductor size 2 AWG (with a voltage drop of 2.8 volts) was substituted; this substitution is acceptable for this installation. The conductor resistances are taken from Table 8 in Chapter 9 of the *NEC* and are given for conductors at 75°C.

The PV conductors are protected with a 45-amp (1.25 x 36) single-pole circuit breaker on this grounded system. The circuit breaker should be rated to accept 2 AWG conductors and have terminals rated for use with 75°C-insulated conductors.

Cable size 6 AWG THHN cable is used in the control center and has an ampacity of 65 amps at 30°C when evaluated with 75°C insulation. Wire size 2 AWG from the negative dc input is used to the point where the grounding electrode conductor is attached instead of the 6 AWG conductor used elsewhere to comply with grounding requirements.

The 75-watt peak load draws about 6.25 amps and 10-2 with ground (w/gnd) nonmetallic sheathed cable (type NM) was used to wire the cabin for the pump and a few lights. DC-rated circuit breakers rated at 20 amps were used to protect the load wiring, which is in excess of the peak load current of 7.8 amps (1.25 x 6.25) and less than the cable ampacity of 30 amps.

Current-limiting fuses in a fused disconnect are used to protect the dc-rated circuit breakers, which may not have an interrupt rating sufficient to withstand the short-circuit currents from the battery under fault conditions. RK-5 fuses were chosen with a 45-amp rating in the charge circuit and a 30-amp rating in the load circuit. The fused disconnect also provides a disconnect for the battery from the charge controller and the dc load center.

The equipment grounding conductors should be 10 AWG and the grounding electrode conductor should be 2 AWG. A smaller grounding electrode conductor (as small as 8 AWG) may be acceptable to the local inspector.

All components should have a voltage rating of at least 1.25 x 22 = 27.5 volts.

EXAMPLE 5 Small Residential Stand-Alone System

Array Size: 10, 12-volt, 51-watt modules; I_{sc} = 3.25 amps, V_{oc} = 20.7 volts

Batteries: 800 amp-hours at 12 volts

Loads: 5 amps dc and 500-watt inverter with 90% efficiency

Description

The PV modules are mounted on the roof. Single-conductor cables are used to connect the modules to a roof-mounted junction box. Potential reverse fault currents indicate that a PV combiner be used with a series fuse for each PV module. UF two-conductor sheathed cable is used from the roof to the control center. Physical protection (wood barriers or conduit) for the UF cable is used where required. The control center, diagrammed in Figure E-5, contains disconnect and overcurrent devices for the PV array, the batteries, the inverter, and the charge-controller.

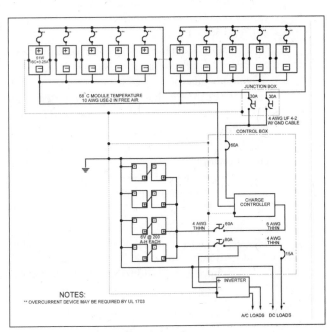

Figure E-5. Small Residential Stand-Alone System

Calculations

The module short-circuit current is 3.25 amps.

CONTINUOUS CURRENT: 1.25 x 3.25 = 4.06 amps

80% OPERATION: 1.25 x 4.06 = 5.08 amps per module

The maximum estimated module operating temperature is 68°C.

From *NEC* Table 310.17:

The derating factor for USE-2 cable is 0.58 at 61-70°C.

Cable 14 AWG has an ampacity at 68°C of 20.3 amps (0.58 x 35) (max fuse is 15 amps—see notes at bottom of Tables 310-16 & 17).

Cable 12 AWG has an ampacity at 68°C of 23.2 amps (0.58 x 40) (max fuse is 20 amps).

Cable 10 AWG has an ampacity at 68°C of 31.9 amps (0.58 x 55) (max fuse is 30 amps).

Cable 8 AWG has an ampacity at 68°C of 46.4 amps (0.58 x 80).

The array is divided into two five-module subarrays. The modules in each subarray are wired from module junction box to the PV combiner for that subarray and then to the array junction box. Cable size 10 AWG USE-2 is selected for this wiring, because it has an ampacity of 31.9 amps under these conditions, and the requirement for each subarray is 5 x 4.06 = 20.3 amps. Evaluated with 75°C insulation, a 10 AWG cable has an ampacity of 35 amps at 30°C, which is greater than the actual requirement of 20.3 amps (5 x 4.06) [Table 310.16 must be used]. In the array junction box on the roof, two 30-amp fuses in pullout holders are used to provide overcurrent protection for the 10 AWG conductors. These fuses meet the requirement of 25.4 amps (125% of 20.3) and have a rating less than the derated cable ampacity.

In this junction box, the two subarrays are combined into an array output. The ampacity requirement is 40.6 amps (10 x 4.06). A 4 AWG UF cable (4-2 w/gnd) is selected for the run to the control box. It operates in an ambient temperature of 40°C and has a temperature-corrected ampacity of 86 amps (95 x 0.91). This is a 60°C cable with 90°C conductors and the final ampacity must be restricted to the 60°C value of 70 amps, which is suitable in this example. Appropriately derated cables **must** be used when connecting to fuses that are rated for use only with 75°C conductors.

A 60-amp circuit breaker in the control box serves as the PV disconnect switch and overcurrent protection for the UF cable. The minimum rating would be 10 x 3.25 x 1.56 = 51 amps. The *NEC* allows the next larger size; in this case, 60 amps, which will protect the 70 amp rated cable. Two single-pole, pullout fuse holders are used for the battery disconnect. The charge circuit fuse is a 60-amp RK-5 type.

The inverter has a continuous rating of 500 watts at the lowest operating voltage of 10.75 volts and an efficiency of 90% at this power level. The continuous current calculation for the input circuit is 64.6 amps ((500 / 10.75 / 0.90) x 1.25).

The cables from the battery to the control center **must** meet the inverter requirements of 64.6 amps plus the dc load requirements of 6.25 amps (1.25 x 5). A 4 AWG THHN has an ampacity of 85 amps when placed in conduit and evaluated with 75°C insulation. This exceeds the requirements of 71 amps (64.6 + 6.25). This cable can be used in the custom power center and be run from the batteries to the inverter.

The discharge-circuit fuse **must** be rated at least 71 amps. An 80-amp fuse should be used, which is less than the cable ampacity.

The dc-load circuit is wired with 10 AWG NM cable (ampacity of 30 amps) and protected with a 15-amp circuit breaker.

The grounding electrode conductor is 4 AWG and is sized to match the largest conductor in the system, which is the array-to-control center wiring. This size would be appropriate for a concrete-encased grounding electrode.

Equipment-grounding conductors for the array and the charge circuit can be 10 AWG based on the 60-amp overcurrent devices. The equipment ground for the inverter **must** be an 8 AWG conductor based on the 80-amp overcurrent device. [Table 250.122]

All components should have at least a dc voltage rating of 1.25 x 20.7 = 26 volts.

EXAMPLE 6 Medium Sized Residential Hybrid System

Array Size: 40, 12-volt, 53-watt modules; I_{sc} = 3.4 amps, V_{oc} = 21.7 volts

Batteries: 1000 amp-hours at 24 volts

Generator: 6 kW, 240-volt ac

Loads: 15 amps dc and 4000-watt inverter, efficiency =.85

Description

The 40 modules (2120 watts STC rating) are mounted on the roof in five subarrays consisting of eight modules mounted on a single-axis tracker. The eight modules are wired in series and parallel for this 24-volt system. Five source circuits are routed to a custom power center. Single-conductor cables are used from the modules to roof-mounted PV combiners for each source circuit. The fuse for each series string of modules is rated at least 1.56 times the module I_{sc}, but less than or equal to the maximum module protective fuse marked on the back of the module. From the combiners, UF sheathed cable is run to the main power center.

Blocking diodes are not required or used to minimize voltage drops in the system.

A ground-fault protection device provides compliance with the requirements of *NEC* Section 690.5.

The charge controller is a relay type.

DC loads consist of a refrigerator, a freezer, several telephone devices, and two fluorescent lamps. The maximum combined current is 15 amps.

The 4000-watt sine-wave inverter supplies the rest of the house.

The 6-kW natural gas fueled, engine-driven generator provides back-up power and battery charging through the inverter. The 240-volt output of the generator is fed through a 5-kVA transformer to step it down to 120 volts for use in the inverter and the house. The transformer is protected on the primary winding by a 30-amp circuit breaker [450.3(B)]. Figure E-6 presents the details.

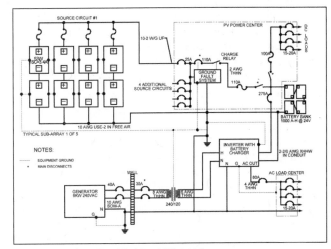

Figure E-6. Medium Sized Residential Hybrid System

Calculations

The subarray short-circuit current is 13.6 amps (4 x 3.4).

CONTINUOUS CURRENT: 1.25 x 13.6 = 17 amps

80% OPERATION: 1.25 x 17 = 21.25 amps

The temperature derating factor for USE-2 cable at 61-70°C is 0.58.

The ampacity of 10 AWG USE-2 cable at 70°C is 31.9 amps (55

x 0.58). [310.17]

The temperature derating factor for UF cable at 36-40°C is 0.91 for the 90°C conductors [310.16].

The ampacity of 10-2 w/gnd UF cable at 40°C is 36.4 amps (40 x 0.91), but is restricted to use with an overcurrent device of no more than 30 amps.

The source-circuit circuit breakers are rated at 25 amps (requirement is 125% of 17 amps = 21.25).

The PV array short-circuit current is 68 amps (5 x 13.6).

CONTINUOUS CURRENT: 1.25 x 68 = 85 amps

80% OPERATION: 1.25 x 85 = 106 amps

A 110-amp circuit breaker is used for the main PV disconnect after the five source circuits are combined.

A 110-amp RK5 current-limiting fuse is used in the charge circuit of the power center, which is wired with 2 AWG THHN conductors (115 amps with 75°C insulation).

The dc-load circuits are wired with 10-2 w/gnd NM cable (30 amps) and are protected with 20- or 30-amp circuit breakers. A 100-amp RK-5 fuse protects these breakers and the load circuits from excess current from the batteries.

Inverter

The inverter can produce 4000 watts ac at 22 volts with an efficiency of 85%.

The inverter input current ampacity requirements are 267 amps ((4000 / 22 / 0.85) x 1.25). See Appendix F for more details.

Two 2/0 AWG USE-2 cables are paralleled in conduit between the inverter and the batteries. The ampacity of this cable (rated with 75°C insulation) at 30°C is 280 amps (175 x 2 x 0.80). The 0.80 derating factor is required because there are four current-carrying cables in the conduit.

A 275-amp circuit breaker with a 25,000-amp interrupt rating is used between the battery and the inverter. Current-limiting fusing is not required in this circuit.

The output of the inverter can deliver 4000 watts ac (33 amps) in the inverting mode. It can also pass up to 60 amps through the inverter from the generator while in the battery charging mode.

Ampacity requirements, ac output: 60 x 1.25 = 75 amps. This reflects the *NEC* requirement that circuits are not to be operated continuously at more than 80% of rating.

The inverter is connected to the ac load center with 4 AWG THHN conductors in conduit, which have an ampacity of 85 amps when used at 30°C with 75°C overcurrent devices. An 80-amp circuit breaker is used near the inverter to provide a disconnect function and the overcurrent protection for this cable.

Generator

The 6-kW, 120/240-volt generator has internal circuit breakers rated at 27 amps (6500-watt peak rating). The *NEC* requires that the output conductors between the generator and the first field-installed overcurrent device be rated at least 115% of the nameplate rating ((6000 / 240) x 1.15 = 28.75 amps). Since the generator is connected through a receptacle outlet, a 10-4 AWG SOW-A portable cord (30 amps) is run to a NEMA 3R exterior circuit breaker housing. This circuit breaker is rated at 40 amps and provides overcurrent protection for the 8 AWG THHN conductors to the transformer. These conductors have an ampacity of 44 amps (50 x 0.88) at 40°C (75°C insulation rating). The circuit breaker also provides an exterior disconnect for the generator. Since the transformer isolates the generator conductors from the system electrical ground (separately derived system), the neutral of the generator is grounded at the exterior disconnect. The generator equipment-grounding conductors and grounding electrode conductor are bonded to the main system equipment-grounding conductors and the grounding electrode conductor.

A 30-amp circuit breaker is mounted near the PV Power Center in the ac line between the generator and the transformer. This circuit breaker serves as the interior ac disconnect for the generator and is grouped with the other disconnects in the system.

The output of the transformer is 120 volts. Using the rating of the generator, the ampacity of this cable **must** be 62.5 amps ((6000 / 120) x 1.25). A 6 AWG THHN conductor was used, which has an ampacity of 65 amps at 30°C (75°C insulation rating).

Grounding

The module and dc-load equipment grounds **must** be 10 AWG conductors. Additional lightning protection will be afforded if a 6 AWG or larger conductor is run from the array frames to ground. The inverter equipment-grounding conductor **must** be a 4 AWG conductor based on the size of the overcurrent device for this circuit. [250.122] The grounding electrode conductor **must** be 2-2/0 AWG or a 500 kcmil conductor, unless there are no other conductors connected to the grounding electrode and that electrode is a ground rod; then this conductor may be reduced to 6 AWG [250.50].

DC Voltage Rating

All dc circuits should have a voltage rating of at least 55 volts (1.25 x 2 x 22).

EXAMPLE 7 Rooftop Utility-interactive System

Array Size: 24, 50-volt, 240-watt modules;
$I_{sc} = 5.6$; $V_{oc} = 62$

Inverter: 200-volt nominal dc input

240-volt ac output at 5000 watts with an efficiency of 0.95.

Description

The rooftop array consists of six parallel-connected strings of four modules each. A PV combiner contains a fuse for each string of modules and a surge arrestor. All wiring is RHW-2 in conduit. The inverter is located adjacent to the service entrance load center where PV power is fed to the grid through a back-fed circuit breaker. Figure E-7 shows the system diagram.

Calculations

The string short-circuit current is 5.6 amps.

CONTINUOUS CURRENT: 1.25 x 5.6 = 7 amps

80% OPERATION: 1.27 x 7 = 8.75 amps

The array short-circuit current is 33.6 amps (6 x 5.6).

CONTINUOUS CURRENT: 1.25 x 33.6 = 42 amps

80% OPERATION: 1.25 x 42 = 52.5 amps

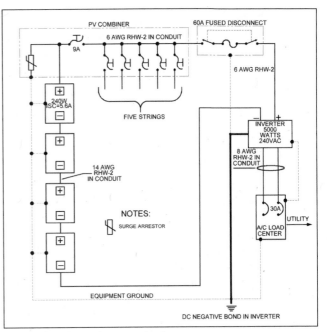

Figure E-7. Rooftop Utility-interactive System

The modules in each string are connected in series. The modules and attached conductors operate at 63°C. The temperature-derating factor for RHW-2 at this temperature is 0.58. The required 30°C ampacity for this cable is 15 amps (8.75 / 0.58). RHW-2 14 AWG cable has an ampacity of 25 amps with 90°C insulation and 20 amps with 75°C insulation so there is no problem with the end of the cable connected to the fuse (with 75°C terminals) since the 7 amps is below either ampacity. Even with 60°C fuse terminals, the ampacity of a 14 AWG conductor would be 20 amps at 30°C. If the PV combiner were operating at 63°C, the fuse would have to be temperature corrected according to the manufacturer's instruction and the use of 14 AWG conductors would still be acceptable when evaluated at 7 amps.

This cable is protected with a 9-amp fuse.

The cable from the PV combiner to the main PV disconnect operates at 40°C. The temperature derating factor for RHW-2 with 90°C insulation is 0.91. This yields a 30°C ampacity requirement of 58 amps (52.5 / 0.91). RHW-2 6 AWG meets this requirement with an ampacity of 75 amps (90°C insulation), and a number 6 AWG cable with 75°C insulation has an ampacity of 65 amps, which also exceeds the 42 amp requirement for overcurrent devices with 75°C terminals.

Overcurrent protection is provided with a 60-amp fused disconnect. Since the negative dc conductor of the array is grounded, only a single-pole disconnect is needed.

The inverter output current is 21 amps (5000 / 240).

80% OPERATION: 1.25 x 21 = 26 amps.

The cable from the inverter to the load center operates at 30°C. Cable size 8 AWG RHW-2 (evaluated with 75°C insulation) has an ampacity of 50 amps.

A back-fed 30-amp, two-pole circuit breaker provides an ac disconnect and overcurrent protection in the load center.

The equipment-grounding conductors for this system should be at least 10 AWG conductors. The ac and dc grounding electrode conductors should be a 6 AWG conductor. An 8 AWG grounding electrode conductor might be allowed if provided with physical protection by installing in conduit.

Although not shown on the diagram, there will be a dc grounding electrode conductor from the inverter to a separate dc grounding electrode (or system). The dc grounding electrode must be bonded to the ac grounding electrode. Alternatively, the dc grounding electrode conductor may be connected directly to the ac grounding electrode.

All dc circuits should have a voltage rating of at least 310 volts (1.25 x 4 x 62). Typically, 600-volt rated conductors, fuses, and related dc equipment would be used.

EXAMPLE 8 Integrated Roof Module System, Utility-Interactive

Array Size: 192, 12-volt, 22-watt thin-film modules; I_{sc} = 1.8 amps; V_{mp} = 15.6 volts; V_{oc} = 22 volts

Inverter: ±180-volt dc input; 120-volt ac output; 4000 watts; 95% efficiency

Description

The array is integrated into the roof as the roofing membrane. The modules are connected in center-tapped strings of 24 modules each. Eight strings are connected in parallel to form the array. Strings are grouped in two sets of four and a series fuse protects the module and string wiring as shown in Figure E-8. The bipolar inverter (not currently in production) has the center tap dc input and the ac neutral output grounded. The 120-volt ac output is fed to the service entrance load center (fifty feet away) through a back-fed circuit breaker.

The manufacturer of these thin-film modules has furnished data that show that the maximum V_{oc} under worst-case low temperatures is 24 volts. The multiplication factor of 1.25 on V_{oc} does not apply [690.7(A)]. The design voltage will be 24 x 24 = 576 volts. The module manufacturer has specified (label on module) 5-amp module protective fuses that **must** be installed in each (+ and -) series string of modules.

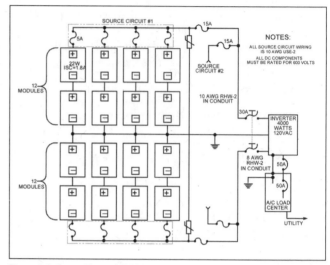

Figure E-8. Center-Tapped PV System

Calculations

Each string short-circuit current is 1.8 amps.

CONTINUOUS CURRENT (estimated for thin-film modules): 1.25 x 1.8 = 2.25 amps

80% OPERATION: 1.25 x 2.25 = 2.8 amps

Each source circuit (4 strings) short-circuit current is 7.2 amps (4 x 1.8).

CONTINUOUS CURRENT: 1.25 x 7.2 = 9 amps

80% OPERATION: 1.25 x 9 = 11.25 amps

The array (two source circuits) short-circuit current is 14.4 amps (2 x 7.2).

CONTINUOUS CURRENT: 1.25 x 14.4 = 18 amps

80% OPERATION: 1.25 x 18 = 22.5 amps

USE-2 cable is used for the module connections and operates at 75°C when connected to the roof-integrated modules. The temperature-derating factor in the wiring raceway is 0.41. For the strings, the 30°C ampacity requirement is 5.5 amps (2.25 / 0.41)[310.16].

Each source circuit conductor is also exposed to temperatures of 75°C. The required ampacity for this cable (at 30°C) is 22.0 amps (9 / 0.41).

Wire size 10 AWG USE-2 is selected for moisture and heat resistance. It has an ampacity of 40 amps at 30°C (90°C insulation) and can carry 35 amps when limited to a 75°C insulation rating (used for evaluating terminal temperature limitations on the fuses). This cable is used for both string and source-circuit wiring. Fifteen-amp fuses are used to protect the string and source-circuit conductors.

The array wiring is inside the building and RHW-2 is used in metal conduit (2005 *NEC* 690.31(E)). It is operated at 50°C when passing through the attic. The temperature derating factor is 0.82, which yields a 30°C ampacity requirement of 22 amps (18 / 0.82). Cable size 10 AWG has an ampacity of 40 amps (90°C insulation) or 35 amps (evaluated with 75°C insulation). Both of these ampacities exceed the 22-amp requirement. Twenty-five amp fuses are required to protect these cables, but 30-amp fuses are selected for better resistance to surges. Since the inverter has high voltages on the dc-input terminals (charged from the ac utility connection), a load-break rated, pullout fuse holder is used.

The inverter is rated at 4000 watts at 120 volts and has a 33-amp output current. The ampacity requirement for the cable between the inverter and the load center is 42 amps ((4000 / 120) x 1.25) at 30°C. Wire size 8 AWG RHW-2 in conduit connects the inverter to the ac-load center, which is fifty feet away and, when evaluated at with 75°C insulation, has an ampacity of 50 amps at 30°C. A 50-amp circuit breaker in a small circuit-breaker enclosure is mounted next to the inverter to provide an ac disconnect for the inverter that can be grouped with the dc disconnect. Another 50-amp circuit breaker is back-fed in the service entrance load center to provide the connection to the utility.

The modules have no frames and, therefore, no equipment grounding requirements. The inverter and switchgear should have 10 AWG equipment grounding conductors. The dc system grounding electrode conductor (GEC) should be an 8 AWG conductor installed in conduit for mechanical protection. This dc GEC is connected to the existing ac GEC.

All dc components in the system should have a minimum voltage rating of 600 volts (24 x 24 = 576).

EXAMPLE 9 Residential Utility-Interactive, Multiple-Inverter System

PV array: 3, 12-module strings of 185 watt, 24V modules; V_{oc} = 42V; I_{sc} = 6.2A

Inverters: 3, 2500-watt, 240Vac output

Residential Service Entrance/Load Center: 200A with 200A main circuit breaker

The PV modules are connected in three series strings of 12 modules each. The coldest ambient temperature is 15°F. Maximum system voltage is 570V (12 x 42 x 1.13) [690.7]. Each series string is connected to one pole of a Square D HU361RB heavy-duty safety switch with a special listing suitable for this application (each pole rated at 600 volts dc). The three outputs of the disconnect are connected to three 2500-watt inverters.

The modules are connected in series with the attached 14 AWG USE-2 conductors and attached connectors.

At a 75°C operating temperature, the 14 AWG USE-2 conductors in free air have an ampacity of 14 amps (35 x 0.41) which is higher than the 10 amps needed (6.2 x 1.56). At the ends of each string of 12 modules, the 14 AWG conductors are spliced (soldered and covered with listed, outdoor-rated heat-shrink tubing [110.14(B)]) to 10 AWG USE-2/RHW-2 conductors which are run in conduit to the readily accessible Square D disconnect located on the outside of the residence near the utility meter [690.14(C)].

The inverter has been certified by the manufacturer as having no capability to backfeed ac current from the utility grid into faults in the dc PV wiring

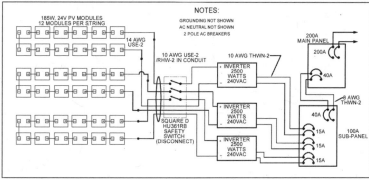

Figure E-9. Utility-Interactive Three-Inverter System

and therefore no overcurrent devices are required in the dc PV disconnect [690.9(A)EX]. The local inspector must accept or reject this certification until the UL Standard 1741 for inverters includes a test for back feeding from the utility.

Inverter output current is 10.4 amps (2500/240)

Ampacity requirements: 13 amps (10.4 x 1.25)

Circuit breaker for each inverter: 15 amps

The ac output conductors of the inverter could be 14 AWG THWN-2 that meets ampacity requirements at 45°C and with 75°C insulation (circuit breaker terminal temperature limitations). However, 10 AWG THWN-2 conductors were used to minimize voltage rise between the inverter outputs and the utility point of connection.

NEC section 690.64(B)(2) imposes a 40-amp maximum PV backfed circuit breaker rating limitation on the main panel (1.2 x 200 – 200). Connecting three double-pole 15-amp circuit breakers from the inverters would total 45 amps exceeding the limitation of 40 amps. Each bus of the 120/240 load center should be analyzed separately, but they are identical in this example. Therefore, a subpanel is used to combine the output of the three inverters before sending the combined output to the main panel.

Subpanel Main Breaker: 3 x 10.4 x 1.25 = 39 amps, round to 40 amps

Subpanel rating from 690.64(B)(2) where X is minimum subpanel rating:

1.20 X = 3 x 15 + 40 = 85, X= 70.8, round up to 100 amp panel size.

Conductors between subpanel and main panel must be rated for 39 amps.

8 AWG THHN in conduit is rated at 46.2 amps (55 x 0.91) at 45°C. With 75°C insulation, the ampacity is 50 amps at 30°C and 41 amps at 45°C. Both are adequate for the required 40 amps.

A 40-amp backfed breaker is installed in the main panel for the residence and meets NEC 690.64(B)(2) at this location.

APPENDIX F: DC Currents on Single-Phase Stand-alone Inverters

When the sinusoidal ac output current of a stand-alone inverter goes to zero 120 times per second, the input dc current also goes nearly to zero. With a resistive ac load connected to the inverter, the dc current waveform resembles a sinusoidal wave with a frequency of 120 Hz. The peak of the dc current is significantly above the average value of the current, and the lowest value of dc current is near zero.

An example of this is shown in the Figure F1. This is an example of a single-phase stand-alone inverter operating with a 4000-watt resistive load. The input battery voltage is 22 volts. The figure shows the dc current waveform. The measured average dc current is 254 amps. The RMS value of this current is 311 amps.

The calculated dc current for this inverter (as was done in Example 6 in Appendix E) is 214 amps (4000/22/0.85) when using the manufacturer's specified efficiency of 85%.

The RMS value of current is the value that causes heating in conductors and is the value of current that causes overcurrent devices to trip. In this case, if the inverter were operated at 100% of rated power and at a low battery voltage, the conductors and overcurrent devices would have to be rated to carry 311 amps, not the calculated 214 amps. Code requirements would increase the cable ampacity requirements and overcurrent device ratings to 388 amps (1.25 x 311).

Loads that have inductive components may result in even higher RMS values of dc currents.

The systems designer should contact the inverter manufacturer in cases where it is expected that the inverter may operate at loads approaching the full power rating of the inverter. The inverter manufacturer should provide an appropriate value for the dc input current under the expected load conditions.

Some inverters may employ topologies that filter the dc input current resulting in less ripple

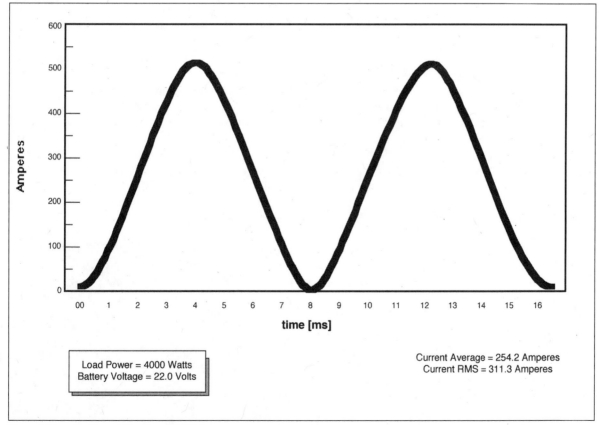

Figure F-1. Inverter Current Waveform (dc side)

APPENDIX G: Grounding PV Modules

Grounding PV modules to reduce or eliminate shock and fire hazards is necessary but difficult. Copper conductors are typically used for electrical connections, and the module frames are generally aluminum. It is well known that copper and aluminum do not mix as was discovered from numerous fires in houses wired with aluminum wiring in the 1970's. PV modules generally have aluminum frames. Many have mill finish, some are clear coated, and some are anodized for color. The mill finish aluminum and any aluminum surface that is scratched quickly oxidizes. This oxidation and any clear coat or anodizing form an insulating surface that makes for difficult long-lasting, low resistance electrical connections (e.g. frame grounding). The oxidation/anodizing is not a good enough insulator to prevent electrical shocks, but it is good enough to make good electrical connections difficult.

Underwriters Laboratories (UL) which tests and lists all PV modules sold in the US requires very stringent mechanical connections between the various pieces of the module frame to ensure that these frame pieces remain mechanically and electrically connected over the life of the module. These low-resistance connections are required because a failure of the insulating materials in the module could allow the frame to become energized at up to 600 volts (depending on the system design). The *National Electrical Code* (*NEC*) requires that any exposed metal surface be grounded if it could be energized. The installer of a PV system is required to ground each module frame. The code and UL Standard 1703 require that the module frame be grounded at the point where a designated grounding provision has been made. The connection must be made with the hardware provided using the instructions supplied by the module manufacturer.

The designated point marked on the module must be used since this is the only point tested and evaluated by UL for use as a long-term grounding point. UL has established that using other points such as the module structural mounting holes coupled with typical field installation "techniques" do not result in low-resistance, durable connections to aluminum module frames. If each and every possible combination of nut, bolt, lock washer and star washer could be evaluated for electrical properties and installation torque requirements AND the installers would all use these components and install them according to the torque requirements, it might be possible to use the structural mounting holes for grounding.

Most US PV module manufacturers are providing acceptable grounding hardware and instructions. Japanese module manufactures are frequently providing less-than-adequate hardware and unclear instructions. Future revisions of UL 1703 should address these issues.

In the meantime, installers have to struggle with the existing hardware and instructions, even when they are poor. SWTDI has identified suitable grounding hardware and provides that information when installers ask about grounding—a frequent topic.

For those modules that have been supplied with inadequate or unusable hardware or no hardware at all, here is a way to meet the intent of the code and UL Standard 1703.

For those situations requiring an equipment-grounding conductor larger than 10 AWG, a thread-forming stainless steel 10-32 screw can be used to attach an ILSCO GBL4 DBT lug to the module frame at or adjacent to the point marked for grounding. A #19 drill is required to make the proper size hole for the 10-32 screw. The 10-32 screw is required so that at least two threads are cut into the aluminum (a general UL requirement for connections of this kind). The thread-forming screw is required so that an airtight, oxygen-free mating is assured between the screw and the frame to prevent the aluminum from reoxidizing. It is not acceptable to use the hex-head, green grounding screws (even when they a have 10-32 threads) because they are not listed for outdoor exposure and will eventually corrode. The same can be said for other screws, lugs, and terminals that have not been listed for outdoor applications. Hex-head stainless steel "tech" screws and sheet metal screws do not have sufficiently fine threads to make the necessary low resistance, mechanically durable connection. The only thread-forming, 10-32 stainless steel screws that have been identified so far have Phillips heads; not the fastest for installation. (See Figure G-1.)

Figure G-1. ILSCO GBL4-DBT Lug

Figure G-2. Improper Module Grounding

The ILSCO GBL4 DBT lug is a lay-in lug with a stainless steel screw made of solid copper and then tin plated. It accepts a 4 AWG to 14 AWG copper conductor. It is listed for direct burial use (DB) and outdoor use and can be attached to aluminum structures (the tin plate). The much cheaper ILSCO GBL4 lug looks identical but is tin plated aluminum, has a plated screw, and is not listed for outdoor use. An alternative to the GBL4 DBT has not been identified, but the search continues.

If the module grounding is to be done with a 14 AWG to 10 AWG conductor, then the ILSCO lug is not needed. Two number 10 stainless steel flat washers would be used on the 10-32 screw, and the copper wire would be wrapped around the screw between the two flat washers that would isolate the copper conductor from the aluminum module frame.

What size conductor should be used? The minimum code requirement is for the equipment grounding conductor for PV source and output circuits to be sized to carry 1.25 times the short-circuit currents at that point. While this may allow a 14 AWG conductor between modules, a conductor this small would require physical protection between the grounding points. Some inspectors will allow a 10 AWG bare conductor to be routed behind the modules from grounding point to grounding point if the conductors are well protected from damage, as they would be in a roof-mounted array. If needed, an 8 AWG or 6 AWG sized conductor may be required (to meet the code or to satisfy the inspector) and then the ILSCO lugs should be used.

It is desirable to use the module mounting structure for grounding. Rack manufacturers have been urged to get their products listed as field-installable grounding devices, but they may be running up against that aluminum oxidation problem also as well as the lack of consistency in tightening nuts and bolts in the field.

The code allows metal structures to be used for grounding and even allows the paint or other covering to be scraped away to ensure a good electrical contact. Numerous types of electrical equipment are grounded with sheet metal screws and star washers. This works on common metals like steel, but not on aluminum due to the oxidation.

Module manufacturers are being encouraged to make that aluminum connection in the factory and to provide a copper-compatible terminal in the j-box or on the frame as is done with the 300-watt RWE-Schott modules.

Unfortunately many PV systems are being grounded improperly even when the proper hardware has been supplied. Figure G-2, a photo taken in March 2004, illustrates that even the proper hardware can be misused. Here, the stainless-steel isolation washer has been installed in the wrong sequence and the copper grounding wire is being pushed against the aluminum frame; a condition sure to cause corrosion and loss of electrical contact in the future.

APPENDIX H: PV Ground Fault Protection Devices and The *National Electrical Code*, Section 690.5

Section 690.5, Ground Fault Protection, of the 1987 *National Electrical Code (NEC)* added new requirements for photovoltaic (PV) systems mounted on the roofs of dwellings. The requirements are intended to reduce fire hazards resulting from ground faults in PV systems mounted on the roofs of dwellings. There is no intent to provide any shock protection since the 5ma level of protection would not be possible on a PV array with distributed leakage currents, and the requirement is not to be associated with a direct current (dc) GFCI. The ground fault protection device (GFPD) is intended to deal only with ground faults and not line-to-line faults.

The requirements for the ground-fault protection device have been modified in subsequent revisions of the Code. The requirements for the device in the 2002 *NEC* are as follows.

1. Detect a ground fault
2. Interrupt the fault current
3. Indicate that there was a ground fault
4. Open the ungrounded PV conductors

As the 1990 *NEC* was published, no hardware had been developed to meet these requirements. Under a two-year contract (1990-1991) from the Salt River Project, a Phoenix, Arizona utility, John Wiles at the Southwest Technology Development Institute at New Mexico State University developed prototype designs and hardware to meet the requirements. The designs were released to the PV industry and GFPDs based on these designs and other concepts began appearing in PV equipment and subsystems in the late 1990's. Listed equipment is now available for both stand-alone and utility-interactive systems.

To understand how these GFPDs work, it must be understood that nearly all currently available inverters, both stand-alone and utility-interactive, employ a transformer that isolates the dc grounded circuit conductor (usually the negative) from the ac grounded circuit conductor (usually the neutral). With this transformer isolation, the dc side of a PV system may be considered similar to a separately derived system and, as such, must have a single dc bonding connection that connects the dc grounded circuit conductor to a common grounding point where the dc equipment-grounding conductors and the dc grounding electrode are connected. Like grounded ac systems, only a single dc bonding connection is allowed. If more than one bonding connection (a.k.a. bonding jumper) were allowed on either the ac side of the system or on the dc side of the system, unwanted currents would circulate in the equip-ment-grounding conductors and would violate *NEC* Section 250.6.

Currently available GFPDs as both separate devices for adding to stand-alone PV systems and as internal circuits in most utility-interactive inverters serve as the dc bonding connection.

In any ground-fault scenario on the dc side of the PV system, ground-fault currents from *any* source (PV modules or batteries in stand-alone systems) must eventually flow through the dc bonding connection on their way from the energy source through the fault and back to the energy source. This includes single ground faults involving the positive conductor faulting to ground or in the negative conductor faulting to ground. In negative-conductor (a grounded conductor) ground faults, parallel paths for the negative currents are created by the fault path and they will flow through the dc bonding connection. Double ground faults are beyond the ability of any equipment to deal with and are not required to be addressed by the *NEC* or standards established by Underwriters Laboratories (UL).

To meet the *NEC* Section 690.5 requirements, a typical GFPD has a 1/2 amp to 1 amp and sometimes 5 amp overcurrent device installed in the dc bonding connection. When the dc ground-fault currents exceed the current rating of the device, it opens. By opening, the overcurrent device interrupts the ground-fault current as required in *NEC* Section 690.5. If a circuit breaker is employed as the overcurrent device, the tripped position of the breaker handle provides the indicating function. When a fuse is used, an additional electronic monitoring circuit in the inverter provides an indication that there has been a ground-fault. The indication function is also an *NEC* 690.5 requirement. There is no automatic resetting of these devices.

In the GFPD using a circuit breaker as the sensing device, an additional circuit breaker is mechanically connected (common handle/common trip) to the sensing circuit breaker. These types of GFPDs may be found in both stand-alone and utility-interactive systems. This additional circuit breaker (usually rated at 100 amps and used as a switch rather than an overcurrent device) is connected in series with the ungrounded circuit conductor from the PV array. In this manner, when a ground-fault is sensed and interrupted, the added circuit breaker disconnects the PV array from the rest of the circuit providing an additional indication that something has happened that needs attention.

Even though the GFPD uses a 100-amp circuit breaker in the ungrounded PV conductor, the 100-amp circuit breaker *may not* be used as the PV disconnect because in normal use of the system, turning off this breaker would unground the system and this is undesirable in non-fault situations.

In the GFPD installed in utility-interactive inverters using a fuse as the sensing element, the electronic controls in the inverter that indicate that there has been a fault, also turn the inverter off and open the internal connections to the ac line. In listing these inverters, UL had indicated that this method of turning off the inverter to provide an additional indication of trouble meets the requirements of 690.5(B) for disconnecting the ungrounded PV conductor.

It should be noted that the dc GFPD detects and interrupts ground faults **anywhere** in the dc wiring and the GFPD may be **located anywhere** in the dc system. Because the normal location for the dc bonding connection is at or near the dc disconnect, this bonding connection is usually made at the dc power center where there is ready access to the dc grounding electrode. GFPDs installed in the utility-interactive inverters or installed in dc power centers on stand-alone systems are the most logical places for these devices. There is no significant reason to install them at the PV module location. This configuration would significantly increase the length of the dc grounding electrode conductor and complicate its routing. To achieve significant additional safety enhancements would require a GFPD at every module. Equipment to do this does not exist and there are no requirements for such equipment.

The diagram on the next page shows both positive (red) and negative (blue) ground faults and the paths that the fault currents take. As noted above, all ground fault currents must pass through the dc bonding connection where the GFPD sensing device is located.

These devices are fully capable of interrupting ground faults occurring anywhere in the dc system including faults at the PV array or anywhere in the dc wiring from the PV module to the inverter and even to the battery in stand-alone systems. All of this can be done from any location on the dc circuit. Fire reduction and increased safety are achieved by having these GFPD on residential PV systems. Keeping the PV source and output conductors outside the dwelling until the point of first penetration and requiring the readily accessible dc disconnect also enhance the safety of the system. See Sections 690.14 of the *2002 NEC* for details.

During a ground-fault, the dc system bonding connection is opened, and if the ground fault cures itself for some reason (e.g., an arc extinguishes), the dc system remains ungrounded until the system is reset. A positive-to-ground fault may allow the negative conductor (now ungrounded) to go to the open-circuit voltage with respect to ground. This is addressed by the marking requirements of Section 690.5(C). A very high value resistance is usually built into the GFPD and this resistance bleeds off static electric charges and keeps the PV system loosely referenced to ground (but not solidly grounded) during ground-fault actions. The resistance is selected so that any fault currents still flowing are only a few milliamps—far too low to be a fire hazard.

APPENDIX I: Selecting Overcurrent Devices and Conductors in PV Systems

1. Define Continuous Currents

The unique nature of PV power generators dictate that all ac and dc calculated currents are continuous and are based on the worst-case conditions. There are no non-continuous currents and all currents are treated as continuous.

   A. DC currents in PV source and PV output circuits are calculated as 125% of the short-circuit current (I_{sc}) (690.8(A)(1)).

   B. AC inverter (stand-alone or utility-interactive) output currents are calculated at the rated output of the inverter (690.8(A)(3)).

   C. DC inverter input currents from batteries are calculated based on the rated output power of the inverter at the lowest battery voltage that can maintain that output (690.8(A)(4)). Inverter dc to ac efficiency must also be factored into the calculation.

2. Select Overcurrent Device

   A. The overcurrent device will be rated at 125% of continuous current (690.8(B)(1)).

   1.) If the overcurrent device is in a listed assembly and the combined assembly is listed for 100% duty, then use 100% continuous current to size the overcurrent device (690.8(B)(1) EX)).

   2.) The calculated value of the overcurrent device may be rounded up to next standard rating (where the rating is less than or equal to 800A (240.4(B). Standard values of overcurrent devices in PV source and output circuits are 1-15 amps in 1-amp increments (690.9(C)).

   In PV source circuits, the value should be less than or equal to the value of the maximum series protective fuse marked on the back of the module. If desired (for unforeseeable reasons), this selected value could be increased to the size of the maximum protective fuse found on the back of the module. However, this will impact conductor sizing and other overcurrent device requirements.

   B. If the overcurrent device is exposed to temperatures (operating conditions) greater than 40°C and/or less than 25°C, temperature correction factors must be applied to the device rating (110.3(B)).

3. Select Conductor

   A. A conductor should be selected with a 30°C ampacity *not less than* 125% of continuous current (215.2(A)(1)).

   B. The conductor selected must have 30°C ampacity after corrections for conditions of use (ambient temperature and conduit fill) *not less than* the continuous currents (no 125% used at this time).

   1.) Apply the conductor selection requirements at all points of different temperatures and or conduit fill.

   2.) Use the 10%/10-foot rule where appropriate (310.15(A)(2) EX).

   C. Select the larger conductor from 3.A. or 3.B (310.15(A)(2)).

4. Evaluate conductor temperature at each termination

   A. Ampacities for conductor size selected in 3.C should be selected from Table 310.16 using 60°C or 75°C ampacity columns depending on conductor temperature rating of the device terminals (110.14(C)).

   B. Ampacity must *not be less than 125%* of continuous current.

   C. Increase the conductor size, if necessary, to meet 4.B at all terminations.

5. Verify that the Overcurrent Device Protects Conductors

   A. The rating of the overcurrent device (after any corrections for conditions of use—2.B.) selected in 2 must *not be more than* the ampacity of the conductor selected in 4.C. The ampacity used for the conductor is that found under the conditions of use (3). Rating round up is allowed (240.4(B)).

   B. A larger conductor size should be selected if the conductor selected in 4.C is not protected by the overcurrent device.

APPENDIX J: Module Series Fuse Requirements

PV Modules and the Series Overcurrent Device

A properly sized overcurrent device is generally required in each string of PV modules to protect the individual modules from reverse currents and to protect the conductor from overloads.

The issue is: Can PV modules or strings of modules be connected in parallel and still meet the UL/NEC requirements (marked on the back of each module) for a series fuse (or breaker, both known as overcurrent protective devices (OCPD)) for each module/string of modules? UL marks the modules based on the reverse-current tests. The NEC requires that the manufacturer's instructions and labels be followed. The intent of the module marking is to protect the module at the marked level from reverse currents. This is a maximum value for the OCPD. Lesser values can be used as long as they meet the NEC requirement of $1.56 * I_{sc}$ to protect the conductor associated with the module or string of modules.

Many installers of 12, 24, and 48-volt PV systems ignore the module OCPD requirement and connect modules/strings in parallel. Can it be done and how? Tests have been run indicating that the module OCPD requirement is valid.

Consider n modules or strings connected in parallel. The NEC requires that an OCPD be installed in the combined output to protect the cable from reverse currents from batteries or back feed of ac currents through an inverter. The OCPD will have a minimum rating of $1.56 * n * I_{sc}$ amps. It is sized at this value to allow forward currents from the array to pass through without interruption and to keep the overcurrent device from operating at more than 80% of rating.

Examine the circuit where there are n modules/strings connected in parallel. Place a ground-fault in one module/string. Examine the sources of fault current that would affect that module string. Ignore current from the faulted module/string itself since the wiring in that string is already sized to carry all currents generated in the string.

First there is the back feed current from the battery or the inverter in those systems with these components. It is limited to the NEC required OCPD of $1.56 * n * I_{sc}$. This current is added to the current from the remaining modules connected in parallel. In this case, the current is $(n-1) * 1.25 * I_{sc}$. The 1.25 is required because of daily-expected irradiance values that are greater than the STC-rated I_{sc}. With a little algebra, the resulting fault current is:

I fault $= (2.81 * n - 1.25) * I_{sc}$ amps. (Fault Current Equation)

Note that this equation does not account for OCPD roundup, so each system must be checked with the actual OCPD values. OCPDs for PV module source and output circuits are rated in one- (1) amp increments from 1 amp to 15 amps [690.9(C)]. Above 15amps, the standard rating increments apply [240.6].

If the module can pass the UL reverse current test at this level and be so marked (the maximum protective series fuse on the label), then is it possible to parallel modules (pick an "n") without a series OCPD for each module/string?

For example, Solarex (now BP Solar) at one time marked their MSX 60 modules with a 20-amp OCPD requirement. The I_{sc} for the MSX 60 is 3.8 amps. Here are the required calculations and checks for two strings in parallel.

The paralleled circuit OCPD will be $2 * 1.56 * 3.8 = 11.8$ amps (assume 12 amp OCPD is used—The NEC now requires module/string OCPDs in one-amp increments up to 15 amps). This OCPD will allow 12 amps of fault current to reach the faulted module/string. Actually OCPDs are tested to pass 110% of rating at 25° C, but that is a complicating factor that does not need to be addressed. Another $1.25 * 3.8 = 4.75$ amps will come from the parallel-connected module for a total of 16.75 amps. This is acceptable since the MSX 60 module is marked for 20 amps.

However, if we try to parallel 3 MSX-60s, the fault current equation yields a fault current of 29+ amps that exceeds the 20-amp limit on the module. The single OCPD is $3 * 1.56 * 3.8 = 17.8$ amps (since OCPDs at this rating are not common, a 20-amp OCPD must be used). The two parallel-connected modules contribute $2 * 1.25 * 3.8 = 9.5$ amps for a total potential fault current of 29.5 amps. This is significantly above the maxim series protective fuse of 20 amps.

In most cases, it is not possible to parallel many more than two modules with a single OCPD unless the marked maximum series OCPD is very large in relation to I_{sc} for the module. Some of the thin-film technologies may be able to do this and that could be an installation benefit for them.

The faults can occur anywhere in the module/string, so a single cell could be responsible for driving voltages over 1 volt that could produce the reverse currents.

What about currents generated within the faulted module string? In the portion of the module/string below the fault (toward the grounded end of the mod-

NEC SUGGESTED PRACTICES

ule/string), the currents flow in the forward direction toward the fault and may or may not cause problems. As far as the contribution to the fault current is concerned, the contribution only appears in the fault path/arc and does not affect the ampacity of the cable. Above the fault, the currents in that portion of the module/string appear to oppose the external fault currents that are trying to reverse the flow of current. Since the location of the fault cannot be controlled ahead of time, worst-case currents must be assumed.

The increased marking value of 20 amps on the example Solarex MSX 60 allows for two modules to be connected in parallel and it does make it easier for the installer to use a single OCPD with larger cable to meet both the *NEC* required cable protection and the UL-required module protection with one large OCPD instead of a smaller OCPD plus a larger OCPD.

Conductor ampacity must also be addressed if modules are going to be paralleled on a single OCPD. The conductors for each string must be able (under fault conditions) to carry the current from the other parallel strings (modules) plus the current that may be back fed from the inverter or battery. In the case with n strings in parallel and a single OCPD in the combined output, the conductor ampacities would be as follows:

Each of the string conductors would have to have an ampacity of $1.25\,(n-1)*I_{sc} + 1.56\,n\,I_{sc}$. If the equation is factored the required ampacity becomes $A=(2.81*n-1.25)*I_{sc}$. As before, OCPD roundup is not considered and the values should be recalculated with actual OCPD values.

The combined output circuit conductors would require an ampacity of $1.56*n*I_{sc}$

Some utility-interactive inverters on the market have redundant internal circuitry that prevents currents from being backfed through the inverter. This removes one source of currents in the above equation. With these products, it is possible to have two and sometimes more strings of modules in parallel with no OCPDs in the dc circuits. The inverter manufacturers should be contacted for information in this area. The above equations can be modified by deleting the combined circuit OCPD and then solved to determine both the requirements for OCPDs and the necessary ampacity of the conductors.

APPENDIX K: Flexible, Fine-Stranded Cables: Incompatibilities with Set-Screw Mechanical Terminals and Lugs

Reports have been received over the last several years about field-made connections that have failed when flexible, fine-stranded cables have been used with mechanical terminals or lugs that use a set screw to hold the wire in the terminal. See Figure K-1 for examples of such terminals.

These terminals are used on nearly all circuit breakers (except those with stud-type terminals), fuse holders, disconnects, PV inverters, charge controllers, power distribution blocks, some PV modules, and many other types of electrical equipment.

Fine-stranded conductors and cables are considered as those cables having stranding more numerous than Class B stranding. Class B stranding (the most common) will normally have 7 strands of wire per conductor in sizes 18-2 AWG, 19 strands in sizes 1-4/0 AWG, and 37 strands in sizes 250-500 kcmil. Conductors having more strands than these are widely available and are in different classes such as K and M used for portable power cords and welding cables. Commonly used building-wire cables such as USE, THW, RHW, THHN and

Figure K-1. Examples of Mechanical Terminals

the like are most commonly available with Class B stranding, but are also *readily available with higher stranding*. Fine-stranded cables are frequently used by PV installers to ease installation and are used in PV systems for battery cables, power conductors to large utility-interactive inverters and elsewhere.

Some modules are supplied with fine-stranded interconnecting cables with attached connectors. While the crimped-on connectors listed with the module are suitable for use with the fine-stranded conductors, an end-of-string conductor with mating connector may also be supplied with the fine-stranded conductor, and the unterminated end of that conductor will not be compatible with mechanical terminals.

According to UL Standard 486 A-B, a terminal/lug/connector must be listed and marked for use with conductors stranded in other than Class B. With no marking or factory literature/instructions to the contrary, the terminal may *only* be used with conductors with the most common Class B stranded conductors. They are not suitable and should not be used with fine-stranded cables. UL engineers have said that few (if any) of the normal screw-type mechanical terminals that the PV industry commonly uses have been listed for use with fine stranded wires. **The terminal must be marked or labeled specifically for use with fine-stranded conductors.**

UL suggests two problems, both of which have been experienced in PV systems. First, the turning screw tends to break the fine wire strands, reducing the amount of copper available to meet the listed ampacity. Second, the initial torque setting does not hold and the fine strands continue to compress after the initial tightening. Even

Figure K-2. Destroyed Mechanical Terminal From PV Syste

after subsequent retorquing, the connection may still loosen. The loosening connection creates a higher-than-normal resistance connection that heats and may eventually fail. See Figure K-2 for a failed mechanical terminal from a PV system.

Solutions

All electrical equipment listed to UL Standards has:
- Terminals rated for the required current and sized to accept the proper conductors
- Sufficient wire bending space to accommodate the Class B stranded conductors in a manner that meets the wire bending requirements of the *NEC*
- Provisions to accept the appropriate conduit size for these conductors where conduit is required.

It is therefore unnecessary to use the fine-stranded cables except possibly when dealing with conductors 4/0 AWG and larger.

In those cases where a fine-stranded cable must be used, a few manufacturers make a limited number of crimp-on compression lugs in various sizes that are suitable for use with fine-stranded cables. See Figure K-3.

Figure K-3. Typical Compression Lug

Factory-supplied markings and literature indicate which lugs are suitable. An example is the ILSCO FE series of lugs in sizes 2/0 AWG and larger. Burndy makes a YA series of lugs in sizes 14 AWG and up. In both cases the lugs are solid copper. It should be emphasized: *Most crimp-on lugs are not listed for use with fine-stranded wire.* Where the crimp-on compression lugs can be used, they *must* be installed using the tools recommended by the manufacturer and, of course, they must be attached to a stud with a nut and washer.

Burndy and others make pin adapters (a.k.a. pigtail adapters) that can be crimped on fine-stranded cables. These pin adapters provide a protruding pin that can be inserted into a standard screw-type mechanical connector. Again, not all pin adapters/pigtail adapters are listed for use with fine-stranded conductors; some are intended for use with aluminum wire and others provide only a conversion to a smaller AWG size for B Class conductor or a pin adapter for Class B conductors.

It is suggested that the use of fine-stranded conductors be avoided wherever possible. Where such cables must be used, they should only be terminated with the appropriate connectors/lugs. Previously installed systems should be revisited and the cables replaced where possible or terminated properly.

APPENDIX L: Ungrounded PV Systems

The 2005 *NEC* will permit (not require) the installation of PV systems that do not have one of the dc PV source and PV output conductors grounded. This new go-ahead for ungrounded systems will be in addition to the existing allowance for ungrounded PV systems operating below a systems voltage of 50 volts.

There are a number of additional requirements for these ungrounded PV systems. These additional requirements were established to ensure the safety of the system. Since the United States has over a 100-year tradition of installing, inspecting, and servicing grounded electrical systems, the training and the infrastructure for installing and inspecting ungrounded systems will need to be established.

Most equipment in common use in the United States is designed for use only on grounded electrical systems. Much of the existing PV balance of systems equipment such as power centers and charge controllers, and even some inverters are designed today for use in grounded systems. Radio frequency (RF) filters, required to meet FCC emissions requirements are frequently installed from only the positive conductor to chassis assuming that the negative conductor is grounded. Disconnects in power centers are installed only in the positive conductor and the negative conductors are routed through grounding blocks bonded to the chassis. . The transition to ungrounded PV systems will necessitate new hardware designs and new thinking for surge protection, overcurrent protection, and disconnects.

Electricians and PV installers are trained to install grounded systems. Inspectors are trained and experienced in inspecting grounded, not ungrounded, electrical systems.

Europe, on the other hand, has many years of experience installing not only ungrounded PV systems but also ungrounded ac electrical systems.

Unfortunately, the installation practices and available equipment on each side of the Atlantic have few commonalities. The few items that are common in both arenas, such as the availability of conduit, are used in entirely different ways. The ungrounded European PV systems have as good a safety record as the grounded US PV systems. This addition to the *2005 NEC* was made to permit the US PV industry to utilize European experience while using US equipment and still meet all safety requirements. The resulting requirements are as follows:

1. Ground-fault detectors will be required on all ungrounded PV arrays for fire protection purposes, not shock protection.

2. Disconnects and overcurrent protection will be required on each circuit conductor unless the system design requires no overcurrent protection in that circuit.

3. The PV source and PV output circuit conductors will be required to be in a raceway or be part of a multi-conductor sheathed cable. This requirement emulates the European use of "double insulated" cable, which is not yet available in the US. When such "double-insulated" cables become available, are tested and listed to an appropriate UL safety standard, then it is anticipated that they too would meet the intent of this requirement.

4. A warning label shall be placed on any termination or location where the ungrounded conductors in raceways may be exposed stating the following:

WARNING: ELECTRIC SHOCK HAZARD. THE DIRECT CURRENT CIRCUIT CONDUCTORS OF THIS PHOTOVOLTAIC POWER SYSTEM ARE UNGROUNDED BUT MAY BE ENERGIZED WITH RESPECT TO GROUND DUE TO LEAKAGE PATHS AND/OR GROUND FAULTS.

5. The inverters or charge controllers used in systems with ungrounded photovoltaic source and output circuits shall be listed for the purpose.

Product Index

PRODUCT INDEX

PRODUCT INDEX

PRODUCT INDEX

PRODUCT INDEX

Subject Index

SUBJECT INDEX

SUBJECT INDEX

WIRE AND ADAPTER PRODUCTS

SPECIALTY WIRE

Meter Wire

Special 18-gauge wire for meters. Has 4 twisted pairs, a total of 8 conductors. Each twisted pair is shielded for an ultra-clean signal. Poly jacketed bundle. Custom cut and sold by the foot.

26659   18 AWG 4-Pair Meter Wire $1.50/ft.

Sub Pump Cable

This flat 2-conductor, 10-gauge jacketed submersible cable is required for the SHURflo Solar Sub splice kit, and is sometimes difficult to find locally. Cut to length. Specify how many feet you need.

26540   10-2 Sub Pump Cable $1/ft.

DC ADAPTERS

12-Volt Fused Replacement Plug

This fused plug has a unique polarity reversing feature. It includes a 5-amp fuse and four sizes of snap-in strain reliefs, which accommodate wire gauges from 24 to 16 AWG. It is rated at 8 amps continuous duty and can be fused to 15 amps for protecting any electric device. USA.

26103   Fused Replacement Plug $3.95

12-Volt Plug Adapter

This simple, durable plastic adapter provides an elegant conversion from any standard 110-volt fixture. Simply insert your 110-volt plug into the connector end of this adapter (no cutting!), insert a 12-volt bulb into the light socket of your lamp, and you're in 12-volt heaven for cheap! USA.

26104   Plug Adapter $5

12-Volt Extension Cords

These low-voltage extension cords have a cigarette lighter receptacle on one end and a cigarette lighter plug on the other end. They are available in 10-foot lengths. The maximum recommended current is 4 amps for the 10-foot cord. USA.

26113   Extension Cord, 10-ft. $13.95

Mobile DC Power Adapter

Use a 12-volt source to run and recharge any battery-powered device using less than 12 volts DC. Our adapter delivers stable DC power at 3, 4.5, 6, 7.5, 9, or 12 volts at up to 2.0 amps. Voltage varies less than 0.5 volt in our tests. Perfect for your camera, DVD or MP3 player, camcorder, Game Boy®, cell phone, or whatever. The 6-foot fused input cord with LED power indicator plugs into any lighter socket, the 1-foot output cord has six adapter plugs. A storage compartment in the power adapter holds your spare plugs to prevent loss. Slide switch selects voltage. China.

53-0105 Mobile DC Power Adapter $24

Double Outlet Adapter

The double outlet power adapter permits the use of two 12-volt products in a single outlet. Two receptacles connect to a single adapter plug. Rated at 10 amps. USA.

26107   Double Outlet Adapter $9.95

ADDENDUM

12-Volt Extension Cord Receptacle

This receptacle, made of all brass parts in a break-resistant plastic housing, connects with 12-volt cigarette lighter plugs. Includes 6-inch lead wire. USA.

26108   Extension Cord Receptacle $6.95

12-Volt Extension Cord with Battery Clips

These unique cords have spring-loaded clips to attach directly to battery terminals. They increase the use and value of 12-volt products like lights, TV sets, radios, or appliances. The maximum amperage that can be put through the 10-foot cord is 4 amps. USA.

26112   Extension Cord and Clips, 10-ft. $14.95

12-Volt Receptacle

Our basic 12-volt receptacle is identical to an automobile cigarette lighter. It is made of a chrome-plated all metal housing with brass internal parts and fits a standard single gang electrical junction box. Handles up to 15-amp surge, but 7 to 8 amps is the maximum continuous current from any cigarette lighter plug. USA.

For ease of installation, use 6" of flexible lamp zip cord between receptacle terminals and incoming romex wire.

26660   DC Outlet, Chrome, Single $6.50

Real Goods Renewables: We Are Experienced!

Real Goods has provided reliable solar power systems to over 60,000 homes and villages worldwide since 1978.

You don't want your project to be a test bed for unproven technology or system design. You want to go to a source of information, products, and service that are of the highest quality. You want to leave the details to a company that has the experience and capabilities to do the job right the first time. You want a company with experience.

Real Goods has been providing clean, reliable, renewable energy to people all over the planet for almost thirty years. If you have a facility where utility power is unavailable or of poor quality, consider our resources:

◆ *Renewable Energy Systems Design and Installation (solar, wind, and hydro)*
◆ *Large-Scale Uninterruptible Backup Power*
◆ *Solar Architecture*
◆ *High-Efficiency Appliances*
◆ *Biological Waste Treatment Systems*
◆ *Green Building Techniques*
◆ *Power Quality Enhancement*
◆ *Whole Systems Integration*
◆ *Utility Intertie with Renewable Energy*
◆ *Water Quality and Management*

Our eight full-time technicians with decades of combined experience in renewable energy can work with you to make your project a success. Our consultation and experience allow you to make informed decisions on which technologies are best suited for your application. Our solar, wind, and micro-hydro electric products can be coupled with diesel generators to make a hybrid electrical system scaled to your needs, offering the best environmental advantages and cost.

Gaiam Real Goods is the most experienced provider of energy conservation, renewable energy, and other sustainable products in the world. Our commitment to excellence has resulted in our being chosen for numerous special

Yucatan Peninsula, Mexico: The solar power system for this resort was designed and provided by Real Goods.

projects. Our commitment to each unique situation results in exceptionally high levels of customer satisfaction.

We have helped millions of people become aware of the positive actions they can take to minimize environmental impact. We can introduce you to technologies and techniques that can save you thousands of dollars and free you from utility company power forever. Our experience comes from dealing with people's unusual needs every day.

"I'm writing to say thank you to both you and the staff at Real Goods who made our dream of an independent energy resort a reality. I realize that our time constraint deadline taxed everyone to the maximum. Yet even under the most adverse conditions, including working with a foreign government and regulations, and a limited skilled labor pool, Real Goods never missed a beat.

Your thoroughness, attention to detail and ability to keep all the players "rowing in the same direction" is certainly commendable. I was also very impressed by your on-going support and commitment to being there for us during our precarious start-up period.

It's not every day that an organization invests $2 million dollars in an energy system. It's nice to know that we made the right choice of a supplier.

Thank you from everyone here at The Essene Way, we look forward to seeing you on the beach again."

Sincerely, Tom Ciola,
Director of the Essene Way Resort

101-Kilowatt Photovoltaic Array, The Essene Way Resort, Belize, Central America.

Real Goods Renewables provides solar electric power for vital communications on Mt. Everest expeditions and many other "outback" applications.

REAL GOODS®

For credit card orders, call toll-free:
1-800-919-2400

360 Interlocken Blvd., #200, Broomfield, CO 80021-3492

SLSB12

Regular business hours: Monday–Friday: 7:30 am–5 pm (PST)
Saturday: 7:30 am–4 pm (PST)

Order Fax toll-free: 1-800-482-7602   Tech Fax: 303-222-3599
International Fax: 707-744-2104   Overseas Orders: 707-744-2100
Customer Service: 1-800-919-2400   Email: techs@realgoods.com

ORDERED BY:

A Name _____

Address _____

City _____ State ____ Zip ____

DAYTIME PHONE: _____
(In case we have questions about your order)

Email Address _____

ALTERNATIVE SHIPPING OR GIFT ADDRESS

B Name _____

Address _____

City _____ State ____ Zip ____

Message _____

ITEM #	SIZE/ COLOR	QTY.	PAGE	DESCRIPTION	CIRCLE ADDRESS	PRICE	TOTAL
					A B		
					A B		
					A B		
					A B		
					A B		
					A B		
					A B		
					A B		
					A B		
					A B		
					A B		
					A B		
					A B		
					A B		
					A B		
					A B		
					A B		
					A B		
					A B		
					A B		
					A B		
					A B		

SHIPPING AND HANDLING CHARGES (PER ADDRESS)

SUBTOTAL	CHARGE	SUBTOTAL	CHARGE
Up to $25.00	$5.95	$125.01-$150.00	$14.95
$25.01-$50.00	$7.95	$150.01-$180.00	10% of subtotal
$50.01-$70.00	$9.95	$180.01-$200.00	9% of subtotal
$70.01-$100.00	$11.95	$200.00+	8% of subtotal
$100.01-$125.00	$13.95		

Does not include freight collect items. For AK, HI, U.S. territories, and Canada, please call for rates.

EXPRESS DELIVERY–Order by 12:30 pm Eastern Standard Time (EST) and for only $9.00 additional, your order will be delivered within 2 business days of the time you placed your order. For next business day delivery, add $16.50 to your shipping charges. These rates apply to all packages up to 10 lbs. and to in-stock merchandise only. Not available on products shipped directly from the manufacturer (❖). Call for rates on heavy or bulky packages, international shipments or deliveries to Alaska, Hawaii, Virgin Islands, or Puerto Rico.

PAYMENT METHOD:

❑ Check #_____

❑ Credit Card (MasterCard, Visa, Discover, or Amex)
Account Number (please include all numbers)

Expiration Date: [ ] – [ ]

CID #: [   ]

Signature _____

TOTAL OF GOODS	
SALES TAX (CA, CO & OH ONLY)	
SHIPPING (each address— see chart at left)	
EXPRESS or Additional Shipping Charges	
YES I'd like to make a donation to the Institute	
TOTAL ENCLOSED (U.S. dollars only)	

THANK YOU FOR YOUR ORDER
Prices subject to change.

SOLAR LIVING SOURCEBOOK

561

Ordering Information

Order Toll-Free 800-919-2400
Monday–Friday 7:30 am– 5:00 pm (Pacific Time)
Saturday 7:30 am– 4:00 pm (Pacific Time)

Fax Toll-Free 800-482-7602
Please include return phone number.

Order Through Our Secure Website
www.realgoods.com

Order by Mail
Use attached order form.

International Orders
Phone Orders: 707-744-2100
Fax: 707-744-2104
(Please call or fax for shipping quote. Funds must be in U.S.
dollars, drawn from a U.S. bank.) We charge $5 for catalogs
mailed outside the U.S.

Toll-Free Customer Service
If you have a problem with your order, our Customer Service
Representatives are available between 7:30 am and 5:00 pm
(Pacific Time) Monday through Friday. Call toll-free at
800-919-2400.

Product Information
Our representatives can provide you with detailed information
about any of our products. If you need to know more, please call
us at 800-919-2400.

Technical Assistance
For technical assistance with a renewable energy system,
please call 800-919-2400, Monday through Friday, 7:30 am–
5:00 pm and Saturday, 7:30 am–4:00 pm Pacific Time,
or fax 303-222-3599.

Shipping
Allow 7 to 10 working days for delivery, depending on your
distance from our Distribution Center in West Chester, Ohio.

Overweight or bulky items may require additional shipping
charges.

There are a few items that ship directly from the manufacturer;
these items are designated on your invoice. Some large items
(like refrigerators) are shipped **freight collect** from the
manufacturer. We will be happy to provide you with a shipping
estimate. Allow extra time for delivery of these items.

Price Change
Prices, specifications, and availability are subject to change.

Freight Damage, Returns, and Adjustments
Inspect all shipments upon arrival. If you discover damage,
please notify the carrier immediately. Be sure to save all shipping
cartons in case you have to return merchandise. Ship to us via
UPS or USPS insured. We charge a stocking fee of 10% to 25%
for merchandise not in original condition, damaged, or
incomplete. Please return items to: Real Goods, 9107 Meridian
Way, West Chester, OH 45069-6534. *Please note—if the item you
ordered is shipped directly from manufacturer, there may be
additional instructions; please call prior to returning.*

REAL GOODS®

For credit card orders, call toll-free:
1-800-919-2400

360 Interlocken Blvd., #200, Broomfield, CO 80021-3492

SLSB12

Regular business hours: Monday–Friday: 7:30 am–5 pm (PST)
Saturday: 7:30 am–4 pm (PST)

Order Fax toll-free: 1-800-482-7602   Tech Fax: 303-222-3599
International Fax: 707-744-2104   Overseas Orders: 707-744-2100
Customer Service: 1-800-919-2400   Email: techs@realgoods.com

ORDERED BY:

A Name _____

Address _____

City _____ State ____ Zip _____

DAYTIME PHONE: _____
(In case we have questions about your order)

Email Address _____

ALTERNATIVE SHIPPING OR GIFT ADDRESS

B Name _____

Address _____

City _____ State ____ Zip _____

Message _____

ITEM #	SIZE/COLOR	QTY.	PAGE	DESCRIPTION	CIRCLE ADDRESS	PRICE	TOTAL
					A B		
					A B		
					A B		
					A B		
					A B		
					A B		
					A B		
					A B		
					A B		
					A B		
					A B		
					A B		
					A B		
					A B		
					A B		
					A B		
					A B		
					A B		
					A B		
					A B		
					A B		
					A B		
					A B		
					A B		

SHIPPING AND HANDLING CHARGES *(PER ADDRESS)*

SUBTOTAL	CHARGE	SUBTOTAL	CHARGE
Up to $25.00	$5.95	$125.01-$150.00	$14.95
$25.01-$50.00	$7.95	$150.01-$180.00	10% of subtotal
$50.01-$70.00	$9.95	$180.01-$200.00	9% of subtotal
$70.01-$100.00	$11.95	$200.00+	8% of subtotal
$100.01-$125.00	$13.95		

Does not include freight collect items. For AK, HI, U.S. territories, and Canada, please call for rates.

EXPRESS DELIVERY–Order by 12:30 pm Eastern Standard Time (EST) and for only $9.00 additional, your order will be delivered within 2 business days of the time you placed your order. For next business day delivery, add $16.50 to your shipping charges. These rates apply to all packages up to 10 lbs. and to in-stock merchandise only. Not available on products shipped directly from the manufacturer (❖). Call for rates on heavy or bulky packages, international shipments or deliveries to Alaska, Hawaii, Virgin Islands, or Puerto Rico.

PAYMENT METHOD:

❑ Check #_____

❑ Credit Card (MasterCard, Visa, Discover, or Amex)
  Account Number *(please include all numbers)*

Expiration Date: ☐☐ — ☐☐

CID #: ☐☐☐☐

Signature _____

TOTAL OF GOODS	
SALES TAX (CA, CO & OH ONLY)	
SHIPPING (each address— see chart at left)	
EXPRESS or Additional Shipping Charges	
YES I'd like to make a donation to the Institute	
TOTAL ENCLOSED (U.S. dollars only)	

THANK YOU FOR YOUR ORDER
Prices subject to change.

Our Pledge to You

We strive for your total satisfaction.
If any product does not meet your expectations,
for any reason, at any time, give us a call and
we'll work with you. You may return most items
(and any unused items in original packaging)
within 30 days of purchase date for a full refund
or replacement. Past this period, normal
manufacturers' warranties apply.

Ordering Information

Order Toll-Free 800-919-2400
Monday–Friday 7:30 am–5:00 pm (Pacific Time)
Saturday 7:30 am–4:00 pm (Pacific Time)

Fax Toll-Free 800-482-7602
Please include return phone number.

Order Through Our Secure Website
www.realgoods.com

Order by Mail
Use attached order form.

International Orders
Phone Orders: 707-744-2100
Fax: 707-744-2104
(Please call or fax for shipping quote. Funds must be in U.S.
dollars, drawn from a U.S. bank.) We charge $5 for catalogs
mailed outside the U.S.

Toll-Free Customer Service
If you have a problem with your order, our Customer Service
Representatives are available between 7:30 am and 5:00 pm
(Pacific Time) Monday through Friday. Call toll-free at
800-919-2400.

Product Information
Our representatives can provide you with detailed information
about any of our products. If you need to know more, please call
us at 800-919-2400.

Technical Assistance
For technical assistance with a renewable energy system,
please call 800-919-2400, Monday through Friday, 7:30 am–
5:00 pm and Saturday, 7:30 am–4:00 pm Pacific Time,
or fax 303-222-3599.

Shipping
Allow 7 to 10 working days for delivery, depending on your
distance from our Distribution Center in West Chester, Ohio.

Overweight or bulky items may require additional shipping
charges.
There are a few items that ship directly from the manufacturer;
these items are designated on your invoice. Some large items
(like refrigerators) are shipped **freight collect** from the
manufacturer. We will be happy to provide you with a shipping
estimate. Allow extra time for delivery of these items.

Price Change
Prices, specifications, and availability are subject to change.

Freight Damage, Returns, and Adjustments
Inspect all shipments upon arrival. If you discover damage,
please notify the carrier immediately. Be sure to save all shipping
cartons in case you have to return merchandise. Ship to us via
UPS or USPS insured. We charge a stocking fee of 10% to 25%
for merchandise not in original condition, damaged, or
incomplete. Please return items to: Real Goods, 9107 Meridian
Way, West Chester, OH 45069-6534. *Please note—if the item you
ordered is shipped directly from manufacturer, there may be
additional instructions; please call prior to returning.*

**Would You Like to Support
Our Education Efforts?**

Feel free to add a tax-deductible
donation to your order and
support our nonprofit
Solar Living Institute.

www.solarliving.org